Ecological and Human Health Impacts of Contaminated Food and Environments

This book discusses linkages between the natural and disturbed chemical composition of the earth's surface and ecological and human health. It reviews the environmental geochemical cycles of natural elements and persistent toxic substances (PTS) in the environment, highlighting the degradation of soil and water resources due to human activities such as extraction and usage of minerals. There is an attempt to provide evidence concerning the health effects of consuming contaminated food, due to frequent consumption of mercury-laden fish. Lastly, sources, fates, and ecological effects of various PTS are presented, including microplastics and associated chemicals.

This book:

- Details linkages between the natural and disturbed chemical composition of the earth's surface, and environmental and human health, focusing on food contamination
- Discusses emerging pollutants with potential widespread hazardous effects such as bisphenol A and phthalates
- Reviews safe food production and quality, as well as the management, regulation, and policies concerning toxic chemicals
- Contains cutting-edge knowledge on safe food production and remediating technologies
- Describes how geochemical cycling results in food contamination.

Urbanization, Industrialization, and the Environment
Series Editor: Viduranga Waisundara

Crisis Management of Chronic Pollution: Contaminated Soil and Human Health
Edited by Magalie Lesueur Jannoyer, Philippe Cattan, Thierry Woignier, and Florence Clostre

Biochar: Production, Characterization, and Applications
Edited by Yong Sik Ok, Sophie M. Uchimiya, Scott X. Chang, and Nanthi Bolan

Phytotechnology: A Sustainable Platform for the Development of Herbal Products
Edited by Wanderley Pereira Oliveira

Waste Technology for Emerging Economies
Edited by Dr. T.C. Bamunuarachchige and Professor H.K.S. de Zoysa

Advanced Materials towards Energy Sustainability: Theory and Implementations
Edited by Samsul Ariffin Abdul Karim and Poppy Puspitasari

Acknowledging Indigenous Knowledge: Voices of Tropical Forest People
By Purabi Bose

Ecological and Human Health Impacts of Contaminated Food and Environments
Edited by Ming Hung Wong

Ecological and Human Health Impacts of Contaminated Food and Environments

Edited by
Ming Hung Wong

CRC Press
Taylor & Francis Group
Boca Raton London New York

CRC Press is an imprint of the
Taylor & Francis Group, an **informa** business

First edition published 2026
by CRC Press
2385 NW Executive Center Drive, Suite 320, Boca Raton FL 33431

and by CRC Press
4 Park Square, Milton Park, Abingdon, Oxon, OX14 4RN

CRC Press is an imprint of Taylor & Francis Group, LLC

ISBN: 9781032787060 (hbk)
ISBN: 9781032794617 (pbk)
ISBN: 9781003492115 (ebk)

DOI: 10.1201/9781003492115

Typeset in Times
by codeMantra

Contents

About the Editor

Professor Wong is a Member of the European Academy of Sciences and Arts and Chang Jiang Chair Professor of the Ministry of Education, China. He served as the Editor-in-Chief of 'Environmental Geochemistry and Health' for 20 years (2002–2023). Professor Wong was the Coordinator of Central and North-East Asia of 'Regionally Based Assessment of Persistent Toxic Substances' and a Panel Member (of 3 experts) of 'Chemicals Management Issues of Developing Countries and Countries with Economies in Transition', sponsored by UNEP/GEF, during 2001–2003 and 2010–2012, respectively.

His research areas included 'Environmental toxicology', 'Ecological restoration', and 'Resource reuse'. He has published over 850 SCI papers. In addition to his PhD (Durham), he was awarded two higher Doctoral Degrees based on published papers in 1992 and 2004: DSc (Durham) and DSc (Strathclyde). Recently, he has been awarded an Honorary DSc from the Southern Federal University, Russia (31 May 2024), in recognition of his involvement as the Lead Scientist in the Mega Project on Bioremediation of Polluted Ecosystems sponsored by the Russian Government.

Professor Wong has been ranked 6th for 3 years and 8th for 2 years (career-long ranking) globally in Environmental Science according to the World's Top 2% Scientists (Stanford University, 2020–2024). Based on Research.com's Best Researchers in Various Disciplines (2023, 2024), he ranked No. 1 in China under Environmental Science.

Preface

Food and environmental contaminants' ecological and human health impacts are pressing issues that demand urgent attention and action. This book brings together a comprehensive collection of research and insights from leading experts in the field. The 21 chapters in this volume address a wide array of contaminants, their sources, exposure pathways, and the resulting health implications, providing a critical resource for researchers, policymakers, and practitioners alike.

Below are the descriptions of the chapters organized under these parts:

Part 1: Environmental Contaminants and Their Management

Chapter 1 discusses the threats posed by persistent organic pollutants (POPs) and emerging chemicals, highlighting findings from UNEP/GEF projects that emphasize the need for sustainable management practices to mitigate their impacts on health and the environment. Chapter 2 explores the connections between ecological challenges and human health within global governance frameworks, using global mercury governance as a case study to illustrate the need for integrated approaches in addressing these intertwined issues. Chapter 3 reviews the diverse laws and regulations to ensure food safety across the Asia Pacific, emphasizing the need to harmonize standards to protect public health from food contaminants. Focusing on organic tea production in China, Chapter 4 examines the environmental and health risks associated with organic farming practices and advocates for improved regulatory frameworks to enhance safety and sustainability.

Part 2: Sources and Exposure Pathways

Chapter 5 investigates the formation of persistent toxic substances from biomass combustion, detailing their health risks and the need for cleaner energy practices to mitigate these hazards. Chapter 6 highlights the severe threat of heavy metals in agriculture, discussing their sources, impacts on food security, and necessary preventive measures to protect human health. Chapter 7 presents microbiome engineering as a novel approach to restoring soil health, emphasizing the importance of microbial dynamics in sustainable agricultural practices. Chapter 8 examines the pervasive issue of per- and polyfluoroalkyl substances (PFAS), detailing exposure pathways and associated health risks and calling for improved regulatory measures. Chapter 9 deals with the potential health impacts of microplastics and associated chemicals, calling for stringent control of the misuse and overuse of plastics. Chapter 10 explores the environmental and health risks associated with persistent toxic substances from molasses-based distilleries, advocating for better detection and regulation.

Part 3: Case Studies in Vietnam

Chapter 11 provides an overview of persistent bioaccumulative toxic (PBT) substances in Vietnam, discussing their sources, contamination profiles, and health risks and emphasizing the need for collaborative management efforts. Chapter 12 assesses the pollution status of emerging contaminants in Vietnam, highlighting exposure risks and the need for targeted research and policy interventions.

Part 4: Biogeochemical Cycles and Health Risks

Chapter 13 discusses the global mercury cycle, its sources, and the factors influencing mercury accumulation in fish, stressing the long-term health risks associated with dietary exposure.

Chapter 14 highlights the application of copper fungicide in farmland, which results in the accumulation of copper in crops and soils, posing human and environmental risks. Chapter 15 highlights the health risks associated with arsenic contamination in rice, emphasizing the importance of dietary management and public awareness. Chapter 16 discusses various mitigation strategies for reducing arsenic levels in rice, highlighting the need for further research and innovative approaches. Chapter 17 explores the contrasting behaviors of arsenic and cadmium in paddy soils, discussing their environmental impacts and potential mitigation strategies.

Part 5: Impacts on Human Health and Development

Chapter 18 details the health risks associated with cadmium-contaminated crops, emphasizing the need for dietary changes to protect human health. Chapter 19 discusses the impact of trace element exposure on male reproductive health, highlighting the delicate balance required for optimal fertility. Chapter 20 reviews the effects of endocrine-disrupting chemicals on maternal and child health, emphasizing the need for sound health policies to mitigate risks. Chapter 21 provides an overview of endocrine disruptors' prevalence and health impacts, particularly in children, and calls for further research and protective measures.

This book is vital for understanding the complex interactions between environmental contaminants and human health, offering insights and recommendations for future research and policy development.

Ming Hung Wong
BSc (CUHK), MSc, PhD, DSc (Durham), MBA, DSc (Strathclyde), Hon DSc (SFedU)
10 December 2024

List of Contributors

Qumber Abass
Department of Environmental Biotechnology, Faculty of Energy and Environmental Engineering, Silesian University of Technology, Gliwice, Poland

Muhammad Adnan
State Key Laboratory of Environmental Geochemistry Institute of Geochemistry, Chinese Academy of Sciences Guiyang, Guizhou, China

Muhammad Ubaid Ali
College of Chemical Engineering, Huaqiao University, Xiamen, China

Sajid Ali
College of Chemical Engineering, Huaqiao University, Xiamen, China

Hoang Quoc Anh
Faculty of Chemistry, Vietnam National University, Hanoi, Vietnam

Yimin Cai
College of Environmental and Resource Sciences, Zhejiang Agriculture and Forestry University, Hangzhou, China

Ram Chandra
Department of Environmental Microbiology, School of Earth and Environmental Sciences, Babasaheb Bhimrao Ambedkar University (A Central University), Lucknow, Uttar Pradesh, India

Xin Chen
School of Ecology, Environment and Resources, Guangdong University of Technology, Guangzhou, China

Xun Wen Chen
Department of Ecology, College of Life Science and Technology, Jinan University, Guangzhou, China

Yujing Chen
The Jockey Club School of Public Health and Primary Care, The Chinese University of Hong Kong Hong Kong, China

Yihan Chi
School of Environmental Science and Engineering, College of Engineering, Southern University of Science and Technology, Shenzhen, China

Le Van Chieu
Vietnam National University, Hanoi, VNU Town in Hoa Lac, Thach That, Hanoi, Vietnam

Kenneth Chow
Department of Statistics and Data Science, University of California, Los Angeles, USA

Ming Kei Chung
The Jockey Club School of Public Health and Primary Care; Institute of Environment, Energy and Sustainability; and Li Ka Shing Institute of Health Sciences, The Chinese University of Hong Kong, Hong Kong, China

Wen-Jing Deng
Department of Science and Environmental Studies, The Education University of Hong Kong, Hong Kong, China

Nicholas Dickinson
Faculty of Agriculture and Life Science, Lincoln University, Canterbury, New Zealand

Ksenia Gerasimova
Higher School of Economics, University of Moscow, Russia

Maria Gertsen
Laboratory of Biogeochemistry, Tula State Lev Tolstoy Pedagogical University (Tolstoy University), Tula, Russia

Denis Ivanov
Department of Geography, Shenzhen MSUBIT University, Shenzhen, China

Dasom Jeon
Dasom Jeon, Rural Development Administration (RDA), Jeonju, Republic of Korea

Zhoujia Jiang
College of Environmental and Resource Sciences, Zhejiang Agriculture and Forestry University, Hangzhou, China

Xiangyi Jin
College of Environmental and Resource Sciences, Zhejiang Agriculture and Forestry University, Hangzhou, China

Kyoung-Woong Kim
Department of Environment and Engineering, Gwangju Institute of Science and Technology (GIST), Gwangju, South Korea

Ekaterina Kravchenko
Academy of Biology and Biotechnology, Southern Federal University, RostovonDon, Russia

Hui Li
Department of Ecology, College of Life Science and Technology, Jinan University, Guangzhou, China

Wai Chin Li
Department of Science and Environmental Studies, The Education University of Hong Kong, Hong Kong, China

Yixin Li
Institute of Agricultural Resources and Environment, Guangdong Academy of Agricultural Sciences Guangzhou, China

Yichun Li
Institute of Agricultural Resources and Environment, Guangdong Academy of Agricultural Sciences, Guangzhou, China

Ziying Li
Shenzhen Academy of Metrology and Quality Inspection, Shenzhen, China

Li-Zi Lin
Joint International Research Laboratory of Environment and Health, Ministry of Education, Guangdong Provincial Engineering Technology Research Center of Environmental Pollution and Health Risk Assessment, Department of Occupational and Environmental Health, School of Public Health, Sun Yat-sen University, Guangzhou, China

Yating Luo
College of Environmental and Resource Sciences, Zhejiang Agriculture and Forestry University, Hangzhou, China

Tu Binh Minh
Faculty of Chemistry, Vietnam National University Hanoi, Hanoi, Vietnam

Tatiana Minkina
Academy of Biology and Biotechnology, Southern Federal University, Rostov-on-Don, Russia

Mehr Ahmed Mujtaba Munir
College of Environment, Zhejiang University of Technology, Hangzhou, China.

Vu Duc Nam
Center for High Technology Research and Development, Vietnam Academy of Science and Technology Hanoi, Vietnam

Da Ouyang
College of Environmental and Resource Sciences, Zhejiang Agriculture and Forestry University, Hangzhou, China

Irina Perelomova
Department of Human Anatomy and Physiology, Medical Institute, Tula State University, Tula, Russia

Leonid Perelomov
Laboratory of Biogeochemistry, Tula State Lev Tolstoy Pedagogical University (Tolstoy University), Tula, Russia

Diane Purchase
Faculty of Science and Technology, Department of Natural Sciences, Middlesex University, London, UK

MD Rokonuzzaman
Department of Science and Environmental Studies, The Education University of Hong, Kong Hong Kong, China

Cheng Shen
Zhejiang Province Key Laboratory of Recycling and Eco-Treatment of Waste Biomass, Zhejiang University of Science and Technology, Hangzhou, China.

Jiaxin Shen
Department of System Engineering, University of Virginia, VA, USA

Jiping Sheng
School of Agriculture and Rural Development (SARD), Renmin University, Beijing, China

Ling Shu
College of Environmental and Resource Sciences Zhejiang Agriculture and Forestry University Hangzhou, China

Kshitij Singh
Department of Environmental Microbiology, School of Earth and Environmental Sciences, Babasaheb Bhimrao Ambedkar University (A Central University), Lucknow, Uttar Pradesh, India

Lingcheng Su
Department of Chemistry, Hong Kong Baptist University, Hong Kong, China

Wenfan Su
Department of System Engineering, University of Virginia, VA, USA

Yuanyuan Tang
School of Environmental Science and Engineering, Southern University of Science and Technology Shenzhen, China

Trinh Thi Tham
Faculty of Environment Hanoi University of Natural Resources and Environment Hanoi, Vietnam

Le Minh Thuy
Faculty of Chemistry, University of Science, Vietnam National University, Hanoi, Vietnam

Thanh Son Tran
Laboratory of Quality Assurance National Institute for Food Control (NIFC), Hanoi, Vietnam

Tran Manh Tri
Faculty of Chemistry Vietnam National University, Hanoi, Vietnam

Martin Tsz-Ki Tsui
Department of Earth and Environmental Sciences, School of Life Sciences, and Institute of Environment, Energy and Sustainability, The Chinese University of Hong Kong, Hong Kong, China

Yulia Venevtseva
Department of Propaedeutic of Internal Diseases, Medical Institute, Tula State University, Tula, Russia

Sidan Wang
Institute of International Relations, China Foreign Affairs University, Beijing, China

Yemeng Wang
The Jockey Club School of Public Health and Primary Care, The Chinese University of Hong Kong, Hong Kong, China

James Tsz-Fung Wong
Department of Civil and Environmental Engineering, The Hong Kong University of Science and Technology, Hong Kong, China

Ming Hung Wong
Consortium on Health, Environment, Education, and Research (CHEER), The Education University of Hong Kong, Hong Kong, China

Anwen Xiao
Institute of Agricultural Resources and Environment, Guangdong Academy of Agricultural Sciences, Guangzhou, China

Mengyao Xing
College of Environmental and Resource Sciences, Zhejiang Agriculture and Forestry University, Hangzhou, China

Yi Yang
State Key Laboratory of Marine Pollution and Department of Chemistry City University of Hong Kong, Hong Kong, China

Zhihong Ye
School of Life Sciences, Sun Yat-sen University, Guangzhou, China

Qifeng Yin
College of Environmental and Resource Sciences, Zhejiang Agriculture and Forestry University Hangzhou, China

Balal Yousaf
Department of Environmental Biotechnology, Faculty of Energy and Environmental Engineering, Silesian University of Technology, Gliwice, Poland

Lihang Yu
The Jockey Club School of Public Health and Primary Care, The Chinese University of Hong Kong, Hong Kong, China

Ruilian Yu
College of Chemical Engineering Huaqiao University, Xiamen, China

Mengwei Yuan
School of Life Sciences, Faculty of Science, The Chinese University of Hong Kong, Hong Kong, China

Haibo Zhang
College of Environmental and Resource Sciences, Zhejiang Agriculture and Forestry University, Hangzhou, China

Jin Zhang
School of Environment and Natural Resources, Zhejiang University of Science and Technology Hangzhou, China

Miao Miao Zhang
Department of Ecology, College of Life Science and Technology, Jinan University, Guangzhou, China

Puqiu Zhao
School of Economics, University of East Anglia, Norwich, UK; and Institute of International Relations, China Foreign Affairs University, Beijing, China

Section I

Environmental Contaminants and Their Management

1

Managing Legacy POPs and Emerging Chemicals: Insights from UNEP/GEF Projects

Ming Hung Wong

1.1 Introduction

Sustainability is guided by three interconnected pillars: social, environmental, and economic. Environmental sustainability emphasizes resource management and pollution control, while social sustainability focuses on equity, justice, and community well-being. Economic sustainability requires businesses to balance profit with environmental and social responsibilities. Among these pillars, ecological sustainability is paramount, as it ensures the maintenance of biological system properties and preserves biodiversity and productivity indefinitely (Clune & Zehnder, 2020).

The rapid socio-economic development over recent decades has exacerbated environmental pollution, posing significant ecological and human health threats. The Lancet Commission on Pollution and Health highlighted that pollution is the leading environmental cause of disease and premature death globally (Landrigan et al., 2018).

Persistent toxic substances (PTS), including heavy metals and persistent organic pollutants (POPs), pose significant environmental and human health threats. Heavy metals, such as mercury, cadmium, and lead, are non-biodegradable and can accumulate in living organisms, leading to toxic effects. POPs, including organic compounds like DDT and dioxins, resist degradation and can bioaccumulate in the food chain. Emerging chemicals of concern, such as flame retardants and bisphenol A (BPA), have not been fully assessed for their environmental and health impacts.

Sources and pathways of various PTS include (1) Industrial emissions: Manufacturing processes, particularly those involving chlorine, release POPs into the atmosphere and waterways (UNEP, 2021). Heavy metals often enter the environment through mining, industrial discharges, and improper waste disposal. (2) Agricultural runoff: Pesticides and fertilizers, particularly those containing POPs, can leach into soil and water systems, impacting aquatic ecosystems (Gilliom et al., 2006). (3) Urban runoff: Contaminants from urban areas, including stormwater runoff, introduce POPs into local waterways (Furlong et al., 2017). (4) Atmospheric deposition: POPs can evaporate and travel long distances, ultimately depositing in remote regions affecting wildlife and ecosystems (Hageman et al., 2015).

The ecological consequences of PTS are alarming, including (1) Biodiversity loss: POPs disrupt food webs and reduce species populations, particularly in birds and fish. (2) Ecosystem disruption: These harmful substances can alter habitat quality and the overall health of ecosystems, making them less resilient to environmental changes (Ghosh et al., 2019). The human health implications of exposure to PTS are profound: (1) Cancer: Numerous studies link POP exposure to increased cancer risks, particularly for breast and prostate cancers. (2) Reproductive disorders: Dioxins and other POPs are associated with reproductive and developmental issues, affecting hormonal balance and fetal development. (3) Neurological problems: Heavy metals like lead and mercury are neurotoxins, leading to cognitive deficits and behavioral issues in children (Landrigan et al., 2018).

The persistent nature and widespread occurrence of various PTS necessitate urgent action to mitigate their impacts on human health and the environment. Enhanced regulatory frameworks and public awareness are essential to address the challenges posed by these hazardous chemicals.

DOI: 10.1201/9781003492115-2

This review examines the critical issues surrounding PTS, including POPs and persistent toxic metals, emerging chemicals of concern, and potential toxic ingredients in personal and pharmaceutical products, drawing insights from the United Nations Environment Programme (UNEP) and Global Environment Facility (GEF) initiatives. The review underscores the importance of international cooperation and sustainable management practices to mitigate these threats. By integrating findings from key literature and ongoing projects, this work aims to provide a comprehensive roadmap for addressing the multifaceted challenges of legacy and emerging contaminants, emphasizing the need for global collective action and sustainable practices.

1.2 Global Initiatives on POPs

The impacts and management of POPs have been a significant focus of international environmental initiatives, particularly following the influential work of Rachel Carson in her book *Silent Spring*. Carson's work highlighted the dangers of chemical pesticides like DDT, which have severe ecological and health impacts due to their persistence, bioaccumulation, and biomagnification in food chains, leading to declines in species such as bald eagles and peregrine falcons due to thin-shelled eggs caused by DDT exposure (Carson, 1962; Derryberry et al., 2020).

1.2.1 Stockholm Convention on POPs

The Stockholm Convention on POPs represents a significant global effort to address these hazardous substances' environmental and health risks. The rationale for the Convention stems from recognizing the severe impacts that POPs have on human health and the environment, necessitating coordinated international action.

The emergence of POPs as a global concern began in the 1970s and 1980s, culminating in the 1996 Intergovernmental Forum on Chemical Safety's conclusion that urgent action was required. This led to establishing the POPs Intergovernmental Negotiating Committee (INC) by the UNEP in 1997, which aimed to develop a comprehensive treaty. The Convention was adopted in May 2001 and entered into force on May 17, 2004, with the initial focus on the 'Dirty Dozen,' a group of 12 harmful POPs, including DDT, PCBs, and dioxins/furans. Table 1.1a lists the 12 chemicals included initially in the Stockholm Convention (2004: https://pops.int/).

The Convention's rationale is based on several key characteristics of POPs: (1) Persistence: POPs resist environmental degradation, remaining in ecosystems for decades or longer. (2) Toxicity: Many POPs are highly toxic, posing significant health risks, including immune system suppression

TABLE 1.1
The Stockholm Convention on POPs- Chemicals Under Control (Stockholm Convention: https://pops.int/)

Pesticides	Industrial Chemicals	
(a) The Dirty Dozen		
1. Aldrin (A)	1. Hexachlorobenzene (HCB) (A) (C)	1. Polychlorinated dibenzo-p-dioxins (PCDD) (C)
2. Chlordane (A)	2. Polychlorinated biphenyls (PCB) (A) (C)	2. Polychlorinated dibenzofurans (PCDF) (C)
3. DDT (B)		
4. Dieldrin (A)		
5. Endrin (A)		
6. Heptachlor (A)		
7. Hexachlorobenzene (HCB) (A) (C)		
8. Mirex (A)		
9. Toxaphene (A)		

(*Continued*)

TABLE 1.1 *(Continued)*

The Stockholm Convention on POPs- Chemicals Under Control (Stockholm Convention: https://pops.int/)

Pesticides	Industrial Chemicals	Unintended By-Products
(b) Newly Added Chemicals (11th Meeting, May 1–12, 2023)		
1. Alpha hexachlorocyclohexane (A)	12. Methoxychlor (A)	
2. Beta hexachlorocyclohexane (A)	13. Pentachlorobenzene (A) (C)	
3. Chlordecone (A)	14. Pentachlorophenol and its salts and esters (A)	
4. Decabromodiphenyl ether (commercial mixture, c-decaBDE) Plus (A)	15. Perluorohexane sulfonic acid (PFHxS), its salts and PFHxS-related compounds (A)	
5. Dechlorane Plus (A)	16. Perfluorohexane sulfonic acid (PFOS), its salts and perfluorooctane sulfonyl fluoride (PFOSF) (B)	
6. Dicofol (A)	17. Perfluorooctanoic acid (PFOA), its salts and PFOA-related compounds (A)	
7. Hexabromobiphenyl (A)	18. Polychlorinated naphthalenes (PCNs)(A)(C)	
8. Hexabromocyclododecane (A)	19. Short-chain chlorinated paraffins (SCCPs) (A)	
9. Hexabromodiphenyl ether and heptabromodiphenyl ether (commercial octabromodiphenyl ether) (A)	20. Technical endosulfan and its related isomers (A)	
10. Hexachlorbutadiene (A) (C)	21. Tetrabromodiphenyl ether and pentabromodiphenyl ether (A)	
11. Lindane (A)	22. UV-328 (UV filter used as an antioxidant for plastics) (A)	

Annex A—*Intentionally produced chemicals that need to be eliminated.*
Annex B—*Intentionally produced chemicals with restrictions.*
Annex C—Unintentionally produced chemicals.

and reproductive issues. (3) Bioaccumulation: These substances accumulate in the fatty tissues of organisms, leading to increased concentrations in food chains, particularly affecting apex predators, including humans. (4) Long-range transport: POPs can be transported over vast distances through atmospheric processes, resulting in contamination far from their sources (Stockholm Convention, 2024: https://pops.int/; Galloway et al., 2017).

The Convention mandates that each party undertake specific actions to mitigate the risks associated with POPs: (1) Prohibition and elimination: Parties must prohibit and eliminate the production and use of intentionally produced POPs. (2) Restriction: The production and use of particular POPs (such as DDT) listed in Annex B must be restricted. (3) Release reduction: Parties must reduce or eliminate releases from unintentionally produced POPs (such as dioxin/furans), utilizing the best available techniques and environmental practices. (4) Management of stockpiles and wastes: Safe and environmentally sound management of stockpiles and wastes containing POPs (such as DDT) is mandated. (5) Identification of additional POPs: The Convention provides procedures for identifying and controlling additional POPs, with the POPs Review Committee responsible for assessing new substances for inclusion.

The Convention mandates specific actions for its parties, including the prohibition and elimination of intentionally produced POPs, restrictions on certain substances, and the safe management of stockpiles and wastes containing POPs. The ongoing scientific review process allows for adding new POPs to the treaty, ensuring that emerging threats are addressed promptly (Derryberry et al., 2020).

Table 1.1b lists the additional chemicals added to the control list (on the 11th meeting, May1–12, 2023) (Stockholm Convention, 2024: https://pops.int/)

The newly added chemicals included perfluoroalkyl and polyfluoroalkyl substances (PFAS), certain polybrominated diphenyl ethers (PBDEs), and highly hazardous pesticides. These additions stemmed from growing concerns about these substances' environmental and health impacts, which are known for their persistence in the environment, potential to bioaccumulate, and adverse effects on human health and ecosystems. The decision reflects a global commitment to phase out harmful chemicals and promote safer alternatives, addressing urgent environmental challenges and protecting vulnerable populations.

The effectiveness of the Stockholm Convention in controlling legacy POPs can be observed through various national and international initiatives that have emerged since its adoption. Many countries have enacted legislation to comply with the Convention's requirements, leading to significant reductions in the production and use of POPs, resulting in a marked decrease in environmental contamination from these substances.

Moreover, the Convention has facilitated international cooperation and knowledge sharing, enabling countries to adopt best practices for managing POPs. The ongoing scientific review process allows for adding new POPs to the treaty, ensuring that emerging threats are addressed promptly. In conclusion, the Stockholm Convention on POPs serves as a critical framework for global action against POPs, demonstrating a commitment to protecting human health and the environment from these hazardous substances.

1.2.2 Regionally Based Assessment of Persistent Toxic Substances (PTS) (2000–2003)

The Regionally Based Assessment of PTS, conducted from 2000 to 2003 under the auspices of the UNEP and the GEF, aimed to elucidate the global impact of PTS through a systematic regional approach.

The project addressed several critical objectives: (1) Assessing the damages and threats: The initiative sought to measure the adverse effects of PTS on human health and the environment. (2) Establishing priorities: It provided a rationale for GEF and UNEP to prioritize future chemical safety and management actions. (3) Regional disparities: The project aimed to identify differences in PTS priorities among various regions. (4) Source identification: It focused on identifying the sources of PTS within the defined regions (Whylie et al., 2003).

The project was particularly significant in assessing the transboundary transport of PTS and the root causes contributing to PTS-related challenges. Additionally, it aimed to identify regional and global priority issues concerning PTS and their management capacities (Dachs & Méjanelle, 2010). This comprehensive assessment complemented international frameworks, including the Stockholm Convention, by evaluating PTS's sources, fates, and ecological and health impacts across 12 world regions. The UNEP Chemicals Division spearheaded the initiative, appointing regional coordinators to enhance management efficiency. These coordinators were drawn from governmental, industrial, and academic sectors, ensuring a multidisciplinary approach.

Data collection on POPs during the project presented numerous challenges. Non-standardized sampling methods, analytical techniques, and data presentation complicate comprehensive efforts to understand PTS impacts. Despite these hurdles, the project yielded 12 regional reports detailing the sources, fates, and effects of various POPs and other PTS, including mercury (Hg), lead (Pb), tributyltin (TBT), PBDEs, and polycyclic aromatic hydrocarbons (PAHs). Each regional report also prioritized additional PTS not listed in the Stockholm Convention's control regime.

For instance, the Central and Southeast Asia region (in which the author served as the coordinator) highlighted specific chemicals of concern, such as HCH, PAH, PBDE, PPCP, organotins, and organic mercury (Wong et al., 2002). The compilation of these regional reports culminated in a global assessment report, which underscored the significance of PTS in environmental policy (Whylie et al., 2003). Notably, it was gratifying to observe that several of these chemicals were eventually included in the Convention's control list (Table 1.1b).

The assessment identified several root causes contributing to the persistence of PTS globally: (1) Chemical persistence: Many PTS are resistant to degradation, leading to long-range transport and disposal challenges. (2) Low water solubility: These substances often exhibit high solubility in fats and oils, resulting in bioaccumulation along food chains (3) High toxicity: The detrimental effects of low concentrations in the environment are often overlooked, as they can be 'invisible' to the public and policymakers, delaying necessary action. (4) Unsustainable production and consumption: Current economic pressures promote using low-cost chemicals, exacerbating global environmental degradation. (5) Cost considerations: The appeal of inexpensive chemical options can overshadow environmental and health considerations. (6) Perceived effectiveness: Some PTS have gained a reputation for ease of use and efficacy, complicating efforts to phase them out. (7) Lack of awareness: There remains a significant gap in public understanding regarding the risks associated with PTS and their potential impacts on health and the environment.

Between 2002 and 2006, workshops were organized to enhance environmental awareness and build capacity to monitor and assess POPs. These included: (1) Workshop on Environmental Monitoring of POPs in East Asia (Tokyo, December 2–4, 2002): Focused on regional collaboration and standardization of methodologies. (2) East Asian Trial POPs Air Monitoring (2004): Established monitoring stations in Beijing, Seoul, Hanoi, Medan, and Jakarta. (3) UNEP Workshop to Develop Global POPs Monitoring Program (Geneva, March 24–27, 2003): Aimed at supporting the effectiveness evaluation of the Stockholm Convention. (4) Workshop on Environmental Monitoring and Risk Assessment of POP Pollutants (Hong Kong, November 23–25, 2006): Addressed ongoing monitoring and risk assessment strategies. (5) Assessment of Existing Capacity and Capacity Building Needs (Beijing, December 13–16, 2005): Focused on identifying gaps in analytical capacity in developing countries.

These initiatives played crucial roles in addressing the global challenges of POPs, demonstrating a commitment to protecting human health and the environment, fostering regional collaboration, and enhancing the overall capacity to manage and mitigate the risks associated with these toxic and persistent substances.

1.3 Emerging Chemicals of Concern

Emerging chemicals encompass many substances increasingly recognized for their potential health and environmental risks, posing threats to wildlife and humans. Emerging chemicals of concern, such as flame retardants, per- and poly-fluoroalkyl substances (PFAS), phthalates, and heavy metals, pose severe health and environmental challenges. Some of these chemicals have been used for a lengthy period and only found out they are not so safe. A typical example is flame retardants (PBDEs). Including specific flame retardants in the Stockholm Convention on POPs marks a critical step toward mitigating their impact on human health and the environment. This indicates the importance of managing various emerging chemicals for protecting ecological and human health.

1.4 Emerging Chemicals Management Issues

Emerging chemicals pose significant health and environmental risks and are increasingly recognized in regulatory frameworks. The project on the Emerging Chemicals Management Issues in Developing Countries and Countries with Economies in Transition (CEIT) sponsored by UNEP/GEF supported the sound management of chemicals throughout their life cycle to minimize significant adverse effects on human health and the global environment (Bouwman et al., 2012).

The Chemical Abstracts Service (CAS) (https://www.cas.org/ja/cas-data/cas-registry) notes that there are 60 million registered organic and inorganic substances globally, but only a small fraction has been evaluated by regulatory bodies. For instance, the USEPA tracks about 84,000 commercially significant chemicals, with only a limited number assessed for toxicity, particularly among high-volume substances.

Countries included in the study included Central and South America, Africa, Asia, Eastern Europe, and Oceania. Table 1.2 lists the 22 Emerging Chemicals Management Issues Identified (with no order of precedence), grouped under process-based, product-based, effect-based, and compound class-based. The most concerning issue is heavy metals in all regions surveyed except Oceania. If excluding Oceania, the ranking of all the 22 issues: heavy metals, PAHs, mixture effects, open burning, endocrine disruption, sewage, inorganic fertilizer, arsenic, E-waste, PPCPs, mine waste, lead in paints, ellict drugs, cadmium in fertilizer, food additives, phthalates, BPA, organotins, marine debris, alkylphenols, ammunition/conflict, and nanoparticle/material.

The increasing complexity of emerging chemical management issues (ECMIs) poses significant risks to human health and the environment, particularly in CEIT, which often lacks adequate regulation and enforcement. Issues such as heavy metals and PAHs are becoming more pressing, with their impacts better understood than in the past. Interventions targeting ECMIs are complicated by their interactions, making isolated solutions ineffective. For instance, regulating harmful E-waste processing in one region may shift the problem elsewhere.

TABLE 1.2

The Emerging Chemicals Management Issues Identified (with No Order of Precedence) (https://www.thegef.org/sites/default/files/publications/GEF_Guidance_on_Emerging_Chemicals-Full-Report_0.pdf)

Processed-Based		Product-Based		Effect-Based		Compound Class-Based	
1	Electronic waste	1	Pb in paints	1	Endocrine disruption	1	PAHs
2	Ammunition, propellants, military equipment, and environ chem legacy of war and conflict	2	Artificial fertilizers			2	Arsenic
3	Mine wastes/drainage	3	Cd fertilizers			3	Bisphenol A
4	Sewage sludge/biosolids—for land application	4	Pharmaceuticals and personal care products			4	Alkylphenols
5	Open burning—with emphasis on open burning of biomass	5	Illicit drugs			5	Parabens
		6	Food additives—melamine in milk			6	Phthalates
						7	PBDEs
						8	TBTs
						9	PFOA/PFOS
						10	Heavy Metals

While localized responses can yield global benefits, copying regulations from developed countries may be counterproductive for developing countries and CEIT, which face unique challenges. Most existing knowledge on ECMIs stems from developed country research, which may not apply effectively to other contexts. Consequently, there is a pressing need for tailored research and monitoring in developing countries, alongside exploring existing multilateral interventions. Effective management options exist but require local adaptation to ensure successful implementation and mitigate the threats posed by ECMIs (Jones, 2019).

In addition to health and environmental risks, the report emphasizes the growing transboundary movement of chemicals and the need for comprehensive regulation. It identifies and prioritizes emerging chemical management issues (ECMIs), particularly concerning the needs of developing countries and CEIT (Garcia et al., 2021).

1.5 Significant Chemical-Related Issues that Pose Risks to Human Health and the Environment (UNEP, 2022)

The Assessment Report on Issues of Concern, published by the UNEP on December 23, 2022, provides a critical overview of 19 significant chemical-related issues that pose risks to human health and the environment (UNEP, 2022: https://www.unep.org/resources/report/assessment-report-issues-concern-chemicals-and-waste-issues-posing-risks-human).

These issues are categorized into two main groups: eight emerging policy issues identified under the Strategic Approach to International Chemicals Management (SAICM) and 11 chemicals with emerging evidence of risks highlighted in the Global Chemicals Outlook II (GCO-II) (Table 1.3).

SAICM and GCO-II are two critical frameworks addressing the complex challenges of chemical management worldwide. SAICM provides a comprehensive strategy for the sound management of chemicals throughout their life cycle, while GCO-II assesses emerging risks and trends related to hazardous chemicals. These initiatives inform policymakers and stakeholders about chemical safety, guiding actions to protect human health and the environment.

TABLE 1.3
Eight Emerging Issues Identified Under the Strategic Approach to International Management (SAICM) and Eleven Chemicals with Emerging Evidence of Risks Highlighted in the GLOBAL CHEMICALS Outlook II (GCO-II) (https://www.unep.org/resources/report/assessment-report-issues-concern-chemicals-and-waste-issues-posing-risks-human)

	Emerging Policy Issues		Emerging Evidence of Risks
1	Chemicals in products	1	Arsenic
2	Endocrine-disrupting chemicals (EDCs)	2	Cadmium
3	Environmentally persistent pharmaceutical pollutants (EPPP)	3	Glyphosate
4	Hazardous substances in the life cycle of electrical & electronic products (HSLEEP)	4	Lead
5	Highly hazardous pesticides (HHPs)	5	Microplastics
6	Lead in paint	6	Neonicotinoids
7	Nanotechnology and manufactured nanomaterials	7	Organotins
8	Per- and polyfluoroalkyl substances (PFASs)	8	Phthalates
		9	Polycyclic aromatic hydrocarbons (PAHs)
		10	Triclosan
		11	Bisphenol A

1.5.1 Strategic Approach to International Chemicals Management (SAICM)

SAICM was established in 2006 during the International Conference on Chemicals Management (ICCM) held in Dubai, United Arab Emirates. It emerged from the recognition that chemicals pose significant risks to human health and the environment, necessitating a comprehensive global framework for their management. The primary objective of SAICM is to promote the sound management of chemicals throughout their life cycle, from production and use to disposal. It aims to ensure that all chemicals are produced and used in ways that minimize adverse effects on human health and the environment by 2020 and beyond. SAICM encourages countries to develop national and regional strategies for the management of chemicals. It facilitates collaboration among stakeholders, including governments, industry, and civil society, to share best practices and enhance capacities. The framework also emphasizes integrating chemical management into broader sustainable development and public health initiatives.

Emerging Policy Issues under SAICM included the following (Table 1.3):

1. Chemicals in products: This issue addresses the presence of hazardous chemicals in consumer products, which can lead to exposure during manufacturing, use, and disposal. Effective management strategies are needed to minimize risks to health and the environment.
2. Endocrine-disrupting chemicals (EDCs): EDCs interfere with hormonal systems and can result in developmental, reproductive, and neurological disorders. The growing evidence of their prevalence in various environments necessitates global action to regulate and reduce exposure.
3. Environmentally persistent pharmaceutical pollutants (EPPP): Pharmaceuticals can persist in the environment, leading to contamination of water sources and impacts on aquatic life. This issue underscores the need for improved waste management practices and more environmentally safer pharmaceuticals.
4. Hazardous substances in the life cycle of electrical and electronic products (HSLEEP): The electronics lifecycle often involves hazardous substances that pose risks during production, use, and disposal. Addressing these risks requires a comprehensive approach to product design and recycling.

5. Highly hazardous pesticides (HHPs): HHPs are known for their severe health and environmental impacts. Their regulation is crucial to protect vulnerable populations, particularly in agricultural communities where exposure is high.
6. Lead in paint: Lead exposure from paint is a significant public health issue, particularly in children. Efforts to ban lead-based paints and promote safer alternatives are essential for preventing lead poisoning.
7. Nanotechnology and manufactured nanomaterials: While nanotechnology offers innovative solutions, the potential health and environmental risks associated with manufactured nanomaterials are not fully understood. Ongoing research and regulation are needed to ensure safety.
8. Per- and polyfluoroalkyl substances (PFASs): Often referred to as 'forever chemicals,' PFASs are resistant to degradation and accumulate in the environment and human bodies. Their widespread use raises concerns about long-term health effects, prompting calls for regulatory action.

1.5.2 Global Chemicals Outlook II (GCO-II)

GCO-II is the second edition of the Global Chemicals Outlook report, published by the UNEP in 2019. It builds on the first GCO published in 2013 and responds to the growing concern about chemicals' impacts on health and the environment. The GCO-II aims to comprehensively assess the global status of chemicals management and highlight emerging issues and risks associated with hazardous chemicals. It is a crucial resource for policymakers, industry leaders, and civil society to understand the challenges and opportunities in managing chemicals sustainably. The report identifies critical areas requiring attention and action, including specific chemicals of concern and broader chemical production and use trends. It underscores the need for enhanced regulatory frameworks, better data collection, and increased stakeholder engagement. GCO-II emphasizes transitioning to safer alternatives and promoting green chemistry practices.

Emerging Evidence of Risks Identified by GCO-II (Table 1.3) included the following:

1. Arsenic: Exposure to arsenic, primarily through contaminated water and food, poses serious health risks, including cancer and developmental issues. Mitigation strategies are urgently needed in affected regions.
2. Cadmium: This toxic metal accumulates in the body and can cause kidney damage and bone weakening. Its presence in batteries and certain agricultural products demands stringent regulations.
3. Glyphosate: As a widely used herbicide, glyphosate has raised concerns regarding its carcinogenic potential. Ongoing research and regulatory scrutiny are necessary to assess its safety.
4. Lead: Beyond paint, lead exposure occurs through various channels, including contaminated soil and water. Comprehensive strategies are required to reduce exposure and remediate affected areas.
5. Microplastics: These tiny plastic particles are pervasive in the environment and can enter the food chain, posing risks to human health and ecosystems. Research into their sources and impacts is critical for effective management.
6. Neonicotinoids: This class of pesticides has been linked to declines in pollinator populations. Their regulation is essential to protect biodiversity and food systems.
7. Organotins: Organotins are used in various industrial applications. They are toxic and can disrupt endocrine function. Their use in antifouling paints and other products requires careful management.
8. Phthalates: Commonly found in plastics, phthalates are associated with reproductive and developmental issues. Regulatory efforts are needed to limit their use in consumer products.
9. Polycyclic aromatic hydrocarbons (PAHs): These harmful compounds are produced from burning fossil fuels and can cause cancer and other health problems. Efforts to reduce emissions and exposure are vital.

10. Triclosan: An antibacterial agent found in many consumer products, triclosan has been linked to antibiotic resistance and endocrine disruption. Its use should be carefully evaluated and regulated.
11. BPA: Commonly used in plastics, BPA exposure has raised concerns about its effects on reproductive health and development. Alternatives should be promoted to minimize risks.

Both SAICM and GCO-II play vital roles in addressing the challenges of chemicals in our world. While SAICM sets a strategic framework for sound management, GCO-II provides essential information on emerging risks and opportunities for improvement. Together, they aim to foster a safer and healthier environment for current and future generations.

The Assessment Report on Issues of Concern is crucial for informing international discussions about chemical management. Highlighting these emerging policy issues and risks encourages stakeholders to take action toward safer chemical practices and better protection of human health and the environment. Addressing these challenges requires collaboration among governments, industries, and civil society to promote sustainable solutions and safeguard future generations.

1.6 Plastics and Associated Toxic Chemicals

Recently, much attention has been paid to the misuse and overuse of plastics and the ecological and human implications worldwide. The predominant linear model of plastic use—characterized by a sequence of 'take, make, use, and dispose'—has become a significant driver of natural resource depletion, environmental degradation, and health crises.

The production and disposal of plastics are linked with our circular economy with several significant environmental issues, including (1) Releasing toxic chemicals and waste: Some POPs are used as additives in some plastics, and dioxins/furans are by-products of PVC production. (2) Affecting climate change: Producing plastics using fossil fuels, open burning, and incineration of plastic wastes emit greenhouse gases (390m t of CO_2 in 2012). (3) Infringing on International waters: widespread plastics pollution. (4) Lowering biodiversity: It is the 2nd most crucial threat to coral reefs and affects climate change due to entanglements and ingestion of plastics by aquatic animals. (5) Accelerating land degradation and damaging food systems: Microplastics induce land degradation and food contamination (Barra et al., 2018).

Plastics enter ecosystems through various pathways: (1) Improper disposal: Landfills and open burning of plastic waste release harmful chemicals into the atmosphere. (2) Microplastics: Microplastics are tiny plastic particles (<2 mm) resulting from the degradation of more oversized plastic items. They can be ingested by marine organisms or taken up by crops, leading to bioaccumulation and transferring to humans. MPs and associated chemicals have potential health hazards. The proliferation of plastics has led to the release of associated toxic chemicals into the environment. BPA, phthalates, and adsorbed POPs accumulate in bivalves and are transferred to humans, causing health problems (Zhang et al., 2020).

Microplastics may induce inflammation, genotoxicity, oxidative stress, apoptosis, and necrosis in human tissues. MPs may cause a shift in microbial composition (e.g., lung and gastrointestinal tract). In addition to heavy metals and POPs attached to the small particles, there are two major toxic chemicals associated with microplastics: (1) BPA (a constituent of plastics) is found in various consumer products; it is related to reproductive toxicity. (2) Phthalates are plasticizers that can leach into the environment, accumulate in living organisms, and are linked to reproductive health issues, particularly in children. Both are recognized as endocrine disrupters, which lower the reproductive success of all animals, including humans. The chemicals associated with plastics may impose longer-term health impacts: birth defects, increased risk of breast cancer, increased risk of prostate cancer, skin diseases, lowered IQs of developing fetuses, increased infertility in males, memory loss, deafness, etc. (Wright & Kelly, 2017; Dueñas-Moreno et al., 2023).

The contamination of aquatic environments with microplastics raises concerns about the safety of drinking water and food sources. The World Health Organization (WHO) is assessing the potential health risks of microplastics, underscoring the urgency of addressing this issue.

TABLE 1.4

Key International Conventions Focused on Controlling Toxic Chemicals

	Enforcement Year	Convention	Toxic Chemicals
1	2006 (ongoing)	Strategic approach to international chemicals management (SAICM)	A broad range of chemicals promoting sound management
2	5 May 1992	Basel Convention	Hazardous wastes, including toxic chemicals
3	24 Feb 2004	Rotterdam Convention	Certain hazardous chemicals and pesticides
4	17 May 2004	Stockholm Convention	Persistent organic pollutants (POPs)
5	16 Aug 2017	Minamata Convention	Mercury and mercury compounds
6	March 2022	United Nations Environment Assembly (UNEA) resolution on plastic pollution	It aims to create a legally binding treaty to address plastic pollution globally
7	Ongoing (under negotiation)	Global plastics treaty	A proposed treaty addresses the entire lifecycle of plastics, from production to disposal

1.7 International Conventions and Management Strategies

1.7.1 International Conventions

International conventions are essential for the sound management of toxic chemicals worldwide. Table 1.4 lists the critical international conventions for such a purpose.

1. Strategic Approach to International Chemicals Management (SAICM) (https://www.saicm.org/)

   SAICM is a broad framework that promotes the sound management of chemicals throughout their lifecycle. It encourages national governments to implement policies and practices for chemical safety, supports capacity building, and encourages stakeholder engagement. It covers a wide range of chemicals, focusing on promoting sustainable practices and minimizing risks.
2. Basel Convention (https://www.basel.int/)

   Basel Convention is a treaty that controls the transboundary movement of hazardous wastes and their disposal. It requires countries to minimize hazardous waste generation and ensure environmentally sound management. It includes a variety of hazardous wastes, particularly those containing toxic chemicals like heavy metals and certain industrial by-products.
3. Rotterdam Convention (https://www.pic.int/)

   Rotterdam Convention focuses on promoting shared responsibilities in the international trade of certain hazardous chemicals. It implements a prior informed consent (PIC) procedure, requiring exporting countries to obtain consent from importing countries before shipping toxic chemicals. It covers specific hazardous chemicals and pesticides identified in the annexes of the Convention.
4. Stockholm Convention (https://pops.int/)

   Stockholm Convention is a global treaty aimed at eliminating or reducing the release of POPs. It establishes measures for reducing and removing POPs, including restrictions on production and use. It targets chemicals known for their persistence in the environment and potential harm to human health, such as DDT and PCBs.
5. Minamata Convention (https://minamataconvention.org/en)

   Minamata Convention focuses on protecting human health and the environment from the adverse effects of mercury and mercury compounds. It includes provisions for reducing mercury emissions, phasing out mercury-containing products, and managing mercury waste. It primarily addresses mercury and its compounds, which are known for their toxic effects.
6. United Nations Environment Assembly (UNEA) Resolution on Plastic Pollution (https://www.unep.org/inc-plastic-pollution)

   It aimed at creating a legally binding treaty to address the global issue of plastic pollution. It advocates for international cooperation to manage plastic pollution, including reducing plastic production and enhancing recycling efforts. It primarily addresses plastics and associated chemicals that contribute to environmental pollution.

7. Global Plastics Treaty (https://www.worldwildlife.org/pages/global-plastics-treaty)
   It is a proposed treaty addressing the entire lifecycle of plastics, from production to disposal. It seeks to establish binding commitments for countries to manage plastic waste effectively and reduce plastic pollution. It encompasses many plastics and their chemical additives, which can harm the environment and human health.

These conventions collectively aim to manage and mitigate the risks associated with toxic chemicals and promote safer practices globally.

1.7.2 Management Strategies

Managing POPs and emerging chemicals of concern is a pressing challenge requiring a multifaceted approach. Several vital strategies are essential for effective management (Zhang et al., 2022; UNEP, Strategic Approach to International Chemicals Management: https://www.saicm.org/):

1. Robust regulatory frameworks: Strengthening regulations, such as those established by the Stockholm Convention, is vital for ensuring compliance and protecting public health and the environment. Regulatory frameworks provide the essential guidelines to combat the risks associated with these pollutants.
2. Public awareness and education: Raising awareness about the dangers posed by POPs and emerging chemicals is crucial. Educational initiatives empower consumers to make informed choices, fostering a society that values health and sustainability. Engaging the public can lead to behavioral changes and support for necessary policy reforms.
3. Sustainable practices: Adopting sustainable practices, including reducing plastic use and improving waste management, can significantly mitigate the environmental impacts of these pollutants. Transitioning toward more sustainable production and consumption patterns is essential for long-term ecological health.
4. Research and monitoring: Continued research and monitoring are critical for understanding the complex interactions among multiple pollutants. This knowledge is necessary for effective risk assessments and for informing future regulatory decisions. Innovation in identifying new pollutants and developing safer alternatives is vital to advancing public safety.

1.8 General Conclusion

Managing PTS, including POPs and emerging chemicals of concern, is a critical global challenge that requires immediate and coordinated action. The ecological and health impacts of these substances are profound, with evidence linking exposure to serious health issues such as cancer, reproductive disorders, and neurological problems. The persistence and bioaccumulation of these chemicals in the environment exacerbate their risks, necessitating robust regulatory frameworks and international cooperation.

The Stockholm Convention on POPs serves as a pivotal framework for global action, mandating the prohibition and restriction of harmful substances while facilitating the identification of new chemicals of concern. Recent additions to the Convention, including perfluoroalkyl and PFAS, reflect a growing recognition of the need to address emerging threats. However, the effectiveness of such international agreements hinges on the commitment of individual nations to implement and enforce regulations that protect public health and the environment.

Moreover, the increasing complexity of ECMIs, particularly in developing countries and countries with economies in transition (CEIT), underscores the necessity for tailored approaches that consider local contexts and challenges. Public awareness and education play crucial roles in fostering a culture of sustainability and encouraging responsible consumption practices.

In conclusion, addressing the challenges posed by PTS and emerging chemicals requires a multifaceted strategy that includes strengthening regulatory frameworks, enhancing public awareness, promoting sustainable practices, and fostering international collaboration. By prioritizing these actions, we can mitigate the risks associated with toxic chemicals and work toward a healthier, more sustainable future for all.

Acknowledgments

The author thanks his former postgraduate students who have been involved in various research projects. Financial support from UNEP/GEF, the National Science Foundation of China, the Research Grants Council of Hong Kong, the Croucher Foundation of Hong Kong, and the British Council is gratefully acknowledged.

REFERENCES

Barra R, Leonard SA, Whaley C, Bierbaum R (2018). *Plastics and The Circular Economy*. UNEP/GEF, Washington, DC.

Bouwman H, Wong MH, Barra R (2012). *GEF Guidance on Chemicals Management Issues in Developing Countries & Countries with Economy in Transition*. UNEP/GEF, Washington, DC.

Carson R (1962). *Silent Spring*. Houghton Mifflin Internet Archive, Boston, MA.

Clune WH, Zehnder AJB (2020). The evolution of sustainability models, from descriptive, to strategic, to the three pillars framework for applied solutions. *Sustain Sci* 15, 1001–1006.

Dachs J, Méjanelle L (2010). Global fate of persistent organic pollutants. *Environ Sci Technol* 44(15), 5635–5641.

Derryberry EP, et al. (2020). The legacy of Rachel Carson: A review of the impacts of DDT on wildlife. *Environ Sci Policy* 112, 1–10.

Dueñas-Moreno J, Mora A, Kumar M, Meng XZ, Mahlknecht J (2023). Worldwide risk assessment of phthalates and bisphenol A in humans: The need for updating guidelines. *Environ Int* 181, 108294.

Furlong ET, et al. (2017). Urban runoff and water quality. *Environ Monit Assess* 189(4), 1–15.

Galloway TS, et al. (2017). Bioaccumulation of persistent organic pollutants in marine organisms. *Mar Pollut Bull* 114(1), 1–10.

Garcia M, et al. (2021). Transboundary movement of chemicals: Implications for developing countries and CEIT. *J Chem Manage Regul* 10(4), 287–301.

Ghosh U, et al. (2019). Impact of persistent organic pollutants on biodiversity. *Environ Sci Technol* 53(12), 6789–6798.

Gilliom RJ, et al. (2006). *Pesticides in the Nation's Streams and Ground Water, 1992–2001*. U.S. Geological Survey Circular 1291, Reston, VA.

Hageman KJ, Bogdal C, Scheringer M (2015). Long-range and regional atmospheric transport of POPs and implications for global cycling. *Comprehen Anal Chem* 67, 363–87.

Jones A (2019). Challenges in environmental chemical management: Insights from developing countries. *Inter J Environ Res* 27(2), 67–81.

Landrigan PJ, et al. (2018). The lancet commission on pollution and health. *Lancet* 391(10119), 462–512.

UNEP (2021). *Global Chemicals Outlook II: From Legacies to Innovative Solutions*. United Nations Environment Programme, Washington, DC.

Whylie P, Albaiges A, Barra R, Bouwman H, Dyke D, Wania F, Wong MH (2003). *Global Report*. UNEP/GEF, Washington, DC.

Wong MH, Choi K, Grosheva E, Saki S, Shibata Y, Suzuki N, Wang J, Zhou H, Leung A (2002). *Regional Report, Central & NE Asia*. UNEP/GEF, Washington, DC.

Wright SL, Kelly FJ (2017). Plastics and human health: A micro issue? *Environ Sci Technol* 51(12), 6634–6647.

Zhang F, Man YB, Mo WY, Man KY, Wong MH (2020). Direct and indirect effects of microplastics on bivalves, with a focus on edible species: A mini-review. *Crit Rev Environ Sci Technol* 50(20), 2109–2143.

Zhang Z, Malik MZ, Khan A, Ali N, Malik S, Bilal M (2022). Environmental impacts of hazardous waste, and management strategies to reconcile circular economy and eco-sustainability. *Sci Tot Environ* 807(2), 150856.

2

International Institutional Linkages between Ecological Challenges and Human Health

Sidan Wang and Puqiu Zhao

2.1 Introduction

There is a growing awareness of the ecological, food, and public health challenges. The contaminated topsoil and increasing demand for food require rethinking global food systems and sustainable agricultural practices. The UN Secretary-General Antonio Guterres urgently needed immediate action to address the global food security emergency in 2020 (UN 2020). Undoubtedly, food-related security stands as the primary agenda for public health. Like many international environmental issues, because of their centrality to human and societal well-being, pressing global health challenges always spill over the national borders (Davies et al. 2014). Thus, international institutions and cooperation on ecological challenges and human health issues are crucial.

However, it should be noted that public health can be influenced by various factors that interact in complex ways and vary in different contexts (Davies 2010). Food safety can be considered a highly political issue, more than a scientific issue, including the fragmented authority and the conflicts between the corporate and public interests from an economic perspective (Nestle 2010). Generally speaking, there is a relatively small body of discussions on food safety being structured into the frameworks of political analysis. The existing studies have mainly focused on the actors addressing human health issues, driving motivations behind global food safety issues, and international relations approaches to food safety. Observing how the actors have been constructing ideas to raise the linkages between ecological challenges and human health issues is also essential. The following sections will review the existing studies.

2.1.1 Actors: Constructing Human Health Governance

Research on actors in global health governance and food safety is critical for understanding how human health problems can be linked with international institutions and global politics. As it addresses, an accurate depiction of global politics must consider how states affect their citizens and engage with other actors in addressing the issues of impacting the well-being of those citizens (Davies 2010). As the main actors, the state and the national government could not be the only actors in global health issues because the global health community has already extended far beyond government (Davies 2010; Novotny 2007). Besides, a more comprehensive range of international societal norms and political and economic considerations should be pursued (Davies 2010). This requires observing how different actors have been engaged in global health governance. In this sense, beyond environmental, biological, and social factors, political and organisational factors play an essential role in the sustainability of public health problems (Hardon and Blume 2005; Le Loup et al. 2010). International public health governance actors should include foreign governments, non-governmental organisations (NGOs), pharmaceutical companies, private donors, and global organisations (Davies 2010). They can be the World Bank, the International Monetary Fund (IMF), the World Heart Federation, the World Medical Association, the World Health

DOI: 10.1201/9781003492115-3

Organization (WHO), the International Federation of Pharmaceutical Manufacturers Association (IFPMA), the United Nations International Children's Emergency Fund (UNICEF) and the Food and Agricultural Organization (FAO) (Harris 1991; Lee 2003; Brugha 2008).

Many international actors can have the power to shape the agenda of global health. In addition to the actors, institutions can be the rules of the game in society, and the constraints that are humanly devised shape human interaction (North 1990). For example, the World Bank has been evolving to be a large financial contributor to global public health, and it even plays an important role in producing knowledge of global health and influencing the agenda of global health governance (Tichenor et al. 2021). From the perspective of actors, it is apparent that international institutions are crucial actors and have been interacting with the global health agenda. While global public health governance has witnessed its links to various international institutions, it is crucial to discover whether the health agenda and the WHO have been engaged in global environmental governance. This work discusses how international environmental institutions have seen human health issues.

2.1.2 Ideas: Driving Motivations Behind Different Actors

The driving motivations for establishing the linkages between the international institutions and the global health issue can be, to some extent, explained by political and institutional considerations. The members of the WHO have their obligations to be consistent with international agreements and provisions for the protection of human life and health, and they thus need to take sanitary actions regarding food and ecological challenges. However, the driving motivation can be the power competition in making rules. One significant change in the global health arena is the shift from the traditional concepts of social justice and equity in public health provision towards markets and efficiency (Poku and Whiteside 2002). From the international trade and economic view, food companies can use their financial power to influence governmental regulations and thus impact public interest (Nestle 2010). Due to this, governments have a dilemma between encouraging the trade of foodstuff and protecting the consumer (Gray 1999). While economic growth seems essential to improving global health governance, the political agenda can be prioritised, including social mobilisation and political participation. For instance, the consequences of adopting social mobilisation as a vehicle for fulfilling the right to freedom from diseases have far-reaching effects on national politics and international relations (Mills 2005; Le Loup et al. 2010).

Moreover, the collaboration of states between the member states in the international institutions, especially those developed and developing countries, is crucial for implementing the regulations and strengthening the capacity and network in terms of global health governance. International institutions have the potential to facilitate cooperation and have witnessed the importance of international collaboration (Keohane 1988). The global scope of the problems indicates that transnational institutions are essential (Harris 1991). As the critical international relations theory, the green theory focuses on the ecological challenges and argues that the state system is locked up to deal with the ecological crisis (Burchill 2013). Though decentralisation, in principle, can make health systems more responsive to local needs, it always causes governance problems. It exacerbates local institutional weaknesses, undermining the right to health and the capabilities of addressing global health issues (Stubbs and Kentikelenis 2017). The transformation of international organisations, such as the new international health regulations in the WHO, provides institutional frameworks to enhance international cooperation in global health, strengthen the national health systems, and boost horizontal and vertical health governance (Fidler and Gostin 2006). Beyond the human health issues, a range of ecological challenges, including climate change, biodiversity, desertification, and plastic pollution, have been observed and integrated into this work to understand the linkages between ecological and human health issues.

2.1.3 Practice: International Relations Approach

Based on the actors and driving motivations, the interaction approach and the power-orientated framework can be used to explain the linkages between ecological challenges and global health issues through international institutions. Global health issues have been overlooked because they do not fit into the dominant state-centric approach of the discipline of international politics. This challenges the conventional wisdom of international relations and requires understanding, beyond the state-centric paradigm, social, economic, and political linkages between local, national, and international activities (Thomas 1989). This is where international relations can contribute profound insights into how and why these actors have accumulated such power and how this blurring of the domestic and international realms of health concerns has affected areas of global health (Davies 2010).

On the one hand, the international relations scholarship can improve understanding of the politics around global health issues and explain why attempts to improve global health are always a bit messier and even 'dirtier' than they appear on paper. On the other hand, it can deepen the understanding of global health governance by exposing the contrasting logics and rationalities underpinning international politics and health (Davies et al. 2014). Previous studies have argued that global health issues should be analysed using international relations approaches and frameworks (Thomas 1989; Lee 2003; Davies 2010). The international politic of health can be understood as the statist and primarily security-focused approach and the globalist approach to focusing on the well-being and rights of individuals (Davies 2010). The globalist perspective starts with individual health needs and then considers how global actors and structures impact individuals. It also observes the intensity of industry pressure and the degree of public demand for governmental intervention in social policy.

In addition to the potential approaches to international institutions regarding global health issues, Keohane (1988) offers two main approaches to epistemology: rationalistic theory and reflective theory. Rationalist theory can be employed to understand the commitment problem and the direction of change in world politics through empirical work and actual practice in international institutions. The reflective theory has been developed based on the weakness of the rationalist theory. Notably, the green theory can be seen as the reflective theory in the studies on international relations around ecological challenges. This theory shows a critical stance towards the traditional theories of international relations like classical realism and neo-realism. It argues that eco-centrism and prioritising the environment from the social perspective can be, more than the power, states, and resistance, recognised as a core of the green theory (Eckersley 2021). However, Laferrière and Stoett (1999) argue that theories, like green theory, focusing on respecting the environment and creating an ecological society, fail to be achieved in the modern, technologically advanced world. The key should be the relevance of ecological considerations in determining the order of the political community (Laferrière and Stoett 1999).

Another approach to epistemology is examining the effectiveness of international regimes and institutions. Jackson and Bührs (2015) discuss institutional and ecological effectiveness in terms of the effectiveness framework of main environmental regimes. Institutional effectiveness refers to the ability of international organisations to implement and enforce their rules and regulations effectively. Ecological effectiveness refers to the extent to which an international environmental regime and agreement successfully achieves positive ecological outcomes and impacts. They argue that ecological factors should be integrated into definitions and evaluations of international institutions based on empirical data and indicators. Other studies discuss the institutional effectiveness of addressing abrupt ecological crises. Galaz et al. (2011) argued that the ecological crisis challenges the monitoring system, the coordination across multi-level actors, and their decision-making capacity. Their research also sketched the framework of the effective international institutions dealing with the ecological crisis, which should include the warning system, decision-making system, and post-crisis communication and learning system.

2.1.4 Research Gaps

While the existing studies focus on the role of the WHO and global health issues in international relations, there are theoretical, methodological, and empirical gaps in international institutions regarding ecological challenges and human health. First, there is a knowledge gap in the research on the actors of global health and ecological politics. While much research on the actors in global ecological and health governance can be identified, there is a lack of research on how and why the various international actors can interact. Although several studies see global health governance as a multi-level governance system (Amaya and De Lombaerde 2019; Ferreira 2022), the traditional framework of understanding global health politics has been challenged and outdated, especially during and after the Covid-19 (Davies and Wenham 2020; Papamichail 2023). This also implies global environmental governance. Thus, new legal and institutional structures are required to maintain the resilience of the policy-making process of ecological governance (Woolley 2014). Discussing and exploring the new perspective of political analysis and international relations is vital and meaningful for addressing global human health issues. Second, there is a theoretical and analytical gap in the research. The existing studies focus on the approach to structural international relations. The globalist, rationalistic, and effectiveness approaches employ the structural approach to regard the nations and other actors as the same functional unit.

However, domestic policies cannot be ignored by the global health governance issues. Bernstein and Cashore (2012) argue that global environmental governance has been involved in the external behaviour of actors, domestic practices, policies, and policy-making processes. For instance, Haas (1989) identifies the epistemic community as combining power and knowledge for the ecological policy-making system. Finally, there is an empirical gap in the international institutional linkages between ecological challenges and human health. While the existing studies on global health governance have been focusing on infectious disease regulations and treaties (Davies 2008; Davies and Wenham 2020; Rubin and Saidel 2016; Fraundorfer 2022), this chapter thus has the potential to observe and assess the international institutional linkages between the ecological governance and human health. Following discussions on the linkages between the institutional frameworks under United Nations Environment Programme (UNEP) and WHO, this work selects the Minamata Convention on Mercury as a case to observe the relationships.

2.2 Environment-Based International Institutions and Their Implications for the Linkages between Global Environmental Challenges and Public Health

While international environmental institutions can be identified widely across a range of global environmental challenges, this work has to focus on a small number of cases to observe the linkages between global environmental challenges and public health. The UNEP has seen its key role in raising the international agenda on environmental protection and supporting the related institutional institutions. Thus, it is identified as a significant case to discuss and observe the linkages. Beyond the UNEP, other international environmental arrangements deserve receiving attention. In 1992, Rio Earth Summit achieved the United Nations Framework Convention on Climate Change (UNFCCC), the Convention on Biological Diversity (CBD), and the United Nations Convention to Combat Desertification (UNCCD). Inter-governmental platforms for scientific research and evidence, such as the Intergovernmental Panel on Climate Change (IPCC), have also been engaged in constructing global environmental governance.

This critical moment is a milestone in global environmental governance and paves the way for international negotiations. Following these institutional arrangements and topics, this work discusses an ongoing framework of negotiating the international plastics instrument and discovers its links to public health. Therefore, to understand the interlinks between global environmental governance and public health issues, this work observes how public health issues have been constructed in the documents concerning the given international institutions and related environmental challenges. The documents were collected via the Earth Negotiations Bulletin (ENB) database, which recorded key statements from various global environmental negotiations (International Institute for Sustainable Development 2024). Based on this, the documents used in the work are called the ENB documents.

2.2.1 United Nations Environment Programme (UNEP)

Regarding the environment-based international institutions, the primary attention is given to UNEP. UNEP has recognised the risks of damaging ecosystems to health and the benefits of conserving ecosystems for public health (UNEP 2020). This shows a close link between health and the environment. And beyond this link, UNEP also recognises the importance of establishing a link between pollution and health. It raises environmental challenges, including air pollution, water and ocean contamination in public health fields (food chain, environmental impacts on human health, and biological and chemical contributions to antimicrobial resistance) (UNEP 2024b).

TABLE 2.1
Environment-Based International Institutions and Their Implications for Human Health Issues

Environment-Based International Institutions	Their Implications for Human Health Issues
UNFCCC (and IPCC)	Impacts of climate change over diseases including waterborne diseases, non-communicable diseases, premature deaths, malnutrition, and mental health Climate-induced conflicts and violence Climate adaptation and climate-resilient food and food supply
CBD	Biodiversity-related impacts on public health around freshwater, wetlands, air pollution, agriculture, food security, nutrition, infectious diseases, microbial diversity, biomedical resources, pharmaceuticals, traditional medicine and green spaces for public health
UNCCD	Climate change, food security, environmental pollution, water supply and pandemics Sand and dust storms and their impacts on health and the environment Impacts of land degradation on food security
Plastics negotiations	Plastic pollution and public health This linkage requires more research to support the institutional construction

UNEP has been collaborating with the Food and Agriculture Organization (FAO) of the United Nations, the WHO, and the World Organisation for Animal Health (WOAH) to address health challenges by integrating environmental crises, human health, planetary boundaries, and animals. They issued *One Health Joint Plan of Action (2022–2026)* in 2022, requiring actions on improving health systems, controlling diseases and pandemics, mitigating risks of food safety, managing antimicrobial resistance, and raising the importance of environmental challenges (FAO et al. 2022).

UNEP has implemented a range of institutional arrangements, including the Ministerial Declaration on Health, Environment, and Climate Change achieved around the 2016 UN climate conference, the new Memorandum of Understanding with WHO in 2018, and the Implementation Plan 'Towards a Pollution-free Planet' adopted by the fourth session of the United Nations Environment Assembly in 2019 (UNEP 2020). They have implications for UNEP to play a role in constructing the linkages between ecological challenges and human health, and they also show the contributions from various international environmental institutions to strengthen the linkages.

Beyond UNEP, various international environmental institutions have been involved in constructing the linkages between ecological challenges and human health. Table 2.1 shows that these international institutions have seen human health issues from their distinctive functional perspectives. The detailed linkages across international institutions are summarised in the following few sections.

2.2.2 Global Climate Governance and UNFCCC

The linkages between climate change and public health have been confirmed by scientific research. The IPCC Sixth Assessment Report offered a chapter on health, well-being, and the changing structure of communities. It revealed a close link between climate impacts and a wide range of public health challenges with high confidence in scientific certainty (Intergovernmental Panel on Climate Change 2023). According to this report, climate impacts can be linked to infectious diseases such as waterborne diseases, non-communicable diseases (cardiovascular diseases, respiratory diseases, cancer, and diabetes), premature deaths, malnutrition, and mental health. Beyond these health challenges, climate-induced conflicts and violence have been categorised as challenges to public health and human well-being. The IPCC report shows a very close link between climate change and public health and indicates the impacts of climate crises on health challenges.

To some extent, the scientific dimension of the linkage between climate change and health has been translated into political discussions. Before the 2015 Paris Climate Conference, the expected international climate agreement was linked to the importance of public health issues, and it was thus called a public health agreement (UNFCCC 2015). Institutionally, WHO and UNFCCC have collaborated to address climate and health challenges (WHO 2020b). These arrangements reflect a positive trend in global climate and health governance. However, the linkages between climate and health fail to be substantially identified in international climate negotiations.

Health was linked to adaptation during the 2023 UN climate change conference in Dubai. According to the ENB documents, the negotiations discussed the global goal of adaptation and improving resilience to climate-related health impacts. Also, the Global Goal on Adaptation was required to consider climate-resilient food and food supply problems. In this sense, the negotiations focused on health issues and food safety in the discussions around adaptation to climate change. In contrast to the interlinks between climate change and public health identified in the UNEP and UNFCCC documents, the climate negotiations have paid much less attention to the health challenges.

2.2.3 Global Biodiversity Governance and CBD

Like UNFCCC, CBD has established institutional connections with WHO. CBD has established a joint work programme with WHO and released the Connecting Global Priorities: Biodiversity and Human Health Report 2015, CBD and WHO established the Interagency Liaison Group on Biodiversity and Health. They invited other international stakeholders, including FAO and UNFCCC, to support the construction of this group (CBD 2017). CBD has been accelerating collective actions to address biodiversity and health issues and raising the construction of a global action plan to mainstream linkages into national actions (CBD 2021).

Addressing health issues seems to be an essential field in global biodiversity governance. UNEP, WHO, and the secretariat of CBD have been working on identifying the main biodiversity-related impacts on public health around freshwater, wetlands, air pollution, agriculture, food security, nutrition, infectious diseases, microbial diversity, biomedical resources, pharmaceuticals, traditional medicine, and green spaces for public health (WHO and CBD 2015). In other words, biodiversity loss negatively impacts public health in the environmental, agricultural, food, microbial, and medical dimensions. This shows the linkages between biodiversity and health have a wide range of specific actions to be considered.

Based on the existing institutional arrangements, the importance of health in biodiversity governance has been reflected in the UN biodiversity negotiations. According to the ENB document, the 2022 UN biodiversity conference raised linkages between biodiversity and public health in negotiations. The Intergovernmental Science-Policy Platform on Biodiversity and Ecosystem Services (IPBES) required employing the nexus assessments on biodiversity, water, food, and health. This conference discussed the cooperation with other international organisations and particularly emphasised the role of the WHO in the negotiations on access to pathogens and utilisation of genetic resources. It is important to note that the linkage between biodiversity and health has been categorised as an essential theme of the negotiations. The conference called for strengthening, mainstreaming, and mobilising financial support to the linkages between biodiversity and health. In contrast to receiving less attention from the climate negotiations, health issues have been seen as influential on the agenda of the biodiversity negotiations.

2.2.4 Global Land Governance and UNCCD

The UNCCD has identified the linkages between desertification and public health and recognises the impacts of changes in the land on human health via climate change, food security, environmental pollution, water supply, and pandemics (UNCCD 2024). While human health has been identified in the UNCCD, institutional linkages between WHO and UNCCC remain unclear (WHO 2012). This is very different from UNFCCC and CBD, which are constructing institutional collaboration between their functional works and WHO's agenda on human health.

This weak institutional linkage can be observed in the negotiations under the framework of UNCCD. As the ENB document shows, the term 'health' around the human community has rarely been mentioned in negotiations. The role of WHO was not identified in the document. The health issues were narrowly linked to the discussions on sand and dust storms. For example, Iraq raised the importance of their impacts on health and the environment. Saint Lucia also mentioned the transboundary impacts of Saharan dust on health across the Atlantic and Caribbean regions. While this statement raises the health impact as an international issue, it fails to show more detailed information and suggestions.

Food security can be identified in UNCCD's negotiations. The problems of drought were seen as an essential threat to food security. The document identified FAO, urging pressure on land and calling for land degradation neutrality. This demonstrates a close link between food security and land degradation. In this sense, food security can be more or less identified on the agenda of global land governance.

2.2.5 Global Plastics Governance and Negotiations

In addition to the established mechanisms of the international environmental negotiations, constructing a new international agreement for managing and controlling global plastics remains ongoing in 2024. UNEP has raised the impacts of plastic pollution on human health very generally (UNEP 2024a). However, UNEP has been investing its attention in the environmental dimension of global plastics governance. Plastics issues have been seen as challenges of marine pollution. To address the root cause of marine plastic pollution, global governance has been diverting its attention towards controlling the production of plastics by employing the life-cycle approach (Nielsen et al. 2020; Wang 2023).

While the international community and stakeholders have been negotiating a new global agreement on ending plastic pollution, they have seen a close link between plastic pollution and health in their discussions. According to the ENB document, in April 2024, the international plastic pollution negotiations constructed the term 'health' very clearly, thus seeing protecting health as an essential objective. WHO was engaged in discussions and negotiations, and its role in plastics governance received much support from various stakeholders. Based on this, defining and researching the linkage between plastic and health has been required. Compared to other international environmental negotiations, the plastic negotiations attach importance to its links to human health. For example, Cameroon raised human health as a priority of plastic governance. Therefore, this shows a positive expectation towards achieving an institutional link between addressing plastic pollution and improving public health.

2.3 WHO: Health-Based International Institution and Its Implications for the Linkages

Being very different from international environmental institutions, WHO has a distinctive understanding of global environmental challenges and impacts on human health. While WHO substantially raises the linkage between climate change and human health, it has been, to some extent, paying attention to other global environmental impacts on public health.

WHO and UNFCCC have closely monitored the interlinks between climate change and public health, mainly via WHO–UNFCCC Health and Climate Change Country Profile Project. This project is engaged primarily in mobilising national health agencies to raise their attention to the impacts of climate change on health, improving health systems, and health-related climate actions (WHO 2020b). This case demonstrates a robust institutional collaboration between the two international organisations. Beyond the interagency work, WHO sees its role in cooperation with the key actors of global climate governance and negotiations. In 2017, WHO raised a unique initiative to address the climate and health challenges in the Small Island Developing States. It called for raising their strong voice, strengthening investments in climate and health fields, improving the climate-related health system, and mobilising international financial resources (WHO 2017).

WHO has seen its influence on UN climate negotiations and conferences. It issued a special report for the UN Climate Conference (COP-24) in 2018 and offered recommendations. This report required controlling climate pollutants, carbon emissions, and air pollution, assessing health issues in the climate actions, raising investments in health-related climate adaptation, mobilising local stakeholders to address climate and health issues, enhancing governmental roles in climate and health actions, and integrate the health challenges into a global climate and public health governance (WHO 2018). This case demonstrates the role of WHO in enhancing the linkages between climate change and human health institutionally, practically, and discursively.

WHO also looks at other international environmental challenges beyond climate change issues. It focuses on the impacts of biodiversity on public health, human needs, traditional medicine, local communities, and infectious diseases (WHO 2015). Also, WHO (2020a) reveals the impacts of desertification on public health, recognising land degradation and desertification contributing to malnutrition caused by the shortage of food and water, diseases caused by contaminated water, respiratory diseases caused by air pollution, and infectious diseases. However, in contrast to its emphasis on the linkages between climate change and public health, WHO has been paying less attention to other environmental challenges. This is not to say that WHO neglects a wide range of environmental impacts on public health. Instead, this is because climate change has emerged as a powerful theme in global governance, and it has moved beyond climate politics to other political and environmental affairs (Aykut and Maertens 2021).

WHO also plays a vital role in constructing the link between public health and environmental pollution. Global plastics governance has evolved into a new international agreement for controlling plastic pollution. To better understand plastic's challenges, WHO (2019) called for academic and scientific work to assess the impacts of microplastics on human health, as their potential risks are everywhere, even in the drinking water system. It is thus essential to observe how WHO's contributions can be translated into plastics negotiations and governance.

2.4 Case Study: Global Mercury Governance and the Minamata Convention on Mercury

2.4.1 Environment Dimension of Establishing International Mercury Institutions

Global mercury governance and the Minamata Convention on Mercury are exciting cases demonstrating the link between mercury pollution and human health. The so-called Minamata disease caused by mercury pollution was identified in Japan in the 1950s. However, it had not yet received attention from global environmental governance until the 1972 Stockholm Human Environment Conference (Allan 2021; Selin and Selin 2022). While this triggered substantial global attention to mercury pollution and its impacts on human health, it had been trapped in the division of global solutions between an international treaty and a voluntary approach (Allan 2021). These different attitudes towards global solutions can be used to explain why the international mercury institutions have been constructing quite slowly.

The Minamata Convention on Mercury was adopted in 2013, explicitly raising the linkages between the use of mercury and human health and recalling the historical lessons and experiences of the so-called Minamata disease (Selin 2014). However, it is imperative to note that this is an international environmental institution. Before the emergence of the Minamata Convention, UNEP initiated the Global Mercury Partnership in 2005 and supported a wide range of stakeholders, such as governments, international organisations, and non-governmental organisations, to take global collective action on addressing mercury pollution and its impacts on human health (UNEP 2024c). Like controlling hazardous chemicals and transboundary wastes, managing mercury is a concern of global environmental governance. Beyond UNEP, the Global Environment Facility provided funding for phasing out the use of mercury in the healthcare sector and medical devices (WHO 2024). This case demonstrates how international environmental institutions have been engaged in health-related issues.

However, while a range of international institutions such as UNEP and WHO had been working towards addressing global mercury issues, the process of constructing international institutions of governing mercury was somewhat constrained by other international environmental negotiations. Because several essential stakeholders (including developed and developing countries) had invested their attention towards the international negotiations around the Rotterdam Convention adopted in 1998 and the Stockholm Convention adopted in 2001 to address hazardous chemicals and pesticides and persistent organic pollutants, respectively. They thus failed to focus on the negotiations on the global mercury governance (Selin 2014). This significantly contributes to the slow establishment of the international mercury agreement.

Compared to the challenging journey of reflecting the pollution of mercury into international institutions, the linkages between the pollution of mercury and human health have been identified in the mercury negotiations. According to ENB documents, the talks and the Conferences of the Parties to the Convention have primarily been focused on the environmental dimension of managing and controlling mercury. For instance, the negotiations covered various issues, including mercury supply, artisanal and small-scale gold mining, and emissions and releases of mercury. Not surprisingly, the mercury negotiations have been defined and constructed around international environmental institutions, particularly under UNEP. However, human health has frequently been identified in mercury negotiations. This is in very contrast to other global environmental negotiations. It is expected that stakeholders will call for contributions from the public health and environmental sectors to be integrated to address mercury issues. For instance, Brazil said addressing mercury-contaminated sites should invite health and environmental experts to the 2017 negotiations. Beyond the state actors, international organisations have seen their attitudes towards the linkages between mercury pollution and human health. For instance, the International Labor Organization emphasised the importance of ensuring a healthy working environment and addressing mercury-related risks in the 2023 negotiations.

2.4.2 Health Dimension of Establishing International Mercury Institutions

While the Convention refers to mercury as a critical environmental and political issue, it establishes clear guidelines for international organisations like WHO to serve their functions in mercury-related works (Selin 2014). WHO plays a professional role in the health dimension of the Convention's progress compared to UNEP, particularly in the executive phases. The institutional linkages between WHO and other international institutional arrangements were explicitly highlighted in WHO's report by the Secretariat in 2013 (WHO 2013). First, international mercury institutions should be working with other health-related international organisations. Article 16 of the Convention focuses on health issues. The Conference of the Parties should consult, collaborate, and exchange relevant information with WHO about addressing health-related issues and activities (UNEP 2023). Hence, the successful adoption of the Convention at the practical level can be attributed to the regular consultations with WHO (Eriksen and Perrez 2014).

Second, WHO is in a prominent position to offer relevant scientific evidence, call on plausible guidelines, and improve public awareness of the use of mercury. For instance, WHO has positively supported the use of thimerosal, recognised as an organomercury compound, in multi-dose vaccines based on scientific evidence, and it has eliminated the doubts and concerns about its potential adverse health effects. While there were different stances on using thimerosal, negotiators finally followed the existing WHO policies (Selin 2014). Moreover, WHO published training materials for health workers, issued guidance for identifying populations at risk from mercury exposure, and designed protocols for monitoring human exposure to mercury (WHO 2013). These institutional arrangements demonstrate the critical role of WHO in global mercury governance.

Third, in partnership with other international institutions, WHO initiated concrete actions to manage mercury-related products. For instance, WHO and Health Care Without Harm have jointly launched a global initiative to offer choices to mercury-containing thermometers and sphygmomanometers since 2008 (WHO 2013). This case demonstrates that health-related social groups and organisations recognise mercury's negative environmental impacts and their roles in addressing relevant actions. Another case is that WHO has been working with UNEP to gradually phase down the use of dental amalgam (WHO 2013). Dentistry uses approximately 340 tonnes of mercury annually, and around 20%–30% of it finally enters the waste stream (Selin 2014). These actions show that a range of health-related sectors and actors have not only emphasised the health impacts of mercury but also recognised the environmental impacts of the use of mercury within the health sector.

Finally, WHO also emphasises the engagement with national health sectors in managing mercury and other chemicals like lead and pesticides, which harm human health. To enhance the information exchange concerning the linkages between risky chemical exposure and human health, WHO invited its member states to participate in the Chemical Risk Assessment Network formulated in 2013 (WHO 2021b). Moreover, for instance, WHO provided a case study on Brazil to share the lessons on essential integration among diverse sectors and stakeholders and outlined guidelines at the national level (WHO 2021a). Meanwhile, WHO has also been involved in national programmes, such as product partnerships with the US government (Sun 2017). These national institutional arrangements demonstrate that the linkages between mercury and public health have been, to some extent, translated and embedded at the national level.

Generally speaking, the case of global mercury governance shows how the linkages between mercury pollution and the health sector have been adopted in the existing international institutions. While the Minamata Convention on Mercury refers to a global environmental institution, it has also been linked to health-related institutions, particularly WHO. A range of stakeholders must recognise both the environmental impacts of mercury in the health sector and the health impacts of environmental pollution.

2.5 Discussion

The linkages between ecological challenges and human health have been emerging on the agenda of global governance. They can also be identified in existing international institutions improving ecological and human health governance. Climate change, biodiversity, desertification, and plastic pollution have been linked to human health. Particularly, according to the negotiations recorded in the ENB documents, biodiversity loss and land degradation impact food security and threaten human health and well-being. However, international institutions have the potential to develop a broader range of projects to translate the linkages into practice. For example, there are three challenges in addressing climate change and public health issues: asking for collective actions on reducing emissions and improving public health, constructing climate-resilient health systems, and mitigating the impacts of climate change on human health (Campbell-Lendrum et al. 2023). Addressing them requires an integrated approach to institutional construction across different international organisations. Similarly, the World Bank identifies the linkage between climate change and public health by calling for improving health systems, constructing low-carbon health infrastructure, and addressing the climate impacts on health (World Bank 2024). This demonstrates a huge potential to build global ecological and human health governance. Particularly, the case of global mercury governance shows the mutual coordination between environmental and human health institutions.

However, the institutional gaps between ecological challenges and human health remain clear. The International Institute for Sustainable Development (IISD) issued a report on the role of health in the global environmental agenda, identifying biodiversity, climate change, pollution, and food systems as main environmental issues in the health field and seeing water-related diseases, women's health, biodiversity-related technologies, mental health and ecosystem, and traditional knowledge and medicines as the health topics in the environmental governance (Willetts et al. 2022). However, a limited range of these issues have narrowly been embedded in the international institutions and the negotiations. This triggers two crucial points to be considered and discussed. First, it is essential to research how innovative ideas for enhancing ecological and public health governance can be translated into institutional arrangements and policy actions. Determining the agenda of global governance depends on the power of constructing ideas within international institutions and politics. Second, it remains unclear how the linkages can be categorised into the existing international institutions. Who should be playing a leading role in governing the interdisciplinary challenges? WHO or UNEP (and other international environmental institutions)? For example, environment-based international organisations focus on environmental challenges more than pandemics and diseases.

Similarly, WHO raises, rather than prioritising, environmental dimensions of human health challenges. International organisations have different functions and capabilities for addressing global challenges.

However, it is essential to note that the Minamata Convention on Mercury demonstrates the possibility of combining international environmental institutions and human health agencies.

The health-driven international institutions cannot enhance the role of human health in global environmental governance. This chapter employs WHO as a case to observe the linkages between ecological challenges and human health in the global public health agenda. WHO has been working with several international environmental institutions to construct joint projects. For example, WHO and UNFCCC have seen their joint efforts to raise public awareness of climate and human health challenges. However, the extent to which WHO has translated its agenda and policies into implementing global climate governance remains unclear. And within the climate negotiations, human health issues have been narrowly linked to climate adaptation. Climate mitigation, a critical theme in negotiations and governance, has rarely been identified and linked to human health. Based on this, not surprisingly, the role of WHO has not been substantially identified in international environmental negotiations. Following these findings, it is recommended that WHO has the potential to develop its role in the talks and raise its voice on international platforms to address global challenges.

The linkages between ecological challenges and human health have been translated differently between international institutional arrangements and their related negotiations. Human health is reflected in a wide range of climate-related documents and joint projects between WHO and UNFCCC, which are substantially determined by the critical role of climate change in global governance. Climate change dominates global environmental governance, while many ecological challenges fail to attract international attention. This situation can be called climatisation, seeing climate change as a significant theme on the agenda of global governance (Maertens 2021; Aykut and Maertens 2021). This determines that many challenges and issues, including human health, have been linked to climate actions in the existing international institutions and related documents. However, human health issues have been weakly identified in the climate negotiations. This is very sharply contrasted to the linkages between biodiversity and human health. As biodiversity has been widely linked with human health, food security, and medicines, its close links to human health have been identified in the international biodiversity negotiations. Functional interactions between biodiversity loss and human health challenges determine this linkage type. Similarly, the linkages between mercury and human health can be identified in international institutions mainly because they have collective demands and functions of addressing mercury pollution.

Global environmental and public health governance has been fragmented across many stakeholders. First, a nexus approach can be employed to address functional fragmentation of the international institutions, understanding and explaining the complicated picture of mapping the various challenges (Liu et al. 2018; Lalawmpuii and Rai 2023). The water–energy–food–environment–climate nexus can be used to integrate the different environmental challenges and have implications for international institutions (Kalvani and Celico 2023; Mperejekumana et al. 2024; Nasrollahi et al. 2021). Beyond the environmental nexus, for example, biodiversity and health issues can be interlinked by making the biodiversity and health nexus (Gautier, Gardon, and Déprés 2023). Thus, it is vital to integrate the nexus into global public health governance rather than locating a single ecological challenge in human health institutions. It is also very innovative to create the water–food–environment–climate–health nexus and map the networks of the obstacles. This nexus may support the institutional linkages between WHO and international environmental institutions beyond UNFCCC. The case of global mercury governance has witnessed this nexus by linking water contamination, chemical pollution, food safety, and environmental pollution in terms of its impacts on human health.

Second, a multi-level governance approach can be employed to explain the vertical fragmentation of global environmental and public health governance. First, setting the linkages across the international institutions on the global governance agenda requires the national actors to translate them into their policy-making processes. The nature of domestic political and administrative structures mainly determines this. For example, whether or not the health agency plays a vital role in the governmental administrative system matters. It is challenging to see the administrative coordination across public health agencies, food and agriculture, and environmental protection at a national level. Second, translating the linkages into their implementation at subnational levels might encounter difficulties. Local interests might not be consistent with national motivations. Local governments might also have an understanding of the linkages and policy priorities. In this sense, it is essential to raise the voices of subnational actors over the linkages.

2.6 Conclusion

This chapter has discussed the linkages between ecological challenges and human health across international institutions. Regarding global environmental governance, climate change has been substantially linked to human health issues. Human health issues have been identified in international biodiversity negotiations and have the potential to be an important topic in international plastics negotiations. However, it is essential to note that human health has been narrowly embodied in the agenda of global environmental governance. On the other hand, WHO has seen its proactive role in constructing the linkages between ecological challenges and human health by participating in international environmental negotiations, issuing theme-focused reports, and establishing joint institutions with other international environmental agencies, particularly UNFCCC. Also, the Minamata Convention on Mercury focuses on constructing international environmental institutional arrangements and involving the role of WHO in its global governance.

However, this work identifies the institutional gaps in the horizontal, functional, and vertical linkages. It requires the nexus approach and the multi-level governance to improve global environmental and public health governance. Based on this, future works can focus on the nature of domestic policy governance in terms of addressing ecological challenges and human health. This shows an excellent opportunity for observing and explaining the interdisciplinary approach to global ecological and public health governance.

Acknowledgements

The authors thank Professor Ming Hung Wong for his valuable and constructive comments on this chapter. The authors declare no competing interests.

REFERENCES

Allan, J. 2021. 'The Incremental Approach to Governing Mercury'. International Institute for Sustainable Development. 1 November 2021. https://www.iisd.org/articles/governing-mercury#:~:text=When%20the%20UNEP%20Governing%20Council%20agreed%20to%20form%20the%20Global

Amaya, AB and P De Lombaerde. 2019. 'Multi-Level Health Governance and Health Diplomacy: Regional Dimensions'. *Regions and Cohesion* 9 (1): 86–92. https://doi.org/10.3167/reco.2019.090107

Aykut, SC and L Maertens. 2021. 'The Climatization of Global Politics: Introduction to the Special Issue'. *International Politics* 58 (4): 501–518. https://doi.org/10.1057/s41311-021-00325-0

Bernstein, S and B Cashore. 2012. 'Complex Global Governance and Domestic Policies: Four Pathways of Influence'. *International Affairs* 88 (3): 585–604. https://doi.org/10.1111/j.1468-2346.2012.01090.x

Brugha, R. 2008. 'Global Health Initiatives and Public Health Policy'. In K. Heggenhougen and S. Quah (Eds.), *International Encyclopedia of Public Health* (pp. 72–81). San Diego, CA: Academic Press. https://doi.org/10.1016/B978-012373960-5.00239-2

Burchill, S. 2013. *Theories of International Relations [Electronic Resource]*. London, UK: Macmillan Education.

Campbell-Lendrum, D, T Neville, C Schweizer, and M Neira. 2023. 'Climate Change and Health: Three Grand Challenges'. *Nature Medicine* 29 (7): 1631–1638. https://doi.org/10.1038/s41591-023-02438-w

CBD. 2017. *Report of the First Meeting of the Interagency Liaison Group on Biodiversity and Health*. Geneva: CBD.

———. 2021. *Biodiversity and Health*. Geneva: CBD.

Davies, SE. 2008. 'Securitizing Infectious Disease'. *International Affairs* 84 (2): 295–313. https://doi.org/10.1111/j.1468-2346.2008.00704.x

———. 2010. 'What Contribution Can International Relations Make to the Evolving Global Health Agenda?' *International Affairs* 86 (5): 1167–1190. https://doi.org/10.1111/j.1468-2346.2010.00934.x

Davies, SE, S Elbe, A Howell, and C McInnes. 2014. 'Global Health in International Relations: Editors' Introduction'. *Review of International Studies* 40 (5): 825–834. https://doi.org/10.1017/S0260210514000308

Davies, SE and C Wenham. 2020. 'Why the COVID-19 Response Needs International Relations'. *International Affairs* 96 (5): 1227–1251. https://doi.org/10.1093/ia/iiaa135

Eckersley, R. 2021. 'Greening States and Societies: From Transitions to Great Transformations'. *Environmental Politics* 30 (1–2): 245–265. https://doi.org/10.1080/09644016.2020.1810890

Eriksen, HH and FX Perrez. 2014. 'The Minamata Convention: A Comprehensive Response to a Global Problem'. *Review of European, Comparative and International Environmental Law* 23 (2): 195–210. https://doi.org/10.1111/reel.12079

FAO, UNEP, WHO, and WOAH. 2022. *Global Plan of Action on One Health (2022–2026). One Health Joint Plan of Action, 2022–2026*. Rome, Italy: FAO, UNEP, WHO, and WOAH.

Ferreira, MM. 2022. 'European Multilevel Governance in the Health Sector: The Discursive Construction of a European Health Union'. *International Journal of Interdisciplinary Civic and Political Studies* 18 (1): 59–76. https://doi.org/10.18848/2327-0071/CGP/v18i01/59-76

Fidler, DP and LO Gostin. 2006. 'The New International Health Regulations: An Historic Development for International Law and Public Health'. *Journal of Law, Medicine and Ethics* 34 (1): 85–94. https://doi.org/10.1111/j.1748-720X.2006.00011.x

Fraundorfer, M. 2022. 'Global Governance of Infectious Disease Outbreaks'. In *Global Governance in the Age of the Anthropocene* (pp. 117–159). Cham: Springer. https://doi.org/10.1007/978-3-030-88156-6_4

Galaz, V, F Moberg, EK Olsson, E Paglia, and C Parker. 2011. 'Institutional and Political Leadership Dimensions of Cascading Ecological Crises'. *Public Administration* 89 (2): 361–380. https://doi.org/10.1111/j.1467-9299.2010.01883.x

Gautier, A, S Gardon, and C Déprés. 2023. 'The Emergence of the Biodiversity/Health Nexus: Making Biodiversity a Health Issue'. *Review of Agricultural, Food and Environmental Studies* 104 (1): 27–46. https://doi.org/10.1007/s41130-023-00189-3

Gray, PS. 1999. 'The Regulation of Food Safety—General Aspects'. In K. Van der Heijden (Ed.), *International Food Safety Handbook: Science, International Regulation, and Control*. New York: Routledge.

Haas, PM. 1989. 'Do Regimes Matter? Epistemic Communities and Mediterranean Pollution Control'. *International Organization* 43 (3): 377–403. https://doi.org/10.1017/S0020818300032975

Hardon, A and S Blume. 2005. 'Shifts in Global Immunisation Goals (1984–2004): Unfinished Agendas and Mixed Results'. *Social Science and Medicine* 60 (2): 345–356. https://doi.org/10.1016/j.socscimed.2004.05.008

Harris, JM. 1991. 'Global Institutions and Ecological Crisis'. *World Development* 19 (1): 111–122. https://doi.org/10.1016/0305-750X(91)90042-G

Intergovernmental Panel on Climate Change. 2023. 'Health, Wellbeing and the Changing Structure of Communities'. In *IPCC Sixth Assessment Report – Impacts, Adaptation and Vulnerability*, 1041–1170. Cambridge University Press. https://doi.org/10.1017/9781009325844.009

International Institute for Sustainable Development. 2024. 'Negotiations | IISD Earth Negotiations Bulletin'. IISD Earth Negotiations Bulletin. 2024. https://enb.iisd.org/negotiations

Jackson, W and T Bührs. 2015. 'International Environmental Regimes: Understanding Institutional and Ecological Effectiveness'. *Journal of International Wildlife Law and Policy* 18 (1): 63–83. https://doi.org/10.1080/13880292.2014.957030

Kalvani, SR and F Celico. 2023. 'The Water–Energy–Food Nexus in European Countries: A Review and Future Perspectives'. *Sustainability (Switzerland)* 15 (6): 4960. https://doi.org/10.3390/su15064960

Keohane, RO. 1988. 'International Institutions: Two Approaches'. *International Studies Quarterly* 32 (4): 379–396. https://doi.org/10.2307/2600589

Laferrière, E and PJ Stoett. 1999. *International Relations Theory and Ecological Thought: Towards a Synthesis*. London: Routledge. https://doi.org/10.4324/9780203169407

Lalawmpuii, B and PK Rai. 2023. 'Role of Water-Energy-Food Nexus in Environmental Management and Climate Action'. *Energy Nexus* 11, 100230. https://doi.org/10.1016/j.nexus.2023.100230

Lee, K. 2003. *Globalization and Health: An Introduction*. London: Palgrave Macmillan.

Liu, J, V Hull, H Charles J Godfray, D Tilman, P Gleick, H Hoff, C Pahl-Wostl, et al. 2018. 'Nexus Approaches to Global Sustainable Development'. *Nature Sustainability* 1: 466–476. https://doi.org/10.1038/s41893-018-0135-8

Loup, GL, S Fleury, K Camargo, and B Larouzé. 2010. 'International Institutions, Global Health Initiatives and the Challenge of Sustainability: Lessons from the Brazilian AIDS Programme'. *Tropical Medicine and International Health* 15 (1): 5–10. https://doi.org/10.1111/j.1365-3156.2009.02411.x

Maertens, L. 2021. 'Climatizing the UN Security Council'. *International Politics* 58 (4): 640–660. https://doi.org/10.1057/s41311-021-00281-9

Mills, A. 2005. 'Mass Campaigns versus General Health Services: What Have We Learnt in 40 Years about Vertical versus Horizontal Approaches?' *Bulletin of the World Health Organization* 83 (4): 315–316. https://doi.org//S0042-96862005000400017

Mperejekumana, P, L Shen, S Zhong, F Muhirwa, MS Gaballah, and JMV Nsigayehe. 2024. 'Integrating Climate Change Adaptation into Water-Energy-Food-Environment Nexus for Sustainable Development in East African Community'. *Journal of Cleaner Production* 434, 140026. https://doi.org/10.1016/j.jclepro.2023.140026

Nasrollahi, H, R Shirazizadeh, R Shirmohammadi, O Pourali, and M Amidpour. 2021. 'Unraveling the Water-Energy-Food-Environment Nexus for Climate Change Adaptation in Iran: Urmia Lake Basin Case-Study'. *Water (Switzerland)* 13 (9): 1282. https://doi.org/10.3390/w13091282

Nestle, M. 2010. *Safe Food : The Politics of Food Safety. California Studies in Food and Culture*. Vol. 5. Berkeley, CA: University of California Press.

Nielsen, TD, J Hasselbalch, K Holmberg, and J Stripple. 2020. 'Politics and the Plastic Crisis: A Review throughout the Plastic Life Cycle'. *Wiley Interdisciplinary Reviews: Energy and Environment* 9 (1): e360. https://doi.org/10.1002/wene.360

North, DC. 1990. *Institutions, Institutional Change and Economic Performance*. Cambridge University Press. https://doi.org/10.1017/cbo9780511808678

Novotny, TE. 2007. 'Global Governance and Public Health Security in the 21st Century'. *California Western International Law Journal* 38 (1): 19–40.

Papamichail, A. 2023. 'Reinscribing Global Hierarchies: COVID–19, Racial Capitalism and the Liberal International Order'. *International Affairs* 99 (4): 1673–1691. https://doi.org/10.1093/ia/iiad091

Poku, NK and A Whiteside. 2002. 'Global Health and the Politics of Governance: An Introduction'. *Third World Quarterly* 23 (2): 191–195. https://doi.org/10.1080/01436590220126586

Rubin, H and N Saidel. 2016. 'Global Governance Structure for Infectious Disease: An Enforceable Strategy'. *Orbis* 60 (2): 279–295. https://doi.org/10.1016/j.orbis.2016.02.003

Selin, H. 2014. 'Global Environmental Law and Treaty-Making on Hazardous Substances: The Minamata Convention and Mercury Abatement'. *Global Environmental Politics* 14 (1): 1–19. https://doi.org/10.1162/GLEP_a_00208

Selin, H and NE Selin. 2022. 'From Stockholm to Minamata and beyond: Governing Mercury Pollution for a More Sustainable Future'. *One Earth* 5 (10): 1109–1125. https://doi.org/10.1016/j.oneear.2022.09.001

Stubbs, T and A Kentikelenis. 2017. 'International Financial Institutions and Human Rights: Implications for Public Health'. *Public Health Reviews* 38: 27. https://doi.org/10.1186/s40985-017-0074-3

Sun, Y. 2017. 'Transnational Public-Private Partnerships as Learning Facilitators: Global Governance of Mercury'. *Global Environmental Politics* 17 (2): 21–44. https://doi.org/10.1162/GLEP_a_00399

Thomas, C. 1989. 'On the Health of International Relations and the International Relations of Health'. *Review of International Studies* 15: 273–280. https://doi.org/10.1017/S0260210500112884

Tichenor, M, J Winters, KT Storeng, J Bump, JP Gaudillière, M Gorsky, M Hellowell, et al. 2021. 'Interrogating the World Bank's Role in Global Health Knowledge Production, Governance, and Finance'. *Globalization and Health* 17, 110. https://doi.org/10.1186/s12992-021-00761-w

UN. 2020. '"We Need to Act Now to Avoid the Worst Impacts of Our Efforts to Control the Pandemic" | United Nations'. 9 June 2020. https://www.un.org/en/coronavirus/we-need-act-now-avoid-worst-impacts-our-efforts-control-pandemic

UNCCD. 2024. 'Health | UNCCD'. 2024. https://www.unccd.int/land-and-life/health/overview

UNEP. 2020. 'Health & Environment'. 24 December 2020. https://www.unep.org/resources/factsheet/health-environment

———. 2023. 'Minamata Convention on Mercury: Text and Annexes -2023 Edition'. Geneva. www.minamataconvention.org

———. 2024a. 'Life-Cycle Approach & Plastic Pollution | UNEP'. 2024. https://www.unep.org/news-and-stories/story/what-life-cycle-approach-and-how-can-it-help-tackle-plastic-pollution

———. 2024b. 'Pollution and Health | UNEP - UN Environment Programme'. 2024. https://www.unep.org/topics/chemicals-and-pollution-action/pollution-and-health

———. 2024c. 'Who We Are | Global Mercury Partnership'. 2024. https://www.unep.org/globalmercurypartnership/who-we-are

UNFCCC. 2015. 'Climate Change Agreement Is Public Health Agreement'. 3 June 2015. https://unfccc.int/news/climate-change-agreement-is-public-health-agreement

Wang, S. 2023. 'International Law-Making Process of Combating Plastic Pollution: Status Quo, Debates and Prospects'. *Marine Policy* 147 (January), 105376. https://doi.org/10.1016/j.marpol.2022.105376

WHO. 2012. *Our Planet, Our Health, Our Future Human Health and the Rio Conventions: Biological Diversity, Climate Change and Desertification*. Geneva: WHO

———. 2013. *Public Health Impacts of Exposure to Mercury and Mercury Compounds: The Role of WHO and Ministries of Public Health in the Implementation of the Minamata Convention on Mercury*. Geneva: WHO.

———. 2015. 'Biodiversity and Health'. 3 June 2015. https://www.who.int/news-room/fact-sheets/detail/biodiversity-and-health

———. 2017. 'Launch of Special Initiative to Address Climate Change Impact on Health in Small Island Developing States'. World Health Organization News. 12 October 2017. https://www.who.int/news/item/12-11-2017-launch-of-special-initiative-to-address-climate-change-impact-on-health-in-small-island-developing-states

———. 2018. 'Health Benefits Far Outweigh the Costs of Meeting Climate Change Goals'. World Health Organization News. 5 December 2018. https://www.who.int/news/item/05-12-2018-health-benefits-far-outweigh-the-costs-of-meeting-climate-change-goals

———. 2019. 'WHO Calls for More Research into Microplastics and a Crackdown on Plastic Pollution'. 2019. https://www.who.int/news/item/22-08-2019-who-calls-for-more-research-into-microplastics-and-a-crackdown-on-plastic-pollution

———. 2020a. 'Climate Change: Land Degradation and Desertification'. 26 October 2020. https://www.who.int/news-room/questions-and-answers/item/climate-change-land-degradation-and-desertification

———. 2020b. 'Promoting and Protecting Health through the UNFCCC'. World Health Organization News. 1 January 2020. https://www.who.int/news/item/01-01-2020-promoting-and-protecting-health-through-the-unfccc

———. 2021a. 'Case Study: Implementation of the Minamata Convention by the Health Sector in Brazil'. Overview. 3 June 2021. https://www.who.int/publications/i/item/9789240024533

———. 2021b. *Framework for the Use of Systematic Review in Chemical Risk Assessment*. Geneva: WHO and CBD.

———. 2024. 'NationsUnitetoEliminateMercury-ContainingMedicalDevices'. 15May2024. https://www.who.int/news/item/15-05-2024-nations-unite-to-eliminate-mercury-containing-medical-devices#:~:text=Led%20by%20the%20United%20Nations%20Environment%20Programme%20(UNEP),%20with

WHO and CBD. 2015. *Connecting Global Priorities: Biodiversity and Human Health A State of Knowledge Review*. Geneva: WHO and CBD.

Willetts, E, L Grant, J Bansard, PM Kohler, T Rosen, P Bettelli, and M Schröder. 2022. 'Health in the Global Environmental Agenda: A Policy Guide'. https://www.iisd.org/ publications/health-global-environment-agenda-policy-guide

Woolley, O. 2014. *Ecological Governance: Reappraising Law's Role in Protecting Ecosystem Functionality*. Cambridge: Cambridge University Press. https://doi.org/10.1017/CBO9781107447080

World Bank. 2024. 'Health and Climate Change'. The World Bank Group Brief. 5 April 2024. https://www.worldbank.org/en/topic/health/brief/health-and-climate-change

3

Laws and Regulations of Food Contaminants Pertinent to Public Health and Food Safety in the Asia Pacific

James Tsz-Fung Wong and Ming Hung Wong

3.1 Introduction

Food contamination is a critical issue that affects public health and food safety worldwide. Food safety is taking the frontal stage in production, processing, and distribution. The World Health Organization (WHO) estimates that food-borne bacteria, parasites, toxins, and allergens cause more than 20 million cases of illness and about 5000 deaths in Europe annually (WHO, 2015). The presence of harmful substances in food can pose risks to consumers, resulting in various health problems and even fatalities in many cases. Understanding the background and significance of food contamination is crucial for implementing effective laws and regulations to protect consumers and ensure food safety.

The economic impact of food contamination is also substantial, with expenditures associated with health care, productivity loss of the society, and impact on the food industry. By highlighting the significance of food contamination, we can underscore the urgency and necessity of robust laws and regulations.

The primary objectives of this chapter are to (1) discuss the importance of food contamination from the public health point of view, (2) delve deeper into the impact of food contamination on public health and food safety, and (3) explore the laws and regulations associated with contaminants, such as chemical contaminants and microbial pathogens. It is hoped that by understanding the specific risks posed by food contaminants, policymakers, and regulatory authorities can develop targeted strategies to mitigate them.

3.1.1 Importance of Laws and Regulations for Public Health and Food Safety

Formulating rules and regulations to ensure food safety is particularly crucial for the Asia Pacific region. The Asia Pacific region encompasses various cultures, cuisines, and agricultural practices. This diversity necessitates tailored food safety regulations that address specific regional hazards and practices. With some of the most densely populated countries in the world, ensuring food safety is vital to prevent widespread outbreaks of food-borne illnesses that can affect large numbers of people quickly. In addition, many countries in the region are experiencing rapid economic growth, leading to increased food production and trade. Effective food safety regulations are therefore essential to support this growth while protecting public health and maintaining consumer confidence at the same time.

The Asia Pacific region is a significant player in the global food trade. Harmonised food safety standards can facilitate trade relations, ensuring that products meet international safety benchmarks and thus enhance market access. Meanwhile, the region is particularly vulnerable to climate change, which can impact food production and safety. Regulations can help mitigate risks associated with climate-related disruptions and promote sustainable agricultural practices. Many countries in the region face challenges in public health infrastructure. Clear food safety regulations can strengthen monitoring and enforcement

DOI: 10.1201/9781003492115-4

mechanisms, ultimately improving health outcomes. The importance of formulating food safety regulations in the Asia Pacific region is underscored by its unique demographic, economic, and environmental challenges, as well as the need for global competitiveness and consumer protection.

This section will highlight the crucial role of laws and regulations in addressing food contamination. Effective regulatory frameworks provide the legal basis for controlling and managing food contaminants throughout production and distribution. The subsection will discuss the objectives of food safety regulations, such as setting safety standards, establishing monitoring systems, and enforcing compliance. Additionally, it will explore the role of regulatory authorities, industry stakeholders, and consumers in ensuring the implementation and effectiveness of these laws and regulations.

3.2 International Perspectives on Food Safety

3.2.1 Global Organisations and Initiatives Promoting Food Safety

The Codex Alimentarius Commission (CAC) is an international group established by the Food and Agriculture Organization (FAO) of the United Nations and the WHO under the Joint FAO/WHO Food Standards Programme in 1963 to develop food standards for protecting consumers' health (FAO/WHO, 2024). Codex Alimentarius, established by CAC, is an assemblage of food standards, guidelines, and codes of practice to protect consumers' health and ensure fair practices. Requirements for food listed in the texts are food products for consumers are safe and free from adulteration.

Texts established by Codex include a wide range of provisions related to food hygiene, food additives, residues of pesticides and veterinary drugs, contaminants, labelling and presentation, methods of analysis and sampling, as well as import and export inspection, etc. Independent expert bodies would, while setting the standard, provide professional and scientific advice to enhance the credibility of the standards. The said expert bodies perform safety assessments that are based on the literature (C.A., 2024)

3.2.2 International Standards and Guidelines for Food Contaminants

International food safety authorities such as the Joint Food and Agriculture Organization of the United Nations (FAO)/WHO Expert Committee on Food Additives (JECFA) have conducted an in-depth assessment of the potential adverse health impacts of a wide range of metallic contaminants and established appropriate Health-Based Guidance Values (HBGVs) based on the toxicological data published in the literature.

There are also different internationally recognised standards and guidelines developed by various organisations and authorities, including but not limited to the United Nations FAO, the WHO, the CAC, the World Organization for Animal Health (WOAH), International Plant Protection Convention (IPPC), European and Mediterranean Plant Protection Organization (EPPO), International Association of Chemical Analysts (AOAC), International Collaborative Council for Pesticide Analysis (CIPAC), European Organization for Cooperation and Development (OECD), European Food Safety Authority (EFSA), and International Organization for Standardisation (ISO).

3.2.3 Role of International Cooperation in Ensuring Food Safety

At the international level, an integrated safety system has been adopted for all food systems, focusing on prevention, proactive actions, and control from "farm to table" to ensure better food safety, hygiene, and quality (Lin et al., 2021). For instance, the European Commission (EU) implements an integrated policy approach to build safety into the food and feed chain. A system with three layers of control has been implemented, namely, the audit of the control systems of Member States and countries by the EU, the responsibility of Member States, and the allocation of responsibility to food-handling businesses and farmers. The EU also implemented a monitoring system to keep contaminants from food and animal feed. Maximum acceptable limits apply to food and feed products and materials that come in contact with food to exclude contamination of mycotoxins, plant toxins, metals, halogenated persistent organic pollutants, and processing contaminants (European Commission, 2017).

3.3 Food Safety Regulations in Different Jurisdictions, with a Focus on the Asia Pacific

Governments worldwide adopt various policy interventions to mitigate food safety risks. Mandatory policy tools are the most effective, followed successively by guiding policy and voluntary policy tools (Sha et al., 2024). Prabhakar et al. (2010) reported that the Asia-Pacific region's policy environment does not have a significant tendency towards food safety-related issues. Different laws and regulations under various ministries and even different departments within the same jurisdictions lead to the formulation and implementation of food safety laws and regulations within the region. The food-related laws and regulations in the Asia Pacific will be elaborated in the following sessions.

3.3.1 Metallic Contaminants

Metallic contaminant is regulated in various jurisdictions. For example, the Legislative Council of Hong Kong completed its scrutiny of the Food Adulteration (Metallic Contamination) (Amendment) Regulation 2018 ("the Amendment Regulation") in October 2018. There are 14 metallic contaminants covered by the Amendment Regulation, namely, antimony, arsenic, cadmium, chromium, lead, mercury, tin, barium, boron, copper, manganese, nickel, selenium, and uranium. The Amendment Regulation stipulates 144 maximum levels (MLs) for the metallic contaminants in respect of different foods, provides where necessary definitions for individual food/food groups, and replaces/deletes obsolete clauses.

The MLs are summarised in Table 3.1. According to the amendment Regulation, a person must not import, consign, deliver, manufacture, or sell for human consumption any specified food or compounded food containing a specified metal more than the ML (FEHD, 2024). Various factors, such as the toxicity of the metal, usual daily intake amount of the food, and anticipated consumer, determine the MLs

TABLE 3.1

Maximum Levels (ML) of Metals in Food or Food Groups in the HKSAR

Metal	Selected Food or Food Groups	Maximum Level (mg/kg)
Antimony, Sb	Vegetables, cereals, meat of animal and poultry, fish, crabs, prawns and shrimps, oysters	1
	Bottled or packed drinking waters, other than natural mineral waters	0.02
	Natural mineral waters	0.005
Arsenic, As (expressed as total arsenic)	Vegetables, cereals (other than rice), meat of animal, editable offal of animal, meat of poultry, editable offal of poultry, food grade salt	0.5
	Editable fats and oils (other than fish oil), fat spreads and blended spreads	0.1
	Bottled or packaged drinking waters, including natural mineral waters	0.01
Arsenic, As (expressed as inorganic arsenic)	Seaweed	1
	Aquatic animals other than fish	0.5
	Husked rice	0.35
	Polished rice	0.2
	Fish, Fish oil	0.1
Barium, Ba	Bottled or packed drinking waters, other than natural mineral waters	1.3
	Natural mineral waters	0.7

(Continued)

TABLE 3.1 (*Continued*)

Maximum Levels (ML) of Metals in Food or Food Groups in the HKSAR

Metal	Selected Food or Food Groups	Maximum Level (mg/kg)
Boron, B	Natural mineral waters	5
	Bottled or packed drinking waters, other than natural mineral waters	2.4
Cadmium, Cd	Crustaceans, Bivalve molluscs, Cephalopods, Gastropods	2
	Kidney of cattle, pigs, goats and sheep	1
	Liver of cattle, pigs, goats and sheep, liver of poultry, food grade salt	0.5
	Leafy vegetables (including Brassica leafy vegetables), wheat, husked rice, polished rice	0.2
	Legume vegetables, pulses, root and tuber vegetables, stalk and stem vegetables, vegetables unless otherwise specified, cereals (other than buckwheat, cañihua, quinoa, wheat, and rice), fish	0.1
	Bulb vegetables, brassica vegetables (other than brassica leafy vegetables), fruiting vegetables (cucurbits), fruiting vegetables (other than cucurbits and tomatoes), the meat of cattle, pigs, goats and sheep, meat of poultry	0.05
	Bottled or packed drinking waters, including natural mineral waters	0.003
Chromium, Cr	Pulses, cereals, meat of animals, meat of poultry, fish, crabs, prawns and shrimps, oysters	1
	Vegetables other than pulses	0.5
	Bottled or packed drinking waters, including natural mineral waters	0.05
Copper, Cu	Bottled or packed drinking waters other than natural mineral waters	2
	Natural mineral waters	1
Lead, Pb	Tea (green and black), dried chrysanthemum	5
	Broadleaf holly leaves, food-grade salt	2
	Bivalve molluscs	1.5
	Mango chutney, editable fungi, aquatic animals (other than fish, crustaceans, and bivalve molluscs)	1
	Coffee beans, edible offal of cattle, editable offal of pig, edible offal of poultry, lime preserved eggs, crustaceans	0.5
	Jams, jellies and marmalades, table olives	0.4
	Leafy vegetables (including Brassica leafy vegetable, other than spinach), fish	0.3
	Cranberry, currants, elderberry, cereal grains (other than buckwheat, cañihua and quinoa), coffee beverages, poultry eggs, tea beverages made from green or black tea, carbonated beverages, wine	0.2
	Fruits (other than cranberry, currants, and elderberry), canned fruits, bulb vegetables, Brassica vegetables (other than Brassica leafy vegetables), legume vegetables, pulses, root and tuber vegetables, canned vegetables, pickled cucumbers, meat of cattle, pig, goat and sheep, meat of poultry, editable fats and oils, fat spreads and blended spreads	0.1

(*Continued*)

TABLE 3.1 (*Continued*)

Maximum Levels (ML) of Metals in Food or Food Groups in the HKSAR

Metal	Selected Food or Food Groups	Maximum Level (mg/kg)
	Fruit juices exclusively from berries and other small fruits, fruiting vegetables (Cucurbits), fruiting vegetables (other than Cucurbits), tomatoes (preserved by heat treatment and hermetically sealed), canned chestnuts and canned chestnuts puree	0.05
	Fruit juices (other than fruit juices exclusively from berries and other small fruits)	0.03
	Milk, secondary milk products	0.02
	Infant formula and follow-up formula, bottled or packaged drinking waters, including natural mineral waters	0.01
Manganese, Mn	Natural mineral waters	0.4
Mercury, Hg (expressed as methyl-mercury)	Fish	0.5
Mercury, Hg (expressed as total mercury)	Aquatic animals other than fish	0.5
	Edible fungi, food-grade salt	0.1
	Meat of animals, edible offal of animals, meat of poultry, edible offal of poultry, poultry eggs	0.05
	Rice, husked rice, polished rice, maize, maize flour, wheat, wheat flour	0.02
	Milk, secondary milk products	0.01
	Natural mineral waters	0.001
Mercury, Hg (expressed as inorganic mercury)	Bottled or packaged drinking waters other than natural mineral waters	0.006
Nickel, Ni	Bottled or packaged drinking waters other than natural mineral waters	0.07
	Natural mineral waters	0.02
Selenium, Se	Bottled or packaged drinking waters other than natural mineral waters	0.04
	Natural mineral waters	0.01
Tin, Sn	Canned foods, other than canned beverages	250
	Canned beverages	150
Uranium, U	Bottled or packaged drinking waters other than natural mineral waters	0.03

Source: HKSARG (2018).

of different selected food and food products. For instance, the ML of Pb in tea could be up to 5 mg/kg, while it is only 0.01 mg/kg for infant formula and follow-up formula milk.

3.3.2 Pesticide Residues

As the global food supply chain becomes increasingly interconnected, managing pesticide residues in food products has become a pressing concern, particularly in Asia Pacific. Many Asian countries, such as mainland China, Thailand, and Vietnam, major agricultural producers and exporters, rely on pesticides to boost crop yields and protect against pests. However, the improper or excessive application of these chemicals can lead to the accumulation of harmful residues in food commodities, posing possible health risks to consumers. Regulatory bodies across the Asia Pacific have responded by implementing stricter controls, including establishing maximum residue limits, and enhancing monitoring and testing programs.

The key regulations for pesticide residues in food in different jurisdictions in the Asia Pacific are summarised in Table 3.2. Most jurisdictions implement the regulation by imposing laws and standards, with the allowable amount of pesticide residues directly stated in the law or the standard. However, there is no unified standard for the region, and it is incoherent for food-related industry, especially for food products produced, processed, and consumed in different Asian Pacific jurisdictions.

3.3.3 Mycotoxins and Other Harmful Substances

Mycotoxins are potentially hazardous secondary metabolites produced by fungi. They can enter the food chain through contaminated plant-based food components or toxigenic fungal development associated with the food. Mycotoxins can build up in various commonly cultivated crops, such as corn, soybeans, and peanuts, in the field and during transportation or storage (El-Sayed et al., 2022). Mycotoxin contamination of foods by mycotoxigenic fungi such as *Aspergillus*, *Alternaria*, and *Penicillium spp.* is a notable risk to food safety.

Implementation of the prerequisite programs, such as the hazard analysis and critical control point (HACCP)-based procedures, can minimise contamination of mycotoxin. HACCP system identifies, evaluates, and controls hazards crucial for food safety. At the same time, traditional methods can be employed for detoxification after the food is contaminated, such as extrusion and irradiation (Hamad

TABLE 3.2

Regulations on Pesticide Residues at Different Jurisdictions in Asia Pacific

Jurisdictions	Key Regulation/Standard	References
Australia	Agricultural and Veterinary Chemicals Code Act 1994	Australian Pesticides and Veterinary Medicines Authority (2021)
Cambodia	Proclamation No. 002 of the Ministry of Agriculture, Forestry, and Fisheries (MAFF)	Kingdom of Cambodia (2007)
Hong Kong SAR	Public Health and Municipal Services Ordinance (Chapter 132)	HKSARG (2012)
Indonesia	Law No. 18/2012 ("The Food Law")	United States Department of Agriculture (2019)
Japan	Food Sanitation Act (Act No. 233 of 1947)	Ministry of Health, Labour and Welfare of Japan (2019)
People's Republic of China	National Food Safety Standard - Maximum Residue Limit of Pesticides in Foods (GB2763—2021)	National Health Commission of the People's Republic of China (2021)
Malaysia	Pesticides Act 1974 (No. 149 of 1974)	The Government of Malaysia (2015)
Philippines	Presidential Decree No. 1144 creating the Fertiliser and Pesticide Authority and abolishing the Fertiliser Industry Authority	Department of Agriculture (2020)
Republic of Korea	Food Sanitation Act, Ministry of Food and Drug Safety Notification No. 2017-57 and administrative Advance Notice No. 2017-289, 2017-290, 2017-366	Ministry of Food and Drug Safety (2017)
Singapore	Food Regulations (Cap. 283)	Singapore Statutes Online (2024)
Taiwan	Article 15 of the Food Safety and Hygiene Act	Ministry of Health and Welfare (2023)
Thailand	B.E.2554 (2011)	Ministry of Public Health (2011)
Viet Nam	Section 30.3(1) and 30.5(1) of the Food and Drugs Act	Vietnam Sanitary and Phytosanitary (2021)

et al., 2023; Shanakhat et al., 2018). Some other harmful substances may affect food safety, such as radionuclides, synthetic chemical dyes (such as Sudan red dyes), and petroleum products (such as diesel).

3.4 Comparative Analysis of Jurisdictions, with a Focus on Asia Pacific

3.4.1 Similarities and Differences in Regulatory Approaches

Governments of different jurisdictions have adopted various yet similar regulatory approaches to control food safety, such as laws and regulations, import control, product recall mechanisms, and mandatory or voluntary reporting of illness or injury caused by the use or foreseeable misuse of foods.

Table 3.3 summarises the food safety regulatory approaches in different Asian Pacific jurisdictions. Most jurisdictions implement mandatory reporting of illness or injury caused by the use or foreseeable

TABLE 3.3

Food Safety Regulatory Approaches in Different Jurisdictions in the Asia Pacific

Jurisdiction/ Regulatory Approach	Mandatory Reporting[a]	Licensing and Approval Requirements to Import/ Export Food	Enforcement Authorities	Major Laws/ Regulations
Australia	✓	✓	• Australian competition and consumer commission ("ACCC") • National measurements institute ("NMI") • Australian border force • Local food authorities • (Food Standards Australia New Zealand, 2023)	• National measurement act 1960
Mainland China	• No mandatory reporting regime	✓	• State market supervision administration • General administration of customs • National health commission (NHFPC)	• PR China food safety law
Hong Kong SAR	• No mandatory reporting regime	✓	• Food and environmental hygiene department • Customs & excise department	• Part V of the public health and municipal services ordinance (Cap. 132) • Food and drugs (composition and labelling) regulations (Cap. • 132W) ("FDR") • Food safety ordinance (Cap. 612) ("FSO")
Indonesia	✓	✓	• Food & drugs supervisory agency • Directorate general of standardisation and consumer protection ministry of trade	• The food law

(Continued)

TABLE 3.3 (*Continued*)

Food Safety Regulatory Approaches in Different Jurisdictions in the Asia Pacific

Jurisdiction/ Regulatory Approach	Mandatory Reporting[a]	Licensing and Approval Requirements to Import/ Export Food	Enforcement Authorities	Major Laws/ Regulations
Japan	✓	✓	• Prefectural governor • Food sanitation inspectors at local health centres • Minister of health, labour and welfare • Secretary general of the CAA	• Food sanitation act • Health promotion act • Food labelling act
Malaysia	• No mandatory reporting regime	✓	• Ministry of agriculture and agro-based industry Malaysia • Customs • Department of agriculture • Ministry of domestic trade, co-operatives and consumerism	• Malaysian quarantine and inspection services act 2011 • Consumer protection act 1999
Philippines	✓	✓	• Food safety regulatory agencies	• The food safety act of 2013
Singapore	• No mandatory reporting regime	✓	• National environment agency • Agri-food & veterinary authority of Singapore	• Food (amendment) regulations 2017 • The sale of food act (Cap 283) • Environmental public health (food hygiene) regulations
Taiwan	✓	✓	• Taiwan food and drug administration • Department of health of each municipal government	• Act governing food safety and sanitation (Taiwan FDA, 2024)
Thailand	✓	✓	• Food and drug administration • Royal Thai police • Customs department	• Food act BE 2522 (AD 1979)
Vietnam	✓	✓	• Ministry of health, Vietnam food administration under ministry of health, departments of health • Vietnam competition authority under the ministry of industry and trade	• Decree No. 178/2013/ND-CP

[a] Mandatory reporting regime for the reporting of illness or injury caused by the use or foreseeable misuse of foods.

misuse of foods, which is an effective administrative tool to protect consumers. Multiple parties within the same jurisdiction often monitor and enforce food safety-related matters, such as the authority overseeing consumer's rights, customs, and health commission. It depends on the jurisdiction's administrative hierarchy.

In the Hong Kong Special Administrative Region (HKSAR), the fundamental food law is laid down in Part V of the Public Health and Municipal Services Ordinance (Cap. 132). It covers general protection for food purchasers, offenses concerning the sale of adulterated food, food hygiene, destruction of unfit food, etc. Various secondary regulations are under Part V of Cap. 132 apply to specific categories of food products, such as Food Adulteration (Metallic Contamination) Regulations (Cap. 132V) and Harmful Substances in Food Regulations (Cap. 132AF) (HKSARG, 2023).

Australia imposes a mandatory reporting system for food safety issues. The Australian Consumer Law imposes a duty to report on food suppliers who become aware of the death, severe injury, or illness of any person and consider that the death or serious injury or illness was caused, or may have been caused, by the use or foreseeable misuse of the relevant food (Authority, 2023).

The People's Republic of China (PRC) Food Safety Law sets forth statutory requirements for producing, circulating, recalling, and importing food products. In general, it applies to different categories of food products. There are additional regulations that apply to other specific food types. For example, the product formula of infant formula dairy products should be registered with the China Food and Drug Administration (CFDA). Some food product-specific mandatory national safety standards, such as the GB 7101-2015, apply to beverages (PRC, 2009). The PRC Food Safety Law also imposes other obligations on food businesses, such as verifying that the raw materials, additives, and associated products used in food production meet relevant food standards, keeping records for testing raw materials and additives of food products, etc.

In Indonesia, a person must report any known incidents resulting from contaminated food to the medical service unit, which shall take timely follow-up actions. Depending on the scale of the incident, the medical service unit may collect food samples and report the case to the unit that oversees health affairs and to Badan Pengawas Obat dan Makanan (BPOM), also known as the National Agency of Drug and Food Control. However, no specific regulation requires importers or manufacturers to report illness incidents to the BPOM (B.P.O.D.M.I., 2022).

According to the Japanese Food Sanitation Act (FSA), importation, manufacturing, processing, and distribution of foods or food additives that are not hygienic, may pose a health hazard, and livestock contracting certain diseases, etc., are prohibited (M.O.H.L.A.W., 2024, Japan MECSST, 2024).

The Food Act 1983 ("Act") and its subsidiary legislation primarily govern food quality and integrity in Malaysia. However, there is no mandatory reporting requirement of known cases of sickness or death due to the consumption of foods. Any such voluntary reporting could be submitted to the Ministry of Health Malaysia, the State Health Department's Food Safety and Quality Division, or directly to the Ministry of Health (P.R.K.K.M., 2024).

3.4.2 Strengths and Weaknesses of Different Regulatory Systems

As elaborated in the previous session, mandatory reporting systems, import food controls, and comprehensive food safety regulations have unique strengths and weaknesses within the broader landscape of regulatory frameworks. Mandatory or encouraged voluntary reporting requirements can enhance transparency and provide valuable data for risk assessment and response by the managing authorities but may also create administrative burdens for businesses, especially for small enterprises.

Import controls can help intercept contaminated or adulterated products at the border, but they can also be resource-intensive and potentially disrupt trade flows. Timely handling of imported food by customs is crucial for frozen or fresh food. Comprehensive food safety regulations offer robust protections but can also be criticised for their complexity and potential to stifle food-associated businesses. Ultimately, the most effective regulatory systems will strategically combine elements of these different approaches, balancing public health priorities with the needs of industry and consumers. By understanding the nuances and trade-offs, policymakers can work to strengthen food safety frameworks worldwide.

3.5 Emerging Challenges and Future Trends

The global food system and food type continue to evolve, and new and complex challenges are emerging that require a proactive and adaptive approach to food safety regulation. The rapid pace of technological advancement, the globalisation of supply chains, and the growing consumer demand for diverse and innovative food products all contribute to an increasingly dynamic regulatory landscape. Policymakers in different jurisdictions must grapple with issues such as integrating advanced traceability systems, regulating novel food processing methods, and harmonising standards across international or regional borders. Furthermore, the rise of e-commerce and the proliferation of online food platforms present unique challenges in ensuring the safety of food products delivered directly to consumers. To meet these emerging challenges, future food safety regulations must be flexible, evidence-based, and responsive to the changing needs of the industry and the public. We can build a more resilient and trustworthy food system that safeguards public health by anticipating and addressing these trends.

3.5.1 Emerging Food Contaminants and Associated Risks

Research has been conducted over the past decade on a wide variety of widely identified food contaminants, comparatively less on those emerging contaminants including but not limited to microplastics, metallic nanoparticles, endocrine disruptors, pharmaceuticals, hormones, and toxins, as well as industrially related synthetic dyes and dye-containing hazardous pollutants. Some are reported to be present in various types of food, such as freshwater fish, honey, milk, and beer (Diaz-Basantes et al., 2020; Picó et al., 2019). Even though human exposure to most emerging pollutants through ingesting contaminated food is inevitable and poses a risk to human health, associated guidelines or regulations are not yet well established for food safety.

For example, microplastics serve as vehicles for heavy metals and bisphenols, increasing health risks. However, the presence of microplastics in food has been overlooked in policy formulation, and there is no microplastic regulation in food or drinking water (Usman et al., 2022). For microplastic-associated chemicals such as bisphenol A and phthalates, most human exposures occur mainly through packaged food and beverages contaminated by migrating from packaging material (Baluka and Rumbeiha, 2016). Many jurisdictions have already restricted or prohibited BPA use in food contact materials, such as mainland China, Japan, Malaysia, and South Korea (SGS Kazakhstan Ltd., 2024). On the other hand, in Australia, certain retailers have voluntarily phased out the use of bisphenol A in plastic feeding bottles made of polycarbonate (Food Standards Australia New Zealand, 2024). These would be effective regulatory tools to avoid the presence of bisphenol A and phthalates in food products from the source.

Apart from emerging contaminants, climate change may also have a role to play in food safety, which enhances the occurrence and virulence of food-borne pathogens and subsequently results in a rise in food-borne infection and intoxication (Duchenne-Moutien and Neetoo, 2021). The spread and survival of zoonotic pathogens are affected by various environmental factors such as precipitation, humidity, and temperature. The change of these factors by climate change, for instance, the irreversibly raised ambient temperature and more frequent rainfall, would likely flavour the spread and survival of pathogens (Mora et al., 2022). Climate change may also affect food-related organisms or toxins. For instance, foods may be contaminated with pathogens to a level that imposes human health risks.

Progressively developing climate change is anticipated to positively affect infections and intoxications, resulting in a growing public health concern (Dietrich et al., 2023). Extreme weather events induced by climate change impact the emission and transformation of contaminants. For example, wildfires facilitated the emission of various contaminants, such as PAHs, VOCs, and particulate matter. Both floods and droughts affect contaminant dynamics across various phases (Bolan et al., 2023). That said, climate change plays a role in spreading infectious diseases, resulting in a thread of food safety.

3.5.2 Evolving Regulatory Strategies and Technologies

The food regulatory system provides a framework for the development and implementation of food standards, with the priority of protecting public health and safety. Innovations within the food industry have seen the introduction of new science and technologies applied to food production, which has implications for its safety. Emerging environmental issues, such as climate change, are encouraging more research and development on the environmental effects of how food is grown, manufactured, and processed (Witherspoon and Donse, 2023).

For instance, there has been a growing emphasis on using predictive modelling to assess emerging food-related risks, including food-borne pathogens and contaminants, and therefore mitigate potential risks in the food industry. Predictive modelling aligns with HACCP principles, contributing to a systematic approach to food safety. Integration of predictive modelling optimises preventive measures in food safety. Predictive mathematical modelling comprehends the behaviour of food microorganisms (Kumar et al., 2024).

3.5.3 Anticipated Changes in Food Safety Regulations

There is a lack of information on how food security is affected by emerging contaminants. Apart from regulating downstream, it is anticipated that the managing authorities could take a more proactive role in eliminating the source of pollution. For instance, to improve plastic waste management along microplastic legislation, to reduce the microplastic pollution in marine environments and, therefore, the presence of microplastics in seafood and foodstuff and associated potential health risks.

3.6 Conclusion

3.6.1 Importance of Ongoing Efforts in Ensuring Food Safety

Figure 3.1 shows the collaboration of different parties for food safety protection to improve human health. It illustrates the role and responsibility of different parties and various levels, from local to international. At the regional level, it involves legislation, enforcement of local laws, and other regulatory approaches. At the internal level, it consists of setting up international standards or approaches by

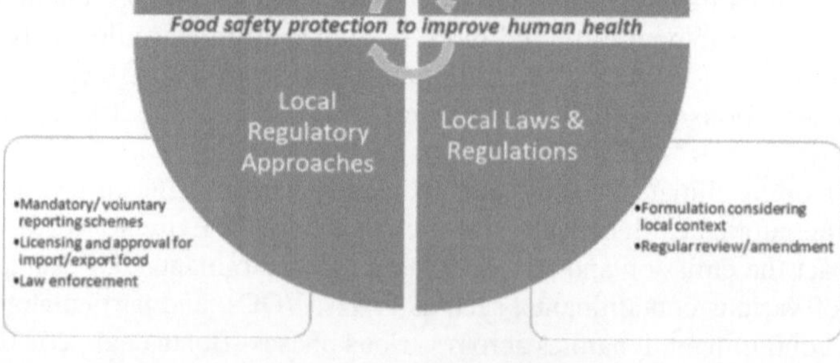

FIGURE 3.1 Food safety protection to improve human health (Fung et al., 2018).

global organisations or initiatives. Ensuring food safety is an ongoing and critical endeavour. Each party, from farm to fork, bears responsibility, and each step should adhere to food safety standards within its justification. We can protect consumers from food-borne illnesses and uphold public health and trust. Continued oversight, research, and education are crucial.

Governments, businesses, and consumers shall remain vigilant, quickly detecting and addressing emerging food safety risks. Technological innovations can also enhance traceability and safety. Through unrelenting efforts by all, we can ensure everyone safely enjoys the food. This is a never-ending task, but we must persist, for health and safety are paramount. Further research should focus on monitoring and eliminating emerging contaminants not currently regulated along the food supply chain and determining how much food safety is affected.

3.6.2 Call to Action for Strengthening Food Safety Regulations

Outdated regulations and lax enforcement would have left consumers vulnerable to unnecessary risks. Different justifications shall demand higher standards from producers, distributors, and retailers to protect public health. Rigorous inspections, transparent supply chain monitoring, and swift penalties for incompliance are essential. Continuous funding for research and development of innovative detection and control technologies is also critical. The above would drive the reforms necessary to ensure that food for human consumption is wholesome and secure.

REFERENCES

Australian Pesticides and Veterinary Medicines Authority. (2021). Agricultural and Veterinary Chemicals Code (MRL Standard) Instrument 2019. Retrieved Jul 20, 2024, from https://faolex.fao.org/docs/pdf/aus205405.pdf

Authority, A. N. Z. F. (2023). *Safe Food Australia: A Guide to the Food Safety Standards*. Australia New Zealand Food Authority, Australia.

Baluka, S. A. & Rumbeiha, W. K. (2016). Bisphenol A and food safety: Lessons from developed to developing countries. *Food and Chemical Toxicology*, 92, 58–63.

Bolan, S., Padhye, L. P., Jasemizad, T., Govarthanan, M., Karmegam, N., Wijesekara, H., Amarasiri, D., Hou, D., Zhou, P., Biswal, B. K., Balasubramanian, R., Wang, H., Siddique, K. H. M., Rinklebe, J., Kirkham, M. B., & Bolan, N. (2023). Impacts of climate change on the fate of contaminants through extreme weather events. *Science of the Total Environment*, 909, 168388.

B. P. O. D. M. I. (2022). Profil BPOM. Retrieved May 30, 2024, from https://www.pom.go.id/profil

C. A. (2024). About Codex Alimentarius. Codex Alimentarius. Retrieved May 30, 2024, from https://www.fao.org/fao-who-codexalimentarius/about-codex/en/

Department of Agriculture. (2020). Pesticide Regulatory Policies and Implementing Guidelines. Retrieved Jul 20, 2024, from https://www.cirs-group.com/files/aty8b0q8or9c/content/file/20240105/1704424493330283623010050.pdf

Diaz-Basantes, M. F., Conesa, J. A., & Fullana, A. (2020). Microplastics in honey, beer, milk and refreshments in Ecuador as emerging contaminants. *Sustainability*, 12(14), 5514.

Dietrich, J., Hammerl, J. A., Johne, A., Kappenstein, O., Loeffler, C., Nöckler, K., Rosner, B., Spielmeyer, A., Szabo, I., Richter, M. H., & Richter, M. H. (2023). Impact of climate change on food-borne infections and intoxications. *Journal of Health Monitoring*, 8(Suppl 3), 78.

Duchenne-Moutien, R. A. & Neetoo, H. (2021). Climate change and emerging food safety issues: A review. *Journal of Food Protection*, 84(11), 1884–1897.

El-Sayed, R. A., Jebur, A. B., Kang, W., & El-Demerdash, F. M. (2022). An over view on the major mycotoxins in food products: Characteristics, toxicity, and analysis. *Journal of Future Foods*, 2(2), 91–102.

European Commission, Directorate-General for Agriculture and Rural Development, Directorate-General for Trade. (2017). Ensuring Food is Safe: The Veterinary and Phytosanitary System of the European Union Explained, Publications Office. https://data.europa.eu/doi/10.2762/969146

FAO WHO. (2024). About Codex Alimentarius, FAO/WHO Food Standards Programme. https://www.fao.org/fao-who-codexalimentarius/about-codex/en/"

FEHD. (2024). Guideline on Food Adulteration (Metallic Contamination) (Amendment) Regulation 2018, Food and Environmental Hygiene Department. https://www.cfs.gov.hk/english/whatsnew/whatsnew_fstr/files/Metal%20guidelines-eng.pdf

Food Standards Australia New Zealand. (2024). Bisphenol A (BPA). Retrieved Jul 20, 2024, https://www.foodstandards.gov.au/consumer/chemicals/bpa

Fung, F., Wang, H. S., & Menon, S. (2018). Food safety in the 21st century. *Biomedical Journal*, 41(2), 88–95.

Hamad, G. M., Mehany, T., Simal-Gandara, J., Abou-Alella, S., Esua, O. J., Abdel-Wahhab, M. A., & Hafez, E. E. (2023). A review of recent innovative strategies for controlling mycotoxins in foods. *Food Control*, 144, 109350.

HKSARG. (2012). Pesticide Residues in Food Regulation. Retrieved Jul 20, 2024, from https://faolex.fao.org/docs/pdf/hk121711.pdf

HKSARG. (2018). Food Adulteration (Metallic Contamination) (Amendment) Regulation 2018. Gazette of the Government of the Hong Kong Special Administrative Region. Retrieved Jul 17, 2024, from https://www.gld.gov.hk/egazette/pdf/20182223/es220182223113.pdf

HKSARG. (2023). Part V (Food and Drugs) of the Public Health and Municipal Services Ordinance (Cap. 132) (Highlights). Center for Food Safety. Retrieved May 26, 2024, from https://www.cfs.gov.hk/english/food_leg/food_leg_list.html

Japan – MECSST. (2024). Ministry of Education, Culture, Sports, Science and Technology (2015). Standard Tables of Food Composition in Japan, Seventh Revised Edition. Available from: URL: https://www.mext.go.jp/en/policy/science_technology/policy/title01/detail01/sdetail01/sdetail01/1385122.htm

Kingdom of Cambodia. (2007). Proclamation No.002 MAFF, 03/01/2007 on the List of Maximum Residue Limit of Pesticides in Agricultural Products of Plant Origin. Retrieved Jul 20, 2024, from https://faolex.fao.org/docs/pdf/cam81986.pdf

Kumar, V., Ahire, J. J., & Taneja, N. K. (2024). Advancing microbial food safety and hazard analysis through predictive mathematical modeling. *The Microbe*, 2, 100049.

Lin, X., Chang, S. C., Chou, T. H., Chen, S. C., & Ruangkanjanases, A. (2021). Consumers' intention to adopt blockchain food traceability technology towards organic food products. *International Journal of Environmental Research and Public Health*, 18(3), 912.

Ministry of Food and Drug Safety. (2017). Pesticide MRLs in Food 2017. Retrieved Jul 20, 2024, from https://www.fao.org/faolex/results/details/en/c/LEX-FAOC190507/

Ministry of Health, Labour and Welfare of Japan. (2019). Guideline for Application for Establishment and Revision of Maximum Residue Limits for Agricultural Chemicals used outside Japan. Retrieved Jul 20, 2024, from https://faolex.fao.org/docs/pdf/jap189799.pdf

Ministry of Health and Welfare. (2023). Standards for Pesticide Residue Limits in Foods. Retrieved Jul 20, 2024, from https://faolex.fao.org/docs/pdf/tw137925.pdf

Ministry of Public Health. (2011). Food Containing Pesticide Residues. Retrieved Jul 20, 2024, from https://faolex.fao.org/docs/pdf/tha159851.pdf

M. O. H. L. A. W. (2024). Food Safety. Retrieved May 30, 2024, from https://www.mhlw.go.jp/stf/seisakunitsuite/bunya/kenkou_iryou/shokuhin/index_00006.html

Mora, C., McKenzie, T., Gaw, I. M., Dean, J. M., von Hammerstein, H., Knudson, A. T., Setter, R. O., Smith, C. Z., Webster, K. M., Patz, J. A., & Franklin, E. C. (2022) Over half of known human pathogenic diseases can be aggravated by climate change. *Nature Climate Change*, 12(9), 869–875.

National Health Commission of the People's Republic of China. (2021). National Food Safety Standard - Maximum Residue Limit of Pesticides in Foods (GB2763—2021). Retrieved Jul 20, 2024, from https://faolex.fao.org/docs/pdf/chn215904.pdf

P. R. K. K. M. (2024). Food Act 1983. Retrieved May 30, 2024, from https://www.moh.gov.my/index.php/database_stores/attach_download/317/16

Picó, Y., Belenguer, V., Corcellas, C., Díaz-Cruz, M. S., Eljarrat, E., Farré, M., Gago-Ferrero, P., Huerta, B., Navarro-Ortega, A., Petrovic, M., Rodríguez-Mozaz, S., Sabater, L., Santín, G., & Barceló, D. (2019). Contaminants of emerging concern in freshwater fish from four Spanish Rivers. *Science of the Total Environment*, 659, 1186–1198.

Prabhakar, S. V. R. K., Sano, D., & Srivastava, N. (2010). Food safety in the Asia-Pacific Region: current status, policy perspectives and a way forward. In *Sustainable Consumption and Production in the Asia-Pacific Region: Effective Responses in a Resource Constrained World*, 215–238. Institute for Global Environmental Strategies, Hayama, Japan.

PRC. (2009). Order of the President of the People's Republic of China (No. 9). Food Safety Law of the People's Republic of China. Retrieved May 26, 2024, from http://www.npc.gov.cn/zgrdw/englishnpc/Law/2011-02/15/content_1620635.htm

SGS Kazakhstan Ltd. (2024). Food Contact Material Regulations – Asia Pacific Region. Retrieved Jul 20, 2024, from https://www.sgs.com/en-kz/services/food-contact-material-regulations-asia-pacific-region

Sha, D., Du, P., & Wu, L. (2024). Classification and prediction of food safety policy tools in China based on machine learning. *Journal of Food Protection*, 87, 100276.

Shanakhat, H., Sorrentino, A., Raiola, A., Romano, A., Masi, P., & Cavella, S. (2018). Current methods for mycotoxins analysis and innovative strategies for their reduction in cereals: An overview. *Journal of the Science of Food and Agriculture*, 98(11), 4003–4013.

Singapore Statutes Online. (2024). Food Regulations. Retrieved Jul 20, 2024, from https://sso.agc.gov.sg/SL/SFA1973-RG1?DocDate=20180327&ProvIds=Sc9-

Taiwan Food and Drug Administration, Taiwan FDA. (2024). Food Nutrients & Composition Database (New Edition) (available in Chinese). Available from: URL: https://consumer.fda.gov.tw/Food/TFND.aspx?nodeID=178

The Government of Malaysia. (2015). Pesticides Act 1974 (No. 149 of 1974). Retrieved Jul 20, 2024, from https://www.fao.org/faolex/results/details/en/c/LEX-FAOC013354/

United States Department of Agriculture. (2019). Food and Agricultural Import Regulations and Standards Report. Retrieved Jul 20, 2024, from https://apps.fas.usda.gov/newgainapi/api/report/downloadreportbyfilename?filename=Food+and+Agricultural+Import+Regulations+and+Standards+Report_Jakarta_Indonesia_3-18-2019.pdf

Usman, S., Abdull Razis, A. F., Shaari, K., Azmai, M. N. A., Saad, M. Z., Mat Isa, N., & Nazarudin, M. F. (2022). The burden of microplastics pollution and contending policies and regulations. *International Journal of Environmental Research and Public Health*, 19(11), 6773.

Vietnam Sanitary and Phytosanitary. (2021). Vietnam Sanitary and Phytosanitary Notification Authority and enquiry point. Notice - Notification of WTO Members: Proposal to amend the List of Maximum Residue Limits (MRLs) for Veterinary Drugs in Foods (Proposed MRL 2021-1). Retrieved Jul 20, 2024, from http://www.spsvietnam.gov.vn/en/gspsncan1418

Witherspoon, B. & Donse, E. (2023). *Food Safety—Regulations and Policy in Australia and New Zealand*. Australia New Zealand Food Authority, Australia.

World Health Organisation (WHO). (2015). *Estimates of the Global Burden of Food-Borne Diseases*. 255p. WHO, Switzerland.

4

Organic Agriculture as a Tool to Promote, Assess and Communicate Health and Environmental Risks: The Case of Organic Tea in China

Jiping Sheng, Wenfan Su, Jiaxin Shen, Denis Ivanov, and Ksenia Gerasimova

4.1 Introduction

Organic certification is a communication system confirming that the products do not contain harmful substances. It is obtained using methods that preserve natural resources and improve the condition of ecosystems. Such labeling is assigned only to products from farms, animal husbandry, poultry farming, and beekeeping. It serves as a monitoring and evaluation instrument and a tool of communication with consumers willing to pay a premium to have assurance that the product they buy is safe and the process is controlled. In this chapter, we discuss how organic production and regulation can help to alleviate ecological and environmental risks in Chinese tea production.

A definition for organic agriculture given by FAO (Food and Agriculture Organization) is multidimensional: it is "one of several approaches to sustainable agriculture," "a holistic production management system which promotes and enhances agro-ecosystem health, including biodiversity, biological cycles, and it is "one of several approaches to sustainable agriculture," and "it is opposed to using synthetic inputs, such as synthetic fertilizers and pesticides, veterinary drugs, genetically modified seeds and breeds, preservatives, additives and irradiation" (FAO, 1999).

Organic agriculture is also an ideology supported by global environmental movements such as IFOAM (International Federation of Organic Agriculture Movement): "Organic Agriculture is a production system that sustains the health of soils, ecosystems, and people. It relies on ecological processes, biodiversity, and cycles adapted to local conditions rather than using inputs with adverse effects. Organic Agriculture combines tradition, innovation, and science to benefit the shared environment and promote fair relationships and a good quality of life for all involved" (IFOAM, 2005a, 2005b). The IFOAM has formulated four principles of organic agriculture: the principle of health refers to preserving the health of soil, plants, animals, humans, and planet as a whole; the principle of ecology promotes respect of living ecological systems and cycles; the principle of fair is about social justice and food sovereignty; and the principle of care is the main principle of sustainability promoting responsibility toward current and future generations (IFOAM, 2005b). Since organic agriculture is a Western concept, it is difficult to translate the word "organic" in Chinese (有机), and it can be interpreted in different ways: The idea of organic food comes from abroad, and it is translated into "有机." It is a confusing word in Chinese. You can say it is ecological, natural, etc. However, some people compare it to an inorganic mineral (无机物). Last but not least, the Chinese often use sentences like "形成一个有机整体," which means a system or an institution that works well just like a human body, as a natural system. Therefore, when consumers read the word organic, they often struggle to understand what is organic (Gerasimova et al., 2021). This creates a case of being lost in translation "when an obscure and vague meaning leads to the situation when organic product is not completely recognized by consumers in a commodity market" (Gerasimova et al., 2021). Even people who work with organic tea, let alone tea consumers, have a slightly different understanding of what it means.

DOI: 10.1201/9781003492115-5

Camellia sinensis, commonly known as tea, is a traditional crop grown in the tropics and subtropics and requires acid soil. It is a commodity crop that significantly contributes to the economies of several countries in Asia and Africa. Environmental constraints include water deficiency due to inadequate and/or poorly distributed rainfall, application of fertilizers for higher production, and high susceptibility to pathogens and pest attacks in hot, humid climates. Application of pesticides and fungicides presents health risks to tea consumers. Organic cultivation with lower and safer usage of pesticides and fertilizers is a more sustainable and environmentally friendly approach to producing this popular crop. Despite having a marginal share in the global tea industry (less than 1.3% in 2017) (Hajra, 2017), the organic tea sector constantly rises. In 2023, the global organic tea market reached 1.3 billion USD, and the experts are expecting the market to reach US$ 2.6 Billion by 2032, exhibiting a compound annual growth rate (CAGR) of 7.94% during 2024–2032 (IMARC, 2023).

While the numbers speak for themselves, organic production and certification do not entirely rule out risks. It could be a good example to discuss alleviating and communicating health and environmental risks. This chapter will review ecological and human health risks related to the production and consumption of Chinese tea and discuss how organic production and certification serve as a risk management tool in the contemporary complex risk environment.

For empirical data, we rely on our series of socio-anthropological studies of organic tea producers in the Huangshan Mountain, which is privately owned, and Xinanyuan, located in the Liukou area, the Xiuning County, Anhui Province and interviews with tea masters in Beijing and Shenzhen in 2017–2024. To categorize risks, we refer to the study conducted by a research team from Renmin University that summarized and weighted the risks present in the organic tea industry (Sheng et al., 2024). This analysis determined that three indicators—soil conditions in the tea garden, soil management methods, and pest, disease, and weed prevention measures—exhibit "Critical risk" levels. If mismanaged, these factors could precipitate severe negative outcomes, such as diminished tea quality, decreased crop yield, and substantial ecological harm to the tea garden. Consequently, these findings emphasize the necessity of careful management and preventive strategies to mitigate potential threats and maintain the sustainable productivity of the tea garden. Another indicator was establishing an internal management system and product recall mechanism associated with a "Serious risk" level. Unaddressed could result in considerable negative impacts, including operational inefficiencies, financial repercussions, and reputation damage. Below we will illustrate these risks (soil conditions, soil management, pest and weed control, and internal management and control) and discuss how current Chinese organic agriculture regulations can address them. We also briefly review health risk management in the organic tea industry and related consumer behavior.

4.2 Risk Management in Organic Tea Production

4.2.1 Environmental and Human Risks in Tea Organic Production

Organic tea was the first organic food in China, certified in full (SKAL) in Zhejiang Province in 1990 (Sheng et al., 2024). From then on, more organic tea is being produced and certified in China. A steady growth in the country's organic tea acreage is shown in Figure 4.1.

Tea production consists of several stages: plucking, when tea pickers pick the tea leaves manually; withering; rolling; fermentation/oxidization; firing/drying; and sorting and grading. There are no additional risks in comparison to non-organic tea production. On the contrary, some environmental risks will be lower due to avoiding most harmful pesticides. Since tea is a natural product, its quality and risks are associated with the environment where it grows. Tea plants have specific characteristics determined by their cultivation's geography. These include temperature, light, humidity, and soil. The combination of these particular factors limits the scope of suitable farmlands. Tea gardens are being lifted or cultivated near urban areas to increase production acreage (Huang, 2017, cited in Gerasimova, 2021).

A hot temperature is required during harvest (spring—early autumn), not lower than the daily average of 20°C. Moreover, daily temperature differences are also desirable for most varieties, such as a hot day and a cool night. When a plant needs to work hard to survive, it grows less intensively, but the content of

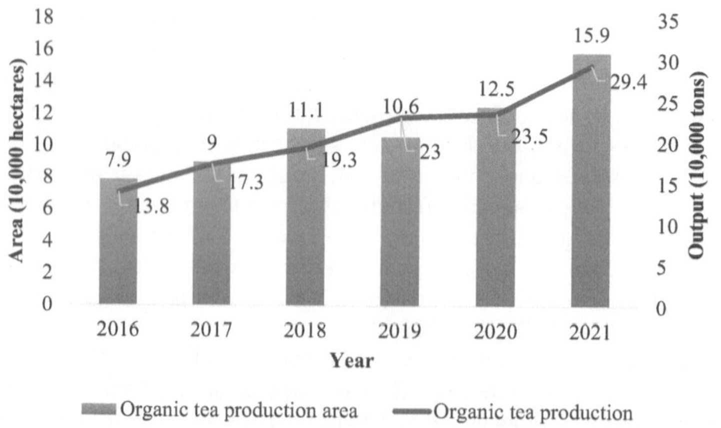

FIGURE 4.1 Production area and output of organic tea (including products during the organic conversion period) in China from 2016 to 2021 (*Source*: State Administration for Market Regulation, China Agricultural University 2022).

various substances in branches and leaves increases. Therefore, the best tea varieties are obtained where the plant is on the edge of the comfort zone (Interview with the tea master, Shenzhen, November 2024). There should be plenty of sun for tea production, but it is better if the light is diffused rather than direct. For example, when growing some lowland tea varieties, tall acacias, and oak shade plantations, frequent fogs perform this function in high-altitude areas, giving the air the necessary humidity. In addition to the drainage factor, the degree of saturation of the soil with minerals is also essential. Tea plants require medium loamy soils with high iron content and deep groundwater (Daraselia et al., 1989). The quality of tea and many environmental risks come from soil management.

Two particular matters, as mentioned above, should be addressed in reviewing environmental risks arising from tea production—the application of fertilizers and pesticides.

4.2.1.1 Risks from the Application of Fertilizers

The soils of tea plantations (red earth, yellow earth, and podzolic) are characterized by low natural fertility. The high air temperature and abundance of precipitation provide optimal conditions for the normal growth and development of the tea plant and contribute to the intensive decomposition of the organic part of the soil with subsequent washing and flushing of nutrients. Thus, the average tea plants grown from 1 hectare consume 160 kg of nitrogen, 20 kg of phosphorus, and 86 kg of potassium. To maintain high yielding capacity of plantations, the nutrients consumed must be replenished by applying fertilizers (Didmanizde, 2021).

Nitrogen is the main component of protein; therefore, its availability in agriculture is one of the determining factors of yield. For most plants, the source of nitrogen is ammonia and soil nitrates. Only some plants can absorb the molecular nitrogen of the atmosphere and convert it into organic nitrogenous substances. Therefore, it is vital to biologically bind atmospheric nitrogen as much as possible and direct it to the nutrition of plants and soil microorganisms (Kretovich, 1980). In the soil, bound nitrogen is represented by four compounds: ammonium salt nitrogen, nitrate nitrogen, organic nitrogen of proteins, and their cleavage products—amino acids, peptides, amines, amides, and humus nitrogen. In natural conditions, soil microorganisms are of great importance for the nutrition of plants with nitrogen, which mineralizes the organic nitrogen contained in the soil, eventually converting it into ammonia (Kretovich, 1972), which is the starting compound used by plants for the synthesis of amino acids and proteins. The content of nitrogen compounds absorbed by the plant, even in the same soil, is subject to significant fluctuations. Large amounts of nitrogen are removed from the soil with the harvest, which is not replenished by nitrogen fixation of the atmosphere with symbiotic and free-living nitrogen fixators and decomposition of plant and animal residues in the soil. Therefore, nitrogen fertilizers are needed for high and stable yields (Koren'kov et al., 1979). Out of the three most essential elements of plant mineral nutrition—nitrogen, phosphorus, and potassium—nitrogen is the most deficient.

Nitrogenous substances—proteins, amino acids, and caffeine—play a significant role in improving the quality of tea. With enzymatic properties, proteins regulate metabolic processes in the plant. Amino acids are also involved in the formation of flavor during the processing of green tea leaves (Didmanidze, 2021).

Early studies, such as Elliot and Whitehead (1934), showed that tea producers relied on fertilizers from the beginning of commercial cultivation since tea was an intense monocrop. In most tea-producing countries, fertilization's efficiency reached about 50%–60% of the total efficiency of all agro-technical techniques. Contemporary studies show that global crops use only 50% of the applied nitrogen effectively (Govindasamy et al., 2023).

While it is impossible to exclude nitrogen application from yield production, its unrestricted use causes water, soil, and air pollution. For example, excessive nitrogen content in the ground causes the growth of weeds or other harmful crops that can compete with valuable plants and reduce yields. An increased amount of nitrogen in the air causes acid precipitation, harming the vegetation and living organisms. Up to 30% is lost through nitrogen application through surface runoff (Chen & Lin, 2016). Contemporary tea studies revealed that average fertilizer inputs generally surpassed the tea plant's nutrient demands (Sitienei et al., 2013).

In the meantime, parallel techniques can partially compensate for the deficit in nitrogen and thus help reduce the application of nitrogen fertilizers, and they are aligned with organic cultivation. In the 1970s, Rahman F. summarized the results of the first extended field experiments with nitrogen fertilizers in commercial tea production (cited in Daraselia, 1989). They showed that with widely varying indicators of soil fertility, the curves of responsiveness to nitrogen did not change significantly. However, they noted that tea varieties significantly influenced the nitrogen demand for leaf tea. Equally important was the presence of other factors of agricultural technology, such as shading, the method of pruning, and the presence of optimal amounts of other nutrients, especially potassium. While shading of the tea plots could partially compensate for the lack of nitrogen, with age, the need for tea plants for nitrogen decreased, and in heavily pruned plantations, the need for nitrogen was lower. It was found that the best tea yields are achieved using organic and inorganic nutrient sources (Gunapala & Ulpathakumbura, 2021).

To cultivate new tea plantations in a suitable natural environment, farmers interfere in the natural landscape and clearance of natural and secondary forests in subtropical China. The conversion of forest hillslopes into tea gardens changes the slope structure, soil properties, and vegetation, which include such soil nutrient losses as nitrogen and phosphorous through the surface runoff along the hillslopes, and thus strongly contributes to the loss of soil nutrients and local water quality deterioration (Wang et al., 2023).

In a tea garden, the land plot between rows of tea plants has the most concentration of nitrogen fertilizers. This is seen in the correlation between the higher use of nitrogen fertilizers and the higher content of theanine, which has increased due to the use of nitrogen fertilizers. This affects the taste, and tea producers must address this during the processing stage. Organic tea has a low level of theanine but is rich in polyphenols, which makes tea tasty and beneficial for human health (Interview with the tea master, Shenzhen, November 2024).

Increasing soil fertility levels by other means decreased the need for nitrogen in tea plants. Thus, fertilizing the soil with organic compost can help to reduce the use of nitrogen fertilizers necessary for the healthy growth of tea plants (Gerasimova et al., 2021). Under the application of organic fertilizers (labeled 15N), plants use less nitrogen fertilizer. Less of it is lost and firmly fixed in the soil than nitrogen mineral fertilizers. However, laboratory tests have shown that variations in 15 N values start to appear only after three years of continuous organic cultivation, and organic tea did not always have higher 15N values than commercial nonorganic teas (Hayashi et al., 2011).

Another related matter to mention is soil acidification. It directly affects the growth of tea trees and tea yield. Jia et al. (2023) found that increasing soil pH (3.29~5.32) helps to increase the nutrient content of the rhizosphere soil of tea trees, thus improving the antioxidant enzyme activity and photosynthesis capacity of tea tree leaves and, as a result, promoting plant growth.

4.2.1.2 Risks from Weed and Pest Management

Untreated weeds contribute to any tea-producing country and can lose 50%–70% of tea yields (Deka & Barua, 2015). The general risks from weed to tea plants include restricted branching and frame development and harboring pathogens and pests as alternative hosts. Despite dislikes from environmental activists, herbicide application remains a cheap and effective means of weed control. Manual weeding is a difficult task and also damages soil fertility. According to tea farmers, when manual weeding is done, weeds physically remove 32 kg N, 40 kg K, and 320 kg Carbon from the field, reducing the organic carbon and nutrients" (Rajaduraj, 2021).

Farmers' observations confirmed that mature tea has a 20% yield loss after 6 months without regular weeding. If manual weeding is resorted to, yield loss after 4 months is 15%. Herbicide spraying creates a mulch of dead weeds on the surface, improves water retention, adds organic matter, and recycles nutrients weeds remove. In contrast, when manual weeding is done, soil fertility and water holding capacity of the soil are lost due to erosion, and 30–45 cm of topsoil is lost, as a result leading to a 30%–50% loss of crop productivity (Rajaduraj, 2021).

The presence of weeds significantly reduces harvester productivity when weed foliage reaches the plucking table, interfering in harvesting operations, reducing harvester output and income, and reducing quality of life. Harvester's safety and well-being at work are compromised by leeches and reptiles in the fields because of profuse weed growth (Rajaduraj, 2021).

The application of pesticides means the presence of pesticide residue in food products, including tea. General health risks associated with pesticides are toxicity, carcinogenicity, negative fertility, and neurological effects (Ahmad et al., 2024).

International legislation sets the limits for the permitted maximum of pesticide residues. The strictest levels are set by the European Union (Fernandez et al., 2023). Detection of pesticide residues in tea is challenging due to the co-extraction of highly complex components in tea matrices and targeted compounds. The matrix effect can disturb the accuracy of the final residue assessment results (Meng et al., 2022).

Recent studies have shown that proper management with organic fertilizer replacement can effectively mitigate soil acidification and nutrient deficiency in tea plantations and increase soil organic matter. Replacing pesticides with organic alternatives is also beneficial for mitigating environmental risks. Still, differences in pesticide residue between various treatments were not evident due to the rapid degradation of pesticides (Xie et al., 2019).

4.2.1.3 Health Risks and Benefits of Organic Tea

Drinking tea is beneficial for human health. For example, organic acids in tea leaves contribute to health benefits such as antioxidation, promotion of digestion and absorption, acceleration of gastrointestinal transit, and regulation of intestinal flora (Chen et al., 2023). It also has hepaprotective effects (Yilmaz et al., 2024). Organic food is attributed to many health benefits resulting from its lower pesticide, synthetic fertilizer, and heavy metal exposure. In Table 4.1, we summarized the upper limits of undesirable toxic elements, such as heavy metals, which come from soil and water and can be found in organic food, as set by the Chinese government.

A significant problem, however, is the estimation of its actual health benefits. In their review of 35 published papers as of January 2019, Vigar et al. (2019) state that, while most of the studies are predominantly observational, they show health benefits of consuming organic food such as "reduced incidence of infertility, birth defects, allergic sensitization, otitis media, pre-eclampsia, metabolic syndrome, high BMI, and non-Hodgkin lymphoma."

It is necessary to indicate that no studies have been conducted to demonstrate the health benefits of organic tea versus conventional tea. In this case, we can only hypothesize that, with rising health risks arising from conventional pesticide residues in tea, organic tea has fewer such risks, and drinking tea has medicinal effects.

However, the limitations of observational studies are quite severe since organic food consumers are more likely to be better educated, have higher socio-economic status, and are health-conscious in

TABLE 4.1
The Upper Limits of Undesirable Elements of the Organic Tea Garden

Quality Standard	Item	The Upper Limits of Undesirable Elements
Organic tea garden soil	pH value ≦	4.0–6.5
	Cadmium/(mg/kg) ≦	0.20
	Mercury/(mg/kg) ≦	0.15
	Arsenic/(mg/kg) ≦	40
	Lead/(mg/kg) ≦	50
	Chromium/(mg/kg) ≦	90
	Copper/(mg/kg) ≦	50
Organic tea garden irrigation water	pH value ≦	5.5–7.5
	Mercury/(mg/L) ≦	0.001
	Cadmium/(mg/L) ≦	0.005
	Arsenic/(mg/L) ≦	0.05
	Lead/(mg/L) ≦	0.1
	Chromium (hexavalent)/(mg/L) ≦	0.1
	Cyanide/(mg/L) ≦	0.5
	Chloride chloride/(mg/L) ≦	250
	Fluoride compounds/(mg/L) ≦	2.0
	Petroleum/(mg/L) ≦	5

Source: Jie (2016); State Administration for Market Regulation, China Agricultural University (2022).

general, which might limit the opportunities to interpret the estimates causally. On the other hand, randomized control trials have their limitations: frequently, just a single food item is substituted. Even studies of whole diet interventions are often carried out within a short time frame and measure only intermediate outcomes such as pesticide excretion. Their ability to examine long-term clinically relevant health outcomes is still problematic.

A recent review of evidence by Jiang et al. (2024) included 23 observational and 27 interventional studies. However, while making broad conclusions on the beneficial health effects of organic food consumption, it also stressed the importance of studying its long-term health consequences. It noted the limitations of observational studies, such as organic food consumers being younger and thinner, having diets of higher nutritional quality, and healthier lifestyles.

Specifically, Chinese consumers who perceive organic tea as a healthy option and status symbol are likelier to state their intentions to purchase organic tea (James et al., 2019). Organic tea has lower risks than conventional tea but still requires proper research and process compliance with organic regulations.

4.2.2 Risk Assessment in Organic Tea Certification in China

The unique characteristics of tea necessitate more specific operational norms for its production, processing, and certification. There are two types of agricultural certification for compliance with the requirements that speak about the product's naturalness and environmental friendliness. These are ecological and organic certifications. In eco-certification, the requirements are softer and more straightforward (but this does not mean that a product with eco-labeling is worse than a product with organic labeling). Complete transfer to organic cultivation under all organic requirements is a very long-term and expensive process; thus, many farms choose a softer version—ecological certification, at least as a temporary transition.

The first organic certification in China was introduced in 1989 by the Rural Ecological Research Centre of the Nanjing Institute of Environmental Sciences, a branch of the State Environmental Protection Agency, which became the first member from China to join the IFOAM. The first organic inspectors came from Europe and the United States (Gould, 2017). However, the current list of the top ten organic certifiers, which is provided in Table 4.2, includes only Chinese organizations, with Beijing Continental

TABLE 4.2

List of the Top Ten Certification Bodies Based on the Number of Organic Product Certificates Issued in China in 2021

Certification Authority	Number of Certified Enterprises (Number)	Number of Certificates (Sheets)	Number of Organic Certificates (Sheets)	Number of Conversion Certificates (Sheets)
Beijing Continental Hengtong Certification Co., Ltd.	2577	3883	2934	949
Nanjing Guohuan Organic Product Certification Center Co., Ltd.	1055	2544	2127	417
China Organic Food Certification Center (COFCC)	1068	1711	1392	319
Hangzhou Wantai Certification Co., Ltd	941	1487	1241	246
Zhonglv Guozheng (Beijing) Certification Center Co., Ltd.	1084	1434	589	845
Hangzhou Gelu Certification Co., Ltd.	596	796	569	227
OXIT Certification Co., Ltd.	567	786	534	252
Hangzhou Zhongnong Quality Certification Center Co., Ltd.	409	709	562	147
Beijing Zhong'an Quality and Environment Certification Center Co., Ltd.	370	565	443	122
Fangyuan Certification Group Co., Ltd.	308	533	363	170

Certification Authority

Hengtong Certification Co., ltd (CHTC), Nanjing Guohuan Organic Product Certification Center Co., Ltd., and China Organic Food Certification Center (COFCC) being top three current leading certifying organizations. For export, a Chinese organic food product must meet international organic regulations by IFOAM, the United Nations, the European Union, the USA, and Japan (Table 4.3).

Only 7% of China's tea acreage can meet all export requirements for organic tea imposed by Japan, the European Union, and the U.S. National Organics Program by USDA. This is known as Certified Organic (Tier 1) (Bolton, 2019). However, 50% of Chinese tea, or around 1.4 billion kg, meets Ecological Certification requirements. In February 2015, China adopted a "zero-growth action plan" that regulates fertilizer and pesticide use to eliminate "excessive application of fertilizer and blind application, which brings about cost increase and environmental pollution" (Bolton, 2019). This explains the recent high number of eco-labeled tea farms in China.

Organic and eco-certification are not risk-free themselves. Risk factors associated with organic product certification can be categorized into five key areas: certification and inspection arrangements, inspector deficiencies, sampling and inspection methods, post-inspection supervision, and customer-related risks. International comparison has identified similar risk factors within the European Union's organic certification process, focusing on compliance with Regulation 834/2007 (Gambelli et al., 2011). Zhen et al. (2020) developed a risk-based indicator for organic crop production, finding production, management, and environment as the highest concern. Therefore, a comprehensive and consistent approach to risk assessment still needs to be developed for the organic tea certification process.

Maintaining the integrity of the certification process is complex and laden with risks at various stages, from environmental conditions to processing operations. This complexity is further compounded by discrepancies in the interpretation of organic standards across various jurisdictions (Zhen et al., 2020). Such inconsistencies can undermine consumer confidence and corporate credibility, emphasizing the need for risk assessment within the organic tea certification process. Establishing a risk assessment index system soon would facilitate organic tea gardens or companies passing organic tea certification, thereby fostering consistency.

TABLE 4.3
The Organic Tea Standards

National/Regional/Relevant Organization	Standards or Regulations
Alliance of organic farming movements	Basic standards for organic farming
Codex alimentarius commission	Guide to the production, processing, marketing, and labeling of organic food
The European Union	Regulations on organic production and processing
The United States	Organic Food Act and its implementing regulations (NOP regulations)
Japan	JAS Act on organic farming
China	Food safety law of the People's Republic of China, agricultural product quality safety law of the People's Republic of China.
	Regulations on certification and accreditation of the People's Republic of China.
	Organic product certification management measures.
	GB/T 19630-2019 requirements for production, processing, labeling, and management system of organic products, NY 5196-2002 organic tea, NY 5199-2002 environmental conditions for organic tea production, NY 5197-2002 technical regulations for organic tea production, NY 5198-2002 technical regulations for organic tea processing

China's current effective legal and regulatory basis has established a relatively complete legal and regulatory system for certification and accreditation, certification rules, and technical standard systems, providing regulatory safeguards for the effectiveness and orderly development of organic agricultural product certification. The current legal, regulatory, and standard system in China mainly includes the following levels:

The legal level includes national food safety law and administrative regulations for organic certification, listed in Table 4.3 under subsection of China. Through analysis of these regulations, the potential key risk points in Chinese organic tea certification can be preliminarily identified, which will be explained below.

4.2.2.1 Identification and Analysis of Risk Factors in Environmental Conditions for Organic Tea Production

Environmental factors play a significant role in tea product safety, primarily due to the potential presence of environmental toxins or chemical hazards. These can include a variety of toxic metals, non-metals, and organic and inorganic compounds that may contaminate tea. These pollutants can persist in the product at trace levels, measured in parts per million (mg/kg) or parts per billion (μg/kg). While the concentrations may be low, they can still induce acute or chronic poisoning in humans. Numerous harmful chemicals prevalent in industrial waste, such as mercury, arsenic, lead, chromium, cadmium, and other metal and non-metal toxins like fluorides, can contaminate natural environments like water bodies, the atmosphere, and soil. Prolonged exposure to these toxins can lead to their accumulation in tea plants, resulting in contaminated products.

Upon comparing domestic and international standards for organic agriculture and products, it is evident that Chinese standards are more stringent, particularly concerning the production site environment, with broader definitions in terms of types and scope (Qiao et al., 2016; Han et al., 2018). This is manifested in explicit, strict requirements for the production site environment and clear guidelines on soil, irrigation water, and atmospheric and environmental quality. Provisions for buffer zones have also been established, further ensuring the integrity of the organic production process.

4.2.2.2 Identification and Analysis of Risk Factors in Production and Planting Conditions

In China's organic product standards, the primary stipulations for production and cultivation encompass soil fertility management and pest and weed control. Detailed guidelines for these areas are elaborated in GB/T 19630-2019 National Standard for Organic Products. The technical regulations for organic tea production delineate the planning and establishment of the organic tea production base, soil management and fertilization, pest and weed control, tea tree pruning and harvesting, conversion techniques, testing methods, and identification of organic tea gardens. The organic tea production base must adhere to the nine standards specified in the technical regulations for organic tea production. These include base planning, road and water system design, tea garden development, tea tree variety selection and planting, tea garden ecological construction, soil management, types and methods of fertilization, disease, insect, and weed control, and criteria for pesticide usage. Based on these regulations, the planning and construction of the tea garden, soil management and fertilization techniques, pest and weed control measures, and the management and operation of tea garden production and transportation constitute key risk points that influence and potentially jeopardize the safety of tea products (Saikia et al., 2014; Xie et al., 2021).

4.2.2.3 Identification and Analysis of Risk Factors in Processing Conditions

The certification of organic tea is governed by the GB/T 19630-2019 National Standard for Organic Products (Table 4.3). It necessitates adherence to several key considerations in the processing conditions. These encompass the appropriate use of ingredients, additives, and processing aids, along with the execution of processing procedures that maintain nutritional content, prevent contamination, and comply with water sanitation standards. Additionally, radiation treatment and harmful filtering materials must be avoided. These regulations further define six essential areas for scrutiny: raw materials, auxiliary materials, processing facilities, processing equipment, processing personnel, and processing technology.

Traditionally categorized as a primary agricultural product processing activity, tea processing has often been subjected to low sanitary standards. This, coupled with using outdated facilities and equipment, usually handled by local farmers with limited hygiene awareness, can result in biological contamination and variable quality.

The lengthy cycle and multifaceted conditions inherent in tea tree production amplify the necessity for educational training for tea growers or processing personnel (Deka & Goswami, 2020; Deka et al., 2022). Certification protocols for raw material bases, routine inspections, and the emphasis on mitigating microbial hazard risks are fundamental to this process. Therefore, the processing facilities, equipment, personnel, and technologies all represent critical risk points that must be addressed for the successful certification of organic tea.

4.2.2.4 Identification of Risk Factors in Labeling and Business Management

Analyzing the risk factors in labeling and business management of organic products according to Chinese national standards reveals six critical control points: labeling and certification logos, management systems and documentation, resource management, internal inspection systems, traceability, product recall mechanisms, complaint handling, and continuous improvement.

Misrepresentation of labeling and certification logos, which serve as primary organic identifiers, can significantly misdirect consumers (Charlebois et al., 2016), underscoring the importance of accurate organic labeling. Quality management systems and their associated documentation form the backbone of effective business operations, with deficiencies potentially impairing enterprise functionality. Resource management, encompassing human resources, infrastructure, and the work environment, is instrumental for seamless enterprise operation, with inadequacies potentially impacting product safety and quality. Internal inspection systems serve as quality control gatekeepers; subpar implementation can hinder the detection of product quality and safety risks. Traceability systems and product recall mechanisms are indispensable tools for managing product quality and safety risks (Wang & Yang, 2015); their absence can impair risk control efficacy during product issues. Complaint management and continuous improvement are integral components of a quality management system. Inefficiency in these areas can breed recurring quality and safety issues. The Technical Regulations for Organic Tea Production also mandate clear labeling and accurate certification markings. Thus, an effective organic tea quality management system is established. In summary, the safety of tea products hinges on key risk points, including accurate labeling, efficient management systems, resource management, internal inspections, traceability, complaint handling, and continuous improvement.

4.3 Conclusion

The risk management process consists of risk cognition or initial assessment, risk communication, and risk alleviation actions, which are actual risk management. Risk is inherent to human life and has existed as a given condition since the dawn of humanity. The concept of risk characterizes a peculiar, intermediate state between security and destruction, where the perception of threatening risks determines thought and action (Beck, 1992). Contemporary risks are difficult to communicate to the public. Policy-makers must provide accurate assessments and develop adequate regulation frameworks since scientific research cannot provide straightforward answers. There should be channels of communication and control with the public. Organic agriculture does not offer the surest guarantee that all risks will be eliminated from food production. However, many traditional and new techniques present steps to reduce these risks. The Chinese policy of reducing the excessive application of fertilizers and pesticides exemplified in the "zero growth action plan" indicates the political will to promote sustainable agriculture and support national food safety. Considering that transferring a conventional tea garden into an organic enterprise is time-consuming and costly, additional subsidies can be advised. Another area of risk management is communication with consumers.

In this case, organic labeling helps promote rules of organic cultivation, for compliance with which consumers pay the premium and oblige labeling agencies to provide control over the production process. In fact, the Chinese organic tea industry has managed to follow regulations, avoid public mistrust crises, and maintain its good reputation.

Through the literature review analysis, empirical observations, and interviews with tea producers, we have systematically sorted and summarized the key risk points encountered in organic tea production and certification in China. The most environmental risks come from the natural environment's conditions and soil management. Reviewing the Chinese national standards and relevant laws and regulations, we established the main factors affecting the risk assessment of the organic tea certification process: original environment' conditions, production and planting status, processing status, and production management.

To sum up, soil conservation and soil fertility maintenance are two basic requirements for any crop, especially for a perennial vegetative harvest crop such as tea, and without fertile soil or depleted soil, plant life cannot be economically sustained. Compliance with strict organic standards ensures consumers' trust, provides them with safe and nutritious food, and obliges producers to cultivate land sustainably. Thus, organic cultivation contributes to achieving the Sustainable Development Goals (N3, 12, and 15).

Acknowledgments

This work was supported by the Potential and Policy of Diversified Food Supply in China (RKX2024007) and the Tracking of Public Consumption Demand for Green Food (High-Quality Green Agricultural Products) (GFJY20240027).

REFERENCES

Ahmad, M.F., Ahmad, F.A., Alsayegh, A., Zeyaullah, M., Alshahrani, A., Muzammil, K., Ali Saati, A., Elbendary, E., Kambal, N., Abdelrahman, M.H., Hussain, S. (2024) Pesticides impacts on human health and the environment with their mechanisms of action and possible countermeasures. *Heliyon* 10(7): e29128. doi: 10.1016/j.heliyon.2024.e29128

Beck, U. (1992) *Risk Society. Towards a New Modernity*. University of Munich: Sage L.

Bolton, D. (2019) Chinese Tea Quality Director Explains "Zero-Growth" Action Plan for Pesticides and Fertilizer. *World Tea News*. June 18 2019. Available at: https://www.worldteanews.com/Features/chinese-tea-quality-director-explains-zero-growth-action-plan-pesticides-and-fertilizer

Charlebois, S., Schwab, A., Henn, R., Huck, C.W. (2016) Food fraud: An exploratory study for measuring consumer perception towards mislabeled food products and influence on self-authentication intentions. *Trends in Food Science & Technology* 50: 211–218.

Chen, C.F., Lin, J.Y. (2016) Estimating the gross budget of applied nitrogen and phosphorus in tea plantations. *Sustainable Environment Research* 26(3): 124–130.

Chen, H., Yu, F., Kang, J., Li, Q., Warusawitharana, H.K., Li, B. (2023) Quality chemistry, physiological functions, and health benefits of organic acids from tea (Camellia sinensis). *Molecules* 28(5): 2339. doi: 10.3390/molecules28052339

Daraselia, M.K., Vorontsov, V.V., Gwasalia, V.P., Tsunava, V.P. (1989) *Kultura tchaya*. Akademia nauk Gruzinskoy SSR. Metzniereba: Tbilisi.

Deka, J., Barua, I. (2015) Weeds of tea fields and their control. *Conference Paper. Conference: Plant Protection in Tea: Recent Advances*. Jorhat, Assam. doi: 10.13140/RG.2.1.5150.9845

Deka, N., Goswami, K. (2020) Organic cultivation and farm entrepreneurship: A case of small tea growers in rural Assam, India. *Agroecology and Sustainable Food Systems* 44(4): 446–466.

Deka, N., Goswami, K., Anurupam, K. (2022) What will drive the small tea growers towards environment-friendly cultivation? Implications from the tea sector in Assam, India. *Climate and Development* 14(5): 443–458.

Didmanizde, R.N. (2021) Teknologiia zakladki I uhoda za tchainymiy plantaziami. *Agroingeneria* 6(106): 38–42.

Elliot, E.C., Whitehead, F.I. (1934) Tea planting in Ceylon. *Colonial Times*. V.94.

FAO (1999) Organic Agriculture. COAG/99/9 Rev.1 Available at: https://www.fao.org/docrep/meeting/X0075e.htm as viewed 01.09.2017.

Fernandez, I.A.A., Maciel, G.M., Gonsalves Bortolini, D., Pedro, A.C., Rubio, F.T.V., Querne de Carvalho, K., Windson, C., Haminiuk, I. (2023) The bitter side of teas: Pesticide residues and their impact on human health. *Food and Chemical Toxicology* 179: 113955.

Gambelli, D., Solfanelli, F., Zorn, A., Lippert, C., Dabbert, S., Zanoli, R. (2011) Modelling risk based inspections in EU organic certification: Data requirements and analysis tools. *Organic is Life* 2: 207727.

Gerasimova, K., Sheng, J., Zhao, J. (2021) COVID-19 and other challenges: A case study of certified organic green tea producers in China. *Critical Sociology* 47(4–5): 591–607.

Gould, D. (2017) *Introduction Speech to the 2nd Organic Asia Congress*. 19 May 2017, Nanchong, Xichong.

Govindasamy, P., Muthusamy, S.K., Bagavathianan, M., Mowrer, J., Kumar Jagannadham, P.T., Maity, A., Halli, H.M., Sujayanand, G.K., Vadivel, R., Das, T.K., Babu S. Rathore, S.S. (2023) Nitrogen use efficiency—a key to enhance crop productivity under a changing climate. *Frontiers in Plant Science* 14: 1121073. doi: 10.3389/fpls.2023.1121073

Gunapala, P.K.R.N., Ulpathakumbura, C. (2021) Effects of Organic and Inorganic Fertilizer Application for Tea Cultivation in Sri Lanka. Available online: https://www.researchgate.net/publication/351943232_EFFECTS_OF_ORGANIC_AND_INORGANIC_FERTILIZER_APPLICATION_FOR_TEA_CULTIVATION_IN_SRI_LANKA

Hajra, N.G. (2017) Organic tea: Global market and forecast sales. *Journal of Tea Science Research* 7: 11.

Han, W.Y., Wang, D.H., Fu, S.W., Ahmed, S. (2018) Tea from organic production has higher functional quality characteristics compared with tea from conventional management systems in China. *Biological Agriculture & Horticulture* 34(2): 120–131.

Hayashi, N., Ujihara, T., Kishi, Y., Matsuo, H. (2011) Annual variation of natural 15N abundance in tea leaves and its practicality as an organic tea indicator. *Journal of Agricultural Food Chemistry* 59(18): 10317–10321. doi: 10.1021/jf202215z

IFOAM (2005a) *Definition of Organic Agriculture*. Available at: https://www.ifoam.bio/en/organic-landmarks/definition-organic-agriculture as viewed 01.09.2017

IFOAM (2005b) *Principles of Organic Agriculture*. Available at: https://www.ifoam.bio/sites/default/files/poa_english_web.pdf/

IMARC (2023) *Organic Tea Market Report by Product (Camelia Sinesis Based, Herbal), Taste (Flavored, Plain), Form (Tea Bag, Leaf and Powder, Liquid), Distribution Channel (Supermarkets and Hypermarkets, Convenience Stores, Online Stores, and Others), and Region 2024–2032*. Report ID: SR112024A3832. IMARC: L.

James, M.X., Hu, Z., Leonce, T.E. (2019) Predictors of organic tea purchase intentions by Chinese consumers. *Journal of Agribusiness in Developing and Emerging Economies* 9(3): 202–219. doi: 10.1108/JADEE-03-2018-0038

Jia, X., Zhang, Q., Wang, Y., Zhang, Y. (2023) Changes of physiological characteristics, element accumulation and hormone metabolism of tea leaves in response to soil pH. *Frontiers in Plant Science Sections Plant Abiotic Stress* 14: 26. doi: 10.3389/fpls.2023.1266026

Jiang, B., Pang, J., Li, J., Mi, L., Ru, D., Feng, J., Cai, L. (2024) The effects of organic food on human health: a systematic review and meta-analysis of population-based studies. *Nutrition Reviews* 82(9): 1151–1175.

Jie, L. (2016) Construction and Empirical Study of Risk Assessment System for Organic Tea Whole Process Certification ---Based on the Analytic Hierarchy Process and Risk Matrix Method. Master's dissertation. Beijing: Renmin University of China.

Koren'kov, D.A., Lavrova, I.A., Kharchenko, N.V. (1979) Ispolzovanie rasteniyami I transformatzia azota v pochve pri vnesenii mineralnyh udobreniy y organicheskogo veshestva. *Agrohimia* 7: 3–7.

Kretovich, V.L. (1972) *Obmen azota v rasteniyah*. Nauka: M.

Kretovich, V.L. (1980) *Biohyimia rasteniy*. Vyshaya shkola: M.

Meng, X., Song, W., Zheng, P., Cui, C., Hou, R. (2022) Rapid determination of 134 pesticides in tea through multi-functional filter cleanup followed by UPLC-QTOF-MS. *Food Chemistry* 370: 130846.

Rajaduraj, R. (2021) Sri Lanka tea under threat from pests, fungal outbreaks, weeds due to agro-chemical ban. Economynext. https://www.historyofceylontea.com/ceylon-publications/ceylon-tea-articles/sri-lanka-tea-under-threat-from-pests-fungal.html

Saikia, G.K., Barbora, A.C., Deka, M.K. (2014) Pest, disease and weed incidence and crop yield as influenced by organic culture in tea [Camellia sinensis (L.) O. Kuntze]. *Agricultural Science Digest-A Research Journal* 34(2): 119–122.

Sheng, J., Su, W., Li, J., Qiao, Y. (2024) *Developing Risk Assessment Index System for Organic Tea Certification Based on China Organic Regulations*. Unpublished paper. Beijing: Renmin University.

Sitienei, K., Home, P.G., Kamau, D.M., Wanyoko, J.K. (2013) The influence of fertilizer type and application rates in tea cultivation on nitrogen and potassium efficiencies. *African Journal of Agricultural Research* 8: 3770–3777.

State Administration for Market Regulation, China Agricultural University. (2022) *Certification of Organic Products and Development of the Organic Industry*. Beijing: China Agricultural Science and Technology Press.

Qiao, Y., Halberg, N., Vaheesan, S., Scott, S. (2016) Assessing the social and economic benefits of organic and fair trade tea production for small-scale farmers in Asia: A comparative case study of China and Sri Lanka. *Renewable Agriculture and Food Systems* 31(3): 246–257.

Vigar, V., Myers, S., Oliver, C., Arellano, J., Robinson, S., Leifert, C. (2019) A systematic review of organic versus conventional food consumption: is there a measurable benefit on human health? *Nutrients* 12(1): 7.

Wang, M.C., Yang, C.Y. (2015) Analyzing organic tea certification and traceability system within the Taiwanese tea industry. *Journal of the Science of Food and Agriculture* 95(6): 1252–1259.

Wang, Y., Lu, X., Gan, Y., Li, Y. (2023) Conversion of forest hillslopes into tea fields increases soil nutrient losses through surface runoff. *Land* 12(2): 448.

Xie, S., Feng, H., Yang, F., Zhao, Z. (2019) Does dual reduction in chemical fertilizer and pesticides improve nutrient loss and tea yield and quality? A pilot study in a green tea garden in Shaoxing, Zhejiang Province, China. *Environmental Science and Pollution Research* 26(2): 2464–2476. doi: 10.1007/s11356-018-3732-1

Xie, S., Yang, F., Feng, H., Yu, Z., Liu, C., Wei, C., Liang, T. (2021) Organic fertilizer reduced carbon and nitrogen in runoff and buffered soil acidification in tea plantations: Evidence in nutrient contents and isotope fractionations. *Science of the Total Environment* 762: 143059.

Yilmaz, A., Dizman, F., Akyildiz, K., Karakas, S.M., Mercantepe, T., Uydu, H.A., Tumkaya, L., Ozturk, K. (2024) The hepatoprotective effects of camellia sinensis on cisplatin-induced acute liver injury. *Life* (Basel) 14(9): 1077. doi: 10.3390/life14091077

Zhen, H., Qiao, Y., Meng, F., Li, H., Liu, Y., Jia, Y., Zanoli, R., Gambelli, D., Francesco Solfanelli, F. (2020) Developing risk indicator system of non-compliance for organic crop farms based on China organic regulations. *Ecological Indicators* 114: 106317.

Section II

Sources and Exposure Pathways

5

Persistent Toxic Substances Emitted from Biomass Combustion and Their Health Impacts

Ekaterina Kravchenko and Tatiana Minkina

5.1 Introduction

Using biomass for w and lighting has been integral to the development of civilization. Since the first torch was lit, biomass has been crucial in providing energy for human societies. In developed countries, the use of biomass declined with the advent of the Industrial Revolution, as high-energy-density fuels such as peat, coal, gas, and oil became available. Nevertheless, due to the rise of global warming and energy security concerns, biomass is once again becoming an increasingly important part of the energy mix.

Biomass is the world's third largest natural energy source (Wielgosiński et al., 2017), contributing 55% of the renewable energy mix and over 6% of the global energy supply (IEA, 2023). According to the Net Zero Emissions by 2050 Scenario, the utilization of bioenergy is projected to escalate significantly by 2030, replacing a substantial portion of fossil fuels. From 2010 to 2022, the use of modern bioenergy has grown at an average annual rate of approximately 3% and continues to follow an upward trajectory (IEA, 2023). Historically, using biomass for power generation has been limited and primarily focused on utilizing residues from sawmills or other industrial processes. Over the past decade, biomass has gained popularity as it is viewed as the only sizeable renewable fuel source available for power generation when needed. The initial large-scale deployment of biomass involved co-firing in coal-fired power plants. However, as the issue of climate change has gained prominence, policies have shifted away from co-firing toward total plant conversions or new builds, as co-firing still involves burning coal.

Despite its potential benefits, biomass combustion raises significant concerns regarding the emission of various toxic substances, including particulate matter and persistent toxic substances (PTS). This review aims to thoroughly investigate these emissions and their impacts on both the environment and human health. In addition to identifying these pollutants, this chapter includes several case studies from different regions, such as Southeast Asia, Africa, Brazil, and China, where biomass burning has significantly contributed to airborne particulate matter and polycyclic aromatic hydrocarbons (PAHs). These case studies will provide a comparative analysis of the regional variations in biomass-derived air pollutants and their implications for public health and ecosystems. Furthermore, this review will assess existing biomass combustion technologies, offering insights into how they can be optimized to reduce toxic emissions and enhance the sustainability of biomass energy production.

5.1.1 Definition and Types of Biomass

Biomass encompasses all living matter on Earth and typically refers to material derived from growing plants or animal manure (Demirbas, 2007). It is a broad term encompassing organic materials sourced from plants, trees, crops, and algae, all of which harness solar energy through photosynthesis to sustain life. This solar energy stored in plants, animals, or their by-products is biomass energy, which can be extracted by burning biomass as fuel. Biomass can be economically and environmentally beneficially produced, primarily through perennial crops. Waste biomass streams hold promise as potential sources for diverse products, including fuel, polymers, and building materials (Tripathi et al., 2019).

DOI: 10.1201/9781003492115-7

While biomass can be categorized in various ways, it generally falls into three primary classifications: natural crops (including aquatic and terrestrial plantations), energy crops, and biomass waste (Darmawan & Aziz, 2022). Determining the extent of crop-generated biomass presents challenges, as it involves distinguishing between what constitutes a "loss" (from production, post-harvesting, and processing) and what qualifies as "waste" (such as retail or consumer loss) (Saracoglu & Gündüz, 2009). Biomass waste results from biomass processing, from agricultural and industrial activities and is driven by the high demand for food and other human necessities. The substantial increase in these industrial activities has resulted in a significant accumulation of biomass waste. Agricultural biomass wastes and residues primarily consist of crop stalks, leaves, roots, fruit peels, and seed/nut shells, typically discarded or burned. However, they represent a valuable potential source of feedstock material. Cereal crops constitute a significant portion of the annual global generation of agricultural residues. Globally, 66% of residual plant biomass comes from cereal straw, with over 60% of these residues produced in low-income countries (Tripathi et al., 2019). Sugarcane stems and leaves are the second-largest contributors, alongside residual biomass from other sources such as oil crops, roots and tubers, nuts, fruits, and vegetables.

Utilizing biomass waste for energy offers several advantages, including avoiding conflicts with essential human needs for food and materials and reducing environmental impacts. Due to its economic viability, effectively harnessing biomass waste as an energy source is believed to meet financial requirements and align with the demand for industrial waste treatment to mitigate further environmental harm. Biomass waste can be further categorized into human-made and natural biomass waste, and its utilization can mitigate environmental issues caused by improper treatment methods like open burning or toxic waste disposal (Elhacham et al., 2020). Refer to Figure 5.1 for an overview of biomass classification.

5.1.2 Methods of Utilization of Biomass-Combustion Technologies

Various technologies exist for managing solid residues, including waste minimization or reuse, recycling discarded materials into new products, treatment of residues, and landfilling (Zheng & Koziński, 2000). Turning these solid residues into valuable products would reduce or eliminate their presence, optimizing natural resource utilization.

There is a growing global interest in harnessing biomass for energy production. The primary driving factors behind biomass combustion typically involve the carbon neutrality of sustainably cultivated biomass or using biomass residues and waste materials (Nussbaumer, 2003). Significant reserves of natural biomass and biomass waste remain untapped, presenting substantial potential for increasing sustainable bioenergy utilization. Presently, combustion stands as the predominant and well-established technology for biomass utilization. However, there is a pressing need for efficiency, emissions reduction, and cost-effectiveness enhancements to capitalize on this resource fully. Additionally, alternative approaches such as gasification warrant exploration, and the synergistic combination of different processes, such as using gasification as a precursor for co-combustion, holds promise for further advancements in biomass energy utilization.

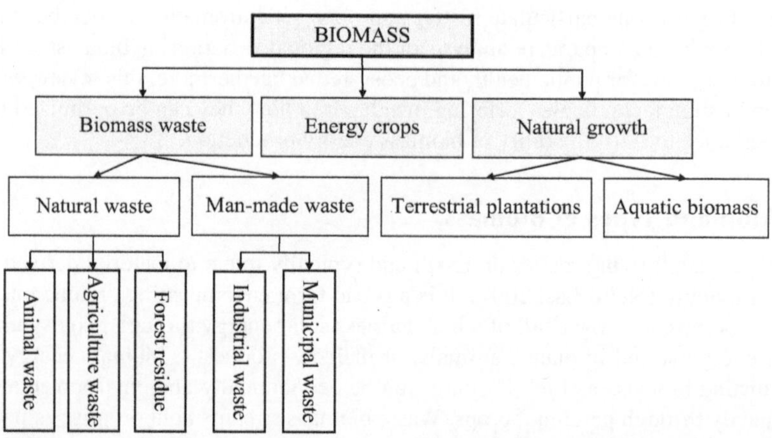

FIGURE 5.1 Biomass classification (Darmawan et al., 2023).

Thermal treatment methods, such as waste combustion, are introduced as alternatives to landfilling, but they are restricted in many areas due to stringent regulations. Thermal treatment offers a viable disposal option for numerous waste streams, as it can effectively eliminate hazardous components like organics and pathogens. This process significantly reduces the mass and volume of waste and allows for energy recovery. However, there are limitations to consider. A major concern is the public perception that combustion may release solid and gaseous toxic substances into the environment. For instance, toxic heavy metals cannot be destroyed during combustion, generating metal particles or vapors when wastes containing these metals are incinerated. Volatile metal compounds, such as lead, mercury, cadmium, and chromium, are emitted as vapors or fine particles, while non-volatile compounds are released with the ash and require safe disposal. Consequently, while most solid wastes are combustible or suitable for incineration, only a tiny portion is combusted. Hence, combustion processes must be enhanced to generate safely landfilled or reused products.

5.1.3 Biomass Fuel

Biofuels encompass a range of alternative fuels sourced from biomaterials, including grain crops (such as corn, wheat, sugarcane, sugar beet, and cassava), crop residues (such as rice straw, rice husk, corn cobs, wheat straw, and corn stover), and various waste biomass materials (such as food waste, livestock waste, and other by-products). These renewable resources are utilized worldwide in gaseous forms like biogas and liquid forms like bioethanol and biodiesel, offering sustainable alternatives to traditional fossil fuels.

Biogas, a gas produced through organic digestion in anaerobic conditions by various microorganisms, is an alternative energy source. It is generated both on small farm scales and at significant industrial levels. Biogas comprises methane (55%–65%), carbon dioxide (35%–40%), nitrogen (0%–3%), hydrogen (0%–1%), and hydrogen sulfide (0%–1%). It has been utilized since 1958, derived from diverse feedstock such as agricultural waste, animal manures, sewage sludge, and industrial organic waste (Hirani et al., 2018). Bioethanol production involves the fermentation and distilling of sugars extracted directly from sugarcane crops or starch-rich crops such as corn, wheat, potatoes, and cassava. During fermentation by yeast, around 10% of sugars are converted into glycerol and succinic acid, typical by-products in bioethanol plants. The United States emerged as the leading bioethanol producer globally, with an output of 59.73 billion liters in 2019, accounting for approximately 54% of the world's bioethanol production (Green, 2024). Brazil followed with 32.44 billion liters, representing around 29% of the global production in the same year (Aruwajoye et al., 2020). Nevertheless, both countries primarily rely on corn and sugarcane as their primary crops for bioethanol production. Biodiesel is a renewable and eco-friendly alternative to fossil fuels in diesel engines without requiring modifications. Biodiesel production commenced in the early 1990s, gaining widespread adoption as a renewable energy source for the transportation sector. Since then, its production has steadily risen (Hirani et al., 2018).

Biomass fuel encompasses various types, spanning first, second, and third generations. First-generation biomass fuel is derived from food industry by-products like corn and soy. However, its production competes with food cultivation, leading to environmental concerns and price inflation (Hirani et al., 2018). Second-generation biomass fuel utilizes non-edible plant residues, offering a more environmentally friendly alternative as it does not compete with food production. Third-generation biomass fuel is sourced from microorganisms like algae, holding significant potential as a future energy source. Technologies for biomass fuel production vary. Methods like transesterification, hydrothermal decomposition, or pyrolysis are employed for biodiesel, while anaerobic digestion processes organic waste for biogas production.

Implementing biofuels as a sustainable energy source has seen notable success stories across various countries, showcasing its potential as a viable alternative to traditional fossil fuels. In Russia, biofuel has been integrated into energy production for both heating and electricity. Notably, in 2017, "Tatenergo" inaugurated the first production line for pellets derived from forest biomass in Tatarstan. Furthermore, Russia's agricultural sector benefits from a support program to facilitate the transition to biofuels, thereby reducing reliance on conventional fuels (Tveritinova, 2008). Across the Atlantic, the United States is another prominent player in the biofuel landscape. With facilities spanning 27 states, "Poet," a leading biodiesel producer, exemplifies the nation's commitment to bioenergy (Winters, 2011).

Additionally, the USA's Renewable Fuel Standard program mandates the inclusion of biofuels in gasoline and diesel production, fostering a more sustainable energy mix. In Brazil, biofuel production, mainly derived from sugarcane, holds a significant share of the country's energy consumption (Antunes et al., 2019; Karp et al., 2021), reaching approximately 18% in 2018. Brazil's Proálcool program promotes ethanol as a viable alternative to traditional fuels like gasoline and diesel, further solidifying the nation's position as a biofuel pioneer (Stolf & Oliveira, 2020). Germany emerges as a leader in biofuel utilization within Europe. The country boasts biogas, biodiesel, and bioethanol facilities, supported by initiatives such as the Renewable Energy Sources Act (EEG) (Yang et al., 2021). This legislation channels funds into projects to expand renewable energy sources, including biofuels, thereby contributing to Germany's sustainable energy transition. In Sweden, biofuels play a pivotal role, with approximately 30% of the total fuel consumption stemming from bio-based sources. The country actively pursues projects that produce biogas, biodiesel, and bioethanol, underscoring its commitment to a low-carbon future (Amiandamhen et al., 2020). These examples highlight the global momentum behind biofuel adoption, driven by a collective recognition of its environmental benefits and potential to mitigate climate change.

5.1.4 Release of Ash

According to some estimates, as biomass production rises, it is anticipated that by 2050, 33%–50% of the global energy demand could be met through biomass combustion (Vassilev et al., 2013). However, with the escalation in biomass combustion comes a corresponding increase in the annual production of waste ash. A critical parameter for assessing biomass is the magnitude of its ash residue. This characteristic helps determine potential biomass contamination, the affinity of elements and compounds to inorganic or organic matter, and predict some technological and environmental issues. The ash content of biomass varies, with estimates ranging from 0.1% to 67.0%, averaging 7.2% (Vassilev et al., 2017). High ash content is characteristic of algae, solid household waste, paper sludge, and wastewater, ranging from 30% to 70% (Vassilev et al., 2015). The high ash values in aquatic species are associated with their ability to accumulate minerals from the water they grow. Additionally, increased ash content in some biomass types may result from inclusions of fragmented sand or soil and metallic pollutants Nunes et al. (2017).

Minerals present in biomass ash residue originate from natural and human-induced processes. They can be categorized based on their origin into primary, secondary, and tertiary minerals. Primary minerals refer to those that existed previously or are inherent components of the biomass, remaining unchanged during combustion. These typically include refractory silicates, oxyhydroxides, and phosphates (Tabakaev et al., 2023; Tabakaev et al., 2019). Secondary minerals are newly formed phases that arise during biomass combustion, representing various mineral classes. Tertiary minerals are formed as a result of the transformation of the ash residue. These typically consist of carbonates and hydroxides and are influenced by the corresponding phase's primary formation process and mechanism.

Studies show that major and minor elements in biomass ash typically include Al, Ca, Cl, Fe, O, K, Mg, P, Na, S, Mn, Si, and Ti, usually expressed in oxide form (Vassilev et al., 2012). Wood biomass tends to have low concentrations of Cl, K, Na, S, and Si, with relatively high levels of Ca compared to other types of biomass. Agricultural biomass yields higher ash content and thus contains more ash-forming elements than the predominant portion of forest biomass. Agricultural residues primarily consist of maize, rice, wheat, sugarcane, and soybean residues. Among these, maize, rice, and wheat, as cereal crops, contribute 68% to the total agricultural residues (Zhai et al., 2021). Ash compositions of cereal crop residues are predominantly composed of SiO_2 and K_2O, with moderate levels of CaO and Cl_2O, and minor amounts (<3%) of other elements. Sugarcane residue ashes exhibit a similar SiO_2 content to cereal crop residue ashes but slightly lower levels of K_2O and CaO, with relatively higher Al_2O_3 and Fe_2O_3. Soybean residue ash contains notably lower SiO_2 content compared to cereals and sugarcane (all of which are grass types), but higher levels of K_2O, CaO, MgO, and P_2O_5, with minor amounts (<3%) of other elements (Chaw et al., 2004).

The ash composition resulting from biomass combustion varies based on the biomass's characteristics and the incineration conditions (Odzijewicz et al., 2022). Typically, this waste is disposed of in landfills, but it is evident that ashes resulting from biomass combustion represent one of the earliest forms of mineral fertilizer. According to Zhai et al. (2021), agricultural residue ashes contain abundant potassium and useful phosphate for fertilizer. Owing to their high concentrations of calcium and magnesium, these by-products serve as substitutes for calcium fertilizers, exhibiting soil de-acidifying properties (Wojcik et al., 2020). Additionally, ashes from biomass combustion contain various other fertilizing elements (potassium, phosphorus, sulfur, silicon, and micronutrients), making them suitable for soil reclamation and enhancing conditions for plant growth (Albuquerque et al., 2021; Odzijewicz et al., 2022).

Materials scientists' research demonstrates the promising potential of utilizing biomass ash in construction materials. Thomas et al. (2021) have confirmed the viability of employing biomass ashes from various sources, such as bamboo leaves, date palms, elephant leaves, banana leaves and plantain peels, rice straw, olive waste, wheat straw, and corn cob. These ashes can serve as pozzolanic materials in cement concrete or as alternative activators in geopolymer concrete, exhibiting the necessary properties for building materials.

5.1.5 Formation of Combustion Aerosols

The growing trend of biomass combustion raises concerns about the health effects associated with inhaling emissions from combustion, which have been linked to adverse health outcomes in humans. Aerosols from biomass combustion result from natural wildfires and anthropogenic activities, such as incinerating wood or agricultural by-products. Globally, smoke from biomass combustion is recognized as a significant environmental risk factor, with the deposition of smoke particles in the lungs playing a crucial role in toxicological effects (Londahl et al., 2008).

Combustion aerosols (CAs), particularly those containing PAHs, have been implicated in these health effects, with PAHs being emitted during ignition and intense combustion in traditional wood stoves (Martinsson et al., 2015). Additionally, CA affects the Earth's radiative balance by absorbing and scattering incoming solar radiation. Specific fractions of CA, known as brown carbon, are highly light-absorbing in the UV region. It is estimated that biomass burning contributes to approximately 40% of the global emissions of black carbon (Adler et al., 2011). These aerosols comprise various substances, including nitrates, ammonium, sulfur, organic materials, and black carbon. The presence of biomass-burning aerosols can significantly influence local air quality, visibility, the behavior of clouds, human health, and global and regional climate conditions.

Primary organic aerosols, which biomass burning emits, constitute a substantial portion (ranging from 20% to 90%) of submicron particulate matter in the atmosphere, according to findings from Zhang et al. (2007). Once in the atmosphere, these aerosols transform the photochemical oxidation processes. However, the atmospheric aging of burning aerosols refers to their changes over time, and the consequences of this process are poorly understood. Global model simulations suggest that brown carbon contributes significantly to the total absorption by atmospheric aerosols, yet the climate effects of CA remain uncertain.

5.2 Types of PTS Formed

5.2.1 Heavy Metals

Several trace elements within the biomass's elemental composition are classified as heavy metals, including B, Be, Cd, Cr, Cu, Mn, Ni, Rb, Se, and Zn. Heavy metals are known for their toxic effects on living organisms (Co, Ni, Cu, Zn, Sn, As, Rb, Cd, Hg, Ag, Au, Pb, Mn, and Mo), with four metals (As, Cd, Pb, and Ni) being hazardous (Zhang et al., 2014). The primary routes of heavy metal entry into the human body include respiratory, oral, and dermal pathways. Once inside the body, they can accumulate, causing various toxic effects, such as nephrotoxicity, neurotoxicity, hepatotoxicity, skin toxicity, cardiovascular toxicity, and others. The main potential health risks of heavy metal exposure are presented in Table 5.1.

The transformation of heavy metals during biomass combustion depends on the specific forms of these trace elements in the raw material (Chen et al., 2019). In the combustion process, heavy metals may be present in their elemental form or as compounds, such as oxides, chlorides, and sulfates. Metal compounds often exhibit greater volatility than the individual metals themselves. Consequently, biomass combustion poses a risk of causing secondary environmental pollution (Hunce et al., 2019). Additionally, alkali metals play a role in forming slags and solid particles and, via evaporation-condensation reactions, may also be released into the atmosphere (Delplanque et al., 2013).

During biomass combustion, part of the ash volatilizes, transitioning into the gas phase. The chemical composition of the fuel, the surrounding gaseous environment, the temperature of biomass combustion, and the method of combustion used to determine the size of this fraction. Jagustyn et al. (2017) found that in wood pellets and chips, which are intended for commercial use and can be burned in boilers not equipped with exhaust gas cleaning systems, there are exceedances of permissible levels of As, Cr, Pb. Wiinikka et al. (2013) discovered that during the combustion of wood and fossil fuels, fly ash particles

TABLE 5.1

The Most Hazardous Toxic Metals and Their Impact on Human Health

Heavy Metal	Anthropogenic Primary Sources	Ways of Entry into the Human Body	Target Organs	Health Hazards	References
Arsenic (As)	Pesticides, wood preservatives and industrial, mining and smelting wastes	Gastrointestinal and respiratory tract, skin contact	Circulatory and nervous system, skin	Hyperkeratosis, hyperpigmentation, high blood pressure, heart attacks and other circulatory disease, respiratory cancer, neurodegenerative diseases, reducing the weights of the testes and the number of sperm in the epididymis.	Abernathy et al. (2003), Garza-Lombó et al. (2019), Huang et al. (2019), Kim and Kim (2015), Mahurpawar (2015)
Cadmium (Cd)	Coal mines, stack dust, metal industry, wastewater, fertilizers	Gastrointestinal and respiratory tract	Kidney, hepatic, lung, skeleton, nervous system, cell cycle	Cancer mutations and genotoxic abnormalities, neurodegenerative defects: myotrophic lateral sclerosis, Parkinson's disease, Alzheimer's disease, multiple sclerosis, glucosuria, fanconi-like syndrome, phosphaturia, aminoaciduria, liver failure.	Charkiewicz et al. (2023), Mishra and Bharagava (2016)
Chromium (Cr)	Electroplating, leather, metallurgic, chromates, pigments, cement, weapons, ammunition, welding and agricultural fertilizers industries	Skin contact, respiratory and gastrointestinal tract	Liver and kidneys, skin, immune system	Steatosis of hepatocytes, parenchymatous degeneration and necrosis, contact dermatitis, systemic contact dermatitis, and skin cancer, nasal septum perforation, bronchitis, asthma, pneumonitis, pharyngitis.	Alvarez et al. (2021), Hasanein and Emamjomeh (2019)
Lead (Pb)	Electroplating, paint industries, industrial dust, road transports and vehicles	gastrointestinal and respiratory tract	Bones, kidney, liver, reproductive, nervous, urinary, immune and cardiovascular system	Encephalopathy, peripheral, neuropathy, central nervous disorders, anemia, chronic cirrhosis, allergies, infectious diseases, and autoimmunity, cancer.	Hegazy and Fouad (2014), Hsiao et al. (2011), Su (2014)
Nickel (Ni)	Metal plating industries, combustion of fossil fuels, Ni mining, and electroplating industries	Gastrointestinal and respiratory tract, skin contact	Respiratory system, skin	Cancer: lung, nose, larynx and prostate; respiratory failure, lung embolism, birth defects, asthma and chronic bronchitis, allergic reactions, kidney, and cardiovascular diseases	Kumar and Trivedi (2016), Seilkop and Oller (2003)
Tin (Sn)	Burning of fossil fuels (coal or oil) or waste incineration, and the production of tin, organotins, iron, steel, and nonferrous metals, pesticides	Gastrointestinal and respiratory tract, skin contact	Lungs, liver, and kidneys, lymphatic tissues, brain	Skin and eye irritation, respiratory irritation, gastrointestinal effects, smell disturbance, hearing loss, visual disturbance, seizures, neural necrosis, encephalopathy, cerebral edema, vasodilatation, hypotension, heart failure, inflammation of the liver, genitourinary effects.	Cima (2011)

(*Continued*)

TABLE 5.1 (*Continued*)
The Most Hazardous Toxic Metals and Their Impact on Human Health

Heavy Metal	Anthropogenic Primary Sources	Ways of Entry into the Human Body	Target Organs	Health Hazards	References
Mercury (Hg)	Artisanal and small-scale gold mining, combustion of coal and other fossil fuels for energy and heat production in power plants and in industrial and residential boilers, primary non-ferrous metals production, cement production, large-scale gold production and waste from consumer products, pig iron production, chlor-alkali industry	Gastrointestinal and respiratory tract	Brain and kidneys	Memory loss, epilepsy, depression, dementia, Parkinson, visual dysfunction, kidney dysfunction, immunostimulant, immunosuppressant, lymphoproliferation, myocardial infarction, arrhythmias, cardiomyopathy, menstrual cycle disorder, reduction in libido, impotency, necrotizing bronchitis, pneumonitis, respiratory failure, inhibition of protein synthesis, microtubule disruption, teratogenesis.	Okpala et al. (2018), Sundseth et al. (2017)
Copper (Cu)	Mining, smelting, and refining of copper, industries producing products from copper such as wire, pipes, and sheet metal, and fossil fuel combustion	Gastrointestinal and respiratory tract, skin contact	Liver and kidneys, brain, nerves	Abdominal pain, cramps, nausea, diarrhea, and vomiting, chronic anemia, neurological symptoms: exhaustion, depression, irritability, heart and renal failure, hepatic necrosis, encephalopathy.	Alengebawy et al. (2021), Mishra and Bharagava (2016)
Zinc (Zn)	Mining, smelting, and sewage irrigation, fertilizer	Gastrointestinal and respiratory tract, skin contact	Immune system, liver, kidney	Epigastric pain, vomiting, nausea, and fatigue, impaired immune function, decreased levels of high-density lipoprotein and increased levels of low-density lipoprotein, toxic and dystrophic changes in the heart, and reduced filtration capacity of the kidneys.	Mishra and Bharagava (2016), Schoofs et al. (2024)
Cobalt (Co)	Burning of fossil fuels and sewage sludge, phosphate fertilizers, mining and the transformation of cobalt-containing ores, and industries that produce or process cobalt	Gastrointestinal and respiratory tract	Respiratory system, cardiovascular system, myocardium, the thyroid gland, and the nervous system	Reversible hearing and vision impairment, cardiomyopathy, skin rashes, pimples, dermatitis and dermal flares, asthma, parenchymal lung disease.	Packer (2016), Paustenbach et al. (2013)
Manganese (Mn)	Iron and steel foundries, power plants, coke ovens, and entrainment of soils containing Mn	Gastrointestinal and respiratory tract, skin contact	Nervous system	Alzheimer and Parkinson disease, mitochondrial dysfunction.	Mahurpawar (2015), Santamaria and Sulsky (2010)

contained significant amounts of Cd, Cr, Mn, Ni, and Zn. It was revealed that Cd, Zn, and Pb volatilization is enhanced at high temperatures in a reducing environment. Apparently, the reducing environment prevents the formation of less volatile metal oxides, limiting the access to O_2. These heavy metals can be present in the aerosol fraction due to their high volatility.

Heavy metals that form less volatile and non-volatile compounds migrate into charcoal and ash (e.g., Co, Ni, Cr, V). The presence of chlorine in the combustion chamber leads to the formation of metal chlorides with low melting points and their release (Xie et al., 2020). Thus, the volatilization of Cd, Pb, Zn, and Cu, which have a greater tendency to form chlorides rather than oxides, occurs. Wobst et al. (2001) also confirmed that a significant presence of HCl in the gas phase leads to increased evaporation of potentially hazardous elements. Meanwhile, the presence of silicates contributes to the binding of metals in stable compounds in the ash (Zha et al., 2021). Zheng and Koziński (2000) noted that ~77% of heavy metals were retained in the ash during pulp and paper bio-sludge combustion. Despite the retention of heavy metals in the ash, they can often be more mobile and toxic to the environment.

As the combustion gases cool down, homogeneous and heterogeneous condensation can occur—homogeneous condensation results in new particles with diameters less than a micrometer. Heterogeneous condensation occurs on the surfaces of entrained ash particles. In such particles, the concentration of heavy metals will be higher than in the original biomass.

The release of heavy metals and their compounds into the gas phase depends on these elements' melting and boiling temperatures. These temperatures do not exceed 1000°C for most elements and are less commonly below 500°C. These metals will likely be emitted at combustion temperatures. Chen et al. (2019) demonstrated that the evaporation of Pb varies from 19% to –49% at a temperature of 600°C to 70%–96% at 800°C–1000°C for all types of studied biomass (agricultural waste, tree bark, and coal). Jiménez et al. (2008) found that with an increase in temperature, the concentration of As, Cr, Zn, Pb, Cu, Co, Fe, Ni, and Mn increases in the fine fraction of fly ash. This trend is also explained by the fact that as the temperature increases, the vapor pressure of typical metal vapors increases, leading to higher emissions of heavy metals.

The relative enrichment factor (REF) represents the ratio of the concentration of an element in the ash formed in the combustion chamber to its ash content in the original fuel (Nalbandian, 2012). Metals are then categorized into three classes: Class I is the least volatile, and Class III is the most volatile. Class II is subdivided into a, b, and c. Metals from the first class (e.g., Mg, Co, Fe, and Ti) will be present in equal concentrations of ash residue and volatile ash and almost not contain fine particles (REF~1). The second class is characterized by an increase in the content of these elements with a decrease in particle size. Class IIa, which includes manganese and chromium, has REF< 0.7 for bottom ash and ~1.3–2 for fly ash. Metals in class IIc (As, Mo, Pb, Zn) have a REF< 0.7 for bottom ash and>4 for fly ash. The intermediate group has a REF<0.7 for bottom ash and 2–4 for fly ash. The ash content of metals in the third class is insignificant. These elements are emitted into the gas phase (Hg and Se) (Nalbandian, 2012; Nzihou & Stanmore, 2013). Mercury is particularly notable for its volatility, which can be entirely emitted as vapor in its oxidized form ($HgCl_2$) or as elemental mercury (Hg^0) (Nzihou & Stanmore, 2013).

In the study conducted by Zha et al. (2021), the mass fractions of calcium, aluminum, and iron were almost indistinguishable from the values in the original biomass due to their low volatility at high temperatures. The volatility of various heavy metals varies little when burning different types of fuels. The relatively high volatility of Pb and Zn (> 50 mass %) is confirmed by Wiinikka et al. (2013). For As, Cd, Cr, Cu, and Hg, volatility was 10–50 mass %. Kovacs et al. (2016) found that Zn, Cu, Cd, Pb, Cr, Ni, Co, Fe, and their compounds are sufficiently volatile at a temperature of 800°C.

Emissions of heavy metals also depend on factors such as boiler design, combustion conditions, cleaning systems, and the properties of the fuel itself. A valuable tool for managing emissions of heavy metals is the calculation of an emission factor during the biomass combustion process. This factor is based on the initial concentration of trace elements in the biomass being tested and the calorific value of the fuel. Economically and environmentally, fuels with a higher calorific value are preferable as they require less quantity and potentially result in lower emissions of toxic compounds (Jagustyn et al., 2017). Implementing dust precipitation and cleaning the condensates formed in the chimney is recommended to minimize heavy metal emissions.

5.2.2 Polycyclic Aromatic Hydrocarbons

PAHs are a class of chemical compounds that contain two or more fused benzene rings. They are one of the main products of incomplete combustion of biofuel. The formation of PAHs during biomass combustion occurs due to the thermal decomposition of organic materials in conditions of limited oxygen access.

During pyrolysis, organic materials decompose at high temperatures without oxygen, forming various gases, liquids, and solid residues.

PAHs are highly persistent in the environment and can cause significant harm to human health due to their carcinogenic properties. This makes controlling their emissions during biomass combustion critically important, requiring the application of modern purification technologies and optimization of combustion processes. Light PAHs such as naphthalene, phenanthrene, pyrene, and fluorine predominate in biomass combustion emissions. Heavy PAHs, including benzo[a]pyrene and indeno[1,2,3-cd]pyrene, are usually emitted in smaller quantities. Once released into the atmosphere, PAHs can settle in the soil, accumulate in plants, and enter the animal food chain. Most PAHs exhibit toxic, mutagenic, and/or carcinogenic effects on living organisms. Due to their excellent solubility in lipids, they easily penetrate epithelial cells (Bao et al., 2023) and are well absorbed in the gastrointestinal tract (Abdel-Shafy & Mansour, 2016). The potential health risks of 16 PAHs are detailed in Table 5.2. However, there is limited data on the specific effects of individual PAHs.

It is crucial to control numerous combustion parameters to prevent the formation of PAHs. Critical factors include (1) Biomass moisture: High moisture impedes complete combustion, facilitating the accumulation of PAHs; (2) Biomass composition: The different content of cellulose, lignin, and other organic components affects the type and amount of PAHs formed. Lignin, for example, can contribute to the formation of more complex PAHs (Chen et al., 2017); (3) Temperature and duration of combustion: High temperatures promote complete combustion, whereas low temperatures can increase PAH emissions due to incomplete combustion; (4) Heat distribution: Uneven heat distribution creates zones of inefficient combustion, which also contributes to the formation of PAHs.

Effective control and management of biomass combustion are vital to minimizing PAH emissions. Understanding the mechanisms of PAH formation and the impact of various combustion factors allows for

TABLE 5.2

PAHs and Health Risks

PAH	Number of Rings	Health Risk	References
Naphthalene	2	Irritates eyes, skin, and respiratory tract. Causes kidney and liver damage, skin allergies, dermatitis, cataracts, and retinal damage, and may affect the central nervous system. Can destroy red blood cells, causing hemolytic anemia. Carcinogen.	USEPA (2003), Yost et al. (2021)
Anthracene	3	Skin irritation; Embryotoxic effects.	IPCS (2010)
Phenanthrene		Induces cytotoxicity, oxidative stress, and apoptosis in human primary alveolar epithelial cells. Causes kidney and liver damage. Causes cardiovascular diseases: electrical dysfunction, atherosclerosis, myocardial infarction, and stroke	Ke et al. (2018)
Acenaphthene		Asthma	Agudelo-Castañeda et al. (2017)
Acenaphthylene		No epidemiological evidence yet of their carcinogenic or non-carcinogenic effects	Ugochukwu and Ochonogor (2018)
Fluorene		Decreased red blood cells	USEPA (2003)
Fluoranthene	4	High genotoxicity. Nervous system, heart, and cerebral-vascular diseases, allergic diseases.	Li et al. (2016)
Benzo[a]anthracene		Carcinogens. Could induce cell apoptosis via ROS or mitochondrial-mediated pathway.	Lin et al. (2018), Mujtaba et al. (2015)
Chrysene		Carcinogens. High genotoxicity. Photosensitizer in the presence of UVB irradiation. Produced reactive oxygen species, oxidative stress, apoptosis, and DNA damage in human skin epidermal cell line. May be phototoxic as well as photogenotoxic under UVB irradiation. Causes oxidative stress to hepatocytes, resulting in cytotoxicity and inflammation	Ali et al. (2011), Zhu et al. (2023)
Pyrene		DNA damage	Fullove and Yu (2013)

(Continued)

TABLE 5.2 (*Continued*)

PAHs and Health Risks

PAH	Number of Rings	Health Risk	References
Dibenzo[a,h] anthracene	5	Carcinogens. Nervous system, heart, and cerebral-vascular diseases. Exerts toxic effects on the liver, enhanced CD4+ and CD8+ T-cell proliferation and cytokine production, has immunosuppressive potential	He et al. (2020), Okona-Mensah et al. (2005)
Benzo[a]pyrene		Carcinogens: tumors in the forestomach, oral cavity, liver, and lungs. High genotoxicity. DNA damage. Induces apoptosis via ROS or mitochondrial pathways. Alters neurotransmitter levels, BaP-DNA adducts, impaired cognition, learning, and memory. Skin irritation. Ischemic heart disease, myocardial infarction, atherosclerosis. Prenatal exposure: mental retardation, decreased IQ, anxiety, and attention issues.	Chepelev et al. (2015), Moffat et al. (2015), Perera et al. (2012)
Benzo[b] fluoranthene		Carcinogens. DNA damage. Inhibited cell growth, increased ROS production. Altered oxidative stress indicators, decreased mitochondrial membrane potential, disrupted mitochondrial regulators. Reproductive damage via apoptosis and autophagy.	Bao et al. (2023), Guo et al. (2020)
Benzo[k] fluoranthene		Induce DNA damage and steatosis. Carcinogens. Changes in miRNA expression.	Burgoon et al. (2017), Song et al. (2012)
Indeno[1,2,3-c,d] pyrene	6	Carcinogens. Could produce many mutagenic substances in the early stages of metabolism, which may cause necrosis and apoptosis of cells. Promote ferroptosis sensitivity in airway epithelial cells, promote Th17 differentiation, and affect Tfh cell differentiation, exacerbating asthma	Liu et al. (2023), Luo et al. (2015), Sun et al. (2020)
Benzo[g,h,I] perylene		DNA damage. Formation of stable adducts in the DNA. Cytotoxic and genotoxic compounds to a bronchial cell line.	Castro-Gálvez et al. (2019), Zaragoza-Ojeda et al. (2016)

the development of technologies and strategies to reduce their formation. This can include biomass pretreatment, optimization of combustion parameters, and application of post-combustion gas cleaning technologies.

5.2.3 Particulate Matter

The presence of particulate matter in the atmosphere primarily results from biomass combustion. These elements have significant impacts on global climate and human health. Evidence strongly links the increase in cardiovascular and respiratory diseases to high concentrations of particulate matter in the atmosphere. For instance, wood smoke causes acute inflammation of the respiratory system, increases the frequency of asthma and allergic diseases, and raises mortality rates.

Understanding particulate matter sources is vital for effectively controlling harmful emissions and protecting public health. Two main sources of particulate matter are identified:

1. Primary subtype particles appear directly in the atmosphere from construction, agricultural activities, unpaved roads, and the burning of various types of wood and waste.
2. Secondary subtype particles are formed due to gas-chemical changes from emissions from internal combustion engines and industrial activities.

Particulate matter is classified into three types based on diameter: PM1: Particles up to 1 micron in size, including various viruses, bacteria, and pathogens; PM2.5: Particles 2.5 microns or smaller, encompassing smoke, fog, and haze; PM10: Particles 10 microns or smaller, including pollen, spores, and dust.

Smaller particulate matters pose a greater danger because their size allows them to penetrate deeper into the lungs. Table 5.3 provides a comparative characterization of particulate matter.

The chemical composition of particulate matter is categorized primarily into organic and inorganic components. Organic components encompass substances such as black carbon, composed of pure carbon in multiple bonded forms. This form of carbon is generated through the incomplete combustion of biomass, fossil fuels, or biofuels. Common sources include household heating systems, transportation, agriculture and off-road vehicles, industrial processes, and burning forests and savannahs. Research suggests that black carbon is responsible for 15%–30% of the global warming effect. These particles can absorb solar energy and emit thermal radiation, which in turn causes additional heating of the Earth's surface. Another significant organic component is brown carbon, which comprises a mixture of polar and non-polar organic molecules, often including PAHs and their derivatives. Brown carbon originates from biomass combustion and can include substances from smoldering fires, coal burning, and organic compounds released by soil and vegetation. It has a strong capacity to absorb ultraviolet light, significantly contributing to atmospheric heating. Inorganic components of particulate matter include mineral particles such as magnesium, potassium, and calcium, typically linked to ash content. Additionally, metals such as mercury, lead, cadmium, and nickel are prevalent, posing significant toxicity risks to human health and the environment.

Studies on particulate matter formation indicate that the nature and volume of these emissions are heavily influenced by factors such as the temperature of heating, and aeration, the conditions under which smoldering and combustion occur, and how frequently the temperature regime changes. It is essential to conduct a systematic and precise analysis of the sources of these pollutants and their components to mitigate the detrimental effects of particulate matter on the atmosphere.

Currently, several analytical models are employed to assess these sources. One-dimensional models, such as the Chemical Mass Balance (Chen et al., 2017), need detailed source profiles and are typically used alongside other methods. Multidimensional models like Positive Matrix Factorization (Reff et al., 2007) or Principal Component Analysis (Onat et al., 2012) use the least squares method to analyze factor profiles from extensive data. The Unmix model by the Environmental Protection Agency provides precise source contributions but requires specific conditions for best results. Each model presents unique features and limitations, necessitating careful selection based on specific research needs and the nature of the particulate matter being analyzed.

TABLE 5.3
Particulate Matters and their Associated Adverse Health Effects

Particulate Matter	Localization in the Respiratory System	Toxic Effect	References
PM1	Can undergo deeper in the lung system, being the smallest fraction capable of reaching the bloodstream	Oxidative stress, local pulmonary, systemic inflammatory responses, chronic bronchitis, cardiovascular diseases, carcinogenic risk (for example, lung cancer)	Pateraki et al. (2019), Samoli et al. (2016)
PM2.5	Absorbed by alveoli, or can enter other organs via lung ventilation	Asthma, bronchitis, pneumonia, influenza and acute respiratory tract infection, cardiovascular diseases, skin diseases, lung cancer	Xing et al. (2016), Yang et al. (2000)
PM10	Able to enter into the respiratory tract and reach the lungs	Immune responses, irritate the lungs, lung constriction, producing shortness of breath, damage cells, aggravated coughing, aggravated asthma, chronic bronchitis, cardiovascular diseases: arrhythmia, cerebrovascular diseases, high blood pressure, ischemic heart disease, cancer	Misiukiewicz-Stepien and Paplinska-Goryca (2021), Zhang et al. (2021)

5.3 Case Studies

5.3.1 Biomass Burning-Derived Airborne Particulate Matter in Southeast Asia

In Southeast Asia, the interaction between anthropogenic activities and natural phenomena has led to recurrent vegetation and peatland fires, predominantly in Indonesia's regions of Sumatra and Borneo. These fires are significant sources of smoke haze emissions that affect local air quality and have far-reaching impacts across Southeast Asia. The particulate matter emitted from these biomass-burning activities undergoes complex physical and chemical transformations in the atmosphere, posing substantial challenges in characterizing its composition and understanding its environmental and health impacts.

Recent research findings indicate that vegetation and peatland fires release large quantities of particulate matter in various sizes and complex chemical compositions, including trace gases (Adam et al., 2021). These emissions contribute to elevated pollution levels and reduced visibility near the sources and in densely populated areas further away due to the transboundary migration of smoke. The impact of biomass smoke varies annually and depends on factors such as the type of fire, combustion conditions, and prevailing weather patterns.

Notably, severe haze episodes in 1997/1998, 2006, 2013, 2015, and 2019 were marked by high levels of PM2.5 (Adam et al., 2021). During these episodes, the composition of particulate matter at observational sites predominantly consisted of primary and secondary carbonaceous aerosols, with a notable increase in toxic trace elements, raising concerns over adverse health effects. Research has highlighted that respiratory, cardiovascular, and all-cause mortality risks are heightened, particularly affecting vulnerable groups such as children and older people.

Despite the focus on health impacts, less attention has been given to immediate measures for exposure mitigation. Protecting public health from inhaling harmful particulate matter should be prioritized in future research endeavors. The dialogue on air quality in Southeast Asian countries has intensified, urging a science-driven policy approach to address the negative consequences of biomass-burning-derived particulate matter effectively.

5.3.2 On the Contribution of Biomass Burning to PAHs in Air in Africa

Wildfires across Africa's diverse forests liberate many pollutants, including PAHs. Due to its carcinogenic properties, Benzo(a)pyrene (BaP) is of particular concern. The dynamics of BaP emissions are deeply influenced by the frequency of wildfires and the nature of the burned biomass. Research from 2001 to 2014 reveals that Africa contributes significantly to global BaP emissions from wildfires, accounting for nearly half of the global totals from similar sources (Wu et al., 2022). These emissions, while predominantly affecting local air quality, also have transcontinental consequences, influencing air quality far beyond the African continent.

The health implications of BaP exposure in Africa are alarming. Particularly in Sub-Saharan Africa, where annual average BaP concentrations can reach as high as 5–8 ng/m^3, and even 10 ng/m^3 in some regions, lifetime cancer risks associated with BaP exposure often exceed the World Health Organization's acceptable risk levels. In some areas, the lifetime cancer risks surpass the high-risk threshold, posing significant health risks to local populations.

The international community has recognized the severity of PAH emissions from African wildfires. Under the auspices of initiatives like the Task Force on Hemispheric Transport of Air Pollution, efforts are underway to enhance the understanding of these emissions' sources, the environmental conditions influencing them, and their health impacts. Such knowledge is critical for developing effective strategies to mitigate these emissions.

Despite global efforts to reduce PAH emissions from fossil fuels, the role of wildfire biomass burning as a source of PAH contamination is expected to increase. Therefore, reducing the incidence of wildfires and improving forest management in Africa are crucial steps forward.

5.3.3 Air Pollution from Sugarcane Biomass Burning in Brazil

Sugarcane burning before harvest is common in Brazil, primarily to facilitate manual harvesting and reduce the threat from venomous animals. However, this method significantly deteriorates air quality and impacts public health in areas heavily involved in sugarcane agriculture.

Research focusing on the nasal mucociliary system, which serves as the first line of defense against airborne pollutants in the upper airways, has shown acute effects from exposure to particulate matter released during sugarcane burning. In a specific study, the saccharin transit time test evaluated both non-smokers and light smokers working as sugarcane cutters for lung function, exhaled carbon monoxide (CO) levels, nasal symptoms, and mucociliary clearance. The tests were conducted before the harvest season and again shortly after the harvest commenced. The findings were telling; although exhaled CO levels increased in smokers, they remained unchanged post-harvest. Significantly, the saccharin transit time demonstrated a marked decrease in mucociliary clearance in both groups after exposure to the smoke, indicating impairment of this crucial defense mechanism without corresponding nasal symptoms.

Moreover, an ecological time-series study conducted in the city of Araraquara, São Paulo, sought to correlate the levels of total suspended particles from sugarcane burning with hospital admissions for asthma (Arbex et al., 2007). The study established a clear relationship using data collected from local hospital records and daily measurements of total suspended particles. Increases in total suspended particle levels were closely followed by the rise in asthma hospital admissions, with effects lasting several days after the initial spike in particle concentration. A 10 $\mu g/m^3$ rise in the 5-day moving average concentrations of total suspended particles corresponded to an 11.6% increase in asthma admissions, underscoring the direct impact of sugarcane burning on respiratory health. These studies collectively highlight the significant health risks associated with pre-harvest sugarcane burning.

5.3.4 Impact of Biomass Burning on PAH Emissions in China

In China, the use of biomass for cooking and heating is prevalent in rural and semi-urban areas, making the study of PAH emissions particularly relevant for public health and environmental policy.

In a detailed examination conducted across three Chinese provinces by Du et al. (2020), researchers evaluated the emission of PAHs from indoor biomass burning in residential cooking stoves. This study provided a real-world perspective on PAH emissions, revealing that emission factors varied considerably, ranging from 42 to 370 mg/kg. These variations were influenced mainly by the type of biomass fuel and stove technology used. Notably, the emission factors observed in real-world settings were consistently higher than those reported in controlled laboratory studies, suggesting that actual exposure to PAHs might be underrepresented in previous assessments.

The results showed significant differences in the profiles of emitted PAHs compared to those recorded in earlier studies. Particularly concerning was the high presence of non-priority PAHs, which, despite their high toxicity, are less frequently monitored. The study also found that the absorption of PAHs by organic carbon in particulate matter was the dominant mechanism for their transport in the gas-particle mix. A notable regional disparity in PAH emissions was observed, with northern China showing higher emission levels with more tremendous toxic potential, primarily due to burning crop residues. This contrasted with southern China, where wood combustion was more prevalent. The impact of these emissions is further compounded in northern regions by the additional burning of biomass for heating during the colder winter months.

Despite ongoing efforts to reduce biomass use through a nationwide shift toward cleaner energy sources, the persistence of solid fuel use continues to pose significant health risks. The government's push to replace solid fuels with cleaner alternatives is crucial, but the remaining impact of PAH emissions from solid fuels continues to be a critical concern for air quality and public health.

5.4 Conclusion

Biomass combustion poses significant challenges due to the formation of PTS like heavy metals, PAHs, and particulate matter, which pose risks to human health and the environment. This chapter examines biomass types, combustion technologies, and emissions, highlighting the multifaceted implications of PTS. Through this investigation, it has been established that the composition of the biomass heavily influences the types and quantities of PTS generated, the conditions under which it is burned, and the efficiency of the combustion process. The mechanisms of PTS formation were explored, revealing how these toxic substances emerge during biomass burning and identifying the various pathways through which humans and wildlife can be exposed. These pathways include inhalation of contaminated air, ingestion of polluted food and water, and dermal contact, highlighting the pervasive nature of exposure risks.

The health risks associated with PTS are substantial, encompassing respiratory issues, cardiovascular diseases, and various forms of cancer. In light of these findings, it is clear that while biomass offers significant benefits as a renewable energy source, the associated environmental and health risks cannot be overlooked. Mitigation strategies are essential to address these risks. Improvements in combustion technologies are necessary to ensure complete combustion and minimize PTS formation. Additionally, stringent emission controls and monitoring systems must be implemented to effectively detect and manage PTS emissions.

Acknowledgments

The study was financial support of the Ministry of Science and Higher Education of the Russian Federation, agreement no. 075-15-2022-1122 and by the Strategic Academic Leadership Program of the Southern Federal University ("Priority 2030").

REFERENCES

Abdel-Shafy, H. I., & Mansour, M. S. (2016). A review on polycyclic aromatic hydrocarbons: Source, environmental impact, effect on human health and remediation. Egyptian Journal of Petroleum, 25(1), 107–123. https://doi.org/10.1016/j.ejpe.2015.03.011

Abernathy, C. O., Thomas, D. J., & Calderon, R. L. (2003). Health effects and risk assessment of arsenic. The Journal of Nutrition, 133(5), 1536S–1538S. https://doi.org/10.1093/jn/133.5.1536S

Adam, M. G., Tran, P. T., Bolan, N., & Balasubramanian, R. (2021). Biomass burning-derived airborne particulate matter in Southeast Asia: A critical review. Journal of Hazardous Materials, 407, 124760.

Adler, G., Flores, J., Abo Riziq, A., Borrmann, S., & Rudich, Y. (2011). Chemical, physical, and optical evolution of biomass burning aerosols: A case study. Atmospheric Chemistry and Physics, 11(4), 1491–1503.

Agudelo-Castañeda, D. M., Teixeira, E. C., Schneider, I. L., Lara, S. R., & Silva, L. F. (2017). Exposure to polycyclic aromatic hydrocarbons in atmospheric PM1. 0 of urban environments: Carcinogenic and mutagenic respiratory health risk by age groups. Environmental Pollution, 224, 158–170. https://doi.org/10.1016/j.envpol.2017.01.075

Albuquerque, A. R., Angélica, R. S., Merino, A., & Paz, S. P. (2021). Chemical and mineralogical characterization and potential use of ash from Amazonian biomasses as an agricultural fertilizer and for soil amendment. Journal of Cleaner Production, 295, 126472.

Alengebawy, A., Abdelkhalek, S. T., Qureshi, S. R., & Wang, M.-Q. (2021). Heavy metals and pesticides toxicity in agricultural soil and plants: Ecological risks and human health implications. Toxics, 9(3), 42. https://doi.org/10.3390/toxics9030042

Ali, D., Verma, A., Mujtaba, F., Dwivedi, A., Hans, R., & Ray, R. (2011). UVB-induced apoptosis and DNA damaging potential of chrysene via reactive oxygen species in human keratinocytes. Toxicology Letters, 204(2–3), 199–207.

Alvarez, C. C., Gómez, M. E. B., & Zavala, A. H. (2021). Hexavalent chromium: Regulation and health effects. Journal of Trace Elements in Medicine and Biology, 65, 126729. https://doi.org/10.1016/j.jtemb.2021.126729

Amiandamhen, S. O., Kumar, A., Adamopoulos, S., Jones, D., & Nilsson, B. (2020). Bioenergy production and utilization in different sectors in Sweden: A state of the art review. BioResources, 15(4), 9834.

Antunes, F. A., Chandel, A. K., Terán-Hilares, R., Milessi, T. S., Travalia, B. M., Ferrari, F. A., Hernandez-Pérez, A. F., Ramos, L., Marcelino, P. F., & Brumano, L. P. (2019). Biofuel production from sugarcane in Brazil. In Khan, M. T., Khan, I. A. (eds) Sugarcane Biofuels: Status, Potential, and Prospects of the Sweet Crop to Fuel the World, 99–121. Springer, Chem.

Arbex, M. A., Martins, L. C., de Oliveira, R. C., Pereira, L. A. A., Arbex, F. F., Cançado, J. E. D., Saldiva, P. H. N., & Braga, A. L. F. (2007). Air pollution from biomass burning and asthma hospital admissions in a sugar cane plantation area in Brazil. Journal of Epidemiology & Community Health, 61(5), 395–400.

Aruwajoye, G. S., Kassim, A., Saha, A. K., & Gueguim Kana, E. B. (2020). Prospects for the improvement of bioethanol and biohydrogen production from mixed starch-based agricultural wastes. Energies, 13(24), 6609.

Bao, Q., Zhao, S., Liu, Y., Yao, Y., You, J., & Xiong, J. (2023). Benzo [b] fluoranthene induced oxidative stress and apoptosis in human airway epithelial cells via mitochondrial disruption. Journal of Applied Toxicology, 43(7), 1083–1094. https://doi.org/10.1002/jat.4445

Burgoon, L. D., Druwe, I. L., Painter, K., & Yost, E. E. (2017). Using in vitro high-throughput screening data for predicting benzo [k] fluoranthene human health hazards. Risk Analysis, 37(2), 280–290. https://doi.org/10.1111/risa.12613

Castro-Gálvez, Z., Garrido-Armas, M., Palacios-Arreola, M. I., Torres-Flores, U., Rivera-Torruco, G., Valle-Rios, R., Amador-Muñoz, O., Hernández-Hernández, A., & Arenas-Huertero, F. (2019). Cytotoxic and genotoxic effects of Benzo [ghi] perylene on the human bronchial cell line NL-20. Toxicology in Vitro, 61, 104645. https://doi.org/10.1016/j.tiv.2019.104645

Charkiewicz, A. E., Omeljaniuk, W. J., Nowak, K., Garley, M., & Nikliński, J. (2023). Cadmium toxicity and health effects—a brief summary. Molecules, 28(18), 6620. https://doi.org/10.3390/molecules28186620

Chaw, S.-M., Chang, C.-C., Chen, H.-L., & Li, W.-H. (2004). Dating the monocot–dicot divergence and the origin of core eudicots using whole chloroplast genomes. Journal of Molecular Evolution, 58, 424–441.

Chen, C., Luo, Z., & Yu, C. (2019). Release and transformation mechanisms of trace elements during biomass combustion. Journal of Hazardous Materials, 380, 120857.

Chen, J., Li, C., Ristovski, Z., Milic, A., Gu, Y., Islam, M. S., Wang, S., Hao, J., Zhang, H., & He, C. (2017). A review of biomass burning: Emissions and impacts on air quality, health and climate in China. Science of the Total Environment, 579, 1000–1034.

Chepelev, N. L., Moffat, I. D., Bowers, W. J., & Yauk, C. L. (2015). Neurotoxicity may be an overlooked consequence of benzo [a] pyrene exposure that is relevant to human health risk assessment. Mutation Research/Reviews in Mutation Research, 764, 64–89. https://doi.org/10.1016/j.mrrev.2015.03.001

Cima, F. (2011). Tin: Environmental pollution and health effects. Encyclopedia of Environmental Health, 10, 351–359.

Darmawan, A., Asyhari, T., Dunggio, I., Salmahaminati, & Aziz, M. (2023). Energy harvesting from tropical biomasses in Wallacea region: Scenarios, technologies, and perspectives. Biomass Conversion and Biorefinery, 14, 1–19. https://doi.org/10.1007/s13399-023-04223-8

Darmawan, A., & Aziz, M. (2022). *An Overview of Biomass Waste Utilization*. Elsevier, Amsterdam, The Netherlands.

Delplanque, M., Collet, S., Del Gratta, F., Schnuriger, B., Gaucher, R., Robinson, B., & Bert, V. (2013). Combustion of Salix used for phytoextraction: The fate of metals and viability of the processes. Biomass and Bioenergy, 49, 160–170.

Demirbas, A. (2007). Combustion of biomass. Energy Sources, Part A: Recovery, Utilization, and Environmental Effects, 29(6), 549–561. https://doi.org/10.1080/009083190957694

Du, W., Yun, X., Chen, Y., Zhong, Q., Wang, W., Wang, L., Qi, M., Shen, G., & Tao, S. (2020). PAHs emissions from residential biomass burning in real-world cooking stoves in rural China. Environmental Pollution, 267, 115592.

Elhacham, E., Ben-Uri, L., Grozovski, J., Bar-On, Y. M., & Milo, R. (2020). Global human-made mass exceeds all living biomass. Nature, 588(7838), 442–444. https://doi.org/10.1038/s41586-020-3010-5

Fullove, T. P., & Yu, H. (2013). DNA damage and repair of human skin keratinocytes concurrently exposed to pyrene derivatives and UVA light. Toxicology Research, 2(3), 193–199.

Garza-Lombó, C., Pappa, A., Panayiotidis, M. I., Gonsebatt, M. E., & Franco, R. (2019). Arsenic-induced neurotoxicity: A mechanistic appraisal. JBIC Journal of Biological Inorganic Chemistry, 24, 1305–1316. https://doi.org/10.1007/s00775-019-01740-8

Green, Y. (2024). The application of ethanol fuel: Taking the United States as an example. Biological Evidence, 14 1.

Guo, J., Huang, J., Zhang, L., Li, C., Qin, Y., Liu, W., Li, J., & Huang, G. (2020). Benzo [b] fluoranthene impairs mouse oocyte maturation via inducing the apoptosis. Frontiers in Pharmacology, 11, 1226. https://doi.org/10.3389/fphar.2020.01226

Hasanein, P., & Emamjomeh, A. (2019). Beneficial effects of natural compounds on heavy metal–induced hepatotoxicity. In Watson, R. R., Preedy, V. R. (eds) Dietary Interventions in Liver Disease, 345–355. Elsevier, Amsterdam, The Netherlands. https://doi.org/10.1016/B978-0-12-814466-4.00028-8

He, C., Shen, M., Morita, K., Wang, D., Kanazawa, T., & Yoshida, Y. (2020). Nuclear factor of activated T cells as a marker of in vivo low-dose dibenzo [a, h] anthracene exposure. Journal of Applied Toxicology, 40(9), 1239–1247. https://doi.org/10.1002/jat.3981

Hegazy, A. M., & Fouad, U. A. (2014). Evaluation of lead hepatotoxicity; histological, histochemical and ultrastructural study. Forensic Medicine and Anatomy Research, 2(03), 70. https://doi.org/10.4236/fmar.2014.23013

Hirani, A. H., Javed, N., Asif, M., Basu, S. K., & Kumar, A. (2018). A review on first-and second-generation biofuel productions. In Kumar, A., Ogita, S., Yau, Y. Y. (eds) Biofuels: Greenhouse Gas Mitigation and Global Warming: Next Generation Biofuels and Role of Biotechnology, 141–154. Springer, New Delhi. https://doi.org/10.1007/978-81-322-3763-1_8

Hsiao, C.-L., Wu, K.-H., & Wan, K.-S. (2011). Effects of environmental lead exposure on T-helper cell-specific cytokines in children. Journal of Immunotoxicology, 8(4), 284–287. https://doi.org/10.3109/1547691X.2011.592162

Huang, H.-W., Lee, C.-H., & Yu, H.-S. (2019). Arsenic-induced carcinogenesis and immune dysregulation. International Journal of Environmental Research and Public Health, 16(15), 2746. https://doi.org/10.3390/ijerph16152746

Hunce, S. Y., Clemente, R., & Bernal, M. P. (2019). Energy production potential of phytoremediation plant biomass: Helianthus annuus and Silybum marianum. Industrial Crops and Products, 135, 206–216.

IEA. (2023). *Tracking Clean Energy Progress 2023.* IEA, Paris. Licence: CC BY 4.0. https://doi.org/https://www.iea.org/reports/tracking-clean-energy-progress-2023

IPCS. (2010). *Polycyclic Aromatic Hydrocarbons, Selected Non-Heterocyclic*. IPCS, Geneva.

Jagustyn, B., Kmieć, M., Smędowski, Ł., & Sajdak, M. (2017). The content and emission factors of heavy metals in biomass used for energy purposes in the context of the requirements of international standards. Journal of the Energy Institute, 90(5), 704–714.

Jiménez, S., Perez, M., & Ballester, J. (2008). Vaporization of trace elements and their emission with submicrometer aerosols in biomass combustion. Energy & Fuels, 22(4), 2270–2277.

Karp, S. G., Medina, J. D., Letti, L. A., Woiciechowski, A. L., de Carvalho, J. C., Schmitt, C. C., de Oliveira Penha, R., Kumlehn, G. S., & Soccol, C. R. (2021). Bioeconomy and biofuels: The case of sugarcane ethanol in Brazil. Biofuels, Bioproducts and Biorefining, 15(3), 899–912.

Ke, S., Liu, Q., Yao, Y., Zhang, X., & Sui, G. (2018). An in vitro cytotoxicities comparison of 16 priority polycyclic aromatic hydrocarbons in human pulmonary alveolar epithelial cells HPAEpiC. Toxicology Letters, 290, 10–18. https://doi.org/10.1016/j.toxlet.2018.03.005

Kim, Y.-J., & Kim, J.-M. (2015). Arsenic toxicity in male reproduction and development. Development & Reproduction, 19(4), 167. https://doi.org/10.12717/DR.2015.19.4.167

Kovacs, H., Szemmelveisz, K., & Koós, T. (2016). Theoretical and experimental metals flow calculations during biomass combustion. Fuel, 185, 524–531.

Kumar, S., & Trivedi, A. (2016). A review on role of nickel in the biological system. International Journal of Current Microbiology and Applied Sciences, 5(3), 719–727.

Li, X., Yang, Y., Xu, X., Xu, C., & Hong, J. (2016). Air pollution from polycyclic aromatic hydrocarbons generated by human activities and their health effects in China. Journal of Cleaner Production, 112, 1360–1367. https://doi.org/10.1016/j.jclepro.2015.05.077

Lin, S., Ren, A., Wang, L., Huang, Y., Wang, Y., Wang, C., & Greene, N. D. (2018). Oxidative stress and apoptosis in benzo [a] pyrene-induced neural tube defects. Free Radical Biology and Medicine, 116, 149–158. https://doi.org/10.1016/j.freeradbiomed.2018.01.004

Liu, K.-Y., Gao, Y., Xiao, W., Fu, J., Huang, S., Han, X., Hsu, S.-H., Xiao, X., Huang, S.-K., & Zhou, Y. (2023). Multidimensional analysis of lung lymph nodes in a mouse model of allergic lung inflammation following PM 2.5 and indeno [1, 2, 3-cd] pyrene exposure. Environmental Health Perspectives, 131(3), 037014. https://doi.org/10.1289/EHP11580

Londahl, J., Pagels, J., Boman, C., Swietlicki, E., Massling, A., Rissler, J., Blomberg, A., Bohgard, M., & Sandström, T. (2008). Deposition of biomass combustion aerosol particles in the human respiratory tract. Inhalation Toxicology, 20(10), 923–933. https://doi.org/10.1080/08958370802087124

Luo, Y., Li, S., She, Y., Zhong, R., Zhang, S., & An, X. (2015). Metabolic transformation of indeno [1, 2, 3–cd] pyrene and toxicities of its metabolites. Bulletin of Environmental Contamination and Toxicology, 94, 112–117. https://doi.org/10.1007/s00128-014-1411-6

Mahurpawar, M. (2015). Effects of heavy metals on human health. International Journal of Research Granthaalayah, 530(516), 1–7.

Martinsson, J., Eriksson, A., Nielsen, I. E., Malmborg, V. B., Ahlberg, E., Andersen, C., Lindgren, R., Nystrom, R., Nordin, E., & Brune, W. (2015). Impacts of combustion conditions and photochemical processing on the light absorption of biomass combustion aerosol. Environmental Science & Technology, 49(24), 14663–14671.

Mishra, S., & Bharagava, R. N. (2016). Toxic and genotoxic effects of hexavalent chromium in environment and its bioremediation strategies. Journal of Environmental Science and Health, Part C, 34(1), 1–32. https://doi.org/10.1080/10590501.2015.1096883

Misiukiewicz-Stepien, P., & Paplinska-Goryca, M. (2021). Biological effect of PM10 on airway epithelium-focus on obstructive lung diseases. Clinical Immunology, 227, 108754. https://doi.org/10.1016/j.clim.2021.108754

Moffat, I., Chepelev, N. L., Labib, S., Bourdon-Lacombe, J., Kuo, B., Buick, J. K., Lemieux, F., Williams, A., Halappanavar, S., & Malik, A. I. (2015). Comparison of toxicogenomics and traditional approaches to inform mode of action and points of departure in human health risk assessment of benzo [a] pyrene in drinking water. Critical Reviews in Toxicology, 45(1), 1–43. https://doi.org/10.3109/10408444.2014.973934

Mujtaba, S. F., Dwivedi, A., Yadav, N., Ch, R., Kushwaha, H. N., Mudiam, M. K., Singh, G., & Ray, R. S. (2015). Superoxide mediated photomodification and DNA damage induced apoptosis by Benz (a) anthracene via mitochondrial mediated pathway. Journal of Photochemistry and Photobiology B: Biology, 142, 92–102. https://doi.org/10.1016/j.jphotobiol.2014.11.006

Nalbandian, H. (2012). Trace element emissions from coal. IEA Clean Coal Centre, 601.

Nunes, L., Matias, J., & Catalão, J. (2017). Biomass in the generation of electricity in Portugal: A review. Renewable and Sustainable Energy Reviews, 71, 373–378. https://doi.org/10.1016/j.rser.2016.12.067

Nussbaumer, T. (2003). Combustion and co-combustion of biomass: Fundamentals, technologies, and primary measures for emission reduction. Energy & Fuels, 17(6), 1510–1521. https://doi.org/10.1021/ef030031q

Nzihou, A., & Stanmore, B. (2013). The fate of heavy metals during combustion and gasification of contaminated biomass—a brief review. Journal of Hazardous Materials, 256, 56–66.

Odzijewicz, J. I., Wołejko, E., Wydro, U., Wasil, M., & Jabłońska-Trypuć, A. (2022). Utilization of ashes from biomass combustion. Energies, 15(24), 9653.

Okona-Mensah, K., Battershill, J., Boobis, A., & Fielder, R. (2005). An approach to investigating the importance of high potency polycyclic aromatic hydrocarbons (PAHs) in the induction of lung cancer by air pollution. Food and chemical Toxicology, 43(7), 1103–1116. https://doi.org/10.1016/j.fct.2005.03.001

Okpala, C. O. R., Sardo, G., Vitale, S., Bono, G., & Arukwe, A. (2018). Hazardous properties and toxicological update of mercury: From fish food to human health safety perspective. Critical Reviews in Food Science and Nutrition, 58(12), 1986–2001. https://doi.org/10.1080/10408398.2017.1291491

Onat, B., Şahin, Ü. A., & Bayat, C. (2012). Assessment of particulate matter in the urban atmosphere: Size distribution, metal composition and source characterization using principal component analysis. Journal of Environmental Monitoring, 14(5), 1400–1409.

Packer, M. (2016). Cobalt cardiomyopathy: A critical reappraisal in light of a recent resurgence. Circulation: Heart Failure, 9(12), e003604.

Pateraki, S., Fameli, K.-M., Assimakopoulos, V., Bougiatioti, A., Maggos, T., & Mihalopoulos, N. (2019). Levels, sources and health risk of PM2. 5 and PM1-bound PAHs across the greater athens area: The role of the type of environment and the meteorology. Atmosphere, 10(10), 622. https://doi.org/10.3390/atmos10100622

Paustenbach, D. J., Tvermoes, B. E., Unice, K. M., Finley, B. L., & Kerger, B. D. (2013). A review of the health hazards posed by cobalt. Critical Reviews in Toxicology, 43(4), 316–362. https://doi.org/10.3109/10408444.2013.779633

Perera, F. P., Tang, D., Wang, S., Vishnevetsky, J., Zhang, B., Diaz, D., Camann, D., & Rauh, V. (2012). Prenatal polycyclic aromatic hydrocarbon (PAH) exposure and child behavior at age 6–7 years. Environmental Health Perspectives, 120(6), 921–926. https://doi.org/10.1289/ehp.1104315

Reff, A., Eberly, S. I., & Bhave, P. V. (2007). Receptor modeling of ambient particulate matter data using positive matrix factorization: Review of existing methods. Journal of the Air & Waste Management Association, 57(2), 146–154.

Samoli, E., Atkinson, R. W., Analitis, A., Fuller, G. W., Beddows, D., Green, D. C., Mudway, I. S., Harrison, R. M., Anderson, H. R., & Kelly, F. J. (2016). Differential health effects of short-term exposure to source-specific particles in London, UK. Environment International, 97, 246–253. https://doi.org/10.1016/j.envint.2016.09.017

Santamaria, A. B., & Sulsky, S. I. (2010). Risk assessment of an essential element: Manganese. Journal of Toxicology and Environmental Health, Part A, 73(2–3), 128–155. https://doi.org/10.1080/15287390903337118

Saracoglu, N., & Gündüz, G. (2009). Wood pellets—tomorrow's fuel for Europe. Energy Sources, Part A: Recovery, Utilization, and Environmental Effects 31(19), 1708–1718.

Schoofs, H., Schmit, J., & Rink, L. (2024). Zinc toxicity: Understanding the limits. Molecules, 29(13), 3130. https://doi.org/10.3390/molecules29133130

Seilkop, S. K., & Oller, A. R. (2003). Respiratory cancer risks associated with low-level nickel exposure: An integrated assessment based on animal, epidemiological, and mechanistic data. Regulatory Toxicology and Pharmacology, 37(2), 173–190. https://doi.org/10.1016/S0273-2300(02)00029-6

Song, M.-K., Song, M., Choi, H.-S., & Ryu, J.-C. (2012). Benzo [k] fluoranthene-induced changes in miRNA-mRNA interactions in human hepatocytes. Toxicology and Environmental Health Sciences, 4, 143–153. https://doi.org/10.1007/s13530-012-0129-2

Stolf, R., & Oliveira, A. P. R. D. (2020). The success of the Brazilian alcohol program (Proálcool)-a decade-by-decade brief history of ethanol in Brazil. Engenharia Agrícola, 40, 243–248.

Su, C. (2014). A review on heavy metal contamination in the soil worldwide: Situation, impact and remediation techniques. Environmental Skeptics and Critics, 3(2), 24.

Sun, L., Fu, J., Lin, S.-H., Sun, J.-L., Xia, L., Lin, C.-H., Liu, L., Zhang, C., Yang, L., & Xue, P. (2020). Particulate matter of 2.5 μm or less in diameter disturbs the balance of TH17/regulatory T cells by targeting glutamate oxaloacetate transaminase 1 and hypoxia-inducible factor 1α in an asthma model. Journal of Allergy and Clinical Immunology, 145(1), 402–414. https://doi.org/10.1016/j.jaci.2019.10.008

Sundseth, K., Pacyna, J. M., Pacyna, E. G., Pirrone, N., & Thorne, R. J. (2017). Global sources and pathways of mercury in the context of human health. International Journal of Environmental Research and Public Health, 14(1), 105. https://doi.org/10.3390/ijerph14010105

Tabakaev, R., Kalinich, I., Mostovshchikov, A., Dimitryuk, I., Asilbekov, A., Ibraeva, K., Gaidabrus, M., Shanenkov, I., Rudmin, M., & Yazykov, N. (2023). Microwave pyrolysis of cattle manure: Initiation mechanism and product characteristics. Biomass Conversion and Biorefinery, 4, 1–12.

Tabakaev, R., Kanipa, I., Astafev, A., Dubinin, Y., Yazykov, N., Zavorin, A., & Yakovlev, V. (2019). Thermal enrichment of different types of biomass by low-temperature pyrolysis. Fuel, 245, 29–38. https://doi.org/10.1016/j.fuel.2019.02.049

Thomas, B. S., Yang, J., Mo, K. H., Abdalla, J. A., Hawileh, R. A., & Ariyachandra, E. (2021). Biomass ashes from agricultural wastes as supplementary cementitious materials or aggregate replacement in cement/geopolymer concrete: A comprehensive review. Journal of Building Engineering, 40, 102332.

Tripathi, N., Hills, C. D., Singh, R. S., & Atkinson, C. J. (2019). Biomass waste utilisation in low-carbon products: Harnessing a major potential resource. NPJ Climate and Atmospheric Science, 2(1), 35. https://doi.org/10.1038/s41612-019-0093-5

Tveritinova, E. (2008). Analysis of Organic-Waste-Based Bioenergy Potential in the Russian Federation: Potential of Substitution of Conventionally Produced Energy with Bioenergy Considering Current Level of Forest, Agricultural and Municipal Waste in Different Federal Districts of the Russian Federation. https://api.semanticscholar.org/CorpusID:113141102

Ugochukwu, U. C., & Ochonogor, A. (2018). Groundwater contamination by polycyclic aromatic hydrocarbon due to diesel spill from a telecom base station in a Nigerian City: Assessment of human health risk exposure. Environmental Monitoring and Assessment, 190, 1–14. https://doi.org/10.1007/s10661-018-6626-2

USEPA. (2003). *Health Effects Support Document for Naphthalene*. USEPA, Washington, DC.

Vassilev, S. V., Baxter, D., Andersen, L. K., & Vassileva, C. G. (2013). An overview of the composition and application of biomass ash. Part 1. Phase–mineral and chemical composition and classification. Fuel, 105, 40–76. https://doi.org/10.1016/j.fuel.2012.09.041

Vassilev, S. V., Baxter, D., Andersen, L. K., Vassileva, C. G., & Morgan, T. J. (2012). An overview of the organic and inorganic phase composition of biomass. Fuel, 94, 1–33.

Vassilev, S. V., Vassileva, C. G., Song, Y.-C., Li, W.-Y., & Feng, J. (2017). Ash contents and ash-forming elements of biomass and their significance for solid biofuel combustion. Fuel, 208, 377–409. https://doi.org/10.1016/j.fuel.2017.07.036

Vassilev, S. V., Vassileva, C. G., & Vassilev, V. S. (2015). Advantages and disadvantages of composition and properties of biomass in comparison with coal: An overview. Fuel, 158, 330–350. https://doi.org/10.1016/j.fuel.2015.05.050

Wielgosiński, G., Łechtańska, P., & Namiecińska, O. (2017). Emission of some pollutants from biomass combustion in comparison to hard coal combustion. Journal of the Energy Institute, 90(5), 787–796. https://doi.org/10.1016/j.joei.2016.06.005

Wiinikka, H., Grönberg, C., & Boman, C. (2013). Emissions of heavy metals during fixed-bed combustion of six biomass fuels. Energy & fuels, 27(2), 1073–1080.

Winters, P. (2011). Current status of cellulosic biofuel commercialization in the United States. *Industrial Biotechnology*, 7(5), 365–374. https://doi.org/10.1089/ind.2011.7.365

Wobst, M., Wichmann, H., & Bahadir, M. (2001). Distribution behavior of heavy metals investigated in a laboratory-scale incinerator. Chemosphere, 44(5), 981–987.

Wojcik, M., Stachowicz, F., & Masłoń, A. (2020). The use of wood biomass ash in sewage sludge treatment in terms of its agricultural utilization. Waste and Biomass Valorization, 11, 753–768.

Wu, M., Luo, J., Huang, T., Lian, L., Chen, T., Song, S., Wang, Z., Ma, S., Xie, C., & Zhao, Y. (2022). Effects of African BaP emission from wildfire biomass burning on regional and global environment and human health. Environment International, 162, 107162.

Xie, K., Hu, H., Xu, S., Chen, T., Huang, Y., Yang, Y., Yang, F., & Yao, H. (2020). Fate of heavy metals during molten salts thermal treatment of municipal solid waste incineration fly ashes. Waste Management, 103, 334–341.

Xing, Y.-F., Xu, Y.-H., Shi, M.-H., & Lian, Y.-X. (2016). The impact of PM2. 5 on the human respiratory system. Journal of Thoracic Disease, 8(1), E69. https://doi.org/10.3978/j.issn.2072-1439.2016.01.19

Yang, F., Ma, Y., & He, K. (2000). A brief introduction to PM2. 5 and related research. World Environ, 4, 33–35.

Yang, X., Liu, Y., Thrän, D., Bezama, A., & Wang, M. (2021). Effects of the German renewable energy sources act and environmental, social and economic factors on biogas plant adoption and agricultural land use change. Energy, Sustainability and Society, 11, 1–22.

Yost, E. E., Galizia, A., Kapraun, D. F., Persad, A. S., Vulimiri, S. V., Angrish, M., Lee, J. S., & Druwe, I. L. (2021). Health effects of naphthalene exposure: A systematic evidence map and analysis of potential considerations for dose–response evaluation. Environmental Health Perspectives, 129(7), 076002. https://doi.org/10.1289/EHP7381

Zaragoza-Ojeda, M., Eguía-Aguilar, P., Perezpeña-Díazconti, M., & Arenas-Huertero, F. (2016). Benzo [ghi] perylene activates the AHR pathway to exert biological effects on the NL-20 human bronchial cell line. Toxicology Letters, 256, 64–76. https://doi.org/10.1016/j.toxlet.2016.05.023

Zha, J., Huang, Y., Clough, P. T., Xia, Z., Zhu, Z., Fan, C., Yu, M., Yan, Y., & Cheng, H. (2021). Green production of a novel sorbent from kaolin for capturing gaseous $PbCl_2$ in a furnace. Journal of Hazardous Materials, 404, 124045.

Zhai, J., Burke, I. T., & Stewart, D. I. (2021). Beneficial management of biomass combustion ashes. Renewable and Sustainable Energy Reviews, 151, 111555.

Zhang, Q., Jimenez, J. L., Canagaratna, M., Allan, J. D., Coe, H., Ulbrich, I., Alfarra, M., Takami, A., Middlebrook, A., & Sun, Y. (2007). Ubiquity and dominance of oxygenated species in organic aerosols in anthropogenically-influenced Northern Hemisphere midlatitudes. Geophysical Research Letters, 34(13), L13801.

Zhang, X., Zhang, X., Gao, B., Li, Z., Xia, H., Li, H., & Li, J. (2014). Effect of cadmium on growth, photosynthesis, mineral nutrition and metal accumulation of an energy crop, king grass (Pennisetum americanum×P. purpureum). Biomass and Bioenergy, 67, 179–187.

Zhang, Y., Ma, Y., Feng, F., Cheng, B., Wang, H., Shen, J., & Jiao, H. (2021). Association between PM10 and specific circulatory system diseases in China. Scientific Reports, 11(1), 12129. https://doi.org/10.1038/s41598-021-91637-x

Zheng, G., & Koziński, J. (2000). Thermal events occurring during the combustion of biomass residue. Fuel, 79(2), 181–192. https://doi.org/10.1016/S0016-2361(99)00130-1

Zhu, Z., Chen, Z., Sakurai, T., Chiba, H., & Hui, S.-P. (2023). Adverse effects of chrysene on human hepatocytes via inducement of oxidative stress and dysregulation of xenobiotic metabolism. Polycyclic Aromatic Compounds, 43(2), 1030–1041. https://doi.org/10.1080/10406638.2021.2023200

6

Heavy Metals in Food Products: An Insight into Food Security as a Global Problem

Muhammad Ubaid Ali, Ruilian Yu, Sajid Ali, Mehr Ahmed Mujtaba Munir, Balal Yousaf, Qumber Abass, and Muhammad Adnan

6.1 Introduction

Heavy metal (HMs) pollution is a global issue that leads to serious environmental concerns and severe health hazards. Some of the HMs, such as iron (Fe), copper (Cu), selenium (Se), and zinc, are considered essential elements as these elements help regulate body metabolism (Wuana & Okieimen, 2011). However, others, such as arsenic (As) (a metalloid), lead (Pb), chromium (Cr), cadmium (Cd), and Hg, have severe carcinogenic and non-carcinogenic health impacts even at very low dosages (Ali et al., 2019). Rapid increases in population, urbanization, industrialization, and land use systems are key influencing factors (Briffa et al., 2020). The diversity of these HMs has increased significantly due to anthropogenic activities such as the industrial revolution, economic globalization, mining, chemical fertilizer use, sewage sludge discharge, and pesticides (Ali et al., 2019). However, excessive accumulation of these HMs in food products such as agricultural products, poultry products, and livestock is a significant concern in emerging food security issues globally and especially in human health (Chowdhury & Alam, 2024).

The presence of HMs in agricultural products is mainly attributed to contaminated soil (Ahogle et al., 2022). Soil is the primary requirement for crops; however, HMs from point sources such as coal mines, chemical industries, energy industries, and thermal power plants, and non-point sources such as soil and sediment erosion, runoff, and freight storage exert significant impacts on their quality (Lin et al., 2022; Rai et al., 2019). Plants uptake these HMs through roots and are stored in different parts, such as stems, leaves, and grains (Tőzsér et al., 2023). Many crops have been shown to bioaccumulate different HMs when grown on contaminated soils, indicating the entrance of these HMs into the food chain (FC). The use of these agricultural feedstocks and grains in poultry and animal husbandry poses a significant threat not only to the livestock and poultry directly but also represents a major human health risk through consumption of these products, leading to serious concern about food security (Afonne & Ifediba, 2020; Rai et al., 2019).

The population worldwide should have physical, economic, and social access to sufficient, nutritious, and healthy food at all times that can fulfill their dietary and active health needs (FAO, 1996). Previously, many studies have reported higher concentrations of HMs in commonly used agricultural products (cereals, root and tubers, fruits, and vegetables), livestock products (meat, milk, and milk products), and poultry (meat and eggs). Human dietary exposure can lead to the accumulation of HMs in human fatty tissues and bone matrix (Elbagermi et al., 2012; Raikwar et al., 2008; Rajaganapathy et al., 2011; Rathebe & Mosoeu, 2023). This can lead to depletion of essential elements, endocrine disruption, growth retardation, cardiovascular problems, renal, gastrointestinal, neurological, reproductive, and immune system problems. This chapter aims to provide insight into HMs' impact on food security and the negative impacts these HMs can pose on crops and livestock. The potential and effective management strategies are also highlighted.

DOI: 10.1201/9781003492115-8

6.2 Heavy Metal Sources in Different Food Products

6.2.1 Soil-Crop System

The key sources of HMs in the soil and agricultural environment include wastewater irrigation, metallic-herbicides or pesticides, the application of sewage sludges, and phosphate-based fertilizers (El-Kady & Abdel-Wahhab, 2018; Rai et al., 2019). Pollutants found in food crops, both from anthropogenic and natural sources, concern human health. These contaminants can be transferred due to soil-to-plant tissue incorporation or direct atmospheric deposition on leaves or other parts of plants (Gan et al., 2017; Zhuang et al., 2009). Particulate matter (PM), along with HMs emitted due to industrial and vehicular emissions, ultimately deposits and accumulates in soil and then in the FC (Luo et al., 2019). The concentrations of HMs (Fe, Cu, Cr, Pb, Ni, and Mn) are significantly high in the distilleries, sludge, chemicals, electroplating, textile, and leather industries (Rathebe & Mosoeu, 2023). Coal-based power plants are considered the critical source of higher Hg concentrations in soil (Rodríguez Martín & Nanos, 2016). Electroplating effluent includes Cd, Cu, Ni, Cr, and Zn, which can lead to stunted development, chlorosis, and leaf necrosis, leading to plant mortality (França et al., 2017). It was found in China that the production of lead-acid-based batteries leads to the emission of PM-bound HMs, which subsequently accumulate in the agro-ecosystem (soil and crops) (Zhang et al., 2019).

The application of phosphate fertilizer (Phospho-gypsum) leads to higher concentrations of different HMs in soil and crops, including significant levels of Cd and Cr, with tomatoes and green peppers having the highest Pb bioavailability (Saeed et al., 2023). Leachates from municipal solid waste and landfills include significant quantities of HMs and metalloids, posing a risk to soil-crop systems. Polluted soils have been linked to elevated HM levels in plant-based foods (Nyiramigisha et al., 2021; Yousaf et al., 2016).

Sewage sludge and improperly treated wastewater are frequently used to increase agricultural output. However, the sites irrigated with improperly treated water cannot assure the safety and quality of crops grown (Minhas et al., 2015; Mishra et al., 2023). The application of nanotechnology in many industries, including agriculture, energy, medicine, and the environment, has an impact on the environment because it releases hazardous metals and nanoparticles (NPs) into the environment uncontrollably (Rai et al., 2018; Xiong et al., 2017).

6.2.2 Livestock

Livestock are a significant source of protein for millions of people around the globe, making it the world's leading food production industry (Aljohani, 2023). HMs (Cd, Pb, Hg, As, and Cr), which are harmful compounds, can negatively affect livestock and poultry. Additionally, essential HMs (Fe, I, Co, Cu, Mn, Mo, and Se) can be harmful if they exceed the safety standards (Baltenweck et al., 2020; Mehmood et al., 2023). These metals can enter the FC through agricultural products that domestic animals eat, posing a severe risk to human health. The common HM routes of exposure on these farms are through ingestion (Khan et al., 2008; Witkowska et al., 2021). Domestic animals coexist with humans in the same habitat. They are mostly exposed to HMs through the consumption of plants and grains, which are often derived from agricultural sources, and, to a lesser extent, exposed to air pollution through inhalation (Amadi & Venkataraman, 2022; Letuka et al., 2023; Mehmood et al., 2023). Drinking water is another primary source of livestock exposure to HMs. This exposure is hard to estimate, especially when the animal drinks water from random sources when grown in an extensive poultry or grazing system (Naveedullah et al., 2013).

The feed provided to broilers has been linked to health complications in humans (Aljohani, 2023). The principal components of chicken feed consist of grain seed, soybean meal, grass meal, biscuits, meat, blood, fish meal, fat, oil, cottonseed bran, maize gluten bran, guar bran, peanut cake minerals, subtherapeutic antibiotics, etc. (Awuchi, 2020; Suleman et al., 2022). The concentration of many hazardous HMs is high, as some metals are added as food additives to increase production. Elements such as Mn, Cu, Fe, Se, and Zn are added as food supplements to enhance livestock's physiological and biochemical functions and overcome the deficiency of essential elements (Aljohani, 2023; Jamil et al., 2023). Selenium is used to minimize oxidative cell damage, while Ca is used for bone formation and to avoid blood clotting

(David et al., 2023; Surai et al., 2018). Zinc and Mn are also used in proper eggshell and feather formation (Alagawany et al., 2020). When assessing the exposure levels of hazardous HM toxic dosages, several aspects that can affect dietary intake, such as the handling of food, the type and quality of raw materials, additives, the ingestion of soil, and unintentional contamination, should be taken into account (Rather et al., 2017). Previously, a study found that the possible way of HM contamination in pig tissues is through feed that is loaded with several additives with high levels of HMs (Khalef et al., 2022). In sheep, higher levels of HMs were attributed to an extensive grazing system, especially near or around industrial zones (Ghosh & Singh, 2014). In the case of poultry, the higher levels of pollution were attributed to backyard rarities, particularly farm waste in landfills, animal feces, and polluted road ditches (Aljohani, 2023).

6.3 Levels of Heavy Metals in Different Crops and Livestock

6.3.1 Crops

6.3.1.1 Cereals

Cereal grains constitute a significant type of staple food used worldwide as a preliminary source of carbohydrates, and among these cereals, 60% of the energy intake worldwide is attributed to rice, wheat, and maize (Afonne & Ifediba, 2020; FAO, 2017). Contamination of these products with HMs is a public concern owing to their high consumption, especially in underdeveloped countries. Rice is considered one of the most essential human foods; however, it has received significant attention concerning HM pollution. Concentrations of HMs in different agricultural products are presented in Table 6.1. Research has shown that rice can uptake As from soil and store it in straw and grains in organic and inorganic As (FAO/WHO, 2017). In central China, it was revealed that brown rice grown near three metal mines had higher levels of As than the regulatory safety limits. Furthermore, the study showed that rice consumption could be a significant source of HM exposure in the local population, especially for As (Fan et al., 2017). Maximum permissible limits of different HMs in cereals and vegetables by different agencies are presented in Table 6.2.

Another study in China near an e-waste recycling site studied the HM contents of polished rice and its hull. The results revealed that the concentrations of HMs for most of the studied HMs apart from Co were higher in hull than in rice. The average amount of Pb (0.69 mg/kg) was three to five times higher

TABLE 6.1

Concentrations of Heavy Metals in Different Crops

Country	Food Crops	HMs Content (mg/kg)	Source of HMs	References
Pakistan	Wheat	Cd 2.8, Cr 4.2, Cu 28, Ni 7.5, Pb 15.6 and Fe 270	Wastewater irrigation	Hassan et al. (2013)
Northern China	Wheat	Cd 0,647, Pb 0.645, Zn 37, and Cu 5.18	Lead smelting	Li et al. (2021)
USA	White rice	As 131, Pb 2.8, and Cd 6.5	//	Tatahmentan et al. (2020)
Pakistan	Basmati rice	Cd (2.7–9.8), Zn (5–9.9), Fe (2–13), Cu (1.84–10.86) and Ni (3.2–13.3)	//	Tariq et al. (2021)
Pakistan	Spinach	Pb 0.34, Cu 1.4, Cd 0.11 and Zn 0.15	Sanitary water irrigation	Noor et al. (2023)
Turkey	Potato	Fe 49–73, Cu 3.1–5.4, Zn 13.8–19, Mn 6.9–13.1, Pb 0.5–0.8, Ni 2–3.6 and Cd 0.08–0.3		Öztürk et al. (2011)
China	Radish	Cu 0.34, Cr 0.03, Pb 0.07, Cd 0.012, Zn 2.48, Ni 0.07	Treated wastewater irrigation	Song et al. (2009)
Kenya	Maize grains	Zn 0.113, Cd 0.225, Cu 0.122, Co 0.042 and Pb 0.324	//	Akenga et al. (2018)
Australia	French beans, beetroot, and kale	Cr (0.00078–0.049), Pb (0.001–0.11) Cu (0.016–0.66) and Zn (0.038–0.145)	Urban stormwater	Tom et al. (2014)
Iran	Soybean	Cd 42.0 ± 19.06, Cu 13440.1 ± 1437.0, Pb 34.44 ± 14.82, and Zn 32566.8 ± 4965	//	Sobhanardakani et al. (2016)

TABLE 6.2
Maximum Permissible Limits of Different Heavy Metals in Cereals and Vegetables by Different Agencies

Heavy Metals	FAO/WHO	Chinese Standards (GB 2762-2005)	USDA	European Union (EC:No.629/2008)
As	0.15	0.15	0.00001	0.1
Cd	0.2	0.1–0.2	0.0026	0.2
Cu	40	10	0.01	//
Cr	1.3	≤0.05	0.03	//
Hg	0.02	0.02	0.01	0.02
Pb	0.03	0.02	0.02	0.02
Zn	60	50	//	//

Sources: EU (2015), Lukuman (2018), USDA (2018), Wang et al. (2015), Clever and Jie (2015), Elafify et al. (2023) and FAO/WHO (2017).

than the maximum permissible levels (MPLs) (China's national standard for safety milled rice criteria (NY5115-2002)). For Cd, 31% of samples were higher than the MPL, and the level was higher in local rice than in commercial rice (Fu et al., 2008). At three industrial sites in Bangladesh, it was found that the amounts of HMs like Cu, Zn, Fe, Cr, and Co were higher than the levels allowed by WHO/FAO. They suggested strict rules to ensure crops are safe for food safety (Hasan et al., 2022).

Maize is considered the second most important cereal after wheat; however, the point concern is that it can also accumulate higher HM levels. The content of HMs in maize seeds collected from the Ilesha gold mining area, Osun State, Nigeria, revealed that the mean contents of As, Cd, and Pb were in the range of 0.6–2.0, 1.5–4.6, and 0.90–2.5 mg/kg, respectively, which were higher than the MPL set by WHO/FAO (Adekiya et al., 2018). In Delta State, Nigeria revealed that the concentrations of Cd (29.1%), Cr (21%), and Pb (44.4%) exceeded the permissible limits of food safety set by the European Union (EU) (Ifie et al., 2022). A study near the Zn smelting site in China revealed that the mean concentrations of Cd, Cr, Pb, and Zn in maize exceeded the maximum permissible limits of Chinese food standards (MHPRC, 2012). It revealed that the concentration of Zn (45.1 ± 1.5 mg/kg) was twice the value of MPL, while in the case of Cd, 95.2% of the samples exceeded MPL (0.2 mg/kg). For Cr, the maximum concentration was fivefolds higher than the limits (Wei et al., 2020). In Pakistan, maize crops irrigated with industrial wastewater showed the highest levels of contamination and posed significant carcinogenic risks (Iqbal et al., 2019).

Wheat has shown high vulnerability in HM accumulation, especially in the case of Cd, Pb, and Cu (Zhou & Zheng, 2022). Asgari and Cornelis (2015) revealed that the Cd levels in wheat grains were significantly higher than those recommended by the United States Environmental Protection Agency (US EPA). In contrast, the concentration of Ni in wheat and maize was several times higher than the recommended guidelines. A significantly higher carcinogenic risk was associated with Cu, Cd, and Cr through the consumption of wheat grains. In Abbottabad, Pakistan, the concentration of most of the survey HMs was higher in wheat grain samples in areas irrigated with municipal wastewater and exceeded the recommended dietary levels of the United States (US). The highest accumulation levels were noticed in the cases of Fe, Cr, and Ni (Hassan et al., 2013). Xue et al. (2019) revealed that among wheat grown in soil irrigated with sewage water in Northern China, 24% of the grain samples collected exceeded China's national food safety limits.

6.3.1.2 Fruits and Vegetables

Due to their high nutritional value and rich sources of essential elements, fruit and vegetable consumption is exceptionally high worldwide. However, using HM-contaminated fruits and vegetables is considered one of the critical routes of dietary exposure to HMs (Gupta et al., 2022). Recently, many studies have reported high levels of HMs in fruits and vegetables, which is alarming to both food security and human health. A study in Maharashtra, India, assessed the HM levels in fruits and vegetables. The study reported that the mean levels of HMs such as Pb, Hg, Cd, and As were extremely high in vegetables as compared to fruits, with the highest levels in garlic and potatoes. Higher values of Health

Risk Index (HRI) and Hazard Index (HI) were found to be associated with vegetable consumption (Mawari et al., 2022). A meta-analysis revealed that Hg contamination in the soil can pose a significant health risk by consuming HM-contaminated vegetables (Afonne & Ifediba, 2020). A study showed that HM levels in fruits and leafy vegetables from Bangladesh exceeded the safety limits set by FAO/WHO, and the primary threat was Pb, Cd, and As (Shaheen et al., 2016).

In Saudi Arabia, the concentration of HMs in different vegetables was studied in four urban and industrial zones. The study reported that the levels of all studied HMs (Hg, Cu, Pb, Cd, Fe, and Zn) exceeded the maximum permissible limits of the Joint FAO and WHO. The highest concentrations were noticed in leafy vegetables such as parsley, Jews mallow, and spinach. The highest Zn concentrations were found in legumes and peas, while in cucumber, the highest Pb levels were recorded (Ali & Al-Qahtani, 2012). İslamoğlu et al. (2021) investigated fruit and vegetables in Istanbul, Turkey, and showed that the levels of Pb exceeded the Turkish Food Codex for safety standards in all samples. In São Paulo State, Brazil, the concentration of Cr and Pb in 44% of the fruits and vegetable samples exceeded the safety standards set by Brazilian legislation (Guerra et al., 2012).

6.3.2 Livestock

6.3.2.1 *Milk and Derived Products*

Milk is considered one of the most nearly complete daily diets due to beneficial nutrients such as fats, carbohydrates, proteins, and minerals. However, increased urbanization and industrialization resulted in higher concentrations of physical, biological, and chemical hazards in the environment, which eventually find their way into the FC and harm the quality and quantity of the final products (Aggarwal et al., 2022). Concentrations of HMs in buffalo's milk and dairy products (cheese, dry milk, and processed cheese) collected from different vendors and supermarkets in Mansoura, Egypt, were investigated. The results revealed that the concentrations of Pb and Cd exceeded 44%, and 68% of the tested samples exceeded the safety standards of USDA (2018) (Elafify et al., 2023). It was revealed that in fresh milk collected from Pemuliahan, Indonesia, 83% of the samples were contaminated with Pb, including 57% exceeding the MRL set by the EPA. In the case of Cd, 90% of samples exceeded MRL, while 67% of pasteurized and 60% of milk samples had higher levels of Pb and Cd above MPL (Harlia et al., 2018).

Milk and dairy products collected from Beni-Suef governorate, Egypt, were investigated for HMs pollution. The results revealed that the levels of Pb in all 77 collected samples exceeded the permissible limit of 0.02 mg/kg by FAO/WHO. Furthermore, the study reported that dairy products were poor sources of essential elements such as Fe, Cu, and Zn (Meshref et al., 2014). It was reported that in Arak, Iran, 27% of the collected dairy products (yogurt, cheese, and pasteurized milk) had higher levels of Pb than the values recommended by the European Union and national standards of Iran (20 mg/kg) (Rezaei et al., 2014). In southwest Nigeria, it was also reported that the Pb, Fe, Cu, and Cr levels were higher than those stipulated by the WHO (Onibon et al., 2019). Milk and its products are commonly used in a day-to-day routine, and the contamination of these products is alarming to human health. Therefore, government agencies should upgrade the regulation and monitoring systems and enforce precautionary measures to protect consumer health.

6.3.2.2 *Meat and Derived Products*

As a rich source of proteins, vitamins, and essential minerals such as Fe, Zn, Ca, and Se, meat plays a vital role in human nutrition. Although meat has several advantages, it can act as a source of toxic pollutants through the bioaccumulation of HMs and trace elements at hazardous levels, which can enhance the prevalence of different diseases (Emami et al., 2023). Concentrations of different HMs in livestock products are presented in Table 6.3. In Algeria, the levels of Cd, Pb, Cu, Fe, and Zn were evaluated in fresh meat from sheep and camels and found that the levels of Cd and Pb were higher than the MPL set by the FAO for fresh meat (Badis, 2014). Alturiqi and Albedair (2012) conducted a study in major cities of Saudi Arabia (Tabouk, Riyadh, Damamm, and Jazan) and observed the highest levels of HMs in meat

TABLE 6.3
Concentration of Heavy Metals in Different Livestock Products

Country	Food Crops	HM Content (mg/kg)	References
Eastern Uganda	Beef	Cd 0.41, Cu 1.45, Cr 19.37, Co 1.63, Fe 164.3, Ni 15, Pb 5.4, and Zn 44.4	Kasozi et al. (2021)
Saudi Arab	Meat products	Fe 242.4, Cu 18.51, Pb 15.43 and Hg 0.125	Alturiqi and Albedair (2012)
Nigeria	Goat, Chicken, and cattle	Cr 0.23 to 1.22, Pb 0.1 to 1.34, Cu 0.10 to 1.44, Fe 0.98 to 4.65, Ni 0.01 to 1.09, Mn 0.45 to 4.11, Cd 0.07 to 0.76 and As 0.01 to 0.34	Akan et al. (2010)
China	Pork, beef, mutton and chicken	As, 0.018, Cd 0.002, Cr 0.061, Cu 0.801, Hg 0.0038, Ni 0.055, and Pb 0.029	Han et al. (2022)
Italy	Pork and beef	Pork As 0.15, Pb 0.019, Zn 45 Beef As 0.012, Pb 0.019, Zn 49	Di Bella et al. (2020)
Ghana	Grilled Beef	Mn 0.19–0.80, Zn 1.45–2.24, Cu 0.05–0.09 and Pb 0.63–1.63	Adzitey et al. (2015)
Egypt	Raw milk, Kareish cheese, processed cheese, and milk powder	Hg 0.0014, 0.0378, 0.0022, and 0.0204, As 0.012, 0.0764, 0.052, and 0.120 Pb 0.1016, 0.2924, 0.1256, and 0.3352, Cd 0.07, 0.1172, 0.0576, and 0.1252, Cr 0.1044, 0.174, 0.1016, and 0.208	Elafify et al. (2023)
Bangladesh	Cow milk	Cd 0.053, Cr 0.67, Cu 1.78, Fe 0.9, Mn 0.09 and Pb 0.033	Muhib et al. (2016)
Pakistan	Cow and buffalo milk	Pb 19.65 ± 43.86, Cr 2.10 ± 2.33 and As 0.48 ± 0.73	Abid et al. (2023)
Iran	Chicken meat	Pb 0,03, Cd 0.008, As 0.007, Ni 0.008, Cu 0.59 and Fe 6.15	Kia et al. (2024)
Bangladesh	Edible parts of chicken	Pb 0.33 to 4.6, Cd 0.004to 0.125, Cr 0.006 to 0.94, Fe 4.05 to 92.31, Cu 0.67 to 4.15, and Zn 4.45 to 23.75	Hossain et al. (2023)
Egypt	Chest Thigh Live	Al 11.535, Cd 0.054, Pb 2.56, Ni 0.23 and Co 0.06	Kamaly and Sharkawy (2023)
India	Eggs	Co 0.07, Cu 14.6, Cr 0.07, Fe 85.7, Ni 1.21, Pb 0.99 and Se 0.94	Giri and Singh (2019)

products such as sausage and luncheon and reported that the levels of all the studied HMs (Hg, Cd, Pb, Cu, Zn, and Fe) in these products exceeded the MPL limits set by FAO/WHO. Permissible limits for HMs in meat and derived products are present in Table 6.4.

About 600 sheep samples, including muscle, liver, and kidney samples, were analyzed for Hg, As, Pb, Cd, and Cr contamination in Kuwait. The results revealed that HMs were detected in 100% of the collected samples. The detected levels for all the HMs except Cr were found to be high compared to

TABLE 6.4
Maximum Permissible Limits of Heavy Metals in Meat and Derived Products by Different Agencies

Heavy Metals	FAO/WHO	Chinese Standards	USDA
Cd	0.5	0.1	0.1
Cr	0.05	1.0	1.0
Cu	0.05–0.5	//	//
Fe	0.01	//	//
Hg	0.1	0.05	0.1
Pb	0.1	0.02	0.5
Zn	0.3–1	//	//

Source: Amfo-otu and Joseph (2014), FAO/WHO (2000), USDA (2018), and MHPRC (2017).

the MPL set by different international agencies (Abd-Elghany et al., 2020). In the southeast region of Nigeria, the EDI for all the studied HMs, including As, Pb, and Cd, in 450 goat samples exceeded the recommended safety limits set by USEPA (Njoga et al., 2021). In Mosul city, Iraq, the HM concentrations in 20% of local meat samples, 12.5% of imported cattle meat, and 22.5% of imported buffalo meat exceeded the MPL limits for Pb set by the United States Department of Agriculture.

In the case of Cr, the values were 2.5% for imported cattle and 5% each for local and imported buffalo meat (Al-Naemi et al., 2020). A study on HM levels in game animals such as wild boars revealed that the levels of Cd and Pb in all the studied samples exceeded the MPL set by the EU and recommended against the consumption of such meat (Gašparík et al., 2017). The levels of Pb in cow and buffalo meat exceeded the MPL level of 0.1 mg/kg set by the WHO in Peshawar, Pakistan (Nawaz et al., 2015). Although meat and related products have high nutritional values, most of the previous literature showed utmost concern for consuming excessive amounts due to the higher levels of different HMs in these products.

6.3.2.3 Poultry Meat and Eggs

The poultry industry is one of the largest industries worldwide. Since 2000, egg production has increased by 63% (83 Mt) in 2019, while 40% of world meat production is attributed to the poultry industry (Aljohani, 2023). In three districts (Malakand, Dir lower, and Peshawar) of Khyber Pakhtunkhwa, Pakistan, it was found that the concentration of Pb in the liver and Cd and Cr in the breast and thighs in Peshawar, Pb, and Cr in the liver and thighs in Dir lower, and Cr in the liver, thighs, and breast in Malakand exceeded the MPL of FAO/WHO. The highest concentration of most of the elements was found in the liver (Khan et al., 2016). In Dhaka, Bangladesh, it was found that 75% of the poultry bone and 60% of the egg yolk and albumens sample had Pb levels higher than FAO/WHO limits. While in the case of Cd, poultry bones (40%) and yolk and albumen (89%) exceeded the FAO/WHO limits (Hossain et al., 2024). Another study conducted in Hamedan, Iran, studied the concentrations of Pb, Cr, Cd, and Cu in hen eggs. The results showed that the mean content of Cd (0.18 ± 0.04 mg/kg) and Cr (0.31 ± 0.03 mg/kg) exceeded the safety standards set by FAO/WHO (Branch, 2017).

In Thailand, 98% of the studied duck egg samples were contaminated with Pb, and the levels were above FAO/WHO standards (Aendo et al., 2019). In Assiut, Egypt, the concentration of Pb in 33% and Fe in 67% of the chicken meat and liver samples exceeded the levels set by the Egyptian Organization for Standardization. Regarding WHO standards, 94% of samples in the case of Pb and 17% for Cd exceeded the safety standards (Kamaly & Sharkawy, 2023). The mean concentrations of HMs in poultry, chicken liver, meat, and blood were in the range of 0.136–31.83, 0.120–26.15, and 0.62–21.15 mg/kg, respectively, in Thi-Qar, Iraq, and most of the studied elements exceeded the safety standards set by FAO/WHO (Talib Jawad et al., 2021).

6.4 Influence of Heavy Metals on Food Security

6.4.1 Soil-Crop System

The growth and development of cereals are significantly impacted by HM contamination due to the disruption of different plant physiological processes. HMs such as Cd and Pb are well known for their detrimental impacts on plant physiology (Vasilachi et al., 2023). The sources, exposure, and impacts on crops are presented in Figure 6.1. Crucial enzymatic activities such as photosynthesis and respiration can be altered due to high levels of these HMs (Houri et al., 2020). A main enzyme termed Rubisco, which plays a crucial role in carbon fixation during photosynthesis, can be interfered with by Cd (Amaral et al., 2024). The disruption of this enzyme due to Cd reduces photosynthesis, leading to decreased growth and biomass production (Aslam et al., 2021). Lead plays a key role in disrupting the mitochondrial electron transport chain, resulting in cellular respiration impairment and energy production, thereby posing a detrimental effect on plant health (Aslam et al., 2021). Identifying the impacts of HMs on different cereals could be challenging due to the difference in HM uptake capacity between different cereals with different specific responses and tolerances (Pandey et al., 2021; Rane et al., 2021).

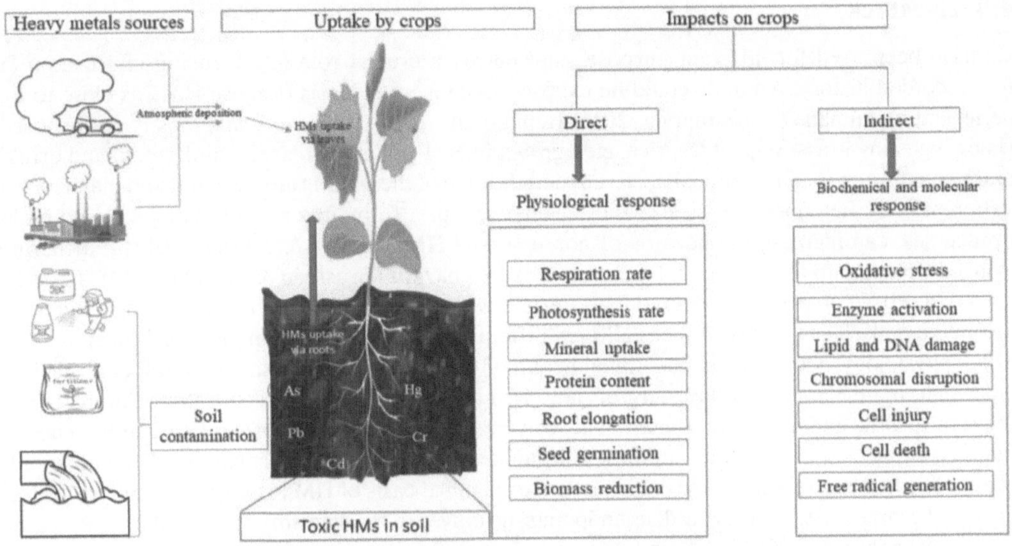

FIGURE 6.1 Heavy metal sources and impacts on crops.

These HMs can also challenge plant physiology by interfering with the uptake of essential nutrients. The uptake of essential elements such as Fe and Zn by roots can be hindered by Cd, causing an imbalance of nutrients and disruption of cellular phenomena, leading to symptoms like chlorosis (reduction in chlorophyll production) (Li et al., 2023). High levels of soil Cu and Ni can cause toxicity in cereals, leading to stunted growth, a reduction in crop yield, cell oxidative stress, and impairment of enzymatic activities. High levels of Zn may interfere with the uptake of Fe and Mn, leading to nutrient imbalance. While in cases of high Cr levels, the cereals are likely to have symptoms such as decreased metabolic processes, leaf damage, growth, and yield reduction (Vasilachi et al., 2023).

6.4.2 Crop Yield

Regarding food security, the reduction in quality and yield is of great concern due to HM exposure. High levels of HMs in soil can compromise the yield and nutritional values of harvested crops by posing detrimental impacts on several aspects of crop production (Arif et al., 2019). The degree of impact on yield depends on several factors, such as types of HMs types and crops, exposure degree, and duration (Hagemeyer, 1999). HMs such as Hg, Pb, and Cd can disrupt plants' crucial physiological problems, as discussed above, significantly reducing overall productivity. Exposure to HMs can also affect plant reproductive processes, which results in smaller grain sizes and lower yields (Emamverdian et al., 2015). Previously, Cd has been linked to reduced pollen grain generation and seed production (Tavarez et al., 2022).

For a successful plant life cycle, reproduction is one of the key stages that ensures the continuation of seeds and species production. However, the biochemical and physiological phenomena involved in flower and fruit production are highly impacted by HM pollution. One of the significant ways HMs disrupt the flowering process is by interfering with and disrupting plant hormone signaling. Hormones such as cytokinins and gibberellins play a vital role in flower and fruit regulation; however, HMs can lead to the suppression and delay of flowers by altering the synthesis, perception, and transport of these hormones (Groszmann et al., 2020; Ur Rahman et al., 2023). The nutritional composition of the different cereals is also compromised due to HM pollution causing micronutrient deficiency, and this decrease in essential elements such as Zn, Mn, and Fe can negatively impact humans and animals. To address these impacts of HMs on food quality and quantity, a holistic approach is in high demand that includes soil HM remediation technologies, proper soil management, and experimenting with crop variants that are resistant to HMs (González et al., 2017; Tong et al., 2021).

6.4.3 Livestock

HMs have been used for different purposes and played a crucial role in advancing civilization even before recorded history. Animals could be exposed to dangerous levels of these HMs as close to those experienced by humans (Acheampong, 2016; Hejna et al., 2018). These animals mostly get exposed to HMs by ingesting grass, vegetables, rice, etc., grown in HM-polluted areas, feeding stock, and drinking contaminated water (Hejna et al., 2018). Even though some of these HMs are essential and play vital roles in different body functions, the lack of these elements, apart from affecting physiological and biological processes, can increase the hazards of non-essential HMs (Mehri & Marjan, 2015). Furthermore, prolonged exposure to these essential elements can be a hazard (Brusseau & Artiola, 2019). Toxic HMs (such as Cd, Pb, Cr, and As) can potentially interfere with cellular functions by replicating essential elements in the body. The toxicity of these HMs in different animals depends on the HMs' form, poisoning method, physiology, age, sex, and nutritional health of the exposed animal species (Gall et al., 2015).

Upon entering the animal's body, most of these HM are accumulated and concentrated in the liver and kidneys in specific and other vital organs. This can lead to symptoms such as oxidative stress, endocrine disruption, congenital disabilities, cardio- and immune toxicity, and teratogenicity (Acheampong, 2016). However, due to a lack of knowledge on the accurate chemical basis of HM toxicity due to their wide range of chemical characteristics and hazardous endpoints, no universal mechanisms exist for all dangerous metal toxicity (Jaishankar et al., 2014). These HMs are indestructible; however, they can be found as free metal ions, which can change their bioavailability and toxicity. These free metal ions can lead to various health disorders in animals through their interaction with biological systems, as they are chemically reactive (Brusseau & Artiola, 2019). The sources, exposure, and impacts on livestock are presented in Figure 6.2.

6.4.3.1 Influence on Domestic Animals (Cow, Goat, Sheep, and Buffalo)

Young calves are particularly highly vulnerable to HM toxicity among cattle (Gupta et al., 2021). This is mostly attributed to their curious nature, habit of licking, and inability to discriminate between eating habits (Constable et al., 2017). This can lead to accidental Pb poisoning by ingesting Pb-bearing objects present in the surrounding environment as industrial, domestic, or agricultural waste (Constable et al., 2017). Before, it was discovered that eating pasture near secondary Pb and Zn smelters caused acute Pb toxicity in cows and buffalo and subclinical Pb toxicity in goats. This changed the essential trace element profile (Acheampong, 2016).

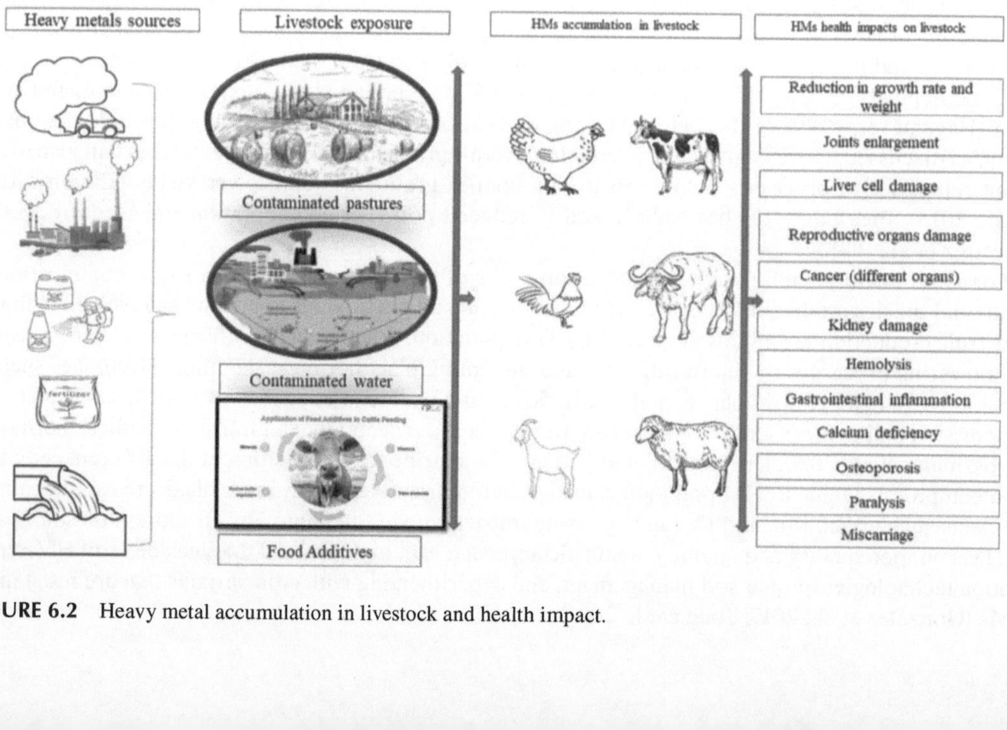

FIGURE 6.2 Heavy metal accumulation in livestock and health impact.

Furthermore, the levels of Pb in the liver and kidney of Pb-poisoned cattle were 0.425 and 8.84 mg/kg, respectively, contributing 72% and 84% of the total concentration in the respective organs. It has been shown that the tolerance levels of goats and sheep to HMs such as Pb and Cd are higher than those of cattle (Abd Ei-Salam et al., 2013; Ogbomida et al., 2018). In sheep, it might be attributed to the fact that they excrete higher levels of Pb, Cr, and Ni in their excrement as compared to goats and cows, and this might also be the reason for lower levels of HMs in sheep milk as compared to others (Hussein et al., 2021). Goats, on the other hand, are more tolerant to Pb than cows and sheep; however, they can show symptoms of lethal toxicity with dominant signs of central nervous system (CNS) involvement with long-term exposure (Acheampong, 2016).

Although exposure to domestic animals is infrequent and uncommon in the case of Hg, ingestion of organic and inorganic Hg can lead to accidental acute and chronic poisoning (Rafati-Rahimzadeh et al., 2014). The most common cause of exposure and chronic poisoning may be attributed to the presence of phenyl Hg, which is present in the form of organic Hg in treated grains (Syversen & Kaur, 2012). Using these grains as feed for an extended period and in large quantities may result in the development of clinical diseases. In horses, Hg poisoning can be attributed to Hg-containing medicines and cutaneous absorption from skin dressings containing HgO (mercuric oxide) (Syversen & Kaur, 2012). Previous studies experimented on cows raised within a 5-km radius of a power plant. The results showed that cows that were exposed had significantly lower hemoglobin levels and higher levels of albumin, serum secretion, and urea nitrogen in their blood, all of which are signs of Hg toxicity. The experiment further concluded that cows exposed to Hg-containing fly ash can have detrimental impacts on the human population that may be direct or indirect via the FC (Mahajan et al., 2012).

Toxicity of Cd is often noticed in sheep and cattle when the concentration in the feed is higher than 50 mg/kg. Chronic neurotoxicity and liver damage have been noticed upon consumption of higher levels of Cd. Chronic intake leads to the accumulation of HMs in the testes, liver, kidneys, and lungs, specifically in animals (Almeer et al., 2019; Balali-Mood et al., 2021). The main symptoms noticed in exposed animals are hair matting, weight loss, dryness, keratosis, brittle horns, and, in addition, necropsy (degenerative changes) in many organs and stomach epithelium hyperkeratosis (Gumasta et al., 2018). Necrotic alterations and vascular degenerations were noticed in the liver, lungs, and kidneys of buffalos and cattle near industrial areas, and the Cd levels in tissues were frequently higher than 2 mg/kg (Sarkar et al., 2013). Experimental poisoning of sheep with 2.5 mg/kg of body weight resulted in demineralization of bones, nephropathy, and anemia. Other than these, health problems, such as stillbirth, congenital flaws, and abortion, could also be potential outcomes of Cd exposure (Hashem et al., 2019; Sarkar et al., 2013).

Among all the livestock species, sheep are considered the most vulnerable to Cu toxicity, and it is reported worldwide (Minervino et al., 2018). Cattles were considered the most tolerant of Cu, and Cu poisoning cases were sporadic; however, Cu toxicity and the mortality rate are on the rise globally. Due to their weak homeostatic control for Cu adsorption, they are more sensitive. Acute Cu poisoning in sheep and calves may occur when they consume 20–110 mg/kg Cu of their body weight, while in mature cattle, the concentration is 200–800 mg/kg Cu of their body weight (Suttle, 2012). Goats have a significantly higher Cu requirement, up to 15–25 mg/kg than sheep, and they can tolerate a higher Cu dietary level that might be attributed to their hepatic adsorption (Feedap et al., 2016). The chronic symptoms of Cu poisoning include stomach aches, salivation, dehydration due to diarrhea, and collapses before death, which generally take place within 24 h. Cu poisoning-affected sheep commonly show symptoms such as jaundice, thirst, anorexia, anemia, and high breathing and heart rate (Newcomer et al., 2021).

6.4.3.2 Poultry

Teratogenic impacts have been noticed in different animals, including chickens, which can lead to gastrointestinal problems, ear abnormalities, and appendage deformations. Upon ingestion of Cd, it binds with a metallothionine protein high in cysteine (Aljohani, 2023). Due to similarity in the oxidation state, Cd can replace Zn in metallothionine and obstruct the protein's ability to act as a scavenger of free radicals. This can lead to liver hypotoxicity, accumulation in renal tissues, and nephrotoxicity (Smirnova et al., 2023). In adult poultry birds, Cd exposure causes oxidative stress and enzymes that increase antioxidant activity in the erythrocytes. Furthermore, it leads to the formation of proximal tubular cells that can cause bone damage or alter kidney function, leading to bone mineralization (Kar & Patra, 2021). Upon adsorption in the digestive track, it can cause essential element (Fe and Ca) deficiencies. Lead poisoning is also common in poultry, and its presence in poultry feed can lead to symptoms such as stunted chicken growth and blood levels of delta-aminolevulinic acid dehydratase (Mukherjee et al., 2022). Previously, many studies have reported the presence of As, Cd, Hg, and Pb residues in chicken liver. Lead is considered the most dangerous element regarding poultry health as it can bind and deactivate vital enzymes. The highest levels of Pb were found in the bones of chickens, followed by skeletal muscle, kidneys, and livers (Mukherjee et al., 2022).

Bio-transformation of As occurs in the kidney and liver of chickens. This results in the production of methylated metabolites distributed throughout the body. Prolonged exposure to As, even at lower levels in poultry, may lead to symptoms of skin discoloration, indigestion, and stomach pain, while higher levels may lead to different types of cancer (skin, liver, lungs, and kidney) (Abu et al., 2022; Sattar et al., 2016). The accumulation concentration varies with the amount ingested, exposure duration, type of element, age, and breed (Acheampong, 2016). As the negative impacts are dose-dependent, prolonged exposure may result in harmful tissue changes and damage. Methyl mercury (MeHg) is one of the most hazardous types of Hg, and it is found abundantly in the environment and poses extremely negative health impacts on poultry (Kojima et al., 2023). Even at very low levels, it can cause extreme damage to the liver and kidneys of poultry, along with anemia and depressed growth. The toxicity is often higher in young, growing chickens than in adults. Among other symptoms, nitric oxide suppression, oxidative stress, and cytokine profile disruption are highly noticeable (Aljohani, 2023).

6.5 Management Strategies of Heavy Metals to Ensure Food Security

6.5.1 Ecological Remediation

Using biochar (BC) is considered a commanding eco-remediation practice that can reduce soil contamination from HMS and provide several advantages. Waste-derived BC has the potential to absorb HMS efficiently by changing the soil's physico-chemical properties and lowering the phyto-availability of dangerous HMs (Omondi et al., 2016; Peng et al., 2018). According to the measure-monitor model, BC and composting can enhance cation exchange capacity, humification, and the nutrient cycle while effectively lowering soil metal concentrations. A global health problem is the intake of foods containing As (Beesley et al., 2013; Qiao et al., 2018). Regulating requirements have not been set for residential gardens or farmlands (de Oliveira et al., 2017). Biochar amendments have grown in favor of an inexpensive, environmentally acceptable method for removing soil connected to food crops and drinking water. However, due to variations in biochar's acidity, its effectiveness in removing As is not constant. Several studies found that biochar had a minimal impact. However, several studies indicate that applying biochar can effectively lower As levels in even the anoxic paddy soil environment (Wang et al., 2017). In plant-soil systems, a combined strategy of phytoremediation along with some chemical and organic modifications may lower the bioavailability of As. BC amendments had a more significant impact on the availability of Cd and Pb in food crops' palatable parts than those of Zn, Ni, Mn, Cr, and Co (Lessl et al., 2014; de Oliveira et al., 2017).

For HMs in subsystems (soil-crop food), green technologies such as phytoremediation (phytoextraction, rhizofiltration, phytodegradation, phytostabilization, and phytovolatilization) are highly effective. Wetland plants are a prime candidate for applying phytotechnologies (Mehmood et al., 2023; Rai et al., 2019). In many agroclimatic zones worldwide, phytoremediation can improve human health and assist in the growth of HM-free vegetables (Zaidi & Khan, 2017). Phytoextracting metals from soil is a capability of terrestrial and wetland plants known as metallophytes or hyperaccumulators (Suman et al., 2018). Furthermore, in contrast to conventional chemical technologies, the use of phyto-technologies for metal remediation is both economical and environmentally sustainable (Wang & Delavar, 2023). Yoon et al. (2006) investigated the function of phyto-technologies in association with food safety related to human health.

Phyla nodiflora and *Gentiana pennelliana* proved to be beneficial in the phyto-stabilization of Cu, Pb, and Zn, among other natural plants that were employed in experimental settings in Florida, USA. Since macrophytes and wetland plants greatly collect HMs from industrial wastewater from sponge iron production, cleaned wastewater would be appropriate for irrigating agricultural lands. After an extensive worldwide investigation and bibliometric study on soil contamination between 1999 and 2012, the most examined research issue was HM pollution and bio/phytoremediation (Guo et al., 2014). Physicochemical treatments (electro-reclamation, leaching, excavation, and thermal treatment) can alter soil's physicochemical and biological aspects and are fast and efficient for highly polluted areas, but chemical and biological aspects of soil are fast and efficient for highly polluted areas. Still, cost-effectiveness and the chances of secondary contamination are the influencing factors. By contrast, other green technologies (biostimulation, bioaugmentation, composting, bioleaching, bioremediation, land aeration, bioventing, and bio/phytoremediation) preserve natural soil properties. Biological remediation, which uses plants or bacteria, is an eco-friendly, solar-powered, and reasonably priced technique.

6.5.2 Source Reduction

Minimizing the origins of HMs is a highly efficient approach to enhancing human well-being. Significantly reducing the levels of HMs in food crops can be achieved by avoiding improperly treated wastewater and sewage sludge. The presence of Cd in the environment can be significantly reduced by decreasing the pollution source. Additionally, plant breeding and agronomy advancements can help control the HMs soil for food crops (Mclaughlin et al., 1999). Implementing air-quality management measures could decrease the deposition of PM in the soil, thereby reducing food crop contamination. A comparative evaluation of locations (clean water and ineffectively treated wastewater irrigated) showed that the levels of HMs in soil, especially Cd, in hair samples of individuals who consumed products cultivated in soil irrigated with wastewater were notably higher compared to clean water irrigated areas (Massaquoi et al., 2015). Source reduction can be achieved, by decreasing treated sewage usage, or by implementing proper treatment before its utilization for irrigation purposes. In Algeria, by refraining from using poorly treated wastewater, the level of HMs in vegetables in agricultural products (tomatoes, potatoes, and cucumbers) was minimized by around 85% (Cherfi et al., 2015). Optimizing land usage enhances food safety. Crops grown by the roadside are susceptible to HM contamination, such as the presence of Pb in the *Amaranthus dubius* leaves, due to the deposition of PM (Nabulo et al., 2006).

Roadside dust, also known as PM, comes from dense urban traffic and serves as a conduit for the transportation of HMs into crops. Therefore, it is not advisable to cultivate crops within 30 m (Liao et al., 2016). The consumption of crop products (such as sugarcane, rice, and 30 vegetables) grown near an acidic mine drainage zone and subsequently intoxicated by HMs is expected to pose significant dangers to human health. Deposition of PM is a key factor in higher levels of As, Cd, and Pb in those crops (Liao et al., 2016). Ensuring proper parameters of environmental aspects, like pH and organic matter, is crucial, even inside greenhouse settings. Tubers and fruits exhibited minimal susceptibility to HM contamination, making them more viable for greenhouse cultivation than leafy vegetables.

6.5.3 Physiochemical Strategies

HM-contaminated sites were previously restored using chemical methods; however, they are not as good as biological methods. According to Udom et al. (2004), the complexation of metals in soil reduces their availability for food crops and mitigates associated health hazards. Studies have concentrated on creating environmentally friendly compounds for soil remediation, such as ferrate (Rai, 2018a, 2018b). Remedial techniques involving practical chemicals, such as alkaline clay in addition to synthetic zeolites, are beneficial in cleaning up soil contaminated with HMs (Senila & Cadar, 2024). Several studies have shown that soil HMs can be remediated using silicon calcium fertilizer, magnetite, red mud, hematite zeolite, modified biochar, hydrous manganese oxide, and maghemite ((Balakhnina et al., 2015; Senila & Cadar, 2024; Yao et al., 2017). Cutting-edge tools like scanning electron microscopy and X-ray powder diffraction showed that the decreased As and Cd uptake in plants was due to the deactivated metals in the soil, such as silicates, phosphates, and hydroxides (Yao et al., 2017). Madhaiyan et al. (2007) conducted a three-year field study on *Hordeum sativum* (spring barley), showing that soil and plant systems can endure Zn and Pb. HMs and weakly mobile elements in the soil and *H. sativum* decreased with chalk and manure amendments (which contained chalk, natural zeolite, and manure). Wheat is a vital grain in the world's population, and its susceptibility to HM contamination will pose a severe human health risk.

In soils amended with single superphosphate, triple superphosphate, and calcium magnesium phosphate sepiolite in conjunction with ZnSO4, the phyto-availability of Cd and Pb was significantly decreased by ≥54.13% and ≥42.14%, respectively (Guo et al., 2018). Phosphate comprising hydroxyapatite and diammonium phosphate reduced the phytoavailability of Pb and Hg in green plants but was lower in the case of Cd (Waterlot et al., 2017). In a study on the productivity of cereals, no significant health risks were found in the case of cereal grains grown in chemically treated (acid-washed) industrial soil (Kim et al., 2017). The toxicity of HMs in plants and the population consumed decreased significantly with the addition of calcium and EDTA. Metalloids (i.e., As) are a worldwide concern as they cause a reduction in chlorophyll and lipid peroxidation enhancement in the *Spinacia oleracea* seedlings, causing severe toxicity (Shahid et al., 2017). Adding chemicals to soil may result in a neutral reaction when it comes to HM cleanup, which is the opposite of the positive studies that were previously discussed.

6.6 Conclusions

Anthropogenic activities such as urbanization, industrialization, and agricultural advancements are the key sources of environmental HM contamination. These HMs can easily pass into organisms through the FC and sometimes they end up in excessive accumulation in tissues and organs, such as in the case of food crops in grains and leaves and the case of livestock in parts such as muscles, blood, liver, and hair. Apart from that, HMs can also be found in different livestock products such as milk, meat, and eggs, which is alarming regarding food security and human health through the consumption of these products. High levels of HMs in crops and their products are primarily attributed to irrigation with improperly treated sewage sludge and industrial effluents, excessive use of fertilizers, and, to some extent, atmospheric deposition. While in livestock, high levels are mostly attributed to grazing on HMs, contaminated grass and pastures, polluted water bodies, and food additives.

These HMs can have a significant influence on crops and livestock in terms of crop and livestock health, quality, and quantity. HM contamination in plants causes a range of physiological and biological problems, including stunted growth, lower crop yield, cell oxidative stress, impairment of enzymatic activities, decreased photosynthesis, and more. In animals, they have a big effect on reproduction in the kidneys, liver, brain, and skin, which can lead to several health problems. According to the literature presented in the current chapter, HMs such as Cd, As, and Pb were found to be high in crops and livestock compared to other essential elements, which is worth noticing as these elements can be extremely dangerous even at very low levels.

The use of advanced technologies may be useful to minimize exposure to these HMs. Launching proper monitoring programs along with advanced policy interventions may be helpful in the subsequent reduction of HM exposure in food crops, livestock, and the environment. To minimize the levels of HMs in urgent basses in the environment, advanced scientific approaches such as biochar, chemical amendments, bioremediation using plant technologies, and changes in physiochemical parameters could help reduce the negative impacts of HMs on crops, livestock, and human health, along with economic losses. Developing collaborations at national and international levels is crucial to cope sufficiently with HM pollution problems and implement proper toxicity prevention strategies.

REFERENCES

Abd Ei-Salam, N. M., Ahmad, S., Basir, A., Kalsum Rais, A., Bibi, A., Ullah, R., Ali Shad, A., Muhammad, Z., & Hussain, I. (2013). Distribution of heavy metals in the liver, kidney, heart, pancreas and meat of cow, buffalo, goat, sheep and chicken from Kohat market Pakistan. *Life Science Journal*, *10*(Suppl. 7), 937–940.

Abd-Elghany, S. M., Mohammed, M. A., Abdelkhalek, A., Saud Saad, F. S., & Sallam, K. I. (2020). Health risk assessment of exposure to heavy metals from sheep meat and offal in Kuwait. *Journal of Food Protection*, *83*(3), 503–510. https://doi.org/10.4315/0362-028X.JFP-19-265

Abid, K., Shams, Z. I., Tahir, M. S., & Zubair, A. (2023). Determination of heavy metals in cow's and buffalo's fresh raw milk from different areas of Pakistan. *Arab Gulf Journal of Scientific Research*, *42*(3), 757–770. https://doi.org/10.1108/AGJSR-11-2022-0266

Abu, S. M., Monira, N., Asmaul, H., Mahbub, E. E. A. T. M., Hena, K. A., & Masudur, R. M. (2022). Seroprevalence of Newcastle disease in layer chickens and pathology in clinically affected chickens at Gazipur, Bangladesh. *Continental Veterinary Journal*, *2*(1), 35–41.

Adekiya, A. O., Oloruntoba, A. P., Ojeniyi, S. O., & Ewulo, B. S. (2018). Heavy metal composition of maize and tomato grown on contaminated soils. *Open Agriculture*, *3*(1), 414–426. https://doi.org/10.1515/opag-2018-0046

Adzitey, F., Kumah, A., & Seddoh, B. K. M. (2015). Assessment of the presence of selected heavy metals and their concentration levels in fresh and grilled beef/guinea fowl meat in the tamale metropolis, ghana. *Research Journal of Environmental Sciences*, *3*, 152–158. https://doi.org/10.1201/9781420033564-12

Aendo, P., Thongyuan, S., Songserm, T., & Tulayakul, P. (2019). Carcinogenic and non-carcinogenic risk assessment of heavy metals contamination in duck eggs and meat as a warning scenario in Thailand. *Science of the Total Environment*, *689*, 215–222. https://doi.org/10.1016/j.scitotenv.2019.06.414

Afonne, O. J., & Ifediba, E. C. (2020). Heavy metals risks in plant foods – need to step up precautionary measures. *Current Opinion in Toxicology*, *22*, 1–6. https://doi.org/10.1016/j.cotox.2019.12.006

Aggarwal, A., Verma, T., & Ghosh, S. (2022). Heavy metal residues in milk and milk products and their detection method. *Trends and Innovations in Food Science*, *2*, 100261.

Ahogle, A. M. A., Letema, S., Schaab, G., Ngure, V., Mwesigye, A. R., & Korir, N. K. (2022). Heavy metals and trace elements contamination risks in peri-urban agricultural soils in Nairobi city catchment, Kenya. *Frontiers in Soil Science*, *2*, 104805. https://doi.org/10.3389/fsoil.2022.1048057

Akan, J. C., Abdulrahman, F. I., Sodipo, O. A., & Chiroma, Y. A. (2010). Distribution of heavy metals in the liver, kidney and meat of beef, mutton, caprine and chicken from Kasuwan Shanu Market in Maiduguri Metropolis, Borno State, Nigeria. *Research Journal of Applied Sciences, Engineering and Technology*, *2*(8), 743–748.

Akenga, T., Kerich, E., Machuka, W., & Ronoh, E. (2018). Selected heavy metals concentration in maize grains and leaves in Homahills, Homabay County. *African Journal of Education, Science and Technology*, *4*(4), 50–58.

Alagawany, M., Elnesr, S. S., Farag, M. R., Tiwari, R., Yatoo, M. I., Karthik, K., Michalak, I., & Dhama, K. (2020). Nutritional significance of amino acids, vitamins and minerals as nutraceuticals in poultry production and health–a comprehensive review. *Veterinary Quarterly*, *41*(1), 1–29. https://doi.org/10.1080/01652176.2020.1857887

Ali, M. H. H., & Al-Qahtani, K. M. (2012). Assessment of some heavy metals in vegetables, cereals and fruits in Saudi Arabian markets. *Egyptian Journal of Aquatic Research*, *38*(1), 31–37. https://doi.org/10.1016/j.ejar.2012.08.002

Ali, M. U., Liu, G., Yousaf, B., Ullah, H., Abbas, Q., Munir, M. A. M., & Irshad, S. (2019). Biomonitoring and health risks assessment of trace elements in various age- and gender-groups exposed to road dust in habitable urban-industrial areas of Hefei, China. *Environmental Pollution*, *244*, 809–817. https://doi.org/10.1016/j.envpol.2018.10.084

Aljohani, A. S. M. (2023). Heavy metal toxicity in poultry: a comprehensive review. *Frontiers in Veterinary Science*, *10*(June), 1–12. https://doi.org/10.3389/fvets.2023.1161354

Almeer, R. S., AlBasher, G. I., Alarifi, S., Alkahtani, S., Ali, D., & Abdel Moneim, A. E. (2019). Royal jelly attenuates cadmium-induced nephrotoxicity in male mice. *Scientific Reports*, *9*(1), 1–12. https://doi.org/10.1038/s41598-019-42368-7

Al-Naemi, H. S., Al-Sanjary, R. A., Faraj, R. A., & Saadi, A. (2020). Detection of lead, chromium and cobalt in meats of cattle and buffalo from retails of mosul city. *Iraqi Journal of Veterinary Sciences*, *34*(2), 447–451. https://doi.org/10.33899/ijvs.2019.126069.1224

Alturiqi, A. S., & Albedair, L. A. (2012). Evaluation of some heavy metals in certain fish, meat and meat products in Saudi Arabian markets. *Egyptian Journal of Aquatic Research*, *38*(1), 45–49. https://doi.org/10.1016/j.ejar.2012.08.003

Amadi, E. V., Venkataraman, A., & Papadopoulos, C. (2022). Nanoscale self-assembly: concepts, applications and challenges. *Nanotechnology*, *33*:132001.

Amaral, J., Lobo, A. K. M., & Carmo-Silva, E. (2024). Regulation of Rubisco activity in crops. *New Phytologist*, *241*(1), 35–51. https://doi.org/10.1111/nph.19369

Amfo-otu, R., & Joseph, B. A. S. A. (2014). Meat contamination through singeing with scrap tyres in akropongakuapem. *Applied Research Journal*, *1*(1), 12–19.

Arif, N., Sharma, N. C., Yadav, V., Ramawat, N., Dubey, N. K., Tripathi, D. K., Chauhan, D. K., & Sahi, S. (2019). Understanding heavy metal stress in a rice crop: toxicity, tolerance mechanisms, and amelioration strategies. *Journal of Plant Biology*, *62*(4), 239–253. https://doi.org/10.1007/s12374-019-0112-4

Asgari, K., & Cornelis, W. M. (2015). Heavy metal accumulation in soils and grains, and health risks associated with use of treated municipal wastewater in subsurface drip irrigation. *Environmental Monitoring and Assessment*, *187*(7), 410. https://doi.org/10.1007/s10661-015-4565-8

Aslam, M., Aslam, A., Sheraz, M., Ali, B., Ulhassan, Z., Najeeb, U., Zhou, W., & Gill, R. A. (2021). Lead toxicity in cereals: mechanistic insight into toxicity, mode of action, and management. *Frontiers in Plant Science*, 11(February), 587785. https://doi.org/10.3389/fpls.2020.587785

Awuchi, C. G. (2020). Nutritional diseases and nutrient toxicities : a systematic review of the diets and nutrition. *International Journal of Advanced Academic Research*, *6*(1), 1–46.

Badis, B. (2014). Levels of selected heavy metals in fresh meat from cattle, sheep, chicken and camel produced in Algeria. *Annual Research & Review in Biology*, *4*(8), 1260–1267. https://doi.org/10.9734/arrb/2014/7430

Balakhnina, T. I., Bulak, P., Matichenkov, V. V., Kosobryukhov, A. A., & Włodarczyk, T. M. (2015). The influence of Si-rich mineral zeolite on the growth processes and adaptive potential of barley plants under cadmium stress. *Plant Growth Regulation*, *75*(2), 557–565. https://doi.org/10.1007/s10725-014-0021-y

Balali-Mood, M., Naseri, K., Tahergorabi, Z., Khazdair, M. R., & Sadeghi, M. (2021). Toxic mechanisms of five heavy metals: mercury, lead, chromium, cadmium, and arsenic. *Frontiers in Pharmacology*, *12*(April), 1–19. https://doi.org/10.3389/fphar.2021.643972

Baltenweck, I., Enahoro, D., Frija, A., & Tarawali, S. (2020). Why is production of animal source foods important for economic development in Africa and Asia? *Animal Frontiers*, *10*(4), 22–29. https://doi.org/10.1093/af/vfaa036

Beesley, L., Marmiroli, M., Pagano, L., Pigoni, V., Fellet, G., Fresno, T., Vamerali, T., Bandiera, M., & Marmiroli, N. (2013). Biochar addition to an arsenic contaminated soil increases arsenic concentrations in the pore water but reduces uptake to tomato plants (Solanum lycopersicum L.). *Science of the Total Environment*, *454–455*, 598–603. https://doi.org/10.1016/j.scitotenv.2013.02.047

Branch, H. (2017). Assessment of levels and health risk of heavy metals (Pb, Cd, Cr, and Cu) in commercial hen's eggs from the city of Hamedan. *Pollution*, *3*(4), 669–677. https://doi.org/10.7508/pj.2017.0

Briffa, J., Sinagra, E., & Blundell, R. (2020). Heavy metal pollution in the environment and their toxicological effects on humans. *Heliyon*, *6*(9), e04691. https://doi.org/10.1016/j.heliyon.2020.e04691

Brusseau, M., & Artiola, J. (2019). Chemical contaminants. In: M. L. Brusseau, I. L. Pepper & C. P. Gerba (Eds.) Environmental and Pollution Science. New York: Academic Press, pp. 175–190.

Cherfi, A., Achour, M., Cherfi, M., Otmani, S., & Morsli, A. (2015). Health risk assessment of heavy metals through consumption of vegetables irrigated with reclaimed urban wastewater in Algeria. *Process Safety and Environmental Protection*, *98*, 245–252. https://doi.org/10.1016/j.psep.2015.08.004

Chowdhury, A. I., & Alam, M. R. (2024). Health effects of heavy metals in meat and poultry consumption in Noakhali, Bangladesh. *Toxicology Reports*, 12(October 2023), 168–177. https://doi.org/10.1016/j.toxrep.2024.01.008

Clever, J., & Jie, M. (2015). China - Peoples Republic of China announces standards for Food Contact Metal Materials and Articles as WTO SPS Notification 1014. GAIN Report No. CH15079. https://apps.fas.usda.gov/newgainapi/api/report/downloadreportbyfilename?filename=China%20announces%20standards%20for%20Food%20Contact%20Metal%20Materials%20_Beijing_China%20-%20Peoples%20Republic%20of_1-12-2016.pdf

Constable, P., Hinchcliff, K., Done, S., & Grunberg, W. (2017). *Veterinary Medicine: A Textbook of Diseases of Cattle, Horses, Sheep, Pigs and Goats*. 11th ed. Missouri: Elsevier Ltd, pp. 421–425.

David, L. S., Anwar, M. N., Abdollahi, M. R., Bedford, M. R., & Ravindran, V. (2023). Calcium nutrition of broilers: current perspectives and challenges. *Animals*, *13*(10), 1–26. https://doi.org/10.3390/ani13101590

de Oliveira, L. M., Suchismita, D., Gress, J., Rathinasabapathi, B., Chen, Y., & Ma, L. Q. (2017). Arsenic uptake by lettuce from As-contaminated soil remediated with Pteris vittata and organic amendment. *Chemosphere*, *176*, 249–254. https://doi.org/10.1016/j.chemosphere.2017.02.124

Di Bella, C., Traina, A., Giosuè, C., Carpintieri, D., Lo Dico, G. M., Bellante, A., Del Core, M., Falco, F., Gherardi, S., Uccello, M. M., & Ferrantelli, V. (2020). Heavy metals and PAHs in meat, milk, and seafood from augusta area (Southern Italy): contamination levels, dietary intake, and human exposure assessment. *Frontiers in Public Health*, *8*(July), 273. https://doi.org/10.3389/fpubh.2020.00273

Elafify, M., EL- Toukhy, M., Sallam, K. I., Sadoma, N. M., Mohammed Abd-Elghany, S., Abdelkhalek, A., & El-Baz, A. H. (2023). Heavy metal residues in milk and some dairy products with insight into their health risk assessment and the role of Lactobacillus rhamnosus in reducing the lead and cadmium load in cheese. *Food Chemistry Advances*, *2*(March), 100261. https://doi.org/10.1016/j.focha.2023.100261

Elbagermi, M. A., Edwards, H. G. M., & Alajtal, A. I. (2012). Monitoring of heavy metal content in fruits and vegetables collected from production and market sites in the Misurata Area of Libya. *ISRN Analytical Chemistry*, *2012*, 1–5. https://doi.org/10.5402/2012/827645

El-Kady, A. A., & Abdel-Wahhab, M. A. (2018). Occurrence of trace metals in foodstuffs and their health impact. *Trends in Food Science & Technology*, *75*, 36–45.

Emamverdian, A., Ding, Y., Mokhberdoran, F., & Xie, Y. (2015). Heavy metal stress and some mechanisms of plant defense response. *Scientific World Journal*, 2015, 7–9. https://doi.org/10.1155/2015/756120

Euorpean Union (EU) (2015). Commission Regulation (EU) 2015/1005 of 25 June 2015 amending Regulation (EC) No 1881/2006 as regards maximum levels of lead in certain foodstuffs (Text with EEA relevance). https://data.europa.eu/eli/reg/2015/1005/oj.

Fan, Y., Zhu, T., Li, M., He, J., & Huang, R. (2017). Heavy metal contamination in soil and brown rice and human health risk assessment near three mining areas in Central China. *Journal of Healthcare Engineering*, *2017*, 4124302. https://doi.org/10.1155/2017/4124302

FAO. (1996). *Rome Declaration on World Food Security and World Food Summit Plan of Action.* Rome: World Food Summit.

FAO. (2017). *Staple foods: what do people eat?* www.fao.org/4/U8480E/U8480E07.htm

FAO/WHO. (2000). Report of the 32nd Session of the Codex Committee of the Food Additives Contaminants, Beijing People's Republic of China, FAO Rome, 36–40.

FAO/WHO. (2017). Joint Food Standards Programme Codex Alimentarius Commission: Report of the 11th Session of the Codex Committee on Contaminants in Foods. Rio de Janeiro, Brazil 3–7, https://www.fao.org/fao-who-codexalimentarius/en/. Accessed 12 March 2023.

Feedap, A. F., Ramos, F., Rychen, G., Saarela, M., Villa, R. E., Wallace, R. J., & Wester, P. (2016). Revision of the currently authorised maximum copper content in complete feed. *EFSA Journal*, *14*(8), 4563. https://doi.org/10.2903/j.efsa.2016.4563

França, F. C. S. S., Albuuerque, A. M. A., Almeida, A. C., Silveira, P. B., Filho, C. A., Hazin, C. A., & Honorato, E. V. (2017). Heavy metals deposited in the culture of lettuce (Lactuca sativa L.) by the influence of vehicular traffic in Pernambuco, Brazil. *Food Chemistry*, *215*, 171–176. https://doi.org/10.1016/j.foodchem.2016.07.168

Fu, J., Zhou, Q., Liu, J., Liu, W., Wang, T., Zhang, Q., & Jiang, G. (2008). High levels of heavy metals in rice (Oryza sativa L.) from a typical E-waste recycling area in southeast China and its potential risk to human health. *Chemosphere*, *71*(7), 1269–1275. https://doi.org/10.1016/j.chemosphere.2007.11.065

Gall, J. E., Boyd, R. S., & Rajakaruna, N. (2015). Transfer of heavy metals through terrestrial food webs: a review. *Environmental Monitoring and Assessment*, *187*(4), 3. https://doi.org/10.1007/s10661-015-4436-3

Gan, Y., Wang, L., Yang, G., Dai, J., Wang, R., & Wang, W. (2017). Multiple factors impact the contents of heavy metals in vegetables in high natural background area of China. *Chemosphere*, *184*(27), 1388–1395. https://doi.org/10.1016/j.chemosphere.2017.06.072

Gašparík, J., Binkowski, Ł. J., Jahnátek, A., Šmehýl, P., Dobiaš, M., Lukáč, N., Błaszczyk, M., Semla, M., & Massanyi, P. (2017). Levels of metals in kidney, liver, and muscle tissue and their influence on the fitness for the consumption of wild boar from Western Slovakia. *Biological Trace Element Research*, *177*(2), 258–266. https://doi.org/10.1007/s12011-016-0884-z

Ghosh, M., & Singh, S. P. (2014). A review on phytoremediation of heavy metals and utilization of it's by products M.Ghosh. *The Asia Pacific Journal of Energy and Environment*, *6*(4), 214–231. https://doi.org/10.1007/s10681-014-1088-2

Giri, S., & Singh, A. K. (2019). Heavy metals in eggs and chicken and the associated human health risk assessment in the mining areas of Singhbhum copper belt, India. *Archives of Environmental and Occupational Health*, *74*(4), 161–170. https://doi.org/10.1080/19338244.2017.1407284

González, Á., del Mar Gil-Diaz, M., & del Carmen Lobo, M. (2017). Metal tolerance in barley and wheat cultivars: physiological screening methods and application in phytoremediation. *Journal of Soils and Sediments*, *17*(5), 1403–1412. https://doi.org/10.1007/s11368-016-1387-4

Groszmann, M., Chandler, P. M., Ross, J. J., & Swain, S. M. (2020). Manipulating gibberellin control over growth and fertility as a possible target for managing wild radish weed populations in cropping systems. *Frontiers in Plant Science*, *11*(March), 1–27. https://doi.org/10.3389/fpls.2020.00190

Guerra, F., Ricardo Trevizam, A., Muraoka, T., Chaves Marcante, N., & Guidolin Canniatti-Brazaca, S. (2012). Heavy metals in food chain heavy metals in vegetables and potential risk for human health. *Scientia Agricola*, *69*(1), 54–60. https://www.scielo.br/pdf/sa/v69n1/v69n1a08.pdf

Gumasta, P., Dubey, A., Swamy, M., & Verma, Y. (2018). Study on cadmium associated microscopic lesions in cattle and buffaloes. *Indian Journal of Veterinary Pathology*, *42*(4), 303. https://doi.org/10.5958/0973-970x.2018.00069.x

Guo, G., Lei, M., Chen, T., & Yang, J. (2018). Evaluation of different amendments and foliar fertilizer for immobilization of heavy metals in contaminated soils. *Journal of Soils and Sediments*, *18*(1), 239–247. https://doi.org/10.1007/s11368-017-1752-y

Guo, K., Liu, Y. F., Zeng, C., Chen, Y. Y., & Wei, X. J. (2014). Global research on soil contamination from 1999 to 2012: a bibliometric analysis. *Acta Agriculturae Scandinavica Section B: Soil and Plant Science*, *64*(5), 377–391. https://doi.org/10.1080/09064710.2014.913679

Gupta, A. R., Bandhyopadhyay, S., Sultana, F., & Swarup, D. (2021). Heavy metal poisoning and its impact on livestock health and production system. *Indian Journal of Animal Health*, *60*(2-Spl), 1–23. https://doi.org/10.36062/ijah.2021.spl.00421

Gupta, N., Yadav, K. K., Kumar, V., Prasad, S., Cabral-Pinto, M. M. S., Jeon, B. H., Kumar, S., Abdellattif, M. H., & Alsukaibia, A. K. D. (2022). Investigation of heavy metal accumulation in vegetables and health risk to humans from their consumption. *Frontiers in Environmental Science*, *10*(February), 1–12. https://doi.org/10.3389/fenvs.2022.791052

Hagemeyer, J. (1999). Ecophysiology of plant growth under heavy metal stress. In: M. N. V. Prasad, & J. Hagemeyer (Eds.) Heavy Metal Stress in Plants. Berlin/Heidelberg, Germany: Springer, pp. 157–181.

Han, J. L., Pan, X. D., & Chen, Q. (2022). Distribution and safety assessment of heavy metals in fresh meat from Zhejiang, China. *Scientific Reports*, *12*(1), 1–8. https://doi.org/10.1038/s41598-022-07214-3

Harlia, E., Rahmah, K. N., & Suryanto, D. (2018). Food safety of milk and dairy product of dairy cattle from heavy metal contamination. *IOP Conference Series: Earth and Environmental Science*, *102*(1), 12050. https://doi.org/10.1088/1755-1315/102/1/012050

Hasan, G. M. M. A., Das, A. K., & Satter, M. A. (2022). Accumulation of heavy metals in rice (Oryza sativa. L) grains cultivated in three major industrial areas of Bangladesh. *Journal of Environmental and Public Health*, *2022*, 1836597. https://doi.org/10.1155/2022/1836597

Hashem, M., El-Dein, I., & Abd ELgelil, S. (2019). Clinicopathological studies on the ameliorative effects of selenium and vitamin E against cadmium toxicity in chickens. *Zagazig Veterinary Journal*, *47*(3), 277–287. https://doi.org/10.21608/zvjz.2019.13192.1045

Hassan, N. U., Mahmood, Q., Waseem, A., Irshad, M., Faridullah, & Pervez, A. (2013). Assessment of heavy metals in wheat plants irrigated with contaminated wastewater. *Polish Journal of Environmental Studies*, *22*(1), 115–123.

Hassan. Emami, M., Saberi, F., Mohammadzadeh, S., Fahim, A., Abdolvand, M., Ali Ehsan Dehkordi, S., Mohammadzadeh, S., & Maghool, F. (2023). A review of heavy metals accumulation in red meat and meat products in the Middle East. *Journal of Food Protection*, *86*(3), 100048. https://doi.org/10.1016/j.jfp.2023.100048

Hejna, M., Gottardo, D., Baldi, A., Dell'Orto, V., Cheli, F., Zaninelli, M., & Rossi, L. (2018). Review: nutritional ecology of heavy metals. *Animal*, *12*(10), 2156–2170. https://doi.org/10.1017/S175173111700355X

Hossain, A., Ahmed, M. W., Rabin, M. H., Kaium, A., Razzaque, M. A., & Zamil, S. S. (2024). Heavy metal quantification in chicken meat and egg: an emerging food safety concern. *Journal of Food Composition and Analysis*, 126(November 2023), 105876. https://doi.org/10.1016/j.jfca.2023.105876

Hossain, E., Nesha, M., Chowdhury, M. A. Z., & Rahman, S. H. (2023). Human health risk assessment of edible body parts of chicken through heavy metals and trace elements quantitative analysis. *PLoS One*, 18(3 March), 1–19. https://doi.org/10.1371/journal.pone.0279043

Houri, T., Khairallah, Y., Zahab, A. A., Osta, B., Romanos, D., & Haddad, G. (2020). Heavy metals accumulation effects on the photosynthetic performance of geophytes in mediterranean reserve. *Journal of King Saud University - Science*, *32*(1), 874–880. https://doi.org/10.1016/j.jksus.2019.04.005

Hussein, A. A., Omer, S. S., Ali, I. S., & Suleiman, B. F. (2021). Assessment of hepatic mineral composition in sheep, cattle, chicken, and fish in Erbil City, Kurdistan Region-Iraq. *Kurdistan Journal of Applied Research, 2021*, 46–55. https://doi.org/10.24017/science.2021.1.6

Ifie, I., Igwebuike, C. G., Imasuen, P., Akalamudo, W., Oghenebrorhie, O., Akpodiete, J. O., & Eze, U. A. (2022). Assessment of aflatoxin and heavy metals levels in maize and poultry feeds from Delta State, Nigeria. *International Journal of Environmental Science and Technology, 19*(12), 12551–12560. https://doi.org/10.1007/s13762-022-03996-1

Iqbal, Z., Abbas, F., Ibrahim, M., Ayyaz, M. M., & Shafaqat Ali, A. M. (2019). Surveillance of heavy metals in maize grown with wastewater and their impacts on animal health in peri-urban areas of Multan, Pakistan. *Pakistan Journal of Agricultural Research, 56*(2), 321–328.

İslamoğlu, A. H., T.Kahvecioğlu, T., Bönce, G., Gedik, E., & Güneş, F.E. (2021). Determination of Heavy Metals in Some Fruits, Vegetables and Fish by ICP-MS. *Eurasian Journal of Food Science & Technology, 5*(1), 67–76.

Jaishankar, M., Tseten, T., Anbalagan, N., Mathew, B. B., & Beeregowda, K. N. (2014). Toxicity, mechanism and health effects of some heavy metals. *Interdisciplinary Toxicology, 7*(2), 60–72. https://doi.org/10.2478/intox-2014-0009

Jamil, M., Khatoon, A., Saleemi, M., Gul, A. M. I., & Majeed, W. (2023). An overview of copper toxicity and public health concerns with mitigation strategies In: R. Z. Abbas, N. M. Saeed, M. Younus, L. Aguilar Marcelino, & A. Khan (Eds.) One Health Triad. Faisalabad, Pakistan: Unique Scientific Publishers, Vol. 2, pp. 162–167.

Kamaly, H. F., & Sharkawy, A. A. (2023). Health risk assessment of metals in chicken meat and liver in Egypt. *Environmental Monitoring and Assessment, 195*(7), 9. https://doi.org/10.1007/s10661-023-11365-9

Kar, I., & Patra, A. K. (2021). Tissue bioaccumulation and toxicopathological effects of cadmium and its dietary amelioration in poultry—a review. *Biological Trace Element Research, 199*(10), 3846–3868. https://doi.org/10.1007/s12011-020-02503-2

Kasozi, K. I., Hamira, Y., Zirintunda, G., Alsharif, K. F., Altalbawy, F. M. A., Ekou, J., Tamale, A., Matama, K., Ssempijja, F., Muyinda, R., Kawooya, F., Pius, T., Kisakye, H., Bogere, P., Matovu, H., Omadang, L., Etiang, P., Mbogua, J., Ochieng, J. J., Otim, O. (2021). Descriptive analysis of heavy metals content of beef from Eastern Uganda and their safety for public consumption. *Frontiers in Nutrition, 8*(February), 1–10. https://doi.org/10.3389/fnut.2021.592340

Khalef, R. N., Hassan, A. I., & Saleh, H. M. (2022). Heavy metal's environmental impact. In: A. I. Hassan, & H. M. Saleh (Eds.) Environmental Impact and Remediation of Heavy Metals (Issue September 2023). London: IntechOpen Limited. https://doi.org/10.5772/intechopen.103907

Khan, S., Cao, Q., Zheng, Y. M., Huang, Y. Z., & Zhu, Y. G. (2008). Health risks of heavy metals in contaminated soils and food crops irrigated with wastewater in Beijing, China. *Environmental Pollution, 152*(3), 686–692. https://doi.org/10.1016/j.envpol.2007.06.056

Khan, Z., Sultan, A., Khan, R., Khan, S., And, I., & Kamran, F. (2016). Concentrations of heavy metals and minerals in poultry eggs and meat produced in Khyber Pakhtunkhwa, Pakistan. *Meat Sciences and Veterinary Public Health, 1*(May 2021), 307363528. https://www.researchgate.net/publication/307363528

Kia, S. A., Aslani, R., Khaniki, G. J., Shariatifar, N., & Molaee-Aghaee, E. (2024). Determination and health risk assessment of heavy metals in chicken meat and edible giblets in Tehran, Iran. *Journal of Trace Elements and Minerals, 7*(January), 100117. https://doi.org/10.1016/j.jtemin.2024.100117

Kim, H. S., Kim, K. R., Kim, W. Il, Owens, G., & Kim, K. H. (2017). Influence of road proximity on the concentrations of heavy metals in Korean Urban agricultural soils and crops. *Archives of Environmental Contamination and Toxicology, 72*(2), 260–268. https://doi.org/10.1007/s00244-016-0344-y

Kojima, L. V., Tuberville, T. D., & Parrott, B. B. (2023). Integrating mercury concentrations in American alligators (Alligator mississippiensis) with hunter consumption surveys to estimate exposure risk. *Environmental Toxicology and Chemistry, 42*(2), 525–534. https://doi.org/10.1002/etc.5524

Lessl, J. T., Luo, J., & Ma, L. Q. (2014). Pteris vittata continuously removed arsenic from non-labile fraction in three contaminated-soils during 3.5 years of phytoextraction. *Journal of Hazardous Materials, 279*, 485–492. https://doi.org/10.1016/j.jhazmat.2014.06.056

Letuka, P., Nkhebenyane, J., & Tywabi-Ngeva, and Z. (2023). Heavy metal contamination in food: the perspective of the sub-saharan informal food trade. *Health Risks of Food Additives - Recent Developments and Trends in Food Sector, i*, 13. https://doi.org/10.5772/57353

Li, L., Zhang, Y., Ippolito, J. A., Xing, W., & Tu, C. (2021). Lead smelting alters wheat flour heavy metal concentrations and health risks. *Journal of Environmental Quality*, *50*(2), 454–464. https://doi.org/10.1002/jeq2.20198

Li, Y., Ur, S., Qiu, Z., Muhammad, S., Farrakh, M., Huang, J., Naveed, S., Li, L., Wang, X., & Cheng, H. (2023). Toxic effects of cadmium on the physiological and biochemical attributes of plants, and phytoremediation strategies : a review. *Environmental Pollution*, 325(November 2022), 121433. https://doi.org/10.1016/j.envpol.2023.121433

Liao, J., Wen, Z., Ru, X., Chen, J., Wu, H., & Wei, C. (2016). Distribution and migration of heavy metals in soil and crops affected by acid mine drainage: public health implications in Guangdong Province, China. *Ecotoxicology and Environmental Safety*, *124*, 460–469. https://doi.org/10.1016/j.ecoenv.2015.11.023

Lin, S., Ali, M. U., Zheng, C., Cai, Z., & Wong, M. H. (2022). Toxic chemicals from uncontrolled e-waste recycling: exposure, body burden, health impact. *Journal of Hazardous Materials*, *426*(June), 127792. https://doi.org/10.1016/j.jhazmat.2021.127792

Lukuman, A. (2018). Heavy metal contents in soil and plants at dumpsites: a case study of awotan and ajakanga dumpsite ibadan, Oyo State, Nigeria Kehinde Gbola Adewuyi. *Journal of Environment and Earth Science*, *7*, 11–24. https://www.researchgate.net/publication/324088079

Luo, X., Bing, H., Luo, Z., Wang, Y., & Jin, L. (2019). Impacts of atmospheric particulate matter pollution on environmental biogeochemistry of trace metals in soil-plant system: a review. *Environmental Pollution*, *255*, 113138. https://doi.org/10.1016/j.envpol.2019.113138

Madhaiyan, M., Poonguzhali, S., & Sa, T. (2007). Metal tolerating methylotrophic bacteria reduces nickel and cadmium toxicity and promotes plant growth of tomato (Lycopersicon esculentum L.). *Chemosphere*, *69*(2), 220–228. https://doi.org/10.1016/j.chemosphere.2007.04.017

Mahajan, V. E., Yadav, R. R., Dakshinkar, N. P., Dhoot, V. M., Bhojane, G. R., Naik, M. K., Shrivastava, P., Naoghare, P. K., & Krishnamurthi, K. (2012). Influence of mercury from fly ash on cattle reared nearby thermal power plant. *Environmental Monitoring and Assessment*, *184*(12), 7365–7372. https://doi.org/10.1007/s10661-011-2505-9

Massaquoi, L. D., Ma, H., Liu, X. H., Han, P. Y., Zuo, S. M., Hua, Z. X., & Liu, D. W. (2015). Heavy metal accumulation in soils, plants, and hair samples: an assessment of heavy metal exposure risks from the consumption of vegetables grown on soils previously irrigated with wastewater. *Environmental Science and Pollution Research*, *22*(23), 18456–18468. https://doi.org/10.1007/s11356-015-5131-1

Mawari, G., Kumar, N., Sarkar, S., Daga, M. K., Singh, M. M., Joshi, T. K., & Khan, N. A. (2022). Heavy metal accumulation in fruits and vegetables and human health risk assessment: findings from Maharashtra, India. *Environmental Health Insights*, *16*, 11786302221119151. https://doi.org/10.1177/11786302221119151

Mclaughlin, M. J., Parker, D. R., & Clarke, J. M. (1999). Metals and micronutrients ± food safety issues. *Field Crops Research*, *60*(1–2), 143–163.

Mehmood, K., Khan, A., Zhang, H., Ali Alkheraije, K., & Tahir, I. (2023). A review of important heavy metals toxicity with special emphasis on nephrotoxicity and its management in cattle. *Frontiers in Veterinary Science*, *9*, 1149720.

Mehri, A., & Marjan, R. F. (2015). Trace elements in human nutrition : a review. *International Journal of Medical Investigation*, *2*(3), 115–128.

Meshref, A. M. S., Moselhy, W. A., & Hassan, N. E. H. Y. (2014). Heavy metals and trace elements levels in milk and milk products. *Journal of Food Measurement and Characterization*, *8*(4), 381–388. https://doi.org/10.1007/s11694-014-9203-6

MHPRC. (2012). *Maximum Levels of Contaminants in Foods*. Beijing, China: Ministry of Health of the People's Republic of China.

MHPRC. (2017). *Ministry of Health of the People's Republic of China Maximum Levels of Contaminants in Foods (GB2762–2017)*. Beijing, China: Ministry of Health of the People's Republic of China.

Minervino, A. H. H., López-Alonso, M., Barrêto Júnior, R. A., Rodrigues, F. A. M. L., Araújo, C. A. S. C., Sousa, R. S., Mori, C. S., Miranda, M., Oliveira, F. L. C., Antonelli, A. C., & Ortolani, E. L. (2018). Dietary zinc supplementation to prevent chronic copper poisoning in sheep. *Animals*, *8*(12), 1–10. https://doi.org/10.3390/ani8120227

Minhas, P. S., Khajanchi, L., Yadav, R. K., Dubey, S. K., & Chaturvedi, R. K. (2015). Long term impact of waste water irrigation and nutrient rates: I. Performance, sustainability and produce quality of peri urban cropping systems. *Agricultural Water Management*, *156*, 100–109. https://doi.org/10.1016/j.agwat.2015.03.012

Mishra, S., Kumar, R., & Kumar, M. (2023). Use of treated sewage or wastewater as an irrigation water for agricultural purposes- environmental, health, and economic impacts. *Total Environment Research Themes*, *6*(May), 100051. https://doi.org/10.1016/j.totert.2023.100051

Muhib, M. I., Chowdhury, M. A. Z., Easha, N. J., Rahman, M. M., Shammi, M., Fardous, Z., Bari, M. L., Uddin, M. K., Kurasaki, M., & Alam, M. K. (2016). Investigation of heavy metal contents in cow milk samples from area of Dhaka, Bangladesh. *International Journal of Food Contamination*, *3*(1), 1–10. https://doi.org/10.1186/s40550-016-0039-1

Mukherjee, A., Kar, I., & Patra, A. (2022). Lead toxicity in livestock and poultry: illuminating the molecular pathomechanism, bioaccumulation and effective mitigation strategies. *Indian Journal of Animal Health*, *6*(2), 103–118. https://doi.org/10.36062/ijah.2022.spl.02322

Nabulo, G., Oryem-Origa, H., & Diamond, M. (2006). Assessment of lead, cadmium, and zinc contamination of roadside soils, surface films, and vegetables in Kampala City, Uganda. *Environmental Research*, *101*(1), 42–52. https://doi.org/10.1016/j.envres.2005.12.016

Naveedullah, N., Hashmi, M. Z., Yu, C., Shen, H., Duan, D., Shen, C., Lou, L., & Chen, Y. (2013). Risk assessment of heavy metals pollution in agricultural soils of siling reservoir watershed in Zhejiang province, China. *BioMed Research International*, *2013*(Cd), 590306. https://doi.org/10.1155/2013/590306

Nawaz, R., Ur Rehman, S., Nawaz, S., & Iftikhar, B. (2015). Analysis of heavy metals in red meat in district Peshawar Khyber Pakhtunkhwa. *Journal of Medical Sciences (Peshawar)*, *23*(3), 166–171.

Newcomer, B. W., Cebra, C., Chamorro, F., Reppert, E., Cebra, M., & Edmondson, M. A. (2021). Diseases of the hematologic, immunologic, and lymphatic systems (multisystem diseases). *Sheep, Goat, and Cervid Medicine*, *2021*, 405–438.

Njoga, E. O., Ezenduka, E. V., Ogbodo, C. G., Ogbonna, C. U., Jaja, I. F., Ofomatah, A. C., & Okpala, C. O. R. (2021). Detection, distribution and health risk assessment of toxic heavy metals/metalloids, arsenic, cadmium, and lead in goat carcasses processed for human consumption in south-eastern Nigeria. *Foods*, *10*(4), 1–17. https://doi.org/10.3390/foods10040798

Noor, A. E., Nisa, Z. un, Sultana, S., Al-Ghanim, K. A., Al-Misned, F., Riaz, M. N., Ahmed, Z., & Mahboob, S. (2023). Heavy metals toxicity in spinach (Spinacia oleracea) irrigated with sanitary wastewater in rural areas. *Journal of King Saud University - Science*, *35*(1), 102382. https://doi.org/10.1016/j.jksus.2022.102382

Nyiramigisha, P., Komariah, A., & Sajidan, M. (2021). Harmful impacts of heavy metal. *Reviews in Agricultural Science*, 9, 1–12.

Ogbomida, E. T., Nakayama, S. M. M., Bortey-Sam, N., Oroszlany, B., Tongo, I., Enuneku, A. A., Ozekeke, O., Ainerua, M. O., Fasipe, I. P., Ezemonye, L. I., Mizukawa, H., Ikenaka, Y., & Ishizuka, M. (2018). Accumulation patterns and risk assessment of metals and metalloid in muscle and offal of free-range chickens, cattle and goat in Benin City, Nigeria. *Ecotoxicology and Environmental Safety*, 151(February 2017), 98–108. https://doi.org/10.1016/j.ecoenv.2017.12.069

Omondi, M. O., Xia, X., Nahayo, A., Liu, X., Korai, P. K., & Pan, G. (2016). Quantification of biochar effects on soil hydrological properties using meta-analysis of literature data. *Geoderma*, *274*, 28–34. https://doi.org/10.1016/j.geoderma.2016.03.029

Onibon, V. O., Tomori, W. B., & Efunwoye, O. O. (2019). Heavy metals content of commercially sold fermented milk products in akure markets, South-West, Nigeria. *International Journal of Scientific & Engineering Research*, *10*(8), 605–617. https://www.ijser.org

Öztürk, E., Atsan, E., Polat, T., & Kara, K. (2011). Variation in heavy metal concentrations of potato (Solanum tuberosum L.) cultivars. *Journal of Animal and Plant Sciences*, *21*(2), 235–239.

Pandey, R., Vengavasi, K., & Hawkesford, M. J. (2021). Plant adaptation to nutrient stress. *Plant Physiology Reports*, *26*(4), 583–586. https://doi.org/10.1007/s40502-021-00636-7

Peng, X., Deng, Y., Peng, Y., & Yue, K. (2018). Effects of biochar addition on toxic element concentrations in plants: a meta-analysis. *Science of the Total Environment*, *616–617*, 970–977. https://doi.org/10.1016/j.scitotenv.2017.10.222

Qiao, J. tao, Li, X. min, & Li, F. bai. (2018). Roles of different active metal-reducing bacteria in arsenic release from arsenic-contaminated paddy soil amended with biochar. *Journal of Hazardous Materials*, *344*(808), 958–967. https://doi.org/10.1016/j.jhazmat.2017.11.025

Rafati-Rahimzadeh, M., Rafati-Rahimzadeh, M., Kazemi, S., & Moghadamnia, A. A. (2014). Current approaches of the management of mercury poisoning: need of the hour. *DARU, Journal of Pharmaceutical Sciences*, *22*(1), 1–10. https://doi.org/10.1186/2008-2231-22-46

Rai, P. K. (2018a). Heavy metal phyto-technologies from Ramsar wetland plants: green approach. *Chemistry and Ecology*, *34*(8), 786–796. https://doi.org/10.1080/02757540.2018.1501476

Rai, P. K. (2018b). *Phytoremediation of Emerging Contaminants in Wetlands*. Boca Raton, FL: CRC Press, Taylor & Francis, pp. 248.

Rai, P. K., Lee, J., Kailasa, S. K., Kwon, E. E., Tsang, Y. F., Ok, Y. S., & Kim, K. H. (2018). A critical review of ferrate(VI)-based remediation of soil and groundwater. *Environmental Research*, *160*(August 2017), 420–448. https://doi.org/10.1016/j.envres.2017.10.016

Rai, P. K., Lee, S. S., Zhang, M., Tsang, Y. F., & Kim, K. H. (2019). Heavy metals in food crops: Health risks, fate, mechanisms, and management. *Environment International*, *125*(November 2018), 365–385. https://doi.org/10.1016/j.envint.2019.01.067

Raikwar, M. K., Kumar, P., Singh, M., & Singh, A. (2008). Toxic effect of heavy metals in livestock health. *Veterinary World*, *1*(1), 28–30. https://doi.org/10.5455/vetworld.2008.28-30

Rajaganapathy, V., Xavier, F., Sreekumar, D., & Mandal, P. K. (2011). Heavy metal contamination in soil, water and fodder and their presence in livestock and products : a review. *International Journal of Environmental Science and Technology*, *4*(3), 234–249.

Rane, J., Singh, A. K., Kumar, M., Boraiah, K. M., Meena, K. K., Pradhan, A., & Vara Prasad, P. V. (2021). The adaptation and tolerance of major cereals and legumes to important abiotic stresses. *International Journal of Molecular Sciences*, *22*(23), 222312970. https://doi.org/10.3390/ijms222312970

Rathebe, P. C., & Mosoeu, L. G. (2023). Fruits and vegetables contaminated with particles of heavy metals: a narrative review to explore the use of electromagnetic fields as an alternative treatment method. *Cogent Food and Agriculture*, *9*(1), 2231686. https://doi.org/10.1080/23311932.2023.2231686

Rather, I. A., Koh, W. Y., Paek, W. K., & Lim, J. (2017). The sources of chemical contaminants in food and their health implications. *Frontiers in Pharmacology*, *8*(NOV), 830. https://doi.org/10.3389/fphar.2017.00830

Rezaei, M., Akbari Dastjerdi, H., Jafari, H., Farahi, A., Shahabi, A., Javdani, H., Teimoory, H., Yahyaei, M., & Malekirad, A. A. (2014). Assessment of dairy products consumed on the Arakmarket as determined by heavy metal residues. *Health*, *06*(05), 323–327. https://doi.org/10.4236/health.2014.65047

Rodríguez Martín, J. A., & Nanos, N. (2016). Soil as an archive of coal-fired power plant mercury deposition. *Journal of Hazardous Materials*, *308*, 131–138. https://doi.org/10.1016/j.jhazmat.2016.01.026

Saeed, M., Ilyas, N., Bibi, F., Shabir, S., Mehmood, S., Akhtar, N., Ali, I., Bawazeer, S., Al Tawaha, A. R., & Eldin, S. M. (2023). Nanoremediation approaches for the mitigation of heavy metal contamination in vegetables: an overview. *Nanotechnology Reviews*, *12*(1), 1–22. https://doi.org/10.1515/ntrev-2023-0156

Sarkar, A., Ravindran, G., & Krishnamurthy, V. (2013). A brief review on the effect of cadmium toxicity: from cellular to organ level. *International Journal of Bio-Technology and Research*, *3*(1), 2249–6858.

Sattar, A., Khan, A., Hussain, H. I., He, C., Hussain, R., Zhiqiang, S., Saleemi, M. K., & Gul, S. T. (2016). Immunosuppressive effects of arsenic in broiler chicks exposed to Newcastle disease virus. *Journal of Immunotoxicology*, *13*(6), 861–869. https://doi.org/10.1080/1547691X.2016.1217105

Selina Acheampong. (2016). Heavy metals' poisoning in farm animals. *Intech*, *11*, 13. https://www.intechopen.com/books/advanced-biometric-technologies/liveness-detection-in-biometrics

Senila, M., & Cadar, O. (2024). Modification of natural zeolites and their applications for heavy metal removal from polluted environments: challenges, recent advances, and perspectives. *Heliyon*, *10*(3), e25303. https://doi.org/10.1016/j.heliyon.2024.e25303

Shaheen, N., Irfan, N. M., Khan, I. N., Islam, S., Islam, M. S., & Ahmed, M. K. (2016). Presence of heavy metals in fruits and vegetables: health risk implications in Bangladesh. *Chemosphere*, *152*, 431–438. https://doi.org/10.1016/j.chemosphere.2016.02.060

Shahid, M., Rafiq, M., Niazi, N. K., Dumat, C., Shamshad, S., Khalid, S., & Bibi, I. (2017). Arsenic accumulation and physiological attributes of spinach in the presence of amendments: an implication to reduce health risk. *Environmental Science and Pollution Research*, *24*(19), 16097–16106. https://doi.org/10.1007/s11356-017-9230-z

Smirnova, E., Moniruzzaman, M., Chin, S., Sureshbabu, A., Karthikeyan, A., Do, K., & Min, T. (2023). A review of the role of curcumin in metal induced toxicity. *Antioxidants*, *12*(2), 12020243. https://doi.org/10.3390/antiox12020243

Sobhanardakani, S., Shiravand, S., & Jameh-Bozorghi, S. (2016). Health risk assessment of heavy metals (Cd, Cu, Pb and Zn) in soybean marketed in Hamedan City, Iran. *Annals of Military and Health Sciences Research*, *14*(3), 11464. https://doi.org/10.5812/amh.11464

Song, B., Lei, M., Chen, T., Zheng, Y., Xie, Y., Li, X., & Gao, D. (2009). Assessing the health risk of heavy metals in vegetables to the general population in Beijing, China. *Journal of Environmental Sciences*, *21*(12), 1702–1709. https://doi.org/10.1016/S1001-0742(08)62476-6

Suleman, S., Qureshi, J. A., & Farooq, W. (2022). Poultry feed contamination and its potential hazards on human health. *Biomedical Letters*, *8*(1), 70–81. https://doi.org/10.47262/bl/8.1.20210901

Suman, J., Uhlik, O., Viktorova, J., & Macek, T. (2018). Phytoextraction of heavy metals: a promising tool for clean-up of polluted environment? *Frontiers in Plant Science*, *871*(October), 1–15. https://doi.org/10.3389/fpls.2018.01476

Surai, P. F., Kochish, I. I., Fisinin, V. I., & Velichko, O. A. (2018). Selenium in poultry nutrition: from sodium selenite to organic selenium sources. *Journal of Poultry Science*, *55*(2), 79–93. https://doi.org/10.2141/jpsa.0170132

Suttle, N. F. (2012). Copper imbalances in ruminants and humans: unexpected common ground. *Advances in Nutrition*, *3*(5), 666–674. https://doi.org/10.3945/an.112.002220

Syversen, T., & Kaur, P. (2012). The toxicology of mercury and its compounds. *Journal of Trace Elements in Medicine and Biology*, *26*(4), 215–226. https://doi.org/10.1016/j.jtemb.2012.02.004

Talib Jawad, S., Abedali, S. T., & Khalef, W. F. (2021). Determination of heavy metals concentration in liver, meat and blood chicken of thi-qar province, Iraq. *Volatiles & Essent. Oils*, *8*(6), 2562–2570.

Tariq, F., Wang, X., Saleem, M. H., Khan, Z. I., Ahmad, K., Malik, I. S., Munir, M., Mahpara, S., Mehmood, N., Ahmad, T., Memona, H., Ugulu, I., Fiaz, S., & Ali, S. (2021). Risk assessment of heavy metals in basmati rice: implications for public health. *Sustainability (Switzerland)*, *13*(15), 13158513. https://doi.org/10.3390/su13158513

Tatahmentan, M., Nyachoti, S., Scott, L., Phan, N., Okwori, F. O., Felemban, N., & Godebo, T. R. (2020). Toxic and essential elements in rice and other grains from the united states and other countries. *International Journal of Environmental Research and Public Health*, *17*(21), 1–12. https://doi.org/10.3390/ijerph17218128

Tavarez, M., Grusak, M. A., & Sankaran, R. P. (2022). Effects of zinc fertilization on grain cadmium accumulation, gene expression, and essential mineral partitioning in rice. *Agronomy*, *12*(9), 12092182. https://doi.org/10.3390/agronomy12092182

Tom, M., Fletcher, T. D., & McCarthy, D. T. (2014). Heavy metal contamination of vegetables irrigated by Urban stormwater: a matter of time? *PLoS One*, *9*(11), 1–21. https://doi.org/10.1371/journal.pone.0112441

Tong, C., Hill, C. B., Zhou, G., Zhang, X. Q., Jia, Y., & Li, C. (2021). Opportunities for improving waterlogging tolerance in cereal crops—physiological traits and genetic mechanisms. *Plants*, *10*(8), 1–22. https://doi.org/10.3390/plants10081560

Tőzsér, D., Horváth, R., Simon, E., & Magura, T. (2023). Heavy metal uptake by plant parts of Populus species: a meta-analysis. *Environmental Science and Pollution Research*, *30*(26), 69416–69430. https://doi.org/10.1007/s11356-023-27244-2

Udom, B. E., Mbagwu, J. S. C., Adesodun, J. K., & Agbim, N. N. (2004). Distributions of zinc, copper, cadmium and lead in a tropical ultisol after long-term disposal of sewage sludge. *Environment International*, *30*(4), 467–470. https://doi.org/10.1016/j.envint.2003.09.004

Ur Rahman, S., Li, Y., Hussain, S., Hussain, B., Khan, W. ud D., Riaz, L., Nadeem Ashraf, M., Athar Khaliq, M., Du, Z., & Cheng, H. (2023). Role of phytohormones in heavy metal tolerance in plants: a review. *Ecological Indicators*, *146*(July 2022), 109844. https://doi.org/10.1016/j.ecolind.2022.109844

USDA. (2018). *USDA-Foreign Agricultural Service (2018).* China: China Releases the Standard for Maximum Levels of Contaminants in Foods. https://www.fas.usda.gov/data/china-china-releases-standard-maximum-levels-contaminants-foods

Vasilachi, I. C., Stoleru, V., & Gavrilescu, M. (2023). Analysis of heavy metal impacts on cereal crop growth and development in contaminated soils. *Agriculture (Switzerland)*, *13*(10), 13101983. https://doi.org/10.3390/agriculture13101983

Wang, J., & Aghajani Delavar, M. (2023). Techno-economic analysis of phytoremediation: a strategic rethinking. *Science of the Total Environment*, *902*(May), 165949. https://doi.org/10.1016/j.scitotenv.2023.165949

Wang, N., Xue, X. M., Juhasz, A. L., Chang, Z. Z., & Li, H. B. (2017). Biochar increases arsenic release from an anaerobic paddy soil due to enhanced microbial reduction of iron and arsenic. *Environmental Pollution*, *220*, 514–522. https://doi.org/10.1016/j.envpol.2016.09.095

Wang, S., Wu, W., Liu, F., Liao, R., & Hu, Y. (2017). Accumulation of heavy metals in soil-crop systems: a review for wheat and corn. *Environmental Science and Pollution Research*, *24*(18), 15209–15225. https://doi.org/10.1007/s11356-017-8909-5

Wang, Z., Zeng, X., Geng, M., Chen, C., Cai, J., Yu, X., Hou, Y., & Zhang, H. (2015). Health risks of heavy metals uptake by crops grown in a sewage irrigation area in China. *Polish Journal of Environmental Studies*, *24*(3), 1379–1386. https://doi.org/10.15244/pjoes/35282

Waterlot, C., Pruvot, C., Marot, F., & Douay, F. (2017). Impact of a phosphate amendment on the environmental availability and phytoavailability of Cd and Pb in moderately and highly carbonated kitchen garden soils. *Pedosphere*, *27*(3), 588–605. https://doi.org/10.1016/S1002-0160(17)60354-0

Wei, X., Zhou, Y., Jiang, Y., Tsang, D. C. W., Zhang, C., Liu, J., Zhou, Y., Yin, M., Wang, J., Shen, N., Xiao, T., & Chen, Y. (2020). Health risks of metal(loid)s in maize (Zea mays L.) in an artisanal zinc smelting zone and source fingerprinting by lead isotope. *Science of the Total Environment*, *742*, 140321. https://doi.org/10.1016/j.scitotenv.2020.140321

Witkowska, D., Słowik, J., & Chilicka, K. (2021). Review heavy metals and human health: Possible exposure pathways and the competition for protein binding sites. *Molecules*, *26*(19), 26196060. https://doi.org/10.3390/molecules26196060

Wuana, R. A., & Okieimen, F. E. (2011). Heavy metals in contaminated soils: a review of sources, chemistry, risks and best available strategies for remediation. *ISRN Ecology*, *2011*, 1–20. https://doi.org/10.5402/2011/402647

Xiong, T., Dumat, C., Dappe, V., Vezin, H., Schreck, E., Shahid, M., Pierart, A., & Sobanska, S. (2017). Copper oxide nanoparticle foliar uptake, phytotoxicity, and consequences for sustainable urban agriculture. *Environmental Science and Technology*, *51*(9), 5242–5251. https://doi.org/10.1021/acs.est.6b05546

Xue, P., Zhao, Q., Sun, H., Geng, L., Yang, Z., & Liu, W. (2019). Characteristics of heavy metals in soils and grains of wheat and maize from farmland irrigated with sewage. *Environmental Science and Pollution Research*, *26*(6), 5554–5563. https://doi.org/10.1007/s11356-018-3997-4

Yao, A., Wang, Y., Ling, X., Chen, Z., Tang, Y., Qiu, H., Ying, R., & Qiu, R. (2017). Effects of an iron-silicon material, a synthetic zeolite and an alkaline clay on vegetable uptake of As and Cd from a polluted agricultural soil and proposed remediation mechanisms. *Environmental Geochemistry and Health*, *39*(2), 353–367. https://doi.org/10.1007/s10653-016-9863-8

Yoon, J., Cao, X., Zhou, Q., & Ma, L. Q. (2006). Accumulation of Pb, Cu, and Zn in native plants growing on a contaminated Florida site. *Science of the Total Environment*, *368*(2–3), 456–464. https://doi.org/10.1016/j.scitotenv.2006.01.016

Yousaf, B., Liu, G., Wang, R., Imtiaz, M., Zia-ur-Rehman, M., Munir, M. A. M., & Niu, Z. (2016). Bioavailability evaluation, uptake of heavy metals and potential health risks via dietary exposure in urban-industrial areas. *Environmental Science and Pollution Research*, *23*(22), 22443–22453. https://doi.org/10.1007/s11356-016-7449-8

Zaidi, A., & Khan, M. S. (2017). Microbial strategies for vegetable production. *Microbial Strategies for Vegetable Production*, 2017, 1–226. https://doi.org/10.1007/978-3-319-54401-4

Zhang, Y., Hou, D., O'Connor, D., Shen, Z., Shi, P., Ok, Y. S., Tsang, D. C. W., Wen, Y., & Luo, M. (2019). Lead contamination in Chinese surface soils: Source identification, spatial-temporal distribution and associated health risks. *Critical Reviews in Environmental Science and Technology*, *49*(15), 1386–1423. https://doi.org/10.1080/10643389.2019.1571354

Zhou, M., & Zheng, S. (2022). Multi-omics uncover the mechanism of wheat under heavy metal stress. *International Journal of Molecular Sciences*, *23*(24), 232415968. https://doi.org/10.3390/ijms232415968

Zhuang, P., McBride, M. B., Xia, H., Li, N., & Li, Z. (2009). Health risk from heavy metals via consumption of food crops in the vicinity of Dabaoshan mine, South China. *Science of the Total Environment*, *407*(5), 1551–1561. https://doi.org/10.1016/j.scitotenv.2008.10.061

7

Soil Health Restoration—A Focus on Microbiome Engineering and Prediction

Xun Wen Chen, Miao Miao Zhang, and Hui Li

7.1 Introduction

7.1.1 Soil Pollution and Land Degradation

Soil pollution and land degradation are still prevalent worldwide [1,2]. The main soil pollutants are heavy metals, persistent organic pollutants (POPs), and emerging contaminants [3], such as polyfluoroalkyl substances (PFAS), antibiotics, and small-sized particles. Additionally, land degradation is of critical concern, including erosion, salinization, compaction, and loss of soil biodiversity [4]. Over ¼ of the Earth's land surface is degraded, and approximately 12 million hectares of productive land is lost annually [1,5,6]. The affected areas are Asia and the Pacific, Europe, Latin America and the Caribbean, Near East and North Africa, North America, and Sub-Saharan Africa. Continuing industrialization, urbanization, and intensive agricultural practices pose a trend of increasing soil pollution and degradation risk. Multifaceted interventions are immediately needed in sustainable remediation techniques, policy measures, and management practices [1,2]. Addressing the critical challenges of soil pollution and land degradation requires a comprehensive understanding of the underlying process and a holistic approach to soil health restoration [7].

7.1.2 Importance of Soil Microbiome on Soil Health

Soil health is a human-perspective term, and it generally describes the state of soil in sustaining microbial, plant, and animal growth, which is associated with nutrient cycling, water and air quality, biodiversity, and human well-being [8]. Soil ecological health and soil health are often used interchangeably. They are slightly different in that soil ecological health is a subset of soil health, emphasizing the ecological attributes from the perspective of ecosystems [9]. Soil microbiome, the complete sets of microbes and genetic materials, is a considerable indicator of soil health because it is soil function-related, interpretable, and responsive to soil management practices [9]. Microbial dynamics is closely related to soil function and health [10–12], and researchers have started to use soil microbial data to evaluate soil health [13]. Essentially, the ability of the microbial community to resist and recover from disturbance (resistance and resilience), known as community stability, is a key indicator of soil health.

Since the number of microbial cells can be up to 10^{10} in 1 g of soil [14,15] and their behaviors are highly complex, stochastic (i.e., random), and context-dependent [16–18], it is challenging to understand and predict microbial behaviors not to say utilization fully. The observability, predictability, and controllability of microbial communities (i.e., complex ecosystems) are the main research directions not only in the soil context [19–21] but also in human microecosystems (e.g., skin and gut) [22–24], attracting continuous research resource allocation [23,25,26]. The Earth Microbiome Project (EMP) [27] and the Human Microbiome Project (HMP) [28] are two significant initiatives in microbiome research. Studying, predicting, and controlling soil microbiomes for ecological and human benefits are particularly challenging. It is a multidisciplinary and demanding task integrating (but not limited to) soil science, soil mechanics, plant biology, microbiology, community ecology, systems biology, network science, chaotic theory, and control theory.

DOI: 10.1201/9781003492115-9

Therefore, identifying and isolating the most important soil bacteria and eliminating the "noise" become a strong motivation. Researchers proposed the concept of the core microbiome [29]. A core microbiome is a collection of microbes that perform essential functions so that we can focus on a substantially smaller subset of soil microbial populations (e.g., in the magnitude of 10^{1-3} instead of 10^{10}), isolate them, and harness most of the beneficial functions [30–33]. Although soil microbes typically vary across environments [19], the soil core microbiome can be ubiquitous, abundant, and relatively stable across different environments or hosts [29,34–36]. Core microbiome has been studied in various ecosystems, such as animal/human gut and plant rhizosphere [30,33]. Mechanistic understanding of community structure dynamics and the ability to predict can provide an essential scientific basis for improving, optimizing, and managing overall soil ecological health, such as sustainable agriculture and biodiversity conservation [37]. This understanding also holds significant theoretical importance for the sustainable use of contaminated and degraded lands.

This chapter briefly reviews soil health indicators and primary approaches for soil health restoration. It then focuses on soil microbiome engineering, exploring the principles of microbial dynamics and the predictability of these dynamics. By acknowledging the complexity and influencing factors, we introduce major ecological models and data-driven approaches for studying microbial dynamics, aiming to enhance soil microbiome engineering for soil health restoration.

7.2 Soil Health and Restoration

7.2.1 Soil Health Indicators

Before discussing soil health restoration, it is essential to define healthy soil. Traditionally, soil assessment is about crop production. But now, soil health is recognized as the continued capacity of soil to sustain microbes, plants, animals, and humans [38]. Soils are living systems with complex and diverse communities of organisms. Soil health encompasses properties and functions such as nutrient and water cycling, biodiversity, organic carbon storage, pathogen suppression, and pollution mitigation [8]. Soil quality measures the soil's ability to perform ecological functions, reflecting its chemical, physical, and biological properties. While inherent properties like soil texture and mineral makeup are unchangeable, management can significantly alter properties such as structure, moisture, and organic matter. These changeable properties indicate soil quality relative to its potential, similar to how the changeable water turbidity or oxygen content indicates water quality. Management practices include applying amendments, growing vegetation, making physical alterations, or removing contaminants.

7.2.2 Traditional Approaches for Soil Health Restoration

Conservation agriculture focuses on minimal soil disturbance and diversified crop rotations [7]. These approaches reduce soil erosion, enhance water retention, improve soil structure, and increase organic matter content. Diverse plant communities can improve pollination and pest suppression, harbor diverse soil microbes, and contribute to soil nutrient cycling, enhancing overall ecosystem functions and services [39]. Similarly, cover cropping and grazing control are also helpful for restoring soil health [40–42]. In addition to plant management, integrated nutrient management is another crucial approach. Organic and inorganic resource inputs enhance soil fertility and nutrient availability for plants. For instance, soil organic carbon (SOC) is critical to soil health and productivity [8]. Practices that promote SOC content, such as organic amendments (e.g., compost) or no-till systems, contribute to soil restoration [43–47]. SOC improves soil structure, water retention, and nutrient availability. By fostering diverse plant communities and managing balanced nutrients and SOC, we can collectively enhance soil health [48,49]. However, since these traditional approaches eventually point to the microbial functions that drive resource cycling, an in-depth understanding of microbial dynamics and possible interventions are essential to applications.

7.2.3 Microbiome Engineering for Soil Health

Microbiome engineering refers to *ex-situ* and/or *in-situ* manipulation of microbial communities favoring desired functions [19]. Humans have a long history of harnessing the benefits of microbes in soils. A classic example is using nitrogen-fixing bacteria (rhizobia) to improve plant growth [50]. This typical early day approach isolates a single microbial strain for crop productivity. Similar cases include using mycorrhizal fungi to improve plant nutrient acquisition and stress resistance [51], *Bacillus* (bacteria) [52] and *Trichoderma* (fungi) [53] to tackle plant pathogens, and plant growth-promoting rhizobacteria (PGPR) to regulate the auxin level stimulating root growth [54], to name a few. Although single inoculants positively promote plant growth, large-scale application faces significant challenges. For example, the benefits often diminish over time, possibly due to interaction between inoculants and native soil microbes [55], and continuous application is required [19]. Inoculants also pose uncertain and hard-to-monitor risks threatening the native microbial community [56]. Notably, a single purpose, such as enhancing crop yield, may not consider the overall soil ecological health and sustainability, which are often multifaceted issues [8].

The simple fact that the soil microbe will hardly be alone lets researchers start to consider the collective behavior of a community [57–59]. For example, a reductionism approach was used to isolate core microbial members that combinedly provide desired outcomes, such as pathogen resistance [60], auxin regulation favoring root elongation [61], and root endodermal differentiation contributing to nutrient uptake [62]. These functions rely on the collective behaviors of the microbial consortia rather than a single species. To construct a microbial consortium, we can isolate potentially beneficial strains with different desired functions, combine them to form a synthetic community (SynCom), and test its collective functions [63]. Since SynCom also receives the impacts from the native community when applying to a specific site, studies have also proposed using a natural community sculptured by natural processes to form a naturally selected community, namely, natural community (NatCom). For instance, isolating microbes from a mine tailing allows screening a NatCom (evolved in the specific environment) that can tolerate metal stress [64,65]. When applying such a NatCom to other metal-contaminated sites, its functions could be retained for soil health improvement [64].

Either single inoculants or consortia, even though they were tested and proven to have the desired structures and functions, we are still facing the long-term sustainability issue. A big hurdle is that we do not know whether they can persist and how they will change over time [20]. This is an ongoing important topic regarding the inoculants' establishment in target sites and their interactions with native biotic and abiotic factors [21,66]. For example, a large-scale field experiment testing the inoculants (mycorrhizal fungi) showed that the factor of pathogen had more weight in predicting the maize growth rather than the expected nutrient availability (mycorrhizal fungi typically improve nutrient uptake) [66].

7.3 Understanding Microbial Dynamics

To harness microbial benefits and predict future events, we first need to understand the principles of microbial dynamics. Microbial dynamics essentially depend on the community context (e.g., what the members are in the first place, i.e., initial community leading to priority effects), microbial interactions (i.e., interaction types and strength levels), and environmental factors (e.g., resource and stress levels) [67].

7.3.1 Community Context

The initial community structure, including species composition and abundance, directly affects the subsequent stable state of community structure and functions [68], known as priority effects [69]. Researchers studied subsequent temporal dynamics by controlling the initial community structure. However, the number of combinations is vast due to the large number of soil bacterial species [14]. For example, 100 species can have up to 2^100 combinations, rendering it a daunting task to survey the dynamics and outcomes of each combination. This poses significant challenges for experiments and predictions [37].

Therefore, researchers used a small number of species. For instance, Callaway et al. [70] and Piccardi et al. [71] used two plants and four bacteria to study their interactions and temporal dynamics. Callaway et al. [70] considered altitude as a stress factor *without* considering resources; Piccardi et al. [71] used sterilants as stress factors and industrial metalworking fluids (petroleum refining products) as resources. The results indicate that species exhibit different behaviors depending on resource availability and stress levels. When resources are abundant (or stress is low), species are more likely to compete. Conversely, when resources are scarce (or stress is high), species are more inclined to cooperate. In other words, unfavorable environments promote cooperation, while favorable environments encourage competition (see a later detailed section). Hu et al. [72] increased the initial number of species to about 50. They found that as the number of species increased (4 → 20 → 50), the abundance of each bacterium changed over time in a pattern of overall stable coexistence → partial stable coexistence → overall continuous fluctuation, similar to the results from Mallmin et al. [73] that species abundance of plankton fluctuates (chaotic regime) when hundreds of species interacting.

In-silico simulation is also possible [37], but the prediction outcome verified using small communities (5–16 species) seems questionable in the real-world soil community. A high-throughput co-culture platform, kChip, has also been established [74,75]. kChip is termed according to the feature that the number of inputs (k) (e.g., strains, media) can be selected by the user. It allows the monitoring of many bacterial combinations with different carbon sources. However, kChip is more focused on bacterial interactions in a more theoretical way, assuming no spatial structure in the growth medium. In contrast, species that can interact under non-structured contexts may not interact in soils (highly structured matrices), hindering direct contact among species [76,77].

The initial community structure and number of species collectively play crucial roles in affecting the subsequent community dynamics. This implies that controlling initial species composition can direct soil microbial dynamics and functions effectively.

7.3.2 Interaction Range

The physical distance between microbes is another essential factor influencing their interactions and temporal dynamics. Microbes disperse and diffuse through various mechanisms in the medium, and this dispersal ability can be quantified using the dispersal rate. In natural soil environments (specifically unsaturated soil, a porous medium consisting of solid, liquid, and gas phases), microbes and other substances are typically immobilized on soil particle surfaces or within water films, with air acting as a barrier [78–81]. As a result, migration is restricted compared with water and air environments. Plant roots and soil fungal mycelium can serve as highways bridging microbial colonies and drastically accelerate dispersal [82–84], although enhanced dispersal does not necessarily suggest later strong interactions [85]. Interaction type and strength, governed by species identity, resources, and stress, determine system stability [85] (see later sections).

When microbes disperse and come close to each other, the interaction range plays a crucial role. Interaction range is an intrinsic characteristic of a microbe, and it refers to the optimal distance at which a microbe can acquire resources from neighboring microbes. Beyond this range, interactions significantly weaken or even disappear. For instance, Dal Co et al. [86] cultivated two interdependent strains of *Escherichia coli* in liquid culture and measured their respective interaction ranges. They found that one strain of *E. coli* had an interaction range of 3 μm, while the other had a range of 12 μm. Bacterial growth was hindered considerably or even halted when the distance exceeded these thresholds. This suggests that bacteria interact primarily with their "neighbors" at the micrometer scale. Different bacterial species exhibit varying interaction ranges (3 μm vs. 12 μm), analogously introverted vs. extroverted personalities in humans. If the initial community contains a specific composition of "introverted" and "extroverted" microbes, it may strongly affect the community's structural distribution and collective behavior.

7.3.3 Interaction Strength and Interaction Diversity

Microbial interactions span a spectrum from cooperation to competition. Six types of interactions are mutualistic symbiosis (+/+), commensalism (0/+), neutral coexistence (0/0), amensalism (–/0), parasitism/exploitation (+/–), and competition (–/–), if the interaction direction is not considered (e.g., we consider

+/– is the same as –/+ here). Interaction strength is the per capita effect of species i on the population growth rate of species j. Interaction strength can be measured in stable states after bacterial monocultures or co-cultures. According to Allesina and Tang [85], reducing interaction strength can stabilize the system because weakened interactions prevent instability as the failure of one species does not easily cascade to others [85]. Additionally, diverse interaction types (a high interaction diversity) are desired for community stability facing disturbance because diverse ecological relationships facilitate a structured and balanced community [87]. Interaction diversity encompasses various relationships (encompassing from competitive to mutualistic) and their respective proportions [87].

7.3.4 Stress-Resource Relationships

Interaction types depend on the coupled effects of stress and resources. Callaway et al. [70] studied plant interactions at high-altitude stress levels to predict interaction types between individuals and proposed the Stress Gradient Hypothesis (SGH). According to SGH, when stress (e.g., high altitude) is severe, cooperation dominates; at lower stress levels, competition prevails. SGH provides a basic framework for predicting interaction types among ecosystem individuals. Piccardi et al. [71] introduced a critical nutrient factor and studied interactions among four bacterial species. They extended the SGH from one dimension (stress alone) to two dimensions: stress → stress + nutrient. They found that strong interactions between bacterial individuals only occurred when stress (e.g., sterilants) and nutrients (e.g., petroleum refining products) were at contrasting levels. Similarly, Ratzke et al. [88] co-cultured eight bacterial species and observed that nutrient-rich environments fostered more competitive interactions. Dominant bacteria grew abundantly, released chemicals, reduced pH, and repressed other bacteria, ultimately reducing biodiversity. These findings provide important implications that the types and levels of stress and resources should not be neglected when aiming at soil microbiome engineering, dynamic prediction, and restoration of soil health.

7.3.5 Predictability

In classical physics, initial conditions and governing equations are enough to predict the future state of an object. A classical case is a pendulum. By knowing the initial conditions (i.e., the length between the fixed point and center of the bob, the initial angle before release, mass, and gravity), the future state of the bob (angle and angular velocity) can be predicted by solving a differential equation. The future state of the object is entirely determined by initial conditions and the governing equation since the system is relatively "simple." However, when the system becomes more complex, such as a double pendulum (consisting of two pendulums, one hanging from the other), the future state becomes hardly predictable [89]. Although the system is still governed by ordinary differential equations, two identical double pendula will exhibit dramatically different trajectories because their future states depend highly on their respective initial conditions. Theoretically, if their initial conditions are the same and down to infinite precision, they yield the same trajectory; however, replicating the infinite precision is practically impossible.

In the context of ecosystems like microbial communities, we can imagine how complex the systems are. To predict the behavior of such systems, we intuitively ask what initial conditions should be considered, what the governing equations are, and how precise the initial conditions can be measured. Answering these three questions is exceptionally challenging, if not impossible. Although stochastic and chaotic processes dominate the dynamics of a microbial community and its collective function, we still have a time window for short-term prediction, which manifests excellent value in facilitating decision making. Typical examples include 7-day weather forecasting that allows preparation for typhoon strikes and evacuating residents in sloping areas vulnerable to heavy rainfall. However, predictions' accuracy diminishes significantly over time due to their sensitivity to initial conditions [90]. In the agricultural context, predicting the one-season behavior of soil microbiome (although challenging) could be extremely helpful in forecasting disease outbreaks, crop productivity, and quality.

7.3.5.1 Concept of Ecological Landscape

Conceptually, the dynamics of complex systems like soil microbiome are guided by a "map" drawn according to the system attractor (imagine a leaf in a river and the river is the attractor guiding the leaf trajectory). A more complex system can have multiple attractors termed attractor manifold (imagine how a golf court settles a ball). An attractor manifold guides a system toward specific outcomes. In ecosystems, the attractor manifold is also known as the ecological landscape (Figure 7.1) [68]. Ecological landscape could also be analogized to the 3D "space" plotted using partial differential equations.

7.3.5.2 Deterministic and Stochastic Processes

Since microbial communities are complex and nonlinear systems, realizing their intrinsic properties is critically important before we attempt to manipulate them. The microbial dynamics is governed by stochastic and deterministic processes [18,95,96] (Figure 7.2). In the stochastic process, environmental factors generate "noise," masking the deterministic process. We can harness the

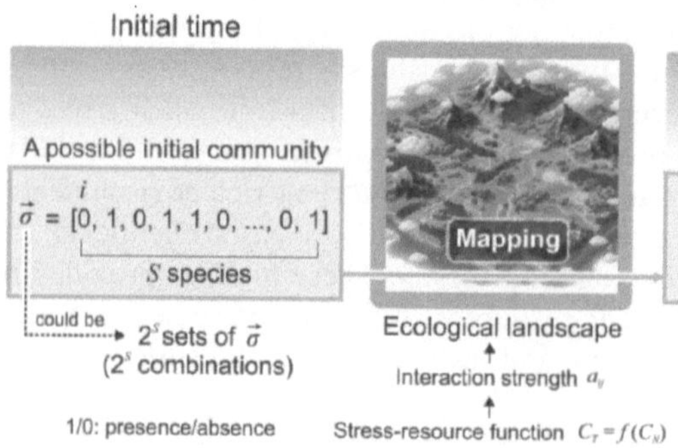

FIGURE 7.1 The concept of ecological landscape in "mapping" the initial community toward the later-time abundance of each species [37,91–94]. The ecological landscape maps the initial species composition to the final steady-state abundances. It captures how individual interactions shape community structures. It also facilitates predicting microbial community structures and functions using limited data, leveraging the sparsity of higher-order interactions. A low level of landscape ruggedness indicates less higher-order interactions, e.g., the growth of a third species depending on the interaction of another two species [92,94]. Stress-resource relationships can also affect interaction strength [71] and influence the ecological landscape.

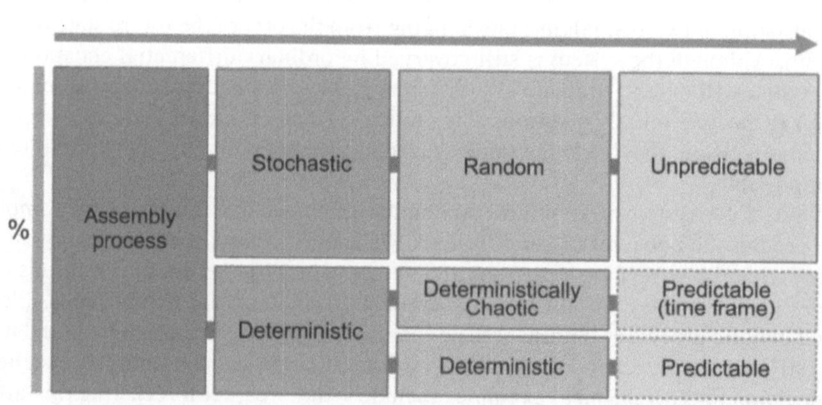

FIGURE 7.2 Theoretical assembly processes and the corresponding predictability. The community assembly process consists of stochastic and deterministic processes. Stochastic processes are random and unpredictable, including dispersal, ecological drift, and chance colonization. The stochastic process may play 60% of the role in governing the microbial assembling process [17]. The deterministic process consists of deterministically chaotic and pure deterministic processes. The deterministically chaotic process is predictable only within specific time frames, while the pure deterministic process is predictable.

deterministic process and, at the same time, recognize the stochastic process and the deterministically chaotic process [97,98]. A deterministic process is one in which specific governing equations can be drawn, although there is still something we cannot predict (see deterministically chaotic process). Collectively, it means there is something we can expect, but some predictions are probably impossible (Figure 7.2).

A deterministically chaotic process refers to a system that exhibits unpredictable and seemingly random behavior, e.g., the abundance changes of individual microbes [72], despite being governed by deterministic rules (e.g., the generalized Lotka–Volterra (GLV) model [99–101], describing abundances of multiple interacting species like preys and predators) without any inherent randomness. Robert May illustrated in the 1970s that even if a simple equation defines the system, the dynamics are unpredictable [102], not because of environmental noise but due to the intrinsic nonlinear dynamics of the system. However, prediction is possible in specific time frames. Although a long-term prediction is practically impossible since the system is susceptible to initial conditions (e.g., initial community context), the system can be relatively ordered and even predictable on short-time scales. Because within a certain time frame, the system is governed by specific rules that determine their near future. However, the accuracy of prediction drastically decreases as time progresses. It raises an interesting question: What is the time frame for us in the context of harnessing core microbiome?

7.3.6 Stability Criteria of a Complex System

Community stability refers to its ability to resist stress and recover after perturbation, which is crucial for the community to sustain functions. Forty years ago, Robert May challenged a central belief in ecology by proving that sufficiently large or complex ecological networks have a probability of persisting close to zero, i.e., chaos with unpredictable future state. His analysis focused on large networks where species interact randomly [102]. However, research has extended May's work to consider well-defined interactions between species, such as predator-prey, mutualistic, or competitive relationships [85]. It shows that reducing complexity favors community stability because it reduces the probability of strong interaction that can lead to cascading failure when certain species are removed. Similarly, higher proportions of weak interaction facilitate stability due to a lower probability of cascading failure [85]. For example, strong mutualistic interactions tend to destabilize ecosystems due to the delicate/fragile balance between costs and benefits, while strong competitive interaction is also destabilizing when one species gains an advantage over others. In addition, diverse interaction types favor stability because a moderate mixture of antagonistic and mutualistic interactions can stabilize population dynamics [87]. Therefore, reducing complexity, weakening interaction, and diversifying interaction types collectively favor community stability.

Environmental stress tends to destabilize ecosystems [103,104] because stress reduces biodiversity and thus the level of redundancy, leading to the loss of species and function. Note that, although stress can reduce biodiversity, a lower biodiversity possibly stabilizes the community due to reduced complexity. Such a contradiction warrants further investigation as maintaining a high biodiversity while reducing complexity seems a dilemma. Importantly, stressful conditions such as pollution can significantly increase the proportion of deterministic processes. For example, the toxicants reduce the biomass and species number of microbes [105], leading to more deterministic microbial assembly processes [17]. Environmental stresses reduce the species pool size to be used to form an initial microbial community. In this regard, referring to Figure 7.2, environmental stresses can reduce the size of the stochastic process and increase the predictability.

On the other hand, decreasing the complexity of the system, such as filtering the core microbiome from the whole population, actually increases system stability, according to Allesina and Tang [85]. Symbiotic fungi associated with plant roots can filter the entire population, leaving a subset of functional bacteria, reducing biodiversity and thus community complexity [35,106,107]. This may effectively establish a stable community across plant species and soil types tackling the context-dependent problem.

In this case, our target seems to be finding a relatively small community (a threshold may be needed to keep necessary levels of biodiversity with enough redundancy) with more weakened interactions among individuals, and providing sufficient desired functions. These fundamental insights help us understand the balance within ecological networks and provide a scientific foundation for microbiome engineering.

7.4 Prediction Approaches

By knowing the principles of microbial dynamics and community stability, we start to model the dynamics, bridging theoretical understanding and practical applications. For engineering and predicting microbial dynamics, mathematical models are the foundation that essentially captures the community's behavior. The existing models are categorized into population-level models (PLMs) and individual-based models (IBMs). IBMs focus more on individual-level characteristics, such as individual heterogeneity and adaptive behavior [108]. In this chapter, we focus on PLMs that emphasize the collective behavior of the microbial community. Statistical and data-driven approaches will also be described.

7.4.1 Ecological Models

7.4.1.1 The Generalized Lotka–Volterra (GLV) Model

The GLV model is a PLM. It is the most straightforward mathematical framework that describes the dynamics of interacting species within a community (Figure 7.3a). The GLV model assumes the system is in a steady state, with constant interaction coefficients, and without spatial structure. It extends the classical Lotka–Volterra model (i.e., predator-prey interaction of two species) and includes multiple species and various interaction types, such as competition, cooperation, and neutralism, but it is less suitable for mutualism (positive feedback leads to unrealistic unlimited growth and instability). The dynamics of species i are described by the following ordinary differential equation.

$$\frac{dN_i}{dt} = N_i\left(r_i + \sum_{j=1}^{n} a_{ij}N_j\right)$$

where N_i and N_j represent the abundances of species i and j, respectively; r_i is the intrinsic growth rate of species i; a_{ij} represents the interaction coefficient between species i and j, with $i=1, 2, 3, \ldots, n$. The term dispersal rate can also be added [72].

The GLV model helps us understand how species coexist and influence each other's abundances. Studies have used the GLV model combined with linear regression (species abundance correlates with concentration of output metabolite) to design synthetic human gut microbiomes for butyrate production. This approach predicted the probability distribution of metabolite concentration from the initial microbial community [30]. Butyrate is a health-relevant metabolite with anti-inflammatory properties. By quantifying microbial interactions, the model can identify and predict high butyrate-producing communities [30]. This approach could be useful when certain microbial compounds are desired, such as antimicrobials to suppress pathogens [52] or auxin [54] to improve root growth.

Real-world interactions involve feedback loops (positive/negative outputs as inputs and so on), time delays, and nonlinearities. Determining accurate and causal interaction coefficients (known as the wiring diagram [23] based on causation not correlation) from empirical data is challenging. When collecting soil samples (typically 0.5–1 g with 3D structure), we do not know whether the bacteria are genuinely interacting. Also, the commonly used approach targeting the 16S rRNA gene obtains compositional data and possibly from inactive cells. It leads to the challenge of reconstructing the interaction strength and interaction types, not to say the actual wiring diagram, which is a causal network rather than a correlation network (e.g., the co-occurrence network). For example, a correlation network would probably not capture the causation of antimicrobials in suppressing other bacteria [110,111]. Also, the model's predictions depend not only on initial community context (internal) but also on environmental factors (external). For example, soil conditions, nutrient availability, and other external factors influence microbial interactions. In summary, the GLV model provides valuable insights into microbial community dynamics, but its simplicity and parameter estimation challenges should be considered when applying it to real-world ecosystems.

In addition to pairwise interactions in the GLV model, if we include higher-order interactions (e.g., the interaction of two species affects the third species and so on), a higher-order interaction model can be drawn (Figure 7.3b). Such a model is very complicated and renders the parameterization extremely difficult [37,112,113]. Studies simulated the higher-order interactions using a relatively smaller community containing 3 [113], 6 [114], or 30 [112] species. But in the real-world context, the number of genuinely (causally) interacting species can be much larger than these, i.e., probably more than a hundred in the soil context [14], making the higher-order interaction model too complicated. Then, deep-learning approaches have been used [115–117]. However, the expense is considerable, and the results are challenging to interpret. Later, researchers borrowed the idea from signal processing that a sparse representation can be obtained from a complex signal. Then, the sparse representation is used to reconstruct the entire signal. Following this, a study demonstrated a way to capture the sparse representation of a higher-order interacting community and then reconstruct and predict the whole community [37]. The method was further tested using other communities containing 5–25 species with known dynamics. The predictions were reasonably good, although the sizes of communities are still relatively small [37]. Finding a sparse representation of the soil microbial communities may be a considerable approach for microbial dynamics prediction facilitating soil health management.

7.4.1.2 Microbial Consumer-Resource Model (MiCRM)

The GLV model considers species only, while the microbial consumer-resource model (MiCRM), including the mediator-explicit model, considers not only species abundance but also the production and release of chemicals by microbial species and the input of external resources into the system (Figure 7.3c

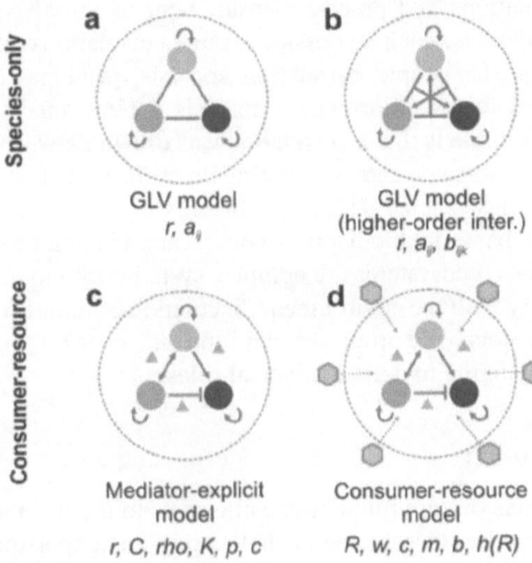

FIGURE 7.3 Schematic illustration of population-level mathematical models that capture microbial community dynamics. (a) The species-only model, such as the GLV model. Only pairwise interactions were considered. Filled circles indicate species. For simplicity, only three species are drawn. Arrows and lines denote interactions. Curved arrows suggest intrinsic properties, such as the intrinsic growth rate of a species (r). a_{ij} is the interaction matrix. (b) The species-only model that considers higher-order interaction. b_{ijk} is the higher-order interaction matrix in that the growth of a third species depends on the interaction of another two species. Due to the complexity of the higher-order interaction model, the method of sparse representation of complexity has been developed [37]. (c) The mediator-explicit model is a special case of microbial consumer-resource (MiCRM). It considers external resources in excess. Triangles are metabolites that facilitate or inhibit the growth of other species. C, resource concentration; rho, strength of interaction (either positive or negative); K, saturation concentration; p, metabolite production rate; c, maximum rate of metabolite consumption. (d) The MiCRM considers metabolites and external resources. Gray hexagons are external resources. R, abundance of resource; w, one unit of resource for the species; c, the capture rate of the species for resource; m, the minimum maintenance energy required by the species; b, a factor converting resource excessive into growth rate; $h(R)$, the intrinsic resource dynamics representing self-inhibition of resource by itself. For IBMs, refer to Refs. [108,109]. More complex models, such as the DEBmicroTrait model, integrate energy budget, PLMs, IBMs, and microbial genome inferred traits, allowing predictions of emergent behavior [26].

and d). MiCRM depicts the dynamics of species *i* using coupled ordinary differential equations dN_i/dt and dR_a/dt (Ni is abundance of species *i* and R_a is the abundance of resource *a*. MiCRM describes the interactions between microbial species within an ecosystem where externally supplied resources are available, either in excess or limited [118,119]. The consumption of these resources can lead to the production of metabolic byproducts, which may facilitate cross-feeding or inhibition among microbes.

MiCRM allows researchers to explore how mechanistic assumptions at the microbial scale influence community-level patterns. By considering factors such as resource availability, cross-feeding, and colonization, MiCRM can qualitatively reproduce observed patterns in survey data, including changes in microbial community composition across environmental gradients. MiCRM has been applied to large-scale microbial surveys, such as EMP and the HMP [120]. Researchers use synthetic data from MiCRM to understand how environmental conditions and taxonomic structure impact observed patterns. For example, harsh environments may affect community composition due to the energetic cost of survival, while environmental filtering shapes competition dynamics.

However, MiCRM simplifies complex microbial ecosystems by assuming homogeneous environments and specific interactions. Real-world microbial communities involve many species and intricate ecological processes. Also, such a consumer-resource model requires accurate understanding of what resources and metabolites are essential. In reality, resource and metabolites are influenced by a wide range of factors, which may not be fully captured by the model. Consequently, MiCRM's predictions are constrained by the accuracy of its assumptions and the available data [120].

7.4.2 Statistical Approach

Regardless of governing equations and precise measurement of initial conditions, researchers could obtain spatial and temporal data as much as possible to find correlations among variables. Data-driven statistical approaches include the simple correlation analysis, principal component analysis (PCA), redundancy analysis (RDA), canonical-correlation analysis (CCA), and structural equation modeling (SEM) [121,122]. A simple example is that a correlation was drawn between soil pH value and bacterial species richness using continental-scale data, suggesting that soils with neutral pH harbor bacterial communities with the highest species richness [123]. If achieving high levels of species richness is desired, aiming for neutral soil pH for bacterial biodiversity conservation is practicable (but also note the biodiversity paradox) [124]. However, ecosystems are complex systems with many intermingling factors, and research outcomes always vary with the environment (fluctuation of initial conditions, let alone precisely identical). The unpredictable stochastic processes can influence over 60% of the microbial community structure [17], making it challenging to derive universal rules.

7.4.3 Data-Driven Approach

Identifying the keystone species or core microbiome allows us to focus on a substantially smaller community. However, finding the core microbiome is challenging and experimentally demanding. Using a data-driven approach based on deep learning, researchers can now learn the assembly rules of microbial communities. The trained deep-learning model can quantify the keystoneness of each species in a microbiome for a habitat [125]. In a more complex context that requires the prediction of plant performance, integration of the omics data from plants and microbes attracts increasing attention, and it requires strong abilities in handling large datasets [20,125].

In terms of studying microbial communities and their interactions involving many species, researchers use Empirical Dynamic Modeling (EDM) to reconstruct attractor manifolds from time-series data [126–128]. EDM does not assume any specific set of equations governing the system (see Section 7.4.1. Ecological Models). Instead, it recovers the system dynamics directly from observed time-series data. EDM reconstructs the underlying state space of a system using lagged coordinate embedding according to the Takens' theorem. The collection of these time-series trajectories forms an attractor manifold. It describes how variables (e.g., genes, microbial abundances) relate over time. EDM acknowledges that relationships among interacting variables can change with different states of the system (state-dependent), meaning that the interaction type and strength change over time. EDM can be applied to microbial

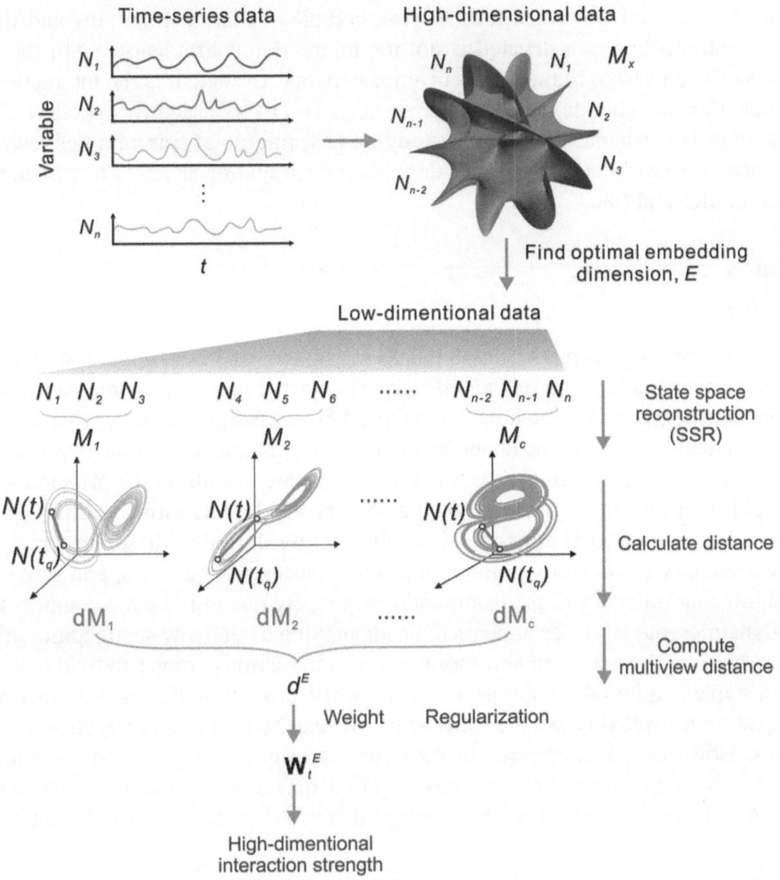

FIGURE 7.4 Schematic illustration of the EDM MDR S-map method. EDM MDR (multiview distance regularized) S-map method to study the high-dimensional interactions among multiple variables (e.g., species) (modified from Ref. [128]). Suppose a microbial community contains *n* species. The species abundances (N_1, N_2, N_3, ..., N_n) change over time and form time-series data, which can be used to reconstruct an overall manifold (M_x) with *n* dimensions. When the data are low-dimensional, such as containing three species, the classic pipeline of EDM can be used to study the system. However, when the data are high-dimensional, such as containing 100 species, the data become too sparse to depict correct neighboring relationships (Euclidean distance) among data points; but capturing correct distances is essentially required when using EDM. So, the high-dimensional M_x needs to be divided into low-dimensional sub-manifolds (M_1, M_2, ..., M_c) after finding the optimal embedding dimension (*E*). *E* is determined by finding the maximum correlation coefficient (namely predictive skill) between forecasts and observations. Here, we suppose $E=3$, which is much smaller than *n*. Now, for each sub-manifold, the distance between $N(t)$ and $N(t_q)$ can be calculated for subsequent analyses.

community data to understand their behavior over time (Figure 7.4). It helps identify patterns, interactions, and causal relationships among variables (e.g., how environmental factors cause the changes in species abundance). Analyzing attractor manifolds gives us insights into community stability, resilience, and response to environmental changes.

However, reconstructing the state space of a system with many species using traditional EDM is challenging due to high dimensionality (curse of dimensionality), data sparsity, complex interactions, noise, and computational limitations. These factors make it difficult to accurately capture and model the dynamics of large ecological systems [128]. A recent model development, the EDM MDR (multiview distance regularized) S-map method (Figure 7.4), can handle many interacting species and reconstruct the interaction network [128]. This approach effectively manages high-dimensional data, allowing for identification of causal relationships and interaction strengths among species. By leveraging advanced computational algorithms, the EDM MDR S-map method overcomes the limitations of traditional models, providing a robust framework for studying complex ecological systems and their underlying interactions. The interaction reconstructed is not the interaction coefficients (a_{ij}) in the GLV model or the correlation coefficients (rho) of the co-occurrence network. Instead, it is the interaction Jacobian that quantifies the net effects of abundance changes in species *j* on the consecutive species *i*. Capturing these complex interactions is particularly useful for studying how microbial communities respond to external disturbances, such as soil pollution and land degradation, providing insights to facilitate resilient soil microbial communities and functions.

7.5 Conclusion

Soil microbiome engineering for plant growth has a long history and is now increasingly focused on soil health restoration, maintenance, and management, particularly under environmental disturbances, such as soil pollution, land degradation, and climate change. This necessitates a comprehensive understanding of which microbial members should be included and how they interact over time, to predict future events. The soil microbiome in a specific context is inherently complex. It follows the principles of chaos, making it unpredictable to a certain extent but allowing for reasonable forecasting within a given timeframe.

Our understanding of microbial dynamics is evolving, revealing that these dynamics are influenced by the initial community composition, interaction range (species-dependent), and interaction strength, all under the significant influence of environmental factors. By recognizing the complexity of microbial assembly and dynamics, our next step appears to be identifying a relatively small community (potentially requiring a threshold) with weakened and more diverse interactions among individuals. This approach aims to create a stable community under perturbation while providing the desired functions.

While we strive for universality, we may still be challenged by the context-dependent nature of microbial interactions. However, advancements in machine learning techniques, the rapid accumulation of high-quality datasets, and enhanced computational power may usher in a new era where context dependency is less of a concern. This aligns with the adage that "change is the only constant."

Acknowledgments

We gratefully acknowledge the support from the National Natural Science Foundation of China (42322711, 42077298, 42477118).

REFERENCES

1. FAO and UNEP (Food and Agriculture Organization of the United Nations and United Nations Environment Programme). Global assessment of soil pollution: Report https://www.fao.org/documents/card/en/c/cb4894en (accessed **Jun 10**, 2021). https://doi.org/10.4060/cb4894en
2. FAO and UNEP (Food and Agriculture Organization of the United Nations and United Nations Environment Programme). Global assessment of soil pollution: Summary for policymakers https://www.fao.org/documents/card/en/c/cb4827en/ (accessed **Sep 23**, 2021). https://doi.org/10.4060/cb4827en
3. Li, X.; Shen, X.; Jiang, W.; Xi, Y.; Li, S. Comprehensive Review of Emerging Contaminants: Detection Technologies, Environmental Impact, and Management Strategies. *Ecotoxicol. Environ. Saf.*, **2024**, *278*, 116420. https://doi.org/10.1016/j.ecoenv.2024.116420
4. IPBES. *The IPBES Assessment Report on Land Degradation and Restoration; Zenodo*, **2018**. https://doi.org/10.5281/zenodo.3237393
5. D'Odorico, P.; Ravi, S. Chapter 11- Land degradation and environmental change. In *Biological and Environmental Hazards, Risks, and Disasters*; Shroder, J. F., Sivanpillai, R., Eds.; Academic Press: Boston, FL, **2016**; pp 219–227. https://doi.org/10.1016/B978-0-12-394847-2.00014-0
6. Kapur, S.; Akça, E. Soil Degradation: Global Assessment. In *Landscape and Land Capacity*; Wang, Y., Eds.; CRC Press: Boston, FL, **2020**.

7. Schneider, R.; Morreale, S. Soil Health Restoration to Address Food and Water Security. In *Life on Land*; Leal Filho, W., Azul, A. M., Brandli, L., Lange Salvia, A., Wall, T., Eds.; Springer International Publishing: Cham, **2021**; pp 951–960. https://doi.org/10.1007/978-3-319-95981-8_147
8. Weil, R. R.; Brady, N. C. *The Nature and Properties of Soils*, 15th ed.; Pearson Education Limited: England, **2016**.
9. Wilhelm, R. C.; Amsili, J. P.; Kurtz, K. S. M.; van Es, H. M.; Buckley, D. H. Ecological Insights into Soil Health According to the Genomic Traits and Environment-Wide Associations of Bacteria in Agricultural Soils. *ISME Commun.*, **2023**, *3* (1), 1–11. https://doi.org/10.1038/s43705-022-00209-1
10. Douglas, G. M.; Beiko, R. G.; Langille, M. G. I. Predicting the Functional Potential of the Microbiome from Marker Genes Using PICRUSt. In *Microbiome Analysis: Methods and Protocols*; Beiko, R. G., Hsiao, W., Parkinson, J., Eds.; Methods in Molecular Biology; Springer: New York, **2018**; pp 169–177. https://doi.org/10.1007/978-1-4939-8728-3_11
11. Langille, M. G. I.; Zaneveld, J.; Caporaso, J. G.; McDonald, D.; Knights, D.; Reyes, J. A.; Clemente, J. C.; Burkepile, D. E.; Vega Thurber, R. L.; Knight, R.; et al. Predictive Functional Profiling of Microbial Communities Using 16S rRNA Marker Gene Sequences. *Nat. Biotechnol.*, **2013**, *31* (9), 814–821. https://doi.org/10.1038/nbt.2676
12. Louca, S.; Parfrey, L. W.; Doebeli, M. Decoupling Function and Taxonomy in the Global Ocean Microbiome. *Science*, **2016**, *353* (6305), 1272–1277. https://doi.org/10.1126/science.aaf4507
13. Wilhelm, R. C.; van Es, H. M.; Buckley, D. H. Predicting Measures of Soil Health Using the Microbiome and Supervised Machine Learning. *Soil Biol. Biochem.*, **2022**, *164*, 108472. https://doi.org/10.1016/j.soilbio.2021.108472
14. Raynaud, X.; Nunan, N. Spatial Ecology of Bacteria at the Microscale in Soil. *PLoS One*, **2014**, *9* (1), e87217. https://doi.org/10.1371/journal.pone.0087217
15. Torsvik, V.; Øvreås, L. Microbial Diversity and Function in Soil: From Genes to Ecosystems. *Curr. Opin. Microbiol.*, **2002**, *5* (3), 240–245. https://doi.org/10.1016/S1369-5274(02)00324-7
16. Fierer, N.; Wood, S. A.; Bueno de Mesquita, C. P. How Microbes Can, and Cannot, Be Used to Assess Soil Health. *Soil Biol. Biochem.*, **2021**, *153*, 108111. https://doi.org/10.1016/j.soilbio.2020.108111
17. Ning, D.; Wang, Y.; Fan, Y.; Wang, J.; Van Nostrand, J. D.; Wu, L.; Zhang, P.; Curtis, D. J.; Tian, R.; Lui, L.; et al. Environmental Stress Mediates Groundwater Microbial Community Assembly. *Nat. Microbiol.*, **2024**, *9* (2), 490–501. https://doi.org/10.1038/s41564-023-01573-x
18. Zhou, J.; Ning, D. Stochastic Community Assembly: Does It Matter in Microbial Ecology? *Microbiol. Mol. Biol. Rev.*, **2017**, *81* (4), e00002-17. https://doi.org/10.1128/MMBR.00002-17
19. Jansson, J. K.; McClure, R.; Egbert, R. G. Soil Microbiome Engineering for Sustainability in a Changing Environment. *Nat. Biotechnol.*, **2023**, *41* (12), 1716–1728. https://doi.org/10.1038/s41587-023-01932-3
20. Berruto, C. A.; Demirer, G. S. Engineering Agricultural Soil Microbiomes and Predicting Plant Phenotypes. *Trends Microbiol.*, **2024**, *32* (9), 858–873. https://doi.org/10.1016/j.tim.2024.02.003
21. Albright, M. B. N.; Louca, S.; Winkler, D. E.; Feeser, K. L.; Haig, S.-J.; Whiteson, K. L.; Emerson, J. B.; Dunbar, J. Solutions in Microbiome Engineering: Prioritizing Barriers to Organism Establishment. *ISME J.*, **2022**, *16* (2), 331–338. https://doi.org/10.1038/s41396-021-01088-5
22. Larsen, P. E.; Field, D.; Gilbert, J. A. Predicting Bacterial Community Assemblages Using an Artificial Neural Network Approach. *Nat. Methods*, **2012**, *9* (6), 621–625. https://doi.org/10.1038/nmeth.1975
23. Liu, Y.-Y. Controlling the Human Microbiome. *Cell Syst.*, **2023**, *14* (2), 135–159. https://doi.org/10.1016/j.cels.2022.12.010
24. Liu, Y.-Y.; Slotine, J.-J.; Barabási, A.-L. Observability of Complex Systems. *Proc. Natl. Acad. Sci. U. S. A.*, **2013**, *110* (7), 2460–2465. https://doi.org/10.1073/pnas.1215508110
25. D'Souza, R. M.; di Bernardo, M.; Liu, Y.-Y. Controlling Complex Networks with Complex Nodes. *Nat. Rev. Phys.*, **2023**, *5* (4), 250–262. https://doi.org/10.1038/s42254-023-00566-3
26. Marschmann, G. L.; Tang, J.; Zhalnina, K.; Karaoz, U.; Cho, H.; Le, B.; Pett-Ridge, J.; Brodie, E. L. Predictions of Rhizosphere Microbiome Dynamics with a Genome-Informed and Trait-Based Energy Budget Model. *Nat. Microbiol.*, **2024**, *9*, 1–13. https://doi.org/10.1038/s41564-023-01582-w
27. Gilbert, J. A.; Jansson, J. K.; Knight, R. The Earth Microbiome Project: Successes and Aspirations. *BMC Biol.*, **2014**, *12*, 69. https://doi.org/10.1186/s12915-014-0069-1
28. Turnbaugh, P. J.; Ley, R. E.; Hamady, M.; Fraser-Liggett, C. M.; Knight, R.; Gordon, J. I. The Human Microbiome Project. *Nature*, **2007**, *449* (7164), 804–810. https://doi.org/10.1038/nature06244

29. Neu, A. T.; Allen, E. E.; Roy, K. Defining and Quantifying the Core Microbiome: Challenges and Prospects. *Proc. Natl. Acad. Sci. U. S. A.*, **2021**, *118* (51), e2104429118. https://doi.org/10.1073/pnas.2104429118
30. Clark, R. L.; Connors, B. M.; Stevenson, D. M.; Hromada, S. E.; Hamilton, J. J.; Amador-Noguez, D.; Venturelli, O. S. Design of Synthetic Human Gut Microbiome Assembly and Butyrate Production. *Nat. Commun.*, **2021**, *12* (1), 3254. https://doi.org/10.1038/s41467-021-22938-y
31. Flint, H. J.; Scott, K. P.; Louis, P.; Duncan, S. H. The Role of the Gut Microbiota in Nutrition and Health. *Nat. Rev. Gastroenterol. Hepatol.*, **2012**, *9* (10), 577–589. https://doi.org/10.1038/nrgastro.2012.156
32. Grieneisen, L.; Dasari, M.; Gould, T. J.; Björk, J. R.; Grenier, J.-C.; Yotpva, V.; Jansen, D.; Gottel, N.; Gordon, J. B.; Learn, N. H.; et al. Gut Microbiome Heritability is Nearly Universal but Environmentally Contingent. *Science*, **2021**, *373* (6551), 181–186. https://doi.org/10.1126/science.aba5483
33. Toju, H.; Peay, K. G.; Yamamichi, M.; Narisawa, K.; Hiruma, K.; Naito, K.; Fukuda, S.; Ushio, M.; Nakaoka, S.; Onoda, Y.; et al. Core Microbiomes for Sustainable Agroecosystems. *Nat. Plants*, **2018**, *4* (5), 247–257. https://doi.org/10.1038/s41477-018-0139-4
34. Bashan, A.; Gibson, T. E.; Friedman, J.; Carey, V. J.; Weiss, S. T.; Hohmann, E. L.; Liu, Y.-Y. Universality of Human Microbial Dynamics. *Nature*, **2016**, *534* (7606), 259–262. https://doi.org/10.1038/nature18301
35. Emmett, B. D.; Lévesque-Tremblay, V.; Harrison, M. J. Conserved and Reproducible Bacterial Communities Associate with Extraradical Hyphae of Arbuscular Mycorrhizal Fungi. *ISME J.*, **2021**, *15*, 2276–2288. https://doi.org/10.1038/s41396-021-00920-2
36. Wang, L.; George, T. S.; Feng, G. Concepts and Consequences of the Hyphosphere Core Microbiome for Arbuscular Mycorrhizal Fungal Fitness and Function. *New Phytol.*, **2023**, *242*, 1529–1533. https://doi.org/10.1111/nph.19396
37. Arya, S.; George, A. B.; O'Dwyer, J. P. Sparsity of Higher-Order Landscape Interactions Enables Learning and Prediction for Microbiomes. *Proc. Natl. Acad. Sci. U. S. A.*, **2023**, *120* (48), e2307313120. https://doi.org/10.1073/pnas.2307313120
38. Lehmann, J.; Bossio, D. A.; Kögel-Knabner, I.; Rillig, M. C. The Concept and Future Prospects of Soil Health. *Nat. Rev. Earth Environ.*, **2020**, *1* (10), 544–553. https://doi.org/10.1038/s43017-020-0080-8
39. Zheng, H.; Wang, L.; Peng, W.; Zhang, C.; Li, C.; Robinson, B. E.; Wu, X.; Kong, L.; Li, R.; Xiao, Y.; et al. Realizing the Values of Natural Capital for Inclusive, Sustainable Development: Informing China's New Ecological Development Strategy. *Proc. Natl. Acad. Sci. U. S. A.*, **2019**, *116*, 201819501. https://doi.org/10.1073/pnas.1819501116
40. Heitschmidt, R. K., Stuth, J. W. (Eds). *Grazing Management: An Ecological Perspective*. Timber Press, Incorporated: Portland, OR, **1991**.
41. Song, S.; Zhu, J.; Zheng, T.; Tang, Z.; Zhang, F.; Ji, C.; Shen, Z.; Zhu, J. Long-Term Grazing Exclusion Reduces Species Diversity but Increases Community Heterogeneity in an Alpine Grassland. *Front. Ecol. Evol.*, **2020**, *8*, 66. https://doi.org/10.3389/fevo.2020.00066
42. Robles, M. D.; Burke, I. C. Legume, Grass, and Conservation Reserve Program Effects on Soil Organic Matter Recovery. *Ecol. Appl.*, **1997**, *7* (2), 345–357. https://doi.org/10.1890/1051-0761(1997)007[0345:LGACRP]2.0.CO;2
43. Balota, E. L.; Machineski, O.; Hamid, K. I. A.; Yada, I. F. U.; Barbosa, G. M. C.; Nakatani, A. S.; Coyne, M. S. Soil Microbial Properties after Long-Term Swine Slurry Application to Conventional and No-Tillage Systems in Brazil. *Sci. Total Environ.*, **2014**, *490*, 397–404. https://doi.org/10.1016/j.scitotenv.2014.05.019
44. Blanco-Moure, N.; Angurel, L. A.; Moret-Fernández, D.; López, M. V. Tensile Strength and Organic Carbon of Soil Aggregates under Long-Term No Tillage in Semiarid Aragon (NE Spain). *Geoderma*, **2012**, *189–190*, 423–430. https://doi.org/10.1016/j.geoderma.2012.05.015
45. Dai, J.; Hu, J.; Zhu, A.; Bai, J.; Wang, J.; Lin, X. No Tillage Enhances Arbuscular Mycorrhizal Fungal Population, Glomalin-Related Soil Protein Content, and Organic Carbon Accumulation in Soil Macroaggregates. *J. Soils Sediments*, **2015**, *15* (5), 1055–1062. https://doi.org/10.1007/s11368-015-1091-9
46. Kou, X.; Ma, N.; Zhang, X.; Xie, H.; Zhang, X.; Wu, Z.; Liang, W.; Li, Q.; Ferris, H. Frequency of Stover Mulching but Not Amount Regulates the Decomposition Pathways of Soil Micro-Foodwebs in a No-Tillage System. *Soil Biol. Biochem.*, **2020**, *144*, 107789. https://doi.org/10.1016/j.soilbio.2020.107789
47. Troy, S. M.; Lawlor, P. G.; O' Flynn, C. J.; Healy, M. G. Impact of Biochar Addition to Soil on Greenhouse Gas Emissions Following Pig Manure Application. *Soil Biol. Biochem.*, **2013**, *60*, 173–181. https://doi.org/10.1016/j.soilbio.2013.01.019

48. Finney, D. M.; Kaye, J. P. Functional Diversity in Cover Crop Polycultures Increases Multifunctionality of an Agricultural System. *J. Appl. Ecol.*, **2016**, *54* (2), 509–517. https://doi.org/10.1111/1365-2664.12765
49. Garland, G.; Edlinger, A.; Banerjee, S.; Degrune, F.; García-Palacios, P.; Pescador, D. S.; Herzog, C.; Romdhane, S.; Saghai, A.; Spor, A.; et al. Crop Cover Is More Important than Rotational Diversity for Soil Multifunctionality and Cereal Yields in European Cropping Systems. *Nat. Food*, **2021**, *2* (1), 28–37. https://doi.org/10.1038/s43016-020-00210-8
50. Masson-Boivin, C.; Sachs, J. L. Symbiotic Nitrogen Fixation by Rhizobia — the Roots of a Success Story. *Curr. Opin. Plant Biol.*, **2018**, *44*, 7–15. https://doi.org/10.1016/j.pbi.2017.12.001
51. Smith, S. E.; Read, D. J. *Mycorrhizal Symbiosis*, 3rd ed.; Academic Press: London, UK, **2008**.
52. Bais, H. P.; Fall, R.; Vivanco, J. M. Biocontrol of *Bacillus Subtilis* against Infection of *Arabidopsis* Roots by *Pseudomonas Syringae* is Facilitated by Biofilm Formation and Surfactin Production. *Plant Physiol.*, **2004**, *134* (1), 307–319. https://doi.org/10.1104/pp.103.028712
53. Tao, C.; Wang, Z.; Liu, S.; Lv, N.; Deng, X.; Xiong, W.; Shen, Z.; Zhang, N.; Geisen, S.; Li, R.; et al. Additive Fungal Interactions Drive Biocontrol of Fusarium Wilt Disease. *New Phytol.*, **2023**, *238* (3), 1198–1214. https://doi.org/10.1111/nph.18793
54. Chieb, M.; Gachomo, E. W. The Role of Plant Growth Promoting Rhizobacteria in Plant Drought Stress Responses. *BMC Plant Biol.*, **2023**, *23* (1), 407. https://doi.org/10.1186/s12870-023-04403-8
55. Bankhead, S. B.; Landa, B. B.; Lutton, E.; Weller, D. M.; Gardener, B. B. M. Minimal Changes in Rhizobacterial Population Structure Following Root Colonization by Wild Type and Transgenic Biocontrol Strains. *FEMS Microbiol. Ecol.*, **2004**, *49* (2), 307–318. https://doi.org/10.1016/j.femsec.2004.04.005
56. Grove, S.; Haubensak, K. A.; Gehring, C.; Parker, I. M. Mycorrhizae, Invasions, and the Temporal Dynamics of Mutualism Disruption. *J. Ecol.*, **2017**, *105* (6), 1496–1508. https://doi.org/10.1111/1365-2745.12853
57. Liu, Y.-X.; Qin, Y.; Bai, Y. Reductionist Synthetic Community Approaches in Root Microbiome Research. *Curr. Opin. Microbiol.*, **2019**, *49*, 97–102. https://doi.org/10.1016/j.mib.2019.10.010
58. Zhang, J.; Liu, Y.-X.; Guo, X.; Qin, Y.; Garrido-Oter, R.; Schulze-Lefert, P.; Bai, Y. High-Throughput Cultivation and Identification of Bacteria from the Plant Root Microbiota. *Nat. Protoc.*, **2021**, *16*, 988–1012. https://doi.org/10.1038/s41596-020-00444-7
59. Bai, Y.; Müller, D. B.; Srinivas, G.; Garrido-Oter, R.; Potthoff, E.; Rott, M.; Dombrowski, N.; Münch, P. C.; Spaepen, S.; Remus-Emsermann, M.; et al. Functional Overlap of the *Arabidopsis* Leaf and Root Microbiota. *Nature*, **2015**, *528* (7582), 364. https://doi.org/10.1038/nature16192
60. Niu, B.; Paulson, J. N.; Zheng, X.; Kolter, R. Simplified and Representative Bacterial Community of Maize Roots. *Proc. Natl. Acad. Sci. U. S. A.*, **2017**, *114* (12), E2450–E2459. https://doi.org/10.1073/pnas.1616148114
61. Finkel, O. M.; Salas-González, I.; Castrillo, G.; Conway, J. M.; Law, T. F.; Teixeira, P. J. P. L.; Wilson, E. D.; Fitzpatrick, C. R.; Jones, C. D.; Dangl, J. L. A Single Bacterial Genus Maintains Root Growth in a Complex Microbiome. *Nature*, **2020**, *587*, 103–108. https://doi.org/10.1038/s41586-020-2778-7
62. Salas-González, I.; Reyt, G.; Flis, P.; Custódio, V.; Gopaulchan, D.; Bakhoum, N.; Dew, T. P.; Suresh, K.; Franke, R. B.; Dangl, J. L.; et al. Coordination between Microbiota and Root Endodermis Supports Plant Mineral Nutrient Homeostasis. *Science*, **2020**, *371* (6525), abd0695. https://doi.org/10.1126/science.abd0695
63. De Roy, K.; Marzorati, M.; Van den Abbeele, P.; Van de Wiele, T.; Boon, N. Synthetic Microbial Ecosystems: An Exciting Tool to Understand and Apply Microbial Communities. *Environ. Microbiol.*, **2014**, *16* (6), 1472–1481. https://doi.org/10.1111/1462-2920.12343
64. Parades-Aguilar, J.; Calderon, K.; Agustin-Salazar, S.; Cerruti, P.; Ambrogi, V.; Gamez-Meza, N.; Medina-Juarez, L. A. Isolation and Identification of Metallotolerant Bacteria with a Potential Biotechnological Application. *Sci. Rep.*, **2024**, *14* (1), 3663. https://doi.org/10.1038/s41598-024-54090-0
65. Mendez, M. O.; Neilson, J. W.; Maier, R. M. Characterization of a Bacterial Community in an Abandoned Semiarid Lead-Zinc Mine Tailing Site. *Appl. Environ. Microbiol.*, **2008**, *74* (12), 3899–3907. https://doi.org/10.1128/AEM.02883-07
66. Lutz, S.; Bodenhausen, N.; Hess, J.; Valzano-Held, A.; Waelchli, J.; Deslandes-Hérold, G.; Schlaeppi, K.; van der Heijden, M. G. A. Soil Microbiome Indicators Can Predict Crop Growth Response to Large-Scale Inoculation with Arbuscular Mycorrhizal Fungi. *Nat. Microbiol.*, **2023**, *8* (12), 2277–2289. https://doi.org/10.1038/s41564-023-01520-w

67. Zhang, Z.; van Kleunen, M.; Becks, L.; Thakur, M. P. Towards a General Understanding of Bacterial Interactions. *Trends Microbiol.*, **2020**, *28* (10), 783–785. https://doi.org/10.1016/j.tim.2020.05.010
68. George, A. B.; Korolev, K. S. Ecological Landscapes Guide the Assembly of Optimal Microbial Communities. *PLoS Comput. Biol.*, **2023**, *19* (1), e1010570. https://doi.org/10.1371/journal.pcbi.1010570
69. Fukami, T. Historical Contingency in Community Assembly: Integrating Niches, Species Pools, and Priority Effects. *Annu. Rev. Ecol. Evol. Syst.*, **2015**, *46*, 1–23. https://doi.org/10.1146/annurev-ecolsys-110411-160340
70. Callaway, R. M.; Brooker, R. W.; Choler, P.; Kikvidze, Z.; Lortie, C. J.; Michalet, R.; Paolini, L.; Pugnaire, F. I.; Newingham, B.; Aschehoug, E. T.; et al. Positive Interactions among Alpine Plants Increase with Stress. *Nature*, **2002**, *417* (6891), 844–848. https://doi.org/10.1038/nature00812
71. Piccardi, P.; Vessman, B.; Mitri, S. Toxicity Drives Facilitation between 4 Bacterial Species. *Proc. Natl. Acad. Sci. U. S. A.*, **2019**, *116* (32), 15979–15984. https://doi.org/10.1073/pnas.1906172116
72. Hu, J.; Amor, D. R.; Barbier, M.; Bunin, G.; Gore, J. Emergent Phases of Ecological Diversity and Dynamics Mapped in Microcosms. *Science*, **2022**, *378* (6615), 85–89. https://doi.org/10.1126/science.abm7841
73. Mallmin, E.; Traulsen, A.; De Monte, S. Chaotic Turnover of Rare and Abundant Species in a Strongly Interacting Model Community. *Proc. Natl. Acad. Sci. U. S. A.*, **2024**, *121* (11), e2312822121. https://doi.org/10.1073/pnas.2312822121
74. Kehe, J.; Ortiz, A.; Kulesa, A.; Gore, J.; Blainey, P. C.; Friedman, J. Positive Interactions Are Common among Culturable Bacteria. *Sci. Adv.*, **2021**, *7* (45), eabi7159. https://doi.org/10.1126/sciadv.abi7159
75. Kehe, J.; Kulesa, A.; Ortiz, A.; Ackerman, C. M.; Thakku, S. G.; Sellers, D.; Kuehn, S.; Gore, J.; Friedman, J.; Blainey, P. C. Massively Parallel Screening of Synthetic Microbial Communities. *Proc. Natl. Acad. Sci. U. S. A.*, **2019**, *116* (26), 12804–12809. https://doi.org/10.1073/pnas.1900102116
76. Engelhardt, I. C.; Patko, D.; Liu, Y.; Mimault, M.; de las Heras Martinez, G.; George, T. S.; MacDonald, M.; Ptashnyk, M.; Sukhodub, T.; Stanley-Wall, N. R.; et al. Novel Form of Collective Movement by Soil Bacteria. *ISME J.*, **2022**, *16* (10), 2337–2347. https://doi.org/10.1038/s41396-022-01277-w
77. Yang, P.; van Elsas, J. D. Mechanisms and Ecological Implications of the Movement of Bacteria in Soil. *Appl. Soil Ecol.*, **2018**, *129*, 112–120. https://doi.org/10.1016/j.apsoil.2018.04.014
78. Bhattacharjee, T.; Datta, S. S. Bacterial Hopping and Trapping in Porous Media. *Nat. Commun.*, **2019**, *10* (1), 2075. https://doi.org/10.1038/s41467-019-10115-1
79. Dechesne, A.; Wang, G.; Gülez, G.; Or, D.; Smets, B. F. Hydration-Controlled Bacterial Motility and Dispersal on Surfaces. *Proc. Natl. Acad. Sci. U. S. A.*, **2010**, *107* (32), 14369–14372. https://doi.org/10.1073/pnas.1008392107
80. Schäfer, A.; Ustohal, P.; Harms, H.; Stauffer, F.; Dracos, T.; Zehnder, A. J. B. Transport of Bacteria in Unsaturated Porous Media. *J. Contam. Hydrol.*, **1998**, *33* (1), 149–169. https://doi.org/10.1016/S0169-7722(98)00069-2
81. Turnbull, G. A.; Morgan, J. A. W.; Whipps, J. M.; Saunders, J. R. The Role of Bacterial Motility in the Survival and Spread of *Pseudomonas Fluorescens* in Soil and in the Attachment and Colonisation of Wheat Roots. *FEMS Microbiol. Ecol.*, **2001**, *36* (1), 21–31. https://doi.org/10.1111/j.1574-6941.2001.tb00822.x
82. Jansa, J.; Hodge, A. Swimming, Gliding, or Hyphal Riding? On Microbial Migration along the Arbuscular Mycorrhizal Hyphal Highway and Functional Consequences Thereof. *New Phytol.*, **2021**, *230* (1), 14–16. https://doi.org/10.1111/nph.17244
83. Kohlmeier, S.; Smits, T. H. M.; Ford, R. M.; Keel, C.; Harms, H.; Wick, L. Y. Taking the Fungal Highway: Mobilization of Pollutant-Degrading Bacteria by Fungi. *Environ. Sci. Technol.*, **2005**, *39* (12), 4640–4646. https://doi.org/10.1021/es047979z
84. Otto, S.; Bruni, E. P.; Harms, H.; Wick, L. Y. Catch Me If You Can: Dispersal and Foraging of *Bdellovibrio Bacteriovorus* 109J along Mycelia. *ISME J.*, **2017**, *11* (2), 386–393. https://doi.org/10.1038/ismej.2016.135
85. Allesina, S.; Tang, S. Stability Criteria for Complex Ecosystems. *Nature*, **2012**, *483* (7388), 205–208. https://doi.org/10.1038/nature10832
86. Dal Co, A.; van Vliet, S.; Kiviet, D. J.; Schlegel, S.; Ackermann, M. Short-Range Interactions Govern the Dynamics and Functions of Microbial Communities. *Nat. Ecol. Evol.*, **2020**, *4* (3), 366–375. https://doi.org/10.1038/s41559-019-1080-2
87. Mougi, A.; Kondoh, M. Diversity of Interaction Types and Ecological Community Stability. *Science*, **2012**, *337* (6092), 349–351. https://doi.org/10.1126/science.1220529

88. Ratzke, C.; Barrere, J.; Gore, J. Strength of Species Interactions Determines Biodiversity and Stability in Microbial Communities. *Nat. Ecol. Evol.*, **2020**, *4* (3), 376–383. https://doi.org/10.1038/s41559-020-1099-4
89. Levien, R. B.; Tan, S. M. Double Pendulum: An Experiment in Chaos. *Am. J. Phys.*, **1993**, *61* (11), 1038–1044. https://doi.org/10.1119/1.17335
90. Kochkov, D.; Yuval, J.; Langmore, I.; Norgaard, P.; Smith, J.; Mooers, G.; Klöwer, M.; Lottes, J.; Rasp, S.; Düben, P.; et al. Neural General Circulation Models for Weather and Climate. *Nature*, **2024**, *632* (8027), 1060–1066. https://doi.org/10.1038/s41586-024-07744-y
91. Chang, C.-Y.; Vila, J. C. C.; Bender, M.; Li, R.; Mankowski, M. C.; Bassette, M.; Borden, J.; Golfier, S.; Sanchez, P. G. L.; Waymack, R.; et al. Engineering Complex Communities by Directed Evolution. *Nat. Ecol. Evol.*, **2021**, *5* (7), 1011–1023. https://doi.org/10.1038/s41559-021-01457-5
92. Sanchez, A.; Bajic, D.; Diaz-Colunga, J.; Skwara, A.; Vila, J. C. C.; Kuehn, S. The Community-Function Landscape of Microbial Consortia. *Cell Syst.*, **2023**, *14* (2), 122–134. https://doi.org/10.1016/j.cels.2022.12.011
93. Sánchez, Á.; Vila, J. C. C.; Chang, C.-Y.; Diaz-Colunga, J.; Estrela, S.; Rebolleda-Gomez, M. Directed Evolution of Microbial Communities. *Annu. Rev. Biophys.*, **2021**, *50*, 323–341. https://doi.org/10.1146/annurev-biophys-101220-072829
94. Skwara, A.; Gowda, K.; Yousef, M.; Diaz-Colunga, J.; Raman, A. S.; Sanchez, A.; Tikhonov, M.; Kuehn, S. Statistically Learning the Functional Landscape of Microbial Communities. *Nat. Ecol. Evol.*, **2023**, *7* (11), 1823–1833. https://doi.org/10.1038/s41559-023-02197-4
95. Dini-Andreote, F.; Stegen, J. C.; van Elsas, J. D.; Salles, J. F. Disentangling Mechanisms That Mediate the Balance between Stochastic and Deterministic Processes in Microbial Succession. *Proc. Natl. Acad. Sci. U. S. A.*, **2015**, *112* (11), E1326–E1332. https://doi.org/10.1073/pnas.1414261112
96. Stegen, J. C.; Lin, X.; Konopka, A. E.; Fredrickson, J. K. Stochastic and Deterministic Assembly Processes in Subsurface Microbial Communities. *ISME J.*, **2012**, *6* (9), 1653–1664. https://doi.org/10.1038/ismej.2012.22
97. Hastings, A.; Hom, C. L.; Ellner, S.; Turchin, P.; Godfray, H. C. J. Chaos in Ecology: Is Mother Nature a Strange Attractor? *Annu. Rev. Ecol. Syst.*, **1993**, *24* (1), 1–33. https://doi.org/10.1146/annurev.es.24.110193.000245
98. May, R. M. *Stability and Complexity in Model Ecosystems.* Princeton University Press: Princeton, NJ, **2019**. https://doi.org/10.1515/9780691206912
99. Lotka, A. J. Analytical Note on Certain Rhythmic Relations in Organic Systems. *Proc. Natl. Acad. Sci. U. S. A.*, **1920**, *6* (7), 410–415. https://doi.org/10.1073/pnas.6.7.410
100. Lotka, A. J. Undamped Oscillations Derived from the Law of Mass Action. *J. Am. Chem. Soc.*, **1920**, *42* (8), 1595–1599. https://doi.org/10.1021/ja01453a010
101. Mühlbauer, L. K.; Schulze, M.; Harpole, W. S.; Clark, A. T. gause R: Simple Methods for Fitting Lotka-Volterra Models Describing Gause's "Struggle for Existence." *Ecol. Evol.*, **2020**, *10* (23), 13275–13283. https://doi.org/10.1002/ece3.6926
102. May, R. M. Will a Large Complex System Be Stable? *Nature*, **1972**, *238* (5364), 413–414. https://doi.org/10.1038/238413a0
103. de Vries, F. T.; Griffiths, R. I.; Bailey, M.; Craig, H.; Girlanda, M.; Gweon, H. S.; Hallin, S.; Kaisermann, A.; Keith, A. M.; Kretzschmar, M.; et al. Soil Bacterial Networks Are Less Stable under Drought than Fungal Networks. *Nat. Commun.*, **2018**, *9* (1), 3033. https://doi.org/10.1038/s41467-018-05516-7
104. Hernandez, D. J.; David, A. S.; Menges, E. S.; Searcy, C. A.; Afkhami, M. E. Environmental Stress Destabilizes Microbial Networks. *ISME J.*, **2021**, *15* (6), 1722–1734. https://doi.org/10.1038/s41396-020-00882-x
105. Srivastava, V.; Sarkar, A.; Singh, S.; Singh, P.; de Araujo, A. S. F.; Singh, R. P. Agroecological Responses of Heavy Metal Pollution with Special Emphasis on Soil Health and Plant Performances. *Front. Environ. Sci.*, **2017**, *5*, 64. https://doi.org/10.3389/fenvs.2017.00064
106. Duan, S.; Feng, G.; Limpens, E.; Bonfante, P.; Xie, X.; Zhang, L. Cross-Kingdom Nutrient Exchange in the Plant–Arbuscular Mycorrhizal Fungus–Bacterium Continuum. *Nat. Rev. Microbiol.*, **2024**, *22*, 773–790. https://doi.org/10.1038/s41579-024-01073-7
107. Wang, L.; Zhang, L.; George, T. S.; Feng, G. A Core Microbiome in the Hyphosphere of Arbuscular Mycorrhizal Fungi Has Functional Significance in Organic Phosphorus Mineralization. *New Phytol.*, **2022**, *238* (2), 859–873. https://doi.org/10.1111/nph.18642

108. Hellweger, F. L.; Clegg, R. J.; Clark, J. R.; Plugge, C. M.; Kreft, J.-U. Advancing Microbial Sciences by Individual-Based Modelling. *Nat. Rev. Microbiol.*, **2016**, *14* (7), 461–471. https://doi.org/10.1038/nrmicro.2016.62
109. Hellweger, F. L.; Bucci, V. A Bunch of Tiny Individuals—Individual-Based Modeling for Microbes. *Ecol. Model.*, **2009**, *220* (1), 8–22. https://doi.org/10.1016/j.ecolmodel.2008.09.004
110. Sugihara, G.; May, R.; Ye, H.; Hsieh, C.; Deyle, E.; Fogarty, M.; Munch, S. Detecting Causality in Complex Ecosystems. *Science*, **2012**, *338* (6106), 496–500. https://doi.org/10.1126/science.1227079
111. Chang, C.-W.; Ushio, M.; Hsieh, C. Empirical Dynamic Modeling for Beginners. *Ecol. Res.*, **2017**, *32* (6), 785–796. https://doi.org/10.1007/s11284-017-1469-9
112. Gibbs, T.; Levin, S. A.; Levine, J. M. Coexistence in Diverse Communities with Higher-Order Interactions. *Proc. Natl. Acad. Sci. U. S. A.*, **2022**, *119* (43), e2205063119. https://doi.org/10.1073/pnas.2205063119
113. Gibbs, T. L.; Gellner, G.; Levin, S. A.; McCann, K. S.; Hastings, A.; Levine, J. M. When Can Higher-Order Interactions Produce Stable Coexistence? *Ecol. Lett.*, **2024**, *27* (6), e14458. https://doi.org/10.1111/ele.14458
114. Sanchez-Gorostiaga, A.; Bajić, D.; Osborne, M. L.; Poyatos, J. F.; Sanchez, A. High-Order Interactions Distort the Functional Landscape of Microbial Consortia. *PLoS Biol.*, **2019**, *17* (12), e3000550. https://doi.org/10.1371/journal.pbio.3000550
115. Fisher, C. K.; Mehta, P. Identifying Keystone Species in the Human Gut Microbiome from Metagenomic Timeseries Using Sparse Linear Regression. *PLoS One*, **2014**, *9* (7), e102451. https://doi.org/10.1371/journal.pone.0102451
116. Michel-Mata, S.; Wang, X.-W.; Liu, Y.-Y.; Angulo, M. T. Predicting Microbiome Compositions from Species Assemblages through Deep Learning. *iMeta*, **2022**, *1* (1), e3. https://doi.org/10.1002/imt2.3
117. Baranwal, M.; Clark, R. L.; Thompson, J.; Sun, Z.; Hero, A. O.; Venturelli, O. S. Recurrent Neural Networks Enable Design of Multifunctional Synthetic Human Gut Microbiome Dynamics. *eLife*, **2022**, *11*, e73870. https://doi.org/10.7554/eLife.73870
118. MacArthur, R. Species Packing and Competitive Equilibrium for Many Species. *Theor. Popul. Biol.*, **1970**, *1* (1), 1–11. https://doi.org/10.1016/0040-5809(70)90039-0
119. Lee, H.; Bloxham, B.; Gore, J. Resource Competition Can Explain Simplicity in Microbial Community Assembly. *Proc. Natl. Acad. Sci. U. S. A.*, **2023**, *120* (35), e2212113120. https://doi.org/10.1073/pnas.2212113120
120. Marsland, R.; Cui, W.; Mehta, P. A Minimal Model for Microbial Biodiversity Can Reproduce Experimentally Observed Ecological Patterns. *Sci. Rep.*, **2020**, *10* (1), 3308. https://doi.org/10.1038/s41598-020-60130-2
121. Bodenhausen, N.; Hess, J.; Valzano, A.; Deslandes-Hérold, G.; Waelchli, J.; Furrer, R.; van der Heijden, M. G. A.; Schlaeppi, K. Predicting Soil Fungal Communities from Chemical and Physical Properties. *J. Sustain. Agric. Environ.*, **2023**, *2* (3), 225–237. https://doi.org/10.1002/sae2.12055
122. Hermans, S. M.; Buckley, H. L.; Case, B. S.; Curran-Cournane, F.; Taylor, M.; Lear, G. Using Soil Bacterial Communities to Predict Physico-Chemical Variables and Soil Quality. *Microbiome*, **2020**, *8* (1), 79. https://doi.org/10.1186/s40168-020-00858-1
123. Fierer, N.; Jackson, R. B. The Diversity and Biogeography of Soil Bacterial Communities. *Proc. Natl. Acad. Sci. U. S. A.*, **2006**, *103* (3), 626–631. https://doi.org/10.1073/pnas.0507535103
124. Clark, J. S.; Dietze, M.; Chakraborty, S.; Agarwal, P. K.; Ibanez, I.; LaDeau, S.; Wolosin, M. Resolving the Biodiversity Paradox. *Ecol. Lett.*, **2007**, *10* (8), 647–659. https://doi.org/10.1111/j.1461-0248.2007.01041.x
125. Wang, X.-W.; Sun, Z.; Jia, H.; Michel-Mata, S.; Angulo, M. T.; Dai, L.; He, X.; Weiss, S. T.; Liu, Y.-Y. Identifying Keystone Species in Microbial Communities Using Deep Learning. *Nat. Ecol. Evol.*, **2024**, *8* (1), 22–31. https://doi.org/10.1038/s41559-023-02250-2
126. Munch, S. B.; Brias, A.; Sugihara, G.; Rogers, T. L. Frequently Asked Questions about Nonlinear Dynamics and Empirical Dynamic Modelling. *ICES J. Mar. Sci.*, **2020**, *77* (4), 1463–1479. https://doi.org/10.1093/icesjms/fsz209
127. Deyle, E. R.; May, R. M.; Munch, S. B.; Sugihara, G. Tracking and Forecasting Ecosystem Interactions in Real Time. *Proc. R. Soc. B Biol. Sci.*, **2016**, *283* (1822), 20152258. https://doi.org/10.1098/rspb.2015.2258
128. Chang, C.-W.; Miki, T.; Ushio, M.; Ke, P.-J.; Lu, H.-P.; Shiah, F.-K.; Hsieh, C. Reconstructing Large Interaction Networks from Empirical Time Series Data. *Ecol. Lett.*, **2021**, *24* (12), 2763–2774. https://doi.org/10.1111/ele.13897

8

Exposure Pathways and Human Health Risks of PFAS

Cheng Shen and Jin Zhang

8.1 Introduction

Per- and polyfluoroalkyl substances (PFAS) are a class of synthetic chemicals extensively used in industrial and consumer products due to their unique properties, including resistance to grease, water, and stains (Evich et al., 2022). These chemicals are defined by their strong carbon-fluorine bonds, which contribute to their high stability and persistence in the environment, leading to their accumulation in ecosystems and human bodies (Bell et al., 2021). The primary exposure pathways for PFAS include ingesting contaminated food and water, inhaling polluted air, and dermal contact with treated products. Once absorbed, PFAS can accumulate in the liver and blood, and due to their long half-life, they can remain in the human body for years, increasing the risk of chronic exposure (Pelch et al., 2019).

Understanding the exposure pathways and associated health effects of PFAS is essential for effective risk assessment, public health protection, regulatory decision-making, targeted remediation efforts, and ongoing research. More accurate risk evaluations and mitigation strategies can be developed by identifying how PFAS enter the human body—whether through water, food, air, or consumer products. Knowledge of the health effects linked to PFAS, including liver damage, immune system disruption, and increased cancer risks, is critical for guiding public health interventions, especially for vulnerable populations. This understanding also informs the development of regulations and policies to minimize exposure and environmental contamination, as highlighted by the European Food Safety Authority (EFSA) (Schrenk et al., 2020). Recognizing specific exposure pathways allows for more effective remediation and contamination control efforts, such as improving water treatment technologies. Continuous research and monitoring, informed by this knowledge, ensure that public health measures remain relevant and effective in addressing emerging risks. Therefore, a comprehensive understanding of PFAS exposure and health impacts is vital for mitigating their adverse effects on human health and the environment.

In this review, we aim to provide a detailed examination of (1) the predominant exposure pathways for PFAS, (2) the factors influencing PFAS exposure, (3) the health impacts associated with exposure, and (4) the limitations of current research and potential solutions. This review seeks to summarize key knowledge gaps and identify future research needs.

8.2 PFAS Exposure Pathways

PFAS exposure occurs primarily through ingestion, inhalation, and dermal contact, with ingestion being the most significant due to contaminated water and food (Figure 8.1). While inhalation and dermal contact contribute less to overall exposure, they still pose risks, particularly in occupational settings and through the use of consumer products, emphasizing the need for more comprehensive studies on these exposure pathways.

DOI: 10.1201/9781003492115-10

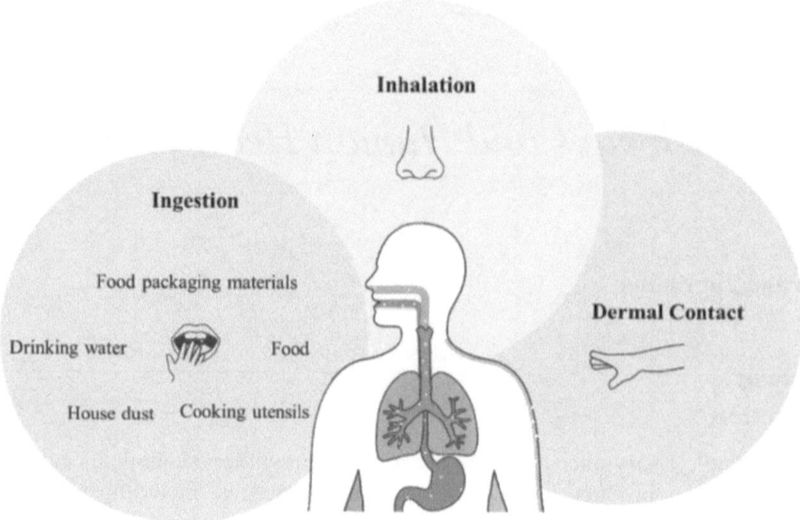

FIGURE 8.1 PFAS exposure pathways.

8.2.1 Ingestion

Ingestion is a critical pathway for PFAS exposure, with several key sources contributing to this risk. Drinking water is a major source of PFAS exposure due to widespread contamination in various water sources, including both surface and groundwater (AbulFadl et al., 2019; Eze et al., 2024; Wang et al., 2024). Significant PFAS contamination has been documented in different regions. For example, the Mid-Ohio Valley in the Americas experienced substantial perfluorooctanoic acid (PFOA) contamination, averaging 3.55 ng/mL due to industrial discharges (Bartell et al., 2010). In Europe, high PFOA levels were reported, particularly in Germany, where concentrations reached up to 598 ng/L in drinking water (Lange et al., 2007). In Asia, tap water in China showed PFOA levels as high as 151 ng/L, exceeding the US EPA's health advisory limit (Tan et al., 2017). Similarly, in Australia, perfluorooctane sulfonate (PFOS) and PFOA were commonly detected in drinking water, with PFOS reaching a maximum concentration of 16 ng/L, contributing to daily intake (Thompson et al., 2011). In response to these health risks, the EPA established advisory limits of 0.004 ng/L for PFOA and 0.02 ng/L for PFOS in 2022 (EPA, 2022), while Chinese standards allow 80 ng/L for PFOA and 40 ng/L for PFOS.

Diet is another critical exposure route; PFAS contamination in food can occur through bioaccumulation in the food chain, food packaging, and using PFAS-containing cookware. The EFSA has evaluated the risks related to the presence of PFAS in food and identified 'Fish meat,' 'Fruit and fruit products,' and 'Eggs and egg products' as the main contributors to PFAS exposure in the diet (EFSA et al., 2020). Among these, fish consumption is often the most significant route of exposure. For instance, an assessment of PFAS-related health risks in Beijing's Miyun Reservoir found that the estimated daily intake (EDI) of high-concern PFAS through drinking water, fish, pork, chicken, and dust ranged from 0.20 to 0.34, 3.44 to 12.61, 0.015 to 0.043, 0.003 to 0.013, and 0.074 to 0.390 ng/kg body weight/day, respectively. The intake of PFAS through fish consumption was significantly higher than in other routes (Meng et al., 2019). This is primarily due to the bioaccumulation and biomagnification of PFAS in aquatic environments. Another study in Belgium's North Sea region used stable isotope and SIBLER analysis to investigate PFAS transfer in the food chain. Still, it could not draw definitive conclusions about PFAS transmission in aquatic food chains (Cara et al., 2022). Environmental exposure and bioaccumulation likely explain the high PFAS content in fish. PFAS are lipophilic, accumulating easily in fatty tissues, and many fish species have high-fat content. Additionally, fish exhibit low PFAS metabolism rates, leading to rapid accumulation in their bodies. In contrast, terrestrial animals like pigs and chickens are primarily exposed through air, and while they may encounter PFAS through drinking water and food, their exposure levels are generally lower.

Food packaging materials (FCMs) and non-stick cookware containing PFAS can also leach these chemicals into food, increasing ingestion risks. PFAS can migrate from FCMs into food through a process influenced by several factors, including the fat content of food, temperature, pH level, and contact time (Eze et al., 2024). Most PFAS currently used in food packaging are short-chain types and fluorotelomer-based derivatives. Even brief interaction between the material and the food can result in PFAS migration. Foods with higher fat content, such as meats, dairy products, and oils, are more prone to this migration. For example, Elizalde et al. (2018) found that the migration of perfluorohexanoic acid (PFHxA), perfluoroheptanoic acid (PFHpA), PFOA, perfluorononanoic acid (PFNA), perfluorodecanoic acid (PFDA), perfluorotridecanoic acid (PFTrDA), and perfluorotetradecanoic acid (PFTeDA) was higher in whole milk than in low-fat milk. Many PFAS compounds are lipophilic, facilitating their transfer from FCMs to food. Additionally, acidic beverages, such as orange juice, have been found to contain higher PFAS levels compared to beverages with a more neutral pH (AbulFadl et al., 2019). Elevated temperatures and prolonged contact time can also increase the rate of PFAS migration from FCMs into food.

Cooking utensils with PFAS coatings, notably non-stick pans, pose a potential risk due to their ability to release PFAS into food during cooking, especially when subjected to high heat. PTFE-coated cookware, commonly known as Teflon, is extensively utilized for its non-stick properties in various industrial and household applications due to its excellent chemical, thermal, and electrical stability and low friction. Despite being non-toxic and inert in its polymeric state, PTFE emits polymer fumes at temperatures above 260°C, which can lead to polymer fume fever in exposed workers at 350°C and pyrolysis at temperatures exceeding 400°C, releasing decomposition products such as tetrafluoroethylene (TFE) and difluorocarbondiradicals (RCF2), with the potential formation of highly toxic perfluoroisobutylene (PFIB) (Sajid and Ilyas, 2017). As PFOA, used in PTFE synthesis, is being phased out due to its toxic effects and replaced by alternatives like GenX, there is an ongoing suspicion that these substitutes may share similar toxicological concerns (Gaballah et al., 2020).

House dust is another important source of PFAS exposure. Dust may contain high concentrations of PFAS, and people ingest these substances through cleaning, touching dust-covered surfaces, or hand-to-mouth contact, especially in children (Poothong et al., 2020). Additionally, Ao et al. (2019) analyzed 168 dust samples and 27 drinking water samples, finding that the total daily intake (TDI) of PFAS for infants, children, adolescents, and adults ranged from 20.67 to 52.97 ng/kg/day, with indoor dust being a greater source than drinking water. This underscores the importance of indoor dust as a source of PFAS, posing potential risks to human health.

8.2.2 Inhalation

Indoor air may contain PFAS precursors, which can enter the body through inhalation and be converted into active PFAS within the body. PFAS in indoor air primarily originates from indoor sources, such as certain consumer products and building materials. For example, a study analyzing 27 commercial paints for volatile and nonvolatile PFAS identified 6:2 FTOH and 6:2 diPAP at concentrations of 0.9–83 μg/g and 0.073–58 μg/g, respectively (Cahuas et al., 2022). These compounds accounted for up to 17% of the total fluorine content in the paints. During paint drying, a significant 40% mass loss of FTOH occurred, indicating volatilization. The ConsExpo model predicted potential human exposure exceeding the reference dose for one commercial paint, highlighting the risk of PFAS inhalation during painting.

These findings suggest that painters are at risk of occupational PFAS exposure. However, research on occupational exposure to airborne PFAS is currently limited, with most studies focusing on ski waxing, textile manufacturing, firefighting, and floor waxing, as well as office workers, retail textile workers, and educational/office buildings (Paris-Davila et al., 2023). Among these occupations, ski waxing shows the highest PFAS air concentrations, reaching 10 μg/m^3, significantly higher than other occupations and residential exposures. Textile manufacturing also shows considerable PFAS levels, particularly during scouring, drying, and heat setting, with PFTetDA air concentrations peaking at approximately 10 μg/m^3 (Paris-Davila et al., 2023). In contrast, floor waxing and office work present lower area air concentrations, often below detection limits, except for PFOA, PFOS, and PFHxA, which have levels around 1E–5 μg/m^3 (Paris-Davila et al., 2023). These findings emphasize the need for tailored protective measures in occupations with high exposure.

Compared to dietary ingestion, there is limited information on the toxicity of PFAS via inhalation. Monnot et al. (2023) examine the appropriateness of route-to-route extrapolation for PFAS to derive guidelines for inhalation exposure, focusing on PFOA and PFOS due to the abundance of their toxicity data. The study suggests an air concentration of 7.0×10^{-5} mg/m^3 (or 0.07 µg/m^3) as an appropriate Reference Concentration (RfC) for PFOA and PFOS, based on the 2016 EPA Reference Dose (RfD) (USEPA, 2016a, 2016b).

8.2.3 Dermal Contact

Dermal exposure to PFAS occurs through direct contact with consumer products, such as clothing, cosmetics, and personal care products that contain PFAS-based coatings or formulations. While dermal absorption of PFAS is generally lower than ingestion and inhalation, prolonged contact with PFAS-contaminated materials can still contribute to cumulative exposure over time.

The potential for dermal contact as a significant pathway for human exposure to PFAS is increasingly recognized but remains insufficiently studied. Although PFAS are found in personal care products and clothing that come into direct contact with the skin, there is a notable lack of data on how these substances are absorbed dermally (Glüge et al., 2020; Schultes et al., 2018). One study by Poothong et al. (2019) used hand wipes to estimate daily dermal uptake of PFOA and PFOS, suggesting substantial exposure via this route. However, these studies often rely on limited sample sizes and do not account for PFAS load variations due to washing frequency or seasonal changes. Further supporting this link, a Spanish study by Serrano et al. (2021) found a correlation between PFAS levels in breast milk and the use of personal care products, underlining the risk posed by dermal exposure to these substances. This growing evidence highlights the need for more comprehensive studies on the long-term effects of dermal exposure to PFAS. The urgent need for more comprehensive dermal permeation and uptake studies is clear to understand better the contribution of dermal exposure to the total PFAS body burden (Ragnarsdóttir et al., 2022).

8.3 Factors Affecting PFAS Exposure

PFAS exposure is a complex issue shaped by various interrelated factors, including biological factors, occupational exposure, socioeconomic factors, and the effectiveness of regulatory and public health measures (Figure 8.2). Understanding these factors is essential for developing effective strategies to reduce PFAS exposure and its associated health risks.

8.3.1 Biological Factors

Age and sex can significantly influence the exposure pathways to PFAS. Children are often at a higher risk due to their frequent hand-to-mouth activities, which can increase the ingestion of PFAS present in the environment. Data from EFSA et al. (2020) indicates that the average exposure to PFAS in adolescents and adults typically falls between 3 and 22 ng/kg body weight per week, with the 95th percentile ranging from 9 to 70 ng/kg body weight per week. In contrast, children's exposure is approximately double that of adults. Brown et al. (2020) assessed the daily dietary intake of individual PFAS from vegetables for different age groups. They discovered that young children (1–2 years) have the highest exposure rates, followed by a gradual decrease with age. This heightened exposure in young children may be due to their higher dietary intake relative to body weight and their developing metabolic systems, which can slow the elimination of PFAS. Play behaviors like playing on the ground can increase the risk of ingesting PFAS from soil. Moreover, the physiological sensitivity and developmental stages characteristic of early childhood might render children more vulnerable to the adverse effects of PFAS.

Sex differences can significantly influence the exposure pathways and health effects associated with PFAS. Biological variations between males and females, including differences in body composition, hormonal balance, and reproductive systems, can lead to distinct absorption, distribution, metabolism, and excretion of PFAS compounds. For instance, studies have shown that women may have a higher body burden of certain PFAS due to differences in fat distribution and hormonal influences that could affect

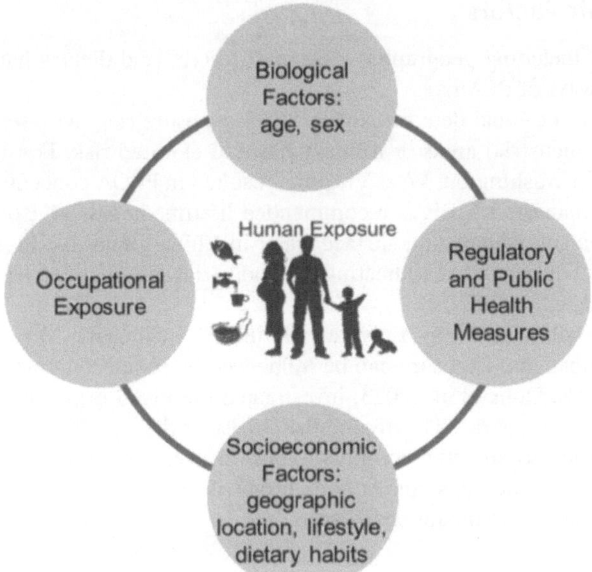

FIGURE 8.2 Factors affecting PFAS exposure.

the storage and elimination of these chemicals (Xie et al., 2021). Additionally, reproductive factors in women, such as menstruation, pregnancy, and breastfeeding, can alter the pharmacokinetics of PFAS, potentially leading to different exposure levels and health risks (Xie et al., 2021). Studies also indicate that PFAS exposure may impact androgenic and estrogenic pathways in a sex-dependent manner. A study by Rivera-Núñez et al. (2023) found that PFAS exposure was associated with higher concentrations of testosterone in males, particularly among those carrying male fetuses. Furthermore, PFAS exposure was associated with increased levels of estradiol and estriol in females, but only when carrying female fetuses, suggesting a complex interaction between PFAS, sex hormones, and fetal sex.

8.3.2 Occupational Exposure

The impact of occupation on exposure pathways to PFAS varies significantly, with different professions facing varying degrees of risk due to their work environment and the substances they handle.

Individuals working in industries that manufacture or use PFAS, such as chemical plants, textile, and leather industries, are at a higher risk of exposure. Occupational exposure to PFAS primarily occurs through inhalation and dermal contact. Human biomonitoring studies have consistently shown that the serum levels of PFOA and PFOS in occupationally exposed individuals are significantly higher than in the general population. For instance, a study in the United States reported geometric mean serum concentrations of PFOS and PFOA in the general population to be 32.0 and 5.6 ng/mL, respectively (Lau et al., 2007). In contrast, workers in fluorochemical production facilities exhibited significantly higher levels, with serum PFOA concentrations ranging up to 500 ng/mL (Olsen et al., 2003). These disparities underscore the impact of occupational exposure on PFAAs body burden. It is crucial to continue monitoring these levels to assess the efficacy of ongoing stewardship programs to reduce PFAA emissions and environmental presence.

Firefighters face unique exposure risks due to their occupational use of PFAS-containing firefighting foams and other materials. Recent studies have indicated that firefighters have elevated levels of PFAS in their blood serum, which may be linked to an increased risk of cancer (Mazumder et al., 2023). Firefighters are exposed to PFAS through various sources, including turnout gear treated with fluoropolymers for water and oil repellency, aqueous film-forming foams (AFFF), and the air and dust at fire scenes and stations. The presence of PFAS in AFFF and the potential for these substances to migrate from treated to untreated layers of turnout gear highlight the need for further research into the health implications for firefighters. Moreover, the review suggests the firefighting community should use PFAS-free alternatives in protective gear and foams to mitigate these risks.

8.3.3 Socioeconomic Factors

Socioeconomic factors, including geographic location, lifestyle, and dietary habits, significantly influence the exposure pathways of PFAS.

Geographic location is a critical determinant of PFAS exposure risk, with residents living near contaminated sites such as industrial areas or military bases at elevated risk. For example, proximity to a fluorochemical facility in Washington, West Virginia, resulted in PFOA concentrations that were alarmingly 190 times higher than the US EPA's recommended lifetime health advisory level of 70 ng/L (Hu et al., 2016). Similarly, a study analyzing surface water in China found that PFAS concentrations were correlated with industrial distributions, indicating that industrial activities can lead to localized increases in PFAS exposure (Wang et al., 2024).

Lifestyle and dietary habits also play a pivotal role in PFAS exposure. Variations in diet, consumer product usage, and occupational exposure can be influenced by socioeconomic status and contribute to differences in PFAS levels. Dong et al. (2023) investigated the PFAS exposure level of residents living near fluorochemical industrial parks. They found that 99.3% of the $\Sigma 36$ PFAS exposure was due to gastrointestinal uptake, while dermal and respiratory exposures were only 0.1% and 0.6%. Moreover, compared to vegetables, fruits, meat, eggs, and aquatic products, cereals represent over 80% of the overall EDI of PFAS, emerging as the dominant contributor.

8.3.4 Regulatory and Public Health Measures

Regulatory and public health measures significantly influence the exposure pathways of PFAS by dictating the production, use, and disposal of these chemicals. The presence and enforcement of regulations can substantially reduce the levels of PFAS in the environment, leading to decreased exposure through various pathways, such as contaminated drinking water, food, and indoor dust (Poothong et al., 2020).

Public health measures, including education and awareness campaigns, are crucial in informing the public about the risks associated with PFAS. These measures can lead to behavioral changes that minimize exposure, such as adopting water filtration systems certified to remove PFAS or selecting food products with lower PFAS content (Vinet and Zhedanov, 2010).

Furthermore, public awareness about the potential health risks of PFAS can encourage the use of safer alternatives and promote practices that reduce the ingestion of PFAS in house dust. The success of these measures is contingent upon the effectiveness of regulation enforcement, the accessibility of alternative products, and the public's ability to access accurate information.

8.4 Human Health Risks of PFAS

PFAS are a group of persistent synthetic chemicals in the environment and the human body. Exposure to PFAS has been associated with various adverse health effects (Figure 8.3). Here are some of the key human health risks related to PFAS exposure.

8.4.1 Cancer

In recent years, epidemiological studies and laboratory analyses have suggested an association between exposure to PFAS and an increased risk of various cancers. For example, a study utilized data from the US National Health and Nutrition Examination Survey (NHANES) and discovered a correlation between PFAS exposure and heightened risks of brain, esophageal, melanomatous skin, prostate, and lung cancers (Moon and Mun, 2024). Additionally, a study focused on the Chinese population unveiled a potential link between PFAS and breast cancer, particularly noting the positive correlation between PFOA and perfluorononanoic acid (PFDA) with breast cancer risk and an inverse association with PFTrDA (Li et al., 2022).

While the carcinogenic mechanisms of PFAS remain not fully understood, research has proposed several pathways. PFAS may activate peroxisome proliferator-activated receptors (PPARs), disrupting normal metabolic pathways and causing lipid and glucose metabolism dysregulation, which could influence

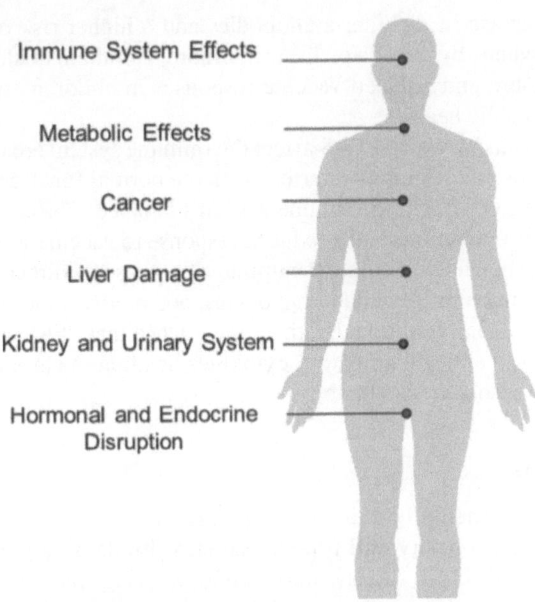

FIGURE 8.3 Human health risks of PFAS.

tumorigenesis and cancer progression (Steenland and Winquist, 2021a). Furthermore, PFAS are known to exhibit endocrine-disrupting properties, capable of mimicking or antagonizing the effects of sex hormones, potentially leading to the development of reproductive system cancers (Steenland and Winquist, 2021b). Exposure to PFAS has also been linked to alterations in DNA methylation, which might lead to the dysregulation of gene expression, affecting cellular processes such as cell proliferation, differentiation, and apoptosis (Steenland and Winquist, 2021b).

8.4.2 Hormonal and Endocrine Disruption

PFASs have garnered considerable attention as endocrine disruptors, with their potential to interfere with the human endocrine system, posing significant risks to reproductive health. Studies have demonstrated that PFAS can disrupt hormonal balance through various mechanisms, impacting reproductive well-being. For instance, research by Zhou et al. (2017) has shown a significant positive correlation between PFAS exposure and menstrual irregularities in women, particularly with compounds such as PFOS, PFOA, PFNA, and PFHxS. Additionally, studies by Aimuzi et al. (2020) and Blake and Fenton (2020) have indicated that PFAS exposure may also be linked to thyroid dysfunction, affecting thyroid hormone levels in pregnant women and neonates. In young adult males, Luo et al. (2021) discovered an inverse relationship between a mixture of PFAS and estradiol (E2) levels, with perfluoroundecanoic acid (PFuDA) identified as the primary contributor.

The mechanisms by which PFAS affect the endocrine system may include direct actions on endocrine organs like the breast and thyroid or indirect effects through disrupting the hypothalamic-pituitary-gonadal (HPGA) axis. Fenton et al. (2009) have shown that PFAS can impact mammary cell culture and accumulate in mammary gland tissue of exposed mice. Zhang et al. (2021) have reported effects on thyroid cell viability and hormone secretion.

8.4.3 Immune System Effects

PFASs are compounds that have been linked to a myriad of adverse health effects, including significant impacts on the immune system. PFAS exposure can impair immune function, reduce immune response, decrease vaccine effectiveness, and increase infection susceptibility. Studies, such as those by Grandjean et al. (2012, 2017), have demonstrated that a doubling of PFOS concentrations in children's serum is associated with a 39% decrease in diphtheria antibodies and a higher risk of falling below protective levels for diphtheria and tetanus by the age of 7. Furthermore, Abraham et al. (2020) identified associations between PFAS exposure and reduced vaccine responses in children and adults, highlighting the extensive implications for public health.

While the exact mechanisms by which PFAS affect the immune system are not fully understood, some hypotheses have been proposed. PFAS may interfere with the normal functioning of immune cells, alter the production of cytokines, or affect the immune system's balance. These disruptions could lead to a diminished ability to fight off infections and a reduced response to vaccination (Ehrlich et al., 2023). For instance, PFASs have been shown to modulate the immune response by directly influencing the behavior of immune cells, including macrophages and lymphocytes, potentially impairing their ability to respond to pathogens (Zhang et al., 2023a). Additionally, there is evidence that PFAS can disrupt the delicate balance of pro-inflammatory and anti-inflammatory cytokines, leading to chronic inflammation and other immune-related disorders (Zhang et al., 2023b).

8.4.4 Metabolic Effects

PFAS are pervasive environmental pollutants that have been linked to metabolic disruptions, notably increasing the risk of obesity and type 2 diabetes. Epidemiological studies, such as those by Attanasio (2019) and Roth et al. (2021), have found correlations between PFAS exposure and increased BMI in adolescents and elevated obesity risk in adults, respectively. The connection to diabetes risk is also supported, with Brosset and Ngueta (2023) and Jensen et al. (2018) showing associations between PFAS exposure and diabetes risk in adults and gestational diabetes in pregnant women. Liu et al. (2020) identified a significant link between higher PFAS levels and apolipoprotein C-III concentrations in lipoproteins. Zheng et al. (2024) furthered this research with a meta-analysis that included 12 studies, revealing a positive association between PFAS exposure and levels of total cholesterol (TC) and low-density lipoprotein (LDL) in children and adolescents and a negative association with the homeostasis model assessment of insulin resistance (HOMA-IR). These findings suggest that PFAS, particularly PFOS, may influence lipid metabolism and insulin resistance during critical developmental periods, highlighting the importance of understanding the long-term health effects of PFAS exposure.

The potential mechanisms by which PFAS may affect metabolism include their structural similarity to PPARs, which could lead to the activation of these receptors and subsequent impacts on lipid and glucose metabolism (Li et al., 2019; Magueresse-Battistoni et al., 2017). PFAS exposure may also disrupt lipid metabolism by influencing bile acid synthesis and enterohepatic circulation (Chen et al., 2020). Animal models have indicated that PFAS exposure is related to insulin resistance and impaired glucose tolerance (Hines et al., 2009; Lv et al., 2013)

8.4.5 Liver Damage

The widespread use of PFAS has resulted in environmental contamination and raised concerns about their harmful effects on the liver. A systematic review and meta-analysis (Costello et al., 2022) evaluated the association between PFAS exposure and liver injury in rodent and epidemiological studies, identifying a consistent link between PFAS exposure and hepatotoxicity. Specifically, higher serum levels of alanine aminotransferase (ALT), a biomarker of liver injury, were associated with exposure to PFOA, PFOS, and PFNA. For example, Darrow et al. (2016) found a positive association between cumulative PFOA exposure and ALT levels in a large cohort, indicating liver cell damage. Long-term exposure to PFAS, particularly PFOA and PFOS, has also been linked to liver diseases such as nonalcoholic fatty liver disease (NAFLD). In children with NAFLD, higher plasma concentrations of PFOS and PFHxS were significantly associated with greater disease severity, including nonalcoholic steatohepatitis (NASH) and liver fibrosis (Jin et al., 2020).

The mechanisms underlying PFAS-induced liver injury are not fully understood. However, it may involve the activation of PPARs alpha (PPARα) and other receptors, leading to disrupted lipid metabolism and inflammation (Fenton et al., 2021). PFAS, structurally similar to fatty acids, can accumulate in the liver, promoting triglyceride accumulation and steatosis (Fenton et al., 2021).

8.4.6 Kidney and Urinary System

PFAS exposure has been consistently associated with reduced kidney function and an increased risk of chronic kidney disease (CKD). Studies have reported a significant association between PFAS exposure and decreased glomerular filtration rates (GFR), a key indicator of kidney function. For instance, studies have demonstrated this link in both adults and children, despite the potential for reverse causation where diminished kidney function might affect PFAS clearance (Park et al., 2021; Wang et al., 2023). A longitudinal study by Blake et al. (2018), spanning 18 years, further confirms the detrimental impact of PFAS on renal glomerular filtration.

Serum PFAS levels exhibit an inverse U-shaped relationship with GFR, indicating an initial accumulation of PFAS with a subsequent decrease in advanced renal disease, potentially due to impaired reabsorption mechanisms (Liang et al., 2023). This relationship is particularly pronounced in the presence of albuminuria, a condition that may exacerbate the underestimation of PFAS exposure's impact on kidney health by cross-sectional studies. The association between PFAS exposure and kidney function is further supported by findings that link these compounds to hypertension and elevated uric acid levels, the latter being a biomarker of renal disease (Jain and Ducatman, 2019).

The incidence of kidney cancer has been on the rise since the 1970s, and PFAS exposure has been implicated in this trend. The C8 Health studies, which are part of the C8 Health Project investigating the health effects of PFOA on residents near the Chemours chemical plant in West Virginia, have reported a hazard ratio of 1.10 for kidney cancer related to PFAS exposure, suggesting a significant association (Nicole, 2013). Stanifer et al. (2018) corroborate these findings, indicating risks ranging from 1.07 to 12.8 for kidney cancer or mortality associated with exposure to long-chain PFAS. Preliminary data from the Veneto region in Italy also point toward a substantial increase in kidney cancer mortality among individuals exposed to PFAS, underscoring the urgency for further research into the role of PFAS in the development of CKD and kidney cancer (Mastrantonio et al., 2018).

While the epidemiological and clinical evidence linking PFAS exposure to kidney disease is growing, the underlying mechanisms are not yet fully understood. PFAS are known to affect various biological pathways, including those related to oxidative stress, PPARs, and NF-E2-related factor 2 (Nrf2), which are crucial for maintaining cellular homeostasis and combating oxidative damage (Liu et al., 2023). Disruption of these pathways can lead to renal tubular damage, fibrosis, and inflammation, potentially contributing to the development of kidney disease.

8.5 Limitations and Possible Solutions

8.5.1 PFAS Mixture Risks

Current risk assessment practices, typically designed for single chemical evaluations, encounter limitations when addressing complex mixtures like PFAS that may interact in combined exposures (Drakvik et al., 2020). These methods may underestimate population exposure levels if they fail to incorporate the cumulative effects of multiple PFAS exposures from various sources and pathways (Anderson et al., 2022). Particularly concerning is the potential for a single-compound approach to underestimate overall risk when multiple PFAS can induce identical adverse health impacts (Bil et al., 2023; EFSA et al., 2019).

In light of these challenges, the scientific community is intensifying research into the toxicity of PFAS mixtures, emphasizing the importance of assessing combined exposures to substances with similar toxicological profiles (Bil et al., 2023). A tiered approach is suggested to tackle these complexities, integrating multiple lines of evidence to address the gaps in analytical and toxicological studies related to PFAS risk assessment (Anderson et al., 2022). This approach necessitates using assumptions that are more likely to overestimate risk, particularly in selecting dose-response models, uncertainty factors, and exposure data, to ensure a more protective assessment strategy.

8.5.2 Limited Data on Detected PFAS

Thousands of PFAS are pervasive in the environment due to their extensive use in various industries and consumer products. However, most human exposure assessments have traditionally focused on a limited number of targeted PFAS compounds, typically up to about 30 substances, representing just a fraction of the total PFAS chemical universe. This targeted approach may not fully capture the breadth of human exposure to the myriad PFAS variants present in the environment, including precursor compounds that can further degrade into known PFAS (McDonough et al., 2022). As a result, there is a growing need for more inclusive and sensitive analytical methods to assess the full range of PFAS exposure in human populations.

Total fluorine quantification, enhanced by suspect screening through high-resolution mass spectrometry, is increasingly recognized as a critical tool for assessing exposure to PFAS. These advanced analytical techniques allow researchers to more accurately identify and quantify the contributions of PFAS precursors that can transform into persistent perfluoroalkyl acids. This comprehensive approach allows for a deeper understanding of the complex exposure pathways and transformation processes associated with PFAS in the environment, thereby enhancing the accuracy and reliability of exposure assessments (De Silva et al., 2021).

8.5.3 Constraints in Health Risk Research

Current research linking PFAS to various health risks, including cancer, endocrine disruption, liver injury, metabolic disorders, and kidney health, faces several limitations. Many studies rely on cross-sectional data, which impedes the determination of causality. Additionally, inconsistencies across studies may stem from variations in exposure levels, durations, study designs, and demographic factors. There is also a need for more comprehensive research on the mechanisms of PFAS toxicity, the effects of PFAS mixtures, early life exposures, and potential sex-specific differences. Moreover, small sample sizes may limit the generalizability of findings, and the molecular mechanisms by which PFAS exposure leads to health outcomes are not yet fully understood. Future research should employ more rigorous experimental designs, including extensive cohort studies with long-term follow-ups and multicenter studies that consider population characteristics, to clarify the mechanisms of PFAS-induced health effects and facilitate accurate risk assessment (Boyd et al., 2022).

8.6 Conclusion

In summary, the intricate exposure pathways of PFAS underscore the urgency for a thorough comprehension of their environmental and biological impact. The ingestion of contaminated water and food, inhalation of PFAS-laden indoor air, and dermal absorption from everyday products are primary routes through which PFAS enters the human body. The complexity is heightened by individual differences such as age, sex, occupation, and socioeconomic status, which can modulate exposure and susceptibility.

The breadth of health implications linked to PFAS exposure is expansive, ranging from cancer to metabolic disorders and potential organ damage. Yet, the current body of research is hindered by methodological limitations, including a reliance on cross-sectional designs and small sample sizes that impede a clear determination of causal relationships.

Looking ahead, there is a pronounced need for robust longitudinal studies with diverse and extensive cohorts to clarify the intricate dynamics between PFAS exposure and health effects. Advanced analytical techniques are also required to detect the myriad of PFAS compounds present in various matrices. Integrating such scientific advancements with strategic regulatory measures and public health initiatives will be pivotal in addressing the pervasive challenge of PFAS. This holistic approach is essential for developing effective interventions that protect public health and mitigate the risks associated with these enduring contaminants.

Acknowledgments

This work was supported by the National Natural Science Foundation of China (42403056), the Natural Science Foundation of Zhejiang Province (LZY22B070002), the National Natural Science Foundation of China (22276174), and The Education University of Hong Kong Seed Funding Grant 2021/22 (FLASS/DRF/IRS-7).

REFERENCES

Abraham, K., Mielke, H., Fromme, H., Völkel, W., Menzel, J., Peiser, M., Zepp, F., Willich, S.N., Weikert, C., 2020. Internal exposure to perfluoroalkyl substances (PFASs) and biological markers in 101 healthy 1-year-old children: associations between levels of perfluorooctanoic acid (PFOA) and vaccine response. *Arch. Toxicol.* 94, 2131–2147. https://doi.org/10.1007/s00204-020-02715-4

AbulFadl, M.M., Sharaf, A.M., Mostafa, M.M., El-Saeid, M.H., 2019. Impact of household cooking on release of fluorinated compounds PFOA and PFOS from Tefal coated cookware to foods. *World J. Adv. Res. Rev.* 3, 024–030. https://doi.org/10.30574/wjarr.2019.3.2.0060

Aimuzi, R., Luo, K., Huang, R., Huo, X., Nian, M., Ouyang, F., Du, Y., Feng, L., Wang, W., Zhang, J., 2020. Perfluoroalkyl and polyfluroalkyl substances and maternal thyroid hormones in early pregnancy. *Environ. Pollut.* 264, 114557. https://doi.org/10.1016/j.envpol.2020.114557

Anderson, J.K., Brecher, R.W., Cousins, I.T., DeWitt, J., Fiedler, H., Kannan, K., Kirman, C.R., Lipscomb, J., Priestly, B., Schoeny, R., Seed, J., Verner, M., Hays, S.M., 2022. Grouping of PFAS for human health risk assessment: findings from an independent panel of experts. *Regul. Toxicol. Pharmacol.* 134, 105226. https://doi.org/10.1016/j.yrtph.2022.105226

Ao, J., Yuan, T., Xia, H., Ma, Y., Shen, Z., Shi, R., Tian, Y., Zhang, J., Ding, W., Gao, L., Zhao, X., Yu, X., 2019. Characteristic and human exposure risk assessment of per- and polyfluoroalkyl substances: a study based on indoor dust and drinking water in China. *Environ. Pollut.* 254, 112873. https://doi.org/10.1016/j.envpol.2019.07.041

Attanasio, R., 2019. Sex differences in the association between perfluoroalkyl acids and liver function in US adolescents: analyses of NHANES 2013–2016. *Environ. Pollut.* 254, 113061. https://doi.org/10.1016/j.envpol.2019.113061

Bartell, S.M., Calafat, A.M., Lyu, C., Kato, K., Ryan, P.B., Steenland, K., 2010. Rate of decline in serum PFOA concentrations after granular activated carbon filtration at two public water systems in Ohio and West Virginia. *Environ. Health Perspect.* 118, 222–228. https://doi.org/10.1289/ehp.0901252

Bell, E.M., De Guise, S., McCutcheon, J.R., Lei, Y., Levin, M., Li, B., Rusling, J.F., Lawrence, D.A., Cavallari, J.M., O'Connell, C., Javidi, B., Wang, X., Ryu, H., 2021. Exposure, health effects, sensing, and remediation of the emerging PFAS contaminants – scientific challenges and potential research directions. *Sci. Total Environ.* 780, 146399. https://doi.org/10.1016/j.scitotenv.2021.146399

Bil, W., Govarts, E., Zeilmaker, M.J., Woutersen, M., Bessems, J., Ma, Y., Thomsen, C., Haug, L.S., Lignell, S., Gyllenhammar, I., Palkovicova Murinova, L., Fabelova, L., Tratnik, J.S., Kosjek, T., Gabriel, C., Sarigiannis, D., Pedraza-Diaz, S., Esteban-López, M., Castaño, A., Rambaud, L., Riou, M., Franken, C., Colles, A., Vogel, N., Kolossa-Gehring, M., Halldorsson, T.I., Uhl, M., Schoeters, G., Santonen, T., Vinggaard, A.M., 2023. Approaches to mixture risk assessment of PFASs in the European population based on human hazard and biomonitoring data. *Int. J. Hyg. Environ. Health* 247, 114071. https://doi.org/10.1016/j.ijheh.2022.114071

Blake, B.E., Fenton, S.E., 2020. Early life exposure to per- and polyfluoroalkyl substances (PFAS) and latent health outcomes: a review including the placenta as a target tissue and possible driver of peri- and postnatal effects. *Toxicology* 443, 152565. https://doi.org/10.1016/j.tox.2020.152565

Blake, B.E., Pinney, S.M., Hines, E.P., Fenton, S.E., Ferguson, K.K., 2018. Associations between longitudinal serum perfluoroalkyl substance (PFAS) levels and measures of thyroid hormone, kidney function, and body mass index in the Fernald Community Cohort. *Environ. Pollut.* 242, 894–904. https://doi.org/10.1016/j.envpol.2018.07.042

Boyd, R.I., Ahmad, S., Singh, R., Fazal, Z., Prins, G.S., Madak Erdogan, Z., Irudayaraj, J., Spinella, M.J., 2022. Toward a mechanistic understanding of poly- and perfluoroalkylated substances and cancer. *Cancers (Basel).* 14, 2919. https://doi.org/10.3390/cancers14122919

Brosset, E., Ngueta, G., 2023. Exposure to per- and polyfluoroalkyl substances and glycemic control in older US adults with type 2 diabetes mellitus. *Environ. Res.* 216, 114697. https://doi.org/10.1016/j.envres.2022.114697

Brown, J.B., Conder, J.M., Arblaster, J.A., Higgins, C.P., 2020. Assessing human health risks from per- and polyfluoroalkyl substance (PFAS)-impacted vegetable consumption: a tiered modeling approach. *Environ. Sci. Technol.* 54, 15202–15214. https://doi.org/10.1021/acs.est.0c03411

Cahuas, L., Muensterman, D.J., Kim-Fu, M.L., Reardon, P.N., Titaley, I.A., Field, J.A., 2022. Paints: a source of volatile PFAS in air—potential implications for inhalation exposure. *Environ. Sci. Technol.* 56, 17070–17079. https://doi.org/10.1021/acs.est.2c04864

Cara, B., Lies, T., Thimo, G., Robin, L., Lieven, B., 2022. Bioaccumulation and trophic transfer of perfluorinated alkyl substances (PFAS) in marine biota from the Belgian North Sea: distribution and human health risk implications. *Environ. Pollut.* 311, 119907. https://doi.org/10.1016/j.envpol.2022.119907

Chen, Z., Yang, T., Walker, D.I., Thomas, D.C., Qiu, C., Chatzi, L., Alderete, T.L., Kim, J.S., Conti, D. V., Breton, C. V., Liang, D., Hauser, E.R., Jones, D.P., Gilliland, F.D., 2020. Dysregulated lipid and fatty acid metabolism link perfluoroalkyl substances exposure and impaired glucose metabolism in young adults. *Environ. Int.* 145, 106091. https://doi.org/10.1016/j.envint.2020.106091

Costello, E., Rock, S., Stratakis, N., Eckel, S.P., Walker, D.I., Valvi, D., Cserbik, D., Jenkins, T., Xanthakos, S.A., Kohli, R., Sisley, S., Vasiliou, V., La Merrill, M.A., Rosen, H., Conti, D.V., McConnell, R., Chatzi, L., 2022. Exposure to per- and polyfluoroalkyl substances and markers of liver injury: a systematic review and meta-analysis. *Environ. Health Perspect.* 130, EHP10092. https://doi.org/10.1289/EHP10092

Darrow, L.A., Groth, A.C., Winquist, A., Shin, H.-M., Bartell, S.M., Steenland, K., 2016. Modeled perfluorooctanoic acid (PFOA) exposure and liver function in a mid-Ohio valley community. *Environ. Health Perspect.* 124, 1227–1233. https://doi.org/10.1289/ehp.1510391

De Silva, A.O., Armitage, J.M., Bruton, T.A., Dassuncao, C., Heiger-Bernays, W., Hu, X.C., Kärrman, A., Kelly, B., Ng, C., Robuck, A., Sun, M., Webster, T.F., Sunderland, E.M., 2021. PFAS exposure pathways for humans and wildlife: a synthesis of current knowledge and key gaps in understanding. *Environ. Toxicol. Chem.* 40, 631–657. https://doi.org/10.1002/etc.4935

Dong, F., Pan, Y., Zhang, J., Hu, J., Luo, Y., Tang, J., Dai, J., Sheng, N., 2023. Comprehensive assessment of exposure pathways for perfluoroalkyl ether carboxylic acids (PFECAs) in residents near a fluorochemical industrial park: the unanticipated role of cereal consumption. *Environ. Sci. Technol.* 57, 19442–19452. https://doi.org/10.1021/acs.est.3c06910

Drakvik, E., Altenburger, R., Aoki, Y., Backhaus, T., Bahadori, T., Barouki, R., Brack, W., Cronin, M.T.D., Demeneix, B., Bennekou, S.H., 2020. Statement on advancing the assessment of chemical mixtures and their risks for human health and the environment. *Environ. Int.* 134, 105267.

EFSA, More, S.J., Bampidis, V., Benford, D., Bennekou, S.H., Bragard, C., Halldorsson, T.I., Hernández-Jerez, A.F., Koutsoumanis, K., Naegeli, H., Schlatter, J.R., Silano, V., Nielsen, S.S., Schrenk, D., Turck, D., Younes, M., Benfenati, E., Castle, L., Cedergreen, N., Hardy, A., Laskowski, R., Leblanc, J.C., Kortenkamp, A., Ragas, A., Posthuma, L., Svendsen, C., Solecki, R., Testai, E., Dujardin, B., Kass, G.E.N., Manini, P., Jeddi, M.Z., Dorne, J.C., Hogstrand, C., 2019. Guidance on harmonised methodologies for human health, animal health and ecological risk assessment of combined exposure to multiple chemicals. *EFSA J.* 17, e05634. https://doi.org/10.2903/j.efsa.2019.5634

EFSA, Schrenk, D., Bignami, M., Bodin, L., Chipman, J.K., del Mazo, J., Grasl-Kraupp, B., Hogstrand, C., Hoogenboom, L. (Ron), Leblanc, J., Nebbia, C.S., Nielsen, E., Ntzani, E., Petersen, A., Sand, S., Vleminckx, C., Wallace, H., Barregård, L., Ceccatelli, S., Cravedi, J., Halldorsson, T.I., Haug, L.S., Johansson, N., Knutsen, H.K., Rose, M., Roudot, A., Van Loveren, H., Vollmer, G., Mackay, K., Riolo, F., Schwerdtle, T., 2020. Risk to human health related to the presence of perfluoroalkyl substances in food. *EFSA J.* 18, 6223. https://doi.org/10.2903/j.efsa.2020.6223

Ehrlich, V., Bil, W., Vandebriel, R., Granum, B., Luijten, M., Lindeman, B., Grandjean, P., Kaiser, A.-M., Hauzenberger, I., Hartmann, C., Gundacker, C., Uhl, M., 2023. Consideration of pathways for immunotoxicity of per- and polyfluoroalkyl substances (PFAS). *Environ. Heal.* 22, 19. https://doi.org/10.1186/s12940-022-00958-5

Elizalde, M.P., Gómez-Lavín, S., Urtiaga, A.M., 2018. Migration of perfluorinated compounds from paperbag to Tenax® and lyophilised milk at different temperatures. *Int. J. Environ. Anal. Chem.* 98, 1423–1433. https://doi.org/10.1080/03067319.2018.1562062

EPA, 2022. *2022 EPA Health Advisory Levels for Four PFAS 331-702*. EPA, Washington, DC.

Evich, M.G., Davis, M.J.B., McCord, J.P., Acrey, B., Awkerman, J.A., Knappe, D.R.U., Lindstrom, A.B., Speth, T.F., Tebes-Stevens, C., Strynar, M.J., Wang, Z., Weber, E.J., Henderson, W.M., Washington, J.W., 2022. Per- and polyfluoroalkyl substances in the environment. *Science* 80, 375. https://doi.org/10.1126/science.abg9065

Ezé, C.G., Okeke, E.S., Nwankwo, C.E., Nyaruaba, R., Anand, U., Okoro, O.J., Bontempi, E., 2024. Emerging contaminants in food matrices: an overview of the occurrence, pathways, impacts and detection techniques of per- and polyfluoroalkyl substances. *Toxicol. Reports* 12, 436–447. https://doi.org/10.1016/j.toxrep.2024.03.012

Fenton, S.E., Ducatman, A., Boobis, A., DeWitt, J.C., Lau, C., Ng, C., Smith, J.S., Roberts, S.M., 2021. Per- and polyfluoroalkyl substance toxicity and human health review: current state of knowledge and strategies for informing future research. *Environ. Toxicol. Chem.* 40, 606–630. https://doi.org/10.1002/etc.4890

Fenton, S.E., Reiner, J.L., Nakayama, S.F., Delinsky, A.D., Stanko, J.P., Hines, E.P., White, S.S., Lindstrom, A.B., Strynar, M.J., Petropoulou, S.-S.E., 2009. Analysis of PFOA in dosed CD-1 mice. Part 2: disposition of PFOA in tissues and fluids from pregnant and lactating mice and their pups. *Reprod. Toxicol.* 27, 365–372. https://doi.org/10.1016/j.reprotox.2009.02.012

Gaballah, S., Swank, A., Sobus, J.R., Howey, X.M., Schmid, J., Catron, T., McCord, J., Hines, E., Strynar, M., Tal, T., 2020. Evaluation of developmental toxicity, developmental neurotoxicity, and tissue dose in zebrafish exposed to genX and other PFAS. *Environ. Health Perspect.* 128, 1–22. https://doi.org/10.1289/EHP5843

Glüge, J., Scheringer, M., Cousins, I.T., DeWitt, J.C., Goldenman, G., Herzke, D., Lohmann, R., Ng, C.A., Trier, X., Wang, Z., 2020. An overview of the uses of per- and polyfluoroalkyl substances (PFAS). *Environ. Sci. Process. Impacts* 22, 2345–2373. https://doi.org/10.1039/D0EM00291G

Grandjean, P., Andersen, E.W., Budtz-Jørgensen, E., Nielsen, F., Mølbak, K., Weihe, P., Heilmann, C., 2012. Serum vaccine antibody concentrations in children exposed to perfluorinated compounds. *JAMA* 307, 391–397.

Grandjean, P., Heilmann, C., Weihe, P., Nielsen, F., Mogensen, U.B., Budtz-Jørgensen, E., 2017. Serum vaccine antibody concentrations in adolescents exposed to perfluorinated compounds. *Environ. Health Perspect.* 125, 77018.

Hines, E.P., White, S.S., Stanko, J.P., Gibbs-Flournoy, E.A., Lau, C., Fenton, S.E., 2009. Phenotypic dichotomy following developmental exposure to perfluorooctanoic acid (PFOA) in female CD-1 mice: low doses induce elevated serum leptin and insulin, and overweight in mid-life. *Mol. Cell. Endocrinol.* 304, 97–105. https://doi.org/10.1016/j.mce.2009.02.021

Hu, X.C., Andrews, D.Q., Lindstrom, A.B., Bruton, T.A., Schaider, L.A., Grandjean, P., Lohmann, R., Carignan, C.C., Blum, A., Balan, S.A., Higgins, C.P., Sunderland, E.M., 2016. Detection of poly- and perfluoroalkyl substances (PFASs) in U.S. drinking water linked to industrial sites, military fire training areas, and wastewater treatment plants. *Environ. Sci. Technol. Lett.* 3, 344–350. https://doi.org/10.1021/acs.estlett.6b00260

Jain, R.B., Ducatman, A., 2019. Perfluoroalkyl substances follow inverted U-shaped distributions across various stages of glomerular function: implications for future research. *Environ. Res.* 169, 476–482. https://doi.org/10.1016/j.envres.2018.11.033

Jensen, R.C., Glintborg, D., Timmermann, C.A.G., Nielsen, F., Kyhl, H.B., Andersen, H.R., Grandjean, P., Jensen, T.K., Andersen, M., 2018. Perfluoroalkyl substances and glycemic status in pregnant Danish women: the odense child cohort. *Environ. Int.* 116, 101–107. https://doi.org/10.1016/j.envint.2018.04.010

Jin, R., McConnell, R., Catherine, C., Xu, S., Walker, D.I., Stratakis, N., Jones, D.P., Miller, G.W., Peng, C., Conti, D. V., Vos, M.B., Chatzi, L., 2020. Perfluoroalkyl substances and severity of nonalcoholic fatty liver in children: an untargeted metabolomics approach. *Environ. Int.* 134, 105220. https://doi.org/10.1016/j.envint.2019.105220

Lange, F.T., Wenz, M., Schmidt, C.K., Brauch, H.-J., 2007. Occurrence of perfluoroalkyl sulfonates and carboxylates in German drinking water sources compared to other countries. *Water Sci. Technol.* 56, 151–158. https://doi.org/10.2166/wst.2007.803

Lau, C., Anitole, K., Hodes, C., Lai, D., Pfahles-Hutchens, A., Seed, J., 2007. Perfluoroalkyl acids: a review of monitoring and toxicological findings. *Toxicol. Sci.* 99, 366–394. https://doi.org/10.1093/toxsci/kfm128

Li, C.-H., Ren, X.-M., Cao, L.-Y., Qin, W.-P., Guo, L.-H., 2019. Investigation of binding and activity of perfluoroalkyl substances to the human peroxisome proliferator-activated receptor β/δ. *Environ. Sci. Process. Impacts* 21, 1908–1914. https://doi.org/10.1039/C9EM00218A

Li, X., Song, F., Liu, X., Shan, A., Huang, Y., Yang, Z., Li, H., Yang, Q., Yu, Y., Zheng, H., Cao, X.-C., Chen, D., Chen, K.-X., Chen, X., Tang, N., 2022. Perfluoroalkyl substances (PFASs) as risk factors for breast cancer: a case–control study in Chinese population. *Environ. Heal.* 21, 83. https://doi.org/10.1186/s12940-022-00895-3

Liang, Y., Zhou, H., Zhang, J., Li, S., Shen, W., Lei, L., 2023. Exposure to perfluoroalkyl and polyfluoroalkyl substances and estimated glomerular filtration rate in adults: a cross-sectional study based on NHANES (2017–2018). *Environ. Sci. Pollut. Res.* 30, 57931–57944. https://doi.org/10.1007/s11356-023-26384-9

Liu, D., Yan, S., Wang, P., Chen, Q., Liu, Y., Cui, J., Liang, Y., Ren, S., Gao, Y., 2023. Perfluorooctanoic acid (PFOA) exposure in relation to the kidneys: a review of current available literature. *Front. Physiol.* 14, 1–13. https://doi.org/10.3389/fphys.2023.1103141

Liu, G., Zhang, B., Hu, Y., Rood, J., Liang, L., Qi, L., Bray, G.A., Dejonge, L., Coull, B., Grandjean, P., Furtado, J.D., Sun, Q., 2020. Associations of Perfluoroalkyl substances with blood lipids and Apolipoproteins in lipoprotein subspecies: the POUNDS-lost study. *Environ. Heal. A Glob. Access Sci. Source* 19, 1–10. https://doi.org/10.1186/s12940-020-0561-8

Luo, K., Liu, X., Nian, M., Wang, Y., Qiu, J., Yu, H., Chen, X., Zhang, J., 2021. Environmental exposure to per- and polyfluoroalkyl substances mixture and male reproductive hormones. *Environ. Int.* 152, 106496. https://doi.org/10.1016/j.envint.2021.106496

Lv, Z., Li, G., Li, Y., Ying, C., Chen, J., Chen, T., Wei, J., Lin, Y., Jiang, Y., Wang, Y., Shu, B., Xu, B., Xu, S., 2013. Glucose and lipid homeostasis in adult rat is impaired by early-life exposure to perfluorooctane sulfonate. *Environ. Toxicol.* 28, 532–542. https://doi.org/10.1002/tox.20747

Magueresse-Battistoni, B. Le, Labaronne, E., Vidal, H., Naville, D., 2017. Endocrine disrupting chemicals in mixture and obesity, diabetes and related metabolic disorders. *World J. Biol. Chem.* 8, 108. https://doi.org/10.4331/wjbc.v8.i2.108

Mastrantonio, M., Bai, E., Uccelli, R., Cordiano, V., Screpanti, A., Crosignani, P., 2018. Drinking water contamination from perfluoroalkyl substances (PFAS): an ecological mortality study in the Veneto Region, Italy. *Eur. J. Public Health* 28, 180–185. https://doi.org/10.1093/eurpub/ckx066

Mazumder, N.-U.-S., Hossain, M.T., Jahura, F.T., Girase, A., Hall, A.S., Lu, J., Ormond, R.B., 2023. Firefighters' exposure to per-and polyfluoroalkyl substances (PFAS) as an occupational hazard: a review. *Front. Mater.* 10, 1–22. https://doi.org/10.3389/fmats.2023.1143411

McDonough, C.A., Li, W., Bischel, H.N., De Silva, A.O., DeWitt, J.C., 2022. Widening the lens on PFASs: direct human exposure to perfluoroalkyl acid precursors (pre-PFAAs). *Environ. Sci. Technol.* 56, 6004–6013. https://doi.org/10.1021/acs.est.2c00254

Meng, J., Liu, S., Zhou, Y., Wang, T., 2019. Are perfluoroalkyl substances in water and fish from drinking water source the major pathways towards human health risk? *Ecotoxicol. Environ. Saf.* 181, 194–201. https://doi.org/10.1016/j.ecoenv.2019.06.010

Monnot, A.D., Massarsky, A., Garnick, L., Bandara, S.B., Unice, K.M., 2023. Can oral toxicity data for PFAS inform on toxicity via inhalation? *Risk Anal.* 43, 1533–1538. https://doi.org/10.1111/risa.14039

Moon, J., Mun, Y., 2024. The association between per- and polyfluoroalkyl substances (PFASs) and brain, esophageal, melanomatous skin, prostate, and lung cancer using the 2003–2018 US National Health and Nutrition Examination Survey (NHANES) datasets. *Heliyon* 10, e24337. https://doi.org/10.1016/j.heliyon.2024.e24337

Nicole, W., 2013. PFOA and cancer in a highly exposed community: new findings from the C8 science panel. *Environ. Health Perspect.* 121, 800379. https://doi.org/10.1289/ehp.121-A340

Olsen, G.W., Burris, J.M., Burlew, M.M., Mandel, J.H., 2003. Epidemiologic assessment of worker serum perfluorooctanesulfonate (PFOS) and perfluorooctanoate (PFOA) concentrations and medical surveillance examinations. *J. Occup. Environ. Med.* 45, 260–270. https://doi.org/10.1097/01.jom.0000052958.59271.10

Paris-Davila, T., Gaines, L.G.T., Lucas, K., Nylander-French, L.A., 2023. Occupational exposures to airborne per- and polyfluoroalkyl substances (PFAS)—A review. *Am. J. Ind. Med.* 66, 393–410. https://doi.org/10.1002/ajim.23461

Park, S.K., Ding, N., Han, D., 2021. Perfluoroalkyl substances and cognitive function in older adults: should we consider non-monotonic dose-responses and chronic kidney disease? *Environ. Res.* 192, 110346. https://doi.org/10.1016/j.envres.2020.110346

Pelch, K.E., Reade, A., Wolffe, T.A.M., Kwiatkowski, C.F., 2019. PFAS health effects database: protocol for a systematic evidence map. *Environ. Int.* 130, 45. https://doi.org/10.1016/j.envint.2019.05.045

Poothong, S., Padilla-Sánchez, J.A., Papadopoulou, E., Giovanoulis, G., Thomsen, C., Haug, L.S., 2019. Hand wipes: a useful tool for assessing human exposure to poly- and perfluoroalkyl substances (PFASs) through hand-to-mouth and dermal contacts. *Environ. Sci. Technol.* 53, 1985–1993. https://doi.org/10.1021/acs.est.8b05303

Poothong, S., Papadopoulou, E., Padilla-Sánchez, J.A., Thomsen, C., Haug, L.S., 2020. Multiple pathways of human exposure to poly- and perfluoroalkyl substances (PFASs): from external exposure to human blood. *Environ. Int.* 134, 105244. https://doi.org/10.1016/j.envint.2019.105244

Ragnarsdóttir, O., Abdallah, M.A.-E., Harrad, S., 2022. Dermal uptake: an important pathway of human exposure to perfluoroalkyl substances? *Environ. Pollut.* 307, 119478. https://doi.org/10.1016/j.envpol.2022.119478

Rivera-Núñez, Z., Kinkade, C.W., Khoury, L., Brunner, J., Murphy, H., Wang, C., Kannan, K., Miller, R.K., O'Connor, T.G., Barrett, E.S., 2023. Prenatal perfluoroalkyl substances exposure and maternal sex steroid hormones across pregnancy. *Environ. Res.* 220, 115233. https://doi.org/10.1016/j.envres.2023.115233

Roth, K., Yang, Z., Agarwal, M., Liu, W., Peng, Z., Long, Z., Birbeck, J., Westrick, J., Liu, W., Petriello, M.C., 2021. Exposure to a mixture of legacy, alternative, and replacement per- and polyfluoroalkyl substances (PFAS) results in sex-dependent modulation of cholesterol metabolism and liver injury. *Environ. Int.* 157, 106843. https://doi.org/10.1016/j.envint.2021.106843

Sajid, M., Ilyas, M., 2017. PTFE-coated non-stick cookware and toxicity concerns: a perspective. *Environ. Sci. Pollut. Res.* 24, 23436–23440. https://doi.org/10.1007/s11356-017-0095-y

Schrenk, D., Bignami, M., Bodin, L., Chipman, J.K., del Mazo, J., Grasl-Kraupp, B., Hogstrand, C., Hoogenboom, L. (Ron), Leblanc, J., Nebbia, C.S., Nielsen, E., Ntzani, E., Petersen, A., Sand, S., Vleminckx, C., Wallace, H., Barregård, L., Ceccatelli, S., Cravedi, J., Halldorsson, T.I., Haug, L.S., Johansson, N., Knutsen, H.K., Rose, M., Roudot, A., Van Loveren, H., Vollmer, G., Mackay, K., Riolo, F., Schwerdtle, T., 2020. Risk to human health related to the presence of perfluoroalkyl substances in food. *EFSA J.* 18, e06223. https://doi.org/10.2903/j.efsa.2020.6223

Schultes, L., Vestergren, R., Volkova, K., Westberg, E., Jacobson, T., Benskin, J.P., 2018. Per- and polyfluoroalkyl substances and fluorine mass balance in cosmetic products from the Swedish market: implications for environmental emissions and human exposure. *Environ. Sci. Process. Impacts* 20, 1680–1690. https://doi.org/10.1039/c8em00368h

Serrano, L., Iribarne-Durán, L.M., Suárez, B., Artacho-Cordón, F., Vela-Soria, F., Peña-Caballero, M., Hurtado, J.A., Olea, N., Fernández, M.F., Freire, C., 2021. Concentrations of perfluoroalkyl substances in donor breast milk in Southern Spain and their potential determinants. *Int. J. Hyg. Environ. Health* 236, 113796. https://doi.org/10.1016/j.ijheh.2021.113796

Stanifer, J.W., Stapleton, H.M., Souma, T., Wittmer, A., Zhao, X., Boulware, L.E., 2018. Perfluorinated chemicals as emerging environmental threats to kidney health. *Clin. J. Am. Soc. Nephrol.* 13, 1479–1492. https://doi.org/10.2215/CJN.04670418

Steenland, K., Winquist, A., 2021a. PFAS and cancer, a scoping review of the epidemiologic evidence. *Environ. Res.* 194, 110690. https://doi.org/10.1016/j.envres.2020.110690

Steenland, K., Winquist, A., 2021b. PFAS and cancer, a scoping review of the epidemiologic evidence. *Environ. Res.* 194, 110690. https://doi.org/10.1016/j.envres.2020.110690

Tan, K.-Y., Lu, G.-H., Piao, H.-T., Chen, S., Jiao, X.-C., Gai, N., Yamazaki, E., Yamashita, N., Pan, J., Yang, Y.-L., 2017. Current contamination status of perfluoroalkyl substances in tapwater from 17 cities in the Eastern China and their correlations with surface waters. *Bull. Environ. Contam. Toxicol.* 99, 224–231. https://doi.org/10.1007/s00128-017-2109-3

Thompson, J., Eaglesham, G., Mueller, J., 2011. Concentrations of PFOS, PFOA and other perfluorinated alkyl acids in Australian drinking water. *Chemosphere* 83, 1320–1325. https://doi.org/10.1016/j.chemosphere.2011.04.017

USEPA, 2016a. *Health Effects Support Document for Perfluorooctane Sulfonate (PFOS).* USEPA, Washington, DC.

USEPA, 2016b. *Health Effects Support Document for Perfluorooctanoic Acid (PFOA).* USEPA, Washington, DC.

Vinet, L., Zhedanov, A., 2010. A "missing" family of classical orthogonal polynomials. *J. Chem. Inf. Model.* 53, 1689–1699.

Wang, J., Shen, C., Zhang, J., Lou, G., Shan, S., Zhao, Y., Man, Y.B., Li, Y., 2024. Per- and polyfluoroalkyl substances (PFASs) in Chinese surface water: temporal trends and geographical distribution. *Sci. Total Environ.* 915, 170127. https://doi.org/10.1016/j.scitotenv.2024.170127

Wang, X., Yan, W., Xie, H., Huang, S., 2023. Associations between perfluoroalkyl substances mixed exposure and uric acid and glomerular filtration rate among adolescents and adults. *Expo. Heal.* 16, 253–262. https://doi.org/10.1007/s12403-023-00551-2

Xie, X., Weng, X., Liu, S., Chen, J., Guo, X., Gao, X., Fei, Q., Hao, G., Jing, C., Feng, L., 2021. Perfluoroalkyl and polyfluoroalkyl substance exposure and association with sex hormone concentrations: results from the NHANES 2015–2016. *Environ. Sci. Eur.* 33, 69. https://doi.org/10.1186/s12302-021-00508-9

Zhang, L., Louie, A., Rigutto, G., Guo, H., Zhao, Y., Ahn, S., Dahlberg, S., Sholinbeck, M., Smith, M.T., 2023a. A systematic evidence map of chronic inflammation and immunosuppression related to per- and polyfluoroalkyl substance (PFAS) exposure. *Environ. Res.* 220, 115188. https://doi.org/10.1016/j.envres.2022.115188

Zhang, L., Louie, A., Rigutto, G., Guo, H., Zhao, Y., Ahn, S., Dahlberg, S., Sholinbeck, M., Smith, M.T., 2023b. A systematic evidence map of chronic inflammation and immunosuppression related to per- and polyfluoroalkyl substance (PFAS) exposure. *Environ. Res.* 220, 115188. https://doi.org/10.1016/j.envres.2022.115188

Zhang, S., Chen, K., Li, W., Chai, Y., Zhu, J., Chu, B., Li, N., Yan, J., Zhang, S., Yang, Y., 2021. Varied thyroid disrupting effects of perfluorooctanoic acid (PFOA) and its novel alternatives hexafluoropropylene-oxide-dimer-acid (GenX) and ammonium 4,8-dioxa-3H-perfluorononanoate (ADONA) in vitro. *Environ. Int.* 156, 106745. https://doi.org/10.1016/j.envint.2021.106745

Zheng, Q., Yan, W., Gao, S., Li, X., 2024. The effect of PFAS exposure on glucolipid metabolism in children and adolescents: a meta-analysis. *Front. Endocrinol. (Lausanne).* 15, 1–10. https://doi.org/10.3389/fendo.2024.1261008

Zhou, W., Zhang, L., Tong, C., Fang, F., Zhao, S., Tian, Y., Tao, Y., Zhang, J., 2017. Plasma perfluoroalkyl and polyfluoroalkyl substances concentration and menstrual cycle characteristics in preconception women. *Environ. Health Perspect.* 125, 1–6. https://doi.org/10.1289/EHP1203

9

Potential Human Health Impacts of Microplastics and Associated Chemicals

Yating Luo, Xiangyi Jin, Mengyao Xing, Zhoujia Jiang, Ling Shu, Qifeng Yin, Da Ouyang, Yimin Cai, and Haibo Zhang

9.1 Introduction

Global plastic production is growing, exceeding 368 million metric tons (Mt) yearly (PlasticsEurope, 2020). With the rapid growth of plastic production, numerous plastics are released into the environment during their life cycle (Li et al., 2021). By 2050, approximately 12,000 Mt of plastic waste will be in landfills or the environment (Geyer et al., 2017). These plastic wastes can be broken into smaller fragments and particles, causing microplastic (MP) contamination. MP was proposed by Thompson et al. (2004) and was defined as plastics with a size of less than 5 mm. MPs were detected in various types of food, such as seafood products (Li et al., 2018; Montero et al., 2023), drinking water and water sources (Acarer, 2023), table salt (Kim et al., 2018; Ruan et al., 2024), air and dust (Zhang et al., 2020; Goßmann et al., 2023), and vegetables and fruits (Oliveri Conti et al., 2020; Pham et al., 2023). The ubiquitous existence of MPs in the environment attracts global attention. Humans cannot escape the fate of ingesting, inhaling, or contacting MPs.

Many studies are reporting on the detection of MPs in human tissues, including the gastrointestinal tract (Ibrahim et al., 2021; Yan et al., 2024), nasal cavity, and lungs (Baeza-Martínez et al., 2022; Jenner et al., 2022; Jiang et al., 2022). These studies provide consistent evidence that MPs in food and air can be ingested or inhaled into the human digestive tract and respiratory tract. Dermal contact is another known way MPs are exposed. Additives in MPs (e.g., brominated flame retardants) may be absorbed by the skin, posing potential risks to the human body (Abafe et al., 2023). Notably, the presence of MPs has been detected in the blood, which may be one of the vital causes of cardiovascular diseases such as thrombi and atherosclerotic plaques (Marfella et al., 2024). With the discovery of more positive correlations between human tissue lesions and the number of detected MPs, scientists have gradually paid attention to the impact of MPs on human health. Recently, there has been an explosion in the number of MP-related articles (Figure 9.1a). Therefore, it is necessary to sort out and summarize them to systematically understand the detection of MPs in various environmental media and their effects on human health.

We collected and analyzed the available articles and review articles primarily published before May 2024 based on the database of Web of Science Core Collection using the keywords "microplastic" or "nanoplastic (NP)," "fish or seafood or food or fruit or vegetable or water or beverage or salt or air or dust," "intestinal or respiratory or blood or genital or placenta or feces or stool," "toxicity," and "human or cell or rat," to summarize the new progress. By analyzing 6988 articles published in 847 journals by authors from 17,958 institutions in 640 countries/regions using Citespace 6.3.1, it was found that the identification of MPs in different environments, especially in the marine, was a hotspot in articles published in 2014–2018 (Figure 9.1b). The influence of MPs on cells and tissues (e.g., oxidative stress and human health) emerged in articles published in 2019–2024 (Figure 9.1c and d). Therefore, this chapter

DOI: 10.1201/9781003492115-11

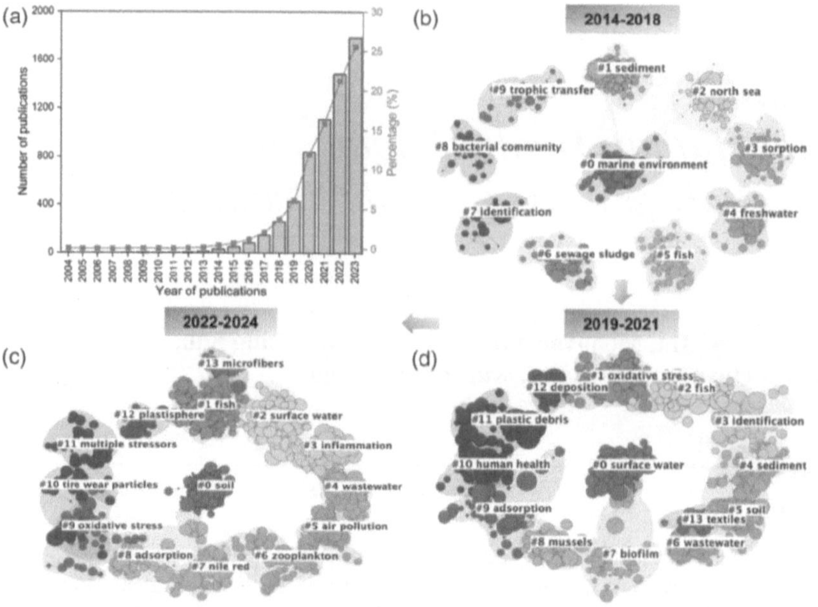

FIGURE 9.1 Statistics (a) and keyword-co-occurrence analysis (b, c, d) of the collected literature. Co-occurrence cluster analysis was performed based on article and citation keywords using the LLR algorithm in CiteSpace 6.3.1.

aims to summarize the exposure pathways of MPs, the uptake-transport-accumulation of MPs in human tissues, and the potential toxicity of MPs to human cells and tissues. Subsequently, we hope to provide the research trends on the impact of MPs on human health.

9.2 Exposure Pathways of Microplastics to Humans

9.2.1 Food Consumption

Seafood accounts for more than 17.0% of global protein consumption (Dawson et al., 2022). MPs have been widely detected in various seafood products, including bivalve molluscs, echinoderms, cephalopods, crustaceans, and finfish, with an abundance ranging from not detected (ND) to 48.5 items/g (Figure 9.2a). High levels of MP contamination were detected in more than 47 seafood species from more than 36 countries (Kibria et al., 2022). Zakeri et al. (2020) detected MPs in the most common commercial fish in the Caspian Sea. MPs were present in 67.6% of the 111 individuals analyzed, with a mean concentration of 2.3 items/fish. MPs (0.7–2.9 items/g) were examined in seawater and mussels from the United Kingdom coastal waters and supermarket sources (Li et al., 2018). It was estimated that consumers consuming 100.0 g processed mussels will ingest 70.0 MP particles. Yulia et al. (2023) detected MP abundance in the whole tissues of green mussels in Jakarta Bay, Indonesia. MPs were identified in all 120 green mussel samples, with fiber (19.0 items/g) > film (15.0 items/g) > fragment (14.5 items/g) being the most common kinds. Furthermore, Fraissinet et al. (2024) used the thermal desorption-proton transfer reaction-mass spectrometry (TD-PTR-MS) technique to quantitatively detect the concentration of MPs in cultured mussel tissues in Italy, which was 10.0–187.0 ng/mg dw.

Further studies found that MP levels in the gastrointestinal tract of seabass were significantly higher than those in muscle. MP levels in seabass followed that in water and feed in aquaculture systems (Matias et al., 2024). It was inferred that MPs in seawater or feed may be ingested by seabass and accumulated in its body. Similarly, Sun et al. (2024d) collected six marine species in the southern Yellow Sea coastal region for MP detection. The results also showed that the high abundance of MPs in demersal and pelagic fish species at different locations was correlated with the high abundance and hotspots of

MPs observed in surface waters near industrial river outputs, suggesting the trophic transfer of MPs in pelagic and benthic habitats, fish, and bivalves. This could exhibit long-term negative impacts on marine ecosystems. A similar phenomenon was also found by Srisiri et al. (2024). It was predicted that in the absence of effective management and intervention measures, the input volume of plastic waste (such as PS foam and PE ropes generated by oyster farming) in Maowei Sea could reach 961.0 tons/year, which is close to the annual input volume of plastics from a river (Tian et al., 2022). Therefore, we should take measures to effectively prevent and control MP pollution in oceans and aquatic life.

In addition to seafood, MPs have also been detected in staple foods. Dessi et al. (2021) bought rice (produced in Australia, India, Pakistan, and Thailand) from Australian supermarkets and found that MPs, mainly PE (95.0%, 45.0–317.0 μg/g dw), were detected in all rice samples. Except for PE, PP, and PET were also detected in rice, with the highest concentrations being 105.0 and 17.0 μg/g dw, respectively. It was assessed that Australians may consume 3.0–4.0 mg MPs through a serving of rice. Gong et al. (2023) found that the transportation process of rice may promote the release of MPs from plastic packaging, and washing rice before cooking can reduce MPs in rice. Shi et al. (2023) also detected the presence of MPs in corn flour. Notably, MPs were commonly found in vegetables and fruits, with abundance ranging from 1.5 to 307.7 items/g (Figure 9.2a). Aydın et al. (2023) detected MPs in tomatoes, cucumbers, pears, apples, onions, and potatoes in Turkey and found that the maximum average MPs was determined in tomato samples (3.6 ± 1.4 items/g). Oliveri Conti et al. (2020) also found that MPs/NPs were commonly found in vegetables (72.2–130.5 items/g; broccoli, lettuce, and carrot) and fruits (52.6–307.7 items/g; apple and pear) in Catania. In addition, MPs were also detected in edible oil (Guo et al., 2023; Battaglini et al., 2024), egg (Liu et al., 2022b), sugar (Afrin et al., 2022), and tea leaves (Li et al., 2022).

MPs have also been detected in beverages, with abundance ranging from ND to 80,000 items/L (Figure 9.2b), and the number of MPs varied greatly among beverages. Socas-Hernández et al. (2024) detected MPs in various common commercial beverages in Spain. Results showed that beverages were ranked from highest to lowest MP content: Beer (95.5 ± 92.0 items/L) > Wine (56.7 ± 73.5 items/L) > Other drinks including juice, energy/sport/tea drink (24.8 ± 27.3 items/L) > Soft drinks (22.5 ± 18.7 items/L) > Bottled water (7.2 ± 10.1 items/L). Among them, the concentration and particle size of MPs in alcoholic beverages were higher than those in non-alcoholic drinks. Both natural and semisynthetic cellulosic particles mainly were found in these MPs in beverages. The average MP concentration in all beverages, including water, natural mineral and mineral water, sparkling soft drinks, cold tea, and some traditional beverages purchased in Türkiye by Basaran et al. (2024), was 2.2 ± 9.9 items/L. The results also indicated that the packaging material would likely be a vital MP source. Moreover, synergistic CO_2 bubbles, pressure, and repeated freeze-thaw cycles significantly increase the release of MPs/NPs (Chen et al., 2023d).

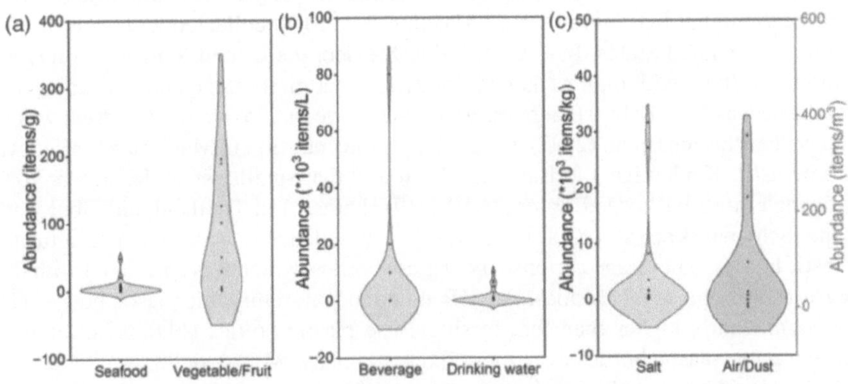

FIGURE 9.2 Statistics on the abundance of MPs in different foods summarized from 91 relevant articles. Key data and references were presented in "Exposure Pathways of Microplastics to Humans."

Notably, Pham et al. (2023) detected MPs in eight food types: salted seafood, fish sauces, seaweed, table salts, soy sauces, honey, beer, and beverage. The median MPs abundances in honey, salt, fish sauce, seaweed, and salted seafood, respectively, were 0.2, 0.2, 0.6, 4.0, and 5.3 items/g, and the median MPs abundances in beverages, beer, and soy sauce were 1.9, 9.0, and 30.0 items/L, respectively. Moreover, deterministic and Monte Carlo calculation results demonstrated that Koreans could potentially ingest 1.4×10^{-4} and 3.1×10^{-4} g MPs/week, respectively. Furthermore, the highest MP intake was 398,520 items/adult/year via tomato consumption. The above studies demonstrated that MP particles are widely present in people's daily diet, posing potential risks to human health.

9.2.2 Drinking Water

MPs have also been detected in drinking water, including tap and bottled water, with an abundance ranging from ND to 10,000 items/L (Figure 9.2b). Acarer (2023) identified the existence of MPs with different polymer types, shapes, and sizes in drinking water sources, influents, and effluents of drinking water treatment plants, tap water, and bottled water. Kosuth et al. (2017) detected MPs in tap water from Indonesia, India, Uganda, Ecuador, Cuba, Lebanon, the United States, and Europe; 83.0% of these samples were contaminated with MP fibers. Based on recommended water intakes, men and women were predicted to consume 14.0 and 10.0 plastic particles daily by drinking the affected water. Zhou et al. (2018) collected tap water samples from 14 countries: Cuba, Ecuador, England, France, Germany, India, Indonesia, Ireland, Italy, Lebanon, Slovakia, Switzerland, Uganda, and the United States. The MP concentration was ND–61.0 items/L, and fibers were the dominant shape. Differently, fragments were found to be the dominant shape of MPs in German tap water and groundwater, and the abundance of MPs was ND–7.0 items/m^3 (Mintenig et al., 2019). In contrast, the MP abundance in tap water in Iran was relatively high (228.0–335.0 items/m^3) (Taghipour et al., 2023).

In tap water samples from different cities in China, 14 MP types were found, among which PE and PP were the main ones, including fragments, fibers, and spheres, with a concentrations range of ND–1247.0 items/L, and MPs smaller than 50 μm were dominated (Tong et al., 2020). In contrast, the concentration of MPs in groundwater was relatively low (4.0–72.0 items/L) and was dominated by fragments (Shi et al., 2022). MPs were also found in South African suburbs in tap water, ranging from 4.7 to 31.0 items/L (Ramaremisa et al., 2024). Furthermore, Vega-Herrera et al. (2022) used an advanced polymer chromatography column coupled to high-resolution mass spectrometry to quantitatively characterize MP concentration, especially for the small size MPs (0.7–20.0 μm), in Spanish tap water. The sum of the total polymers measured in the samples ranged from 514.0 to 1583.0 ng/L. Moreover, Sun et al. (2024c) recompiled baseline data from 34 countries to evaluate the status and drivers of MP pollution in tap water systems worldwide. The results demonstrated that particles smaller than 50 μm dominated global tap water systems, with fragments and polyester (PES) as the most common shape and composition.

In addition to tap water, the use of bottled water has drastically increased over the past few decades (>400 billion L/year in 2022) (IBWA, 2023). Tse et al. (2022) found that the abundance of MPs in bottled water was markedly higher than in tap water. Mason et al. (2018) collected 259 water bottles from 11 countries (China, the United States, Indonesia, India, Mexico, the United Kingdom, France, Germany, Brazil, Lebanon, and Italy) with high packaged drinking water market size and per capita consumption. The MP abundance was ND–10,000.0 items/L with an average of 325.0 items/L, which was higher than that in the tap water. Praveena et al. (2022) detected MP concentration in Malaysian bottled water, which was 8.0–22.0 items/L. Kankanige and Babel (2020) found that smaller-sized MPs (6.5–20.0 μm) contamination in single-use PET-bottled water (140.0 ± 19.0 items/L) in Thailand, and fiber was the dominant MP shape. Schymanski et al. (2018) tested the MP abundance in water from 12 returnable and 10 single-use plastic bottles, 3 beverage cartons, and 9 glass bottles, which was correlated with the materials the bottles/cartons were made of. Besides, the MP abundance in returnable plastic bottles (118.0 ± 88.0 items/L) was significantly higher than that in single-use plastic bottles (14.0 ± 14.0 items/L). It was inferred that recyclable plastic bottles were more likely to release MP particles.

Patil et al. (2024) found that the local bottled water (212.0 ± 100.0 items/L) showed higher MP contamination compared to national bottled water (72.0 ± 36.0 items/L). On the contrary, Samandra et al. (2022) found that the MP abundance in Australian bottled water was lower than in imported bottled water. This indicated that the MP levels varied significantly between national vs. local products. Moreover, Tse et al. (2022) found that the MPs abundance in different particle size ranges in bottled water differed. Small MPs with sizes 1.0–10.0 μm were found to be more abundant in all samples when compared to larger MPs (10.0–50.0 μm, 50.0–5000.0 μm). Small plastic particles pose a higher risk to the human body and should be tested carefully. In addition to MPs themselves, emerging contaminants, including pharmaceuticals and personal care products, bisphenol A, phthalates, alkylphenols, and perfluoroalkyl and polyfluoroalkyl substances were also detected in bottled water from different countries (Akhbarizadeh et al., 2020). To sum up, the existence of MPs in drinking water has become a problem that cannot be ignored.

9.2.3 Table Salt

As an indispensable condiment in daily life, the safety of salt has been widely considered. MPs have been commonly detected in table salt of > 200 brands worldwide, with abundance ranging from ND to 31,680.0 items/kg (Figure 9.2c). Kim et al. (2018) collected 39 brands of commercial salt from supermarkets in 17 different countries on four continents. The results showed that sea salt (675.0 ± 2560.0 items/kg) contained a higher abundance of MPs compared with rock salt (38.0 ± 55.0 items/kg) and lake salt (245.0 ± 307.0 items/kg). Besides, the sea salt samples produced in Asia contained relatively higher MPs than those from other continents. Furthermore, it was also demonstrated that MP pollution level of sea salt in the study area was closely related to that of seawater and marine organisms, tending to increase proportionally to the amounts of plastic debris entering the sea. Similar results were also found by Khuyen et al. (2021). It was inferred that high MP concentrations in sea salt might be attributed to MPs in the sea.

In contrast, the MP concentration of the sea salt purchased from a Malaysian market but made in Australia, France, Iran, Japan, Malaysia, New Zealand, Portugal, and South Africa was relatively low (1.0–10.0 items/kg) (Karami et al., 2017). Lee et al. (2019) found similar results in sea salt from Taiwan (9.8 items/kg). The large variation of MP abundance in table salt may result from different analytical methods. Kim and Song (2021) found that the MP accumulation and composition depended markedly on the measured/targeted minimum cutoff size. Hence, compilation and comparison of different datasets without correction for measured MP size may cause erroneous estimates of human MP intake. When not corrected by size, MP abundance differed 10–600 times among MP identification methods, with the most outstanding value from visual observation, followed by Fourier-transform infrared spectroscopy (FTIR) and Raman spectroscope (Lee et al., 2021). Size-corrected values showed that adults consume up to 19,000 MP particles ≥10 μm yearly through salt.

Ruan et al. (2024) developed a method combining surface-enhanced Raman scattering and stimulated Raman scattering to achieve the filtration, enrichment, and detection of NPs in sea salt. It was estimated that the average person might ingest up to 6 million NPs (0.2–1.0 μm) annually just by consuming sea salt. Sivagami et al. (2021) preliminarily evaluated the cytotoxicity of MPs isolated from sea salt against embryonic kidney–293 cells of humans. The results showed that MP treatment caused cells to form apoptotic bodies via pycnotic nuclei, promoting cell apoptosis. This suggests that these MPs detected in salt may exert lethal effects on human cells, potentially harming human health. The consequences of salt intake in humans require much exploration.

9.2.4 Inhalation

The occurrence of MPs in the air has attracted increasing attention since 2015. Many studies have been conducted on MP detection in air and dust in different regions worldwide (Figure 9.2c). The density of MPs in the air tends to be lower than that of soil minerals. Hence, MP dust can stay in the air for long, facilitating its transport to distant locations (Sridharan et al., 2021). MPs have been detected in the air in the North Atlantic (Goßmann et al., 2023), the west Pacific (Liu et al., 2019), the northwest Pacific (Ding et al., 2022), the South China Sea, and the East Indian Ocean (Wang et al., 2020), even in the free tropospheric (Allen et al., 2021). Liu et al. (2019) found that the MPs abundance in the coastal area

(0.13 ± 0.24 items/m^3) was higher than that in the pelagic area (0.01 ± 0.01 items/m^3). Besides, Western Pacific suspended atmospheric MPs' morphology and composition were consistent with airborne MPs. Hence, it was proposed that suspended atmospheric MP was a vital source of ocean pollution. However, the study by Wang et al. (2020) suggested that atmospheric MP transmission as the primary source of oceanic MPs based on transoceanic studies was not a plausible assumption. The related transmission mechanism remains to be explored.

Many studies attempted to collect and monitor MPs in the air/dust in different regions and under different climate conditions. Zhang et al. (2020) collected 286 indoor dust samples from 12 countries; the abundance of PET- and PC- based MPs was 38.0–120,000.0 and <0.1–1700.0 μg/g, respectively. Notably, economic development (GDP per capita) was found to be one of the crucial factors in the indoor-dust-MPs distribution. Besides, Perera et al. (2022) also found that the abundance and distribution of airborne MPs were related to population density, industrialization level, and human activity. Moreover, the concentration of MPs indoors (0.1–0.9 items/m^3) was higher than that outdoors (ND–0.2 items/m^3). Differently, Zhu et al. (2021) monitored the abundance of MPs in the air of 5 megacities in northern and southeastern China. They found no consistent correlation between airborne MP abundance and typical socioeconomic indices. This might be because the sampling locations were all in megacities. In addition, detection time (Liu et al., 2019; Liu et al., 2025) and rainfall (Li et al., 2023a) also affected the number of MPs in the air.

9.2.5 Miscellaneous

Dermal contact is another known way of exposure to MPs, although it seems less prevalent than ingestion and inhalation. Abbasi and Turner (2021) detected MPs on the surface of the hand (2.1 items/individual/d), face (2.0 items/individual/d), and skin and hair (>3.5 items/individual/d). Since skin pores range from 40.0 to 80.0 μm, NPs (<100.0 nm), synthetic fibers (<25.0 μm), and additives can cross the dermal barrier, causing potential harm to humans (Wu et al., 2022). The German Federal Institute for Risk Assessment found that MPs used in face pack peelings and shower products can lead to the absorption of PE and PP particles in the skin tissues, ultimately causing local inflammation and cytotoxicity (Sharma & Chatterjee, 2017). Moreover, Eom et al. (2024) used cellular spheroids as a 3D cell culture model to examine the penetration of PS MPs. Results showed that MPs (0.1–3.0 μm) adversely affected human dermal fibroblast-derived spheroids and affected gene expression related to the extracellular matrix and integrin-mediated adhesion, which might potentially affect skin disorder development.

The above results demonstrated that MPs can pass through the skin barrier and cause potential harm to the human body, which needs further study. In addition to plastic particles, additives in MPs pass through human skin more easily. Abafe et al. (2023) provided new insights into the skin bioaccessibility of additive brominated flame retardant (BFR) chemicals from different MP types. They assessed subsequent human skin exposure to these chemicals through skin contact with MPs. Results showed that people could get skin exposure to BFRs through contact with MPs in indoor dust (Exposure dose: 0.01–231.0 ng/kg bw/d). Hence, the potential hazards of skin contact with MPs must be taken seriously.

Li et al. (2023b) found that intravenous infusion products could release MPs/NPs during intravenous infusion, which directly entered the human blood circulation system. The results showed that for each intravenous infusion of 250 mL injection solution, about 0.5 μg/10^5–10^{11} items of PVC MPs/NPs were released from intravenous infusion products. Zhu et al. (2023b) also discovered a similar phenomenon, demonstrating that MPs can enter the blood through infusion therapy. Besides, MPs/NPs were also detected in intravenous hypertonic fluids (62.8 ± 72.4 items/L) (Çağlayan et al., 2024) and disposable PVC infusion tubes (PVC NPs: 0.042–0.087 μg/4 h/each tube) (Zheng et al., 2024). Except for intravenous injection, Yang et al. (2023) also found worrisome changes in the distribution of MPs in the blood of patients before (main type: PET, 67.0%) and after (main type: PA, 57.0%) cardiac surgery, as well as the reduction in the predominant size of MPs (Before: 30.0–50.0 μm; After: 20.0–30.0 μm). Moreover, high MP exposure was also detected in surgical environments, reaching 13,048.0 ± 14,283.0 items/m^2/d (Field et al., 2022). Notably, the abundance of MPs in the surgical environment was dynamic. The amounts of

MPs (PET, PP, PE, and PA) for working hours (37,882.0 ± 15,950.0 items/m^2/d) in the operating theatre were significantly higher than that for non-working hours (5332.0 ± 3260.0 item/m^2/d). This implied the influence of surgical procedures on MPs' exposure in the surgical environment. The above studies illustrate a previously ignored but critical pathway of human exposure to MPs, intravenous injection, and surgery.

These studies show that MP pollution has penetrated every aspect of our lives. They exist in water and atmosphere and are widely distributed in our living environment and various scenes. These MPs can enter organisms through the food chain, the air, or direct contact, posing a potential threat to ecosystems and human health.

9.3 Monitoring of Microplastics Accumulation in Human Tissues

Previous studies have shown that MPs are ubiquitous in the environment, food, and consumer products. So, will these foods containing MPs be absorbed after entering the human digestive tract, affecting human health? Many studies have reported the detection of MPs in human tissues, including intestine (7.9–702.7 items/g), lung (0.2–14.2 items/g), placenta (0.3–18.0 items/g), and feces (0.18–138.9 items/g) (Table 9.1). This section reviews relevant literature published recently to help understand the distribution characteristics of MPs in humans.

9.3.1 Digestive Tract

As the first site for MP ingestion, the gastrointestinal tract is critical as the gatekeeper for their bioavailability. Zhang et al. (2022c) found that the abundance of MPs in nasal secretions of people living near plastic factories was significantly higher than that of people living in other areas. It was inferred that the amounts of MPs in the human digestive tract might be related to the environment in which people live. Tonsils are the first line of defense against oral MPs, and multiple types of MPs were detected in tonsils (Zhu et al., 2024), with PVC (46.4%) and PBS (47.3%) being the predominant types. In contrast, the abundance of PBS was relatively reduced in small- (29.8%) and large- (0.6%) intestine tissues. This may be because the PBS is filtered or biodegraded in the digestive tract.

Ibrahim et al. (2021) detected MPs (such as PC, PA, PP) in colon samples, which indicated that MPs are ubiquitously presented in the human digestive tract. Among them, filament or fiber was the most common shape, suggesting that they were difficult to digest and might eventually be excreted from human bodies with feces. Yan et al. (2024) conducted long-term continuous MP exposure experiments on mice, established a quantitative human intestinal MP exposure assessment model, and calculated that the cumulative excretion rate of fecal MPs was stable at around 93.0%. This illustrated that the human digestive tract might not absorb most ingested MPs but eventually excreted with feces. However, the MPs (PE, PMMA, and PA) in tumoral colon tissues were significantly higher than in non-tumoral colon tissues (Cetin et al., 2023). Similarly, Zhang et al. (2024) also found the existence of MPs in gallstones, and significantly high MPs were observed in young patients with cholelithiasis. This indicated a possible connection between colorectal cancer/gallstones and MP exposure.

The above studies show that although most MPs, especially large fragments, are difficult to degrade and absorb by the digestive tract, their impact on human health requires attention. The release of potentially toxic substances during the transfer of MPs in the digestive tract, as well as the effects of the shape and size of ingested MPs on the digestive tract, are desperately in need of attention. Tamargo et al. (2022) simulated and assessed the potential risks of MPs at the gastrointestinal tract level and found that the morphology of PET MPs in the colon was different from that of the original particles, which may be biotransformed in the gastrointestinal tract. On the contrary, Stock et al. (2020) found that after artificial digestion, a unique protein corona formed on the surface of plastic pellets, but the digestive fluids did not break down the pellets or change their shape and size. More studies are needed to fully evaluate the influence of different types of MPs on the human digestive tract.

TABLE 9.1

Presence of Microplastics in Human Tissues or Fluids

Human Body Sample	Detection Method	Sample Number	Type[a]	Abundance	Size (μm)	References
Tonsil	LDIR, FTIR	16	10 types, PVC, PBS	6.0 ± 7.4 items/g	20.3–471.9	Zhu et al. (2024)
Small intestine	LDIR, FTIR	16	10 types. PVC, PBS	9.5 ± 13.1 items/g	20.3–498.1	
Large intestine				7.9 ± 7.0 items/g		
Intestine	LDIR	40	31 types. PU, SR, EVA, PE	425.0–1750.0 items/individual	20.0–500.0	Zhang et al. (2022c)
Colon	ATR-FTIR, Raman spectroscope	31	PE, PMMA, PA	207.8 ± 154.1–702.7 ± 504.3 items/g	1.0–1299.0	Cetin et al. (2023)
Colon	μ-FTIR	11	PC, PA, PP	28.1 ± 15.4 items/g	1100 ± 300	Ibrahim et al. (2021)
Bronchoalveolar lavage fluid	Confocal μ-Raman spectrometer	77	10 types. PP, PE, PET	431.0 ± 277.0 items/L	< 20	Chen et al. (2023a)
Bronchoalveolar lavage fluid	LDIR	18	13 types. PE, PET, PP	13.4 items/g	20–80	Qiu et al. (2023a)
Bronchoalveolar lavage fluid	LDIR, FTIR	32	Smoker: 31 types. PU, PE, silicone; Non-smoker: 17 types. PU, PET, PE, POM	Smoker: 25.9 ± 24.0 items/g; Non-smoker: 13.4 ± 15.3 items/g	/	Lu et al. (2023)
Sputum,	Polarizing microscope	9	/	105.4 items/g	20.5–2476.8	Na et al. (2021)
Nasal lavage fluid				0.8 items/g	50.8–6200.0	
Sputum	LDIR; μ-FTIR	22	21 types. PU, PET, CPE, AV	1.9–9.2 items/mL	20.0–500.0	Huang et al. (2022b)
Sputum	LDIR	16	Couriers: PC, PVC; Office staff: PVC, PA	102.9, 134.3 items/g (Couriers, Office staff)	/	Jiang et al. (2022)
Nasal lavage fluid			Couriers: PA, PE; Office staff: PVC, PA	0.8, 2.6 items/g (Couriers, Office staff)	/	
Lung	μ-FTIR	13	12 types. PP, PET	0.7 ± 0.8 items/g	Length: 223.1 ± 436.2; width: 22.2 ± 20.3	Jenner et al. (2022)
Lung	μ-FTIR, LDIR	50	16 types. AF, PET, PR	0.7 ± 0.7 items/tumor tissue; 0.2 ± 0.5 items/normal tissue	0.0–50.0	Chen et al. (2022b)

(Continued)

TABLE 9.1 (*Continued*)

Presence of Microplastics in Human Tissues or Fluids

Human Body Sample	Detection Method	Sample Number	Type[a]	Abundance	Size (μm)	References
Lung	LDIR	12	12 types. PP, PET, PS	4.3 ± 5.1 items/g	20.0–100.0	Wang et al. (2023b)
Lung	LDIR, FTIR	14	9 types. PVC, PA, PET, PBS	14.2 ± 14.6 items/g	20.3–477.7	Zhu et al. (2024)
Heart	LDIR	15	9 types. PET, PU	14.0–7268.0 items/g	20.0–469.0	Yang et al. (2023)
Gallstone	LDIR, Py-GC/MS	16	PS, PE, PP, PET, EVA	>12.7 items/g >14.6 μg/g	0.0–30.0	Zhang et al. (2024)
Liver	Raman spectroscope	11	6 types. PS, PVC, PET, PMMA, POM, PP	3.2 items/g	4.0–30.0	Horvatits et al. (2022)
Kidneys		3		ND[b]		
Spleen		3		0.9 items/g		
Blood	LDIR	15	8 types. PA, PET	4.0–114.0 items/g	20.0–184.0	Yang et al. (2023)
Blood	ATR-FTIR, Py-GC/MS	22	PET, PS, PE, PMMA	1.6 μg/mL	0.7–500.0	Leslie et al. (2022)
Saphenous vein tissue	μ-FTIR	5	5 types. AR, PVAc, nylon-EVA	15.0 ± 17.2 items/g	7.0–1074.0	Rotchell et al. (2023)
Thrombi	Raman spectrometer, quality control-particle detection system	26	H_2Pc, H-G, PE	1.0–15.0 items/individual	2.1–26.0	Wu et al. (2023)
Thrombi	LDIR, Py-GC/MS	30	PE, PA66, PVC	14.9–221.8 μg/g	20.0–50.0	Wang et al. (2024)
Carotid artery plaque	Py-GC/MS	304	PE, PVC	PE: 21.7 ± 24.5 μg/mg; PVC: 5.2 ± 2.4 μg/mg	<1.0	Marfella et al. (2024)
Atherosclerotic plaque	Py-GC/MS	17	PET, PA-66, PVC, PE	118.7 ± 53.9 μg/g	/	Liu et al. (2024)
Enclosed body fluid	μ-Raman spectroscope	104	23 types. PP, PS, PTFE, PVB, PA, PE, PEAA, PSAN, PVAc	2.0–38.0 items/mL	2.2–103.3	Guan et al. (2023)
Semen	Py-GC/MS, LDIR	6	24 types; PS, PE, PVC	0.2 ± 0.5 items/mL	20.0–100.0	Zhao et al. (2023)
Testis		30		11.6 ± 15.5 items/g		
Semen	μ-Raman spectroscope	10	16 types. PP, PS, PET, PE, POM, PVC, PC, acrylic	ND–2.5 items/mL	2.0–6.0	Montano et al. (2023)
Endometrium	LDIR, questionnaire	20	13 MPs. PEAA, fluororubber, PE	ND–1170.0 items/g	20.0–500.0	Sun et al. (2024b)

(Continued)

TABLE 9.1 (*Continued*)

Presence of Microplastics in Human Tissues or Fluids

Human Body Sample	Detection Method	Sample Number	Type[a]	Abundance	Size (μm)	References
Placenta	μ-Raman spectroscope	43	PE, PS	2.0–38.0 items/individual	Normal: 7.3–27.6; IUGR: 2.9–34.5	Amereh et al. (2022)
Placenta	LDIR	17	14 types. PVC, PP, PBS	0.3–9.6 items/g	20.3–307.3	Zhu et al. (2023a)
Placenta and meconium	LDIR	18 mother–infant pairs	16 types. PA, PU	Placenta: 18.0 items/g; Meconium: 51.4 items/g	20.0–50.0	Liu et al. (2023)
Placenta	Raman XploRA Nano Microspectrometer	4	12 types. PP, paint, coating, dye	12.0 items/sample	5.0–10.0	Ragusa et al. (2021)
Placenta, meconium and feces	μ-FTIR	2	10 types. PE, PP, PS, PU, airborne fallout	Feces: PE (0.096 items/g)	>50.0	Braun et al. (2021)
Placenta	ATR-FTIR, Py-GC-MS	62	12 types. PE, PVC, nylon 66	126.8 ± 147.5 μg/g	/	Garcia et al. (2024)
Feces	FTIR	30	PET, PA, PE, EP	0.2–0.8 items/g	200.0–4900.0	Hasanah et al. (2024)
Feces	Extracted via SPE method, detected by HPLC	19	PET, PC	PET: ND–16.0 μg/g, PC: ND–2.1 μg/g	3000.0–5000.0	Zhang et al. (2021a)
Feces	μ-FTIR	26	8 types. PP, PET, PS, PE, PVC, PC, PA, PU	1.0–36.0 items/g	20.0–800.0	Zhang et al. (2021b)
Feces	FTIR	8	9 types. PP, PET	1.8–17.2 items/g	50.0–500.0	Schwabl et al. (2019)
Feces	Raman spectroscope, intensity contribution	11	PE, PP, PS, PET	3.3–14.0 μg/g	/	Luqman et al. (2021)
Feces	Raman spectroscope, intensity contribution	11	4 types. PP, PE, PS, PET	6.9–16.6 μg/g	/	Wibowo et al. (2021)
Feces	μ-Raman spectroscope	102	15 types, PET, PA	Patients: 41.8 items/g; healthy: 28.0 items/g	1.7–393.8	Yan et al. (2022a)
Feces	μ-Raman spectroscope	8	PS, PP, PE, PET, PVC	20.4–138.9 items/g	< 300.0	Ho et al. (2022)

[a] PES, polyester; PE, polyethylene; PET, polyethylene terephthalate; PVC, polyvinyl chloride; PP, polypropylene; PS, polystyrene; PA, polyamide; PC, polycarbonate; PU, polyurethane; PBS, poly(butylene succinate); PMMA, polymethyl methacrylate; PTFE, polytetrafluoroethylene; PVAc, poly (vinyl propionate/acetate); PVB, polyvinyl butyral; EVA, ethylene vinyl acetate; POM, polyoxymethylene (polyformaldehyde); AV, alkyd varnish; SR, silicon resin; AF, acrylic fiber; AR, alkyd resin; PR, phenoxy resin; PEAA, polyethylene-co-acrylic acid; PSAM, polystyrene-co-acrylonitrile; H_2Pc, phthalocyanine; H-G, hostasol-green.

[b] ND, not detected.

9.3.2 Respiratory Tract

Inhalation is one of the critical ways for humans to contact MPs. Several studies have detected MPs in sputum (Huang et al., 2022b), nasal lavage fluid (Jiang et al., 2022), bronchoalveolar lavage fluid (Baeza-Martínez et al., 2022), and lungs (Jenner et al., 2022). This indicated that MPs in air or dust may enter the human body through the respiratory tract, causing potential harm to cells and tissues. The research results of Jiang et al. (2022) and Na et al. (2021) were consistent. They all found that the abundance of MP in the courier's sputum and nasal lavage fluid was ~100.0 and 0.8 items/g, respectively. Jiang et al. (2022) also found that the abundance of MPs in the nasal lavage of office staff was markedly higher than that of couriers. This may be related to the fact that the MP concentration in indoor air is generally higher than in outdoor air. The results of MPs in the upper respiratory tract of humans demonstrated that MPs in the environment can enter the human body via inhalation.

Bronchoalveolar lavage fluid (BALF) is an excellent medium to represent the condition of the human lower respiratory tract, which can provide helpful information on the environmental status of alveoli and terminal bronchioles (Sartorelli et al., 2020). Baeza-Martínez et al. (2022) first reported the existence of MPs in BALF. They found that the MPs abundance in high-risk occupations (e.g., carpentry, construction) was significantly higher than in low-risk occupations (e.g., salesman, teacher). Moreover, the abundances of total MPs, PU, and SIL in smokers' human BALF were significantly higher than those of non-smokers (Lu et al., 2023). Chen et al. (2023a) also found that the MPs pollution in BALF children in the severe CAP group was considerably higher than in the non-severe group. It was illustrated that occupation and living habits were critical factors affecting the abundance of MPs in BALF, and the inhaled MPs might play a vital role in human respiratory pathology.

Zhu et al. (2024) detected MP distribution in different human body tissues. They found the highest abundance of MPs in lung tissue, followed by small intestine, large intestine, and tonsil, suggesting that inhalation was the primary route of human exposure. Moreover, MPs were identified within all regions (upper-, middle/lingular-, lower- regions) of the lung (Jenner et al., 2022), providing convincing evidence for the presence of MPs within the human lung. Studies explored the correlation between lesions and MP pollution in lung tissues. The results showed that the tumor detection rate in patients with a higher risk history of microfiber exposure was markedly higher than that with a lower one, indicating that microfiber inhalation could be correlated with the formation of ground glass nodules (Chen et al., 2022b). Besides, the MP levels in lung tissues were positively correlated with platelet (PLT), plateletcrit (PCT), and fibrinogen (FIB). They were negatively associated with direct bilirubin (DB), revealing the correlation between MP exposure and blood test indicators (Wang et al., 2023b). Extensive research is needed to elucidate the mechanisms of potential adverse health effects of these MPs in the human respiratory tract and other tissues.

9.3.3 Human Blood

Blood irrigates organs in the human body and is the transport pathway for oxygen, nutrients, and potentially toxic substances (e.g., MPs) around the body to distal tissues and organs. Leslie et al. (2022) detected MPs (including PET, PS, PE, and PMMA) in human blood (1.6 μg/mL). Besides, MPs (AR, PVAc, nylon-EVA) were also detected in human saphenous vein tissues (Rotchell et al., 2023). This indicated that the inhaled or ingested MPs were bioavailable and could be absorbed into the human bloodstream and even vascular tissues. These MPs in blood vessels may be closely linked to human diseases.

Studies detected MPs in thrombi in different parts of the human body (Wu et al., 2023; Wang et al., 2024). Wang et al. (2024) detected fragment-shaped MPs and spherical-shaped MPs in the thrombi in cerebral arteries in the brain, coronary arteries in the heart, and deep veins in the lower limbs. Notably, the abundance of MPs was significantly higher in thrombi in the middle cerebral arteries than in the internal carotid arteries. Furthermore, Chen et al. (2022a) introduced an optically assisted thrombus platform, verified MPs' effect on thrombus formation and structure, and found that MPs invasion led to decreased binding strength of fibrin to platelets. Similarly, Kim et al. (2022) found that NPs promoted procoagulant activation of isolated human red blood cells and thrombosis in rats. This demonstrates that MPs in the thrombi gradually accumulate as blood flows, which may be associated with disease severity.

Liu et al. (2024) found that MPs may be associated with human atherosclerosis. The abundance of MPs in arteries (coronary and carotid arteries) containing atherosclerotic plaques was significantly higher than in aortas without atherosclerotic plaques. In a 34-month follow-up of 304 patients, Marfella et al. (2024) found that patients with detectable MPs/NPs on carotid artery plaques were more likely to have myocardial infarction, stroke, or all-cause death, compared with patients with no detectable MPs/NPs. These results demonstrated that the existence of MPs in blood might be one of the vital causes of cardiovascular diseases such as thrombus and atherosclerotic plaque and requires urgent attention.

Moreover, irregular block-shaped and pigment MPs, andiron compounds, and metallic oxides were detected in thrombus in patients with arterial dissection or acute arterial embolism of lower extremities (Wu et al., 2023). Similarly, iron-free mineral particles, compounds, and carbon or organic groups were also detected in kidney, liver, pelvic, and gall bladder fluids (Guan et al., 2023). Besides, chlorine-containing jagged-edged foreign particles were also detected in carotid artery plaque. It was inferred that these substances might be loaded on MPs and transported to remote tissues along with blood. That is, MPs can act as vectors of potentially toxic chemicals. Liu et al. (2022c) elucidated that co-exposure of MPs and iron disrupted brain iron homeostasis, leading to ferroptosis in cognitive-related brain regions, thereby exacerbating cognitive impairment. Therefore, the coexistence of MPs and potentially toxic chemicals may aggravate the adverse effects of MPs on the human body.

9.3.4 Genital System

There are currently some studies on detecting MPs in the human genital system. Zhao et al. (2023) detected MPs (PS, PE, PVC) in the testis (11.6 ± 15.5 items/g) and semen (0.2 ± 0.5 items/mL) of humans. Moreover, the size of MPs in semen was larger than in testis, and the dominant MPs types and shapes in testicles and semen were different. The MP abundance in semen (0.9 items/mL) detected by Montano et al. (2023) was similar to the previous results. It was indicated that MPs might enter semen via the inflammation-sensitive epididymis and seminal vesicles. In addition, MPs (Up to 1170.0 items/g, including PEAA, fluororubber, and PE) were detected in human endometrium, which played an essential role in human health and female fertility (Sun et al., 2024b). Notably, it was also found that drinking habits might be positively correlated with MP concentration in the human body, and chewing gum was potentially associated with MP exposure. These findings provide critical evidence for the distribution of MPs in the reproductive system.

The above studies demonstrate that ingested or inhaled MPs can enter the human genital system, such as testicles, semen, and endometrium, which may affect male reproductive ability and may even be genetically toxic. Hou et al. (2021) found that sperm cells atrophied, shed, and underwent apoptosis after exposure to PS MPs, and PS MPs led to abnormal testicular spermatogenesis by inducing inflammation. Besides, An et al. (2021) found that PS MPs could enter mice ovaries, causing granulosa cell apoptosis and ovarian fibrosis. Moreover, Wei et al. (2022) found that MP exposure damaged the testes and ovaries and altered the serum hormone levels. There was more PS MP accumulation in the ovaries than in the testes, and female mice were more susceptible to MPs in reproduction and fertility than males. Maternal exposure to PS NPs during gestation and lactation even caused trans-generational toxicity in mice, such as inflammatory response and glycometabolism disorder in the liver. It disrupted the spermatogenesis of mouse offspring (Huang et al., 2022c). These results illustrate that MPs exhibit potential harm to the functions of the human reproductive system. However, more human experimental data are needed to reveal the corresponding mechanisms.

9.3.5 Human Placenta

Continued exposure to MPs has raised concerns about human health, particularly reproductive function. The placenta provides nutrients and oxygen to the fetus via the umbilical cord and is an essential interface in regulating the fetal and maternal environments. Therefore, researchers paid attention to the distribution of MPs in the human placenta. Ragusa et al. (2021) detected MPs, including PP, paint, coating, and dye, in all placental portions: maternal, fetal, and amniotic membranes. Besides, Zhu et al. (2023a) also found a high abundance of MPs (PVC, PP, and PBS) in placentas (2.7 items/g). Furthermore, the concentration of MPs (PE, PVC, and nylon 66) in the placenta was quantitatively evaluated by using ATR-FTIR and py-GC-MS and was 126.8 ± 147.5 μg/g (Garcia et al., 2024). These studies demonstrate the presence of MPs in the human placenta, which may be maternally transmitted.

Intrauterine transmission via the placenta to the fetus is possible, and the potential risks during fetal development must be recognized (Braun et al., 2021). Studies have shown that MPs were found in the placenta in all intrauterine growth restriction (IUGR) pregnancies, with MPs detection rates being markedly higher than that in normal pregnancies (5.3%) (Amereh et al., 2022). Besides, negative associations between MP exposure and birth outcomes (including birth weight, length, head circumference, and 1-min Apgar score) were also observed. Moreover, logistic regression models showed that bottled water, takeaway food, and outdoor work were positively associated with MPs in the placenta. In addition, MPs (mainly CPE) and additives (calcium zinc PVC stabilizer) were also detected in the amniotic fluid of pregnancies complicated with preterm premature rupture of membranes (PPROM) (Halfar et al., 2023).

Similarly, Garcia et al. (2024) also found a negative correlation between MP abundance and gestational age in amniotic fluids. High concentrations of MPs, especially those of 50.0–100.0 μm, significantly impacted the microbial community structure in the placenta and meconium (Liu et al., 2023). These studies enhanced the understanding of MPs transfer through blood-placenta-amniotic fluid-meconium. MPs may be gradually transferred to the placenta after being ingested and inhaled into the human body, potentially affecting newborns' growth and development.

Many studies have revealed the toxicological effects of MPs on the female reproductive system, such as placental dysfunction (Dibbon et al., 2024), inducing miscarriage (Wan et al., 2024), and fetal growth restriction (Aghaei et al., 2022; Chen et al., 2023b; Dibbon et al., 2024), causing placental and fetal metabolic disorders (Chen et al., 2023b) in mice. The ingested PS NPs were shown to breach the intestinal barrier and, subsequently, the placenta's maternal-fetal barrier, entering the fetal circulation and all fetal tissues (Cary et al., 2023). Furthermore, oral (Mortensen et al., 2019) or pulmonary (Fournier et al., 2020) NP exposure can negatively affect placenta function and fetal development. This suggested that MPs/NPs might enter the placenta through inhalation or ingestion, potentially negatively affecting the fetus. Furthermore, PS NPs increased intracellular reactive oxygen species (ROS) production in human placental choriocarcinoma JEG-3 cells, leading to cytotoxicity, cell cycle arrest, and the inhibition of protein kinase A (PKA) activity (Shen et al., 2022). Similarly, NPs could enter the human trophoblastic cytoplasm, decrease cell viability, cause cell cycle arrest, reduce the cell migration and invasion abilities, increase the production of intracellular ROS and proinflammatory cytokines (TNF-α and IFN-γ) (Hu et al., 2022). These findings provide empirical evidence of the potential trophoblast toxicity of NPs in mammals and emphasize the need for risk assessment of MPs/NPs on female reproduction and fetal development.

9.3.6 Human Feces

Many studies have detected MPs in human feces from different regions. Schwabl et al. (2019) detected MPs (PP, PET) in the feces of 8 volunteers residing in Europe and Asia and found an average of about 20.0 items in every 10 g of feces. The study of Ho et al. (2022) showed that the abundance of MPs (PS, PP, PE) in stool samples by Hong Kong residents was approximately 50.0 items/g. Besides, Zhang et al. (2021a) determined PET and PC MPs in the feces of adults and infants from the United States and found that the PET concentration in infant feces (36,000.0 ng/g) was significantly higher than those in adult feces (2600.0 ng/g). Yan et al. (2024) calculated that the cumulative excretion rate of fecal MPs was about 93.0%. It was indicated that intestinal excretion might be vital for the body to eliminate MPs.

The distribution of MPs in human feces was found to be significantly correlated with a variety of variables. A moderate correlation between packaged water and beverage intake and MP abundance in feces was observed by Zhang et al. (2021b), and drinking water could be a source of MP pollution in human feces (Luqman et al., 2021; Wikurendra et al., 2022). Besides drinking water, salt, and oral hygiene products (Wibowo et al., 2021), plastic packaging of drinking water and food and dust exposure (Yan et al., 2022a) may also be a vital source of MP contamination. Yan et al. (2022a) found that the fecal MP abundance in IBD patients (41.8 items/g) was considerably higher than that in healthy individuals (28.0 items/g). This suggests that MP exposure may be related to the occurrence and development of intestinal diseases, thereby enhancing the susceptibility and retention of MPs. Toxicology studies further proved that MPs disrupted the balance between colonic epithelial self-renewal and differentiation, exacerbating colitis (Xie et al., 2023). Because of the potential risks of MPs to human health and the convenience and representativeness of fecal MP detection, we recommend that monitoring of human fecal MPs be conducted to assess the frequency of human exposure to MPs.

9.4 Toxicological Impacts of Microplastics and Associated Chemicals on Humans

The accumulation of MPs in tissues can induce various adverse effects, including oxidative stress and cytotoxicity, imbalance in energy homeostasis and metabolism, disruption of immune function, and neurotoxicity (Deng et al., 2017; Jiang et al., 2022). MPs can also be vectors of microorganisms and potentially toxic chemicals. Therefore, binding MPs to microbes and associated chemicals may enhance their adverse impacts on human tissues (Mammo et al., 2020). This section summarizes MP toxicity research progress based on human and mammalian cells or tissues to illustrate their toxicological effects in humans (Figure 9.3).

9.4.1 Oxidative Stress and Cytotoxicity

Once exposed, MPs may pass across cell membranes and transfer to different tissues, causing damage to the human body (Ragusa et al., 2022). Oxidative stress is an essential indicator in cytotoxicity studies reflecting the degree of disturbance between the production of ROS and antioxidant defenses in cells (Schirinzi et al., 2017; Xie et al., 2020). In addition to ROS, H_2O_2, SOD, GSH-Px, CAT, MDA, T-AOC, and LPO also can be used to assess the degree of oxidative stress in cells exposed to MPs (Deng et al., 2017; Wang et al., 2021b; Sharma et al., 2023). Many studies have shown that once entering cells, MPs/NPs can cause cellular oxidative stress through ROS and affect cell activity, leading to a series of

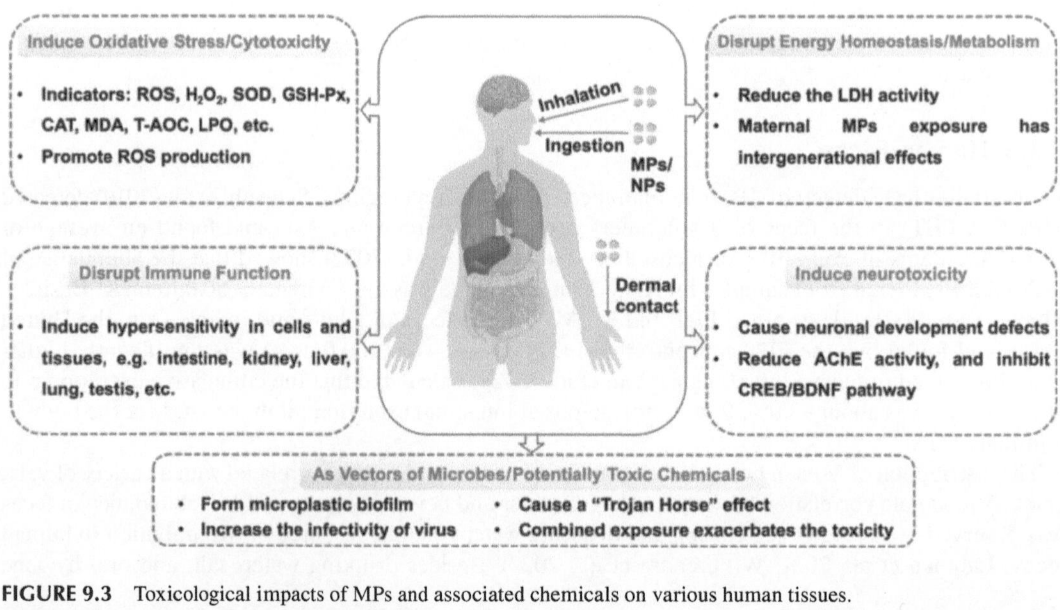

FIGURE 9.3 Toxicological impacts of MPs and associated chemicals on various human tissues.

biological reactions, such as oxidative stress-induced signaling pathways, inflammation, and apoptosis (Liang et al., 2021; Wang et al., 2021b).

The oxidative stress response of human cells, mammalian cells, or tissues to various types of MPs (including PE-, PP-, PET-, PU-, PVC-, and PMMA-NPs/MPs), especially for PS NPs/MPs, has been extensively explored by researchers. Stock et al. (2021) found that the uptake of MPs by human intestinal epithelial and hepatocyte cell lines was material-dependent. A high concentration of PVC MPs (>75,000 μg/mL) significantly suppressed cell viabilities. In contrast, PE-, PP-, and PET-MPs showed less effects on cell viability. Similarly, the addition of PE- and PS-MPs showed no significant reduction of cell viability but markedly promoted ROS generation in human cells (Schirinzi et al., 2017). Interestingly, ROS could only be detected in the low dose of PE MP added group (0.05, 0.1 μg/mL) in the human glioblastoma cell line (T98G). On the contrary, PS MPs significantly promoted ROS generation by the human cervical adenocarcinoma cell line (HeLa) and T98G, especially in the high concentration-treated group (10.0 μg/mL). This demonstrates that the toxicity of PS MPs to human cells may be higher than PE MPs.

The toxicity of MPs to cells is size- and dose-dependent. High dose (1000.0 μg/mL) and small size (20.0 μm) of PP MPs exhibited cell viability-reducing and ROS generation-promoting effects (Hwang et al., 2019). Besides, larger and powdered PP MPs showed low cytotoxicity, indicating that surface property was also one of the main influencing factors. In addition, the surface chemistry of NPs/MPs was also found to affect their toxicity. Multiple studies confirmed that NH_2-modified PS NPs showed more toxic effects on cells than unmodified PS NPs, such as marked oxidative damage and cell death via apoptosis (Chiu et al., 2015; Ruenraroengsak et al., 2015; Inkielewicz-Stepniak et al., 2018; He et al., 2020). NH_2-modified PS NPs increased autophagic flux because of ROS generation and endoplasmic reticulum (ER) stress induced by misfolded protein aggregation, causing autophagic apoptosis (Chiu et al., 2015). Furthermore, exposure to sulfate-modified PS NPs induced foam cell formation, which played a detrimental role in the development of atherosclerosis (Florance et al., 2021).

Cell type is also a factor that cannot be ignored when affecting the toxicity of NPs/MPs. PS NPs reduced the viability of mixed neuronal cells in mice, increased neuronal apoptosis and reactive astrocytosis, but did not affect other cells (Jung et al., 2020). Rubio et al. (2020) also found that the biological effects of PS NPs depended on the leukocyte cell line used; phagocytic cellular (THP-1) likely exhibited high resistance to PS NPs. This indicates that cells in different human body tissues exhibit different tolerances to NPs/MPs, and fragile cells/tissues need more attention during human MP monitoring.

9.4.2 Disruption of Energy Homeostasis and Metabolism

Energy homeostasis is affected by the balance between available energy (intake and reserves) and expenditure. PS MPs can induce insufficient energy metabolism. Deng et al. (2017) found that ATP concentration was significantly reduced, lactate dehydrogenase (LDH) activity was dramatically increased, and the levels of total cholesterol (T-CHO) and triglycerides (TG) were significantly decreased in the liver of mice after exposure to PS MPs. It was demonstrated that the accumulation of MPs may induce energy deficiency and metabolic disorders in the human body. Similar results were also found by Lu et al. (2018), PS MPs induced gut microbiota dysbiosis and hepatic lipid metabolism disorder in mice.

Meanwhile, oral exposure to PS MPs significantly suppressed the secretion of intestinal mucin and induced gut microbiota dysbiosis in mice. This indicated that PS MPs could change the gut microbiota composition and lead to hepatic lipid disorder in mice, the health risks of which to animals cannot be ignored. Maternal MP exposure had intergenerational effects and even caused long-term metabolic consequences in mice's F1 and F2 generations, including metabolic disorders and potential hepatic lipid accumulation in F1 mice (Luo et al., 2019).

Jin et al. (2019) found that PS MPs could alter mice's amino acid and bile acid metabolism. Arginine was significantly elevated after exposure to PS MPs, which might lead to argininaemia. Notably, PS MPs also markedly increased bile acids (TBA) in the liver, suggesting that the enterohepatic circulation of TBA was disturbed, affecting lipids' digestion and absorption and causing metabolic disorders. In addition to enterohepatic cells, dysfunctional lipid homeostasis was detected in tissue-specific and brain-specific macrophages exposed to sulfate-modified PS MPs (Florance et al., 2021). Sulfate-modified PS MPs induced the accumulation of lipids, leading to foam cell formation and contributing to the pathogenesis of atherosclerosis.

Notably, phosphatidylinositol 3-kinase (PI3K) was found to play a pivotal role in the macrocytosis-foam cell formation and autophagy axis. Compared with healthy mice, mice with acute colitis were sensitive to PS MPs. Typical inflammatory factor expression and triglyceride accumulation showed that exposure to PS MPs aggravated dextran sodium sulfate-induced acute colitis and lipid disorders (Zheng et al., 2021). Humans or animals with abnormal functions are more sensitive to MPs and require attention.

9.4.3 Disruption of Immune Function

The immune system is a network of biological systems that protect organisms from diseases, and dysfunction can lead to autoimmune diseases, inflammatory diseases, and cancer. Research has shown that high-dose, small-size PP MPs stimulate the immune system, enhancing potential hypersensitivity to PP MPs by increasing cytokines and histamine levels in cells (Hwang et al., 2019). Besides, the upregulation of inflammatory cytokine transcripts in the distal small intestine and colon in PE MP-exposed mice also indicated the hypersensitivity response of the intestinal immune system to MPs (Djouina et al., 2022). Similarly, the pathways and genes that regulate cellular inflammation in human colorectal adenocarcinoma Caco-2 cells were significantly altered by PE MPs, providing an understanding of the toxicity effects of PE MPs-induced intestinal inflammatory diseases (Wu et al., 2020).

In addition to the digestive system, hypersensitivity reactions also occur in tissues such as kidney (Wang et al., 2021c; Kuang et al., 2024), liver (Zhao et al., 2021), and lung (Xu et al., 2019) during long-term PS MP exposure. Testis is an immune-privileged organ, so testicular immune homeostasis plays a crucial role in spermatogenesis. Jin et al. (2021) found that inflammatory factors were significantly enriched in the testes of mice exposed to PS MPs, indicating that PS MPs induced testicular inflammation. Moreover, the expression of blood-testis barrier (BTB)-associated connexins was significantly downregulated, suggesting that PS MPs disrupted the integrity of BTB. Further research showed that exposure to PS MPs caused reproductive toxicity through oxidative stress and activation of the p38 mitogen-activated protein kinase (MAPK) signaling pathway (Xie et al., 2020). Besides, female mice were more susceptible to MPs than male mice regarding reproduction and fertility (Wei et al., 2022).

Moreover, PS MP consumption can lead to noticeably prolonged proinflammatory responses in joint tissues in mice (Rawle et al., 2022). PS MPs significantly decreased peripheral blood white blood cell (WBC) counts and inhibited the colony-forming ability of bone marrow (BM) cells in mice, causing hematotoxicity to some extent, affecting gene expression, and disturbing related molecular and biological pathways in mouse BM cells (Sun et al., 2021). In addition, after in vivo feeding, PS MPs could accumulate in the spleen, kidneys, heart, liver, lungs, blood, testis and epididymis, brain and thighbone, etc. of mice (Liang et al., 2021; Huang et al., 2022a). It is demonstrated that MPs can cause cytotoxicity to cells or tissues in contact or translocate to distant tissues through blood circulation, thereby causing potential harm to different tissues.

9.4.4 Neurotoxicity and Increasing Incidence of Neurodegenerative Diseases

Exposure to MPs can induce brain damage, causing neurotoxicity and behavioral abnormalities. Jung et al. (2020) isolated mixed neuronal cells and primary astrocytes from mouse brains and explored the effects of PS NPs on the activity and function of these cells. Results showed that PS NPs accumulated in the cytoplasm of mixed neuronal cells, leading to reduced cell viability and defective neuronal development. Besides, PS NP exposure resulted in reactive astrocytosis, which accelerated the apoptosis of nearby damaged neurons. These results demonstrated that PS NPs induced toxicity to cultured neuronal cells. Acetylcholinesterase (AChE) activity is used as a specific biomarker of possible neurotoxic effects of nanomaterials. Multiple studies showed that the activity of AChE in brain tissue was significantly inhibited after exposure to PS MPs/NPs (Estrela et al., 2021; Wang et al., 2022; Sharma et al., 2023).

AChE is responsible for regulating neurotransmission by breaking down acetylcholine. Reduced AChE activity indicated that the cholinergic signaling might be disrupted, affecting various physiological processes such as memory and muscle movement (Sharma et al., 2023).

Moreover, PS MPs/NPs disrupted neuronal arborization and dendritic spine density in the mouse prefrontal cortex, inducing prefrontal cortex-brain-derived neurotrophic factor (PFC-BDNF) neuromorphological toxicity (Suman et al., 2024) and inhibiting the CREB/BDNF pathway (Wang et al., 2022). The PFC acts as the brain's chief executive officer, controlling the highest levels of cognitive and emotional processes. Dysregulation of its synaptic transmission has been linked to social deficits, affective disturbance, and memory loss in brain disorders such as autism, schizophrenia, depression, and Alzheimer's disease (Yan & Rein, 2022b). The BDNF is a target gene of cyclic adenosine monophosphate response element binding protein (CREB) and plays a vital role in the synaptic plasticity of the central nervous system. The dysregulation of CREB/BDNF in the brain may lead to neuronal death and neurodegenerative diseases (Zarneshan et al., 2022).

Furthermore, Liang et al. (2022) mapped cell-specific responses to PS NP exposure in mice brains. They found that PS NPs induced Parkinson's disease (PD)-like neurodegeneration in mice, mainly by causing energy metabolism disorders in the substantia nigra pars compacta (SNc) and striatum. Shan et al. (2022) found that PS NPs penetrated the blood-brain barrier and induced microglia activation in the mice brain (Shan et al., 2022). Paing et al. (2024) also found that microglia in brain cells preferentially uptook PS NPs, causing microglial activation and inflammatory responses, leading to dysregulation of hippocampal neuron activity. This demonstrated that PS NP exposure caused memory impairment in mice and was a potential environmental inducement of cognitive impairment.

Jin et al. (2022) and Lee et al. (2022) also found similar results, showing that exposure to PS MPs led to learning and memory deficits and neurotoxic effects in mice. Interestingly, eliminating the vagus nerve, a modulator of the gut–brain axis, markedly rescued learning and memory in PS MPs-treated mice (Lee et al., 2022). The vagus nerve is one of the regulation pathways between the gut microbiome and its metabolites and the brain, regulating blood-brain barrier integrity, hippocampus-dependent learning and memory, neuroplasticity, and hippocampal inflammation (Suarez et al., 2018; Chen et al., 2021). It would be worth investigating the potentially critical role of the gut-brain axis in PS MPs/NPs-induced neurotoxicity in mouse and human brains in the future.

Notably, besides direct contact with PS NPs, Jeong et al. (2022) found that maternally administered PS NPs could be transmitted to offspring through breast milk and cause brain dysregulation in offspring. Similarly, Shin et al. (2023) found that offspring of PS NPs- exposed dams exhibited signs of anxiety- and depression-like behavior, as well as abnormal social behavior. The above studies illustrate that PS NPs exhibit negative impacts on the memory and cognitive functions, social, emotional, and memory functions of the mouse brain, as well as the neurotoxicity of PS MPs/NPs transmitted from mother to child to the next generation, which cannot be ignored. The relevant mechanisms and prevention strategies remain to be further studied.

9.4.5 Microplastics as Vectors for Microbes and Potentially Toxic Chemicals

MPs can be vectors of microorganisms and potentially toxic chemicals because of their charged rough surface, diverse morphology, and high hydrophobicity. Hence, binding MPs to microbes and chemicals may enhance their harmful effects on human health (Mammo et al., 2020). PS MP consumption led to a significantly prolonged arthritic foot swelling in mice infected with chikungunya virus (CHIKV) (Rawle et al., 2022). Tong et al. (2022) found that MPs could act as a habitat for *Helicobacter pylori* to survive and form a biofilm, helping *H. pylori* to colonize in the mice's stomach quickly, thereby promoting gastric injury and inflammation. Besides, *Pseudomonas aeruginosa* "curled up" to adhere to spherical PS MPs, and PS MPs significantly increased the invasive ability of *P. aeruginosa* into mice epithelial cells (Qi et al., 2022). This might lead to more severe functional impairment and longer inflammatory reactions to the upper respiratory tract. Likewise, SARS-CoV-2 adhered firmly to spherical PS MPs, and the adsorbed virus exhibited higher infectivity for human colon epithelial cells in vitro than free virus, with upregulated expression of apoptosis- and inflammation-related genes (Zhang et al., 2022a). The above studies illustrate that the combinations of MPs and microbial pathogens pose potential health risks, which may alter the pathogenicity of the microorganism and/or the mode/extent of entry/infection.

MPs in environments are quickly colonized by microorganisms, including bacteria, fungi, and algae, thereby forming biofilms on their surfaces (Wang et al., 2021a). The composition of MP biofilm varies with environmental conditions (Sooriyakumar et al., 2022). MP biofilms have unique microbial community structures, which selectively enrich antibiotic-resistance genes and pathogenic bacterial hosts (Wu et al., 2019; Junaid et al., 2022; Kaur et al., 2022). Therefore, the spread of sundry microbes (especially pathogenic microorganisms) present in MP biofilms may exacerbate their risks to human health. The existence of biofilm will promote the accumulation of pollutants in MPs, such as persistent organic pollutants, heavy metals, and antibiotics (He et al., 2022). Toxicological studies on MPs with biofilms must be paid more attention to.

Various additives, such as bisphenols, phthalates, polybrominated diphenyl ethers, polychlorinated biphenyl ethers, organotin, perfluorinated compounds, dioxins, polycyclic aromatic hydrocarbons, organic pollutants, and heavy metals, are added during plastic production (Ullah et al., 2023). MPs/NPs can cause a "Trojan Horse" effect by transferring the additives and other environmental pollutants to mammals, inducing several synergistic, behavioral, histological, and biomolecular alternations (Zhang et al., 2022b). Multiple studies have shown the joint toxicity of MPs and phthalate esters (PAEs), the most common plasticizers, to the human/mouse respiratory system (Shi et al., 2021; Han et al., 2023), digestive system (Deng et al., 2020; Yu et al., 2022), reproductive system (Ma et al., 2023; Sharma et al., 2024), etc. MPs/NPs can transport and release PAEs into mammalian cells and tissues, and the combined exposure of MPs/NPs and PAEs can exacerbate allergic asthma, increasing intestinal permeability and inflammation and blood-testis barrier dysfunction.

Besides, PE MP co-exposure exacerbated the cytotoxicity of tetrabromobisphenol A (TBBPA) on human colon cancer Caco-2 cells, while its effect was weak. Both PE MPs and TBBPA can shift the intestinal microbiota and interfere with the metabolic pathways of intestinal microorganisms, thus disrupting intestinal homeostasis (Huang et al., 2021). Besides, incorporating environmentally relevant concentrations of benzo [a] pyrene altered the physicochemical characteristics of PS MPs. This markedly enhanced the toxicity of PS MPs, including increased ROS levels, intestinal permeability, pathological inflammation, and intestinal bacterial translocation (Shaoyong et al., 2023). Moreover, MP combined with antibiotic (e.g., sulfamethoxazole, tetracycline hydrochloride, doxycycline) exposure perturbed intestinal microbiota homeostasis in mammals (Liu et al., 2022a; Wang et al., 2023a; Sun et al., 2024a). Liu et al. (2022a) especially found that MPs reduced sulfamethoxazole bioaccumulation but aggravated their influences on intestinal microbiota and antibiotic resistance in mice. It was inferred that MPs and antibiotics may exhibit high risks to human health.

In addition, the combined effects of MPs/NPs and heavy metal(loid)s, such as arsenic (As) (Chen et al., 2023c), lead (Pb) (Yuan et al., 2023), cadmium (Cd) (Hu et al., 2024), chromium (Cr) (Liao & Yang, 2020), iron (Fe) (Liu et al., 2022c), and copper (Cu) (Rong et al., 2024), on humans are also very complex (Cao et al., 2021). A study showed that MPs contained and transported considerable levels of adsorbed As, with As abundance on MPs as high as 6.5 mg/kg (Mora et al., 2023). Their co-exposure might have potentially additive effects on mammals. Chen et al. (2023c) found that MPs altered mice's intestinal microbiota and metabolites, thereby increasing the bioavailability of As. Besides, co-exposure to As and PS NPs induced mtROS-dependent ferritin autophagy and ferroptosis, leading to renal fibrosis and structural damage, whereas exposure to As or PS NPs alone did not exhibit nephrotoxicity (Zhong et al., 2024).

Similarly, the expression of carcinogenic biomarkers was exacerbated during long-term co-exposure to NPs and As in vitro (Barguilla et al., 2022). For Pb, the co-exposure of MPs/NPs aggravated Pb deposition liver and Pb-induced hepatotoxicity (Yuan et al., 2023) and aggravated ovarian toxicity in mice (Feng et al., 2022). For Cd, PS MPs/NPs and Cd synergistically destroyed the intestinal barrier and promoted Cd accumulation (Hu et al., 2024), leading to oxidative stress, ferroptosis, and excessive mitophagy in mouse kidneys (Qiu et al., 2023b), and enhanced male reproductive toxicity in mice (Lan et al., 2024).

9.5 Future Research Perspectives

To sum up, MPs were widely detected in various environmental media and even human tissues, causing potential harm to the human body. Based on the above discussion, research on the influences of MPs on human health is still in its early stages. Analyzing the particle size distribution of MPs in samples is crucial as smaller MP particles exhibit significantly higher toxicity. The research on the quantitative and in situ distribution of MPs in different human tissues and its impact mechanisms on human cells/organs/systems also needs to be strengthened. Recommendations for future research are as follows.

1. Establish an MPs/NPs quantitative detection combination system to characterize particle size distribution. The combination of atomic force microscopy (AFM), attenuated total reflectance-Fourier transform infrared spectroscope (ATR-FTIR), and pyrolysis-gas chromatography-mass spectrometry (Py-GC/MS) is suggested. Advanced imaging and analysis techniques for MPs/NPs are also required.
2. Conduct relevant epidemiological investigations and combine them with indoor experiments to make the toxicological research of MPs correspond to actual exposure scenarios and obtain research results with practical guidance. Moreover, developing in vitro models that simulate the complexity of human tissues/organs is needed to better investigate the accumulation and translocation of MPs in the human body.
3. It is critical to distinguish the impact of MP particles themselves on human health from those of additives and other pollutants. Information on additives, biofilms, and potentially toxic chemicals loaded on the surface of MPs also need to be obtained simultaneously when detecting MPs in human tissues. The coexistence of these substances may exacerbate the hazards of MPs. The corresponding mechanisms need to be further elucidated.

REFERENCES

Abafe O A, Harrad S, Abdallah M A E, 2023. Novel insights into the dermal bioaccessibility and human exposure to brominated flame retardant additives in microplastics. *Environmental Science & Technology*, 57: 10554–10562.

Abbasi S, Turner A, 2021. Human exposure to microplastics: a study in Iran. *Journal of Hazardous Materials*, 403: 123799.

Acarer S, 2023. Abundance and characteristics of microplastics in drinking water treatment plants, distribution systems, water from refill kiosks, tap waters and bottled waters. *Science of the Total Environment*, 844: 163866.

Afrin S, Rahman M M, Hossain M N, Uddin M K, Malafaia G, 2022. Are there plastic particles in my sugar? A pioneering study on the characterization of microplastics in commercial sugars and risk assessment. *Science of the Total Environment*, 837: 155849.

Aghaei Z, Sled J G, Kingdom J C, Baschat A A, Helm P A, Jobst K J, Cahill L S, 2022. Maternal exposure to polystyrene micro-and nanoplastics causes fetal growth restriction in mice. *Environmental Science & Technology Letters*, 9: 426–430.

Akhbarizadeh R, Dobaradaran S, Schmidt T C, Nabipour I, Spitz J, 2020. Worldwide bottled water occurrence of emerging contaminants: a review of the recent scientific literature. *Journal of Hazardous Materials*, 392: 122271.

Allen S, Allen D, Baladima F, Phoenix V, Thomas J, Le Roux G, Sonke J, 2021. Evidence of free tropospheric and long-range transport of microplastic at Pic du Midi observatory. *Nature Communications*, 12: 7242.

Amereh F, Amjadi N, Mohseni-Bandpei A, Isazadeh S, Mehrabi Y, Eslami A, Naeiji Z, Rafiee M, 2022. Placental plastics in young women from general population correlate with reduced foetal growth in IUGR pregnancies. *Environmental Pollution*, 314: 120174.

An R, Wang X, Yang L, Zhang J, Wang N, Xu F, Hou Y, Zhang H, Zhang L, 2021. Polystyrene microplastics cause granulosa cells apoptosis and fibrosis in ovary through oxidative stress in rats. *Toxicology*, 449: 152665.

Aydın R B, Yozukmaz A, Şener İ, Temiz F, Giannetto D, 2023. Occurrence of microplastics in most consumed fruits and vegetables from Turkey and public risk assessment for consumers. *Life*, 13: 1686.
Baeza-Martínez C, Olmos S, González-Pleiter M, López-Castellanos J, García-Pachón E, Masiá-Canuto M, Hernández-Blasco L, Bayo J, 2022. First evidence of microplastics isolated in European citizens' lower airway. *Journal of Hazardous Materials*, 438: 129439.
Barguilla I, Domenech J, Rubio L, Marcos R, Hernández A, 2022. Nanoplastics and arsenic co-exposures exacerbate oncogenic biomarkers under an in vitro long-term exposure scenario. *International Journal of Molecular Sciences*, 23: 2958.
Basaran B, Aytan Ü, Şentürk Y, Özçifçi Z, Akçay H T, 2024. Microplastic contamination in some beverages marketed in Türkiye: characteristics, dietary exposure and risk assessment. *Food and Chemical Toxicology*, 189: 114730.
Battaglini E, Miralles P, Lotti N, Soccio M, Fiorini M, Coscollà C, 2024. Analysis of microplastics in commercial vegetable edible oils from Italy and Spain. *Food Chemistry*, 443: 138567.
Braun T, Ehrlich L, Henrich W, Koeppel S, Lomako I, Schwabl P, Liebmann B, 2021. Detection of microplastic in human placenta and meconium in a clinical setting. *Pharmaceutics*, 13: 921.
Çağlayan U, Gündoğdu S, Ramos T M, Syberg K, 2024. Intravenous hypertonic fluids as a source of human microplastic exposure. *Environmental Toxicology and Pharmacology*, 107: 104411.
Cao Y X, Zhao M J, Ma X Y, Song Y W, Zuo S H, Li H H, Deng W Z, 2021. A critical review on the interactions of microplastics with heavy metals: mechanism and their combined effect on organisms and humans. *Science of the Total Environment*, 788: 147620.
Cary C M, DeLoid G M, Yang Z, Bitounis D, Polunas M, Goedken M J, Buckley B, Cheatham B, Stapleton P A, Demokritou P, 2023. Ingested polystyrene nanospheres translocate to placenta and fetal tissues in pregnant rats: potential health implications. *Nanomaterials*, 13: 720.
Cetin M, Demirkaya Miloglu F, Kilic Baygutalp N, Ceylan O, Yildirim S, Eser G, Gul H İ, 2023. Higher number of microplastics in tumoral colon tissues from patients with colorectal adenocarcinoma. *Environmental Chemistry Letters*, 21: 639–646.
Chen C, Liu F, Quan S, Chen L, Shen A, Jiao A, Qi H, Yu G, 2023a. Microplastics in the Bronchoalveolar lavage fluid of Chinese children: associations with age, City development, and disease features. *Environmental Science & Technology*, 57: 12594–12601.
Chen G, Xiong S, Jing Q, van Gestel C A, van Straalen N M, Roelofs D, Sun L, Qiu H, 2023b. Maternal exposure to polystyrene nanoparticles retarded fetal growth and triggered metabolic disorders of placenta and fetus in mice. *Science of the Total Environment*, 854: 158666.
Chen L, Zheng Y, Liu Y, Tian P, Yu L, Bai L, Zhou F, Yang Y, Cheng Y, Wang F, 2022a. Microfluidic-based in vitro thrombosis model for studying microplastics toxicity. *Lab on a Chip*, 22: 1344–1353.
Chen Q, Gao J, Yu H, Su H, Yang Y, Cao Y, Zhang Q, Ren Y, Hollert H, Shi H, 2022b. An emerging role of microplastics in the etiology of lung ground glass nodules. *Environmental Sciences Europe*, 34: 25.
Chen S, Yang J L, Zhang Y S, Wang H Y, Lin X Y, Xue R Y, Li M Y, Li S W, Juhasz A L, Ma L Q, 2023c. Microplastics affect arsenic bioavailability by altering gut microbiota and metabolites in a mouse model. *Environmental Pollution*, 324: 121376.
Chen Y, Xu H, Luo Y, Ding Y, Huang J, Wu H, Han J, Du L, Kang A, Jia M, 2023d. Plastic bottles for chilled carbonated beverages as a source of microplastics and nanoplastics. *Water Research*, 242: 120243.
Chen Y, Xu J, Chen Y, 2021. Regulation of neurotransmitters by the gut microbiota and effects on cognition in neurological disorders. *Nutrients*, 13: 2099.
Chiu H W, Xia T, Lee Y H, Chen C W, Tsai J C, Wang Y J, 2015. Cationic polystyrene nanospheres induce autophagic cell death through the induction of endoplasmic reticulum stress. *Nanoscale*, 7: 736–746.
Dawson A L, Li J Y, Kroon F J, 2022. Plastics for dinner: Store-bought seafood, but not wild-caught from the Great Barrier Reef, as a source of microplastics to human consumers. *Environmental Advances*, 8: 100249.
Deng Y, Yan Z, Shen R, Wang M, Huang Y, Ren H, Zhang Y, Lemos B, 2020. Microplastics release phthalate esters and cause aggravated adverse effects in the mouse gut. *Environment International*, 143: 105916.
Deng Y, Zhang Y, Lemos B, Ren H, 2017. Tissue accumulation of microplastics in mice and biomarker responses suggest widespread health risks of exposure. *Scientific Reports*, 7: 46687.
Dessi C, Okoffo E D, O'Brien J W, Gallen M, Samanipour S, Kaserzon S, Rauert C, Wang X, Thomas K V, 2021. Plastics contamination of store-bought rice. *Journal of Hazardous Materials*, 416: 125778.

Dibbon K C, Mercer G V, Maekawa A S, Hanrahan J, Steeves K L, Ringer L C, Simpson A J, Simpson M J, Baschat A A, Kingdom J C, 2024. Polystyrene micro-and nanoplastics cause placental dysfunction in mice. *Biology of Reproduction*, 110: 211–218.

Ding J, Sun C, He C, Zheng L, Dai D, Li F, 2022. Atmospheric microplastics in the Northwestern Pacific Ocean: distribution, source, and deposition. *Science of the Total Environment*, 829: 154337.

Djouina M, Vignal C, Dehaut A, Caboche S, Hirt N, Waxin C, Himber C, Beury D, Hot D, Dubuquoy L, Launay D, Duflos G, Body-Malapel M, 2022. Oral exposure to polyethylene microplastics alters gut morphology, immune response, and microbiota composition in mice. *Environmental Research*, 212: 113230.

Eom S, Shim W, Choi I, 2024. Microplastic-induced inhibition of cell adhesion and toxicity evaluation using human dermal fibroblast-derived spheroids. *Journal of Hazardous Materials*, 465: 133359.

Estrela F N, Guimarães A T B, Araújo A P d C, Silva F G, Luz T M d, Silva A M, Pereira P S, Malafaia G, 2021. Toxicity of polystyrene nanoplastics and zinc oxide to mice. *Chemosphere*, 271: 129476.

Feng Y, Yuan H, Wang W, Xu Y, Zhang J, Xu H, Fu F, 2022. Co-exposure to polystyrene microplastics and lead aggravated ovarian toxicity in female mice via the PERK/eIF2α signaling pathway. *Ecotoxicology and Environmental Safety*, 243: 113966.

Field D T, Green J L, Bennett R, Jenner L C, Sadofsky L R, Chapman E, Loubani M, Rotchell J M, 2022. Microplastics in the surgical environment. *Environment International*, 170: 107630.

Florance I, Ramasubbu S, Mukherjee A, Chandrasekaran N, 2021. Polystyrene nanoplastics dysregulate lipid metabolism in murine macrophages in vitro. *Toxicology*, 458: 152850.

Fournier S B, D'Errico J N, Adler D S, Kollontzi S, Goedken M J, Fabris L, Yurkow E J, Stapleton P A, 2020. Nanopolystyrene translocation and fetal deposition after acute lung exposure during late-stage pregnancy. *Particle and Fibre Toxicology*, 17: 1–11.

Fraissinet S, De Benedetto G E, Malitesta C, Holzinger R, Materić D, 2024. Microplastics and nanoplastics size distribution in farmed mussel tissues. *Communications Earth & Environment*, 5: 128.

Garcia M A, Liu R, Nihart A, El Hayek E, Castillo E, Barrozo E R, Suter M A, Bleske B, Scott J, Forsythe K, 2024. Quantitation and identification of microplastics accumulation in human placental specimens using pyrolysis gas chromatography mass spectrometry. *Toxicological Sciences*, 199: 81–88.

Geyer R, Jambeck J R, Law K L, 2017. Production, use, and fate of all plastics ever made. *Science Advances*, 3: e1700782.

Gong R, Wang J, Wang X, Liu Y, Shan J, 2023. Analysis of microplastics release from rice package in combination with machine learning and hyperspectral imaging technique. *Food Packaging and Shelf Life*, 39: 101152.

Goßmann I, Herzke D, Held A, Schulz J, Nikiforov V, Georgi C, Evangeliou N, Eckhardt S, Gerdts G, Wurl O, 2023. Occurrence and backtracking of microplastic mass loads including tire wear particles in northern Atlantic air. *Nature Communications*, 14: 3707.

Guan Q, Jiang J, Huang Y, Wang Q, Liu Z, Ma X, Yang X, Li Y, Wang S, Cui W, 2023. The landscape of micron-scale particles including microplastics in human enclosed body fluids. *Journal of Hazardous Materials*, 442: 130138.

Guo X, Dai H, Gukowsky J, Tan X, He L, 2023. Detection and quantification of microplastics in commercially bottled edible oil. *Food Packaging and Shelf Life*, 38: 101122.

Halfar J, Čabanová K, Vávra K, Delongová P, Motyka O, Špaček R, Kukutschová J, Šimetka O, Heviánková S, 2023. Microplastics and additives in patients with preterm birth: the first evidence of their presence in both human amniotic fluid and placenta. *Chemosphere*, 343: 140301.

Han Q, Gao X, Wang S, Wei Z, Wang Y, Xu K, Chen M, 2023. Co-exposure to polystyrene microplastics and di-(2-ethylhexyl) phthalate aggravates allergic asthma through the TRPA1-p38 MAPK pathway. *Toxicology Letters*, 384: 73–85.

Hasanah U, Amqam H, Septami A E, Chalid M, Aris A Z, 2024. Plasticizing pregnancy: microplastics identified in expectant mothers' feces. *Environmental Health Insights*, 18: 11786302241235810.

He S, Jia M, Xiang Y, Song B, Xiong W, Cao J, Peng H, Yang Y, Wang W, Yang Z, 2022. Biofilm on microplastics in aqueous environment: physicochemical properties and environmental implications. *Journal of Hazardous Materials*, 424: 127286.

He Y, Li J, Chen J, Miao X, Li G, He Q, Xu H, Li H, Wei Y, 2020. Cytotoxic effects of polystyrene nanoplastics with different surface functionalization on human HepG2 cells. *Science of the Total Environment*, 723: 138180.

Ho Y W, Lim J Y, Yeoh Y K, Chiou J C, Zhu Y, Lai K P, Li L, Chan P K S, Fang J K H, 2022. Preliminary findings of the high quantity of microplastics in faeces of Hong Kong residents. *Toxics*, 10: 414.

Horvatits T, Tamminga M, Liu B, Sebode M, Carambia A, Fischer L, Püschel K, Huber S, Fischer E K, 2022. Microplastics detected in cirrhotic liver tissue. *EBioMedicine*, 82: 104147.

Hou B, Wang F, Liu T, Wang Z, 2021. Reproductive toxicity of polystyrene microplastics: in vivo experimental study on testicular toxicity in mice. *Journal of Hazardous Materials*, 405: 124028.

Hu J, Zhu Y, Zhang J, Xu Y, Wu J, Zeng W, Lin Y, Liu X, 2022. The potential toxicity of polystyrene nanoplastics to human trophoblasts in vitro. *Environmental Pollution*, 311: 119924.

Hu L, Feng X, Lan Y, Zhang J, Nie P, Xu H, 2024. Co-exposure with cadmium elevates the toxicity of microplastics: Trojan horse effect from the perspective of intestinal barrier. *Journal of Hazardous Materials*, 466: 133587.

Huang D, Zhang Y, Long J, Yang X, Bao L, Yang Z, Wu B, Si R, Zhao W, Peng C, Wang A, Yan D, 2022a. Polystyrene microplastic exposure induces insulin resistance in mice via dysbacteriosis and pro-inflammation. *Science of the Total Environment*, 838: 155937.

Huang S M, Huang X X, Bi R, Guo Q X, Yu X L, Zeng Q H, Huang Z Y, Liu T M, Wu H S, Chen Y L, Xu J L, Wu Y G, Guo P, 2022b. Detection and analysis of microplastics in human sputum. *Environmental Science & Technology*, 56: 2476–2486.

Huang T, Zhang W, Lin T, Liu S, Sun Z, Liu F, Yuan Y, Xiang X, Kuang H, Yang B, 2022c. Maternal exposure to polystyrene nanoplastics during gestation and lactation induces hepatic and testicular toxicity in male mouse offspring. *Food and Chemical Toxicology*, 160: 112803.

Huang W, Yin H, Yang Y, Jin L, Lu G, Dang Z, 2021. Influence of the co-exposure of microplastics and tetrabromobisphenol A on human gut: simulation in vitro with human cell Caco-2 and gut microbiota. *Science of the Total Environment*, 778: 146264.

Hwang J, Choi D, Han S, Choi J, Hong J, 2019. An assessment of the toxicity of polypropylene microplastics in human derived cells. *Science of the Total Environment*, 684: 657–669.

Ibrahim Y S, Tuan Anuar S, Azmi A A, Wan Mohd Khalik W M A, Lehata S, Hamzah S R, Ismail D, Ma Z F, Dzulkarnaen A, Zakaria Z, 2021. Detection of microplastics in human colectomy specimens. *JGH Open*, 5: 116–121.

Inkielewicz-Stepniak I, Tajber L, Behan G, Zhang H, Radomski M W, Medina C, Santos-Martinez M J, 2018. The role of mucin in the toxicological impact of polystyrene nanoparticles. *Materials*, 11: 724.

International Bottled Water Association (IBWA)'s bottled water reporter, 2022 market report findings, 2023.

Jenner L C, Rotchell J M, Bennett R T, Cowen M, Tentzeris V, Sadofsky L R, 2022. Detection of microplastics in human lung tissue using μFTIR spectroscopy. *Science of the Total Environment*, 831: 154907.

Jeong B, Baek J Y, Koo J, Park S, Ryu Y-K, Kim K-S, Zhang S, Chung C, Dogan R, Choi H S, Um D, Kim T K, Lee W S, Jeong J, Shin W H, Lee J R, Kim N S, Lee D Y, 2022. Maternal exposure to polystyrene nanoplastics causes brain abnormalities in progeny. *Journal of Hazardous Materials*, 426: 127815.

Jiang Y, Han J, Na J, Fang J, Qi C, Lu J, Liu X, Zhou C, Feng J, Zhu W, Liu L, Jiang H, Hua Z, Pan G, Yan L, Sun W, Yang Z, 2022. Exposure to microplastics in the upper respiratory tract of indoor and outdoor workers. *Chemosphere*, 307: 136067.

Jin H, Ma T, Sha X, Liu Z, Zhou Y, Meng X, Chen Y, Han X, Ding J, 2021. Polystyrene microplastics induced male reproductive toxicity in mice. *Journal of Hazardous Materials*, 401: 123430.

Jin H, Yang C, Jiang C, Li L, Pan M, Li D, Han X, Ding J, 2022. Evaluation of neurotoxicity in BALB/c mice following chronic exposure to polystyrene microplastics. *Environmental Health Perspectives*, 130: 107002.

Jin Y, Lu L, Tu W, Luo T, Fu Z, 2019. Impacts of polystyrene microplastic on the gut barrier, microbiota and metabolism of mice. *Science of the Total Environment*, 649: 308–317.

Junaid M, Siddiqui J A, Sadaf M, Liu S, Wang J, 2022. Enrichment and dissemination of bacterial pathogens by microplastics in the aquatic environment. *Science of the Total Environment*, 830: 154720.

Jung B K, Han S W, Park S H, Bae J S, Choi J, Ryu K Y, 2020. Neurotoxic potential of polystyrene nanoplastics in primary cells originating from mouse brain. *Neurotoxicology*, 81: 189–196.

Kankanige D, Babel S, 2020. Smaller-sized micro-plastics (MPs) contamination in single-use PET-bottled water in Thailand. *Science of the Total Environment*, 717: 137232.

Karami A, Golieskardi A, Keong Choo C, Larat V, Galloway T S, Salamatinia B, 2017. The presence of microplastics in commercial salts from different countries. *Scientific Reports*, 7: 46173.

Kaur K, Reddy S, Barathe P, Oak U, Shriram V, Kharat S S, Govarthanan M, Kumar V, 2022. Microplastic-associated pathogens and antimicrobial resistance in environment. *Chemosphere*, 291: 133005.

Khuyen V T K, Le D V, Anh L H, Fischer A R, Dornack C, 2021. Investigating the correlation of microplastic pollution between seawater and marine salt using micro-Raman spectroscopy. *Frontiers in Marine Science*, 8: 735975.

Kibria G, Nugegoda D, Haroon A Y, 2022. Microplastic pollution and contamination of seafood (including fish, sharks, mussels, oysters, shrimps and seaweeds): a global overview. In Hashmi M Z (Ed.) *Microplastic Pollution: Environmental Occurrence and Treatment Technologies*, pp. 277–322, Springer: Chem.

Kim E H, Choi S, Kim D, Park H J, Bian Y, Choi S H, Chung H Y, Bae O N, 2022. Amine-modified nanoplastics promote the procoagulant activation of isolated human red blood cells and thrombus formation in rats. *Particle and Fibre Toxicology*, 19: 60.

Kim J S, Lee H J, Kim S K, Kim H J, 2018. Global pattern of microplastics (MPs) in commercial food-grade salts: sea salt as an indicator of seawater MP pollution. *Environmental Science & Technology*, 52: 12819–12828.

Kim S K, Song N S, 2021. Microplastics in edible salt: a literature review focusing on uncertainty related with measured minimum cutoff sizes. *Current Opinion in Food Science*, 41: 16–25.

Kosuth M, Wattenberg E V, Mason S A, Tyree C, Morrison D, 2017. *Synthetic Polymer Contamination in Global Drinking Water*. Orb Media: Washington, DC.

Kuang Q, Gao L, Feng L, Xiong X, Yang J, Zhang W, Huang L, Li L, Luo P, 2024. Toxicological effects of microplastics in renal ischemia–reperfusion injury. *Environmental Toxicology*, 39: 2350–2362.

Lan Y, Hu L, Feng X, Wang M, Yuan H, Xu H, 2024. Synergistic effect of PS-MPs and Cd on male reproductive toxicity: ferroptosis via Keap1-Nrf2 pathway. *Journal of Hazardous Materials*, 461: 132584.

Lee C W, Hsu L F, Wu I L, Wang Y L, Chen W C, Liu Y J, Yang L T, Tan C L, Luo Y H, Wang C C, 2022. Exposure to polystyrene microplastics impairs hippocampus-dependent learning and memory in mice. *Journal of Hazardous Materials*, 430: 128431.

Lee H J, Song N S, Kim J S, Kim S K, 2021. Variation and uncertainty of microplastics in commercial table salts: critical review and validation. *Journal of Hazardous Materials*, 402: 123743.

Lee H, Kunz A, Shim W J, Walther B A, 2019. Microplastic contamination of table salts from Taiwan, including a global review. *Scientific Reports*, 9: 10145.

Leslie H A, van Velzen M J M, Brandsma S H, Vethaak A D, Garcia-Vallejo J J, Lamoree M H, 2022. Discovery and quantification of plastic particle pollution in human blood. *Environment International*, 163: 107199.

Li J, Green C, Reynolds A, Shi H, Rotchell J M, 2018. Microplastics in mussels sampled from coastal waters and supermarkets in the United Kingdom. *Environmental Pollution*, 241: 35–44.

Li J, Zhang J, Ren S, Huang D, Liu F, Li Z, Zhang H, Zhao M, Cao Y, Mofolo S, 2023a. Atmospheric deposition of microplastics in a rural region of North China Plain. *Science of the Total Environment*, 877: 162947.

Li P, Li Q, Lai Y, Yang S, Yu S, Liu R, Jiang G, Liu J, 2023b. Direct entry of micro (nano) plastics into human blood circulatory system by intravenous infusion. *Iscience*, 26: 108454.

Li P, Wang X, Su M, Zou X, Duan L, Zhang H, 2021. Characteristics of plastic pollution in the environment: a review. *Bulletin of Environmental Contamination and Toxicology*, 107: 577–584.

Li Y, Peng L, Fu J, Dai X, Wang G, 2022. A microscopic survey on microplastics in beverages: the case of beer, mineral water and tea. *Analyst*, 147: 1099–1105.

Liang B, Huang Y, Zhong Y, Li Z, Ye R, Wang B, Zhang B, Meng H, Lin X, Du J, 2022. Brain single-nucleus transcriptomics highlights that polystyrene nanoplastics potentially induce Parkinson's disease-like neurodegeneration by causing energy metabolism disorders in mice. *Journal of Hazardous Materials*, 430: 128459.

Liang B, Zhong Y, Huang Y, Lin X, Liu J, Lin L, Hu M, Jiang J, Dai M, Wang B, Zhang B, Meng H, Lelaka J J J, Sui H, Yang X, Huang Z, 2021. Underestimated health risks: polystyrene micro- and nanoplastics jointly induce intestinal barrier dysfunction by ROS-mediated epithelial cell apoptosis. *Particle and Fibre Toxicology*, 18: 20.

Liao Y L, Yang J Y, 2020. Microplastic serves as a potential vector for Cr in an in-vitro human digestive model. *Science of the Total Environment*, 703: 134805.

Liu J, Lv M, Sun A, Ding J, Wang Y, Chang X, Chen L, 2022a. Exposure to microplastics reduces the bioaccumulation of sulfamethoxazole but enhances its effects on gut microbiota and the antibiotic resistome of mice. *Chemosphere*, 294: 133810.

Liu K, Wu T, Wang X, Song Z, Zong C, Wei N, Li D, 2019. Consistent transport of terrestrial microplastics to the ocean through atmosphere. *Environmental Science & Technology*, 53: 10612–10619.

Liu P, Shao L, Guo Z, Zhang Y, Cao Y, Ma X, Morawska L, 2025. Physicochemical characteristics of airborne microplastics of a typical coastal city in the Yangtze River Delta Region, China. *Journal of Environmental Sciences*, 148: 602–613.

Liu Q, Chen Z, Chen Y, Yang F, Yao W, Xie Y, 2022b. Microplastics contamination in eggs: detection, occurrence and status. *Food Chemistry*, 397: 133771.

Liu S, Liu X, Guo J, Yang R, Wang H, Sun Y, Chen B, Dong R, 2023. The association between microplastics and microbiota in placentas and meconium: the first evidence in humans. *Environmental Science & Technology*, 57: 17774–17785.

Liu S, Wang C, Yang Y, Du Z, Li L, Zhang M, Ni S, Yue Z, Yang K, Wang Y, Li X, Yang Y, Qin Y, Li J, Yang Y, Zhang M, 2024. Microplastics in three types of human arteries detected by pyrolysis-gas chromatography/mass spectrometry (Py-GC/MS). *Journal of Hazardous Materials*, 469: 133855.

Liu X, Yang H, Yan X, Xu S, Fan Y, Xu H, Ma Y, Hou W, Javed R, Zhang Y, 2022c. Co-exposure of polystyrene microplastics and iron aggravates cognitive decline in aging mice via ferroptosis induction. *Ecotoxicology and Environmental Safety*, 233: 113342.

Lu L, Wan Z, Luo T, Fu Z, Jin Y, 2018. Polystyrene microplastics induce gut microbiota dysbiosis and hepatic lipid metabolism disorder in mice. *Science of the Total Environment*, 631–632: 449–458.

Lu W, Li X, Wang S, Tu C, Qiu L, Zhang H, Zhong C, Li S, Liu Y, Liu J, 2023. New evidence of microplastics in the lower respiratory tract: inhalation through smoking. *Environmental Science & Technology*, 57: 8496–8505.

Luo T, Wang C, Pan Z, Jin C, Fu Z, Jin Y, 2019. Maternal polystyrene microplastic exposure during gestation and lactation altered metabolic homeostasis in the dams and their F1 and F2 offspring. *Environmental Science & Technology*, 53: 10978–10992.

Luqman A, Nugrahapraja H, Wahyuono R A, Islami I, Haekal M H, Fardiansyah Y, Putri B Q, Amalludin F I, Rofiqa E A, Götz F, Wibowo A T, 2021. Microplastic contamination in human stools, foods, and drinking water associated with Indonesian coastal population. *Environments*, 8: 138.

Ma T, Liu X, Xiong T, Li H, Zhou Y, Liang J, 2023. Polystyrene nanoplastics aggravated dibutyl phthalate-induced blood-testis barrier dysfunction via suppressing autophagy in male mice. *Ecotoxicology and Environmental Safety*, 264: 115403.

Mammo F, Amoah I, Gani K, Pillay L, Ratha S, Bux F, Kumari S, 2020. Microplastics in the environment: interactions with microbes and chemical contaminants. *Science of the Total Environment*, 743: 140518.

Marfella R, Prattichizzo F, Sardu C, Fulgenzi G, Graciotti L, Spadoni T, D'Onofrio N, Scisciola L, La Grotta R, Frigé C, 2024. Microplastics and nanoplastics in atheromas and cardiovascular events. *New England Journal of Medicine*, 390: 900–910.

Mason S A, Welch V G, Neratko J, 2018. Synthetic polymer contamination in bottled water. *Frontiers in Chemistry*, 6: 389699.

Matias R S, Gomes S, Barboza L G A, Almeida C M R, Marques A, Guilhermino L, Valente L M, 2024. Occurrence of microplastics and metals in European seabass produced in different aquaculture systems: implications for human exposure, risk, and food safety. *Science of the Total Environment*, 929: 172535.

Mintenig S M, Löder M G J, Primpke S, Gerdts G, 2019. Low numbers of microplastics detected in drinking water from ground water sources. *Science of the Total Environment*, 648: 631–635.

Montano L, Giorgini E, Notarstefano V, Notari T, Ricciardi M, Piscopo M, Motta O, 2023. Raman microspectroscopy evidence of microplastics in human semen. *Science of the Total Environment*, 901: 165922.

Montero V, Chinchilla Y, Gomez L, Flores A, Medaglia A, Guillen R, Montero E, 2023. Human health risk assessment for consumption of microplastics and plasticizing substances through marine species. *Environmental Research*, 237: 116843.

Mora A, Dueñas-Moreno J, Mahlknecht J, 2023. Microplastics as a vector of arsenic contamination. *Current Opinion in Environmental Science & Health*, 33: 100461.

Mortensen N P, Johnson L M, Grieger K D, Ambroso J L, Fennell T R, 2019. Biological interactions between nanomaterials and placental development and function following oral exposure. *Reproductive Toxicology*, 90: 150–165.

Na J, Geng Y, Jiang Y, Ma R, Cao Z, Zhao Y, Han J, Liu L, Sun W, Pan G, Yan L, Yang Z, 2021. Microplastics detected in sputum and nasal lavage fluid of couriers: a pilot study. *Chinese Journal of Public Health*, 37: 451–454.

Oliveri Conti G, Ferrante M, Banni M, Favara C, Nicolosi I, Cristaldi A, Fiore M, Zuccarello P, 2020. Micro- and nano-plastics in edible fruit and vegetables. The first diet risks assessment for the general population. *Environmental Research*, 187: 109677.

Paing Y M M, Eom Y, Song G B, Kim B, Choi M G, Hong S, Lee S H, 2024. Neurotoxic effects of polystyrene nanoplastics on memory and microglial activation: insights from in vivo and in vitro studies. *Science of the Total Environment*, 924: 171681.

Patil S, Bafana A, Krishnamurthi K, Sivanesan S, 2024. Estimated exposure to microplastics through national and local brands of bottled water in Central India. *Environmental Monitoring and Assessment*, 196: 1–20.

Perera K, Ziajahromi S, Bengtson Nash S, Manage P M, Leusch F D, 2022. Airborne microplastics in indoor and outdoor environments of a developing country in South Asia: abundance, distribution, morphology, and possible sources. *Environmental Science & Technology*, 56: 16676–16685.

Pham D T, Kim J, Lee S H, Kim J, Kim D, Hong S, Jung J, Kwon J H, 2023. Analysis of microplastics in various foods and assessment of aggregate human exposure via food consumption in Korea. *Environmental Pollution*, 322: 121153.

PlasticsEurope, 2020. *Plastics - the Facts 2020. An Analysis of European Latest Plastics Production, Demand and Waste Date*. Plastics Europe: Europe.

Praveena S M, Ariffin N I S, Nafisyah A L, 2022. Microplastics in Malaysian bottled water brands: occurrence and potential human exposure. *Environmental Pollution*, 315: 120494.

Qi Y, Chen Y, Yan X, Liu W, Ma L, Liu Y, Ma Q, Liu S, 2022. Co-exposure of ambient particulate matter and airborne transmission pathogens: the impairment of the upper respiratory systems. *Environmental Science & Technology*, 56: 15892–15901.

Qiu L, Lu W, Tu C, Li X, Zhang H, Wang S, Chen M, Zheng X, Wang Z, Lin M, 2023a. Evidence of microplastics in bronchoalveolar lavage fluid among never-smokers: a prospective case series. *Environmental Science & Technology*, 57: 2435–2444.

Qiu W, Ye J, Su Y, Zhang X, Pang X, Liao J, Wang R, Zhao C, Zhang H, Hu L, 2023b. Co-exposure to environmentally relevant concentrations of cadmium and polystyrene nanoplastics induced oxidative stress, ferroptosis and excessive mitophagy in mice kidney. *Environmental Pollution*, 333: 121947.

Ragusa A, Notarstefano V, Svelato A, Belloni A, Gioacchini G, Blondeel C, Zucchelli E, De Luca C, D'Avino S, Gulotta A, Carnevali O, Giorgini E, 2022. Raman microspectroscopy detection and characterisation of microplastics in human breastmilk. *Polymers*, 14: 2700.

Ragusa A, Svelato A, Santacroce C, Catalano P, Notarstefano V, Carnevali O, Papa F, Rongioletti M C A, Baiocco F, Draghi S, 2021. Plasticenta: first evidence of microplastics in human placenta. *Environment International*, 146: 106274.

Ramaremisa G, Tutu H, Saad D, 2024. Detection and characterisation of microplastics in tap water from Gauteng, South Africa. *Chemosphere*, 356: 141903.

Rawle D J, Dumenil T, Tang B, Bishop C R, Yan K, Le T T, Suhrbier A, 2022. Microplastic consumption induces inflammatory signatures in the colon and prolongs a viral arthritis. *Science of the Total Environment*, 809: 152212.

Rong J, Yuan C, Yin X, Wu X, He F, Wang Y, Leung K S-Y, Lin S, 2024. Co-exposure of polystyrene nanoplastics and copper induces development toxicity and intestinal mitochondrial dysfunction in vivo and in vitro. *Science of the Total Environment*, 930: 172681.

Rotchell J, Jenner L, Chapman E, Bennett R M, Bolanle I O, Loubani M, Sadofsky L, Palmer T, 2023. Detection of microplastics in human saphenous vein tissue using μFTIR: a pilot study. *PLoS One*, 18: e0280594.

Ruan X, Ao J, Ma M, Jones R R, Liu J, Li K, Ge Q, Xu G, Liu Y, Wang T, 2024. Nanoplastics detected in commercial sea salt. *Environmental Science & Technology*, 58: 9091–9101.

Rubio L, Barguilla I, Domenech J, Marcos R, Hernández A, 2020. Biological effects, including oxidative stress and genotoxic damage, of polystyrene nanoparticles in different human hematopoietic cell lines. *Journal of Hazardous Materials*, 398: 122900.

Ruenraroengsak P, Tetley T D, 2015. Differential bioreactivity of neutral, cationic and anionic polystyrene nanoparticles with cells from the human alveolar compartment: robust response of alveolar type 1 epithelial cells. *Particle and Fibre Toxicology*, 12: 1–20.

Samandra S, Mescall O J, Plaisted K, Symons B, Xie S, Ellis A V, Clarke B O, 2022. Assessing exposure of the Australian population to microplastics through bottled water consumption. *Science of the Total Environment*, 837: 155329.

Sartorelli P, Indini S, Bianchi F, d'Alessandro M, Bergantini L, Cameli P, Scancarello G, Antonietta M M, Barabesi L, Bargagli E, 2020. Cytological analysis of bronchoalveolar lavage fluid in asbestos-exposed workers. *La Medicina del Lavoro*, 111: 379.

Schirinzi G F, Pérez-Pomeda I, Sanchís J, Rossini C, Farré M, Barceló D, 2017. Cytotoxic effects of commonly used nanomaterials and microplastics on cerebral and epithelial human cells. *Environmental Research*, 159: 579–587.

Schwabl P, Köppel S, Königshofer P, Bucsics T, Trauner M, Reiberger T, Liebmann B, 2019. Detection of various microplastics in human stool: a prospective case series. *Annals of Internal Medicine*, 171: 453–457.

Schymanski D, Goldbeck C, Humpf H U, Furst P, 2018. Analysis of microplastics in water by micro-Raman spectroscopy: release of plastic particles from different packaging into mineral water. *Water Research*, 129: 154–162.

Shan S, Zhang Y, Zhao H, Zeng T, Zhao X, 2022. Polystyrene nanoplastics penetrate across the blood-brain barrier and induce activation of microglia in the brain of mice. *Chemosphere*, 298: 134261.

Shaoyong W, Jin H, Jiang X, Xu B, Liu Y, Wang Y, Jin M, 2023. Benzo [a] pyrene-loaded aged polystyrene microplastics promote colonic barrier injury via oxidative stress-mediated notch signalling. *Journal of Hazardous Materials*, 457: 131820.

Sharma A, Kaur M, Sharma K, kumar Bunkar S, John P, Bhatnagar P, 2023. Nano polystyrene induced changes in anxiety and learning behaviour are mediated through oxidative stress and gene disturbance in mouse brain regions. *Neurotoxicology*, 99: 139–151.

Sharma K, Sharma A, Bhatnagar P, 2024. Combined effect of polystyrene nanoplastic and di-n-butyl phthalate on testicular health of male Swiss albino mice: analysis of sperm-related parameters and potential toxic effects. *Environmental Science and Pollution Research*, 31: 23680–23696.

Sharma S, Chatterjee S, 2017. Microplastic pollution, a threat to marine ecosystem and human health: a short review. *Environmental Science and Pollution Research*, 24: 21530–21547.

Shen F, Li D, Guo J, Chen J, 2022. Mechanistic toxicity assessment of differently sized and charged polystyrene nanoparticles based on human placental cells. *Water Research*, 223: 118960.

Shi J Y, Dong Y B, Shi Y Y, Yin T T, He W, An T Y, Tang Y L, Hou X W, Chong S J, Chen D N, Qin K J, Lin H, 2022. Groundwater antibiotics and microplastics in a drinking-water source area, northern China: occurrence, spatial distribution, risk assessment, and correlation. *Environmental Research*, 210: 112855.

Shi Q, Tang J, Wang L, Liu R, Giesy J P, 2021. Combined cytotoxicity of polystyrene nanoplastics and phthalate esters on human lung epithelial A549 cells and its mechanism. *Ecotoxicology and Environmental Safety*, 213: 112041.

Shi Y, Yi L, Du G, Hu X, Huang Y, 2023. Visual characterization of microplastics in corn flour by near field molecular spectral imaging and data mining. *Science of the Total Environment*, 862: 160714.

Shin H S, Lee S H, Moon H J, So Y H, Lee H R, Lee E-H, Jung E-M, 2023. Exposure to polystyrene particles causes anxiety-, depression-like behavior and abnormal social behavior in mice. *Journal of Hazardous Materials*, 454: 131465.

Sivagami M, Selvambigai M, Devan U, Velangani A A J, Karmegam N, Biruntha M, Arun A, Kim W, Govarthanan M, Kumar P, 2021. Extraction of microplastics from commonly used sea salts in India and their toxicological evaluation. *Chemosphere*, 263: 128181.

Socas-Hernández C, Miralles P, González-Sálamo J, Hernández-Borges J, Coscollà C, 2024. Assessment of anthropogenic particles content in commercial beverages. *Food Chemistry*, 447: 139002.

Sooriyakumar P, Bolan N, Kumar M, Singh L, Yu Y, Li Y, Weralupitiya C, Vithanage M, Ramanayaka S, Sarkar B, 2022. Biofilm formation and its implications on the properties and fate of microplastics in aquatic environments: a review. *Journal of Hazardous Materials Advances*, 6: 100077.

Sridharan S, Kumar M, Singh L, Bolan N S, Saha M, 2021. Microplastics as an emerging source of particulate air pollution: a critical review. *Journal of Hazardous Materials*, 418, 126245.

Srisiri S, Haetrakul T, Dunbar S G, Chansue N, 2024. Microplastic contamination in edible marine fishes from the upper Gulf of Thailand. *Marine Pollution Bulletin*, 198: 115785.

Stock V, Fahrenson C, Thuenemann A, Dönmez M H, Voss L, Böhmert L, Braeuning A, Lampen A, Sieg H, 2020. Impact of artificial digestion on the sizes and shapes of microplastic particles. *Food and Chemical Toxicology*, 135: 111010.

Stock V, Laurisch C, Franke J, Dönmez M H, Voss L, Böhmert L, Braeuning A, Sieg H, 2021. Uptake and cellular effects of PE, PP, PET and PVC microplastic particles. *Toxicology in Vitro*, 70: 105021.

Suarez A N, Hsu T M, Liu C M, Noble E E, Cortella A M, Nakamoto E M, Hahn J D, de Lartigue G, Kanoski S E, 2018. Gut vagal sensory signaling regulates hippocampus function through multi-order pathways. *Nature Communications*, 9: 2181.

Suman A, Mahapatra A, Gupta P, Ray S S, Singh R K, 2024. Polystyrene microplastics induced disturbances in neuronal arborization and dendritic spine density in mice prefrontal cortex. *Chemosphere*, 351: 141165.

Sun H, Yang B, Zhu X, Li Q, Song E, Song Y, 2024a. Oral exposure of polystyrene microplastics and doxycycline affects mice neurological function via gut microbiota disruption: the orchestrating role of fecal microbiota transplantation. *Journal of Hazardous Materials*, 467: 133714.

Sun J, Sui M, Wang T, Teng X, Sun J, Chen M, 2024b. Detection and quantification of various microplastics in human endometrium based on laser direct infrared spectroscopy. *Science of the Total Environment*, 906: 167760.

Sun R, Xu K, Yu L, Pu Y, Xiong F, He Y, Huang Q, Tang M, Chen M, Yin L, Zhang J, Pu Y, 2021. Preliminary study on impacts of polystyrene microplastics on the hematological system and gene expression in bone marrow cells of mice. *Ecotoxicology and Environmental Safety*, 218: 112296.

Sun T, Teng Y, Ji C, Li F, Shan X, Wu H, 2024c. Global prevalence of microplastics in tap water systems: abundance, characteristics, drivers and knowledge gaps. *Science of the Total Environment*, 929: 172662.

Sun X, Yang R, Ji J, Zhu Z, White J C, Shen Y, 2024d. An evaluation of microplastic contamination in the marine waters and species in the coastal region of the South Yellow Sea, China. *Journal of Hazardous Materials*, 469: 134018.

Taghipour H, Ghayebzadeh M, Ganji F, Mousavi S, Azizi N, 2023. Tracking microplastics contamination in drinking water in Zahedan, Iran: from source to consumption taps. *Science of the Total Environment*, 872: 162121.

Tamargo A, Molinero N, Reinosa J J, Alcolea-Rodriguez V, Portela R, Bañares M A, Fernández J F, Moreno-Arribas M V, 2022. PET microplastics affect human gut microbiota communities during simulated gastrointestinal digestion, first evidence of plausible polymer biodegradation during human digestion. *Scientific Reports*, 12: 528.

Thompson R C, Olsen Y, Mitchell R P, Davis A, Rowland S J, John A W G, McGonigle D, Russell A E, 2004. Lost at sea: where is all the plastic? *Science*, 304: 838–838.

Tian Y C, Yang Z Y, Yu X Y, Jia Z, Rosso M, Dedman S, Zhu J M, Xia Y X, Zhang G P, Yang J Q, Wang J Z, 2022. Can we quantify the aquatic environmental plastic load from aquaculture? *Water Research*, 219: 118551.

Tong H Y, Jiang Q Y, Hu X S, Zhong X C, 2020. Occurrence and identification of microplastics in tap water from China. *Chemosphere*, 252: 126493.

Tong X, Li B, Li J, Li L, Zhang R, Du Y, Zhang Y, 2022. Polyethylene microplastics cooperate with Helicobacter pylori to promote gastric injury and inflammation in mice. *Chemosphere*, 288: 132579.

Tse Y T, Chan S M N, Sze E T P, 2022. Quantitative assessment of full size microplastics in bottled and tap water samples in Hong Kong. *International Journal of Environmental Research and Public Health*, 19: 13432.

Ullah S, Ahmad S, Guo X, Ullah S, Ullah S, Nabi G, Wanghe K, 2023. A review of the endocrine disrupting effects of micro and nano plastic and their associated chemicals in mammals. *Frontiers in Endocrinology*, 13: 1084236.

Vega-Herrera A, Llorca M, Borrell-Diaz X, Redondo-Hasselerharm P E, Abad E, Villanueva C M, Farré M, 2022. Polymers of micro (nano) plastic in household tap water of the Barcelona Metropolitan Area. *Water Research*, 220: 118645.

Wan S, Wang X, Chen W, Wang M, Zhao J, Xu Z, Wang R, Mi C, Zheng Z, Zhang H, 2024. Exposure to high dose of polystyrene nanoplastics causes trophoblast cell apoptosis and induces miscarriage. *Particle and Fibre Toxicology*, 21: 13.

Wang J, Guo X, Xue J, 2021a. Biofilm-developed microplastics as vectors of pollutants in aquatic environments. *Environmental Science & Technology*, 55: 12780–12790.

Wang L, Chen J, Zhang X, Xu M, Zhang X, Zhao W, Cui J, 2023a. Effects of microplastics and tetracycline on intestinal injury in mice. *Chemosphere*, 337: 139364.

Wang S, Han Q, Wei Z, Wang Y, Xie J, Chen M, 2022. Polystyrene microplastics affect learning and memory in mice by inducing oxidative stress and decreasing the level of acetylcholine. *Food and Chemical Toxicology*, 162: 112904.

Wang S, Liu J, Yao Y, Wang Y, Xu S, 2021b. Polystyrene microplastics-induced ROS overproduction disrupts the skeletal muscle regeneration by converting myoblasts into adipocytes. *Journal of Hazardous Materials*, 417: 125962.

Wang S, Lu W, Cao Q, Tu C, Zhong C, Qiu L, Li S, Zhang H, Lan M, Qiu L, 2023b. Microplastics in the lung tissues associated with blood test index. *Toxics*, 11: 759.

Wang T, Yi Z, Liu X, Cai Y, Huang X, Fang J, Shen R, Lu W, Xiao Y, Zhuang W, 2024. Multimodal detection and analysis of microplastics in human thrombi from multiple anatomically distinct sites. *Ebiomedicine*, 103: 105118.

Wang X, Li C, Liu K, Zhu L, Song Z, Li D, 2020. Atmospheric microplastic over the South China Sea and East Indian Ocean: abundance, distribution and source. *Journal of Hazardous Materials*, 389: 121846.

Wang Y L, Lee Y H, Hsu Y H, Chiu I J, Huang C C Y, Huang C C, Chia Z C, Lee C P, Lin Y F, Chiu H W, 2021c. The kidney-related effects of polystyrene microplastics on human kidney proximal tubular epithelial cells HK-2 and male C57BL/6 mice. *Environmental Health Perspectives*, 129: 057003.

Wei Z, Wang Y, Wang S, Xie J, Han Q, Chen M, 2022. Comparing the effects of polystyrene microplastics exposure on reproduction and fertility in male and female mice. *Toxicology*, 465: 153059.

Wibowo A T, Nugrahapraja H, Wahyuono R A, Islami I, Haekal M H, Fardiansyah Y, Sugiyo P W W, Putro Y K, Fauzia F N, Santoso H, Götz F, Tangahu B V, Luqman A, 2021. Microplastic contamination in the human gastrointestinal tract and daily consumables associated with an Indonesian farming community. *Sustainability*, 13: 12840.

Wikurendra E, Aini S, Nagy I, Nurika G, Herdiani N, Shamsudin S, 2022. Source of microplastic pollution within human stool in the Surabaya River Basin Area. *Journal of Chemical Health Risks*, 5: 581.

Wu D, Feng Y D, Wang R, Jiang J, Guan Q Q, Yang X, Wei H C, Xia Y K, Luo Y M, 2023. Pigment microparticles and microplastics found in human thrombi based on Raman spectral evidence. *Journal of Advanced Research*, 49: 141–150.

Wu P, Lin S, Cao G, Wu J, Jin H, Wang C, Wong M H, Yang Z, Cai Z, 2022. Absorption, distribution, metabolism, excretion and toxicity of microplastics in the human body and health implications. *Journal of Hazardous Materials*, 437: 129361.

Wu S, Wu M, Tian D, Qiu L, Li T, 2020. Effects of polystyrene microbeads on cytotoxicity and transcriptomic profiles in human Caco-2 cells. *Environmental Toxicology*, 35: 495–506.

Wu X J, Pan J, Li M, Li Y, Bartlam M, Wang Y Y, 2019. Selective enrichment of bacterial pathogens by microplastic biofilm. *Water Research*, 165: 114979.

Xie S, Zhang R, Li Z, Liu C, Chen Y, Yu Q, 2023. Microplastics perturb colonic epithelial homeostasis associated with intestinal overproliferation, exacerbating the severity of colitis. *Environmental Research*, 217: 114861.

Xie X, Deng T, Duan J, Xie J, Yuan J, Chen M, 2020. Exposure to polystyrene microplastics causes reproductive toxicity through oxidative stress and activation of the p38 MAPK signaling pathway. *Ecotoxicology and Environmental Safety*, 190: 110133.

Xu M, Halimu G, Zhang Q, Song Y, Fu X, Li Y, Li Y, Zhang H, 2019. Internalization and toxicity: a preliminary study of effects of nanoplastic particles on human lung epithelial cell. *Science of the Total Environment*, 694: 133794.

Yan Z, Hao Z, Zhang Z, Liu R, Zhao K, Zhang Y, 2024. A noninvasive quantitative method for evaluating intestinal exposure to microplastics based on the excretion and metabolism patterns of microplastics and their additives. *Environmental Science & Technology*, 58: 7791–7801.

Yan Z, Liu Y, Zhang T, Zhang F, Ren H, Zhang Y, 2022a. Analysis of microplastics in human feces reveals a correlation between fecal microplastics and inflammatory bowel disease status. *Environmental Science & Technology*, 56: 414–421.

Yan Z, Rein B, 2022b. Mechanisms of synaptic transmission dysregulation in the prefrontal cortex: pathophysiological implications. *Molecular Psychiatry*, 27: 445–465.

Yang Y, Xie E, Du Z, Peng Z, Han Z, Li L, Zhao R, Qin Y, Xue M, Li F, 2023. Detection of various microplastics in patients undergoing cardiac surgery. *Environmental Science & Technology*, 57: 10911–10918.

Yu Z, Xia Y, Cheng S, Mao L, Luo S, Tang S, Sun W, Jiang X, Zou Z, Chen C, 2022. Polystyrene nanoparticles aggravate the adverse effects of di-(2-ethylhexyl) phthalate on different segments of intestine in mice. *Chemosphere*, 305: 135324.

Yuan H, Wen S, Zhao Y, Hu L, Xu H, 2023. Polystyrene nanoplastics exacerbated Pb-induced liver toxicity in mice. *Toxicology Research*, 12: 446–456.

Yulia I, Agoes S, Haidar Brabo A, Muhammad Abdilla F, Nisa Indriyasari K, Muthiati Rahmatin N, Catur Putranto T W, Melissa Payus C, 2023. Microplastics in green mussels (Perna viridis) from Jakarta Bay, Indonesia, and the associated hazards to human health posed by their consumption. *Environmental Monitoring and Assessment*, 195: 884–884.

Zakeri M, Naji A, Akbarzadeh A, Uddin S, 2020. Microplastic ingestion in important commercial fish in the southern Caspian Sea. *Marine Pollution Bulletin*, 160: 111598.

Zarneshan S N, Fakhri S, Khan H, 2022. Targeting Akt/CREB/BDNF signaling pathway by ginsenosides in neurodegenerative diseases: a mechanistic approach. *Pharmacological Research*, 177: 106099.

Zhang D, Wu C, Liu Y, Li W, Li S, Peng L, Kang L, Ullah S, Gong Z, Li Z, Ding D, Jin Z, Huang H, 2024. Microplastics are detected in human gallstones and have the ability to form large cholesterol-microplastic heteroaggregates. *Journal of Hazardous Materials*, 467: 133631.

Zhang G, Cao G, Luo R-H, Song Q, Zeng Y, Liu K, Qu J, Lin X, Liu F-L, Wang G, 2022a. Microplastics interact with SARS-CoV-2 and facilitate host cell infection. *Environmental Science: Nano*, 9: 2653–2664.

Zhang J, Wang L, Kannan K, 2020. Microplastics in house dust from 12 countries and associated human exposure. *Environment International*, 134: 105314.

Zhang J, Wang L, Trasande L, Kannan K, 2021a. Occurrence of polyethylene terephthalate and polycarbonate microplastics in infant and adult feces. *Environmental Science & Technology Letters*, 8: 989–994.

Zhang N, Li Y B, He H R, Zhang J F, Ma G S, 2021b. You are what you eat: microplastics in the feces of young men living in Beijing. *Science of the Total Environment*, 767: 144345.

Zhang Q, He Y, Cheng R, Li Q, Qian Z, Lin X, 2022b. Recent advances in toxicological research and potential health impact of microplastics and nanoplastics in vivo. *Environmental Science and Pollution Research*, 29: 40415–40448.

Zhang X, Wang H, Peng S, Kang J, Xie Z, Tang R, Xing Y, He Y, Yuan H, Xie C, 2022c. Effect of microplastics on nasal and intestinal microbiota of the high-exposure population. *Frontiers in Public Health*, 10: 1005535.

Zhao L, Shi W, Hu F, Song X, Cheng Z, Zhou J, 2021. Prolonged oral ingestion of microplastics induced inflammation in the liver tissues of C57BL/6J mice through polarization of macrophages and increased infiltration of natural killer cells. *Ecotoxicology and Environmental Safety*, 227: 112882.

Zhao Q C, Zhu L, Weng J M, Jin Z R, Cao Y L, Jiang H, Zhang Z, 2023. Detection and characterization of microplastics in the human testis and semen. *Science of the Total Environment*, 877: 162713.

Zheng H, Wang J, Wei X, Chang L, Liu S, 2021. Proinflammatory properties and lipid disturbance of polystyrene microplastics in the livers of mice with acute colitis. *Science of the Total Environment*, 750: 143085.

Zheng X, Feng Q, Guo L, 2024. Quantitative analysis of microplastics and nanoplastics released from disposable PVC infusion tubes. *Journal of Hazardous Materials*, 465: 133246.

Zhong G, Qiao B, He Y, Liu H, Hong P, Rao G, Tang L, Tang Z, Hu L, 2024. Co-exposure of arsenic and polystyrene-nanoplastics induced kidney injury by disrupting mitochondrial homeostasis and mtROS-mediated ferritinophagy and ferroptosis. *Pesticide Biochemistry and Physiology*, 201: 105904.

Zhou Z, Kosuth M, Mason S A, Wattenberg E V, 2018. Anthropogenic contamination of tap water, beer, and sea salt. *PLoS One*, 13: e0194970.

Zhu L, Kang Y, Ma M, Wu Z, Zhang L, Hu R, Xu Q, Zhu J, Gu X, An L, 2024. Tissue accumulation of microplastics and potential health risks in human. *Science of the Total Environment*, 915: 170004.

Zhu L, Ma M, Sun X, Wu Z, Yu Y, Kang Y, Liu Z, Xu Q, An L, 2023b. Microplastics entry into the blood by infusion therapy: few but a direct pathway. *Environmental Science & Technology Letters*, 11: 67–72.

Zhu L, Zhu J, Zuo R, Xu Q, Qian Y, Lihui A, 2023a. Identification of microplastics in human placenta using laser direct infrared spectroscopy. *Science of the Total Environment*, 856: 159060.

Zhu X, Huang W, Fang M, Liao Z, Wang Y, Xu L, Mu Q, Shi C, Lu C, Deng H, 2021. Airborne microplastic concentrations in five megacities of northern and southeast China. *Environmental Science & Technology*, 55: 12871–12881.

10

Sources and Impacts of Persistent Toxic Substances in the Environment—The Case of Molasses-Based Distilleries

Kshitij Singh, Diane Purchase, and Ram Chandra

10.1 Introduction

Persistent toxic substances (PTS) represent a class of environmental pollutants characterized by their long-lasting presence in the environment, resistance to degradation, and potential to cause harm to ecosystems and human health (Inobeme et al., 2023). These substances persist in the air, water, and soil and even accumulate in the living system (Krithiga et al., 2022). They can be biomagnified through the food chain in the ecosystem. The primary characteristic of these substances is their ability to endure degradation processes, leading to bioaccumulation in living organisms and posing significant risks to both the environment and human health through biological transformation. Examples of PTS include certain pesticides, industrial chemicals, and unintentional by-products of human activities (Cachada et al., 2018).

The fundamental factors to consider when identifying PTS include the following characteristics:

i. They can be either organic or organometallic substances
ii. They may bioaccumulate in the tissue
iii. They persist in the environment for extended periods
iv. They exert adverse effects on the metabolic activities of organisms
v. Due to their persistent properties, these substances can be transported over long distances

One defining feature of PTS is its ability to bioaccumulate in living organisms. As these substances enter the food chain, they accumulate in tissues at concentrations exceeding those in the surrounding environment (Wong et al., 2012). This bioaccumulation process is further intensified through bio-magnification, wherein the concentration of PTS increases at each trophic level. This phenomenon poses a significant threat to higher-order predators, including humans, as they may experience higher exposure levels.

PTS are broadly categorized into two main groups: persistent organic and inorganic pollutants. Typical examples of persistent organic pollutants (POP) include dichlorodiphenyltrichloroethane (DDT), polychlorinated biphenyls (PCBs), dioxins and furans, Aldrin, Dieldrin, and Endrin. In contrast, persistent inorganic pollutants (PIP) include heavy metals and radionuclides (Bussian et al., 2021). The sources of PTS are diverse, stemming from agricultural practices, industrial activities, and even natural processes. Pesticides, widely used in agriculture to control pests and enhance crop yield, contribute significantly to the PTS load. Industrial processes, especially chemicals, plastics, and electronics, release persistent pollutants into the environment (Rajmohan et al., 2019).

Additionally, specific PTS can form as by-products during combustion processes, such as waste incineration and forest fires. Once released into the environment, PTS undergoes air, water, and soil transport. Long-range atmospheric transport is a notable mechanism leading to the global distribution of these pollutants. Their fate is often determined by factors such as their chemical structure, affinity for organic

DOI: 10.1201/9781003492115-12

matter, and volatility (Scheringer et al., 2009). Some PTS (such as A and B) may adsorb to soil particles, persisting for decades, while others (such as C and D) can bioaccumulate in aquatic organisms, moving up the food chain.

The persistence of these toxic substances in the environment has far-reaching ecological consequences. Aquatic ecosystems are particularly vulnerable, with PTS accumulating in sediments and affecting aquatic life (Ali et al., 2019). The disruption of endocrine systems in wildlife, impaired reproductive success, and developmental abnormalities are expected outcomes. Additionally, PTS can contribute to the decline of certain species and disrupt ecosystems by altering food web dynamics (Nilsen et al., 2019). Humans are exposed to PTS primarily through consuming contaminated food, especially fish, meat, and dairy products. Exposure to these substances can lead to various health effects, including carcinogenicity, neurotoxicity, reproductive disorders, and immune system suppression.

The long half-life of these substances means their impacts will endure for years. Ongoing research is essential to understand emerging PTS and their potential consequences. Moreover, addressing the global nature of the issue requires continued cooperation among nations, industries, and researchers to find sustainable solutions and prevent the introduction of new persistent pollutants. Recognizing the global nature of PTS, international efforts have been made to address their production, use, and release. The Aarhus Protocol 1998, the POPs Convention on Long-range Transboundary Air Pollution (LRTAP), aims to eliminate or restrict the production and use of selected POPs. The Stockholm Convention on POPs, adopted in 2001, is a critical international treaty aimed at eliminating or restricting the production and use of significant PTS. The Convention has listed specific chemicals, including the "dirty dozen," such as DDT, PCBs, and dioxins, for concerted global action.

Despite international efforts to control PTS, challenges persist. Various industries commonly release PTS, including sugarcane molasses-based distilleries, pulp paper mill industry, tanneries, petroleum refineries, and certain chemical plants. Molasses-based distilleries release various PTS following alcohol production (Figure 10.1). The waste produced by sugarcane molasses-based distilleries is highly complex and challenging to manage, comprising heavy metals, organic pollutants such as phenols and sulfides, potentially hazardous volatile organic compounds (VOCs) and persistent organometallic toxic substances. Various studies have demonstrated their toxicity to aquatic and terrestrial ecosystems (Singh et al., 2021).

Moreover, research has shown the presence of diverse androgenic and mutagenic substances in this waste and its wastewater, presenting significant risks to the reproductive system of aquatic life, causing cytotoxic effects on other organisms (Chandra and Kumar, 2017), crops, and soil microbes (Chandra et al., 2008).

These toxic substances pose substantial health risks to humans and wildlife. Heavy metals, for instance, can bioaccumulate in the food chain, resulting in chronic health issues such as neurological disorders, kidney damage, and developmental abnormalities. After prolonged exposure, organic pollutants and VOCs can cause respiratory problems, skin irritation, and cancer (Singh et al., 2021).

This chapter highlights the critical importance of PTS using molasses-based distilleries as a case study to understand and mitigate their environmental and health impacts. By prioritizing pollution control measures, ensuring regulatory compliance, and fostering public engagement, distilleries can minimize their adverse effects on the environment and protect the well-being of current and future generations.

10.2 Case Study—Molasses-based Distilleries

Molasses-based distilleries provide a valuable case study for understanding the impact of PTS on the environment. The global relevance of molasses-based distilleries makes this case study particularly significant. These distilleries are common in many countries, especially in regions where sugarcane is cultivated. The complexity of pollutants produced by these distilleries offers insights into the behavior, transport, and fate of various PTS in the environment. Studying these pollutants helps us comprehend how they interact with different environmental media, including soil, water, and biota. The varied environmental impacts of these distilleries—from soil contamination to water pollution—provide a comprehensive view of the challenges associated with managing PTS. From a public health perspective, understanding the impacts of PTS from molasses-based distilleries is essential. This case study

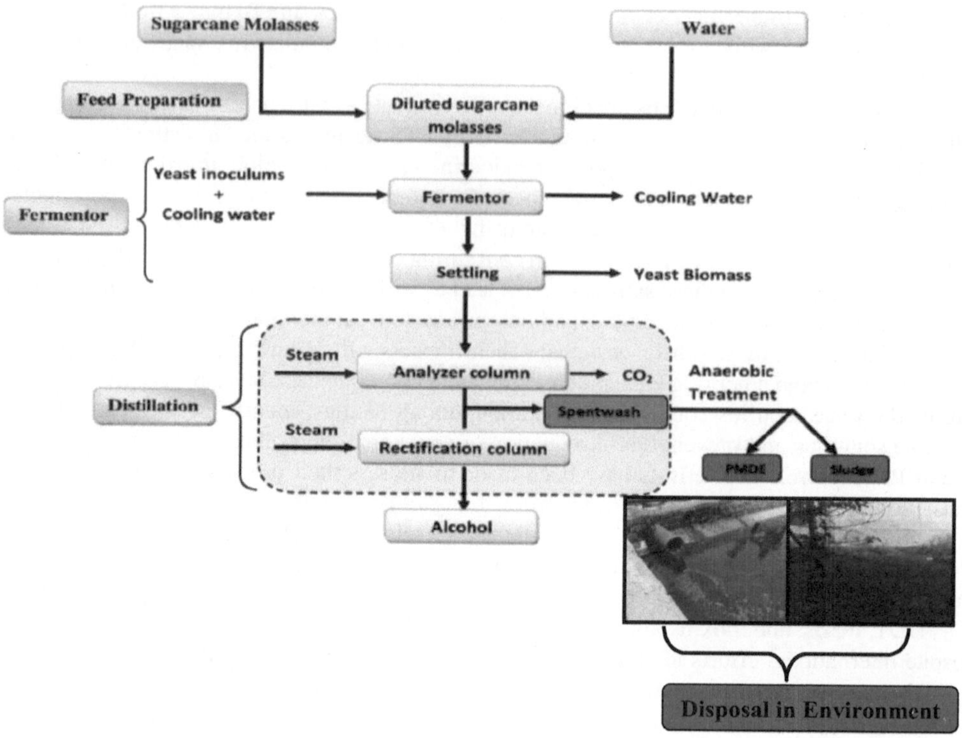

FIGURE 10.1 Detailed alcohol production process and discharge of its waste in the environment after treatment (Kshitij Singh, 2023).

highlights the exposure pathways and health outcomes associated with these pollutants, contributing to efforts to protect human health.

Molasses-based distilleries, integral to producing alcohol and various biofuels, have a significant environmental footprint (Arshad et al., 2019). These industrial facilities, while contributing significantly to economic development, are sources of PTS that pose substantial risks to ecosystems and human health. Figure 10.1 presents the alcohol production processes and the subsequent discharges into the environment. Sugarcane molasses-based distilleries can, therefore, lead to the release of PTS into the environment if not managed properly (Chowdhary et al., 2017).

PTS can be detected through various techniques such as gas chromatography-mass spectrometry (GC-MS), high-performance liquid chromatography (HPLC), inductively coupled plasma mass spectrometry (ICP-MS), X-ray fluorescence (XRF), and nuclear magnetic resonance (NMR) spectroscopy (Durner and Watts, 2021; Flanagan et al., 2008). Table 10.1 presents a range of the physicochemical properties of distillery effluent and sludge. Most of the parameters in the range have been found above the permissible limit of the Central Pollution Control Board (CPCB).

10.2.1 Sources of Persistent Toxic Substances

The primary environmental concern associated with molasses-based distilleries is the wastewater, commonly called spent wash. This effluent is rich in organic compounds, heavy metals, and other toxic substances. The high biochemical oxygen demand (BOD) and chemical oxygen demand (COD) values of spent wash indicate the presence of significant organic pollutants that consume large amounts of oxygen in water bodies, thereby affecting aquatic life. Moreover, this wastewater contains recalcitrant compounds—substances resistant to biodegradation.

TABLE 10.1
Detailed Parameters and Their Ranges in Distillery Effluent and Sludge

Parameter	Range (Effluent)	Range (Sludge)
pH	8.53±0.01	8.1±0.00
BOD	2487.66±110.12	–
COD	6296.66±6.65	–
TS	6940±5.29	–
TDS	10,527.66±6.65	–
VS	1059.33±2.08	–
Chloride	2746±4.35	1272.74±5.13
Total nitrogen	195.33±4.06	
Ammoniacal nitrogen	131.21±1.24	15.70±0.78
Phenol	6892±7.54	501.34±1.22
Sulphate	13,463.33±3.78	145.07±0.68
Phosphate	41.52±1.60	2268.83±1.70
Mn	8.20±0.09	238.47±0.83
Cr	2.97±0.026	02.15±1.07
Zn	16.61±0.00	43.47±1.31
Cu	2.55±0.01	847.46±1.00
Fe	373.95±0.36	5264.49±59.64
Pb	2.59±0.00	31.22±1.14
Cd	BDL	0.510±0.71
Ni	4.18±0.00	15.60±0.54

All values in mg/L except color intensity (Pt–Co unit), pH, and temperature (°C).
BOD, biological oxygen demand; COD, chemical oxygen demand; TS, total solid; TDS, total dissolved solid; VS, volatile solid.

The wastewater from such distilleries, often called vinasse, contains a complex mixture of organic compounds, nutrients, and potentially toxic substances such as cleaning agents, solvents, and additives (Hoarau et al., 2018) (Table 10.2). Distillery-generated spent wash consists of pH between 3.8 and 4.4. It also contains heavy metals such as lead, cadmium, selenium, and other toxic organic compounds. When sugarcane molasses-based distillery, which contains melanoidin as a primary colorant, discharges effluent on the land ecosystem, it affects various aspects of soil, including the biological, physical, and chemical properties such as an organic and inorganic component of soil, soil microflora, soil enzyme, soil water, pH levels, soil texture (Figure 10.2) (Singh et al., 2021). The wastewater can also affect the physiological characteristics of groundwater quality, impacting their pH, TDS, alkalinity, minerals, color, and electrical conductivity (Ratna et al., 2021).

Vinasse is rich in organic matter, which, if not properly treated, can lead to oxygen depletion in water bodies through microbial decomposition. This can result in adverse impacts on aquatic ecosystems and the release of other toxic substances from sediments (Christofoletti et al., 2013). It is also responsible for eutrophication, leading to algal blooms, subsequent depletion of oxygen levels, and releasing toxins produced by certain algae species (Chandra and Kumar, 2017).

10.2.1.1 Heavy Metals

Some PTS, such as heavy metals like mercury, lead, and cadmium, can be present in wastewater as contaminants from raw materials or process inputs. Equipment used in sugarcane molasses distillation, such as tanks, pipes, pumps, and valves, may contain metals such as lead, cadmium, chromium, and nickel. Corrosion or wear of these metal surfaces can release heavy metal contaminants into the processed molasses or surrounding environment (Moutsatsou et al., 2003).

TABLE 10.2

PTS Compounds Detected from Distillery Effluent

Name of Compound	Toxicity
Acetic acid, [(TMS)oxy],-TMS ester	Causes severe eye damage and respiratory toxicity and also causes skin irritation
Silane, (4-ethylphenyl)trimethyl	Hazardous to groundwater
Benzene, 1-ethyl-2-methyl	In pure forms, it may cause skin and severe eye irritation
2,3-D-2-methylsuccinic acid 2TMS	Sympathomimetic toxicity, acute psychosis
β-Eudesmol, TMS ether	May cause eye and skin irritation
2-Butenedioic acid, bis (TMS)ester	May cause genetic disorders
Ethanedioic acid, bis(TMS)ester	Causes neurotoxicity, occupational hepatotoxicity, and methemoglobinemia
Docosane	Causes severe eye damage and respiratory toxicity and also causes skin irritation
Benzoic acid, 2-methyl-, TMS ester	Hazardous to groundwater
n-Pentadioic acid, bis(TMS) ester	Skin irritation
Decanedioc acid	Causes severe eye irritation and skin irritation, and may cause respiratory irritation
Phenol, 2,4-bis(1,1-dimethylethyl)	Causes serious eye irritation and skin irritation
Ethanol, 2(octadecyoxy)	Causes serious eye damage and respiratory toxicity
Phenol, 2, 6-bis(1,1-dimethylethyl)	Aquatic toxicity
Tetradecanoic acid, TMS ester	Causes skin irritation, a possible carcinogen
9,12-Octadecadienoic acid (Z,Z)- TMS ester	Causes severe eye damage and respiratory toxicity and also causes skin irritation
Hexadecanoic acid, TMS ester	Skin irritation
Benzoic acid, 3,4,5 tris(TMS oxy)-TMS ester	Causes severe eye irritation and skin irritation and may cause respiratory irritation
Quercetin 7,3′,4′-trimethoxy	Causes severe eye irritation and skin irritation
Octadecanoic acid, TMS ester	Skin irritation

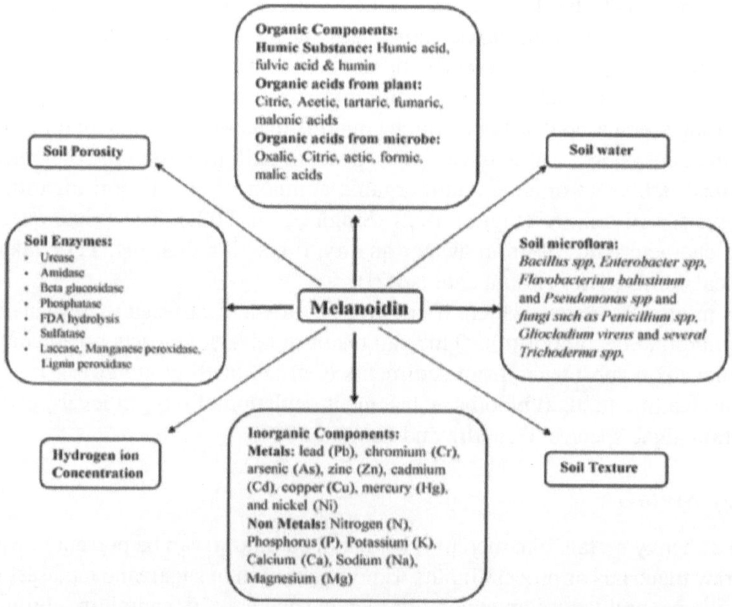

FIGURE 10.2 The release of melanoidin, a major colorant in sugarcane molasses-based distillery wastewater, results in various environmental impacts (Singh et al., 2021).

Certain additives or processing aids used in distillation may contain heavy metals. For example, metal catalysts or clarifying agents may be used to improve fermentation or filtration processes. These additives can contribute to heavy metal contamination in the final product or waste streams if not adequately controlled or removed.

Improper disposal practices can result in these metals leaching into the environment, posing risks to ecosystems and human health through bioaccumulation in food chains (Jagaba et al., 2024). One of the major issues with the wastewater from molasses-based distilleries is its potential for bioaccumulation. Toxic substances in the effluent can enter the food chain, leading to health risks for wildlife and humans (Sonone et al., 2020). Additionally, the wastewater often has high salinity and contains heavy metals such as lead, chromium, and copper. These elements can contaminate soil and water sources, resulting in long-term environmental damage.

10.2.1.2 Organic Toxic Contaminants

Furthermore, sewage from distilleries also contains many organic PTS (Table 10.3) and pathogenic microorganisms, including bacteria, viruses, and parasites, if not adequately treated (Chahal et al., 2016).

PTS emitted from sugarcane molasses-based distilleries, such as PM, VOC, PAHs, and others mentioned earlier, can travel significant distances before depositing onto surfaces or being removed from the atmosphere through processes like rainfall or deposition. This long-range transport of pollutants can have environmental and health implications for regions far beyond the immediate vicinity of the distillery (Silva et al., 2010).

10.2.1.3 Volatile Organic Compounds

In addition, when pollutants are emitted into the air during various stages of the distillation process, combustion of materials, or other activities associated with the distillery operation (Sivaramanan, 2015). VOCs may be released during fermentation, distillation, waste treatment, and other production processes

TABLE 10.3

PTS Compound Detected from Distillery Sludge

Name of Compound	Toxicity
Arachidonic acid	Causes inflammation and mood disorders
2,4-Dimethylbenzene	Skin irritation
Cyclohexanemethanol	Eye irritation
D-Glucopyranoside	Skin corrosion/irritation
1,3-Propanediol	Human carcinogen
1,3-Butanediol	Skin irritation
2,6-Dihydroxyacetophenone	Causes severe eye irritation and skin irritation and may cause respiratory irritation
Pentafluorobenzyloxime	Causes severe eye irritation and skin irritation
5-(p-Aminophenyl)-4-(p-tolyl)-2-thiazolamine	Causes serious eye damage and respiratory toxicity
1-Dodecanol	Aquatic toxicity
1-Cyclohexyldimethylsilyloxydodecane	Causes skin irritation, a possible carcinogen
2-Chloro-4-nitro-N (trimethylsilyl) benzamide	Causes severe eye damage and respiratory toxicity and also causes skin irritation
Cis-11-Eicosenoic acid	Hazardous to groundwater
Trans-13-Octadecenoic acid	In pure forms, it may cause skin and severe eye irritation
Cannabinol	Sympathomimetic toxicity, acute psychosis
10,12-Docosadiynedioic acid	May cause eye and skin irritation
Palmitic acid	May cause genetic disorders
1,2,3,4- Tetrahydronaphthalene	Causes neurotoxicity, occupational hepatotoxicity, and methemoglobinemia
Spironolactone	Hyperkalemia

in sugarcane-based molasses distilleries. These VOCs can include ethanol, acetaldehyde, methanol, and volatile organic acids. While VOC emissions from distilleries can contribute to air pollution and may have health implications, they are not considered POPs in the same sense as POPs (Tripathi et al., 2022). Once released into the atmosphere, these pollutants can undergo various physical and chemical transformations, allowing them to be transported over long distances through atmospheric dispersion, wind patterns, and atmospheric mixing. The extent of transport and dispersion depends on factors such as the properties of the pollutants, meteorological conditions, and atmospheric stability (Holmes and Morawska, 2006). Therefore, distilleries must implement effective pollution control measures and adhere to regulatory standards to minimize the generation and release of PTS into the atmosphere.

10.2.2 Environmental Health Hazards

PTS present significant environmental health risks due to their persistent nature, bioaccumulation potential, and propensity for long-term exposure. These substances accumulate within living organisms, particularly in fatty tissues, as they ingest contaminated food, water, or air. This accumulation intensifies as substances move up the food chain, elevating risks for predators and higher trophic-level organisms (Razinkovas, 2007).

Moreover, PTS undergo biomagnification within food chains, leading to increasingly higher concentrations at each trophic level. This phenomenon results in heightened exposure levels for top predators, including humans, who consume contaminated fish, meat, or dairy products (Ali and Khan, 2018).

The major composition of molasses melanoidin distillery waste contains melanoidin, heavy metals, and phenolic compounds to assess the toxic effect of melanoidin in different combinations on aquatic plants. The study revealed that *Typha angustifolia* grown on a metal solution containing variable concentrations of phenol (200–800 mg/L) and melanoidin (2500–8500 Co–Pt) are well adapted due to increased levels of antioxidants, which minimized the damage caused by reactive oxygen species, high level of color, COD, and BOD. However, toxicity emerges when the concentrations of pollutants exceed the quenching capacity of *T. angustifolia* natural protection system. *T. angustifolia* showed optimum tolerance for various heavy metal bioremediation in the presence of phenol (200–400 mg/L) and melanoidin (3000–4000 Co–Pt). Hence, it can be concluded that *T. angustifolia* is effective for heavy metals bioremediation from metal, melanoidin, and phenol-containing industrial wastewater at optimized conditions.

Several organic and inorganic toxic substances have been detected from molasses-based distillery waste and wastewater as recalcitrant compounds emerging pollutants to the environment. These are persistent and directly impose adverse effects on the ecosystem and indirectly affect all living systems through the food chain.

The study revealed that there was an accumulation and distribution of toxic metals like Cu, Cd, Cr, Zn, Fe, Ni, Mn, and Pb in wheat *Triticum aestivum* and Indian mustard (*Brassica campestris*) irrigated with distillery effluents. Further, analyses of plant samples have indicated the maximum accumulation of Fe (340 mg/kg in wheat root and 560 mg/kg in mustard leaves) followed by Mn and Zn in root > shoot > leaves > seeds. The maximum increase in photosynthetic pigment was observed between 30 and 60 days, while protein content was found to be maximum between 60 and 90 days of the growth period in both plants. An increase in malondialdehyde, cysteine, and ascorbic acid antioxidant content was also observed in the roots and leaves of treated plants up to 60 and 90 days of growth. Hence, wheat and mustard plants irrigated with effluents without adequate treatment are health hazards to the environment, humans, and animals (Chandra et al., 2009).

Furthermore, the effect of anaerobically treated distillery and sludge showed inhibition of amylase activity and seed germination along with the growth parameter of *Phaseolus mungo* L. as shown in Figure 10.3 for the reduction of root and shoot with the increase of distillery effluent concentration during irrigation in the pot experiment (Chandra et al., 2008; Bharagava and Chandra 2010).

The characterization of pollutants in post-methanated (post-methanation?) distillery sludge showed an abundant mixture of androgenic and mutagenic compounds in distillery sludge and leachate. In the GC-MS analysis also showed dodecanoic acid, octadecanoic acid, n-pentadecanoic acid, hexadecanoic acid, b-sitosterol, stigmasterol, b-sitosterol trimethyl ether, heptacosane, dotriacontane, lanosta-8, 24-dien-3-one, 1-methylene-3-methyl butanol, 1-phenyl-1-propanol, 5-methyl-2-(1-methylethyl) cyclohexanol, and 2-ethylthio-10-hydroxy-9-methoxy-1,4 anthraquinone as major organic pollutants along

FIGURE 10.3 Morphological effect of distillery sludge amendment soils at different concentrations (10%–80%) on the growth of root, shoot, and leaves of *P. mungo* after 60 days vs. the control.

with heavy metals (all mg/kg): Fe (2403), Zn (210.15), Mn (126.30, Cu (73.62), Cr (21.825), Pb (16.33), and Ni (13.425). In a simultaneous analysis of bacterial communities using the restriction fragment length polymorphism (RFLP) method, the dominance of Bacillus sp. followed by Enterococcus sp. as autochthonous bacterial communities growing in this extremely toxic environment indicated a primary community for bioremediation. A toxicity evaluation showed a reduction of toxicity in degraded samples of sludge and leachate, confirming the role of autochthonous bacterial communities in the in-situ bioremediation of distillery waste (Chandra and Kumar, 2017). In addition, recent observations showed the decolorization of color-containing organic pollutants present in distillery spent wash leachates by *Arbuscular Mycorrhizas* (Goswami et al., 2023).

Researchers have investigated the nature of distillery wastewater pollutants characterization, and it showed the nature of distillery wastewater pollutants characterization has been examined by researchers and it showed the absorption maxima (λmax) at 295 nm of pollutants present in the spent wash, which indicated that melanoidin is a major colorant along with other organic compounds. GC-MS analysis of the effluent identified several compounds, including 1,2,4,5-tetrahydro-2-methyl-3H-2-benzazepine-3-one, 6-(trifluoromethyl)-1H-imidazole [4,5-c] pyridine, 4,6-di-tert-butyl-1H, 3H-thieno[3,4-c] thiophene, butanoic acid, trimethylsilyl ester, silane, (dodecyloxy)trimethylsilyl ester, 4-styryl [2,2] paracyclophane, 4-(p-cumylphenoxy) phthalonitrile, hexadecanoic acid, trimethylsilyl ester, butyl ester, and squalene. Some of these compounds are known for their androgenic and mutagenic properties. Toxicity tests on post-methanated distillery effluent (PMDE) demonstrated direct toxic effects on catfish (*Heteropneustes fossilis*) even at a concentration as low as 1% (v/v), causing degeneration of the primary and secondary lamellae of the gills and damage to the epithelial layer. At higher concentrations (5%–10%), there was a complete loss of primary and secondary gill lamellae. A significant increase in liver catalase activity was observed in *H. fossilis*, indicating the generation of oxidative free radicals. Similarly, inhibition of α-amylase activity in germinating *Zea mays* (maize) seeds was also observed at 10% PMDE. This study highlights the environmental toxicity of the effluent, as demonstrated by its detrimental effects on freshwater fish and *Zea mays* seeds, as shown in Figure 10.4 (Tripathi et al., 2021).

10.2.3 Regulations and Guidelines for Control of PTS

There are many international guidelines and regulations for controlling PTS. Table 10.4 presents a list of well-known examples.

While economically beneficial, Molasses-based distilleries pose significant environmental and health risks due to the production of PTS. Effective regulatory frameworks are essential for mitigating these risks. Country-specific regulations in India, the United States, the European Union, Brazil, and Australia provided examples of regulations to control potentially toxic substances from molasses-based distilleries.

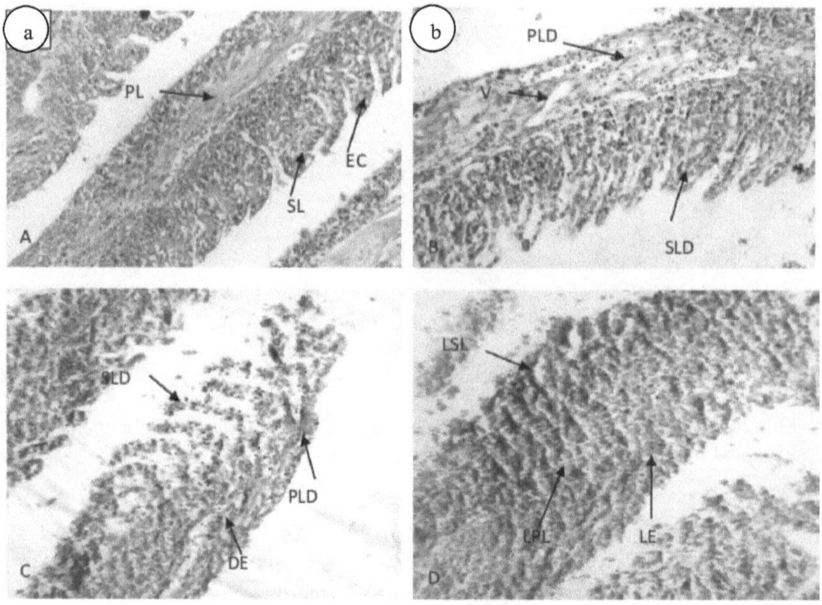

(a)

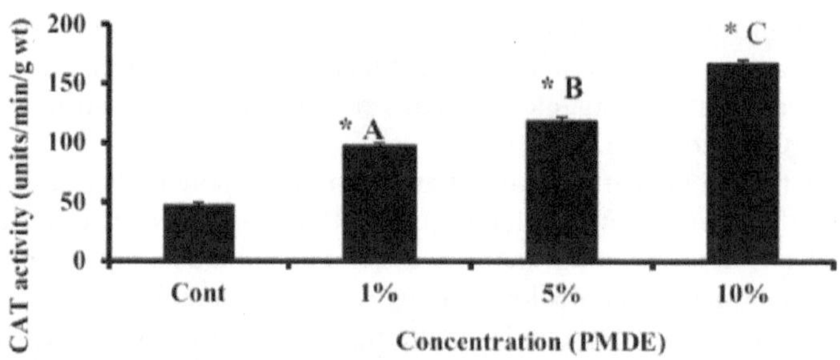

(b)

FIGURE 10.4 **A.** Effect of different concentrations of post-methanated distillery effluent (PMDE; 1%, 5%, 10%) on gill histopathology of the freshwater catfish *Heteropneustes fossilis* after 24h. (a) Control; (b) 1% PMDE; (c) 5% PMDE; (d) 10% PMDE. PL, primary lamellae; SL, secondary lamellae; EC, epithelial cell; PLD, primary lamellae degeneration; SLD, secondary lamellae degeneration; V, vacuolation; DE, damaged epithelium; LPL, loss of primary lamellae; LSL, loss of secondary lamellae; LE, loss of epithelium. [Image captured at 20 ×; hematoxylin and eosin stained. **B.** Catalase activity (units/min/g wt) in liver of *Heteropneustes fossilis* at different concentrations of PMDE (1%, 5%, 10%) and a control. Each group had ten fish in duplicates for 24h. Values are expressed as the mean±SEM. Data were analyzed by one-way ANOVA (*$P<0.001$) and Newman–Keuls test ($P<0.05$; A, B, C). Groups superscripted with different letters significantly differ in intergroup comparison (Tripathi et al., 2021).

10.2.3.1 India

In India, the CPCB under the Ministry of Environment, Forest and Climate Change (MoEFCC) regulates the environmental impact of molasses-based distilleries. The CPCB has established comprehensive guidelines and standards for effluent discharge and emissions, recognizing the severe pollution potential of distillery wastewater, known as spent wash. CPCB mandates stringent limits on BOD, COD, total dissolved solids (TDS), and heavy metals in the effluent. To combat pollution, India has implemented a Zero Liquid Discharge (ZLD) policy for distilleries, requiring them to treat and reuse all wastewater, ensuring no discharge into water bodies. This policy promotes advanced treatment technologies like anaerobic digesters, reverse osmosis, and multi-effect evaporators.

TABLE 10.4
Examples of International Regulations to Control PTS

Legislation	PTS	Notes
Stockholm Convention on Persistent Organic Pollutants	POPs	Adopted in 2001 and entered into force in 2004. Aims to eliminate or restrict the production, use, and release of POPs, which persist in the environment, bioaccumulate through the food chain, and pose risks to human health and the environment. Parties to the Convention are required to take measures to reduce or eliminate the production and release of POPs and manage and dispose of existing POPs in an environmentally sound manner (Lallas, 2001).
Aarhus Protocol on Persistent Organic Pollutants	POPs	A 1998 protocol on POPs adds to the 1979 Geneva Convention on Long-Range Transboundary Air Pollution (LRTAP). Aims "to control, reduce or eliminate discharge, emissions and losses of persistent organic pollutants" in Europe, some former Soviet Union countries, and the United States, to reduce their transboundary fluxes to protect human health and the environment from adverse effects. In May 2013, the protocol has been ratified by 31 states and the European Union
Minamata Convention on Mercury	Hg	Adopted in 2013 and entered into force in 2017. Aims to protect human health and the environment from anthropogenic emissions and releases of mercury and mercury compounds. The Convention includes provisions for controlling mercury emissions from industrial processes, products, and waste and measures for managing and reducing mercury use in artisanal and small-scale gold mining (Kessler, 2013).
Rotterdam Convention on Prior Informed Consent (PIC)	Hazardous chemicals and pesticides	Entered into force in 2004. Aims to promote shared responsibility and cooperation in the international trade of hazardous chemicals and pesticides. The Convention establishes a procedure for the prior informed consent of importing countries before certain hazardous chemicals and pesticides are exported to them.
Basel Convention on the Control of Transboundary Movements of Hazardous Wastes and Their Disposal	Hazardous wastes	Aims to minimize the generation of hazardous wastes, ensure their environmentally sound management, and prevent illegal traffic and dumping. Parties to the Convention are required to take measures to control the generation, export, import, and disposal of hazardous wastes, including those containing persistent toxic substances (Ahmed, 2019)
Toxic Substances Control Act (TSCA) - USA	Chemicals with significant risks	Enacted in 1976. Provides the EPA with the authority to require reporting, record-keeping, testing requirements, and chemical substances and mixture restrictions.
REACH (Registration, Evaluation, Authorization and Restriction of Chemicals) - EU	Carcinogens, mutagens, reproductive toxicants, and substances that are persistent, bioaccumulative, and toxic	Adopted in 2006. Aims to ensure a high level of protection of human health and the environment from the risks posed by chemicals.

The CPCB regulates air emissions from distilleries, limiting VOCs, particulate matter (PM), and other pollutants. Scrubbers, biofilters, and bag filters are encouraged to minimize air pollution.

The Environmental Protection Act (EPA) (MOEF, 2023) governs the handling and disposal of solid and hazardous waste from distilleries. By-products such as spent grains are often utilized as animal feed or for biogas production, promoting waste minimization and resource recovery.

10.2.3.2 United States

In the United States, the Environmental Protection Agency (EPA) enforces regulations under the Clean Water Act (CWA) (U.S. Environmental Protection Agency, 1972) and the Resource Conservation and Recovery Act (RCRA) (U.S. Environmental Protection Agency, 1976) to control pollution from molasses-based distilleries. The CWA sets national standards for effluent discharge, including specific limitations for BOD, COD, and toxic substances. Distilleries are required to obtain permits under the National Pollutant Discharge Elimination System (NPDES), ensuring compliance with these standards through regular monitoring and reporting.

The RCRA regulates the generation, transportation, treatment, storage, and disposal of hazardous waste. Distilleries must adhere to stringent guidelines for managing hazardous by-products, ensuring safe disposal, and minimizing environmental impact.

In addition to federal regulations, individual states may impose more stringent requirements on distilleries. These state-specific regulations often address local environmental conditions and public health concerns, providing additional protection.

10.2.3.3 European Union

The EU adopts a holistic approach to regulating molasses-based distilleries through directives like the Industrial Emissions Directive (IED) (European Parliament, & Council of the European Union, 2000) and the Water Framework Directive (WFD) (European Parliament, & Council of the European Union, 2010). The IED requires distilleries to implement Best Available Techniques (BAT) to minimize environmental impact. This includes advanced wastewater treatment methods, energy-efficient processes, and stringent control of air emissions. The directive emphasizes an integrated approach, considering the effect on air, water, and land. The WFD aims to achieve good status for all EU waters. It sets quality standards for effluent discharge, including limits on nutrients and hazardous substances, to prevent water pollution and protect aquatic ecosystems.

EU regulations also mandate continuous monitoring and transparent reporting of emissions and effluents. Compliance is enforced through regular inspections and penalties for violations, ensuring that distilleries adhere to environmental standards.

10.2.3.4 Brazil

The National Environmental Council (CONAMA) establishes environmental regulations for industries, including molasses-based distilleries in Brazil. CONAMA sets specific limits on pollutants in wastewater and air emissions (Lemos, 2013). Bio-digesters and other advanced treatment technologies are encouraged to meet these standards and reduce environmental impact.

Brazilian regulations promote the utilization of by-products, such as using spent wash for irrigation under controlled conditions and converting waste to energy through biogas production. This approach supports sustainable waste management and resource recovery.

10.2.3.5 Australia

Australian regulations emphasize sustainability, encouraging distilleries to adopt water recycling, energy efficiency, and renewable energy sources. In Australia, environmental protection agencies in each state and territory regulate the environmental impact of molasses-based distilleries (EPA, Victoria, 2018)?). Each state or territory has its environmental protection laws and guidelines. These regulations require distilleries to develop comprehensive environmental plans for wastewater treatment, air emissions control, and solid waste management.

Distilleries must obtain environmental permits and comply with specified conditions. Regular monitoring and reporting are mandated to ensure adherence to environmental standards and promptly address non-compliance.

10.3 Conclusion and Future Strategies

The indiscriminate disposal of PTS (leads to soil, water, and air contamination) has endangered ecosystems and human health. The persistence of these toxins exacerbates their harmful effects, resulting in long-lasting environmental damage. The case study using molasses-based distillery waste demonstrated the need for comprehensive solutions.

Efforts to combat the impact of PTS include extensive monitoring and assessment programs (Gavrilescu et al., 2015). Environmental agencies and research institutions worldwide regularly monitor air, water, soil, and biota to track the levels of PTS. Biomonitoring studies in humans and wildlife provide crucial information on exposure levels and trends, guiding regulatory measures and pollution control strategies. Addressing the legacy of PTS requires advanced technologies for remediation. Innovative techniques such as phytoremediation, bioremediation, and thermal treatment are employed to reduce the concentration of persistent pollutants in contaminated sites (Dadrasnia et al., 2013). Furthermore, implementing source reduction and utilizing alternative, less toxic substances in industrial processes play a crucial role in preventing the continual release of PTS into the environment (Ali et al., 2019).

Molasses-based distillery waste represents a complex environmental challenge with widespread implications for ecosystems and human health. Efforts to mitigate their impact necessitate a multifaceted approach, including international collaboration, stringent regulations, advanced monitoring, and innovative remediation technologies. The ongoing commitment to addressing the legacy of PTS is essential for the well-being of the planet and future generations (Chu &Karr, 2017). Different countries have implemented various regulatory frameworks to control the release of PTS from molasses-based distilleries, with guidelines focusing on advanced wastewater treatment, emissions control, and sustainable waste management. These regulations aim to protect the environment and promote innovative solutions and best practices within the industry.

A combination of immediate and future strategies is required to address this issue effectively. First, strict regulatory enforcement and monitoring mechanisms must be implemented to control the discharge of distillery waste into the environment. This involves imposing stringent effluent standards and penalties for non-compliance to incentivize responsible waste management practices among distilleries. Investing in research and development efforts to develop innovative waste treatment technologies is imperative. Their advanced treatment methods, such as anaerobic digestion, bioremediation, and membrane filtration, promise to mitigate the release of toxic substances from distillery waste while recovering valuable by-products such as biogas and clean water. The distillery industry should promote sustainable practices to reduce waste generation and minimize environmental impacts. Encouraging the adoption of circular economy principles, such as recycling and resource recovery, can help minimize the reliance on single-use resources and reduce the overall environmental footprint of distillery operations. Education and awareness campaigns targeting industry stakeholders and the general public are also essential. Individuals and communities can advocate for stricter regulations and support sustainable practices within the industry by raising awareness about the environmental and health hazards associated with distillery waste.

REFERENCES

Ahmed, I. (*2019*). The basel convention on the control of transboundary movements of hazardous wastes and their disposal: a legal misfit in global ship recycling jurisprudence. *Washington International Law Journal*, *29*, 411.

Ali, H., & Khan, E. (*2018*). Trophic transfer, bioaccumulation, and biomagnification of non-essential hazardous heavy metals and metalloids in food chains/webs—Concepts and implications for wildlife and human health. *Human and Ecological Risk Assessment: An International Journal*, *25*, 1353–1376.

Ali, H., Khan, E., & Ilahi, I. (*2019*). Environmental chemistry and ecotoxicology of hazardous heavy metals: environmental persistence, toxicity, and bioaccumulation. *Journal of Chemistry*, *2019*, 6730305.

Arshad, M., Abbas, M., & Iqbal, M. (*2019*). Ethanol production from molasses: environmental and socioeconomic prospects in Pakistan: feasibility and economic analysis. *Environmental Technology & Innovation*, *14*, 100317.

Bharagava, R. N., & Chandra, R. (*2010*). Effect of bacteria treated and untreated post-methanated distillery effluent (PMDE) on seed germination, seedling growth and amylase activity in Phaseolus mungo L. *Journal of Hazardous Materials*, *180*(1–3), 730–734.

Bussian, B. M., Eugenio, N. R., Wilson, S. C., Ceci, A., Parelho, C., Semenov, D., & Yahyaabadi, M. (*2021*). Main soil contaminants and their fate in the soil environment. *Global Assessment of Soil Pollution: Report*. Food and Agriculture Organization of the United Nations (FAO), Rome.

Cachada, A., Rocha-Santos, T., & Duarte, A. C. (*2018*). Soil and pollution: an introduction to the main issues. In A. Cachada (Ed.), *Soil Pollution* (pp. 1–28). Academic Press, New York.

Chahal, C., Van Den Akker, B., Young, F., Franco, C., Blackbeard, J., & Monis, P. (*2016*). Pathogen and particle associations in wastewater: significance and implications for treatment and disinfection processes. *Advances in Applied Microbiology*, *97*, 63–119.

Chandra, R., & Kumar, V. (*2017*). Detection of Bacillus and Stenotrophomonas species growing in an organic acid and endocrine-disrupting chemical-rich environment of distillery spent wash and its phytotoxicity. *Environmental Monitoring and Assessment*, *189*, 1–19.

Chandra, R., Bharagava, R. N., Yadav, S., & Mohan, D. (*2009*). Accumulation and distribution of toxic metals in wheat (Triticum aestivum L.) and Indian mustard (Brassica campestris L.) irrigated with distillery and tannery effluents. *Journal of Hazardous Materials*, *162*(2–3), 1514–1521.

Chandra, R., Yadav, S., & Mohan, D. (*2008*). Effect of distillery sludge on seed germination and growth parameters of green gram (Phaseolus mungo L.). *Journal of Hazardous Materials*, *152*(1), 431–439.

Chowdhary, P., Yadav, A., Kaithwas, G., & Bharagava, R. N. (*2017*). Distillery wastewater: a major source of environmental pollution and its biological treatment for environmental safety. *Green Technologies and Environmental Sustainability*, *2017*, 409–435.

Christofoletti, C. A., Escher, J. P., Correia, J. E., Marinho, J. F. U., & Fontanetti, C. S. (*2013*). Sugarcane vinasse: environmental implications of its use. *Waste Management*, *33*(12), 2752–2761.

Chu, E. W., & Karr, J. R. (*2017*). Environmental impact: concept, consequences, measurement. *Reference Module in Life Sciences*, *5*, 278–296.

Dadrasnia, A., Shahsavari, N., & Emenike, C. U. (*2013*). Remediation of contaminated sites. *Hydrocarbon*, *65*, 82.

Durner, J., & Watts, D. C. (*2021*). Principles of analytical chemistry for toxicology. *Regulatory Toxicology*, *2021*, 455–497.

Environmental Protection Authority (EPA) Victoria. (*2018*). *Industrial Waste Resource Guidelines: Wastewater Management for Industry*. EPA Victoria, Victoria.

European Parliament, & Council of the European Union. (*2000*). *Directive 2000/60/EC of the European Parliament and of the Council of 23 October 2000 establishing a framework for Community action in the field of water policy*. Official Journal of the European Communities, L 327, 1–73.

European Parliament, & Council of the European Union. (*2010*). *Directive 2010/75/EU of the European Parliament and of the Council of 24 November 2010 on industrial emissions (integrated pollution prevention and control)*. Official Journal of the European Union, L 334, 17–119.

Flanagan, R. J., Taylor, A. A., Watson, I. D., & Whelpton, R. (*2008*). *Fundamentals of Analytical Toxicology*. John Wiley & Sons, New York.

Gavrilescu, M., Demnerová, K., Aamand, J., Agathos, S., & Fava, F. (*2015*). Emerging pollutants in the environment: present and future challenges in biomonitoring, ecological risks and bioremediation. *New Biotechnology*, *32*(1), 147–156.

Goswami, V., Deepika, S., Chandra, R., Babu, C. R., & Kothamasi, D. (*2023*). Arbuscular mycorrhizas accelerate the degradation of colour containing organic pollutants present in distillery spent wash leachates. *Journal of Hazardous Materials*, 452, 131291.

Hoarau, J., Caro, Y., Grondin, I., & Petit, T. (*2018*). Sugarcane vinasse processing: toward a status shift from waste to valuable resource. A review. *Journal of Water Process Engineering*, *24*, 11–25.

Holmes, N. S., & Morawska, L. (*2006*). A review of dispersion modelling and its application to the dispersion of particles: an overview of different dispersion models available. *Atmospheric Environment*, *40*(30), 5902–5928.

Inobeme, A., Adetunji, C. O., Mathew, J. T., Okonkwo, S., Bamigboye, M. O., Ajai, A. I., & Inobeme, J. (*2023*). Advanced nanotechnology for the degradation of persistent organic pollutants. In M. Shah (Ed.), *Microbial Degradation and Detoxification of Pollutants* (pp. 51–72). De Gruyter, Berlin, Boston.

Jagaba, A. H., Lawal, I. M., Birniwa, A. H., Affam, A. C., Usman, A. K., Soja, U. B., & Yaro, N. S. A. (2024). Sources of water contamination by heavy metals. In J. Jaafar, A. A. Zaidi, & M. N. Naseer (Eds.), *Membrane technologies for heavy metal removal from water* (pp. 3–27). CRC Press.

Kessler, R. (*2013*). The minamata convention on mercury: a first step toward protecting future generations. *Environ Health Perspect*, *121*, 304–309.

Krithiga, T., Sathish, S., Renita, A. A., Prabu, D., Lokesh, S., Geetha, R., ... & Sillanpaa, M. (*2022*). Persistent organic pollutants in water resources: fate, occurrence, characterization and risk analysis. *Science of The Total Environment, 831*, 154808.

Lallas, P. L. (*2001*). The Stockholm Convention on persistent organic pollutants. *American Journal of International Law, 95*(3), 692–708.

Lemos, A. F. (*2013*). The CONAMA (In) competence for issue of licensing standards for environmental: analysis of its legality. *Brazilian Journal of Law and Public Policy, 3*, 118.

Moutsatsou, A., Chalarakis, E., & Zarangas, G. (*2003*). Influence of raw materials and distillation equipment on the heavy metal content of waste from an alcoholic anis-type beverage. *Journal of Hazardous Materials, 96*(1), 53–64.

Nilsen, E., Smalling, K. L., Ahrens, L., Gros, M., Miglioranza, K. S., Picó, Y., & Schoenfuss, H. L. (*2019*). Critical review: grand challenges in assessing the adverse effects of contaminants of emerging concern on aquatic food webs. *Environmental Toxicology and Chemistry, 38*(1), 46–60.

Rajmohan, K. V. S., Ramya, C., Viswanathan, M. R., & Varjani, S. (*2019*). Plastic pollutants: effective waste management for pollution control and abatement. *Current Opinion in Environmental Science & Health, 12*, 72–84.

Ratna, S., Rastogi, S., & Kumar, R. (*2021*). Current trends for distillery wastewater management and its emerging applications for sustainable environment. *Journal of Environmental Management, 290*, 112544.

Razinkovas, A. (*2007*). Trophic network models and prediction of toxic substances accumulation in food webs. In I.E. Gonenc, V.G. Koutitonsky, B. Rashleigh, R.B. Ambrose, J.P. Wolflin (Eds.), *Assessment of the Fate and Effects of Toxic Agents on Water Resources* (pp. 279–289). Springer Netherlands, Dordrecht.

Scheringer, M. (*2009*). Long-range transport of organic chemicals in the environment. *Environmental Toxicology and Chemistry: An International Journal, 28*(4), 677–690.

Silva, F. S., Cristale, J., André, P. A., Saldiva, P. H., & Marchi, M. R. (*2010*). PM2. 5 and PM10: the influence of sugarcane burning on potential cancer risk. *Atmospheric Environment, 44*(39), 5133–5138.

Singh, K., Tripathi, S., & Chandra, R. (*2021*). Maillard reaction product and its complexation with environmental pollutants: a comprehensive review of their synthesis and impact. *Bioresource Technology Reports, 15*, 100779.

Singh, K. (*2023*). Study the role of rhizospheric bacterial community during the phytoremediation of distillery waste (Doctoral dissertation). Babasaheb Bhimrao Ambedkar University, Lucknow.

Sivaramanan, S. (*2015*). Air Pollution sources, pollutants and mitigation measures. *Research Gate, 10*(2.1), 5106–8485.

Sonone, S. S., Jadhav, S., Sankhla, M. S., & Kumar, R. (*2020*). Water contamination by heavy metals and their toxic effect on aquaculture and human health through food Chain. Lett. Appl. *NanoBioScience, 10*(2), 2148–2166.

Tripathi, S., Purchase, D., Al-Rashed, S., & Chandra, R. (*2022*). Microbial community dynamics and their relationships with organic and metal pollutants of sugarcane molasses-based distillery wastewater sludge. *Environmental Pollution, 292*, 118267.

Tripathi, S., Singh, K., Singh, A., Mishra, A., & Chandra, R. (*2021*). Organo-metallic pollutants of distillery effluent and their toxicity on freshwater fish and germinating Zea mays seeds. *International Journal of Environmental Science and Technology*, 1–14.

U.S. Environmental Protection Agency. (*1972*). *Clean Water Act (CWA)*. U.S. Government. Retrieved from https://www.epa.gov/laws-regulations/summary-clean-water-act

U.S. Environmental Protection Agency. (*1976*). *Resource Conservation and Recovery Act (RCRA)*. U.S. Government. Retrieved from https://www.epa.gov/rcra

Wong, M. H., Armour, M. A., Naidu, R., & Man, M. (*2012*). Persistent toxic substances: sources, fates and effects. *Reviews on Environmental Health, 27*(4), 207–213.

Section III

Case Studies in Vietnam

11

Monitoring of Persistent Bioaccumulative Toxic (PBT) Substances in Vietnam: Contamination Profiles, Source Characterization, and Health Risk Assessments

Hoang Quoc Anh, Tran Manh Tri, Vu Duc Nam, Trinh Thi Tham, and Tu Binh Minh

11.1 Introduction

Many organic substances produced intentionally and inadvertently have been released into the environment and entered different transportation and transformation processes (Wolff and Crossland, 1991; Ballschmiter, 1992). Organic chemicals are applied in various sectors, such as plastics, resins, synthetic rubber, fibers, dyes, paints, pigments, coatings, industrial fluids, fertilizers, pesticides, pharmaceuticals, medicines, personal care products, and food additives (Wittcoff et al., 2013; Lucattini et al., 2018). Besides, various organic substances are unintentionally formed during high-thermal processes such as fuel combustion, waste incineration, and industrial production (Abdel-Shafy and Mansour, 2016; Liu et al., 2021). The widespread application of organic chemicals and the popularity of thermal sources lead to the ubiquitous occurrence of these compounds in the environment (Lucattini et al., 2018; Kulsoom et al., 2021). Among them, various chemicals have been identified as persistent, bioaccumulative, and toxic (PBT) substances of global concern (Muir and Howard, 2006). The Stockholm Convention on Persistent Organic Pollutants (POPs) has provided a list of typical PBT substances, including pesticides, industrial chemicals, and unintentionally produced compounds, which are primarily organohalogens (Pariatamby and Kee, 2016; Mortimer and Brushett, 2023).

In addition, polycyclic aromatic hydrocarbons (PAHs) and related compounds have received constant attention because of their abundance, bioaccumulation, and high toxicity, including carcinogenicity (Abdel-Shafy and Mansour, 2016; Lawal, 2017). The PBT substances can enter the surrounding environment, such as the atmosphere, soil, and water, through different pathways like evaporation, washout, and leaching (Mill et al., 2019; Thakur and Pathania, 2020). Then, these compounds continue to participate in various natural processes including phase distribution (e.g., gas-particle, air-water, water-particle, and water-sediment partition), transport by atmospheric and riverine flows, decomposition and transformation by abiotic environmental factors (e.g., temperature, sunlight, and oxidizing/reducing agents), and bioaccumulation, biomagnification, and biotransformation in the food web (Gómez et al., 2007; Mamy et al., 2015; Thakur and Pathania, 2020). Human exposure to several PBT substances can lead to increased cancer risk, reproductive disorders, alteration of the immune system, neurobehavioral impairment, endocrine disruption, genotoxicity, and increased congenital disabilities (Thakur and Pathania, 2020).

Vietnam is a developing country in Southeast Asia and is undergoing a process of rapid urbanization and industrialization. However, economic change usually accompanies environmental degradation caused by organic pollutants, especially PBT substances. The complex emission sources of PBT

DOI: 10.1201/9781003492115-14

substances in Vietnam have been attributed to traffic, construction, household, industrial, and inadequate waste processing activities. A clear urban-rural declining trend was observed for contamination degree of several classes of organic pollutants in Vietnam, such as PAHs (Tuyen et al., 2014a, 2014b; Anh et al., 2019a), polychlorinated biphenyls (PCBs) (Anh et al., 2019b), brominated flame retardants (BFRs) (Anh et al., 2018), and current-use chemicals (Anh et al., 2019c, 2019d, 2019e).

Previous studies have indicated the emissions of dioxin-related compounds (DRCs), chlorinated benzenes (CBzs), and PAHs in flue gas, residual ash, and receiving environments from multiple industrial sectors of this country (Hue et al., 2016; Pham et al., 2019a, 2019b; Nguyen et al., 2019a, 2021, 2023a, 2024a). The "localized" pollution by specific organic contaminants was revealed in some informal waste processing areas in northern Vietnam. For example, elevated concentrations of flame retardants and plasticizers were detected in environmental samples from some Vietnamese e-waste recycling sites (Tue et al., 2013; Anh et al., 2017a; Hoang et al., 2022); meanwhile, relatively high concentrations of PAHs, PCBs, and DRCs were found in workplace dust and air from an informal end-of-life vehicle (ELV) dismantling area in northern Vietnam (Takahashi et al., 2017; Anh et al., 2019f, 2020a, 2020b). Human exposure to PBT substances (e.g., PAHs, DRCs, PCBs, and flame retardants) has been evaluated. Higher health risk levels were estimated for urban residents, occupationally exposed persons, children in the waste processing areas, and workers in industrial facilities (Tue et al., 2010a, 2010b, 2013; Tuyen et al., 2014a, 2014b; Hue et al., 2016; Anh et al., 2017a; Takahashi et al., 2017; Pham et al., 2019b). Nevertheless, comprehensive studies on the occurrence and impacts of multiple classes of PBT substances in Vietnam are relatively limited, mainly due to inadequate monitoring proficiency as well as human and financial resources.

This chapter provides a comprehensive and updated overview of the contamination status, spatio-temporal trends, potential emission sources, and impacts of PBT substances in Vietnam. Data from peer-reviewed papers on pollution monitoring and assessment of PBT substances in Vietnam were gathered and assessed, focusing on those published over the last two decades. The targeted groups of PBT substances include organochlorine pesticides (OCPs), DRCs, PCBs, CBzs, BFRs, poly- and perfluoroalkyl substances (PFASs), and PAHs. Information about levels and profiles of these substances in various environmental compartments such as air, dust, soil, water, sediment, biota, and food was presented, followed by a systematic review of their emission sources like legacy agrochemicals, industrial emissions, domestic chemicals, and informal waste processing activities. The impacts of PBT substances on ecological and human health were also reviewed, providing the scientific basis and reference information for developing an effective management and mitigation scheme for these pollutants in Vietnam.

11.2 Contamination Status

11.2.1 Air

Air samples were collected from different locations along Vietnam to characterize the pollution status of PBT substances such as PAHs and their derivatives (Anh et al., 1999; Hien et al., 2007a, 2007b; Kishida et al., 2008; Hong et al., 2016; Anh et al., 2020b; Trung et al., 2021; Vu-Duc et al., 2021; Hoang et al., 2023a; Vo et al., 2024), PCBs (Iwata et al., 1994; Tue et al., 2013; Wang et al., 2016a; Anh et al., 2020b), BFRs (Tue et al., 2013; Anh et al., 2020b), OCPs (Iwata et al., 1994; Wang et al., 2016a), and dioxins (Ngo et al., 2017;Chi et al., 2022). The air samples were mainly collected as particulate matter (PM) or by polyurethane foam-based passive air samplers (PUF-PAS). A summary of PBT levels and distribution characteristics in Vietnamese air samples is provided in Table 11.1.

Concentrations of unsubstituted and alkylated PAHs were generally higher than those of other classes such as nitro PAHs, PCBs, BFRs, OCPs, and polychlorinated dibenzo-*p*-dioxins/furans (PCDD/Fs). Atmospheric levels of PAHs were correlated with urbanization degree and traffic density, showing the abundance of benzo[ghi]perylene (BP)—an indicator of gasoline vehicle emission (Anh et al., 1999; Hien et al., 2007b; Kishida et al., 2008). PAH concentrations in outdoor air PM samples were higher than in indoor samples (Trung et al., 2021; Vu-Duc et al., 2021; Vo et al., 2024). Although a decline in urban-rural trend was found for PAHs, elevated levels of these substances were measured in the air

TABLE 11.1

Levels and Distribution Characteristics of PBT Substances in Vietnamese Air Samples

Pollutants	Sample Type	Location (Sampling Year)	Levels	Distribution Characteristics	References
PAHs	PM	Urban and rural sites, Ho Chi Minh City (1998)	2.15–117 ng/m^3	Higher levels in urban areas than rural sites; BP as a major compound.	Anh et al. (1999)
PAHs	PM	Urban roadside, Ho Chi Minh City (2005–2006)	38.3–72.3 ng/m^3	Higher levels in the rainy season than in the dry season; BP as a major compound.	Hien et al. (2007a)
PAHs and nitro PAHs	PM	Urban sites, Ho Chi Minh City (2005–2006)	7.6±3.8–19±7.6 ng/m^3	Higher levels in traffic sites; BP as a major compound.	Hien et al. (2007b)
PAHs and alkyl PAHs	PM and gas phase	Urban roadside, Hanoi (2005)	PM 63±82 ng/m^3 Gas 480±300 ng/m^3	Higher levels in bus terminals than other roadside sites; 6–7 ring and 3 ring PAHs dominated PM and gas phases, respectively.	Kishida et al. (2008)
PAHs and nitro PAHs	PM	Urban roadside, Hanoi (2010–2011)	PAHs 9.6–63.7 pmol/m^3 Nitro PAHs 0.12–0.96 pmol/m^3	5–6 ring PAHs as major compounds, 9-nitroanthracene, and 1-nitropyrene as major nitro PAHs.	Pham et al. (2012)
PAHs and alkyl PAHs	PUF-PAS	Urban and rural sites (2012–2013)	PAHs 15–87 ng/m^3 Alkyl PAHs 8.0–80 ng/m^3	Higher levels in urban sites than in rural sites.	Hong et al. (2016)
PAHs and methyl PAHs	PUF-PAS	Waste processing and urban sites (2013–2015)	PAHs 5.1–100 ng/m^3 Methyl PAHs 1.0–33 ng/m^3	Higher levels in waste processing sites than in urban sites; 3–4 ring PAHs and methyl PAHs as major compounds.	Anh et al. (2020b)
PAHs and methyl PAHs	PM	Urban roadside sites, Hanoi	210–660 ng/m^3	Higher levels in $PM_{2.5}$ than PM_{10} and $PM_{>10}$; proportions of 5–6 ring PAHs increase with decreasing particle sizes.	Trung et al. (2021)
PAHs	PM	Urban sites, Hanoi (2018)	226–727 ng/m^3	Higher levels in PM_{10} than $PM_{2.5}$; 2–3 ring PAHs as major compounds.	Vu-Duc et al. (2021)
PAHs and methyl PAHs	PUF-PAS	Vehicle processing sites, Bac Giang (2019)	40–95 ng/m^3	Higher levels in workshops than in the control house; methyl PAHs were more abundant in workshops than in the control house.	Hoang et al. (2023a)
PAHs	PM	Urban residential sites (2018–2019)	$PM_{2.5}$ 102.9–155.6 ng/m^3 $PM_{0.1}$ 25.3–52.9 ng/m^3	Higher levels in winter than in summer; Nap and Phe are major compounds.	Vo et al. (2024)

(Continued)

TABLE 11.1 (*Continued*)

Levels and Distribution Characteristics of PBT Substances in Vietnamese Air Samples

Pollutants	Sample Type	Location (Sampling Year)	Levels	Distribution Characteristics	References
PCBs	Low-volume air sampling	Urban and suburban areas (1990)	710–830 pg/m^3	No clear trend was revealed due to the limited sample size (n=3).	Iwata et al. (1994)
PCBs	PUF-PAS	E-waste, urban, and suburban sites (2008)	33–1800 pg/m^3	Higher levels in e-waste workshops than in common houses; tri-CBs were dominant.	Tue et al. (2013)
PCBs	PUF-PAS	Urban, suburban, and rural sites (2012–2013)	135–1518 pg/m^3	Higher levels in suburban sites and sites near urban centers; tri- to penta-CBs were dominant.	Wang et al. (2016a)
PCBs	PUF-PAS	Waste processing and urban sites (2013–2015)	170–1100 pg/m^3	Levels in waste processing sites and urban sites were not significantly different; CB-11 was the most predominant congener.	Anh et al. (2020b)
PBDEs	PUF-PAS	E-waste, urban, and suburban sites (2008)	4.6–720 pg/m^3	Higher levels in e-waste workshops than common houses; BDE-209, -47, -99, and -28 were major congeners.	Tue et al. (2013)
PBDEs	PUF-PAS	Waste processing and urban sites (2013–2015)	5.3–86 pg/m^3	Levels in waste processing sites and urban sites were not significantly different; BDE-209 was the most predominant congener.	Anh et al. (2020b)
OCPs	Low-volume air sampling	Urban and suburban areas (1990)	HCHs 220–12×10^6 pg/m^3 DDTs 1700–2400 pg/m^3 CHLs 31–340 pg/m^3	No clear trend was revealed due to the limited sample size (n=3).	Iwata et al. (1994)
OCPs	PUF-PAS	Urban, suburban, and rural sites (2012–2013)	DDTs 160–3923 pg/m^3 HCHs 59.5–625 pg/m^3 HCB 125–998 pg/m^3 Heptachlor 5.4–200 pg/m^3 Endosulfan 1.0–97.6 pg/m^3 Chlordane 15.3–209 pg/m^3 Mirex 0.39–10.5 pg/m^3	Compositions and spatial variation of almost all OCPs were not significantly different between urban and rural sites; enantiomeric signatures of OCPs exhibited seasonal variation.	Wang et al. (2016a)

(*Continued*)

TABLE 11.1 (*Continued*)
Levels and Distribution Characteristics of PBT Substances in Vietnamese Air Samples

Pollutants	Sample Type	Location (Sampling Year)	Levels	Distribution Characteristics	References
PCDD/Fs	PM and gas phase	Agent Orange hotspot, industrial, and rural sites (2010–2015)	21.3±13–139±84 fg I-TEQ/m^3	Levels increased in the order: industrial > Agent Orange hotspot > rural sites; 2,3,4,7,8-PeCDF was an important contributor to TEQ levels.	Ngo et al. (2017)
PCDD/Fs	PUF-PAS	Dioxin remediation site, Da Nang Airport (2013–2017)	PCDDs 0.38–4.89 pg/m^3 PCDFs 0.14–1.90 pg/m^3 TEQs 0.039–4.37 pg/m^3	Major compounds were 2,3,7,8-TCDD, 1,2,3,4,6,7,8-HpCDD, and OCDD; spatial and temporal trends of PCDD/Fs in the air were related to remediation activities.	Sau et al. (2021)
PCDD/Fs	$PM_{2.5}$	Urban site, Hanoi (2019)	92.4±67.3 fg I-TEQ/m^3	PCDD/F levels may not correlate with $PM_{2.5}$ concentrations; OCDD, 1,2,3,4,6,7,8-HpCDD, 1,2,3,4,6,7,8-HpCDF, and OCDF were major compounds.	Chi et al. (2022)
PCDD/Fs and dioxin-like PCBs	PUF-PAS	Urban, suburban-industrial, and rural sites, Hanoi (2017–2021)	PCDDs 0.076–1.00 pg/m^3 PCDFs 0.25–1.18 pg/m^3 DL-PCBs 2.03–35.4 pg/m^3 TEQs 0.035–0.22 pg/m^3	Higher levels in suburban-industrial and urban sites than in rural sites; dioxin concentrations increased during spring and winter.	Sau (2023)

samples from some rural waste processing workshops (Anh et al., 2020b; Hoang et al., 2023a). Such informal waste recycling sites also detected local air contamination caused by PCBs and BFRs (Tue et al., 2013; Anh et al., 2020b). CB-11, an emerging dichlorinated congener derived from organic pigments, has been identified as the dominant PCB component in the air (Anh et al., 2020b), alongside the contribution of tri- to penta-CBs (Tue et al., 2013; Wang et al., 2016a).

Wang et al. (2016a) reported the widespread occurrence of PCBs and OCPs in the air from several locations along the country with relatively high levels of dichlorodiphenyltrichloroethane and related compounds (DDTs) as compared to other Asian countries. Among BFRs, polybrominated diphenyl ethers (PBDEs) were more abundant than other substances like hexabromocyclododecanes (HBCDs), decabromodiphenyl ethane (DBDPE), 1,2-bis(2,4,6-tribromophenoxy)ethane (BTBPE), hexabromobiphenyl (BB-153), and pentabromoethylbenzene (PBEB) (Tue et al., 2013; Anh et al., 2020b). PCDD/Fs have also been detected in ambient air from different locations in Vietnam, showing higher levels in urban, and industrial hotspot sites than in rural sites.

11.2.2 Dust

Settled dust is a sink of semivolatile compounds, including PBT substances (Lucattini et al., 2018). In Vietnam, indoor and outdoor dust samples have been collected to determine PAHs and methylated PAHs (Tuyen et al., 2014a, 2014b; Anh et al., 2019a, 2019f; Mon et al., 2020), PCBs (Tue et al., 2013; Takahashi et al., 2017; Anh et al., 2019b, 2020a), BFRs (Tue et al., 2013; Anh et al., 2017a; Takahashi et al., 2017; Hoang et al., 2021a, 2022, 2023a; Tran et al., 2022), DRCs (Tue et al., 2010b; Takahashi et al., 2017), and PFASs (Tan et al., 2024). Concentrations of PBT substances in Vietnamese settled dust samples are summarized in Figure 11.1. Analytical results of PBT substances in dust revealed a wide concentration range over 6–7 orders of magnitude, from a few pg/g to hundreds of μg/g. Higher levels of BFRs like PBDEs and DBDPE were found in indoor dust than in outdoor dust, while other groups such as PAHs, methylated PAHs, and PCBs did not show a clear indoor-outdoor distribution trend. Road dust concentrations of PBT groups decreased in the order: PAHs (76–9300 ng/g), methylated PAHs (28–1500 ng/g), PCBs (0.10–120 ng/g), PBDEs (0.90–56 ng/g), and DBDPE (<0.10–73 ng/g) (Tuyen et al., 2014a, 2014b; Anh et al., 2018, 2019a, 2019b). PBT concentrations in indoor dust varied greatly with higher levels of some specific groups like methylated PAHs, PCBs, BFRs, and DRCs found in settled dust samples collected from e-waste and ELV processing workshops (Tue et al., 2010b, 2013; Takahashi et al., 2017; Anh et al., 2019f; Hoang et al., 2022).

Some studies conducted in the same areas over several years did not provide a coherent temporal trend of pollutants such as PAHs in urban road dust (Tuyen et al., 2014a, 2014b; Anh et al., 2019a; Mon et al., 2020) and BFRs in e-waste dust (Tue et al., 2013; Anh et al., 2017a). In Bui Dau e-waste sites, markedly higher levels of PBDEs and DBDPE in floor and equipment dust collected in 2019 than atmospheric deposited dust taken in previous years (2008 and 2014) were primarily attributed to different dust types and sampling strategies (Tue et al., 2013; Anh et al., 2017a; Hoang et al., 2022). Concentrations of typical PBT classes such as PAHs, BFRs, PCBs, and PFASs in Vietnamese dust were within the low to moderate range compared to other countries worldwide.

Accumulation patterns of PBT groups in dust varied between sample types and locations, reflecting different emission sources and environmental behaviors. PAHs were more abundant than methylated PAHs in almost all Vietnamese dust samples, except for some floor dust samples collected from ELV processing workshops (Tuyen et al., 2014b; Anh et al., 2019a, 2019f). High molecular weight (HMW) PAHs accounted for higher proportions in total PAHs than low molecular weight (LMW) compounds. Major PAHs detected in Vietnamese dust were phenanthrene (Phe), fluoranthene (Flt), pyrene (Pyr), chrysene (Chr), and BP. Penta- and hexa-CBs were dominant in dust from urban and waste processing areas, while tri- and tetra-CBs dominated profiles in dust from industrial areas (Tue et al., 2013; Takahashi et al., 2017; Anh et al., 2019b, 2020a). CB-11 was widely detected in both road dust (0.059–1.5 ng/g, accounting for 13% of total PCBs in average; Anh et al., 2019b) and indoor dust (0.52–8.7 ng/g, accounting for 18% of total PCBs in average; Anh et al., 2020a).

BDE-209 was the most predominant PBDE congener detected in dust, suggesting the widespread application of deca-BDE technical mixtures and products containing them (Hoang et al., 2021b).

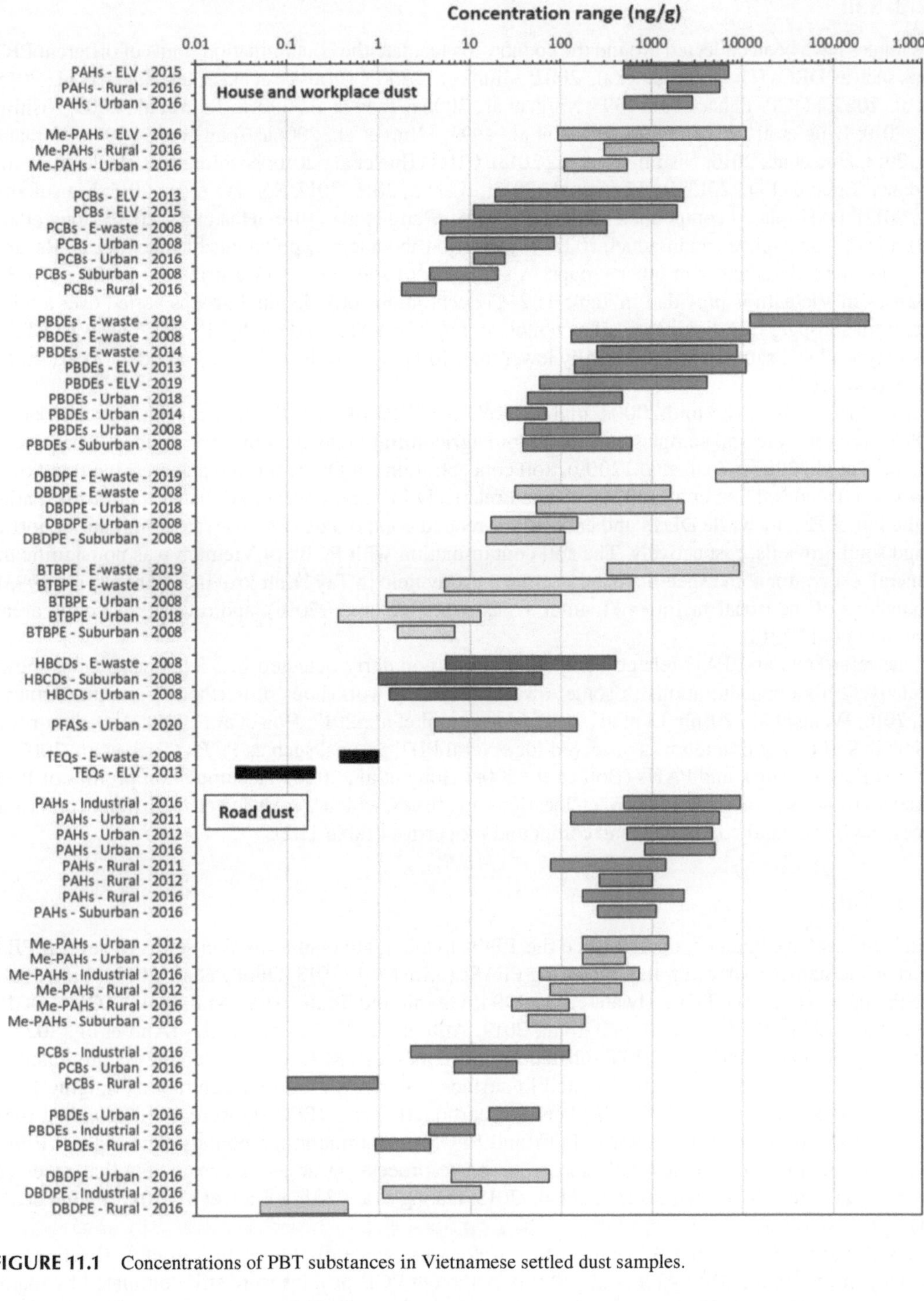

FIGURE 11.1 Concentrations of PBT substances in Vietnamese settled dust samples.

Besides regulated dioxins (i.e., PCDD/Fs and dioxin-like PCBs), brominated dioxins (e.g., PBDD/Fs) and unidentified compounds may contribute significantly to total DRC levels and toxic equivalents in dust (Tue et al., 2010b; Takahashi et al., 2017). Perfluorooctanoic acid (PFOA) and perfluorooctane sulfonate (PFOS) were characterized as major PFASs in urban house dust from Hanoi (Tan et al., 2024).

11.2.3 Soil

Soil samples have been collected around the country to elucidate the contamination status of different PBT groups such as DRCs (Dwernychuk et al., 2002; Minh et al., 2003; Brodsky et al., 2009; Huyen et al., 2015; Le et al., 2022;), OCPs (Thao et al., 1993; Minh et al., 2006a; Toan et al., 2007a; Toan et al., 2009; Nishina et al., 2010; Ding et al., 2024), PCBs (Thao et al., 1993; Minh et al., 2006a; Toan et al., 2007b; Romano et al., 2013; Hue et al., 2016; Nishimura et al., 2018), CBzs (Hue et al., 2016; Nishimura et al., 2018), flame retardants (Eguchi et al., 2013; Romano et al., 2013; Li et al., 2016, 2017; Someya et al., 2016; Matsukami et al., 2017), PAH-related compounds (Boll et al., 2008; Wang et al., 2016b; Hoa et al., 2020; Hong et al., 2020), and PFASs (Nguyen and Baduel, 2023). Information about legacy pollutants like dioxins, PCBs, and OCPs was more abundant than other groups. A summary of soil levels and distribution profiles of PBT substances in Vietnam is provided in Table 11.2. Concentrations of individual groups varied over a wide range, from sub-pg/g (for dioxins) to a few tenths of µg/g (for PAHs and methyl PAHs). The WHO-TEQ levels in typical soil samples were generally lower than 100 pg/g (Minh et al., 2003; Brodsky et al., 2009; Nghiem et al., 2022).

During the 1990s to the mid-2000s, higher levels of OCPs like DDTs and hexachlorocyclohexanes (HCHs) were found in soil samples collected from agricultural areas than in other regions (Thao et al., 1993; Toan et al., 2007a; Toan et al., 2009). Soil concentrations of OCPs showed a decreasing trend over the last two decades. Ding et al. (2024) revealed relatively low soil concentrations with uniform spatial distribution of HCHs, while DDTs and endosulfan-related compounds (ENDs) were dominant in northern and southern soils, respectively. The soil contamination with PCBs in Vietnam was not significant in general, except for some specific locations like a paddy field in Tay Ninh Province (Thao et al., 1993), surroundings of industrial facilities (Toan et al., 2007b; Hue et al., 2016), and e-waste recycling areas (Nishimura et al., 2018).

Flame retardants and PAH-related compounds were popularly detected in Vietnamese soils, showing relatively high concentrations in some e-waste recycling workshops in northern Vietnam (Someya et al., 2016; Wang et al., 2016b; Li et al., 2017; Matsukami et al., 2017; Hoa et al., 2020). An urban-rural gradient in soil contamination was observed for several PBT groups, such as PCBs (Toan et al., 2007b), PBDEs (Li et al., 2016), and PAHs (Boll et al., 2008; Hong et al., 2020). Accumulation profiles of PBT substances in soil samples varied between locations and times, which were affected by different emission sources, environmental conditions, and compound properties (Table 11.2).

11.2.4 Water

Several studies have been conducted since the 1990s to elucidate contamination status of several PBT groups in Vietnamese surface water, including PFASs (Kim et al., 2013; Duong et al., 2015a; Lam et al., 2017; Phung et al., 2021), PCBs (Iwata et al., 1994; Quynh and Toan, 2019; Anh et al., 2021a), OCPs (Iwata et al., 1994), BFRs (Quynh and Toan, 2019; Anh et al., 2021a), and PAHs (Anh et al., 2021a). A summary of concentrations of PBT substances in Vietnamese surface water samples is presented in Table 11.3. PFASs were frequently detected in surface water, with total concentrations ranging from lower than detection limits to 170 ng/L. Perfluorocarboxylic acids (PFCAs) were more abundant than perfluoroalkyl sulfonates (PFSs), with PFOA and PFOS as dominant compounds. The highest PFAS levels were recorded in samples collected from waste processing areas or urban sites that received industrial and domestic discharges (Kim et al., 2013; Duong et al., 2015a; Lam et al., 2017; Phung et al., 2021).

Concentrations of PCBs and BFRs in surface water were generally lower than 10 ng/L (Iwata et al., 1994; Quynh and Toan, 2019; Anh et al., 2021a). Although PCB profiles were still dominated by major congeners of technical PCB mixtures, significant proportions of CB-11, a pigment-derived congener, were detected (Anh et al., 2021a). OCPs were detected in water samples collected along the country in the 1990s, showing a decreasing trend of HCHs > DDTs > chlordane compounds (CHLs) (Iwata et al., 1994); however, updated information about OCPs in water is scarce. Anh et al. (2021a) measured PAHs and methylated PAHs in drainage water of an ELV processing area at concentrations ranging from 290 to 400 ng/L, which were dominated by LMW compounds like naphthalene (Nap) and Phe.

TABLE 11.2

Levels and Distribution Characteristics of PBT Substances in Vietnamese Soil Samples

Pollutants	Location (Sampling Year)	Levels	Distribution Characteristics	References
PCDD/Fs and dioxin-like PCBs	Dumping sites, Hanoi and Ho Chi Minh City (1999–2001)	WHO-TEQs 0.01–59 pg/g	Higher levels in Hanoi dumping sites than in Ho Chi Minh dumping sites and control sites; CB-126 contributed to >95% of TEQs.	Minh et al. (2003)
PCDD/Fs	Cam Lo and Gio Linh, Quang Tri Province (2003)	WHO-TEQs 0.011–10.8 pg/g	TCDD contributed for 56%–99% of TEQs.	Brodsky et al. (2009)
PCDD/Fs	Bien Hoa Airbase, Dong Nai Province (2000)	WHO-TEQs 0.04–1,180,737 pg/g	Soils were heavily contaminated with dioxins, especially TCDD.	Schecter et al. (2001)
PCDD/Fs	Bien Hoa Airbase, Dong Nai Province (2004)	WHO-TEQs 4.6–184 pg/g	Soils were heavily contaminated with dioxins, especially TCDD.	Mai et al. (2007)
PCDD/Fs and dioxin-like PCBs	Bien Hoa Airbase, Dong Nai Province (2014)	WHO-TEQs 32–3300 pg/g	Levels decreased with increasing soil depths; TCDD proportions in TEQs approached 100% in almost all sites.	Huyen et al. (2015)
PCDD/Fs	A Luoi Valley, Thua Thien Hue Province (1996–1999)	I-TEQs 4.9–901 pg/g	Higher levels in military airbases than in Agent Orange sprayed areas; TCDD contributed to 80% to >99% of TEQs.	Dwernychuk et al. (2002)
PCDD/Fs	A So Airbase, Thua Thien Hue Province (2014)	WHO-TEQs 2.7–746 pg/g	Higher levels in zone A (herbicide storage) than in zone B; levels decreased with increasing soil depths.	Le et al. (2019)
PCDD/Fs and dioxin-like PCBs	Bien Hoa Airbase and industrial areas, Dong Nai Province (2020)	WHO-TEQs 0.426–57.4 pg/g	Higher levels in airbase than in industrial areas; major contributors to TEQs were TCDD, 12,378-PeCDD, CB-126, and 23,478-PeCDF.	Nghiem et al. (2022)
OCPs	Paddy field and upland sites, Hanoi, Hue, Tay Ninh, and Ho Chi Minh City (1990)	DDTs 0.73–1300 ng/g HCHs 0.15–55 ng/g	Higher levels in paddy field soils than in upland soils; profiles of OCPs varied between locations.	Thao et al. (1993)
OCPs	Municipal dumping and control sites, Hanoi and Ho Chi Minh City (1999–2001)	DDTs 0.41–83 ng/g HCHs 0.11–2.2 ng/g CHLs <0.01–2.5 ng/g HCB <0.01–2.8 ng/g	Higher levels in dumping sites than in control sites, but the difference was not very high; DDT metabolites were more dominant than DDT; γ-HCH was more abundant than other HCH isomers.	Minh et al. (2006a)
OCPs	Agricultural, industrial, and urban areas, Hanoi (2006)	DDTs <0.02–172 ng/g HCHs <0.05–20.6 ng/g	Higher levels in agricultural areas than in industrial and urban areas.	Toan et al. (2007a)

(Continued)

TABLE 11.2 (*Continued*)

Levels and Distribution Characteristics of PBT Substances in Vietnamese Soil Samples

Pollutants	Location (Sampling Year)	Levels	Distribution Characteristics	References
OCPs	Agricultural areas and villages, Bac Ninh (2006)	DDTs <0.02–161 ng/g HCHs <0.05–9.54 ng/g	Higher levels in agricultural areas than in towns and traditional villages.	Toan et al. (2009)
OCPs	Agricultural fields, Ha Nam Province (2006–2007)	DDTs 2.07–25.4 ng/g	Higher levels in subsurface soils than in surface soils; metabolism and downward movement of DDTs were observed.	Nishina et al. (2010)
OCPs	Five cities along country (2017)	DDTs 0.022–85 ng/g ENDs 0.012–2.4 ng/g HCHs 0.005–0.24 ng/g	Higher levels in surface soils than soil cores, except for HCHs; HCHs showed uniform distribution; DDTs and ENDs were more abundant in north-central and southern areas, respectively.	Ding et al. (2024)
PCBs	Paddy field and upland sites, Hanoi, Hue, Tay Ninh, and Ho Chi Minh City (1990)	0.61–320 ng/g	Higher levels in urban and suburban soils impacted by industrial activities; the highest level of 320 ng/g in a paddy field in Tay Ninh may relate to warfare equipment.	Thao et al. (1993)
PCBs	Municipal dumping and control sites, Hanoi and Ho Chi Minh City (1999–2001)	0.45–40 ng/g	Higher levels in dumping sites than in control sites.	Minh et al. (2006a)
PCBs	Agricultural, industrial, and urban areas, Hanoi (2006)	<0.02–190 ng/g	Higher levels in industrial and urban areas than in agricultural areas; PCB-138 and -153 were major congeners.	Toan et al. (2007b)
PCBs	Thi Nai Lagoon, Binh Dinh Province (2010)	0.67–18.6 ng/g	Higher levels in horticultural and residential areas; PCB profiles varied between locations; CB-11 accounted for 7% of ΣPCBs on average.	Romano et al. (2013)
PCBs	Industrial facilities, northern Vietnam	14.5–130 ng/g	Levels and profiles of PCBs varied between facilities.	Hue et al. (2016)
PCBs	Bui Dau e-waste area, Hung Yen Province (2011–2012)	2.0–7300 ng/g	PCB levels decreased in the following order: open burning sites > near open burning sites > e-waste workshops > rice fields.	Nishimura et al. (2018)
CBzs	Industrial facilities, northern Vietnam	PeCB 0.18–15.5 ng/g HCB <0.05–18.4 ng/g	Higher levels in soils around municipal waste incinerators.	Hue et al. (2016)
CBzs	Bui Dau e-waste area, Hung Yen Province (2011–2012)	17–1400 ng/g	CBz levels decreased in the order: open burning sites > near open burning sites > e-waste workshops > rice fields.	Nishimura et al. (2018)

(Continued)

TABLE 11.2 (*Continued*)

Levels and Distribution Characteristics of PBT Substances in Vietnamese Soil Samples

Pollutants	Location (Sampling Year)	Levels	Distribution Characteristics	References
BFRs	Dumping and reference sites (2000s)	PBDEs 0.02–430 ng/g HBCDs <0.005–1.3 ng/g	Higher PBDE levels in dumping sites than in reference sites; hepta- to deca-BDEs were major compounds.	Eguchi et al. (2013)
PBDEs	Thi Nai Lagoon, Binh Dinh Province (2010)	0.21–4.02 ng/g	Higher levels in horticultural and residential areas; BDE-47 and -99 were major congeners.	Romano et al. (2013)
PBDEs	Urban and rural areas (2012)	0.2–2.4 ng/g	Levels decreased in the order: urban > rural > background sites; BDE-99, -47, and -209 were major compounds.	Li et al. (2016)
Novel BFRs	Urban and rural areas (2012)	Urban/rural 0.68 ± 0.56 ng/g E-waste 21 ng/g	Higher levels in e-waste sites than urban/rural sites; DBDPE was the most dominant compound.	Li et al. (2017)
Emerging flame retardants	Bui Dau e-waste area, Hung Yen Province (2012)	DBDPE n.d.–4200 ng/g BTBPE n.d.–350 ng/g DPs n.d.–65 ng/g HBBz n.d.–16 ng/g	Levels decreased in the order: e-waste workshop > open burning > paddy field sites; DBDPE, BTBPE, DPs, and HBBz were the most frequently detected compounds.	Someya et al. (2016)
Flame retardants	Bui Dau e-waste area, Hung Yen Province (2012–2014)	PBDEs <0.1–9200 ng/g Other BFRs <5–7400 ng/g DPs <0.2–1900 ng/g	Levels decreased in the order: e-waste workshop > open burning > paddy field sites; PBDE levels decreased and total other BFRs and DPs increased over 2012–2014 period.	Matsukami et al. (2017)
PAHs	Urban, suburban, and rural areas, northern Vietnam (2006)	170–1700 ng/g	PAH levels decreased in the order: urban > suburban > rural soils. Major compounds were Nap, Phe, Flt, Pyr, and Chr.	Boll et al. (2008)
PAHs and methyl PAHs	Bui Dau e-waste area, Hung Yen Province (2012)	530–42,000 ng/g	Higher levels in e-waste workshops and open burning sites than in rice fields;	Hoa et al. (2020)
PAHs and alkyl PAHs	Urban and rural sites (2012)	71.9–1290 ng/g	Higher levels in urban sites than in rural sites; PAHs were more abundant than alkyl PAHs.	Hong et al. (2020)
Halogenated PAHs	Bui Dau e-waste area, Hung Yen Province (2012)	0.025–3.96 ng/g	Levels decreased in the order: open burning sites > e-waste workshops > rice paddy fields.	Wang et al. (2016b)
PFASs	Agricultural field, Ho Chi Minh City	PFOS n.d.–661 ng/g	Almost all PFASs were not detected in three soil samples, except for one sample contaminated with an elevated PFOS level of 661 ng/g.	Nguyen and Baduel (2023)

TABLE 11.3

Levels and Distribution Characteristics of PBT Substances in Vietnamese Surface Water Samples

Pollutants	Location (Sampling Year)	Levels	Distribution Characteristics	References
PFASs	Waste recycling/disposal and rural sites (2011)	n.d.–170 ng/L	Higher levels in surface water from e-waste sites; PFOA, PFNA, and PFUnDA were major compounds.	Kim et al. (2013)
PFASs	Rivers and canals in Hanoi, Hue, Da Nang, and Ho Chi Minh City (2013)	0.92–17 ng/L	Higher levels at sites receiving industrial discharge and untreated domestic sewage; rainy season samples had higher levels than dry season ones; PFOA, PFNA, and PFOS were the most frequently detected compounds.	Duong et al. (2015a)
PFASs	9 rivers along the country (2013–2015)	n.d.–14.6 ng/L	Higher levels in water samples from Sai Gon River, PFOA, and PFOS were major compounds.	Lam et al. (2017)
PFASs	4 craft villages in Hanoi, Bac Ninh, and Hung Yen	0.83–58.2 ng/L	PFAS levels decreased in the order: paper recycling, textile, and plastic recycling sites; PFOA, PFHxA, and PFOS were major compounds.	Phung et al. (2021)
PCBs	Hanoi, Hue, and Ho Chi Minh City (1990)	0.57–8.0 ng/L	The highest levels were found in water samples from Nha Be River and Gua Canal, Ho Chi Minh City.	Iwata et al. (1994)
PCBs	Kim Nguu River, Hanoi (2018)	1.89–8.93 ng/L	Higher levels in dry season samples than in wet season ones; CB-52 and -118 were major compounds.	Quynh and Toan (2019)
PCBs	Vehicle processing areas, Bac Giang Province (2015)	3.3–5.0 ng/L	Di- to hexa-CBs were detected with major congeners as CB-11, -28, -31, -44, -52, -70, -95/98, -101, -110/120.	Anh et al. (2021a)
OCPs	Hanoi, Hue, and Ho Chi Minh City (1990)	HCHs 1.9–31 ng/L DDTs 0.29–4.7 ng/L CHLs 0.045–0.55 ng/L	The highest levels found in water samples from Nha Be River, Ho Chi Minh City, and Phu Loc Lake, Hue; α-HCH and *p,p′*-DDD were major compounds.	Iwata et al. (1994)
PBDEs	Kim Nguu River, Hanoi (2018)	1.92–7.08 ng/L	Higher levels in dry season samples than in wet season ones; BDE-47 and -99 were major compounds.	Quynh and Toan (2019)
BFRs	Vehicle processing areas, Bac Giang Province (2015)	0.64–1.7 ng/L	BDE-28, -47, -209, and DBDPE were detected at trace levels.	Anh et al. (2021a)
PAHs and methyl PAHs	Vehicle processing areas, Bac Giang Province (2015)	290–400 ng/L	Nap and Phe were major compounds.	Anh et al. (2021a)

In Vietnam, information about PBT substances in other water types, like groundwater and tap water, is still limited. Duong et al. (2015b) applied a target screening method to semi-quantify the occurrence of 940 organic micro-pollutants in groundwater samples collected from Hanoi and Ho Chi Minh City, indicating low detection frequencies and trace levels of 15 OCPs and 18 PCBs, except for a few DDT compounds. Duong et al. (2015a) detected 7 PFASs in groundwater samples from Hanoi and Ho Chi Minh City at concentrations from <0.1 to 21 ng/L with PFOS, PFOA, and perfluorobutanoic acid (PFBA) as the most frequently observed compounds. Trace levels of PFASs were also measured in Vietnamese tap water (mean 0.31; max 1.19 ng/L) and well water (mean 3.55; max 8.88 ng/L) with major compounds such as PFOA, perfluorononanoic acid (PFNA), perfluorohexane sulfonate (PFHxS), and PFOS (Lam et al., 2017).

11.2.5 Sediment

Because almost all PBT substances are hydrophobic or amphiphilic, they tend to accumulate in the bed sediments through association with sedimentary organic matters, making sediment a sink of pollutants. Therefore, sediment samples have been extensively investigated to characterize the contamination status of multiple organic pollutants such as PAHs, DRCs, PCBs, OCPs, BFRs, and PFASs in Vietnam since the 1990s. A comprehensive summary of PBT levels in Vietnamese sediment samples is provided in Figure 11.2. Ambient concentrations of PBT groups in sediments decreased in the order: PAHs (0.73–6540 ng/g), DDTs (0.01–1100 ng/g), PCBs (0.04–384 ng/g), PBDEs (0.03–170 ng/g), DBDPE (<0.050–59 ng/g), PCDD/Fs (0.2–2.0 ng/g), and PFASs (not detected (n.d.)–0.86 ng/g). In general, levels of PBT substances in Vietnamese sediments were in low to moderate ranges when compared to those from other countries (Frignani et al., 2007; Hong et al., 2008; Saha et al., 2009; Kwan et al., 2013;

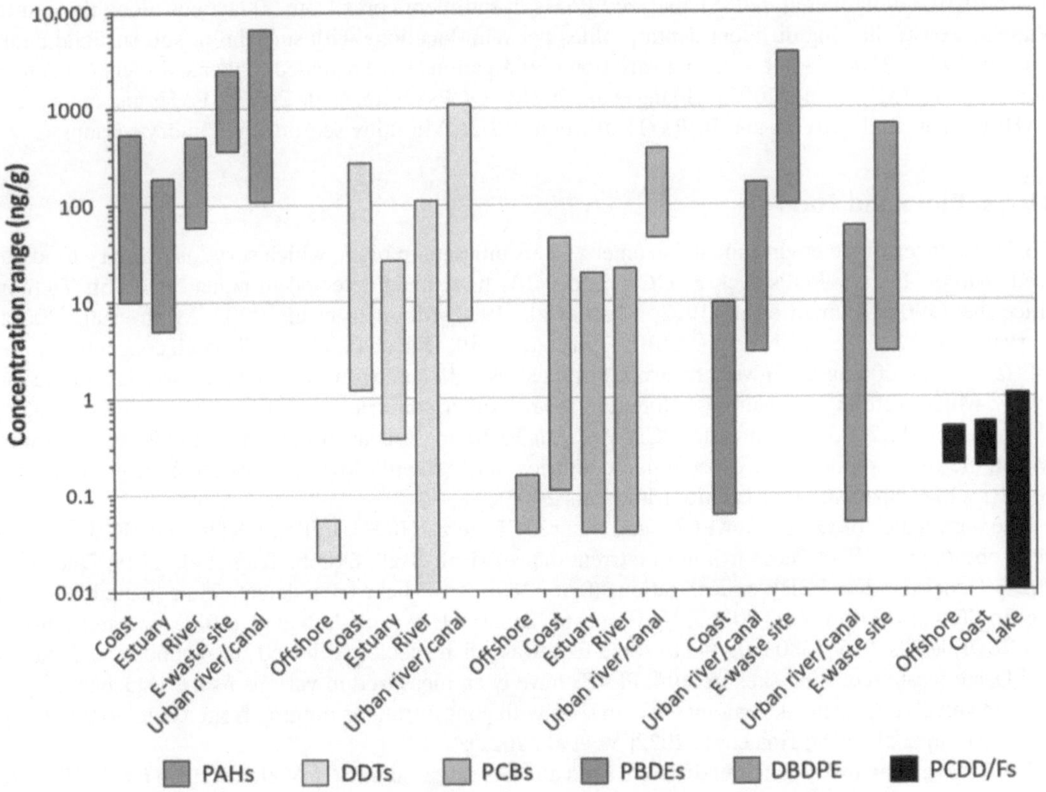

FIGURE 11.2 Spatial distribution of PBT substances in Vietnamese sediment samples.

Hoang et al., 2021c, 2021d). Besides, significant pollution caused by PBDEs and other BFRs was detected in some e-waste recycling sites in northern Vietnam (Anh et al., 2017a; Matsukami et al., 2017).

Sediment samples collected from sewer systems in urban and e-waste recycling sites showed higher concentrations of PAHs than riverine, estuarine, and coastal sediments (Anh et al., 1999; Boll et al., 2008; Babut et al., 2019; Hoa et al., 2020; Hoang et al., 2021c; Thuy et al., 2012). Similar spatial trends of sediment pollution (i.e., city canal > river > estuary) were also observed for PCBs, DDTs, CHLs, and hexachlorobenzene (HCB) (Minh et al., 2007a, 2007b). Markedly low concentrations of PCBs (0.036–0.158 ng/g), HCHs (0.021–0.054 ng/g), and DDTs (0.015–0.079 ng/g) were measured in surface sediments from offshore waters of central Vietnam (Tri et al., 2016).

Levels of legacy POPs like PCBs and OCPs still increased during the 1990s to the mid-2000s in the Red River Delta, northern Vietnam (Nhan et al., 1998, 2001; Hoai et al., 2010), but a significant decrease was found for these substances in southern Vietnam during this period (Minh et al., 2007a, 2007b; Carvalho et al., 2008). Recent studies revealed that current sediment concentrations of PCBs and OCPs in Vietnam were relatively low (Romano et al., 2013; Tran et al., 2019; Hoang et al., 2021d; Le et al., 2022). Levels of PBDEs in Vietnamese sediments decreased in the order: e-waste recycling sites (100–3800 ng/g; Matsukami et al., 2017), urban canals (3.0–170 ng/g; Kwan et al., 2013), and lagoon (n.d.–9.62 ng/g; Romano et al., 2013) and coastal areas (0.06–7.26 ng/g; Nguyen et al., 2024b). DBDPE, a prominent alternative BFR, has been detected in sediments from e-waste sites (3.0–680 ng/g; Matsukami et al., 2017) and urban areas (<0.050–59 ng/g; Hoang et al., 2021d). Trace levels of PFASs were measured in Vietnamese sediments (Lam et al., 2017; Babut et al., 2019).

A few studies investigated temporal trends of PBT substances in Vietnam using sediment core samples (Frignani et al., 2007; Giuliani et al., 2008, 2011; Tri et al., 2016; Tham et al., 2020). An increasing trend toward surface layers has been found for PCBs (Frignani et al., 2007; Giuliani et al., 2011; Tham et al., 2020), PBDEs (Tham et al., 2020), and PCDD/Fs (Tri et al., 2016). Meanwhile, peak concentrations of several OCPs were detected in sediment layers dated from the late 1950s to the early 1980s (Tham et al., 2020). Giuliani et al. (2008) analyzed PAHs in sediment cores from 10 lagoons along the central Vietnam coast, showing different depth profiles between locations with surficial or sub-surficial peaks over the 1970s–2000s. Some seasonal variation investigations have revealed higher sediment concentrations of PCBs (Minh et al., 2007a; Hoang et al., 2021d), OCPs (Minh et al., 2007a), PAHs and methylated PAHs (Hoang et al., 2021c), and BFRs (Hoang et al., 2021d) in rainy seasons than in dry seasons.

11.2.6 Biota and Food

PBT substances in the environment can enter and accumulate in biota, which serve as primary food and feed sources. Legacy POPs such as OCPs and PCBs have been detected in aquatic biota in Vietnam since the 1990s (Kannan et al., 1992; Nhan et al., 1998; Monirith et al., 2003; Minh et al., 2006b; Carvalho et al., 2008; Nishina et al., 2010; Tran et al., 2019; Duc et al., 2024). Elevated concentrations of DDTs (220–34,000 ng/g lw) were measured in mussels collected from the Vietnamese coast in the late 1990s, which were higher than those found in several other countries except for China and Hong Kong (Monirith et al., 2003). Meanwhile, PCB levels in Vietnamese coastal mussels (21–450 ng/g lw) were within the range reported for other Asian countries but markedly lower values recorded in developed countries like Japan and Russia (Monirith et al., 2003).

The dominance of DDTs over PCBs and other OCPs (e.g., HCHs, CHLs, ENDs, and HCB) has also been observed in Vietnamese fish until current dates (Minh et al., 2006b; Tran et al., 2019; Duc et al., 2024). Generally, low PBDE concentrations (0.06–26 ng/g lw) have been detected in aquatic biota samples in Vietnam (Minh et al., 2006b; Nguyen et al., 2024b). However, Anh et al. (2017) reported substantial PBDE levels (216–1380 ng/g lw) in some freshwater fish species collected from ponds and canals in Bui Dau e-waste recycling sites in 2014. PFASs have been measured in various fish, crustacean, gastropod, and bivalve species taken along the countries with concentrations ranging from n.d. to 26.2 ng/g wet weight (Lam et al., 2017; Hoa et al., 2022; Vi et al., 2022).

PAHs have been investigated in different food and beverage samples in Vietnam since the 2000s (Boll et al., 2008; Tran-Lam et al., 2018; Nguyen et al., 2019b; Thi et al., 2020; Nguyen et al., 2023b). Boll et al. (2008) detected PAHs at concentrations ranging from 490 to 650 ng/g dry weight in water spinach from the Yen Duyen aquatic production area, Hanoi, which were somewhat higher than in vegetables from Greece and Thailand. Tran-Lam et al. (2018) measured 18 PAHs in various food types in Hanoi, and their concentrations were higher in tea, coffee, grilled meat, and instant noodle samples than in snack, cake, and dried vegetable ones. In this study, the highest PAH levels were detected in a black tea sample (318 ng/g), a fried noodle sample (183 ng/g), and a coal-grilled meat sample (160 ng/g) (Tran-Lam et al., 2018).

PAHs have been frequently found in different tea types over a concentration range of 75.3–709 ng/g dry weight, with higher levels found in black tea than in other types such as green, oolong, and herbal teas (Nguyen et al., 2019b; Thi et al., 2020). Thi et al. (2020) estimated that 2.6%, 5.0%, 17.2%, and 25.7% of total PAHs in oolong, black, herbal, and green tea can be released into infusion, respectively. Roasted coffee samples collected from Ho Chi Minh City were found to contain six PAHs including Nap (522–1469 ng/g), anthracene (115–299 ng/g), Pyr (11.6–69.0 ng/g), fluorene (12.2–59.6 ng/g), Phe (11.2–49.8 ng/g), and benz[a]anthracene (1.10–4.00 ng/g) (Nguyen et al., 2023b).

11.3 Emission Sources

Emission sources of PBT substances in Vietnam can be classified into four categories: legacy sources (e.g., pesticides used in agriculture and vector controls and imported PCB-containing products), industrial emissions (wastes from industrial enterprises), domestic activities (e.g., vehicle exhaust, municipal wastewater, and emission from consumer products), and informal waste processing activities. Figure 11.3 comprehensively shows the relationship between emission sources, contaminated environmental media, and corresponding PBT groups in Vietnam.

11.3.1 Legacy Sources

As Vietnam is an agricultural country, OCPs have been widely applied since 1960 for crop production and public health protection (Hong et al., 2008; Hoai et al., 2010). Before 1985, about 6000–9000 tonnes of OCPs were utilized in agriculture in Vietnam annually, with DDT as the most intensively used pesticide (Minh et al., 2007a, 2007b). Since the late 1990s, DDT and other OCPs like HCHs, CHLs, ENDs, HCB, and drin compounds were officially banned in Vietnam, but illegal and unintentional uses of these substances have been expected to continue (Hoai et al., 2010; Wang et al., 2016a;

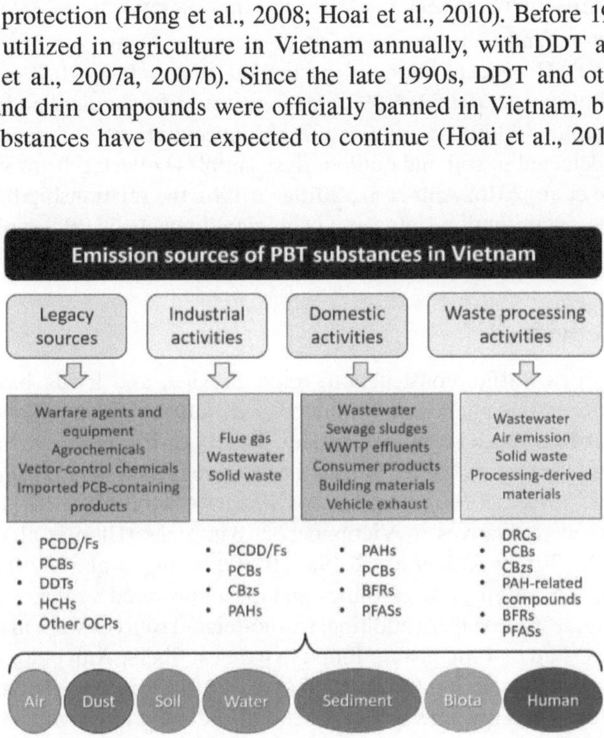

FIGURE 11.3 Emission sources of PBT substances in Vietnam.

Le et al., 2022). The intensive applications of OCPs, especially DDT, in agriculture and vector controls are mainly responsible for the abundance of these pesticide and its metabolites like dichlorodiphenyldichloroethylene (DDE) and dichlorodiphenyldichloroethane (DDD) in environmental media and biota in different areas of Vietnam (Nhan et al., 1998; Monirith et al., 2003; Carvalho et al., 2008; Hoai et al., 2010). Besides, other sources of DDTs in Vietnam have been mentioned by previous studies including the occurrence of DDT in dicofol, the use in antifouling paints for shipping industry, leakages from dumping sites or old stockpiles, and the contribution of long-range transport (Minh et al., 2007b; Hong et al., 2008; Hoai et al., 2010; Nishina et al., 2010; Wang et al., 2016a; Le et al., 2022).

11.3.2 Industrial Emissions

Industrial enterprises (e.g., waste incinerators, boilers, and power, metallurgy, cement, and brick plants) can release considerable amounts of PTS substances like PCDD/Fs, PCBs, CBzs, and PAHs into the surrounding environments through flue gas, wastewater, and solid waste like fly ash and bottom ash. Flue gas samples collected from multiple industrial plants along the country contained a wide range of dioxin concentrations: waste incinerators (0.30–46,800), metallurgy plants (48–4310), boilers (17–1690), cement plants (3.3–837), coal power plants (17–336), and brick plant (1.7) pg WHO-TEQ/Nm^3 (Thuong et al., 2014; Pham et al., 2019a; Thi et al., 2019). Fly ash and bottom ash from these facilities also contained PCDD/Fs and dioxin-like PCBs (DL-PCBs), showing higher levels in samples taken from metallurgy plants (0.174–4290 pg WHO-TEQ/g) and incinerators (7.93–232 pg WHO-TEQ/g) than other facilities like boilers (11.7–78.4), cement plants (0.373–10.9), brick plant (0.12–0.537), and coal power plant (0.001–0.300) pg WHO-TEQ/g (Thuong et al., 2014; Pham et al., 2019b; Thi et al., 2019). The formation mechanism of PCDD/Fs and DL-PCBs in these facilities was estimated to be *de novo* rather than precursor synthesis (Thuong et al., 2014; Pham et al., 2019a, 2019b).

Several studies have detected CBzs in fly ash and bottom ash from different waste incinerators and industrial plants in northern Vietnam (Hue et al., 2016; Nguyen et al., 2019a, 2021, 2024a). Concentrations of CBzs in fly ash and bottom ash ranged from 1.04 to 405 ng/g and from 1.53 to 383 ng/g, respectively (Nguyen et al., 2019a, 2021, 2024a). Levels and profiles of CBzs in the ash samples varied greatly between ash types, sectors, and facilities, implying complicated formation mechanisms. Nguyen et al. (2023a) found elevated PAH concentrations in particulate matter PM_{10} (0.994–27.5 µg/Nm^3), fly ash (0.430–785 µg/g), and bottom ash (0.052–6.09 µg/g) from several municipal and industrial waste incinerators in northern Vietnam. Although relatively high concentrations and specific patterns of PCBs, CBzs, and PAHs were detected in soil and outdoor dust samples collected from some industrial areas in northern Vietnam (Hue et al., 2016; Anh et al., 2019a, 2019b), the relationship between emissions from industrial facilities and contamination status and ecological impacts of PBT substances in surrounding environments has not been well characterized in the country.

11.3.3 Domestic Activities

Domestic activities such as traffic, construction, trade, service, and living have released substantial amounts of PTS substances into indoor and outdoor environments, especially in big cities with high urbanization and modernization degrees in Vietnam. Levels of PAHs, methylated PAHs, PCBs, and BFRs in urban road dust were much higher in Hanoi urban areas than in rural sites (Tuyen et al., 2014a, 2014b, Anh et al., 2018, 2019a, 2019b). Traffic activities have been recognized as principal emission sources of PAHs and their derivatives in Vietnamese urban areas (Hien et al., 2007b; Kishida et al., 2008; Tuyen et al., 2014a, 2014b; Anh et al., 2019a, 2019f). Hoang et al. (2020) detected 19 PAHs and 15 methylated PAHs in soot samples of gasoline- and diesel-powered vehicles in northern Vietnam at 800,000 and 12,000 ng/g, respectively. In addition, traffic-related sources were also described for PBDEs (e.g., elevated PBDE levels up to 4 mg/g were found in car seat fibers; Anh et al., 2017a) and PCBs (e.g., lubricating oils in vehicles; Toan et al., 2007a; Hoai et al., 2010). Road runoff has also been considered a source of PBT substances in Vietnam's urban water and sediment environments (Hoang et al., 2021c, 2021d).

Indoor environments have been contaminated by various organic pollutants released from consumer products, building materials, and furniture. Concentrations of BFRs like PBDEs and DBDPE in Vietnamese house dust were about one order of magnitude higher than those in road dust, suggesting the dominance of their indoor sources over outdoor sources (Anh et al., 2018; Hoang et al., 2021a, 2021b). PBDEs were found in obsolete plastic parts of electronics (e.g., TV rear cover, circuit board, computer keyboards, and laptop fan) in Vietnam at concentrations ranging from 1730 to 97,300 ng/g (Anh et al., 2017a). Several consumer products and old electronics from China, Japan, and South Korea contain moderate to high PBDE levels, which can be imported to Vietnam and serve as potential sources of these substances (Hoang et al., 2021b). Information about other organic pollutants like PCBs and PFASs in consumer products, furniture, and building materials in Vietnam is relatively scarce. Several novel PCB congeners have been inadvertently formed during the production of organic pigments and occurred in various commercial pigments, paints, and other consumer goods worldwide (Anh et al., 2021b).

Although the occurrence of pigment-derived PCBs in their source products has not been reported in Vietnam, CB-11 has been identified as a predominant congener in indoor air and dust samples in Hanoi urban areas, reflecting emission from consumer products and paints (Anh et al., 2020a, 2020b). PFASs were detected in Vietnamese food contact materials (median 0.341; max 624 ng/g; Hoang et al., 2023c) and indoor dust (median 24.1; range 3.99–147 ng/g; Tan et al., 2024); however, current data are not adequate to characterize the source and sink relationship of PFASs in this country. Domestic combustion activities (e.g., cooking, smoking, and incense burning) and outdoor sources (e.g., vehicle exhaust and road dust) are also responsible for the presence of PAHs and Me-PAHs in Vietnamese indoor dust (Anh et al., 2019f).

Untreated domestic wastewater, wastewater treatment plant (WWTP) effluents, sewage sludges, and leachates from dumping sites can cause the pollution of water and sediment environments by PBT substances. Urban canals that receive residential discharges and surface runoff showed higher concentrations of organic pollutants (e.g., PAHs, PCBs, OCPs, PBDEs, and PFASs) than the main rivers in Vietnamese large cities like Hanoi and Ho Chi Minh City (Minh et al., 2007b; Boll et al., 2008; Babut et al., 2019; Hoang et al., 2021c, 2021d). Municipal dumping leachate and discharges of WWTPs were contaminated with markedly higher levels of PFASs than ambient water (Kim et al., 2013; Lam et al., 2017). Sewage sludges from urban canals in Hanoi also contained high concentrations of PAHs (Hung et al., 2015).

11.3.4 Informal Waste Processing Activities

Informal processing activities of different waste types such as e-waste, plastics, papers, scrap metals, and ELVs have been reported as potential sources of various PBT groups like dioxins, PCBs, CBzs, BFRs, PAHs, and PFASs in Vietnam because of primitive technologies and improper waste management (see Figure 11.4 for some examples). Investigations have been conducted in some e-waste and plastic recycling areas in northern Vietnam (e.g., Bui Dau Village in Hung Yen Province and Phu Luu Village in Hai Phong City) since the late 2000s to find out emission sources of various flame retardants and other organic toxicants (Tue et al., 2010b, 2013; Someya et al., 2016; Anh et al., 2017a; Matsukami et al., 2017; Hoa et al., 2020; Hoang et al., 2022).

Critical recycling activities in these areas include: (1) stockpiling of e-waste in both indoor and outdoor areas without emission control measures; (2) manual dismantling and separation of e-waste by simple hand tools; (3) plastic crushing, melting, and recycling into recycled consumer products; (4) open burning of electrical cables and wires to recover metals (notably copper); and (5) open dumping and burning of less valuable materials (Hoang et al., 2023d). The generation of plastic debris, release of dust incorporated in e-waste, and volatility from e-waste have been the causes of air and dust contamination with BFRs and PCBs in recycling workshops and adjacent living areas (Tue et al., 2013; Hoang et al., 2021b, 2022). Source characterization revealed the intensive applications of deca-BDE mixtures and alternative formulations (e.g., DBDPE) in the areas (Tue et al., 2013; Anh et al., 2017a; Hoang et al., 2022). The co-emission of DRCs (especially polybrominated dibenzofurans, PBDFs) as impurities of BFRs from e-waste has also been observed in the workshop-settled dust (Tue et al., 2010b).

FIGURE 11.4 Examples of informal waste processing activities in northern Vietnam (2013–2019): (a) manual dismantling of e-waste; (b) plastic crushing; (c) ELV dismantling by hand tools; (d) e-waste storage on canal banks; (e) open storage of plastic waste; (f) storage of vehicle parts in semi-open workshops; (g) open burning to recover metals from e-waste; (h) open burning of mixed domestic and recycling waste; (i) burning of less valuable materials from vehicles.

The long-term transport of pollutants from e-waste workshops together with outdoor operations (e.g., stockpiling, open burning, and dumping) are responsible for the contamination of BFRs, DRCs, PCBs, CBzs, and unsubstituted, methylated, and halogenated PAHs in soil and sediment in Bui Dau areas (Someya et al., 2016; Suzuki et al., 2016; Wang et al., 2016b; Matsukami et al., 2017; Nishimura et al., 2018; Hoa et al., 2020). Surface water in some craft villages (e.g., Bui Dau e-waste sites and Phong Khe paper recycling sites) also contained PFASs at higher levels than other locations with suspected sources from recycling activities (Kim et al., 2013; Phung et al., 2021).

Informal ELV processing activities have also released PBT substances into the air, dust, and water environments with substantial amounts and/or specific patterns (Takahashi et al., 2017; Anh et al., 2019c, 2019f, 2020a, 2020b, 2021a; Hoang et al., 2022, 2023a, 2023b). The ELV processing areas are in Thuyen Village, Bac Giang Province, where expired vehicles and vehicular parts have been collected and manually dismantled by using primitive tools like hand cutters, hammers, and oxy-fuel torches. The separated components are classified into reusable parts for repairing/resale, recyclable materials (e.g., scrap metals), and less valuable remains for disposal. The ELV processing operations like dismantling, metal scrapping, and stockpiling of vehicular parts have been performed in workshop backyards and semi-open facilities constructed with steel frames and corrugated sheets (Takahashi et al., 2017; Anh et al., 2019f).

Concentrations of PCBs, PAHs, methylated PAHs, and DRCs in floor dust of ELV workshops were higher than those in common house dust, reflecting specific emissions from improper dismantling, waste management, and disposal activities. Oily materials (e.g., fuels, engine oils, and lubricants) have not been well managed during ELV processing, which spills on the workshop floors and can contaminate settled dust by their toxic components like PCBs, PAHs, and especially methylated PAHs (Takahashi et al., 2017; Anh et al., 2019f, 2020a). PCB profiles in the ELV dust samples were dominated by penta- and hexa-CBs, implying the previous applications of technical PCB mixtures like Aroclor 1254, Sovol,

Kanechlor 500, and PCB5 in transportation sectors (Takahashi et al., 2017; Anh et al., 2020a). In particular, methylated PAHs were found at elevated concentrations and remarkable proportions as compared with unsubstituted PAHs in dust and air samples in the ELV workshops, suggesting the sources from petroleum products used in vehicles and engines like crude oils, fuels, lubricants, and other engine oils (Anh et al., 2019f, 2020b; Hoang et al., 2021e, 2023a).

Takahashi et al. (2017) reported that concentrations of DRCs in the ELV dust were higher than those in Vietnamese e-waste dust and common house dust, resulting from mixed sources such as technical PCB mixtures (DL-PCBs), open burning (PCDD/Fs and monobromo-PCDD/Fs), and PBDE-related emissions (PBDFs). Although PBDEs were not identified as major pollutants in the ELV sites, they exhibited specific patterns in ELV dust samples with higher proportions of lower brominated homologs, indicating thermal/photodegradation processes of higher brominated ones (notably BDE-209) and/or emissions from vehicular materials treated with technical penta-BDE mixtures (Hoang et al., 2022, 2023b).

11.4 Ecological Risk Assessments

The occurrence of PBT substances in water and sediment environments can cause negative impacts on aquatic organisms. Although ecotoxicological tests have rarely been performed within pollution monitoring studies in Vietnam, ecotoxicological risks of PBT substances have been estimated by comparing their measured levels with environmental quality guidelines or ecological benchmarks. In general, concentrations of PBT groups like PAHs, PCBs, PBDEs, alternative BFRs, and PFASs in almost all Vietnamese sediment and water samples were lower than critical values such as probable effect levels (PEL), threshold effect levels (TEL), maximum permissible concentrations (MPC), or predicted no-effect concentrations (PNEC), indicating low ecological risks to aquatic biota (Kim et al., 2013; Mai et al., 2018; Toan et al., 2020; Anh et al., 2021a; Hoang et al., 2021c, 2021d; Phung et al., 2021; Le et al., 2022).

The low ecological risks are mainly due to the relatively low degree of contamination of these PBT substances in Vietnamese aquatic environments. Several sediment samples collected about 20 years ago in Hanoi and Ho Chi Minh City showed relatively high concentrations of OCPs and PCBs that exceeded the PEL values (Hoai et al., 2010; Minh et al., 2007a, 2007b), but this situation was rarely observed in estuarine and marine sediments (Hong et al., 2008; Tri et al., 2016). Although significant sediment contamination and bioaccumulation of BFRs have been reported in some Vietnamese e-waste recycling sites (Anh et al., 2017a; Matsukami et al., 2017), the ecological risks of BFRs in these areas have yet to be fully characterized.

11.5 Human Exposure and Health Risk Assessment

11.5.1 Exposure Pathways and Doses

PBT substances in environmental media (e.g., air, dust, and soil), industrial waste (e.g., ash and particulate matter), and food sources can enter human bodies through multiple pathways like inhalation, intentional or unintentional oral ingestion, and dermal contact. Exposure doses of pollutants (expressed as the amount of pollutant per kg body weight per day, e.g., ng/kg/day) have been frequently estimated based on pollutant concentrations in sources, daily exposure rates of media or food, and body weights. The exposure doses were then compared with threshold values (e.g., reference doses RfD, acceptable daily intake ADI, or tolerable daily intake TDI) to estimate exposure degree and to predict health risk. Contributions of exposure pathways in total exposure doses of each pollutant depend on their environmental occurrence and exposed object properties (e.g., age, gender, occupation, and living habits). Generally, acceptable levels of non-carcinogenic and carcinogenic risks were estimated for the Vietnamese population exposed to organic pollutants like PCBs, OCPs, PBDEs, PAHs, and PFASs through soil/dust ingestion, air inhalation, and food consumption. However, these estimations may be incomprehensive due to several limitations such as: (1) different exposure pathways were not collectively evaluated in the same study; (2) information about PBT substances in various food types and drinks was still scarce; (3) bioaccessibility of pollutants was not considered; and (4) chemical-specific approaches were usually applied to derive total toxicity of a pollutant subset instead of effect-based methods (e.g., *in vitro* bioassays) for complex pollutant mixtures.

Median environmental and food concentrations of selected PBT groups (i.e., PCBs, DDTs, PBDEs, and PAHs) and generic exposure scenarios were gathered from previous studies to estimate total daily intake doses and contributions of different pathways (i.e., air inhalation, dust/soil ingestion, and food consumption). Exposure doses were calculated for three age groups, including adults, children, and infants of the general population and in waste recycling areas. Only fish consumption was accounted for by adults and children because information about PBT substances in other food types is limited in Vietnam. At the same time, breast milk was assigned as the principal food for infants. As shown in Figure 11.5, infants have been estimated to receive the highest intake doses of DDTs, PCBs, and PBDEs, mainly contributed by breastfeeding (Tue et al., 2010a, 2010c). Total PBDE intake doses derived for e-waste recycling workers and infants of women involved in e-waste processing were about one to two orders of magnitude higher than those of the general population.

Meanwhile, there is no significant difference in median intake doses of DDTs, PCBs, and PAHs between the general population and waste recycling habitants. Except for PAH intake doses increasing with age, DDTs, PCBs, and PBDEs were the highest for infants, followed by children and adults. Food consumption served as the most critical exposure pathway in almost cases. Proportions of dust ingestion in total PBDE intakes were significant for children and waste processing workers. Air inhalation was also an innegligible exposure source for PCBs and PAHs in children and infants. The contribution of road dust and soil to total intakes of PBT substances was not significant.

Although the intake doses of PBT substances estimated for Vietnamese people were generally lower than respective threshold values, unacceptable levels of risk were reported for some specific cases related to exposure to DRCs (Tue et al., 2010b, 2014; Hue et al., 2014; Hoang et al., 2014; Durant et al., 2015; Tuyet-Hanh et al., 2015; Takahashi et al., 2017; Sau et al., 2021), PBDEs (Tue et al., 2010a), and PAHs (Hoang et al., 2021e; Trung et al., 2021). These observations suggest the need for more detailed and comprehensive exposure and risk assessment of PBT substances in Vietnam, especially in dioxin-contaminated sites and informal waste processing areas.

11.5.2 Internal Exposure and Health Impacts

PBT substances (e.g., PCDD/Fs, PBDFs, PCBs, OCPs, PBDEs, and HBCDs) and metabolites (e.g., hydroxylated PAHs) have been detected in breast milk, serum, adipose tissue, placenta, and urine samples of Vietnamese people. Breast milk has been the most frequently investigated human samples with PBT levels decreasing in the order of OCPs > PCBs > BFRs (Figure 11.6).

Potential dioxin sources, such as open dumping and e-waste recycling sites, are responsible for the accumulation of PCDD/Fs and other DRCs (e.g., PBDFs) in Vietnamese human breast milk (Kunisue et al., 2004; Tue et al., 2014).

The general population in Vietnam has been exposed to legacy POPs such as PCBs and OCPs (Schecter et al., 1997; Minh et al., 2004; Tue et al., 2010c). Levels of PCBs and OCPs in urban residents' serum and breast milk samples were higher than those in suburban and rural ones (Schecter et al., 1997; Tue et al., 2010c). DDT levels in Vietnamese breast milk samples were within the upper range reported for Asian countries (Minh et al., 2004; Tanabe and Kunisue, 2007). Over time, a trend of decreasing OCP and PCB concentration in breast milk was observed (Minh et al., 2004; Tue et al., 2010c). Concentrations of hydroxylated PAHs, biomarkers of PAH exposure, in urine samples of Hanoi urban residents, were 2–10 times higher than those of people in Brisbane, Australia (Thai et al., 2015). People in waste processing areas have also been exposed to some specific PBT groups significantly (Kunisue et al., 2004; Tue et al., 2010a, 2014; Schecter et al., 2018). Breast milk and serum levels of DRCs, PBDEs, and HBCDs measured in e-waste recyclers were much higher than regular residents in the same areas and reference sites

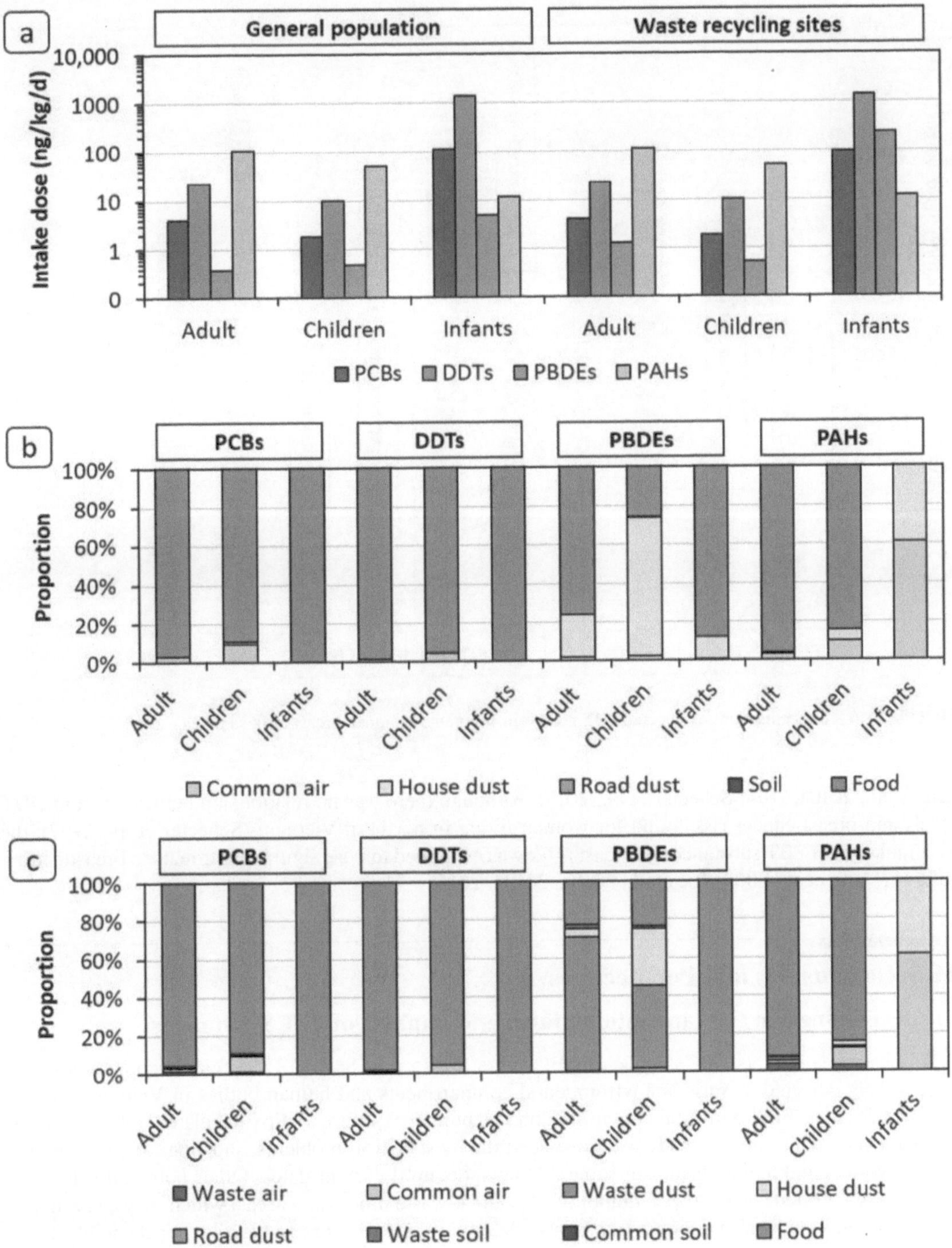

FIGURE 11.5 Daily intake doses of PBT substances estimated for Vietnamese people (a) and exposure pathway profiles in the general population (b) and waste processing areas (c).

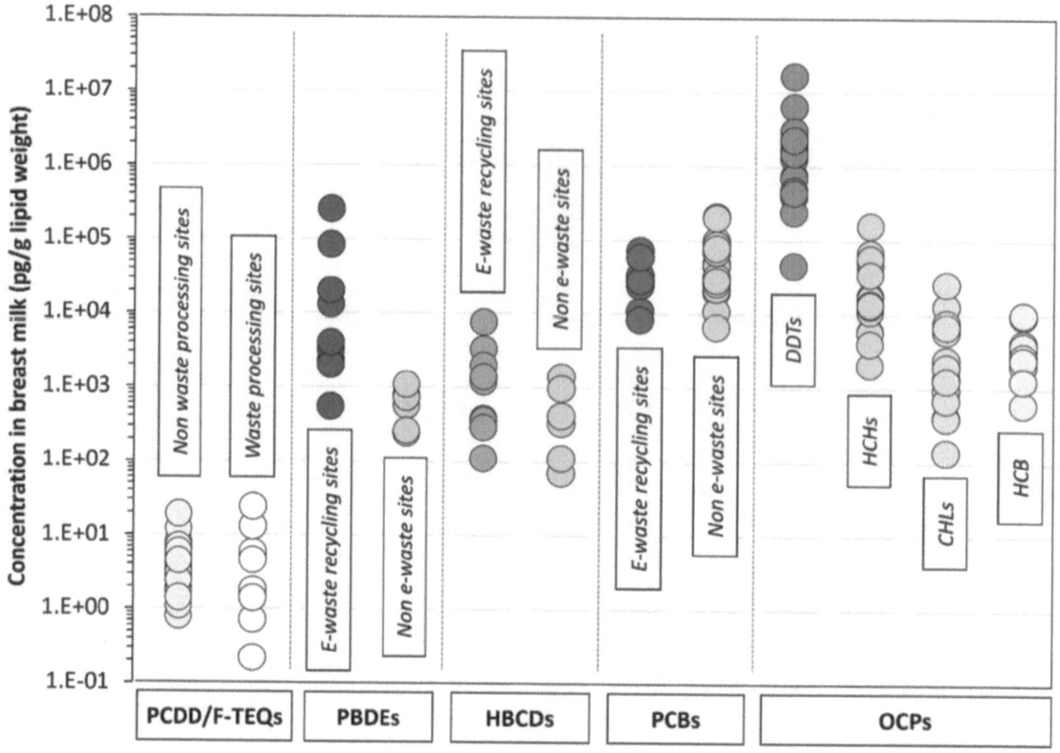

FIGURE 11.6 Concentrations of selected PBT groups in Vietnamese human breast milk samples.

(Tue et al., 2010a, 2014; Schecter et al., 2018). Although there was no relationship between serum DDT levels and breast cancer risk found for women living in northern Vietnam (Schecter et al., 1997), the accumulation of PBT substances in breast milk was estimated to pose significant threats to breastfeeding infants (Minh et al., 2004; Tue et al., 2010a, 2010c, 2014).

11.6 Conclusions and Perspectives

11.6.1 Findings on Contamination Status and Impacts of PBT Substances

PBT substances such as DRCs, OCPs, PCBs, CBzs, BFRs, PFASs, and PAH-related compounds have been widely detected in various environmental compartments and human bodies in Vietnam over the past five decades. Environmental contamination and human exposure to dioxins related to herbicide and defoliant spraying in the Vietnam War have been the most critical problems, showing long-term effects on ecosystems and human health in some AO hotspots until current dates. Other potential sources of DRCs were reported, such as open dumping sites and e-waste processing areas, which may cause dioxin accumulation in indoor/outdoor environments and biotic/abiotic factors and lead to unacceptable dioxin exposure doses in humans.

As an agricultural country with a tropical climate, the past uses and uncontrolled current sources of OCPs for crop production and vector control are responsible for remarkable levels of these pesticides and their metabolites in the environments and human breast milk, especially in the 1990s–2000s. Industrial production and domestic activities have also released considerable amounts of PBT substances into the environment; however, ecological and human health risks caused by these toxicants in Vietnam are estimated to be low and still acceptable. In addition, improper processing activities of modern waste types such as e-waste, obsolete plastics, and ELVs have been identified as potential sources of DRCs, BFRs, PCBs, CBzs, PFASs, and PAH-related compounds, which cause significant local contamination and human exposure events since the late 2000s. Further attention and efforts should be made to characterize pollution status, spatiotemporal trends, emission sources, and adverse impacts of PBT substances in Vietnam.

11.6.2 Future Perspectives

In-depth and comprehensive monitoring, risk assessment, and toxicological and epidemiological studies should be performed for legacy POPs and emerging contaminants. The roles of international/regional collaborations and technical/educational support from other countries have been highlighted by significant results from several collaborative and intercomparison studies. Appropriate management, recycling, and industrial and domestic waste disposal schemes should be implemented, particularly in informal waste processing sites and highly urbanized areas. As a concluding remark, we emphasize the importance of synergistic efforts from the Government, scientific community, industrial and commercial sectors, and all people to safely manage, reduce, and eliminate PBT substances in the environment.

Acknowledgments

Collaborations and support for the VNU Laboratory for Monitoring of Organic Toxicants in the Environments (MOTE) are acknowledged.

REFERENCES

Abdel-Shafy, H. I., and M. S. M. Mansour. 2016. "A review on polycyclic aromatic hydrocarbons: source, environmental impact, effect on human health and remediation". *Egyptian Journal of Petroleum* 25: 107–123. doi:10.1016/j.ejpe.2015.03.011

Anh, M. T., L. M. Triet, J. J. Sauvain, and J. Tarradellas. 1999. "PAH contamination levels in air particles and sediments of Ho Chi Minh City, Vietnam". *Bulletin of Environmental Contamination and Toxicology* 63: 728–735. doi:10.1007/s001289901040

Anh, H. Q., V. D. Nam, T. M. Tri, N. M. Ha, N. T. Ngoc, P. T. N. Mai, D. H. Anh, N. H. Minh, N. A. Tuan, and T. B. Minh. 2017a. "Polybrominated diphenyl ethers in plastic products, indoor dust, sediment and fish from informal e-waste recycling sites in Vietnam: a comprehensive assessment of contamination, accumulation pattern, emissions, and human exposure". *Environmental Geochemistry and Health* 39: 935–954. doi:10.1007/s10653-016-9865-6

Anh, H. Q., K. Tomioka, N. M. Tue, T. M. Tri, T. B. Minh, and S. Takahashi. 2018. "PBDEs and novel brominated flame retardants in road dust from northern Vietnam: levels, congener profiles, emission sources and implications for human exposure". *Chemosphere* 197: 389–398. doi:10.1016/j.chemosphere.2018.01.066

Anh, H. Q., T. B. Minh, T. M. Tri, and S. Takahashi. 2019a. "Road dust contamination by polycyclic aromatic hydrocarbons and their methylated derivatives in northern Vietnam: concentrations, profiles, emission sources, and risk assessment". *Environmental Pollution* 254: 113073. doi:10.1016/j.envpol.2019.113073

Anh, H. Q., I. Watanabe, K, Tomioka, T. B. Minh, and S. Takahashi. 2019b. "Characterization of 209 polychlorinated biphenyls in street dust from northern Vietnam: contamination status, potential sources, and risk assessment". *Science of the Total Environment* 652: 345–355. doi:10.1016/j.scitotenv.2018.10.240

Anh, H. Q., K. Tomioka, N. M. Tue, G. Suzuki, T. B. Minh, P. H. Viet, and S. Takahashi. 2019c. "Comprehensive analysis of 942 organic micro-pollutants in settled dusts from northern Vietnam: pollution status and implications for human exposure". *Journal of Material Cycles and Waste Management* 21: 57–66. doi:10.1007/s10163-018-0745-2

Anh, H. Q., K. Tomioka, N. M. Tue, L. H. Tuyen, N. K. Chi, T. B. Minh, P. H. Viet, and S. Takahashi. 2019d. "A preliminary investigation of 942 organic micro-pollutants in the atmosphere in waste processing and urban areas, northern Vietnam: levels, potential sources, and risk assessment". *Ecotoxicology and Environmental Safety* 167: 354–364. doi:10.1016/j.ecoenv.2018.10.026

Anh, H. Q., T. M. Tri, N. T. T. Thuy, T. B. Minh, and S. Takahashi. 2019e. "Screening analysis of organic micro-pollutants in road dusts from some areas in northern Vietnam: a preliminary investigation on contamination status, potential sources, human exposure, and ecological risk". *Chemosphere* 224: 428–436. doi:10.1016/j.chemosphere.2019.02.177

Anh, H. Q., N. M. Tue, L. H. Tuyen, T. B. Minh, P. H. Viet, and S. Takahashi. 2019f. "Polycyclic aromatic hydrocarbons and their methylated derivatives in settled dusts from end-of-life vehicle processing, urban, and rural areas, northern Vietnam: occurrence, source apportionment, and risk assessment". *Science of the Total Environment* 672: 468–478. doi:10.1016/j.scitotenv.2019.04.018

Anh, H. Q., I. Watanabe, T. B. Minh, N. M. Tue, L. H. Tuyen, P. H. Viet, and S. Takahashi. 2020a. "Polychlorinated biphenyls in settled dusts from an end-of-life vehicle processing area and normal house dusts in northern Vietnam: occurrence, potential sources, and risk assessment". *Science of the Total Environment* 728: 138823. doi:10.1016/j.scitotenv.2020.138823

Anh, H. Q., I. Watanabe, N. M. Tue, L. H. Tuyen, P. H. Viet, N. K. Chi, T. B. Minh, and S. Takahashi. 2020b. "Polyurethane foam-based passive air sampling for simultaneous determination of POP- and PAH-related compounds: a case study in informal waste processing and urban areas, northern Vietnam". *Chemosphere* 247: 125991. doi:10.1016/j.chemosphere.2020.125991

Anh, H. Q., I. Watanabe, L. H. Tuyen, N. M. Tue, N. T. Nghia, P. H. Viet, T. B. Minh, and S. Takahashi. 2021a. "Comprehensive monitoring of more than 1000 organic micro-pollutants in drainage water: case study in a rural village with end-of-life vehicle processing activities in northern Vietnam". *Water, Air, & Soil Pollution* 232: 14. doi:10.1007/s11270-020-04965-5

Anh, H. Q., I. Watanabe, T. B. Minh, and S. Takahashi. 2021b. "Unintentionally produced polychlorinated biphenyls in pigments: an updated review on their formation, emission sources, contamination status, and toxic effects". *Science of the Total Environment* 755: 142504. doi:10.1016/j.scitotenv.2020.142504

Babut, M., B. Mourier, M. Desmet, C. S. Laprade, P. Labadie, H. Budzinski, L. F. D. Alencastro, T. A. Tu, E. Strady, and N. Gratiot. 2019. "Where has the pollution gone? A survey of organic contaminants in Ho Chi Minh city/Saigon River (Vietnam) bed sediments". *Chemosphere* 217: 261–269. doi:10.1016/j.chemosphere.2018.11.008

Ballschmiter, K. 1992. "Transport and fate of organic compounds in the global environment". *Angewandte Chemie International Edition* 31: 487–515. doi:10.1002/anie.199204873

Boll, E. S., J. H. Christensen, and P. E. Holm. 2008. "Quantification and source identification of polycyclic aromatic hydrocarbons in sediment, soil, and water spinach from Hanoi, Vietnam". *Journal of Environmental Monitoring* 10: 261–269. doi:10.1039/B712809F

Brodsky, E.S., A. A. Shelepchikov, D. B. Feshin, V. S. Roumak, N. V. Umnova, A. N. Kuznetsov, T. K. Sau, N. X. Truong, and D. S. Pavlov. 2009. "The current level of dioxin pollution in the area of large-scale spraying of agent orange in Vietnam". *Doklady Biological Sciences* 429: 526–530. doi:10.1134/S0012496609060131

Carvalho, F. P., J. P. Villeneuve, C. Cattini, I. Tolosa, D. D. Thuan, and D. D. Nhan. 2008. "Agrochemical and polychlorobyphenyl (PCB) residues in the Mekong River delta, Vietnam". *Marine Pollution Bulletin* 56: 1476–1485. doi:10.1016/j.marpolbul.2008.04.042

Chi, K. H., Y. T. Huang, H. M. Nguyen, T. T. H. Tran, S. Chantara, and T. H. Ngo. "Characteristics and health impacts of $PM_{2.5}$-bound PCDD/Fs in three Asian countries". *Environment International* 167: 107441. doi:10.1016/j.envint.2022.107441

Ding, Y., S. Qin, H. Huang, X. Tang, X. Li, Y. Zhang, W. Chen, L. P. Nguyen, and S. Qi. 2024. "Selected pesticidal POPs and metabolites in the soil of five Vietnamese cities: sources, fate, and health risk implications". *Environmental Pollution* 342: 123043. doi:10.1016/j.envpol.2023.123043

Duc, N. V., T. L. Minh, H. T. Nguyen, D. A. Dam, X. N. Thi, D. B. Chu, B. V. Hoi, A. Q. Hoang, T. M. Tran, and T. N. M. Pham. 2024. "Halogenated organic pollutants (HOPs) in marine fish collected from central-northern coastal waters in Vietnam". *International Journal of Environmental Analytical Chemistry* 185: 114374. doi:10.1080/03067319.2023.2299740

Duong, H. T., K, Kadokami, H. Shirasaka, R. Hidaka, H. T. C. Chau, L. Kong, T. Q. Nguyen, and T. T. Nguyen. 2015a. "Occurrence of perfluoroalkyl acids in environmental waters in Vietnam". *Chemosphere* 122: 115–124. doi:10.1016/j.chemosphere.2014.11.023

Duong, H. T, K. Kadokami, H. T. C. Chau, T. Q. Nguyen, T. T. Nguyen, and L. Kong. 2015b. "Groundwater screening for 940 organic micro-pollutants in Hanoi and Ho Chi Minh City, Vietnam". *Environmental Science and Pollution Research* 22: 19835–19847. doi:10.1007/s11356-015-5180-5

Durant, J. T., T. G. Boivin, H. R. Pohl, and T. H. Sinks. 2015. "Public health assessment of dioxin-contaminated fish at former US airbase, Bien Hoa, Vietnam". *International Journal of Environmental Health Research* 25: 254–264. doi:10.1080/09603123.2014.938026

Dwernychuk, L. W., H. D. Cau, C. T. Hatfield, T. G. Boivin, T. M. Hung, P. T. Dung, and N. D. Thai. 2002. "Dioxin reservoirs in southern Viet Nam—a legacy of Agent Orange". *Chemosphere* 47: 117–137. doi:10.1016/S0045-6535(01)00300-9

Eguchi, A., T. Isobe, K. Ramu, N. M. Tue, A. Sudaryanto, G. Devanathan, P. H. Viet, R. S. Tana, S. Takahashi, A. Subramanian, and S. Tanabe. 2013. "Soil contamination by brominated flame retardants in open waste dumping sites in Asian developing countries". *Chemosphere* 90: 2365–2371. doi:10.1016/j.chemosphere.2012.10.027

Frignani, M., R. Piazza, L. G. Bellucci, N. H. Cu, R. Zangrando, S. Albertazzi, I. Moret, S. Romano, and A. Gambaro. 2007. "Polychlorinated biphenyls in sediments of the Tam Giang-Cau Hai Lagoon, Central Vietnam". *Chemosphere* 67: 1786–1793. doi:10.1016/j.chemosphere.2006.05.119

Giuliani, S., M. Sprovieri, M. Frignani, N. H. Cu, C. Mugnai, L. G. Bellucci, S. Albertazzi, S. Romano, M. L. Feo, E. Marsella, and D. H. Nhon. 2008. "Presence and origin of polycyclic aromatic hydrocarbon in sediments of nine coastal lagoons in central Vietnam". *Marine Pollution Bulletin* 56: 1504–1512. doi:10.1016/j.marpolbul.2008.04.013

Giuliani, S., R. Piazza, L. G. Bellucci, N. H. Cu, M. Vecchiato, S. Romano, C. Mugnai, D. H. Nhon, and M. Frignani. 2011. "PCBs in Central Vietnam coastal lagoons: levels and trends in dynamic environments". *Marine Pollution Bulletin* 62: 1013–1024. doi:10.1016/j.marpolbul.2011.02.035

Gómez, M. J., F. Pazos, F. J. Guijarro, V. Lorenzo, and A. Valnecia. 2007. "The environmental fate of organic pollutants through the global microbial metabolism". *Molecular Systems Biology* 3: 114. doi:10.1038/msb4100156

Hien, T. T., L. T. Thanh, T. Kameda, N. Takenaka, and H. Bandow. 2007a. "Distribution characteristics of polycyclic aromatic hydrocarbons with particle size in urban aerosols at the roadside in Ho Chi Minh City, Vietnam". *Atmospheric Environment* 41: 1575–1586. doi:10.1016/j.atmosenv.2006.10.045

Hien, T. T., L. T. Thanh, T. Kameda, N. Takenaka, and H. Bandow. 2007b. "Nitro-polycyclic aromatic hydrocarbons and polycyclic aromatic hydrocarbons in particulate matter in an urban area of a tropical region: Ho Chi Minh City, Vietnam". *Atmospheric Environment* 41: 7715–7725. doi: 10.1016/j.atmosenv.2007.06.020

Hoa, N. T. Q, H. Q. Anh, N. M. Tue, N. T. Trung, L. N. Da, T. V. Quy, N. T. A. Huong, G. Suzuki, S. Takahashi, S. Tanabe, P. C. Thuy, P. T. Dau, P. H. Viet, and L. H. Tuyen. 2020. "Soil and sediment contamination by unsubstituted and methylated polycyclic aromatic hydrocarbons in an informal e-waste recycling area, northern Vietnam: occurrence, source apportionment, and risk assessment". *Science of the Total Environment* 709: 135852. doi:10.1016/j.scitotenv.2019.135852

Hoa, N. T. Q., T. T. Lieu, H. Q. Anh, N. T. A. Huong, N. T. Nghia, N. T. Chuc, P. D. Quang, P. T. Vi, and L. H. Tuyen. 2022. "Perfluoroalkyl substances (PFAS) in freshwater fish from urban lakes in Hanoi, Vietnam: concentrations, tissue distribution, and implication for risk assessment". *Environmental Science and Pollution Research* 29: 52057–52069. doi:10.1007/s11356-022-19532-0

Hoai, P. M., N. T. Ngoc, N. H. Minh, P. H. Viet, M. Berg, A.C. Alder, and W. Giger. 2010. "Recent levels of organochlorine pesticides and polychlorinated biphenyls in sediments of the sewer system in Hanoi, Vietnam". *Environmental Pollution* 158: 913–920. doi:10.1016/j.envpol.2009.09.018

Hoang, T. T, W.A. Traag, A. J. Murk, and R. L. A. P. Hoogenboom. 2014. "Levels of polychlorinated dibenzo-p-dioxins, dibenzofurans (PCDD/Fs) and dioxin-like PCBs in free range eggs from Vietnam, including potential health risks". *Chemosphere* 114: 268–274. doi:10.1016/j.chemosphere.2014.05.010

Hoang, A. Q., T. H. Le, M. B. Tu, and S. Takahashi. 2020. "Characterization of unsubstituted and methylated polycyclic aromatic hydrocarbons and screening of potential organic compounds in solid waste and environmental samples by gas chromatography–mass spectrometry". *Journal of Environmental Chemistry* 30: 82–93. doi:10.5985/jec.30.82

Hoang, M. T. T, H. Q. Anh, K. Kadokami, H. T. Duong, H. M. Hoang, T. V. Nguyen, S. Takahashi, G. T. Le, and H. T. Trinh. 2021a. "Contamination status, emission sources, and human health risk of brominated flame retardants in urban indoor dust from Hanoi, Vietnam: the replacement of legacy polybrominated diphenyl ether mixtures by alternative formulations". *Environmental Science and Pollution Research* 28: 43855–43896. doi:10.1007/s11356-021-13822-9

Hoang, A. Q, T. M. Tran, M. B. Tu, and S. Takahashi. 2021b. "Polybrominated diphenyl ethers in indoor and outdoor dust from Southeast Asia: an updated review on contamination status, human exposure, and future perspectives". *Environmental Pollution* 272: 116012. doi:10.1016/j.envpol.2020.116012

Hoang, A. Q., S. Takahashi, N. D. Le, T. T. H. Hoang, T. T. Duong, T. M. H. Pham, T. D. Nguyen, T. X. B. Phung, T. A. H. Nguyen, H. T. Le, M. T. Nguyen, M. B. Tu, Y. T. H. Nguyen, T. M. L. Nguyen, V. P. Phung, and T. P. Q. Le. 2021c. "Unsubstituted and methylated PAHs in surface sediment of urban rivers in the Red River Delta (Hanoi, Vietnam): concentrations, profiles, sources, and ecological risk assessment". *Bulletin of Environmental Contamination and Toxicology* 107: 475–486. doi:10.1007/s00128-021-03174-y

Hoang, A. Q., S. Takahashi, N. D. Le, T. T. Duong, T. M. H. Pham, T. N. M. Pham, T. A. H. Nguyen, T. M. Tran, M. B. Tu, and T. P. Q. Le. 2021d. "Comprehensive determination of polychlorinated biphenyls and brominated flame retardants in surface sediment samples from Hanoi urban area, Vietnam: contamination status, accumulation profiles, and potential ecological risks". *Environmental Research* 197: 111158. doi: 10.1016/j.envres.2021.111158

Hoang, A. Q., G. Suzuki, C. Michinaka, N. M. Tue, L. H. Tuyen, M. B. Tu, and S. Takahashi. 2021e. "Characterization of unsubstituted and methylated polycyclic aromatic hydrocarbons in settled dust: combination of instrumental analysis and in vitro reporter gene assays and implications for cancer risk assessment". *Science of the Total Environment* 788: 147821. doi:10.1016/j.scitotenv.2021.147821

Hoang, A. Q., R. Karyu, N. M. Tue, A. Goto, L. H. Tuyen, H. Matsukami, G. Suzuki, S. Takahashi, P. H. Viet, and T. Kunisue. 2022. "Comprehensive characterization of halogenated flame retardants and organophosphate esters in settled dust from informal e-waste and end-of-life vehicle processing sites in Vietnam: occurrence, source estimation, and risk assessment". *Environmental Pollution* 310: 119809. doi:10.1016/j.envpol.2022.119809

Hoang, A. Q., S. Takahashi, L. H. Tuyen, N. M. Tue, N. M Tu, T. T. T. Nguyen, and M. B. Tu. 2023a. "Polycyclic aromatic hydrocarbons in air and dust samples from Vietnamese end-of-life vehicle processing workshops: contamination status, sources, and exposure risks". *Bulletin of Environmental Contamination and Toxicology* 110: 110. doi:10.1007/s00128-023-03757-x

Hoang, A. Q., S. Takahashi, N. M. Tue, L. H. Tuyen, T. M. Tri, N. T. H. Yen, and M. B. Tu. 2023b. "Occurrence, emission sources, and risk assessment of polybrominated diphenyl ethers and current-use brominated flame retardants in settled dust from end-of-life vehicle processing, urban, and rural areas, northern Vietnam". *Environmental Science and Pollution Research* 30: 2061–2074. doi:10.1007/s11356-022-22396-z

Hoang, A. Q., T. L. Tran, L. H. Tuyen, T. A. H. Nguyen, D. M. Pham, T. C. Nguyen, T. N. Nguyen, D. Q. Phan, M. K. Nguyen, V. Q. Tran, C. T. Pham, Q. D. Bui, and T. Q. H. Nguyen. 2023c. "Perfluoroalkyl substances in food contact materials: preliminary investigation in Vietnam and global comparison". *Environmental Science and Pollution Research* 30: 104181–104193. doi:10.1007/s11356-023-29746-5

Hoang, A. Q., N. M. Tue, M. B. Tu, G. Suzuki, H. Matsukami, L. H. Tuyen, P. H. Viet, T. Kunisue, S. Sakai, and S. Takahashi. 2023d. "A review on management practices, environmental impacts, and human exposure risks related to electrical and electronic waste in Vietnam: findings from case studies in informal e-waste recycling areas". *Environmental Geochemistry and Health* 45: 2705–2728. doi:10.1007/s10653-022-01408-4

Hong, S. H., U. H. Yim, W. J. Shim, J. R. Oh, P. H. Viet, and P. S. Park. 2008. "Persistent organochlorine residues in estuarine and marine sediments from Ha Long Bay, Hai Phong Bay, and Ba Lat Estuary, Vietnam". *Chemosphere* 72: 1193–1202. doi:10.1016/j.chemosphere.2008.02.051

Hong, W. J., H. Jia, W. L. Ma, R. K. Sinha, H. B. Moon, H. Nakata, N. H. Minh, K. H. Chi, W. L. Li, K. Kannan, E. Sverko, and Y. F. Li. 2016. "Distribution, fate, inhalation exposure and lung cancer risk of atmospheric polycyclic aromatic hydrocarbons in some Asian countries". *Environmental Science & Technology* 50: 7163–7174. doi:10.1021/acs.est.6b01090

Hong, W. J., Y. F. Li, W. L. Li, H. Jia, N. H. Minh, R. K. Moon, H. Nakata, K. H. Chi, K. Kannan, and E. Sverko. 2020. "Soil concentrations and soil-air exchange of polycyclic aromatic hydrocarbons in five Asian countries". *Science of the Total Environment* 711: 135223. doi:10.1016/j.scitotenv.2019.135223

Hue, N. T., N. T. T. Thuy, and N. H. Tung. 2016. "Polychlorobenzenes and polychlorinated biphenyls in ash and soil from several industrial areas in North Vietnam: residue concentrations, profiles and risk assessment". *Environmental Geochemistry and Health* 38: 399–411. doi:10.1007/s10653-015-9726-8

Hung, C. V., B. D. Cam, P. T. N. Mai, and B. Q. Dzung. 2015. "Heavy metals and polycyclic aromatic hydrocarbons in municipal sewage sludge from a river in highly urbanized metropolitan area in Hanoi, Vietnam: levels, accumulation pattern and assessment of land application". *Environmental Geochemistry and Health* 37: 133–146. doi:10.1007/s10653-014-9635-2

Iwata, H., S. Tanabe, N. Sakai, A. Nishimura, and R. Tatsukawa. 1994. "Geographical distribution of persistent organochlorines in air, water and sediments from Asia and Oceania, and their implications for global redistribution from lower latitudes". *Environmental Pollution* 85: 15–33. doi:10.1016/0269-7491(94)90234-8

Kannan, K., S. Tanabe, H. T. Quynh, N. D. Hue, and R. Tatsukawa. 1992. "Residue pattern and dietary intake of persistent organochlorine compounds in foodstuffs from Vietnam". *Archives of Environmental Contamination and Toxicology* 22: 367–374. doi:10.1007/BF00212555

Kim, J. W., N. M. Tue, T. Isobe, K. Misaki, S. Takahashi, P. H. Viet, and S. Tanabe. 2013. "Contamination by perfluorinated compounds in water near waste recycling and disposal sites in Vietnam". *Environmental Monitoring and Assessment* 185: 2909–2919. doi:10.1007/s10661-012-2759-x

Kishida, M., K. Imamura, N. Takenaka, Y. Maeda, P. H. Viet, and H. Bandow. 2008. "Concentrations of atmospheric polycyclic aromatic hydrocarbons in particulate matter and the gaseous phase at roadside sites in Hanoi, Vietnam". *Bulletin of Environmental Contamination and Toxicology* 81: 174–179. doi:10.1007/s00128-008-9450-5

Kulsoom, M., V. Shukla, and N. Kumar. 2021. "Environmental dynamics of persistent organic pollutants". In *Persistent Organic Pollutants in the Environment Origin and Role*. CRC Press, Boca Raton, FL. doi:10.1201/9781003053170-1-1

Kunisue, T., M. Watanabe, H. Iwata, A. Subramanian, I. Monirith, T. B. Minh, R. Baburajendran, T. S. Tana, P. H. Viet, M. Prudente and S. Tanabe. 2004. "Dioxins and related compounds in human breast milk collected around open dumping sites in Asian developing countries: bovine milk as a potential source". *Archives of Environmental Contamination and Toxicology* 47: 414–426. doi:10.1007/s00244-004-3172-4

Kwan, C. S., H. Takada, K. Mizukawa, M. Saha, R. Rinawati, and E. C. Santiago. 2013. "Sedimentary PBDEs in urban areas of tropical Asian countries". *Marine Pollution Bulletin* 76: 95–105. doi:10.1016/j.marpolbul.2013.09.023

Lam, N. H., C. R. Cho, K. Kannan, and H. S. Cho. 2017. "A nationwide survey of perfluorinated alkyl substances in waters, sediment and biota collected from aquatic environment in Vietnam: distributions and bioconcentration profiles". *Journal of Hazardous Materials* 323: 116–127. doi:10.1016/j.jhazmat.2016.04.010

Lawal, A. T. 2017. "Polycyclic aromatic hydrocarbons. A review". *Cogent Environmental Science* 3: 1339841. doi:10.1080/23311843.2017.1339841

Le, P. C., D. L. Le, and D. D. Nguyen. 2022. "Accumulation and distribution of persistent organic pollutants in surface sediment of the estuarine region and correlation between genetic and environmental factors: a case study". *Environmental Technology & Innovation* 28: 102908. doi:10.1016/j.eti.2022.102908

Li, W. L., W. L. Ma, H. Jia, W. Hong, H. B. Moon, H. Nakata, N. H. Minh, R. K. Sinha, K. H. Chi, K. Kannan, E. Sverko, and Y. F. Li. 2016. "Polybrominated diphenyl ethers (PBDEs) in surface soils across five Asian countries: levels, spatial distribution and source contribution". *Environmental Science & Technology* 50: 12779–12788. doi:10.1021/acs.est.6b04046

Li, W. L., W. L. Ma, Z. Zhang, L. Liu, W. W. Song, H. Jia, Y. S. Ding, H. Nakata, N. H. Minh, R. K. Sinha, H. B. Moon, K. Kannan, E. Sverko, and Y. F. Li. 2017. "Occurrence and source effect of novel brominated flame retardants (NBFRs) in soils from five Asian countries and its relationship with PBDEs". *Environmental Science & Technology* 51(19): 11126–11135. doi:10.1021/acs.est.7b03207

Liu, X., L. Yang, G. Liu, and M. Zheng. 2021. "Formation of environmentally persistent free radicals during thermochemical processes and their correlations with unintentional persistent organic pollutants". *Environmental Science & Technology* 55: 6529–6541. doi:10.1021/acs.est.0c08762

Lucattini, L., G. Poma, A. Covaci, J. de Boer, M. H. Lamoree, and P. E. G. Leonards. 2018. "A review of semi-volatile organic compounds (SVOCs) in the indoor environment: occurrence in consumer products, indoor air and dust". *Chemosphere* 201: 466–482. doi:10.1016/j.chemosphere.2018.02.161

Mai, P. T. N., N. V. Thuong, T. T. Tham, N. K. Hoang, H. Q. Anh, T. M. Tri, L. S. Hung, D. T. Nhung, V. D. Nam, N. T. M. Hue, N. T. A. Huong, D. H. Anh, N. H. Minh, and T. B. Minh. 2018. "Distribution, accumulation profile, and risk assessment of polybrominated diphenyl ethers in sediment from lake and river systems in Hanoi metropolitan area, Vietnam". *Environmental Science and Pollution Research* 25: 7170–7179. doi:10.1007/s11356-015-5235-7

Mamy, L., D. Patureau, E. Barriuso, C. Bedos, F. Bessac, X. Louchart, F. Martin-laurent, C. Miege, and P. Benoit. 2015. "Prediction of the fate of organic compounds in the environment from their molecular properties: a review". *Critical Reviews in Environmental Science and Technology* 45: 1277–1377. doi:10.1080/10643389.2014.955627

Matsukami, H., G. Suzuki, M. Someya, N. Uchida, N. M. Tue, L. H. Tuyen, P. H. Viet, S. Takahashi, S. Tanabe, and H. Takigami. 2017. "Concentrations of polybrominated diphenyl ethers and alternative flame retardants in surface soils and river sediments from an electronic waste-processing area in northern Vietnam, 2012–2014". *Chemosphere* 167: 291–299. doi:10.1016/j.chemosphere.2016.09.147

Mill, T., J. M. Patel, and C. T. Stevens. 2019. "The environmental fate of synthetic organic chemicals". *Physical Sciences Reviews* 4: 20180075. doi: 10.1515/psr-2018-0075

Minh, N. H., T. B. Minh, M. Watanabe, T. Kunisue, I. Monirith, S. Tanabe, S. Sakai, A. Subramanian, K. Sasikumar, P. H. Viet, B. C. Tuyen, T. S. Tana, and M. S. Prudente. 2003. "Open dumping site in Asian developing countries: a potential source of polychlorinated dibenzo-*p*-dioxins and polychlorinated dibenzofurans". *Environmental Science & Technology* 37: 1493–1502. doi:10.1021/es026078s

Minh, N. H., M. Someya, T. B. Minh, T. Kunisue, H. Iwata, M. Watanabe, S. Tanabe, P. H. Viet, and B. C. Tuyen. 2004. "Persistent organochlorine residues in human breast milk from Hanoi and Hochiminh City, Vietnam: contamination, accumulation kinetics and risk assessment for infants". *Environmental Pollution* 129: 431–441. doi:10.1016/j.envpol.2003.11.012

Minh, N. H., T. B. Minh, N. Kajiwara, T. Kunisue, A. Subramanian, H. Iwata, T. S. Tana, R. Baburajendran, S. Karuppiah, P. H. Viet, B. C. Tuyen, and S. Tanabe. 2006a. "Contamination by persistent organic pollutants in dumping sites of Asian developing countries: implication of emerging pollution sources". *Archives of Environmental Contamination and Toxicology* 50: 474–481. doi:10.1007/s00244-005-1087-3

Minh, N. H., T. B. Minh, N. Kajiwara, T. Kunisue, H. Iwata, P. H. Viet, N. P. C. Tu, B. C. Tuyen, and S. Tanabe. 2006b. "Contamination by polybrominated diphenyl ethers and persistent organochlorines in catfish and feed from Mekong River Delta, Vietnam". *Environmental Toxicology and Chemistry* 25: 2700–2709. doi:10.1897/05-600r.1

Minh, N. H., T. B. Minh, H. Iwata, N. Kajiwara, T. Kunisue, S. Takahashi, P. H. Viet, B. C. Tuyen, and S. Tanabe. 2007a. "Persistent organic pollutants in sediments from Sai Gon–Dong Nai River Basin, Vietnam: levels and temporal trends". *Archives of Environmental Contamination and Toxicology* 52: 458–465. doi:10.1007/s00244-006-0157-5

Minh, N. H., T. B. Minh, N. Kajiwara, T. Kunisue, H. Iwata, P. H. Viet, N. P. C. Tu, B. C. Tuyen, and S. Tanabe. 2007b. "Pollution sources and occurrences of selected persistent organic pollutants (POPs) in sediments of the Mekong River Delta, South Vietnam". *Chemosphere* 67: 1794–1801. doi:10.1016/j.chemosphere.2006.05.144

Mon, E. E., N. Phay, T. Agusa, L. T. Bach, H. M. Yeh, C. H. Huang, and H. Nakata. 2020. "Polycyclic aromatic hydrocarbons (PAHs) in road dust collected from Myanmar, Japan, Taiwan, and Vietnam". *Archives of Environmental Contamination and Toxicology* 78: 34–45. doi:10.1007/s00244-019-00693-y

Monirith, I., D. Ueno, S. Takahashi, H. Nakata, A. Sudaryanto, A. Subramanian, S. Karuppiah, A. Ismail, M. Muchtar, J. Zheng, B. J. Richardson, M. Prudente, N. D. Hue, T. S. Tana, A. V. Tkalin, and S. Tanabe. 2003. "Asia-Pacific Mussel Watch: monitoring contamination of persistent organochlorine compounds in coastal waters of Asian countries". *Marine Pollution Bulletin* 46: 281–300. doi:10.1016/s0025-326x(02)00400-9

Mortimer, M., and A. R. Brushett. 2023. "Persistent organic pollutants (POPs)". In *Marine Pollution – Monitoring, Management and Mitigation*. Springer, Cham. 185–205. doi: 10.1007/978-3-031-10127-4_8

Muir, D. C. G., and P. H. Howard. 2006. "Are there other persistent organic pollutants? A challenge for environmental chemists". *Environmental Science & Technology* 40: 7157–7166. doi:10.1021/es061677a

Nghiem, T. X., A. Q. Hoang, T. D. Nguyen, T. T. Nguyen, P. D. Tran, T. T. Nguyen, and M. B. Tu. 2022. "PCDD/Fs and dioxin-like PCBs in chicken eggs and soils in Dong Nai Province, southern Vietnam: impacts of raising methods and nearby pollution sources". *Bulletin of Environmental Contamination and Toxicology* 108: 136–144. doi:10.1007/s00128-021-03419-w

Nguyen, T. T. N., and C. Baduel. 2023. "Optimization and validation of an extraction method for the analysis of multi-class emerging contaminants in soil and sediment". *Journal of Chromatography A* 1710: 464287. doi:10.1016/j.chroma.2023.464287

Nguyen, C. H., T. Nakahama, T. T. Dang, H. H. Chu, L. V. Hoang, T. Kishimoto, and N. T. Nguyen. 2017. "Expression of aryl hydrocarbon receptor, inflammatory cytokines, and incidence of rheumatoid arthritis in Vietnamese dioxin exposed people". *Journal of Immunotoxicology* 14: 196–203. doi:10.1016/j.chroma.2023.464287

Nguyen, A. T. N., M. Nishijo, T. T. Pham, N. N. Tran, A. H. Tran, L. V. Hoang, H. Boda, Y. Morikawa, Y. Nishino, and H. Nishijo. 2018. "Sex-specific effects of perinatal dioxin exposure on eating behavior in 3-year-old Vietnamese children". *BMC Pediatrics* 18: 213. doi:10.1186/s12887-018-1171-2

Nguyen, H. T., T. T. T. Nguyen, N. H. Tung, A. Q. Hoang, L. H. Pham, and T. B. Minh. 2019a. "Levels, profiles, and emission characteristics of chlorobenzenes in ash samples from some industrial thermal facilities in northern Vietnam". *Environmental Science and Pollution Research* 26: 188–198. doi:10.1007/s11356-018-3591-9

Nguyen, D. T., P. D. Luu, T. D. Doan, Y. H. Dao, and G. T. Le. 2019b. "Polycyclic aromatic hydrocarbon contamination in three tea samples collected in two different areas of Vietnam". *Journal of Food and Nutrition Research* 7: 51–64. doi:10.12691/jfnr-7-1-7

Nguyen, T. T. T., A. Q. Hoang, V. D. Nguyen, H. T. Nguyen, T. V. Vu, X. T. Vuong, and M. B. Tu. 2021. "Concentrations, profiles, emission inventory, and risk assessment of chlorinated benzenes in bottom ash and fly ash of municipal and medical waste incinerators in northern Vietnam". *Environmental Science and Pollution Research* 28(11): 13340–13351. doi:10.1007/s11356-020-11385-9

Nguyen, H. T., V. Q. Pham, T. P. M. Nguyen, T. T. T. Nguyen, B. M. Tu, and P. T. Le. 2023a. "Emission and distribution profiles of polycyclic aromatic hydrocarbons in solid residues of municipal and industrial waste incinerators, northern Vietnam". *Environmental Science and Pollution Research* 30: 38255–38268. doi:10.1007/s11356-022-24680-4

Nguyen, L. K. P, N. P. Nguyen, M. T. Le, Q. M. Bui, and T. S. Cam. 2023b. "Concentrations of polycyclic aromatic hydrocarbons in Vietnamese takeaway coffee: effects of coffee variety, roasting temperature and time". *Food Additives & Contaminants Part A* 40: 346–355. doi:10.1080/19440049.2023.2168067

Nguyen, T. T. T., X. T. Vuong, M. B. Tu, M. H. Trinh, and A. Q. Hoang. 2024a. "Insights into full-congener profiles of chlorinated benzenes in fly and bottom ash: case study in Vietnamese industrial and municipal waste incinerators". *Bulletin of Environmental Contamination and Toxicology* 112: 46. doi:10.1007/s00128-024-03874-1

Nguyen, D. T., X. T. T. Le, T. V. Trinh, and M. V. Do. 2024b. "Evaluation of chemical exposure to the sediment and green mussels (*Perna viridis*) at some coastal sites in northern Vietnam". *Regional Studies in Marine Science* 71: 103413. doi:10.1016/j.rsma.2024.103413

Nhan, D. D., N. M. Am, N. C. Hoi, L. V. Dieu, F. P. Carvalho, J. P. Villeneuve, and C. Cattini. 1998. "Organochlorine pesticides and PCBs in the Red River Delta, North Vietnam". *Marine Pollution Bulletin* 36: 742–749. doi:10.1016/S0025-326X(98)00025-3

Nhan, D. D., F. P. Carvalho, N. M. Am, N. Q. Tuan, N. T. H. Yen, J. P. Villeneuve, and C. Cattini. 2001. "Chlorinated pesticides and PCBs in sediments and molluscs from freshwater canals in the Hanoi region". *Environmental Pollution* 112: 311–320. doi:10.1016/S0269-7491(00)00149-4

Nishimura, C., G. Suzuki, H. Matsukami, T. Agusa, M. Takaoka, S. Takahashi, N. M. Tue, P. H. Viet, S. Tanabe, H. Takigami, and T. Fujimori. 2018. "Soil pollution by chlorobenzenes and polychlorinated biphenyls from an electronic waste recycling area in northern Vietnam". *International Journal of Environment and Pollution* 63: 283. doi:10.1504/IJEP.2018.097863

Nishina, T., C. N. Kien, N. V. Noi, H. M. Ngoc, C. S. Kim, S. Tanaka, and K. Iwasaki. 2010. "Pesticide residues in soils, sediments, and vegetables in the Red River Delta, northern Vietnam". *Environmental Monitoring and Assessment* 169: 285–297. doi:10.1007/s10661-009-1170-8

Pariatamby, A., and Y. L. Kee. 2016. "Persistent organic pollutants management and remediation". *Procedia Environmental Sciences* 31: 842–848. doi:10.1016/j.proenv.2016.02.093

Thuy, P. C., T. Kameda, A. Toriba, N. Tang, and K. Hayakawa. 2012. "Characteristics of atmospheric polycyclic aromatic hydrocarbons and nitropolycyclic aromatic hydrocarbons in Hanoi-Vietnam, as a typical motorbike city". *Polycyclic Aromatic Compounds* 32: 296–312. doi:10.1080/10406638.2012.679015

Pham, M. T. N., H. Q. Anh, X. T. Nghiem, B. M. Tu, T. N. Dao, and M. H. Nguyen. 2019a. "Characterization of PCDD/Fs and dioxin-like PCBs in flue gas from thermal industrial processes in Vietnam: a comprehensive investigation on emission profiles and levels". *Chemosphere* 225: 238–246. doi:10.1016/j.chemosphere.2019.03.024

Pham, M. T. N., H. Q. Anh, X. T. Nghiem, B. M. Tu, T. N. Dao, and D. N. Vu. 2019b. "Residue concentrations and profiles of PCDD/Fs in ash samples from multiple thermal industrial processes in Vietnam: formation, emission levels, and risk assessment". *Environmental Science and Pollution Research* 26: 17719–17730. doi:10.1007/s11356-019-05015-2

Phung, V. T., Nguyen, T. N, Phan, L. A. T, Pham, H. V, and Duong, H. A. 2021. "Determination of per- and polyfluoroalkyl substances in craftvillages and industrial environments of Vietnam". *Journal of Analytical Methods in Chemistry* 2021: 5564994. doi:10.1155/2021/5564994

Quynh, T. X., and V. D. Toan. 2019. "Endocrine disrupting compounds (EDCs) in surface waters of the KimNguu River, Vietnam". *Bulletin of Environmental Contamination and Toxicology* 103: 734–738. doi:10.1007/s00128-019-02710-1

Romano, S., R. Piazza, C. Mugnai, S. Giuliani, L. G. Bellucci, C. N. Huu, M. Vecchiato, S. Zambon, N. D. Hoai, and M. Frignani. 2013. "PBDEs and PCBs in sediments of the Thi Nai Lagoon (Central Vietnam) and soils from its mainland". *Chemosphere* 90: 2396–2402. doi:10.1016/j.chemosphere.2012.10.067

Saha, M., A. Togo, K. Mizukawa, M. Murakami, H. Takada, M. P. Zakaria, N. H. Chiem, B. C. Tuyen, M. Prudente, R. Boonyatumanond, S. K. Sarkar, B. Bhattacharya, P. Mishra, and T. S. Tana. 2009. "Sources of sedimentary PAHs in tropical Asian waters: differentiation between pyrogenic and petrogenic sources by alkyl homolog abundance". *Marine Pollution Bulletin* 58: 189–200. doi:10.1016/j.marpolbul.2008.04.049

Schecter, A., P. Toniolo, L. C. Dai, L. T. B. Thuy, and M. S. Wolft. 1997. "Blood levels of DDT and breast cancer risk among women living in the North of Vietnam". *Archives of Environmental Contamination and Toxicology* 33: 453–456. doi:10.1007/s002449900276

Schecter, A., J. Kincaid, H. T. Quynh, J. Lanceta, H. T. T. Tran, R. Crandall, W. Shropshire, and L. S. Birnbaum. 2018. "Biomonitoring of metals, polybrominated diphenyl ethers, polychlorinated biphenyls, persistent pesticides in Vietnamese female electronic waste recyclers". *Journal of Occupational and Environmental Medicine* 60: 191–197. doi:10.1097/jom.0000000000001200

Someya, M., G. Suzuki, A. C. Ionas, N. M. Tue, F. Xu, H. Matsukami, A. Covaci, L. H. Tuyen, P. H. Viet, S. Takahashi, S. Tanabe, and H. Takigami. 2016. "Occurrence of emerging flame retardants from e-waste recycling activities in the northern part of Vietnam". *Emerging Contaminants* 2: 58–65. doi:10.1016/j.emcon.2015.10.002

Suzuki, G., M. Someya, H. Matsukami, N. M. Tue, N. Uchida, L. H. Tuyen, P. H. Viet, S. Takahashi, S. Tanabe, A. Brouwer, and H. Takigami. 2016. "Comprehensive evaluation of dioxins and dioxin-like compounds in surface soils and river sediments from e-waste-processing sites in a village in northern Vietnam: heading towards the environmentally sound management of e-waste". *Emerging Contaminants* 2: 98–108. doi:10.1016/j.emcon.2016.03.001

Takahashi, S., N. M. Tue, C. Takayanagi, L. H. Tuyen, G. Suzuki, H. Matsukami, P. H. Viet, T. Kunisue, and S. Tanabe. 2017. "PCBs, PBDEs and dioxin-related compounds in floor dust from an informal end-of-life vehicle recycling site in northern Vietnam: contamination levels and implications for human exposure". *Journal of Material Cycles and Waste Management* 19: 1333–1341. doi:10.1007/s10163-016-0571-3

Tan, H., S. Tang, L. Yang, J. Li, Y. Deng, H. Shen, Q. Dai, Y. Gao, P. Wu, L. Zhu, and Z. Cai. 2024. "Global quantification of emerging and legacy per- and polyfluoroalkyl substances in indoor dust: levels, profiles and human exposure". *Science of the Total Environment* 927: 172132. doi:10.1016/j.scitotenv.2024.172132

Tanabe, S., and T. Kunisue. 2007. "Persistent organic pollutants in human breast milk from Asian countries". *Environmental Pollution* 146: 400–413. doi:10.1016/j.envpol.2006.07.003

Thai, P. K., Z. Li, A. Sojodin, A. Fox, N. B. Diep, T. T. Binh, and J. F. Mueller. 2015. "Biomonitoring of polycyclic aromatic hydrocarbons exposure in small groups of residents in Brisbane, Australia and Hanoi, Vietnam, and those travelling between the two cities". *Chemosphere* 139: 358–364. doi:10.1016/j.chemosphere.2015.07.004

Thakur, M., and D. Pathania. 2020. "Environmental fate of organic pollutants and effect on human health". In *Abatement of Environment Pollutants Trends and Strategies*. Elsevier, Amsterdam, The Netherlands. 245–262. doi:10.1016/B978-0-12-818095-2.00012-6

Tham, T. T, H. Q. Anh, B. T. Phuong, L. T. Trinh, N. T. T. Thuy, N. T. H. Yen, T. M. Tri, and T. B. Minh. 2020. "Contamination status and temporal trends of persistent toxic substances in sediment cores from coastal areas of central Vietnam". *Marine Pollution Bulletin* 156: 111222. doi:10.1016/j.marpolbul.2020.111222

Thao, V. D., M. Kawano, and R. Tatsukawa. 1993. "Persistent organochlorine residues in soils from tropical and sub-tropical asian countries". *Environment Pollution* 81: 61–71. doi:10.1016/0269-7491(93)90029-n

Thi, H. N., T. N. T. Thu, L. P. Hai, H. N. Thanh, H. C. Viet, H. C. Thi, T. X. Nghiem, and N. H. Tung. 2019. "Emission of unintentionally produced persistent organic pollutants from some industrial processes in northern Vietnam". *Bulletin of Environmental Contamination and Toxicology* 102: 287–296. doi:10.1007/s00128-018-2519-x

Thi, L. A. P., N. T. Ngoc, N. T. Quynh, N. V. Thanh, T. T. Kim, D. H. Anh, and P. H. Viet. 2020. "Polycyclic aromatic hydrocarbons (PAHs) in dry tea leaves and tea infusions in Vietnam: contamination levels and dietary risk assessment". *Environment Geochemistry and Health* 42: 2853–2863. doi:10.1007/s10653-020-00524-3

Thuong, N. V., V. D. Nam, N. T. M. Hue, L. K. Son, N. V. Thuy, H. D. Tung, N. A. Tuan, T. B. Minh, D. Q. Huy, and N. H. Minh. 2014. "The emission of polychlorinated dibenzo-*p*-dioxins and polychlorinated dibenzofurans from steel and cement-kiln plants in Vietnam". *Aerosol and Air Quality Research* 14: 1189–1198. doi:10.4209/aaqr.2013.03.0099

Thuy, H. T. T., T. T.C. Loan, P. T. Luu, N. V. Dung, L. D. Bao, T. H. Phuong, N. T. Khanh, T. T. H. Yen, and D. X. Huy. 2021. "Spatial and temporal variations of PAHs in surface sediments of estuarine and coast of CanGio wetland, Vietnam". *Environmental Science and Pollution Research* 28: 11962–11975. doi:10.1007/s11356-020-08523-8

Toan, V. D., V. D. Thao, J. Walder, H. R. Schmutz, and C. T. Ha. 2007a. "Contamination by selected organochlorine pesticides (OCPs) in surface soils in Hanoi, Vietnam". *Bulletin of Environmental Contamination and Toxicology* 78: 195–200. doi:10.1007/s00128-007-9149-z

Toan, V. D., V. D. Thao, J. Walder, H. R. Schmutz, and C. T. Ha. 2007b. "Level and distribution of polychlorinated biphenyls (PCBs) in surface soils from Hanoi, Vietnam". *Bulletin of Environmental Contamination and Toxicology* 78: 211–216. doi:10.1007/s00128-007-9120-z

Toan, V. D., V. D. Thao, J. Walder, and C. T. Ha. 2009. "Residue, temporal trend and half-life time of selected organochlorine pesticides (OCPs) in surface soils from Bacninh, Vietnam". *Bulletin of Environmental Contamination and Toxicology* 82: 516–521. doi:10.1007/s00128-008-9629-9

Toan, V. D., T. X. Quynh, and N. T. L. Huong. 2020. "Endocrine disrupting compounds in sediment from KimNguu river, northern area of Vietnam: a comprehensive assessment of seasonal variation, accumulation pattern and ecological risk". *Environment Geochemistry and Health* 42: 647–659. doi:10.1007/s10653-019-00399-z

Tran, T. A. M., G. Malarvannan, T. L. Hoang, V. H. Nguyen, A. Covaci, and M. Elskens. 2019. "Occurrence of organochlorine pesticides and polychlorinated biphenyls in sediment and fish in Cau Hai lagoon of Central Vietnam: human health risk assessment". *Marine Pollution Bulletin* 141: 521–528. doi:10.1016/j.marpolbul.2019.03.006

Tran, L. T., T. C. Kieu, H. M. Bui, N. T. Nguyen, T. T. T. Nguyen, D. T. Nguyen, T. Q. Nguyen, H. T. A. Huong, T. H. Le, S. Takahashi, M. B. Tu, and A. Q. Hoang. 2022. "Polybrominated diphenyl ethers in indoor dusts from industrial factories, offices, and houses in northern Vietnam: contamination characteristics and human exposure". *Environment Geochemistry and Health* 44: 2375–2388. doi:10.1007/s10653-021-01026-6

Tran-Lam, T. T., Y. H. Dao, L. K. T. Nguyen, H. K. Ma, H. N. Tran, and G. T. Le. 2018. "Simultaneous determination of 18 polycyclic aromatic hydrocarbons in daily foods (Hanoi Metropolitan Area) by gas chromatography–tandem mass spectrometry". *Foods* 7: 201. doi:10.3390/foods7120201

Tri, T. M., H. Q. Anh, T. T. Tham, T. V. Quy, N. Q. Long, D. T. Nhung, M. Nakamura, M. Nishida, Y. Maeda, L. V. Boi, and T. B. Minh. 2016. "Distribution and depth profiles of polychlorinated dibenzo-p-dioxins, polychlorinated dibenzofurans, and polychlorinated biphenyls in sediment collected from offshore waters of Central Vietnam". *Marine Pollution Bulletin* 106: 341–346. doi:10.1016/j.marpolbul.2016.03.013

Trung, N. T., H. Q. Anh, N. M. Tue, G. Suzuki, S. Takahashi, S. Tanabe, N. M. Khai, T. T. Hong, P. T. Dau, P. C. Thuy, and L. H. Tuyen. 2021. "Polycyclic aromatic hydrocarbons in airborne particulate matter samples from Hanoi, Vietnam: particle size distribution, aryl hydrocarbon ligand receptor activity, and implication for cancer risk assessment". *Chemosphere* 280: 130720. doi:10.1016/j.chemosphere.2021.130720

Tue, N. M., A. Sudaryanto, T. B. Minh, T. Isobe, S. Takahashi, P. H. Viet, and S. Tanabe. 2010a. "Accumulation of polychlorinated biphenyls and brominated flame retardants in breast milk from women living in Vietnamese e-waste recycling sites". *Science of the Total Environment* 408: 2155–2162. doi:10.1016/j.scitotenv.2010.01.012

Tue, N. M., G. Suzuki, S. Takahashi, T. Isobe, P. T. K. Trang, P. H. Viet, and S. Tanabe. 2010b. "Evaluation of dioxin-like activities in settled house dust from Vietnamese e-waste recycling sites: relevance of polychlorinated/brominated dibenzo-*p*-dioxin/furans and dioxin-like PCBs". *Environment Science & Technology* 44: 9195–9200. doi:10.1021/es102505j

Tue, N. M., A. Sudaryanto, T. B. Minh, B. H. Nhat, T. Isobe, S. Takahashi, P. H. Viet, and S. Tanabe. 2010c. "Kinetic differences of legacy organochlorine pesticides and polychlorinated biphenyls in Vietnamese human breast milk". *Chemosphere* 81: 1006–1011. doi:10.1016/j.chemosphere.2010.09.013

Tue, N.M., S. Takahashi, G. Suzuki, T. Isobe, P. H. Viet, Y. Kobara, N. Seike, G. Zhang, A. Sudaryanto, and S. Tanabe. 2013. "Contamination of indoor dust and air by polychlorinated biphenyls and brominated flame retardants and relevance of non-dietary exposure in Vietnamese informal e-waste recycling sites". *Environment International* 51: 160–167. doi:10.1016/j.envint.2012.11.006

Tue, N. M., K. Katsura, G. Suzuki, L. H. Tuyen, T. Takasuga, S. Takahashi, P. H. Viet, and S. Tanabe. 2014. "Dioxin-related compounds in breast milk of women from Vietnamese e-waste recycling sites: levels, toxic equivalents and relevance of non-dietary exposure". *Ecotoxicology and Environmental Safety* 106: 220–225. doi:10.1016/j.ecoenv.2014.04.046

Tuyen, L. H., N. M. Tue, G. Suzuki, K. Misaki, P. H. Viet, S. Takahashi, and S. Tanabe. 2014a. "Aryl hydrocarbon receptor mediated activities in road dust from a metropolitan area, Hanoi—Vietnam: contribution of polycyclic aromatic hydrocarbons (PAHs) and human risk assessment". *Science of the Total Environment* 491–492: 246–254. doi:10.1016/j.scitotenv.2014.01.086

Tuyen, L. H., N. M. Tue, S. Takahashi, G. Suzuki, P. H. Viet, A. Subramanian, K. A. Bulbule, P. Parthasarathy, A. Ramanathan, and S. Tanabe. 2014b. "Methylated and unsubstituted polycyclic aromatic hydrocarbons in street dust from Vietnam and India: occurrence, distribution and *in vitro* toxicity evaluation". *Environmental Pollution* 194: 272–280. doi:10.1016/j.envpol.2014.07.029

Vi, P. T., N. T. Ngoc, P. D. Quang, N. T. Dam, N. M. Tue, L. H. Tuyen, P. H. Viet, and D. H. Anh. 2022. "Perfluoroalkyl substances in freshwater and marine fish from northern Vietnam: accumulation levels, profiles, and implications for human consumption". *Marine Pollution Bulletin* 182: 113955. doi:10.1016/j.marpolbul.2022.113995

Vo, L. H. T., M. Yoneda, T. D. Nghiem, K. Sekiguchi, Y. Fujitani, D. N. Vu, and T. H. T. Nguyen. 2024. "Characterisation of polycyclic aromatic hydrocarbons associated with indoor $PM_{0.1}$ and $PM_{2.5}$ in Hanoi and implications for health risks". *Environmental Pollution* 343: 123138. doi:10.1016/j.envpol.2023.123138

Vu-Duc, N., L. A. P. Thi, T. L. Minh, L. A. Nguyen, H. N. Thi, L. H. P. Thi, V. A. D. Thi, H. L. Quang, H. Nguyen-Xuan, T. T. Nguyen, P. T. Nguyen, and D. B. Chu. 2021. "Analysis of polycyclic aromatic hydrocarbon in airborne particulate matter samples by gas chromatography in combination with tandem mass spectrometry (GC-MS/MS)". *Journal of Analytical Methods in Chemistry* 2021: 6641326. doi:10.1155/2021/6641326

Wang, W., Y. Wang, R. Zhang, S. Wang, C. Wui, C. Chaemafa, J. Li, G. Zhang, and K. Yu. 2016a. "Seasonal characteristics and current sources of OCPs and PCBs and enantiomeric signatures of chiral OCPs in the atmosphere of Vietnam". *Science of the Total Environment* 542: 777786. doi: 10.1016/j.scitotenv.2015.10.129

Wang, Q., Y. Miyake, T. Amagai, G. Suzuki, H. Matsukami, N. M. Tue, S. Takahashi, S. Tanabe, L. H. Tuyen, P. H. Viet, and H. Takigami. 2016b. "Halogenated polycyclic aromatic hydrocarbons in soil and river sediment from e-waste recycling sites in Vietnam". *Journal of Water and Environment Technology* 14: 166–176. doi:10.2965/jwet.15-053

Wittcoff, H. A., B. G. Reuben, and J. S. Plotkin. 2013. "The evolution of the organic chemicals industry". In *Industrial Organic Chemicals*. John Wiley & Sons, Inc., Hoboken, NJ. 13–47. doi:10.1002/9781118229996.ch1

Wolff, M. J. C, and N. O. Crossland. 1991. "The environmental fate of organic chemicals". In *Controlling Chemical Hazards: Fundamentals of the Management of Toxic Chemicals. The Environmental Fate of Organic Chemicals*. Springer, Dordrecht. 19–46. doi:10.1007/978-94-011-6849-6_2

12

Contaminants of Emerging Concern in Vietnamese Environments: A Recent Assessment of Pollution Status, Emission Sources, and Human Health Risks

Tran Manh Tri, Hoang Quoc Anh, Le Minh Thuy, Vu Duc Nam, Le Van Chieu, and Tu Binh Minh

12.1 Introduction

Endocrine-disrupting chemicals (EDCs) comprise a wide range of substances that can interfere with functions of endogenous hormones and affect various biological systems such as reproductive, immune, central nervous, respiratory, and cardiovascular systems (Metcalfe et al., 2022; Zhu et al., 2023). In recent years, the increasing use of synthetic chemicals in consumer and industrial products has remarkably contributed to the contamination of aquatic and indoor environments, and those chemicals are considered contaminants of emerging concern (CECs). The widespread occurrences of emerging organic chemicals such as phthalates, bisphenol A, tetrabromobisphenol A (TBBPA), bisphenol A diglycidyl ether (BADGE), alkylphenols, and pharmaceuticals have been reported in various environmental media such as water, air, settle dust, sediment and biota samples from industrialized countries (Ma et al., 2014; Kubwabo et al., 2013; Wang et al., 2015; Wang and Kannan, 2016).

Despite this, comprehensive and systematic studies on environmental occurrence, fate, and human exposure to these chemicals in developing countries are still limited. Vietnam is one of the countries in economic transition that have received considerable attention during the past three decades in terms of monitoring and environmental remediation of persistent organic pollutants (POPs). Our research group provided in-depth and systematic reviews of the environmental contamination, sources, and human exposure of the legacy POPs and emerging EDCs in 2016 (Minh et al., 2016, Tri et al., 2016). In recent years, several investigations have been conducted in our laboratory, especially for CECs, which are still being used in a wide range of industrial and consumer products such as phthalate esters (PAEs), linear and cyclic siloxanes, parabens, bisphenol A, which facilitate gaining further in-depth insights into the fate and transport of these chemicals, especially in indoor environments.

In this chapter, recent studies on different groups of CECs such as phthalates esters, siloxanes, parabens, bisphenol A, bisphenol A, diglycidyl ether (BADGE), triclosan and triclocarbans in Vietnamese environment were compiled to elucidate their status of contamination, distribution, and behavior in an ambient and indoor environment. This chapter is based primarily on the studies conducted by our research group in collaboration with international laboratories. In addition, the review was supplemented with data from other studies to provide a complete picture of environmental occurrences and human exposure in Vietnam.

DOI: 10.1201/9781003492115-15

12.2 Contamination Status and Emission Sources

12.2.1 Settled Dust and Air Samples

Settled dust has been considered a simple and effective sample for monitoring organic micro-contaminants. Indoor dust samples were collected from different micro-environments such as homes, hair salons, offices, laboratories, kindergartens, and cars in Hanoi city and some other cities and provinces in northern Vietnam, and residue concentrations of various organic emerging contaminants were examined (Table 12.1). Preliminary screening of indoor dust samples from different micro-environments from Vietnamese revealed an accumulation pattern of different CECs, with PAEs as primary chemicals with the highest concentrations, followed by bisphenol A, siloxanes, BADGEs, and parabens with relatively similar magnitude of contamination (Figure 12.1). This result is consistent with our screening study of a large number of organic contaminants in the settled dust, showing the predominance of phthalates and pharmaceuticals and personal care products (PPCPs) than other groups (Anh et al., 2019).

In general, concentrations of PAEs and siloxanes were in higher ranges than other chemicals, suggesting the extensive use and emission in Vietnam. PAE concentrations were generally higher in workplaces and informal waste recycling areas (such as e-waste and plastic waste) than those in urban micro-environments (homes, offices, laboratories, and supermarkets) (Figure 12.2). Phthalates are used for plasticizers in many consumer and industrial products, and releases from informal plastic waste recycling sites are documented.

Based on the screening results, we targeted in-depth investigations on the potential CEC groups such as PAEs, cyclic and linear siloxanes, and paraben using indoor dust samples as indicators in different indoor micro-environments to elucidate their occurrence, behavior, and possible sources. An interesting result was observed for homes and hair salons, showing remarkably higher concentrations of PAEs and siloxanes than other micro-environments. A subsequent study focusing on hair salons in Hanoi City revealed a similar accumulation pattern of cyclic siloxanes in hair care products and dust samples. In addition, strong positive correlations between siloxane concentrations in dust and indoor air samples were observed. These results indicate that hair care products are primary sources of siloxanes in hair salons. In addition, relatively high levels of phthalates were also encountered in kindergartens, suggesting potential exposure risk for children (Tran et al., 2018).

The informal waste recycling and end-of-life vehicle (ELV) storage areas have been considered potential emission sources of classic POPs such as PCBs, PAHs, and PBDEs (Hoang et al., 2023a, 2023b). To further elucidate the role of these recycling areas in the contamination of emerging contaminants, settle dust samples were collected from various sites in and around these recycling facilities, and residue concentrations of PAEs and cyclic siloxanes were examined (Hoang et al., 2022a, 2022b). Concentrations of both PAEs and cyclic siloxanes were found to be higher in waste recycling than those in urban homes. In waste recycling workshops, large amounts of PAE-containing materials were processed (e.g., manual dismantling, washing, drying, and crushing into small pieces) and stored. Like PBDEs, a legacy brominated flame retardant, these activities can facilitate the PAE and siloxane emission by evaporation, debris formation, and direct transfer through a material-dust interface. As for comparison, PAEs in dust from recycling and industrial workplaces from Vietnam are in a lower range compared to other countries in the region and markedly lower than those in industrialized countries (Hoang et al., 2022a). Similarly, indoor dust from Vietnam contained lower-range concentrations of cyclic siloxanes, higher than those in India and Pakistan but less than most of the countries surveyed (Tran et al., 2017a).

Multimedia monitoring has long been considered a practical approach for a comprehensive and in-depth understanding of the behavior and transport of organic contaminants. Passive air samplers and low-volume air samplers have been proven as useful sampling techniques as they are easy to use, relatively simple and quiet in operation, and suitable for evaluation of long-term human exposures and risk assessment (Tran et al., 2017b, 2017c). Indoor air samples were collected using low-volume samplers from different micro-environments such as homes, offices, hair salons, laboratories, and kindergartens, and concentrations of PAEs and siloxanes were examined (Tran et al., 2017b, 2017c, 2018; Anh et al., 2021; Nguyen et al., 2023). PAE concentrations in air from homes and hair salons were remarkably higher

TABLE 12.1

Concentrations of Several Groups of CECs in Air and Settled Dust Samples from Vietnam

Location, Time	Type of Samples	Number of Samples (*n*)	PAEs[a]	VMSs[b]	PBs[c]	BPA[d]	BADGE[e]	TCS[f]	TCC[g]	Unit	References
4 cities of the North, 2014	**Indoor dust**		**30,200 (3440–106,000)**		**183 (40.0–842)**		**388 (23–1750)**			ng/g	Tran et al. (2016)
	Homes	16	26,800 (3440–79,300)		245 (46.9–842)		324 (27.7–1650)			ng/g	
	Supermarkets/ shops	18	40,200 (8030–106,000)		136 (40.8–431)		491 (23.2–1750)			ng/g	
	Laboratories	7	27,800 (421–53,300)		223 (115–326)		365 (115–666)			ng/g	
	Offices	5	8280 (4500–17,000)		98.4 (40.0–144)		219 (994–353)			ng/g	
4 cities of the North, 2014	Indoor dust	38		291 (n.d.–943)						ng/g	Tran et al. (2017a)
4 cities of the North, 2016–2017	**Indoor air**	**97**	**1040 (106–16,000)**							ng/m^3	Tran et al. (2017b)
	Particle phase		3570 (95.2–13,100)							µg/g	
	Gas phase		775 (57.0–14,900)							ng/m^3	
4 cities of the North, 2016–2017	**Indoor air**	**97**		**469 (41.8–1950)**						ng/m^3	Tran et al. (2017c)
	Particle phase			1880 (141–7220)						µg/g	
	Gas phase			321 (23.8–1580)						ng/m^3	
Hanoi, 2017	Indoor dust	24		671 (178–1240)						ng/g	Tran et al. (2018)
	Indoor air	24		1030 (415–2610)						ng/m^3	
Hanoi, 2018	Air dust		1.92–60.4							ng/m^3	Dung et al. (2019)
4 cities of the North, 2017	Indoor dust	89						33.2* (<5–1090)	19.3* (<3–531)	ng/g	Tran et al. (2020)

(Continued)

TABLE 12.1 (*Continued*)

Concentrations of Several Groups of CECs in Air and Settled Dust Samples from Vietnam

Location, Time	Type of Samples	Number of Samples (*n*)	PAEs[a]	VMSs[b]	PBs[c]	BPA[d]	BADGE[e]	TCS[f]	TCC[g]	Unit	References
Hanoi, 2019–2020	**Indoor air**	**38**									Anh et al. (2021)
	Gas phase		530* (105–1760)	124* (n.d.–647)						ng/m^3	
	Particle phase		1840* (225–9410)	437* (n.d.–5630)						µg/g	
	Bulk air		688* (142–2390)	171* (n.d.–1100)						ng/m^3	
	University laboratories	**16**									
	Gas phase		357* (105–883)	46.7* (n.d.–208)						ng/m^3	
	Particle phase		2170* (227–7130)	252* (n.d.–1560)						µg/g	
	Bulk air		533* (142–1550)	78.7* (n.d.–267)						ng/ m^3	
	University and bank offices	**8**									
	Gas phase		606* (216–1260)	134* (33.7–300)						ng/m^3	
	Particle phase		2690* (637–6070)	324* (70.5–616)						µg/g	
	Bulk air		956* (276–1830)	164* (81.0–347)						ng/ m^3	
	Homes	**14**									
	Gas phase		566* (200–1760)	269* (27.9–647)						ng/m^3	
	Particle phase		1460* (225–9410)	1350* (258–5630)						µg/g	
	Bulk air		688* (265–2390)	479* (64.6–1100)						ng/m^3	

(*Continued*)

TABLE 12.1 (*Continued*)
Concentrations of Several Groups of CECs in Air and Settled Dust Samples from Vietnam

Location, Time	Type of Samples	Number of Samples (*n*)	PAEs[a]	VMSs[b]	PBs[c]	BPA[d]	BADGE[e]	TCS[f]	TCC[g]	Unit	References
	Outdoor air	**38**									
	Gas phase		104* (25.7–460)	33.4* (n.d.–181)						ng/m³	
	Particle phase		429* (44.9–1800)	97.4* (n.d.–1130)						µg/g	
	Bulk air		161* (34.1–515)	43.2* (n.d.–258)						ng/m³	
	University laboratories	**16**									
	Gas phase		72.7* (25.7–240)	13.0* (n.d.–57.5)						ng/m³	
	Particle phase		511* (84.4–1800)	65.8* (n.d.–370)						µg/g	
	Bulk air		127* (34.1–409)	17.6* (n.d.–70.3)						ng/m³	
	University and bank offices	**8**									
	Gas phase		154* (42.1–367)	33.6* (8.01–77.6)						ng/m³	
	Particle phase		820* (186–1580)	88.5* (15.8–200)						µg/g	
	Bulk air		208* (59.5–515)	40.7* (21.2–92.1)						ng/m³	
	Homes	**14**									
	Gas phase		134* (33.7–460)	63.0* (7.41–181)						ng/m³	
	Particle phase		322* (44.9–1500)	327* (n.d.–1130)						µg/g	
	Bulk air		161* (46.2–486)	100* (19.5–258)						ng/m³	
Hanoi, 2017–2019	Indoor dust	58			285 (<MDL–1650)					ng/g	Tran et al. (2021)

(*Continued*)

TABLE 12.1 (*Continued*)

Concentrations of Several Groups of CECs in Air and Settled Dust Samples from Vietnam

Location, Time	Type of Samples	Number of Samples (*n*)	PAEs[a]	VMSs[b]	PBs[c]	BPA[d]	BADGE[e]	TCS[f]	TCC[g]	Unit	References
4 cities of the North, 2019–2020	**Indoor dust**	**60**		**1130 (86.0–5890)**						ng/g	Hoang et al. (2022)a
	Workplaces	32		1560 (126–5890)						ng/g	
	Classrooms	11		576 (86.0–1540)						ng/g	
	Homes	17		873 (115–1680)						ng/g	
5 cities of the North, 2018–2020	**Indoor dust**	**62**	**33,400* (585–153,000)**							ng/g	Hoang et al. (2022b)
	Workplaces	30	49,100* (9210–153,000)							ng/g	
	Homes	32	23,700* (585–83,700)							ng/g	
Hanoi, 2021	Indoor dust	21				626 (n.d.–2230)				ng/g	Thuy et al. (2022)
Bac Ninh, 2021	PM0.1	32	70.4 (1.76–372)	16.7 (MQL–203)						ng/m³	Le et al. (2023)
	PM0.5	32	47.4 (2.24–895)	21.5 (MQL–537)							
Hanoi, 2022	**Air samples**	**90**									Nguyen et al. (2023)
	Indoor air	**78**	**320–4770**							ng/m³	
	Laboratories		757 (402–1380)							ng/m³	
	Homes		1140 (320–3030)							ng/m³	
	Salons		3580 (2570–4470)							ng/m³	
	Ambient air	**12**	**75.1 (35.9–133)**							ng/m³	

Note: *median concentration.

[a] Phthalates; [b]Volatile methyl siloxanes; [c]Parabens; [d]Bisphenol A; [e]Bisphenol A; diglycidyl ether; [f]Triclosan; [g]Triclocarban.

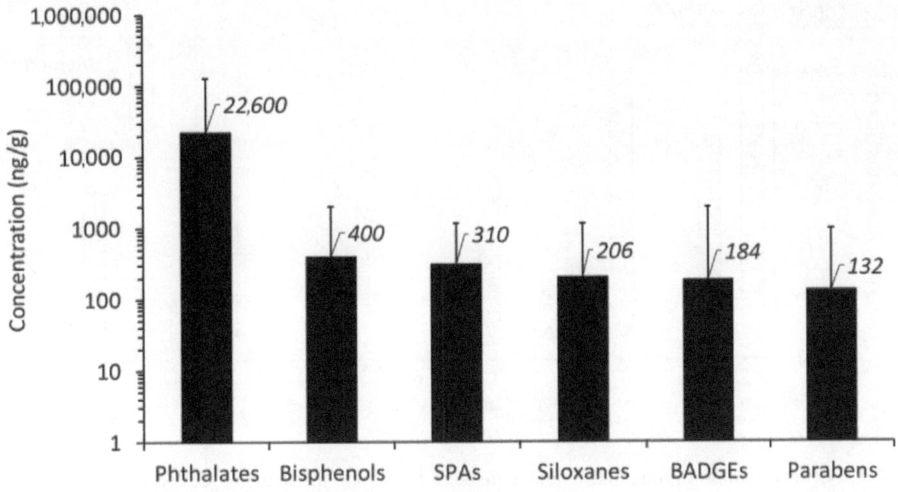

FIGURE 12.1 Concentrations of some CECs in house dust samples from Vietnam.

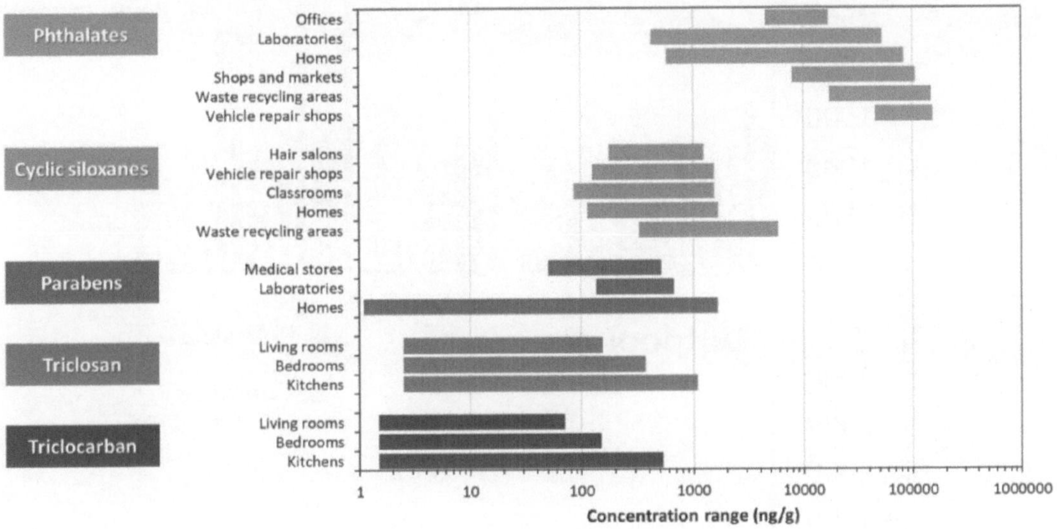

FIGURE 12.2 Concentrations of some CECs in settled dust samples from informal waste recycling areas and indoor micro-environments from Vietnam.

than in other micro-environments (Figure 12.3). Cyclic siloxanes had a lower contamination extent. Still, they showed similar patterns with higher levels in homes and hair salons. Most consumer products used in households and hair salons contain large amounts of PAEs, such as plastic materials (Orecchio et al., 2022), personal care products (Guo et al., 2014; Pagoni et al., 2022), and cosmetics (Koniecki et al., 2011). Our recent findings indicate that such commercial products are the primary sources of PAEs released into the indoor air, whereas the distribution of PAEs in the ambient air is attributed to diffusion through the indoor environment (Ouyang et al., 2019; Szewczyńska et al., 2021; Anh et al., 2021).

Paired indoor and outdoor air were also collected from the exact locations in the Hanoi metropolitan areas to understand the behavior of emerging chemicals in the indoor environment. An interesting result showed higher concentrations of PAEs and cyclic siloxanes in indoor air, about 5–10 times more than those in outdoor air samples (Figure 12.4) (Nguyen et al., 2022; Tran et al., 2017c). These results further

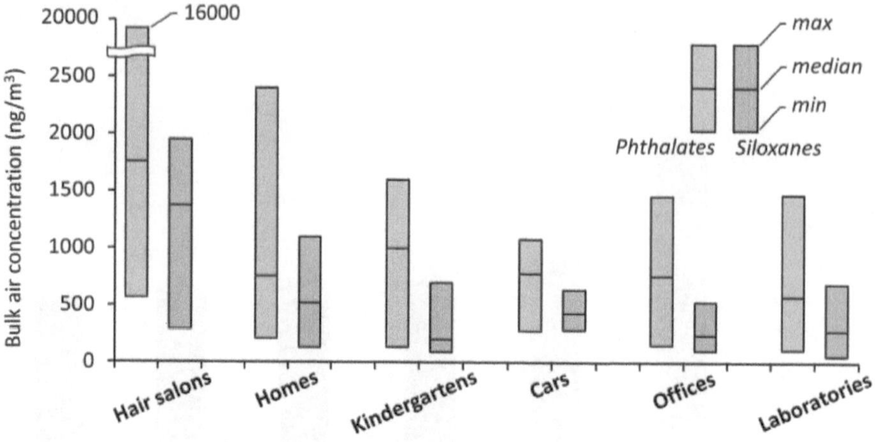

FIGURE 12.3 Concentrations of phthalates and siloxanes in indoor air samples collected from different micro-environments in Vietnam.

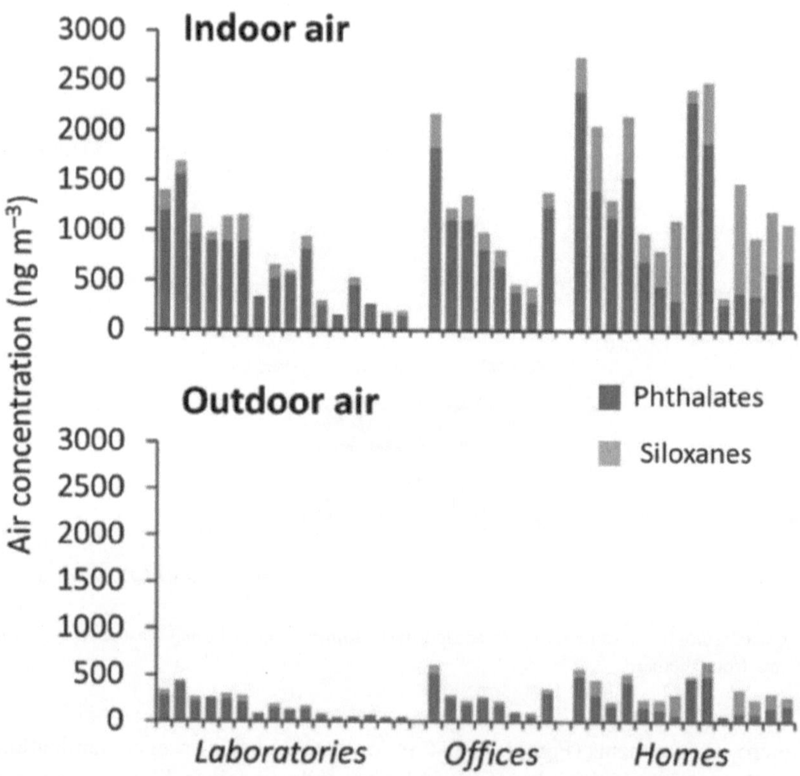

FIGURE 12.4 Comparison of phthalate and siloxane concentrations in indoor and outdoor air samples from the same locations in Vietnam.

support the previous assumption that PAEs originated from indoor environments (Koniecki et al., 2011; Pei et al., 2013; Tran and Kannan, 2015). Moreover, environmental factors and photochemical degradation may be responsible for decreasing PAE levels in the ambient air (Chen et al., 2018; Anh et al., 2021; Zhou, et al., 2021). To further understand the behavior of PAEs in indoor environments, air samples were collected and analyzed in several laboratories during the day and night. PAE levels in nighttime air were significantly higher than in the daytime (Nguyen et al., 2022). These results suggest that several factors may affect the distribution of PAEs in indoor air, such as temperature, pressure, wind speed, or weather. In night conditions, the laboratories are permanently closed, so the circulation of air to the outside is lower than during the daytime.

12.2.2 Water, Sediment, and Biological Samples

Extensive surveys were conducted during August–December 2020 to collect different types of water samples such as tap water, bottled water, lake water, and wastewater in Hanoi Metropolitan areas for understanding of spatial occurrence of PAEs, cyclic siloxanes, and parabens (Le et al., 2021, 2022b; Nguyen et al., 2021) (Table 12.2, Figure 12.5). Tap water samples were collected from domestic water supplies to residents living in Hanoi urban area. Plastic bottled water, including non-carbonated and carbonated water samples, was purchased in Hanoi supermarkets and stores. Lakewater samples were collected from six lakes in Hanoi, including Thien Quang Lake, Ba Mau Lake, Hoan Kiem Lake, Yen So Lake, Truc Bach Lake, and West Lake. Wastewater samples, including influents and effluents, were collected along the lengths of two sewage canals that receive domestic discharges from central Hanoi. The accumulation pattern of all three groups of chemicals (PAEs, cyclic siloxanes, and parabens) was similar and followed the order of wastewater > lake water > tap water > bottled water. As for PAEs, the concentrations were in the order of wastewater (mean: 121,000 ng/L; range: 20,700–405,000 ng/L) > lake water (mean: 51,800 ng/L; range: 19,600–127,000 ng/L) > tap water (mean: 11,200 ng/L; range: 2100–18,000 ng/L) > bottled water (mean: 6400 ng/L; range: 1640–15,700 ng/L). Among PAEs, di-(2-ethylhexyl) phthalate (DEHP) accounted for a major proportion of total concentrations (45%) in wastewater, followed by diisobutyl phthalate (DiBP, 10.3%), and dibutyl phthalate (DBP, 9.53%).

Little geographical variations of PAEs were observed in major lakes throughout Hanoi, suggesting the ubiquitous distribution of these chemicals in surface waters. These lakes receive stormwater from crowded residential areas. High levels of PAEs measured in water samples from lakes can be explained by urban runoff and stormwater discharges (Le et al., 2021). The distribution of siloxanes from lake water was slightly different from PAEs with higher levels in stagnant lakes (Ba Mau and Thien Quang) than in other lakes with flowing and stronger water exchange conditions (Nguyen et al., 2021). Relatively higher levels of parabens were also found in Thien Quang Lake (23.7–244 ng/L) and Truc Bach Lake (31.1–80.5 ng/L), the lakes that have low water exchange and receive discharges from densely populated areas in Hanoi city (Le et al., 2022b). In general, PAEs exhibited higher concentrations and frequency of detection than cyclic siloxanes and parabens in different water samples surveyed. This pattern is consistent with that found in settled dust and air samples, indicating the predominant use of phthalates in various consumer and personal care products and significant emissions sources to indoor and ambient environments.

Extensive surveys were conducted in late 2020 for three rivers and canal systems flowing through Hanoi City to elucidate spatial distributions of PAEs and cyclic siloxanes in sediment samples (Le et al., 2022a; Nguyen et al., 2022, Table 12.2). The extent of contamination is affected by several factors, including water sources, discharging volume, and population density in the vicinities of sampling sites. The highest PAE levels were encountered in sediment from the To Lich River and Nhue River canals. These rivers receive untreated urban wastewater discharges and flow through the higher densely populated areas. In particular, To Lich River received untreated domestic and industrial wastewater with a total volume of about 600,000 m^3/day, and the Nhue River also undergoes about 800,000 m^3/day of wastewater (Nguyen et al., 2019). These results suggest that rapid urbanization and industrialization have increased the PAE pollution levels in sediments. The magnitude of PAE contamination in urban sediment from Vietnam is in a higher range than contamination in urban sediment from Vietnam, which is higher than that of the world. Average PAEs concentrations in three urban rivers in Hanoi, Vietnam, were lower than those in Xi River, Northeast China, but higher than many other rivers and lakes elsewhere (Srivastava et al., 2010; Sun et al., 2013; Gao et al., 2014; 2019; Wang et al., 2014; Li et al., 2016; Lee et al., 2019; Kingsley and Witthayawirasak, 2020).

As for siloxanes, a similar spatial pattern was observed with the highest levels in sediment samples from the To Lich River (range: 260–15,000 ng/g d.w.) and Nhue River (range: 188–6800 ng/g d.w.)

TABLE 12.2

Concentrations of Several Groups of CECs in Water and Sediment Samples from Vietnam

Location, Time	Type of Samples	Number of Samples (n)	PAEs[a]	VMSs[b]	PB[c]	BPA[d]	Unit	References
Hanoi, 2018	**Wastewater**							Quynh and Duc (2019)
	Dry season	6	247.70 (2.95–412.29)				ng/L	
	Wet season	6	122.30 (1.39–204.04)				ng/L	
Hanoi, 2020	Bottled water	21	6400 (1640–15,700)				ng/L	Le et al. (2021)
	Tap water	7	11,200 (2100–18,000)				ng/L	
	Lake water	33	51,800 (19,600–127,000)				ng/L	
	Wastewater	35	121,000 (20,700–405,000)				ng/L	
Hanoi, 2020	Bottled water	20		10.3 (2.31–28.1)			ng/L	Nguyen et al. (2021)
	Tap water	7		12.6 (2.84–19.8)			ng/L	
	Lake water	31		350 (67.9–1100)			ng/L	
	Influent wastewater	25		2380 (221–7400)			ng/L	
	Effluent wastewater	8		155 (63.0–378)			ng/L	
Hanoi, 2021	Bottled water	19			6.92 (1.56–39.9)		ng/L	Le et al. (2022b)
	Tap water	11			28.6 (5.01–54.3)		ng/L	
	Lake water	22			51.7 (18.0–254)		ng/L	

(*Continued*)

TABLE 12.2 (*Continued*)
Concentrations of Several Groups of CECs in Water and Sediment Samples from Vietnam

Location, Time	Type of Samples	Number of Samples (n)	PAEs[a]	VMSs[b]	PB[c]	BPA[d]	Unit	References
	River water	18			32.1 (16.5–52.1)		ng/L	
	Wastewater	30			268 (27.3–1050)		ng/L	
Ho Chi Minh city, 2017–2018	sediment	28				18.56–80.36	ng/g	Tran et al. (2019)
Hanoi, 2020	**Surface sediment**							Le et al. (2022a)
	To Lich river	9	50,000 (11,000–125,000)				ng/g–dw[e]	
	Nhue river	12	29,300 (2140–89,900)				ng/g–dw[e]	
	Day river	15	13,800 (1140–43,100)				ng/g–dw[e]	
Hanoi, 2020	**Surface sediment**	85		2510 (75.4–15,000)				Nguyen et al. (2022)
	To Lich river	25		4760 (260–15,000)			ng/g–dw[e]	
	Nhue river	25		1920 (188–6800)			ng/g–dw[e]	
	Day river	35		1340 (75.4–4600)			ng/g–dw[e]	

[a] Phthalates; [b] Volatile methyl siloxanes; [c] Parabens; [d] Bisphenol A; [e] ng/g-dry weight.

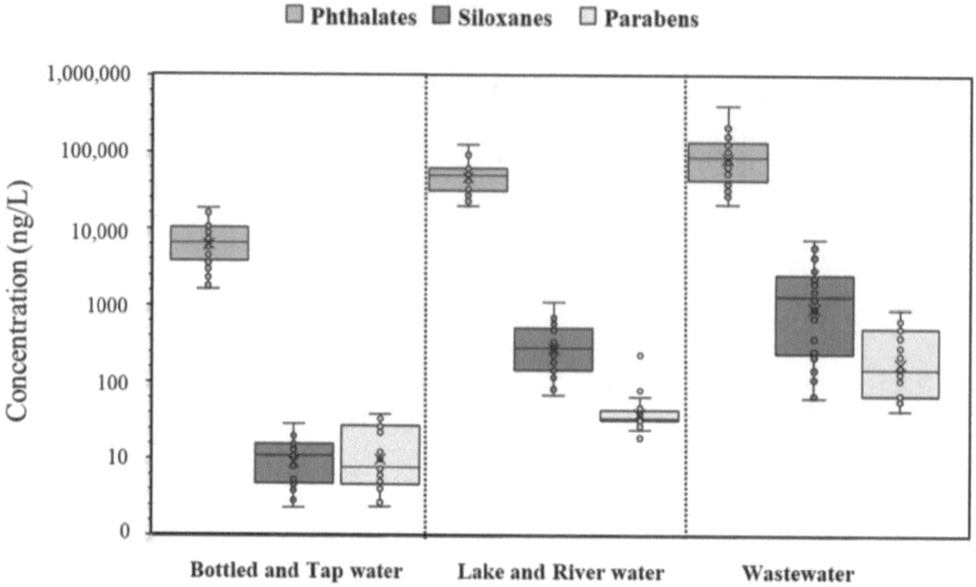

FIGURE 12.5 Concentrations of phthalates, siloxanes, and parabens from different water samples in Hanoi, Vietnam.

(Nguyen et al., 2022). Siloxane concentrations from sampling sites along each river generally exhibited slight fluctuation. Concentrations were lower at the headwater site due to the influx flow and the lack of discharge points. The highest concentrations were observed at the discharge point in a high population-density area. Two sampling locations downstream in the Hanam and Namdinh provinces rivers are lower than other locations, which could be attributed to the dilution downstream, and the area mainly receives agricultural wastewater. Studies of siloxane accumulation in sediment are limited. This is among the first ones in Vietnam. The pattern of cyclic siloxanes in sediment samples from this survey is similar to those reported in previous studies (Zhang et al., 2011a; Lee et al., 2014; Silva et al., 2021).

12.2.3 Biota and Human Samples

Investigations on CECs in biota and human samples from Vietnam are relatively limited. Tran-Lam et al. (2023 and 2024) conducted the most comprehensive studies on the marine fish species collected from coastal areas throughout Vietnam's north, central, and south (Table 12.3). Concentrations of PAEs in five marine fish species ranged from 4.8–39.3 ng/g lipid wt. with DEP as the predominant compound. Feeding habits, tropic levels, and sampling sites with industrial and tourism activities are probable factors affecting the variations in PAEs and other EDCs in fish species examined (Tran-Lam et al., 2024).

The occurrence of emerging compounds such as UV filters, synthetic musk, and pharmaceutically active compounds was also reported, and concentrations of PAEs were in the lower range compared to these chemical groups (Tran-Lam et al., 2024). As for parabens, an interesting result was observed with potential biomagnification along the marine food web with trophic magnification factors ranging from 1.45 to 3.55 (Tran-Lam et al., 2023). From a geographical point of view, fish from central coasts contained higher levels than those in the northern and southern regions, suggesting the role of tourism activities in the central coastal area. Higher contamination by organochlorine pesticides (OCPs), polychlorinated biphenyls (PCBs), and polybrominated diphenyl ethers (PBDEs) in various sites from central coasts of Vietnam was also attributed to the extensive tourism activities in summer seasons (Tham et al., 2019).

As for human samples, surveys were conducted during 2010–2011 to elucidate the geographical distribution of bisphenol S, bisphenol A, PAEs, parabens, triclosan, and triclocarbans in urine samples

TABLE 12.3

Concentrations of Several Groups of CECs in Biota and Human Samples from Vietnam

Locations, Time	Type of Samples	Number of Samples (n)	PAEs[a]	*m*–PAEs[b]	PBs[c]	BPA[d]	BPS[e]	TCS[f]	TCC[g]	Unit	References
Hanoi, 2010	Urine	30		129						ng/mL	Guo et al. (2011)
				115						μg/g creatinine	
Hanoi, 2010	Urine	30				3.32 (0.16–30.1)				ng/mL	Zhang et al., (2011b)
						2.79 (0.10–22.0)				μg/g creatinine	
Hanoi, 2010–2011	Urine	29					0.198 (0.037–0.932)			ng/mL	Liao et al. (2012)
							0.185 (0.050–0.660)			μg/g creatinine	
Vietnam, 2010–2012	Urine	19			3.25*					ng/mL	Honda et al. (2018)
Vietnam, 2010–2012	Urine	19						2.34 (0.08–27.0)	1.08 (0.02–12.3)	ng/mL	Iyer et al., (2018)
Hanoi, 2018	Beverage	8	55.4 (18.2–86.0)							ng/mL	Thuy et al., (2018)
Hanoi, 2018	**Non–alcoholic beverages**										Tran–Lam et al., (2018)
	Mineral water	14	2.9 (1.32–5.77)							μg/L	
	Carbonated drinks	17	128.2 (3.15–339)							μg/L	
	Functional drinks	29	88.2 (2.93–201)							μg/L	
	Juice drinks	32	445 (169–783)							μg/L	

(Continued)

TABLE 12.3 (*Continued*)

Concentrations of Several Groups of CECs in Biota and Human Samples from Vietnam

Locations, Time	Type of Samples	Number of Samples (n)	PAEs[a]	*m*–PAEs[b]	PBs[c]	BPA[d]	BPS[e]	TCS[f]	TCC[g]	Unit	References
	Tea drinks	33	121 (23.1–135)							μg/L	
	Fermented milks	23	150 (23.1–303)							μg/L	
Vietnam, 2022	**Marine fish**										Pham et al. (2024)
	Lizard fish (*Saurida tumbil*)	18			11.8 (6.21–17.2)					ng/g-ww[h]	
	Round scad (*Decapterus punctatus*)	19			11.8 (6.82–25.3)					ng/g-ww[h]	
Vietnam, 2022	**Marine fish**										Tran–Lam et al., (2023)
	Indian halibut (*Psettodes erumei*)	19			29.2 (7.5–61.5)					ng/g–dw[i]	
	Silver pomfret (*Pampus argenteus*)	21			18.5 (4.3–37.9)					ng/g–dw[i]	
	Largehead hairtail (*Trichiurus lepturus*)	16			36.9 (10.8–62)					ng/g–dw[i]	
	Indian mackerel (*Rastrelliger kanagurta*)	26			21.7 (3.4–46)					ng/g–dw[i]	
	yellow–stripe scad (*Selaroides leptolepis*)	32			26.8 (8.5–52.7)					ng/g–dw[i]	

(*Continued*)

TABLE 12.3 (*Continued*)
Concentrations of Several Groups of CECs in Biota and Human Samples from Vietnam

Locations, Time	Type of Samples	Number of Samples (n)	PAEs[a]	*m*–PAEs[b]	PBs[c]	BPA[d]	BPS[e]	TCS[f]	TCC[g]	Unit	References
	Marine fish										
	Indian halibut (*Psettodes erumei*)	19	19.7 (10.5–26.6)							ng/g–lw[j]	
	Silver pomfret (*Pampus argenteus*)	21	20.2 (4.8–39)							ng/g–lw[j]	
	Largehead hairtail (*Trichiurus lepturus*)	16	23.6 (14.4–36.9)							ng/g–lw[j]	
	Indian mackerel (*Rastrelliger kanagurta*)	26	22.1 (10.5–39.3)							ng/g–lw[j]	

Note: *median concentration.
[a] Phthalates diesters; [b]Phthalate metabolites; [c]Parabens; [d]Bisphenol A; [e]Bisphenol S; [f]Triclosan; [g]Triclocarban; [h]wet-weight; [i]dry-weight; [j]lipid-weight.

collected from general populations of different countries, including Vietnam. Among the countries surveyed, PAEs in human urine samples from Vietnam were low to moderate. PAE levels in Vietnam were comparable to those in Japan and Korea, higher than Malaysia, but lower than Kuwait, China, and India (Guo et al., 2011). On the other hand, parabens, triclosan, and triclocarban concentrations in urine samples from Vietnam were the lowest among the countries surveyed (Honda et al., 2018; Iyer et al., 2018), suggesting less exposure degree for Vietnamese general population. This contamination feature was also seen for the settled dust and indoor air samples, showing a greater extent for PAEs and relatively lower levels of parabens and siloxanes, as discussed in earlier sections.

12.3 Human Health Risk Assessment

12.3.1 Inhalation

Air concentrations of phthalates and siloxanes were used to calculate daily intake doses (ID) through inhalation of CECs for different exposure groups, including adults, children, and specific subjects living and working in various micro-environments (such as homes, hair salons, and laboratories). The intake doses of selected compounds such as DEP, DnBP, and DEHP were compared with respective reference doses (RfD) proposed by the US Environmental Protection Agency (US EPA, 1987a, 1987b, 1987c) to estimate hazard quotients (HQ). The equations for intake dose estimation and exposure parameters were generally adopted from the US EPA (US EPA, 2011). The intake doses through air inhalation estimated for various population groups in different micro-environments are summarized in Table 12.4.

The intake dose values estimated for normal adults (e.g., bank and university officers) and university subjects (e.g., laboratory staff, researchers, and students) were comparable due to similar contamination levels of phthalates and siloxanes detected in the air of offices and laboratories. The ID values derived for children were about two to four times higher than those of adults, mainly due to their lower body burdens than adults. The ID values of individual compounds ranged from few ng to several hundreds of ng per kg per day, which were much lower than reference dose of some phthalates such as DEP (800,000 ng/kg/day; US EPA, 1987a), DnBP (100,000 ng/kg/day; US EPA, 1987b), and DEHP (20,000 ng/kg/day; US EPA, 1987c). As a result, these compounds' non-cancer risks were still low (HQ values from 10^{-6} to 10^{-2}). The CR values of DEHP were in the range of 10^{-10} to 10^{-8} and were also smaller than acceptable values of 10^{-6} to 10^{-4}, indicating insignificant extents of cancer risk (Anh et al., 2021).

12.3.2 Dust Ingestion

Intake doses for various CECs through dust ingestion were estimated based on the measured concentrations in the settled dust samples (including road and indoor dust) and dust ingestion rate for adults and children recommended by the US EPA (Tran et al., 2016; Hoang et al., 2002a, 2002b). A general screening survey was conducted in 2014 for indoor dust from four cities in northern Vietnam with different micro-environments, including homes, supermarkets/shops (electric and pharmaceutical shops), laboratories, and offices. The median intakes of PAEs (children: 50.2; adult: 19.4 ng/kg b.w./day) were about two orders of magnitude higher than those of BADGEs and parabens (Tran et al., 2016), indicating the predominant intake dose of PAEs and high usage extent in indoor products compared to other chemicals. A further investigation of house dust collected during 2018–2020 indicated comparable residue concentrations in 2014, suggesting no significant changes in PAE concentrations and human exposure during this period.

Compared to typical houses, human exposure to PAEs through dust ingestion was further investigated in informal waste recycling areas such as vehicle repair shops, waste recycling workshops, and ELV processing workshops (Hoang et al., 2022a). The intake doses estimated for informal occupationally exposed persons were about 3–5 times higher than ordinary residents. However, these intake doses were still several orders of magnitude smaller than reference values proposed by the US Environmental Protection Agency (US EPA, 1987a, 1987b, 1989) and US Consumer Product Safety Commission (US CPSC, 2011), and Korenkamp and Koch (2020) for male developmental toxicity. Based on median concentrations of

TABLE 12.4

Comparison of Intake Dose of Phthalates, Siloxanes, Parabens, Triclosan, and Triclocarban through Air Inhalation, Dust Ingestion, and Food Consumption in Vietnam

		PAEs		VMSs		PBs		TCS		TCC		
Location	**Micro-environment**	**Children**	**Adults**	**Children**	**Adults**	**Children**	**Adults**	**Children**	**Adults**	**Children**	**Adults**	**References**
Air Inhalation (ng/kg body wt./day)												
4 cities of the North, 2016–2017	Homes, cars, kindergartens, laboratories, offices, salons	416	213									Tran et al. (2017b)
4 cities of the North, 2016–2017	Homes, cars, kindergartens, laboratories, offices, hair salons			188	95.9							Tran et al. (2017c)
Hanoi, 2017	Hair salons				Men: 79.5; Women: 103							Tran et al. (2018)
Hanoi, 2019–2020	Laboratories, offices, homes	**EDI** DMP: NA–114 DEP: 6.55–588 DPrP: NA–134 DBP: 10.0–342 DiBP: 7.79–472 DnHP: NA–23.1 BBzP: NA–126 DCHP: NA–176	DMP: NA–56.2 DEP: 2.60–208 DPrP: NA–42.9 DBP: 2.60–111 DiBP: 8.57–174 DnHP: NA–7.46 BBzP: NA–38.4 DCHP: NA–49.0	D4: 14.5–453 D5: 11.7–307 D6: 6.78–207	D4: 5.10–122 D5: 6.68–98.5 D6: 2.18–63.0							Anh et al. (2020)

(Continued)

TABLE 12.4 (*Continued*)

Comparison of Intake Dose of Phthalates, Siloxanes, Parabens, Triclosan, and Triclocarban through Air Inhalation, Dust Ingestion, and Food Consumption in Vietnam

Location	Micro-environment	PAEs		VMSs		PBs		TCS		TCC		References
		Children	Adults	Children	Adults	Children	Adults	Children	Adults	Children	Adults	
		DEHP: 14.5–453	DEHP: 0.93–117									
		DnOP: NA–52.6	DnOP: NA–17.2									
		HQ										
		DEP: 8.19E-06–7.35E-04	DEP: 3.25E-06–2.60E-04									
		DBP: 9.96E-05–3.42E-03	DBP: 2.60E-05–1.11E-03									
		DEHP: NA–1.72E-02	DEHP: 4.63E-05–5.84E-03									
		Cancer risk										
		DEHP: NA–4.38E-08	DEHP: 3.47E-10–4.38E-08									
Bac Ninh, 2021	PM0.1; PM0.5	5.0	2.6	1.8	0.9							Le et al. (2023)
Hanoi, 2022	Laboratories, homes, salons	**EDI**										Nguyen et al. (2023)
		DMP: 0.04–28.7	DMP: 0.03–9.75									
		DEP: 3.94–48.3	DEP: 0.96–23.1									
		DPrP: NA–171	DPrP: NA–42.6									
		DiBP: 4.58–369	DiBP: 1.87–102									
		DBP: 16.3–315	DBP: 5.51–86.6									
		DnHP: NA–28.4	DnHP: NA–8.99									

(*Continued*)

TABLE 12.4 (*Continued*)

Comparison of Intake Dose of Phthalates, Siloxanes, Parabens, Triclosan, and Triclocarban through Air Inhalation, Dust Ingestion, and Food Consumption in Vietnam

Location	Micro-environment	PAEs		VMSs		PBs		TCS		TCC		References
		Children	Adults	Children	Adults	Children	Adults	Children	Adults	Children	Adults	
		BBzP: NA–13.4	BBzP: NA–5.25									
		DCHP: NA–91.2	DCHP: NA–25.8									
		DEHP: 86.9–1140	DEHP: 42.9–357									
		DnOP: NA–403	DnOP: 5.02–144									
		HQ										
		DEP: 4.92E-06–6.04E-05	DEP: 1.19E-06–2.89E-05									
		DBP: 1.63E-04–3.15E-03	DBP: 5.51E-05–8.65E-04									
		DEHP: 4.34E-03–5.72E-02	DEHP: 2.15E-03–1.79E-02									
Indoor Dust Ingestion (ng/kg body wt./day)												
4 cities of the North, 2014	Homes, supermarkets, shops, laboratories, offices	50.2	19.4			0.293	0.113					Tran et al. (2016)
4 cities of the North, 2014	Homes, laboratories, offices			0.39	0.17							Tran et al. (2017a)
Hanoi, 2017	Hair salons				Men: 0.231; women: 0.298							Tran et al. (2018)
4 cities of the North, 2017	Homes							0.266	0.05	0.128	0.024	Tran et al. (2020)

(*Continued*)

TABLE 12.4 (*Continued*)

Comparison of Intake Dose of Phthalates, Siloxanes, Parabens, Triclosan, and Triclocarban through Air Inhalation, Dust Ingestion, and Food Consumption in Vietnam

Location	Micro-environment	PAEs		VMSs		PBs		TCS		TCC		References
		Children	Adults	Children	Adults	Children	Adults	Children	Adults	Children	Adults	
Hanoi, 2017–2019	Homes, laboratories, medical stores					0.685	0.136					Tran et al. (2021)
4 cities of the North, 2019–2020	Work places, classrooms, homes			3.76 (0.287–19.6)	0.376 (0.029–1.96)							Hoang et al. (2022a)
5 cities of the North, 2018–2020	Work places, homes		DEHP: 0.29–44 DBP, DiBP, BBzP: 0.013–18									Hoang et al. (2022b)
Fish Consumption (mg/kg body wt./day)												
Vietnam, 2022	Marine fish					9.34E-04	2.34E-04					Tran-Lam et al. (2023)
Vietnam, 2022	Marine fish	**EDI**										Tran-Lam et al. (2024)
		DBP: 2.7E-08	3.40E-08									
		DEP: 2.1E-07	2.60E-07									
		DEHP: 1.9E-08	2.40E-08									
		HQ										
		DBP: 3.6E-09	4.50E-09									
		DEP: 6.7E-09	8.30E-09									
		DEHP: 9.6E-10	1.20E-09									

(*Continued*)

TABLE 12.4 (*Continued*)

Comparison of Intake Dose of Phthalates, Siloxanes, Parabens, Triclosan, and Triclocarban through Air Inhalation, Dust Ingestion, and Food Consumption in Vietnam

		PAEs		VMSs		PBs		TCS		TCC		
Location	**Micro-environment**	**Children**	**Adults**	**Children**	**Adults**	**Children**	**Adults**	**Children**	**Adults**	**Children**	**Adults**	**References**
Drinking Water Consumption (ng/kg body wt./day)												
Hanoi, 2018	Beverages		DEP: 0.15–100 DBP: 0.27–220.2 BBzP: 0.9–64.5 DEHP: 0.27–1399									Tran-Lam et al. (2018)
Hanoi, 2020	Bottled water	256	254									Le et al. (2021)
Hanoi, 2020	Bottled water			0.412	0.409							Nguyen et al. (2021)
Hanoi, 2021	Bottled water					0.44	0.22–0.27					Le et al. (2022)

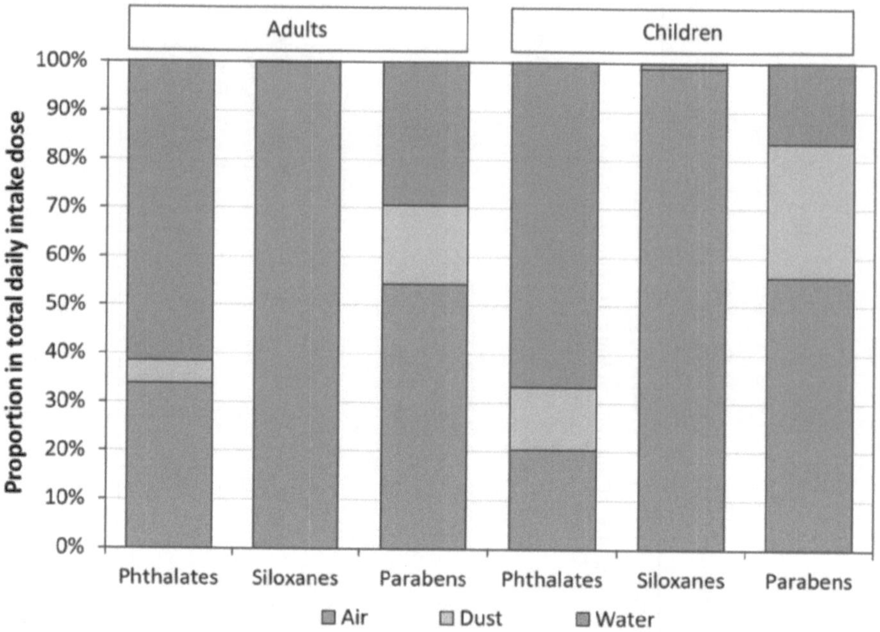

FIGURE 12.6 Contribution of different intake pathways of phthalates, siloxanes, and parabens from Vietnam.

DEHP, a predominant phthalate compound, dust ingestion accounted for about 6.1%–21% of total daily intakes, which was lower than water consumption and air inhalation (Le et al., 2021; Anh et al., 2021). In addition, contributions of dust ingestion pathway in total daily intakes of DnBP, DiBP, and BzBP were minor, accounting for 0.4%–2.3% of the total PAEs.

Cyclic and linear siloxanes in indoor dust samples were first investigated during 2010–2014 in a comprehensive survey of 12 countries, including Vietnam (Tran et al., 2017a). The Median intake doses for Vietnamese people were 0.17 and 0.39 ng/kg b.w./day for adults and children, respectively. These levels were similar to those estimated for India, Pakistan, Colombia, and Saudi Arabia but in the lower range compared to industrialized countries such as the USA, Korea, Japan, and China (Tran et al., 2017a). A more comprehensive survey in workplaces of informal waste recycling areas in some northern provinces during 2019–2020 revealed that daily intake doses of total cyclic siloxanes estimated in children (median: 2.51; range: 0.287–19.6 ng/kg bw/day) were about one order of magnitude higher than adults (median: 0.252; range: 0.029_1.96 ng/kg bw/day) (Hoang et al., 2022b).

The exposure dose pattern was in the order of D5 > D6 > D4 > D3. The exposure doses of D4, D5, and D6 through air inhalation and water consumption were estimated in our previous studies (Anh et al., 2021; Nguyen et al., 2021) and were included for comparison (Figure 12.6). For both adults and children, air inhalation was the most critical pathway accounting for more than 99% of the total exposure doses, while intakes through dust ingestion and water consumption were comparable. The contamination levels of cyclic siloxanes in Vietnamese dust, air, and water samples were generally lower than in other countries. The respective exposure doses estimated for Vietnamese people were also smaller than those reported for people from other parts of the world (Tran et al., 2015; Anh et al., 2021; Nguyen et al., 2021).

12.4 Conclusions

From the updated overview of the occurrence and distribution of CECs in Vietnam since 2016, the following conclusions can be made:

- PAEs remain the most predominant chemicals with the highest detection frequency and residue concentrations among CECs examined in informal waste recycling workplaces and indoor micro-environments. PAE concentrations in settled dust and PAS air samples from recycling areas were higher than in common indoor environments. Indoor dust and air from hair salons and homes of the metropolitan area generally contained higher residue levels of CECs such as PAEs, siloxanes, parabens, triclosan, and triclocarbans than other micro-environments from other provinces, reflecting their widespread use in personal care and household products in densely populated areas. An interesting result was observed on the elevated levels of PAEs and cyclic siloxanes in indoor air compared to the outdoor air samples collected from the same locations, highlighting the role of the indoor environments as a potential source. In addition, PAE levels found in indoor air were comparable to those reported for industrialized countries, indicating a higher risk for indoor human exposure.
- In the aquatic environments, widespread PAEs and cyclic siloxanes were observed in water and sediment from the lake and river systems in the Hanoi metropolitan area with little geographical variations. Similar to settled dust and air, contamination patterns in water and sediment showed high levels of PAEs compared to cyclic siloxanes and parabens. PAE contamination in urban sediment from Vietnam was within the world's higher range, suggesting that rapid urbanization and industrialization have increased the PAE pollution levels in sediments.
- Non-dietary intakes of CECs through various consumption pathways indicate the predominance of PAE and siloxane doses through air inhalation and water consumption. Parabens exhibited intake doses between PAEs and siloxanes with relatively similar contributions of water consumption and air inhalation while having higher dust intake than the other two chemical groups. The exposure doses to CECs through dust ingestion and air inhalation decreased with increasing age, indicating younger groups such as infants, toddlers, and children are at higher risk and deserve further investigations.
- Considering the above findings, the major direction strategies should be continuously targeted to the point sources of pollution, such as informal waste recycling sites and the indoor environments that are home for personal care and consumer products, such as resident houses and cosmetic salons. Since children are sensitive to poisonous chemical exposure, micro-environments that accommodate children, such as kindergartens, schools, and playing grounds, deserve comprehensive studies.

REFERENCES

Anh, H. Q., K. Tomioka, N. M. Tue, G. Suzuki, T. B. Minh, P. H. Viet, and S. Takahashi. 2019. "Comprehensive analysis of 942 organic micro-pollutants in settled dust from northern Vietnam: Pollution status and implications for human exposure". *Journal of Material Cycles and Waste Management* 21: 57–66. doi: 10.1007/s10163-018-0745-2

Anh, H. Q., H. M. N. Nguyen, T. Q. Do, K. Q. Tran, T. B. Minh, and T. M. Tran. 2021. "Air pollution caused by phthalates and cyclic siloxanes in Hanoi, Vietnam: Levels, distribution characteristic, and implication for inhalation exposure". *Science of the Total Environment* 760: 143380. doi: 10.1016/j.scitotenv.2020.143380

Chen, Y., D. Lv, X. Li, and T. Zhu. 2018. "$PM_{2.5}$-bound phthalates in indoor and outdoor air in Beijing: Seasonal distributions and human exposure via inhalation". *Environmental Pollution* 241: 369–377. doi: 10.1016/j.envpol.2018.05.081

Dung, T. A., H. T. Duong, and B. Q. Lap. 2019. "Current status of phthalate contamination of the air dust at some locations in Hanoi and preliminary assessment of the DEHP exposure to human health". *Journal of Water Resources Science and Technology* 64(3): 54–59.

Gao, D., Z. Li, Z. Wen, and N. Ren. 2014. "Occurrence and fate of phthalate esters in full-scale domestic wastewater treatment plants and their impact on receiving waters along the Songhua River in China". *Chemosphere* 95: 24–32. doi: 10.1016/j.chemosphere.2013.08.009

Gao, X., J. Li, X. Wang, J. Zhou, B. Fan, W. Li, and Z. Liu. 2019. "Exposure and ecological risk of phthalate esters in the Taihu Lake basin, China". *Ecotoxicology and Environmental Safety* 171: 564–570. doi: 10.1016/j.ecoenv.2019.01.001

Guo, Y., H Amirah, H. C. Cho, T. B. Minh, M. A. Mohd, H. Nakata, and K. Kannan. 2011. "Occurrence of phthalate metabolites in human urine from several Asian countries". *Environmental Science and Technology* 45: 3138–3144. doi: 10.1021/es103879m

Guo, Y., L. Wang, and K. Kannan. 2014. "Phthalates and parabens in personal care products from China: Concentrations and human exposure". *Archives of Environmental Contamination and Toxicology* 66: 113–119. doi: 10.1007/s00244-013-9937-x

Hoang, A. Q., T. M. Le, H. M. N. Nguyen, H. Q. Le, N. D. Vu, N. C. Chu, G. H. M. Dang, T. B. Minh, S. Takahashi, and T. M. Tran. 2022a. "Phthalic acid esters (PAEs) in workplace and house from Vietnam: Concentrations, profiles, emission sources, and exposure risk". *Environmental Science and Pollution Research* 29: 14046–14057. doi: 10.1007/s11356-021-16851-6

Hoang, A. Q., H. T. Trinh, H. M. N. Nguyen, T. Q. Nguyen, T. X. Nguyen, T. V. Duc, T. T. Nguyen, T. Q. Do, T. B. Minh, and T. M. Tran. 2022b. "Assessment of cyclic volatile methyl siloxanes (CVMSs) in indoor dust from different micro-environments in northern and central Vietnam". *Environmental Geochemistry and Health* 45: 1711–1722. doi: 10.1007/s10653-022-01298-6

Hoang, A. Q., S. Takahashi, L. H. Tuyen, N. M. Tue, N. M. Tu, T. T. T. Nguyen, and M. B. Tu. 2023a. "Polycyclic aromatic hydrocarbons in air dust samples from Vietnamese end-of-life vehicle processing workshops: Contamination status, sources, and exposure risks". *Bulletin of Environmental Contamination and Toxicology* 110: 110. doi: 10.1007/s00128-023-03757-x

Hoang, A. Q., S. Takahashi, N. M. Tue, L. H. Tuyen, T. M. Tran, N. T. H. Yen, and M. B. Tu. 2023b. "Occurrence, emission sources, and risk assessment of polybrominated diphenyl ethers and current-use brominated flame retardants in settled dust from end-of-life vehicle processing, urban, and rural arears, northern Vietnam". *Environmental Science and Pollution Research* 30: 2061–2074.

Honda, M., M. Robinson, and K. Kannan. 2018. "Parabens in human urine from several Asian countries, Greece, and the United States". *Chemosphere* 201: 13–19. doi: 10.1016/j.chemosphere.2018.02.165

Iyer, A. P., J. Xue, M. Honda, M. Robinson, T. A. Kumosani, K. Abulnaja, and K. Kannan. 2018. "Urinary levels of triclosan and triclocarban in several Asian countries, Greece and the USA: Association with oxidative stress". *Environmental Research* 160: 91–96. doi: 10.1016/j.envres.2017.09.021

Kingsley, O., and B. Witthayawirasak. 2020. "Deterministic assessment of the risk of the phthalate esters in sediments of U-Tapao canal, Southern Thailand". *Toxics* 8(4): 93. doi: 10.3390/toxics8040093

Koniecki, D., R. Wang, R. P. Moody, and J. Zhu. 2011. "Phthalates in cosmetic and personal care products: Concentrations and possible dermal exposure". *Environmental Research* 111(3): 329–336. doi: 10.1016/j.envres.2011.01.013

Korenkamp, A., and H. M. Koch. 2020. "Refined reference doses and new procedures for phthalate mixture risk assessment focused on male developmental toxicity". *International Journal of Hygiene and Environmental Health* 224: 113428. doi: 10.1016/j.ijheh.2019.113428

Kubwabo, C., P. E. Rasmussen, X. Fan, I. Kosarac, F. Wu, A. Zidek, and S. L. Kuchta. 2013. "Analysis of selected phthalates in Canadian indoor dust collected using household vacuum and standardized sampling techniques". *Indoor Air* 23(6): 506–514. doi: 10.1111/ina.12048

Le, T. M., H. M. N. Nguyen, V. K. Nguyen, A. V. Nguyen, N. D. Vu, N. T. H. Yen, A. Q. Hoang, T. B. Minh, K. Kannan, and T. M. Tran. 2021. "Profile of phthalic acid esters (PAEs) in bottled water, tap water, lake water, and wastewater samples collected from Hanoi, Vietnam". *Science of the Total Environment* 788, 147831. doi: 10.1016/j.scitotenv.2021.147831

Le, T. M., C. L. T. Pham, H. M. N. Nguyen, T. T. Duong, T. P. Q. Le, D. T. Nguyen, N. D. Vu, T. B. Minh, and T. M. Tran. 2022a. "Distribution and ecological risk assessment of phthalic acid esters in surface sediments of three rivers in Northern Vietnam". *Environmental Research* 209, 112843. doi: 10.1016/j.envres.2022.112843

Le, T. M., P. T. Pham, T. Q. Nguyen, T. Q. Nguyen, M. Q. Bui, H. Q. Nguyen, N. D. Vu, K. Kannan, and T. M. Tran. 2022b. "A survey of parabens in aquatic environments in Hanoi, Vietnam and its implications for human exposure and ecological risk". *Environmental Science and Pollution Research* 29(31): 46767–46777. doi: 10.1007/s11356-022-19254-3

Le, T. M., H. L. Quang, A. H. Tran, M. B. Quang, N. D. Vu, H. N. Thi, H. V. Khanh, K. Kannan, and T. M. Tran. 2023. "Co-occurrence of phthalic acid esters (PAEs) and cyclic volatile methylsiloxanes (cVMSs) in fine particulate matter ($PM_{0.5}$ and $PM_{0.1}$) collected from an industrial area in Vietnam". *Environmental Research* 237(Part 2): 117018. doi: 10.1016/j.envres.2023.117018

Lee, S., H. B. Moon, G. J. Song, K. Ra, W. C. Lee, and K. Kannan. 2014. "A nationwide survey and emission estimates of cyclic and linear siloxanes through sludge from wastewater treatment plants in Korea". *Science of the Total Environment* 497–498: 106–112. doi: 10.1016/j.scitotenv.2014.07.083

Lee, Y. M., J. E. Lee, W. Choe, T. Kim, J. Y. Lee, Y. Kho, K. Choi, and K. D. Zoh. 2019. "Distribution of phthalate esters in air, water, sediments, and fish in the Asan Lake of Korea". *Environment International* 126: 635–643. doi: 10.1016/j.envint.2019.02.059

Li, B., R. Liu, H. Gao, R. Tan, P. Zeng, and Y. Song. 2016. "Spatial distribution and ecological risk assessment of phthalic acid esters and phenols in surface sediment from urban rivers in Northeast China". *Environment Pollution* 219: 409–415. doi: 10.1016/j.envpol.2016.05.022

Liao, C., F. Liu, H Alomirah, V. D. Loi, M. A. Mohd, H. B. Moon, H. Nakata, and K. Kannan. 2012. "Bispheno S in urine from the United States and seven Asian countries: Occurrence and human exposure". *Environmental Science and Technology* 46: 6860–6866. doi: 10.1021/es301334j

Ma, W. L., B. Subedi, and K. Kannan. 2014. "The occurrence of bisphenol A, phthalates, parabens and other environmental phenolic compounds in house dust: A review". *Current Organic Chemistry* 18*(17): 2182–2199. doi: 10.2174/13852728196661404230205

Metcalfe, C. D., S. Bayen, M. Desrosiers, G Muñoz, S Sauvé, V. Yargeau. 2022. "An introduction to the sources, fate, occurrence and effects of endocrine disrupting chemicals released into the environment". *Environmental Research* 207: 112658. doi: 10.1016/j.envres.2021.112658

Minh, N. H., D. H. Anh, T. M. Tran, H. Q. Anh, M. T. N. Pham, N. D. Vu, P. H. Viet, and T. B. Minh. 2016. "Persistent toxic substances in Vietnam: A review of environmental contamination and human exposure". *Persistent Organic Chemicals in the Environment: Status and Trends in the Pacific Basin Countries I Contamination Status Chapter* 3: 55–83. doi: 10.1021/bk-2016-1243.ch003

Nguyen, T. K., H. M. Nguyen, C. K. Truong, M. B. Ahmed, Y. Huang, and J. L. Zhou. 2019. "Chemical and microbiological risk assessment of urban river water quality in Vietnam". *Environmental Geochemistry and Health* 41: 2559–2575. doi: 10.1007/s10653-019-00302-w

Nguyen, H. M. N., H. T. Khieu, N. A. Ta, H. Q. Le, T. Q. Nguyen, T. Q. Do, A. Q. Hoang, K. Kannan, and T. M. Tran. 2021. "Distribution of cyclic volatile methylsiloxanes in drinking water, tap water, surface water, and wastewater in Hanoi, Vietnam". *Environmental Pollution* 285: 117260. doi: 10.1016/j.envpol.2021.117260

Nguyen, H. M. N., H. T. Khieu, H. Q. Le, T. T. Duong, T. Q. Do, T. B. Minh, and T. M. Tran. 2022. "Assessment of distributional characteristics and ecological risks of cyclic volatile methylsiloxanes in sediments from urban rivers in northern Vietnam". *Environmental Science and Pollution Research* 29(20): 29917–29926. doi: 10.1007/s11356-021-18487-y.

Nguyen, A. V., T. V. Vu, C. L. T. Pham, V. N. Nguyen, N. T. Ta, A. Q. Hoang, T. B. Minh, and T. M. Tran. 2023. "Widespread distribution of phthalic acid esters in indoor and ambient air samples collected from Hanoi, Vietnam". *Environmental Science and Pollution Research* 30: 63175–63184. doi: 10.1007/s11356-023-26558-5

Orecchio, S., D. Amorello, R. Indelicato, S. Barreca, and S. Orrechio. 2022. "A short review of simple analytical methods for the evaluation of PAHs and PAEs as indoor pollutants in house dust samples". *Atmosphere* 13(11): 1799. doi: 10.3390/atmos13111799

Ouyang, X., M. Xia, X. Shen, and Y. Zhan. 2019. "Pollution characteristics of 15 gas- and particle-phase phthalates in indoor and outdoor air in Hangzhou". *Journal of Environmental Sciences* 86: 107–119. doi: 10.1016/j.jes.2019.05.008

Pagoni, A., O. S. Arvaniti, and O. L. Klantzi. 2022. "Exposure to phthalates from personal care products: Urinary levels and predictors of exposure". *Environmental Research* 212(Part A): 113194. doi: 10.1016/j.envres.2022.113194

Pei, X. Q., M. Song, M. Guo, F. F. Mo, and X. Y. Shen. 2013. "Concentration and risk assessment of PAE present in indoor air from newly decorated apartments". *Atmospheric Environment* 68: 17–23. doi:10.1016/j.atmosenv.2012.11.039

Pham, P. T., T. C. Quan, Q. T. Le, M. Q. Bui, A. H. Tran, A. T. T. Phung, A. Q. Hoang, T. B. Minh, T. T. T. Lam, H. N. Tran, and T. M. Tran. 2024. "Quantification of parabens in marine fish samples by a rapid, simple, effective sample preparation method". *Environmental Science and Pollution Research* 31(11): 16571–16582. doi: 10.1007/s11356-024-32321-1

Quynh, T. X., and T. V. Duc. 2019. "Endocrine disrupting compounds (EDCs) in surface waters of the KimNguu river, Vietnam". *Bulletin of Environmental Contamination and Toxicology* 103(3). doi: 10.1007/s00128-019-02710-1

Silva, J., F. Bernardo, M. Jesus, T. Faria, A. Alves, N. Ratola, and V. Homem. 2021. "Levels of volatile methylsiloxanes in urban wastewater sludges at various steps of treatment". *Environmental Chemistry Letters* 19: 2723–2732. doi: 10.1007/s10311-021-01191-1

Srivastava, A., V. P. Sharma, R. Tripathi, R. Kumar, D. K. Patel, and P. K. Mathur. 2010. "Occurrence of phthalic acid esters in Gomti River sediment, India". *Environmental Monitoring and Assessment* 169: 397–406. doi: 10.1007/s10661-009-1182-4

Sun, J., J. Huang, A. Zhang, W. Liu, and W. Cheng. 2013. "Occurrence of phthalate esters in sediments in Qiantang River, China and inference with urbanization and river flow regime". *Journal of Hazardous Materials* 248–249: 142–149. doi: 10.1016/j.jhazmat.2012.12.057

Szewczyńska, M., E. Dobrzyńska, and M. Pośniak. 2021. "Determination of phthalates in particulate matter and gaseous phase emitted in indoor air of offices". *Environmental Science and Pollution Research* 28: 59319–59327. doi: 10.1007/s11356-020-10195-3

Tham, T. T., H. Q. Anh, L. T. Trinh, V. M. Lan, N. X. Truong, N. T. H. Yen, N. L. Anh, T. M. Tri, and T. B. Minh. 2019. "Distributions and seasonal variations of organochlorine pesticides, polychlorinated biphenyls, and polybrominated diphenyl ethers in surface sediment from coastal areas of central Vietnam". *Marine Pollution Bulletin* 144: 28–25. doi: 10.1016/j.marpolbul.2019.05.009

Thuy, L. M., N. T. N. Anh, L. T. Trinh, L. Q. Huong, P. T. P. Thao, N. Q. Trung, and T. M. Tri. 2018. "Determination and distribution of phthalate diesters in beverage collected from Hanoi, Vietnam". *VNU Journal of Sciences and Technology* 34(4): 16–20. doi: 10.25073/2588-1140/vnunst.4822

Thuy, L. M., D. T. Lieu, V. D. Nam, N. T. Ngoc, D. H. Anh, and T. M. Tri. 2022. "Determination of Bisphenol in Indoor dust samples by liquid chromatography in combination with quadrupole mass spectrometry". *VNU Journal of Science: Natural Sciences and Technology* 38(3): 113–122.

Tran, B. C., T. T. D. Tran, B. N. Nguyen, and X. V. Le. 2019. "BPA accumulation in the sediment of the area receiving the processed effluents of Phuoc Hiep landfill, Ho Chi Minh city". *Vietnam Journal of Science and Technology* 61(6): 29–33.

Tran, T. M., and K. Kannan. 2015. "Occurrence of phthalates diesters in particulate and vapor phases in indoor air and implications for human exposure in Albany, New York, USA". *Archives of Environmental Contamination and Toxicology* 68: 489–499. doi: 10.1007/s00244-015-0140-0

Tran, T. M., T. B. Minh, T. A. Kumosani, and K. Kannan. 2016. "Occurrence of phthalate diesters (phthalates), p-hydroxybenzoic acid esters (parabens), bisphenol A diglycidyl ether (BADGE) and their derivatives in indoor dust from Vietnam: Implications for exposure". *Chemosphere* 144: 1553–1559. doi: 10.1016/j.chemosphere.2015.10.028

Tran, T. M., K. O. Abualnaja, A. G. Asimakopoulos, A. Covaci, B. Gevao, B. Johnson-Restrepo, T. A. Kumosani, G. Malarvannan, T. B. Minh, H. B. Moon, H. Nakata, R. K. Sinha, and K. Kannan. 2017a. "A survey of cyclic and linear siloxanes in indoor dust and their implications for human exposures in twelve countries". *Environment International* 78: 39–44. doi: 10.1016/j.envint.2015.02.011

Tran, T. M., H. T. Le, T. B. Minh, and K. Kannan. 2017b. "Occurrence of phthalate diesters in indoor air from several Northern cities in Vietnam, and its implication for human exposure". *Science of the Total Environment* 601–602: 1695–1701. doi: 10.1016/j.scitotenv.2017.06.016

Tran, T. M., H. T. Le, N. D. Vu, G. H. M. Dang, T. B. Minh, and K. Kannan. 2017c. "Cyclic and linear siloxanes in indoor air from several northern cities in Vietnam: Levels, spatial distribution and human exposure". *Chemosphere* 184: 1117–1124. doi: 10.1016/j.chemosphere.2017.06.092

Tran, T. M., M. B. Tu, and N. D. Vu. 2018. "Cyclic siloxanes in indoor environments from hair salons in Hanoi, Vietnam: Emission sources, spatial distribution, and implications for human exposure". *Chemosphere* 212: 330–336. doi: 10.1016/j.chemosphere.2018.08.101

Tran, T. M., H. T. Trinh, H. Q. Anh, T. V. Le, S. N. Le, and T. B. Minh. 2020. "Characterization of triclosan and triclocarban in indoor dust from home micro-environments in Vietnam and relevance of non-dietary exposure". *Science of the Total Environment* 732: 139326. doi: 10.1016/j.scitotenv.2020.139326

Tran, T. M., T. T. T. Lam, H. H. T. Mai, L. H. T. Bach, H. M. N. Nguyen, H. T. Trinh, L. T. Dang, T. B. Minh, T. C. Quan, and A. Q. Hoang. 2021. "Parabens in personal care products and indoor dust from Hanoi, Vietnam: Temporal trends, emission sources, and non-dietary exposure through dust ingestion". *Science of the Total Environment* 761: 143274. doi: 10.1016/j.scitotenv.2020.143274

Tran-Lam, T. T., Y. H. Dao, D. T. Nguyen, H. K. Ma, T. Q. Pham, and G. T. Le. 2018. "Optimization of sample preparation for detection of 10 phthalates in non-alcoholic beverages in Northern Vietnam". *Toxics* 6: 69. doi: 10.3390/toxics6040069

Tran-Lam, T. T., A. T. T. Phung, P. T. Pham, M. Q. Bui, Y. H. Dao, and G. T. Le. 2023. "Occurrence, biomagnification, and risk assessment of parabens and their metabolites in marine fish: The case study of Vietnam". *Chemosphere* 344: 140221. doi: 10.1016/j.chemosphere.2023.140221

Tran-Lam, T. T., T. C. Quan, M. Q. Bui, Y. H. Dao, and G. T. Le. 2024. "Endocrine-disrupting chemicals in Vietnamese marine fish: Occurrence, distribution, and risk assessment". *Science of the Total Environment* 908: 168305. doi: 10.1016/j.scitotenv.2023.168305

Tri, T. M., D. H. Anh, P. M. Hoai, N. H. Minh, V. D. Nam, P. H. Viet, and T. B. Minh. 2016. "Emerging endocrine disrupting chemicals and pharmaceuticals in Vietnam: A review of environmental occurrence and fate in aquatic and indoor environments". *Persistent organic Chemicals in the environment: Status and trends in the pacific basin countries II temporal trends Chapter* 10: 223–253. doi: 10.1021/bk-2016-1244.ch010

US Consumer Product Safety Commission (CPSC). 2011. *Toxicity review for dicyclohexyl phthalate; CASRN 84–617.* https://www.cpsc.gov/s3fs-public/ToxicityReviewOfDCHP.pdf.

US EPA. 1987a. *Diethyl phthalate; CASRN 84-66-2.* https://cfpub.epa.gov/ncea/iris/iris_documents/documents/subst/0226_summary.pdf#nameddest=rfd.

US EPA. 1987b. *Dibutyl phthalate; CASRN 84-74-2.* https://cfpub.epa.gov/ncea/iris/iris_documents/documents/subst/0038_summary.pdf#nameddest=rfd.

US EPA. 1987c. *Di(2-ethylhexyl) phthalate; CASRN 117–81–7.* https://cfpub.epa.gov/ncea/iris/iris_documents/documents/subst/0014_summary.pdf#nameddest=rfd.

US EPA. 1989. *Butyl benzyl phthalate; CASRN 85-68-7.* https://cfpub.epa.gov/ncea/iris/iris_documents/documents/subst/0293_summary.pdf#nameddest=rfd

US EPA. 2011. *Exposure factors handbook 2011 edition (final report).* https://cfpub.epa.gov/ncea/risk/recordisplay.cfm?deid=236252

Wang, J., L. Bo, L. Li, D. Wang, G. Chen, P. Christie, and Y. Teng. 2014. "Occurrence of phthalate esters in river sediments in areas with different land use patterns". *Science of The Total Environment* 500–501: 113–119. doi: 10.1016/j.scitotenv.2014.08.092

Wang, W., K. O. Abualnaja, A. G. Asimakopoulos, A. Covaci, B. Gevao, B. Johnson-Restrepo, T. A. Kumosani, G. Malarvannan, T. B. Minh, H. B. Moon, H. Nakata, R. K. Sinha, and K. Kannan. 2015. "A comparative assessment of human exposure to tetrabromobisphenol A and eight bisphenols including bisphenol A via indoor dust ingestion in twelve countries". *Environment International* 83: 183–191. doi: 10.1016/j.envint.2015.06.015

Wang, W., and K. Kannan. 2016. "Fate of parabens and their metabolites in two wastewater treatment plants in New York State, United States". *Environmental Science and Technology* 50(3): 1174–1181. Doi: 10.1021/acs.est.5b05516

Zhang, Z., H. Qi, N. Ren, Y. Li, D. Gao, and K. Kannan. 2011a. "Survey of cyclic and linear siloxanes in sediment from the Songhua River and in sewage sludge from wastewater treatment plants, Northeastern China". *Archives of Environmental Contamination and Toxicology* 60: 204–211. doi: 10.1007/s00244-010-9619-x

Zhang, Z., H. Almirah, H. S. Cho, Y. F. Li, C. Liao, T. B. Minh, M. A. Mohd, H. Nakata, N. Ren, and K. Kannan. 2011b. "Urinary bisphenol A concentrations and their implications for human exposure in several Asian countries". *Environmental Science and Technology* 45: 7044–7050. doi: 10.1021/es200976k

Zhou, X., J. Lian, Y. Cheng, and X. Wang. 2021. "The gas/particle partitioning behavior of phthalate esters in indoor environment: Effects of temperature and humidity". *Environmental Research* 194: 110681. doi: 10.1016/j.envres.2020.110681

Zhu, L., P. Hajeb, P. Fauser, and K. Vorkamp. 2023. "Endocrine disrupting chemicals in indoor dust: A review of temporal and spatial trends, and human exposure". *Science of The Total Environment* 874: 162374. doi: 10.1016/j.scitotenv.2023.162374

Section IV

Biogeochemical Cycles and Health Risks

13

Global Biogeochemical Cycle of Mercury and Critical Factors in Determining Mercury Levels in Fish

Martin Tsz-Ki Tsui

13.1 Introduction

Since the industrial revolution, many metallic elements' environmental levels have been significantly elevated in the surface environments of our planet, leading to widespread contamination at the global scale (Nriagu, 1996). An exception was lead pollution in the Balkans, and high lead pollution appeared as early as the Early Bronze Age (Longman et al., 2018). Mercury (Hg) is widely regarded as an iconic legacy pollutant among these metallic elements. There were three phases of historical emissions since the Industrial Revolution, namely, Phase I—"Gold/Silver Rush" in the North America from 1850 through 1915, Phase II—"War and Depression" from 1915 through 1950, and Phase III—"Coal Surge and Artisanal Gold Mining" since the 1950s until now (Streets et al., 2011). The all-time Hg emission from anthropogenic activities has been estimated to be ~350 Gg, with 39% released before 1850 and 61% after 1850 (Streets et al., 2011).

In recent years, coal combustion and artisanal and small-scale gold mining (ASGM) have become the dominant sources of Hg emissions at the global scale, leading to a continuous increase of global Hg emissions despite efforts in reducing coal-related emissions of Hg by the developed world. Based on the data from "Global Hg Assessment 2018" by the UN Environment Programme (UNEP, 2018), ASGM represented the top Hg emission source at ~837 Mg, coal combustion was ranked the second at ~473 Mg, and non-ferrous metals production was ranked the third at ~327 Mg, with all other sources combined the total anthropogenic Hg emissions in 2018 would be totaled at ~2220 Mg or ~2.2 Gg. It should be noted that this estimate did not consider "natural" Hg emission sources such as volcanoes, Hg re-emissions from various earth surfaces, and wildfires (see Section 13.2).

Despite the temporal increases of Hg emissions at the global scale, there are large variations of Hg emissions among different continents and countries, which are mainly determined by their energy demand (e.g., coal burning), industrial activities and types (e.g., cement production and waste products), and various anthropogenic activities leading to release of Hg (e.g., ASGM and biomass/straw burning). For example, China (~572 Mg) and India (~205 Mg) are the top two Hg-emitting countries due to their large population sizes, expanding industries, and high energy demand (coal burning). The third top Hg emitter belongs to Indonesia (~156 Mg), but most of their Hg emission is due to ASGM activities. A country with much smaller population sizes (e.g., Peru with ~32 million people) would come as the fourth top emitter of Hg (~122 Mg) due mainly to their extremely high ASGM activities. As a result, Asia (east, southeast, and south) represents the most significant Hg emissions (totaling at ~1.08 Gg, or about 50% of total annual global Hg emissions) (UNEP, 2018). In contrast, the former largest emitter of Hg, North America, mainly before World War II, would now only account for slightly less than 2% of global Hg emissions.

With the expected rapid economic developments of many developing worlds collectively referred to as the "Global South", it is anticipated that Hg emissions will continue to rise in the next decade but then may decline in the longer term until 2050, which will also largely depend on the success in combating the illegal ASGM activities across many countries (e.g., government's ban in Ghana, a large Hg emitter at ~47 Mg/year; Osei et al., 2021) as well as other industrial sectors currently with a large uncertainty (Brocza et al., 2024). The Minamata Convention on Mercury is a legally binding treaty with 128

DOI: 10.1201/9781003492115-17

signatories, including large-emitting countries such as China, India, and the United States. Thus, there is a global consensus on reducing anthropogenic Hg emissions.

This chapter aims to provide an overview of environmental sources of Hg, global biogeochemical cycles of Hg especially on the various factors influencing Hg levels in fish, and the implementation of Minamata Convention and the monitoring of its outcome. This chapter is not meant to cover all aspects in this topic but to highlight the most critical factors controlling fish Hg levels and the future outlook of any expected Hg reduction in fish through Minamata Convention.

13.2 Overview of Environmental Cycles of Mercury

Different from organic compounds, which can eventually be degraded and mineralized into inorganic carbon, Hg is a heavy metal that never degrades. In earth's surfaces, Hg can undergo complex environmental transport and transformations, and all these cycling processes can vary widely over space and time, thus making it difficult to generalize global cycling and to understand all complex processes. Most importantly, Hg is considered a global pollutant, contaminating all ecosystem types irrespective of the proximity or remoteness of the point sources (Driscoll et al., 2013).

The global distribution of Hg is mainly attributed to its relatively low boiling point (380°C) and high vapor pressure, leading to a strong gaseous phase of Hg on earth's surface environment. In the atmosphere, the dominant form of Hg is the gaseous elemental Hg [often abbreviated as Hg(0) or GEM], and Hg(0) is relatively inert or unreactive. Therefore, gaseous Hg(0) can travel far distances in the atmosphere (Durnford et al., 2010) (Figure 13.1). The estimated atmospheric residence time of Hg(0) can be up to 1 year, enabling the global distribution of Hg regardless of the distances from point sources.

Occasionally, atmospheric Hg(0) can be oxidized to become Hg(II), and Hg(II) is much more reactive and can dissolve in rain or snow. As such, Hg(II) can be deposited onto the earth's surfaces through wet deposition. However, Hg(0) can also be directly deposited onto the earth's surfaces (Figure 13.1). For example, the stomatal uptake of Hg(0) represents the dominant deposition pathway of Hg onto terrestrial landscapes, termed dry deposition (Obrist et al., 2017). Also, Hg(0) can be deposited onto the oceanic surfaces through direct dissolution into seawater and/or various oxidation processes (e.g., Hg(0) to Hg(II)) between the atmosphere and seawater boundary (Marumoto and Imao, 2015). However, both Hg(0) and Hg(II) are deemed relatively non-toxic due to low uptake capability by animals. Therefore,

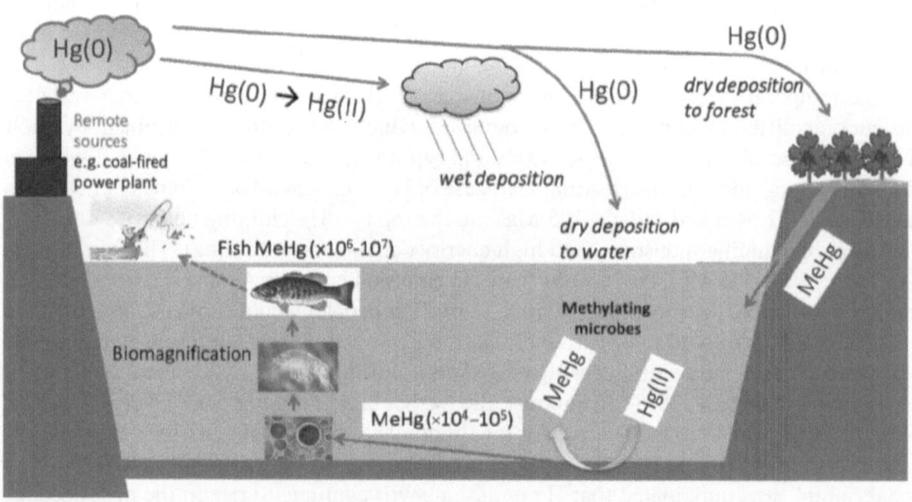

FIGURE 13.1 Diagram illustrates the key biogeochemical processes governing Hg transformations in the atmosphere, forested land, and marine environments. *Note*: Hg(0), gaseous elemental mercury; Hg(II), oxidized inorganic mercury; MeHg, methylmercury. This information was compiled from published studies (Barkay and Gu, 2022; Fitzgerald et al., 2007; Pickhardt and Fisher, 2007; Obrist et al., 2017; Tsui et al., 2010).

their health risks are relatively low compared to their organic counterparts, such as neurotoxic methylmercury (CH_3Hg or MeHg) (Bernhoft, 2012).

In nature, almost all MeHg is formed through microbial methylation processes (Gilmour et al., 2013). We now know that certain anaerobic microbes such as sulfate-reducing bacteria, iron-reducing bacteria, and methanogens possessing the two-gene cluster *hgcAB* can methylate Hg(II) into MeHg (Parks et al., 2013), which occur primarily in hypoxic and anoxic environments including surface sediments in aquatic ecosystems (Figure 13.1). However, microbial MeHg production can also occur in seemingly oxygenated environments including pristine river (Tsui et al., 2010) and the subsurface layers in the open ocean (Blum et al., 2013). Overall, both microbial and geochemical factors control the microbial methylation of Hg(II). The latter would be related to the bioavailability of Hg(II) for the methylating microbes (Regnell and Watras, 2019).

Once produced, the environmental presence of MeHg can also be regulated by at least three pathways (Barkay and Gu, 2022). The first pathway would only apply to environmental media receiving sufficient sunlight, such as surface water and atmosphere (Yu et al., 2024). Interestingly, MeHg photodegradation is much faster in freshwater than in seawater, in which the higher levels of dissolved organic ligands (especially with the presence of sulfur) in freshwater environments have been proposed to enhance the photodegradation rate of MeHg. In contrast, the high chloride levels in seawater would be complex with MeHg and decrease its photodegradation rate (Zhang and Hsu-Kim, 2010).

In contrast to the sunlight-mediated degradation pathway, the second pathway occurs more often in "dark" environments, such as sediments and deep oceans where light is absent. While not all mechanisms are known, at least those associated with the *merB* gene are known to break down MeHg, returning it to Hg(II), while microbes possessing the *merA* gene can convert Hg(II) back to gaseous Hg(0) (Barkay and Gu, 2022). The third pathway is even less understood and is related to *mer*-independent oxidative degradation of MeHg that may involve some strains of anaerobic microbes and aerobic methane-oxidizing bacteria (i.e., methanotrophs) (Lu et al., 2017). Thus, the degradation of MeHg is significant to keep removing this toxic compound from our ecosystems. If not, MeHg is continuously produced by the methylating microbes, then all Hg(0) and Hg(II) would become methylated and very toxic at specific points.

13.3 Methylmercury Entering the Food Webs and Humans

Once produced, MeHg in sediment porewater can be effluxed into the overlying water, MeHg can be produced in situ in the water column, or MeHg can be derived from the terrestrial runoff (Figure 13.1). Before any degradation occurs, dissolved MeHg can be greatly bioconcentrated to the base of aquatic food webs (e.g., phytoplankton) by 10^4–10^5 times (Pickhardt and Fisher, 2007), and then MeHg can be transferred up the food chain by a factor of 2–10 per trophic level, reaching to the highest level at the apex predator (Seco et al., 2021) including sharks (Goyanna et al., 2023), common loons (Schoch et al., 2020), and polar bears (Lippold et al., 2020) targeting aquatic preys. Typically, when comparing MeHg content in the apex predator to those of surrounding water, the ratio of MeHg in the apex predator to water would range from 10^6 to 10^7 times, illustrating the dramatic uptake capability of MeHg into the top animal predators (Wu and Zhang, 2023) (Figure 13.1). Recent studies also demonstrated that terrestrial forest floor food webs can bioaccumulate and biomagnify MeHg to a similar extent as their aquatic counterparts (Tsui et al., 2019), which can transfer MeHg to the top predators such as terrestrial songbirds (Cristol and Evers, 2020) and endangering their population survival.

Besides the threat to wildlife, MeHg bioaccumulation and biomagnification can pose a threat to human health through their dietary consumption through at least two routes (Figure 13.1) (Mergler et al., 2007). First, more than 3 billion people worldwide rely on seafood for essential nutrition, which can include many seafood products and carnivorous marine fish that are high in trophic levels and often tainted with elevated MeHg levels, such as King mackerel, shark, swordfish, and tuna (Zupo et al., 2019). Second, more recent studies from China showed that rice paddy could be a dietary source of MeHg, which can be mediated by anaerobic microbes, especially iron-reducing bacteria (Zhong et al., 2024). MeHg can extensively accumulate in the rice grain, which will be a health concern for about half of the global population (~4 billion) having rice as their staple foods. However, rice MeHg levels in general (except those grown in Hg-contaminated lands, e.g., abandoned Hg mines in Guizhou, China; Qiu et al., 2008) are not high compared to those in marine fish, and thus seafood and fish are still the major routes to which human mainly obtain neurotoxic MeHg (Li et al., 2024).

As a result, to protect the public, the elevated Hg levels in fish can trigger widespread fish consumption advisories or warnings against the consumption of fish due to high Hg levels, which have been issued in freshwater and coastal water bodies such as in the United States (Shimshack et al., 2007), in order to limit the intake of Hg by the populations. In the United States, the U.S. Environmental Protection Agency has derived the fish tissue concentration criterion of MeHg to protect human health through consuming fish and seafood products in "Guidance for Implementing the January 2001 Methylmercury Water Quality Criterion" (USEPA, 2001). The derived tissue concentration criterion for fish fillets is 0.3 mg Hg/kg wet wt. In the United States, individual states deeming their populations high in fish intake may lower the fish tissue concentration criterion. For example, 0.2 mg Hg/kg wet wt. was issued by Minnesota, USA (MPCA, 2009) as there could be ~2 million anglers (out of total state population of ~5.7 million) targeting many freshwater lakes and rivers in this state. However, in other countries or regions, the legal limit of Hg in fish and seafood can be different, for example, 0.5 mg Hg/kg wet wt. was set as the maximum permitted concentration in food by the Centre for Food Safety of Hong Kong SAR Government, China. For sensitive populations such as pregnant women and children, the number of fish meals is recommended to be much less than those of normal adults (USEPA, 2001). Different countries have somewhat different fish consumption guidelines. Still, they often have recommended much-reduced

TABLE 13.1

Fish Consumption Guidelines in the United States, European Union, China, Australia, and New Zealand

Countries	Category	Remarks
United States	**Women in pregnancy and breastfeeding:** ~112 g per meal 2–3 meals per week **Children:** ~28 g per meal (1–3 years old) ~56 g per meal (4–7 years old) ~84 g per meal (8–10 years old) ~112 g per meal (11 years old) 2 meals per week	Highly recommending "best choices" of fish and seafood only Common examples of "best choices": Catfish, crab, lobster, oyster, salmon, shrimp, squid, tilapia, trout, etc.
European Commission	**Women in pregnancy:** 1–4 meals per week <1–2 meals per week (with high Hg content)	Consumption advice should be refined at the national level within the European Union
China	**General adults:** ~150–250 g per meal 2 meals per week	No specified types of fish are recommended concerning Hg content
Australia and New Zealand	**Women in pregnancy and breastfeeding:** ~150 g per meal 1 meal per week **Children:** ~75 g per meal (1–3 years old) 1 meal per week	No specified types of fish are recommended concerning Hg content

Note: United States from U.S. Environmental Protection Agency and U.S. Food and Drug Administration; for China's guideline, it was derived from The Chinese Dietary Guidelines 2022 issued by Chinese Nutrition Society; for Australia and New Zealand, it was derived from Food Standards Australia and New Zealand.

fish consumption amounts for women in pregnancy and breastfeeding, and children (Table 13.1). It appears that the guideline jointly issued by U.S. Environmental Protection Agency and U.S. Food and Drug Administration in 2021 would be the most comprehensive and user-friendly thus far.

13.4 Environmental Factors Governing Mercury Levels in Fish

With the above information, one can speculate that there is a suite of factors controlling environmental Hg transformations and production of MeHg, including (1) ambient factors (e.g., wetland abundance, availability of oxygen, organic matter, and nutrients), (2) ecological factors (e.g., food web structure and trophic position), and (3) physiological factors (e.g., ingestion rate, growth rate, and excretion rate), which can ultimately control MeHg levels in fish and thus determine their risks to humans through dietary consumption and are briefly summarized in Table 13.2.

Regarding the (1) ambient factors, inland studies across geographic gradients have demonstrated that watersheds with increasing abundance of wetland would show increasing aqueous MeHg concentrations (Brigham et al., 2009), which in turn can positively contribute to the rising MeHg content in various aquatic biota types including both macroinvertebrates and fish (Chasar et al., 2009). Thus, wetland abundance in a watershed can be a key factor in elevating MeHg contamination in aquatic ecosystems, which can occur even without any industrial sources. The underlying reason is that MeHg is predominantly produced within these reducing environments (i.e., wetlands) and are key sources of MeHg for downstream streams and lakes. Interestingly, in montane areas with little or no wetland existence, studies have found low MeHg levels in surface water and biota, but macroalgae (e.g., filamentous algae and the associated epiphytes) can serve as the hotspot of in-situ MeHg production and contribute to bioaccumulation and biomagnification of MeHg in the local river food webs (Tsui et al., 2010).

In coastal environments, much of the MeHg production would occur in the surface sediments. Thus, the major geochemical controls would be sedimentary organic matter, microbial communities, and the presence of sulfide (Fitzgerald et al., 2007). Indeed, the organic matter content in surface sediment can determine hypoxia/anoxia and provide energy sources to the Hg-methylating microbial communities. Coastal studies along the salinity gradient found that MeHg production appeared to be reduced with increasing salinity, and one of the explanations is the excessive buildup of sulfide, which can inhibit Hg(II) uptake by the methylating microbes (Ulus et al., 2022). In subtropical and tropical biomes, studies found that coastal environments receiving inputs from major river mouths or coastal wetlands (e.g., mangroves) may have elevated MeHg as these landscape features can export elevated amounts of MeHg to the coastal ecosystems (Bergamaschi et al., 2012; Liu et al., 2021), and in particular mangroves can serve as a biogeochemical hotspot in producing MeHg (Wang et al., 2024).

TABLE 13.2

Summary of Factors that Change Fish Mercury Levels in the Aquatic Environments

Type of Factor	Specific Factor	Effect on Fish Mercury Levels	Main Reasons
Ecological	Increasing food chain length	⇑	Elevated biomagnification of methylmercury
Physiological	Increasing fish growth rate	⇓	Biodiluting methylmercury content in the fast-growing tissues
Environmental	Increasing aqueous methylmercury levels	⇑	Increasing methylmercury entrance into the base of the food chain
Environmental	Increasing water turbidity	⇑	Decreasing methylmercury photodegradation
Environmental	Watershed wetland abundance	⇑	Increasing methylmercury supply to downstream environments
Environmental	Decreasing sea ice thickness	⇓	Increasing methylmercury photodegradation

Source: The information was compiled from published studies (Brigham et al., 2009; Chasar et al., 2009; Lavoie et al., 2013; Point et al., 2011; Sherman and Blum, 2013).

As the largest ecosystem on the planet, the open ocean has long been under debate as to why there would be MeHg in the water column as no surface sediment and continent are "nearby." Some earlier studies have attempted to link the presence of MeHg to different sources in the open ocean, including the terrestrial riverine inputs, the sediments in the deep sea, and the water column (Kraepiel et al., 2003). However, it would later become clear that the terrestrial riverine inputs and deep ocean sediments cannot be the primary sources of MeHg to the surface layers in the open ocean. Therefore, in-situ MeHg production in the water column would likely be the ultimate source. However, the underlying processes regulating microbial Hg(II) methylation in the marine water column are still unknown. Some studies have found that non-microbial processes are important for Hg methylation and MeHg degradation (Munson et al., 2018) while other studies hypothesized that anaerobic methylators in anoxic microenvironments (e.g., sinking marine organic particles) and/or aerobic microbes in oxic seawater could be responsible for direct Hg methylation in the water column (Wang et al., 2022). Along the vertical profile in the open ocean, research has shown the presence of MeHg peaks in deeper water depths (~500 and 1,000 m) in the North Pacific Ocean (Hammerschmidt and Bowman, 2012), while the use of stable Hg isotopes revealed that the MeHg bioaccumulated in the deeper ocean fish would derive from a combination of MeHg originating from the surface (where MeHg was partially photodegraded) and the water column in the aphotic zone (< 200 m) of the open ocean (Blum et al., 2013). Thus, there is still a long way to figure out the exact processes of Hg cycling in vast oceanic environments.

13.5 Biological Factors Governing Mercury Levels in Fish

Besides environmental factors that can be ecosystem-specific, there are important factors related to fish physiology and aquatic food webs that can also govern MeHg accumulation in fish. For example, biokinetic models can reasonably predict MeHg levels in fish by incorporating the food ingestion rate and the growth rate of the fish (Dang and Wang, 2012). For example, if the fish grows slower, one can expect higher MeHg content while the lower MeHg content in fish with a faster growth rate through growth biodilution (Lavigne et al., 2010; Baumann et al., 2017) (Table 13.2). Also, it is possible that the growth biodilution of MeHg also occurs for the prey of the fish, such as macroinvertebrates, which can contribute to decrease in MeHg burden in the fish.

For the aquatic food webs, the food chain length would be the most critical factor in determining the MeHg trophic transfer to the top predatory fish as there is a somewhat similar trophic magnification slope of MeHg (calculated as the slope of $Log_{10}MeHg$ against $\delta^{15}N$) across geographic regions within a biome (e.g., the temperate region) (Lavoie et al., 2013). Across different biomes, a meta-analysis has shown that the trophic magnification slope of MeHg (*defined as above*) would, in general, decrease from polar to temperate and to tropical biomes (Lavoie et al., 2013) due to the increasing average temperature and the growth of biodilution.

The other important factor controlling MeHg content in fish would be the relatively slow efflux (elimination) of MeHg in fish (e.g., 1% per day) (Wang and Wong, 2003). Compared to the uptake processes, we know relatively little regarding the efflux of MeHg from fish or other aquatic animals. In general, we think of this process as compartmentalized and relatively slow compared to other fast-eliminating elements, including its counterpart, Hg(II) (e.g., 5%–10% per day) (Wang and Wong, 2003). Temperature may be another critical factor in controlling the efflux of MeHg, and if so, can add complexity to the biokinetic processes controlling MeHg bioaccumulation in fish (Tsui and Wang, 2004). This is an area that deserves more research.

13.6 Atmospheric Deposition and Linkage to Mercury Uptake in Fish

Compared to natural source emissions, anthropogenic emissions would remain the largest source of modern-world Hg deposition (UNEP, 2018). Theoretically, more atmospheric Hg deposition would lead to more microbial MeHg production in aquatic ecosystems. Such an idea has been tested in a whole watershed experiment at METAALICUS in the Experimental Lakes Area in a remote southwestern Ontario,

Canada region. In this large-scale study, isotopically labeled Hg(II) of different masses (^{198}Hg, ^{200}Hg, and ^{202}Hg) were intentionally added to an experimental lake (^{202}Hg), associated wetland (^{198}Hg), and the upland forest (^{200}Hg), and the researchers followed the isotopic tracers in different ecosystems (lake, forest, and wetland) over time (Harris et al., 2007). The study found that Hg directly deposited into the lake water would be methylated and bioaccumulated in the fish within the first 3 years. However, essentially none of MeHg in fish would be derived from the Hg depositing onto the upland forest in three years. Still, those deposited onto the terrestrial environments would take much longer (decades) to be exported, methylated, and bioaccumulated in the lake's fish (Oswald et al., 2014). Thus, Hg emission, deposition, transport, methylation, and bioaccumulation can have a delayed effect especially when deposited to the terrestrial landscapes that we do not entirely understand currently about the timeline.

In contrast to the above whole-lake experiment, a recent study monitoring the fish Hg levels in China has shown a rather complex relationship between anthropogenic Hg emission and MeHg bioaccumulation in fish over the last four decades (Zhang et al., 2022), in which an increasing Hg emission during the previous four decades since China's economic revolution would correlate with decreasing MeHg bioaccumulation in fish. Many factors were discussed to cause such an unexpected phenomenon, including changes in trophic structure (food chain length, e.g., from wild catch to fish farming) and loss of wetlands on a large scale (i.e., loss of MeHg production hotspots) (Tsui et al., 2023). Therefore, simply cutting down Hg emissions may not be sufficient to solve the MeHg pollution problem since many ecosystem features can mediate the Hg transformations.

13.7 Minamata Convention on Mercury (2013)

Minamata Convention was signed and adopted in the year of 2013 with a total of 128 signatories in order to protect human health from mercury contamination that can compromise human health through regulating Hg-containing products and anthropogenic processes releasing Hg. By the year 2017, the convention entered into force. There has been thorough guidance issued by UNEP in monitoring of Hg and Hg compounds to provide support on the evaluation of the effectiveness of the Minamata Convention (UNEP, 2021). There are different tiers of monitoring recommendations in terms of targeted environments and frequency of monitoring. One of the emphases would be placed on the Hg hotspots and monitoring biological organisms accumulating MeHg because this is our ultimate goal for the Convention—reducing Hg health toxicity worldwide, including wildlife and humans (Evers et al., 2016). Some studies proposed the use of stable Hg isotopes as a novel tool to provide additional information such as source tracking of the bioaccumulated Hg (Kwon et al., 2020).

Upon entering into force for about 7–8 years, there have not been updated reports on the evaluation of the effectiveness of the Minamata Convention regarding Hg reduction observed in fish and other wildlife which, as discussed above, may take much longer time to see. Furthermore, there can be disparity among countries of different development to implement the monitoring programs. For example, the United States has long-run fish monitoring programs (at the state level), national mussel watch program, and national atmospheric deposition network for mercury, etc. for few decades already. All these long-term monitoring networks and data are already in place and vigorously reviewed from time to time, and these data can be used to monitor the outcome of the Convention. However, such rich resources may not exist across many developing worlds and thus the global effectiveness of the Convention would be difficult to know entirely. Therefore, a much larger effort is required to implement environmental monitoring across the globe.

13.8 Conclusions

Many countries across the globe have monitoring programs on persistent pollutants in fish, which are essential sources of contaminants for human consumption and pose a risk to global public health. Due to its strong bioaccumulation and biomagnification nature, MeHg is among fish and seafood's most monitored chemical compounds and a few highly hydrophobic and persistent organic compounds such as polychlorinated biphenyls. Even though the Minamata Convention has entered into force in 2017, we believe that it will take at least decades to see the beneficial effects of Hg emission reduction because the legacy Hg emitted in the last 100–200 years (Streets et al., 2009) would still actively cycle through the surface environments on the planet through deposition and re-emission processes. Thus, more work will be needed to perform sophisticated modeling on Hg bioaccumulation in fish amid the changes in global Hg emissions.

This chapter summarizes our current knowledge of crucial factors determining Hg, especially MeHg, in fish. The ever-changing Hg emissions, climate change, and structures of aquatic food webs and ecosystem habitats make the biogeochemical relationships of Hg continue to be complex, dynamic, and challenging to predict fully. However, the current monitoring of Hg/MeHg in fish by many government bodies appears to be practical and helpful in discovering which fish species and seafood products would be particularly prone to high MeHg contamination, alerting people to consume them cautiously or have them entirely banned for direct consumption. Nevertheless, these regulatory monitoring and cautions issued would help human's intake of Hg but cannot benefit fish-consuming wildlife (e.g., mammals and birds). Thus, a significant reduction in Hg emissions (like carbon emissions) would be necessary to reduce all organismal exposure to highly toxic MeHg on the planet.

Acknowledgments

I am indebted to Professor M.H. Wong, the editor of this book, for his invitation, review comments, and encouragement to write and finish this chapter. This work was financially supported by an award from the Worldwide Universities Network Research Development Fund 2022, a competitive General Research Fund award from the Hong Kong Research Grants Council (14302523), and a Seed Collaborative Research Fund from the State Key Laboratory of Marine Pollution at The City University of Hong Kong (SCRF/0056).

REFERENCES

Barkay, T., and B. Gu. 2022. "Demethylation— the other side of the mercury methylation coin: A critical review." *ACS Environmental Au* 2, 77–97.

Baumann, Z., R. P. Mason, D. O. Conover, P. Balcom, C. Y. Chen, K. L. Buckman, N. S. Fisher, and H. Baumann. 2017. "Mercury bioaccumulation increases with latitude in a coastal marine fish (Atlantic silverside, *Menidia menidia*)." *Canadian Journal of Fisheries and Aquatic Sciences* 74, 1009–1015.

Bergamaschi, B. A., D. P. Krabbenhoft, G. R. Aiken, E. Patino, D. G. Rumbold, and W. H. Orem. 2012. "Tidally driven export of dissolved organic carbon, total mercury, and methylmercury from a mangrove-dominated estuary." *Environmental Science and Technology* 46, 1371–1378.

Bernhoft, R. A. 2012. "Mercury toxicity and treatment: A review of the literature." *Journal of Environmental and Public Health* 2012, 460508.

Blum, J. D., B. N. Popp, J. C. Drazen, C. A. Choy, and M. W. Johnson. 2013. "Methylmercury production below the mixed layer in the North Pacific Ocean." *Nature Geoscience* 6, 879–884.

Brigham, M. E., D. A. Wentz, and G. R. Aiken. 2009. "Mercury cycling in stream ecosystems. 1. Water column chemistry and transport." *Environmental Science and Technology* 43, 2720–2725.

Brocza, F. M., P. Rafaj, R. Sander, F. Wagner, and J. M. Jones. 2024. "Global scenarios of anthropogenic mercury emissions." *Atmospheric Chemistry and Physics* 24, 7385–7404.

Chasar, L. C., B. C. Scudder, A. R. Stewart, A. H. Bell, and G. R. Aiken. 2009. "Mercury cycling in stream ecosystems. 3. Trophic dynamics and methylmercury bioaccumulation." *Environmental Science and Technology* 43, 2733–2739.

Cristol, D. A., and D. C. Evers. 2020. "The impact of mercury on North American songbirds: Effects, trends, and predictive factors" *Ecotoxicology* 29, 1107–1116.

Dang, F., and W. X. Wang. 2012. "Why mercury concentration increases with fish size? Biokinetic explanation." *Environmental Pollution* 163, 192–198.

Driscoll, C. T., R. P. Mason, H. M. Chan, D. J. Jacob, and N. Pirrone. 2013. "Mercury as a global pollutant: Sources, pathways, and effects." *Environmental Science and Technology* 47, 4967–4983.

Durnford, D., A. Dastoor, D. Figueras-Nieto, and A. Ryjkov. 2010. "Long range transport of mercury to the Arctic and across Canada". *Atmospheric Chemistry and Physics*, 10, 6063–6086.

Evers, D. C., S. E. Keane, N. Basu, and D. Buck. 2016. "Evaluating the effectiveness of the minamata convention on mercury: Principles and recommendations for next steps." *Science of the Total Environment* 569–570, 888–903.

Fitzgerald, W. F., C. H. Lamborg, and C. R. Hammerschmidt. 2007. "Marine biogeochemical cycling of mercury." *Chemical Reviews* 107, 641–662.

Gilmour, C. C., M. Podar, A. L. Bullock, A. M. Graham, S. D. Brown, A. C. Somenahally, A. Johs, R. A. Hurt, K. L. Bailey, and D. A. Elias. 2013. "Mercury methylation by novel microorganisms from new environments." *Environmental Science and Technology* 47, 11810–11820.

Goyanna, F. A. A., M. B. Fernandes, G. B. da Silva, and L. D. de Lacerda. 2023. "Mercury in oceanic upper trophic level sharks and bony fishes - A systematic review." *Environmental Pollution* 318, 120821.

Hammerschmidt, C. R., and K. L. Bowman. 2012. "Vertical methylmercury distribution in the subtropical North Pacific Ocean." *Marine Chemistry* 132–133, 77–82.

Harris, R. C., J. W. M. Rudd, M. Amyot, C. L. Babiarz, K. G. Beaty, P. J. Blanchfield, R. A. Bodaly, B. A. Branfireun, C. C. Gilmour, J. A. Graydon, A. Heyes, H. Hintelmann, J. P. Hurley, C. A. Kelly, D. P. Krabbenhoft, S. E. Lindberg, R. P. Mason, M. J. Paterson, C. L. Podemski, A. Robinson, K. A. Sandilands, G. R. Southworth, V. L. St Louis, and M. T. Tate. 2007. "Whole-ecosystem study shows rapid fish-mercury response to changes in mercury deposition." *Proceedings of the National Academy of Sciences of the United States of America* 104, 16586–16591.

Kraepiel, A. M. L., K. Keller, H. B. Chin, E. G. Malcolm, and F. M. M. Morel. 2003. "Sources and variations of mercury in tuna." *Environmental Science and Technology* 37, 5551–5558.

Kwon, S. Y., J. D. Blum, R. Yin, M. T. K. Tsui, Y. H. Yang, and J. W. Choi. 2020. "Mercury stable isotopes for monitoring the effectiveness of the minamata convention on mercury." *Earth-Science Reviews* 203, 103111.

Lavigne, M., M. Lucotte, and S. Paquet. 2010. "Relationship between mercury concentration and growth rates for Walleyes, Northern Pike, and Lake Trout from Quebec Lakes." *North American Journal of Fisheries Management* 30, 1221–1237.

Lavoie, R. A., T. D. Jardine, M. M. Chumchal, K. A. Kidd, and L. M. Campbell. 2013. "Biomagnification of mercury in aquatic food webs: A worldwide meta-analysis." *Environmental Science and Technology* 47, 13385–13394.

Li, M. L., C. P. Thackray, V. W. Y. Lam, W. W. L. Cheung, and E. M. Sunderland. 2024. "Global fishing patterns amplify human exposures to methylmercury." *Proceedings of the National Academy of Sciences of the United States of America* 121, e2405898121.

Lippold, A., J. Aars, M. Andersen, A. Aubail, A. E. Derocher, R. Dietz, I. Eulaers, C. Sonne, J. M. Welker, Ø. Wiig, and H. Routti. 2020. "Two decades of mercury concentrations in Barents Sea polar bears (*Ursus maritimus*) in relation to dietary carbon, sulfur, and nitrogen." *Environmental Science and Technology* 54, 7388–7397.

Liu, M., Q. Zhang, C. Yu, L. Yuan, Y. He, W. Xiao, H. Zhang, J. Guo, W. Zhang, Y. Li, Q. Zhang, L. Chen, and X. Wang. 2021. "Observation-based mercury export from rivers to coastal oceans in East Asia." *Environmental Science and Technology* 55, 14269–14280.

Longman, J., D. Verses, W. Finsinger, and V. Ersek. 2018. "Exceptionally high levels of lead pollution in the Balkans from the Early Bronze Age to the Industrial Revolution." *Proceedings of the National Academy of Sciences of the United States of America* 115, E5661–E5668.

Lu, X., W. Y. Gu, L. Zhao, M. F. Ul Haque, A. A. DiSpirito, J. D. Semrau, and B. Gu. 2017. "Methylmercury uptake and degradation by methanotrophs." *Science Advances* 3, e1700041.

Marumoto, K., and S. Imao. 2015. "Determination of dissolved gaseous mercury in seawater of Minamata Bay and estimation for mercury exchange across air–sea interface." *Marine Chemistry* 168, 9–17.

Mergler, D., H. A. Anderson, L. H. M. Chan, K. R. Mahaffey, M. Murray, M. Sakamoto, and A. H. Stern. 2007. "Methylmercury exposure and health effects in humans: A worldwide concern." *Ambio* 36, 3–11.

MPCA. 2009. *"Implementation Plan for Minnesota's Statewide Mercury Total Maximum Daily Load."* Minnesota Pollution Control Agency, St. Paul, Minnesota, USA. URL: https://www.pca.state.mn.us/sites/default/files/wq-iw4-01p.pdf

Munson, K. M., C. H. Lamborg, R. M. Boiteau, and M. A. Saito. 2018. "Dynamic mercury methylation and demethylation in oligotrophic marine water." *Biogeosciences* 15, 6451–6460.

Nriagu, J. O. 1996. "A history of global metal pollution." *Science* 272, 223.

Obrist, D., Y. Agnan, M. Jiskra, C. L Olson, D. P. Colegrove, J. Hueber, C. W. Moore, J. E. Sonke, and D. Helmig. 2017. "Tundra uptake of atmospheric elemental mercury drives Arctic mercury pollution." *Nature* 547, 201–204.

Osei, L., T. Yeboah, E. Kumi, and E. F. Antoh. 2021. "Government's ban on artisanal and small-scale mining, youth livelihoods and imagined futures in Ghana." *Resources Policy* 71, 102008.

Oswald, C. J., A. Heyes, and B. A. Branfireun. 2014. "Fate and transport of ambient mercury and applied mercury isotope in terrestrial upland soils: Insights from the METAALICUS watershed." *Environmental Science and Technology* 48, 1023–1031.

Parks, J. M., A. Johs, M. Podar, R. Bridou, R. A. Hurt, S. D. Smith, S. J. Tomanicek, Y. Qian, S. D. Brown, C. C. Brandt, A. V. Palumbo, J. C. Smith, J. D. Wall, D. A. Elias, and L. Y. Liang. 2013. "The genetic basis for bacterial mercury methylation." *Science* 339, 1332–1335.

Pickhardt, P. C., and N. S. Fisher. 2007. "Accumulation of inorganic and methylmercury by freshwater phytoplankton in two contrasting water bodies." *Environmental Science and Technology* 41, 125–131.

Point, D., J. Sonke, R. Day, D. G. Roseneau, K. A. Hobson, S. S. Vander Pol, A. J. Moors, R. S. Pugh, O. F. X. Donard, and P. R. Becker. 2011. "Methylmercury photodegradation influenced by sea-ice cover in Arctic marine ecosystems." *Nature Geoscience* 4, 188–194.

Qiu, G., X. Feng, P. Li, S. Wang, G. Li, S. Wang, G. Li, L. Shang, and X. Fu. 2008. "Methylmercury accumulation in rice (*Oryza sativa* L.) grown at abandoned mercury mines in Guizhou, China." *Journal of Agricultural and Food Chemistry* 56, 2465–2468.

Regnell, O., and C. J. Watras. 2019. "Microbial mercury methylation in aquatic environments: A critical review of published field and laboratory studies." *Environmental Science and Technology* 53, 4–19.

Schoch, N., Y. Yang, R. D. Yanai, V. L. Buxton, D. C. Evers, and C. T. Driscoll. 2020. "Spatial patterns and temporal trends in mercury concentrations in common loons (*Gavia immer*) from 1998 to 2016 in New York's Adirondack Park: Has this top predator benefitted from mercury emission controls?" *Ecotoxicology* 29, 1774–1785.

Seco, J., S. Aparício, A. S. Brierley, P. Bustamante, F. R. Ceia, J. P. Coelho, R. A. Philips, R. A. Saunders, S. Fielding, S. Gregory, R. Matias, M. A. Pardal, E. Pereira, G. Stowasser, G. A. Tarling, and J. C. Xavier. 2021. "Mercury biomagnification in a Southern Ocean food web." *Environmental Pollution* 275, 116620.

Sherman, L. S., and J. D. Blum. 2013. "Mercury stable isotopes in sediments and largemouth bass from Florida lakes, USA." *Science of the Total Environment* 448, 163–175.

Shimshack, J. P., M. B. Ward, and T. K. M. Beatty. 2007. "Mercury advisories: Information, education, and fish consumption." *Journal of Environmental Economics and Management* 53, 158–179.

Streets, D. G., M. K. Devane, Z. Lu, T. C. Bond, E. M. Sunderland, and D. J. Jacob. 2011. "All-time releases of mercury to the atmosphere from human activities." *Environmental Science and Technology* 45, 10485–10491.

Streets, D. G., Q. Zhang, and Y. Wu. 2009. "Projections of global mercury emissions in 2050." *Environmental Science and Technology* 43, 2983–2988.

Tsui, M. T. K., S. Liu, R. L. Brasso, J. D. Blum, S. Y. Kwon, Y. Ulus, Y. H. Nollet, S. J. Balogh, S. L. Eggert, and J. C. Finlay. 2019. "Controls of methylmercury bioaccumulation in forest floor food webs." *Environmental Science and Technology* 53, 2434–2440.

Tsui, M. T. K., and W. X. Wang. 2004. "Temperature influences on the accumulation and elimination of mercury in a freshwater cladoceran, *Daphnia magna*." *Aquatic Toxicology* 70, 245–256.

Tsui, M. T. K., J. C. Finlay, S. J. Balogh, and Y. H. Nollet. 2010. "In situ methylmercury production within a stream channel in northern California." *Environmental Science and Technology* 44, 6998–7004.

Tsui, M. T. K., S. Y. Kwon, M. Li, and K. Bishop. 2023. "Revisiting the relationship between mercury emission and bioaccumulation." *Eco-Environment & Health* 2, 1–2.

Ulus, Y., M. T. K. Tsui, A. Sakar, P. Nyarko, N. B. Aitmbarek, M. Ardón, and A. T. Chow. 2022. "Declines of methylmercury along a salinity gradient in a low-lying coastal wetland ecosystem at South Carolina, USA." *Chemosphere* 308, 136310.

UNEP. 2018. "*Global Mercury Assessment 2018*." UN Environment Programme. Geneva, Switzerland.

UNEP. 2021. "*Guidance on Monitoring of Mercury and Mercury Compounds to Support Evaluation of the Effectiveness of the Minamata Convention*." UN Environment Programme. Bali, Indonesia.

USEPA, 2001. "*Guidance for Implementing the January 2001 Methylmercury Water Quality Criterion*." United States Environmental Protection Agency. Office of Science and Technology. Washington, DC, USA.

Wang, K., G. Liu, and Y. Cai. 2022. "Possible pathways for mercury methylation in oxic marine waters." *Critical Reviews in Environmental Science and Technology* 52, 3997–4015.

Wang, S., Tsui, M.T.K., Li, J., and Pan, K. 2024. "Biogeochemical controls on methylmercury distribution in a subtropical wetland ecosystem." *Marine Pollution Bulletin* 207, 116894.

Wang, W. X., and R. S. K. Wong. 2003. "Bioaccumulation kinetics and exposure pathways of inorganic mercury and methylmercury in a marine fish, the sweetlips *Plectorhinchus gibbosus*." *Marine Ecology Progress Series* 261, 257–268.

Wu, P., and Y. Zhang. 2023. "Toward a global model of methylmercury biomagnification in marine food webs: Trophic dynamics and implications for human exposure." *Environmental Science and Technology* 57, 6563–6572.

Yu, C., M. Peng, X. Wang, and X. Pan. 2024. "Photochemical demethylation of methylmercury (MeHg) in aquatic systems: A review of MeHg species, mechanisms, and influencing factors." *Environmental Pollution* 344, 123297.

Zhang, H., W. X. Wang, C. J. Lin, X. B. Feng, J. B. Shi, G. B. Jiang, and T. Larssen. 2022. "Decreasing mercury levels in consumer fish over the three decades of increasing mercury emissions in China." *Eco-Environment & Health* 1, 46–52.

Zhang, T., and H. Hsu-Kim. 2010. "Photolytic degradation of methylmercury enhanced by binding to natural organic ligands." *Nature Geoscience* 3, 473–476.

Zhong, H., W. Tang, Z. Li, C. Sonne, S. S. Lam, X. Zhang, S. Y. Kwon, J. Rinklebe, L. M. Nunes, R. Q. Yu, B. Gu, H. Hintelmann, M. T. K. Tsui, J. Zhao, X. Q. Zhou, M. Wu, B. Liu, Y. Hao, L. Chen, B. Zhang, W. Tan, X. X. Zhang, H. Ren, and Y. R. Liu. 2024. "Soil Geobacteraceae are the key predictors of neurotoxic methylmercury bioaccumulation in rice." *Nature Food* 5, 301–311.

Zupo, V., G. Graber, S. Kamel, V. Plichta, S. Granitzer, C. Gundacker, and K. J. Wittmann. 2019. "Mercury accumulation in freshwater and marine fish from the wild and from aquaculture ponds." *Environmental Pollution* 255, 112975.

14

Copper Fungicide Accumulation in Crops and Soils: Human and Environmental Health Risks

Dasom Jeon and Nicholas Dickinson

14.1 Introduction

This chapter evaluates the consequences of extensive agricultural and horticultural use of copper (Cu) as a pesticide to control various crop diseases. We explore an apparent contradiction between a general tolerance towards existing and continuing widespread Cu contamination of agricultural and horticultural soils in the context of current knowledge of the toxicological significance of this heavy metal to human health and the environment.

14.2 Copper and Human Health

Copper (Cu) is essential for both plant nutrition and animal health and is required for respiration, photosynthesis, and numerous other metabolic processes. Inorganic Cu formulations are used extensively in agriculture and horticulture as a pesticide, defoliant, desiccant, micronutrient fertiliser, and feed additive, but primarily to control a wide range of crop diseases (Table 14.1, Elmer & White, 2016.). This metal is integral to various physiological processes, including brain development, neurotransmitter synthesis, immune response, myelin sheath formation, energy production, and wound healing (Coelho et al., 2020). Exposure routes to this metal arise from historical residues from its numerous industrial and domestic applications.

Excessive exposure to Cu can lead to adverse health effects in humans, including gastrointestinal disturbances and liver damage (Linder & Hazegh-Azam, 1996). Dysregulation of copper homeostasis has also been implicated in the pathogenesis of Alzheimer's disease [AD] (Coelho et al., 2020). The accumulation of copper in the brain with age has been well-documented, with significantly higher levels found in AD patients compared to healthy individuals (Squitti et al., 2002, 2005; Vasudevaraju et al., 2010), contributing to increased oxidative stress and subsequent neurodegeneration (Hsu et al., 2018). Ceruloplasmin is a protein made in the liver that stores and transports most mineral Cu around the body, but elevated levels of non-ceruloplasmin Cu have been observed in the blood of AD patients, suggesting a disruption in copper metabolism correlated with disease severity (Squitti et al., 2002). Genetic factors also played a crucial role in copper susceptibility, with specific genetic variants linked to higher levels of non-ceruloplasmin Cu and increased risk of developing AD (Squitti et al., 2018). Epidemiological evidence has supported these findings, indicating that high dietary intake of copper, particularly when combined with diets high in saturated and trans fats, may accelerate cognitive decline in older adults (Morris et al., 2006). Furthermore, regional studies have observed a higher prevalence of AD in areas with elevated soil copper content, suggesting environmental exposure is a significant factor in disease development (Shen et al., 2014). Excessive intake of copper significantly increases the risk of developing AD among individuals with a predisposition to copper metabolic abnormalities (Arena et al., 2018). Coelho et al. (2020) advocate adopting agroecological approaches to minimise dependence on copper-based fungicides to safeguard human health and the environment.

DOI: 10.1201/9781003492115-18

TABLE 14.1

Copper Formulations Used in Organic Crop Production

Input Class	Substance	AQ/ EU	COR[a]	USDA NOP[b]	Description, Compositional Requirements, Conditions of Use
CP	Bordeaux mixture[c]	R	R	R	Inorganic Cu compounds. Maximum of 3 kg/ha/year. It must be used in a manner that minimises Cu accumulation in soil.
CP	Cu hydroxide	R	R	R	Inorganic Cu compounds. Maximum of 3 kg/ha/year
CP	Cu oxychloride	R	R	R	Inorganic Cu compounds. Maximum of 3 kg/ha/year.
CF CP	Cu sulphate	R	R	R	Used to correct documented soil deficiency. Plant pest control must be used to prevent Cu build-up in the soil. Prohibited for use as defoliant, herbicide, or desiccant.
CF	Trace elements	R	R	R	Use is restricted to cases where soil/plant nutrient deficiency is documented by soil or tissue testing or diagnosed by an independent expert. Micronutrients in either chloride or nitrate forms are prohibited. Micronutrients may not be used as a defoliant, herbicide, or desiccant.

Source: Data from AsureQuality (2018).

[a] The Canadian Std (COR) is available from CAN/CGSB-32.310 https://www.tpsgc-pwgsc.gc.ca/ongc-cgsb/programme-program/normes-standards/internet/032-310/032-310-eng.html. CAN/CGSB-32.311 https://www.tpsgc-pwgsc.gc.ca/ongc-cgsb/programme-program/normes-standards/internet/032-311/032-311-eng.html.

[b] United States Department of Agriculture National Organic Standard (USDA NOP) (Part 205- National Organic program) is available from a link on the AsureQuality website. https://www.ecfr.gov/current/title-7/subtitle-B/chapter-I/subchapter-M/part-205?toc=1.

[c] Copper sulphate.

CF, Crop Fertilisers and Soil Amendments; CP, Crop Pest, Weed, and Disease Control; CT, Crop Management Tools and Production Aids; R, Restricted, Inputs which are allowed with restrictions. All restricted inputs require approval by AsureQuality before use. Approval may be granted if no alternatives are available, and approval will be subject to certain conditions. The use of these materials is discouraged.

14.3 Environmental Impact

Copper is widely used in agriculture for its effective broad-spectrum activity against various pathogens, including bacteria, oomycetes, ascomycetes, and basidiomycetes (Tamm et al., 2022). Despite its frequent application on crops, foliar symptoms of Cu toxicity are rare because there is little surface absorption into leaves. It reaches the soil from spray drift and when it is washed off foliage by rainfall or irrigation or from decomposing plant debris. Copper then binds rapidly to soil colloids and organic matter, which generally results in a lower environmental risk than other more mobile heavy metals (Oorts, 2013). The metal usually accumulates in the surface organic soil horizon and root zone, where it undergoes minimal leaching and has limited uptake by plants beyond the root zone, with little internal transference to above-ground plant parts and into edible tissues. Nevertheless, continuous application of Cu leads to significant accumulation in soil, where it can persist indefinitely.

It is well known that elevated Cu in the soil can inhibit root growth and nutrient uptake, impairing crop productivity and yield (Alloway, 2013), with evidence of disruption of recognised determinants of soil health, including adverse effects on soil fertility, microbial diversity, and nutrient cycling processes (Kabata-Pendias & Szteke, 2015). Copper toxicity poses significant risks to soil-dwelling organisms, including beneficial microbes and earthworms essential for maintaining soil structure and facilitating nutrient cycling (Sauvé et al., 2000). Nonetheless, Cu continues to be widely used and to accumulate in many agricultural and horticultural soils where there is little evidence of reduced crop productivity, contrary to the detrimental impacts that might be expected from toxicological knowledge (Table 14.2).

Clearly, there is a contradiction between the known adverse effects of Cu and its continued use during the production of food crops. It has been argued that the impacts and risks may become evident in the longer term when accumulated Cu may become more labile and more environmentally hazardous (Melendez et al., 2020). Undoubtedly, the implications of its continued use on soil health, ecosystem resilience, and environmental sustainability remain poorly understood and present a significant gap in our understanding of the longer-term environmental impacts of Cu.

TABLE 14.2

Ecological and Human Health Impacts of Copper

Ecological Indicators	Impacts	References
Earthworms	Copper affects earthworm survival and growth. Higher copper concentrations can disrupt earthworm behaviour, leading to decreased earthworm populations, affecting soil structure and nutrient cycling.	Aminah et al. (2020), Eijsackers et al. (2005), Jouquet et al. (2006), Karimi et al. (2021), Van Zwieten et al. (2004)
Microorganisms	Copper impacts the microbial community by inhibiting microbial growth and reducing soil respiration and enzyme activity. This affects nutrient cycling and soil health.	Cao et al. (2020), Chander et al. (2001), Karimi et al. (2021), Li et al. (2017), Merrington et al. (2002), Peña et al. (2018), Wang et al. (2009), Wang et al. (2007), Wyszkowska et al. (2013), Zhang et al. (2016)
Plants	Copper accumulation in soil can lead to phytotoxic effects in plants, inhibiting root growth and affecting plant health and crop yields. High soil copper levels can also lead to plants' reduced uptake of water and nutrients.	Bortoluzzi et al. (2019), Cruz et al. (2022), European Food Safety Authority (EFSA) (2006), López-Bucio et al. (2003), Mitchell et al. (2018), Toselli et al. (2009), Yuan et al. (2013)
Soil ecosystem functionality	Evidence of biological resilience to contamination: Plant root foraging, soil respiration, microbial biomass, earthworm burrowing behaviour, organic matter decomposition.	Jeon (2024)
Humans	Excessive exposure can lead to copper toxicity, affecting liver and kidney function and causing gastrointestinal distress.	Arena et al. (2018), Coelho et al. (2020), Elmer and White (2016), Linder and Hazegh-Azam (1996), Shen et al. (2014), Squitti et al. (2002, 2005, 2018), Vasudevaraju et al. (2010)

Source: From Jeon (2024).

14.4 Extent of Cu Usage in Agriculture and Horticulture

Risks from fungicides generally receive less attention than insecticide exposure, although fungicides account for up to 96% of all pesticide applications (Bereswill et al., 2012). Copper has been used as a fungicide in vineyards in Europe since the 1860s (Helling et al., 2000; Kovačič et al., 2013) and continues to be widely applied to crops in temperate and tropical zones that include potatoes, many fruits, vineyards, coffee, and cocoa (Speiser et al., 2018). In the 1950s, 20–30 kg Cu/ha/year were commonly applied for plant protection in Europe, often with > 80 kg used for tropical crops (IFOAM, 2018). In France, application rates have been reported to have been 15–30 7 kg Cu/ha/year in past decades (Brun et al., 1998), in Italy 0.5–8.7 kg Cu/ha/year (Toselli et al., 2009), in Greece 5–8 applications of 0.5–2.5 kg Cu/ha/year (Vavoulidou et al., 2005), and in Australia 2–14 kg Cu/ha (Pietrzak & McPhail, 2004; Wightwick et al., 2008). In the USA, the annual use of Cu in agricultural production increased over 10 years from an estimated 700 Mg in 1987 to 5200 Mg in 1997 (Aspelin & Grube, 1999). Still, the total amount varies between different countries (Table 14.3).

Copper-based inorganic fungicides continue to be preferentially used in crop production because of their low cost, low mammalian toxicity, prolonged residual effects, and relatively high toxicity to plant pathogens, providing effective control of many fungal and bacterial diseases (Aminah et al., 2020). With contemporary concern about risks from a wide range of more complex formulations of organic pesticides and increased movement towards organic farming practices, continued use of inorganic Cu fungicides appears likely (Bolan et al., 2003). Although Cu is applied to plant foliage, just 15% of most plant protection products reach and are retained at the target, while the remaining 85% reaches the soil surface (Aikpokpodion et al., 2010). Inputs of Cu to vineyards and orchards in America, Australia, Europe, and New Zealand range between 1 and 11 7 kg Cu/ha/year, resulting in Cu accrual rates > 5 mg/kg/year in surface soil (0–0.1 m) (Mirlean et al., 2007; Ruyters et al., 2013; Wightwick et al., 2008).

TABLE 14.3
Maximum Permitted Amounts of Copper, Total Estimated Actual Use, and Percentage Utilisation in Organic Farming in the 12 Surveyed European Countries

	Maximum Permitted Quantities (t/annum)	Total Estimated Real Use (t/annum)	Percentage Utilisation (%)[a]
Belgium	3.9	3.9	100%
Bulgaria	248	67	27%
France	546	473	87%
Germany	91	42	46%
Hungary	34.4	22.1	46%
Italy	3253	1556	48%
Norway	0.6	0.5	78%
Spain	2038	1081	53%
Switzerland	10.4	7.1	68%
United Kingdom	11.5	6.2	54%
Total of 10 countries	**6236**	**3259**	**52%**

Source: Tamm et al. (2021).
[a] Percentage utilisation: (Total estimated actual use)/(Maximum permitted quantities).

The accumulation of Cu in agricultural soil has been widely recorded following repeated application of Cu-based fungicides in horticulture (Chaignon et al., 2003; Gallagher et al., 2001). In an 18-year-old apricot orchard soil in New Zealand, the mean Cu concentration in soil (100 mg/kg) was elevated above adjacent pasture (32.4 mg/kg) and native grassland soil (20.0 mg/kg) (Morgan & Johnston, 1991). Elsewhere, long-term accumulation of Cu in vineyard soils has been extensively reported (Besnard et al., 2001; Fagnano et al., 2020; Komárek et al., 2010; Mirlean et al., 2009; Morgan & Taylor, 2004; Pietrzak & McPhail, 2004; Ruyters et al., 2013; Wightwick et al., 2008; Yang et al., 2005), with soil total Cu concentrations ranging from 200 to 500 mg/kg (Brun et al., 1998). Copper concentrations in 130-year-old vineyard soils in France have been found with 1500 mg/kg (Flores-Vélez et al., 1996), and in vineyards in Brazil (40–100 years) with as much as 3200 mg/kg (Besnard et al., 2001; Mirlean et al., 2007). Rates of accumulation of Cu in soil have also been reported; for instance, apple orchards treated with 16 kg Cu/ha/year showed an increase of 1.37–1.45 mg/kg/year in total soil Cu concentrations (Wang et al., 2015; Wang et al., 2009), annual rates have been reported in apricot orchard soils of 4.23 mg/kg/year (Morgan & Johnston, 1991), and in coffee plantation of 5 mg/kg/year (Lepp et al., 1984). Slow natural attenuation rates of Cu in contaminated soils have emphasised the need for effective management strategies. Dickinson and Lepp (1985) estimated that 80–120 years would be required to achieve a mere 10% reduction in Cu concentrations through natural processes. Additionally, Soteburg and Oijard (cited in Morgan & Johnston, 1991) found that a substantial proportion (76%–86%) of Cu applied to the soil over two to three decades earlier remained in the top 0.2 m of the soil.

14.5 Regulations and Guidance

Copper poses ecological risks as it accumulates in soil and potentially enters the food chain without substantially being degraded or leached from soil (Loska & Wiechuła, 2003). Background concentrations of soil Cu in uncontaminated soil have been reported to range from 20 to 40 mg/kg (Baker, 1990; Kabata-Pendias, 2010; Weng et al., 2003), with an optimal range of Cu in soil for most crops is considered to be 5–30 mg/kg, with deficiency symptoms likely below this threshold (Adrees et al., 2015). Background concentrations vary with soil type; typical total concentrations are 5–80 mg/kg in organic soils, 6–40 mg/kg in mineral soils, and 0.1–3.0 mg/kg for soil extracts. Soil concentrations > 100 mg/kg are often cited to be associated with environmental risks (Schramel et al., 2000). Normal Cu concentrations in plants typically fall within 2.5–25 mg/kg (Allen et al., 1974).

Soil quality guidelines regarding acceptable Cu usage vary between countries (Table 14.4). There was little consistency in regulatory approaches to using Cu products until 1991, the European Union

TABLE 14.4

Comparison of Soil Quality Guidelines (SQG) for Copper (Cu) from International Jurisdictions

	Soil Quality Guideline (SQG)	Soil Total Cu Concentration (mg/kg)	References
Australia	SQG(EC_{50}) (national park/area with high ecological value land use)[a]	80–200	National Environment Protection Council (NEPC) (2011)
	SQG(EC_{50}) (urban residential/public open space land use)[a]	150–400	
Australia and New Zealand	ISQG-Low (Trigger value)[b]	65	Australian and New Zealand Environment and Conservation Council (ANZECC) (2000)
	ISQG-High[b]	270	
	Environmental investigation required value	60	Australian and New Zealand Environment and Conservation Council (ANZECC) (1992)
Canada	Canadian SQG (residential)	63	Canadian Council of Ministers of the Environment (2007)
	Canadian SQG (commercial and industrial)	91	
Denmark	Ecotoxicological soil quality criteria	30 (total)	Carlon (2007)
Europe	EU minimal risk values (residential)	14–70	National Environment Protection Council (NEPC) (2011)
	EU warning risk values (residential)	100–500	
	EU potential risk values (residential)	100–1000	
	EU Cu ecological risk assessment	26–176	
Finland	Threshold value[c]	100	Ministry of the Environment (2007)
	Lower guideline value[d]	150	
	Higher guideline value[e]	200	
Latvia	The precautionary limit value[f]	30–60	Cabinet Regulation No. 804. Republic of Latvia (2005)
	The critical limit value[g]	150	
Netherland	Dutch target value	36	VROM (2000)
	Dutch intervention level	190	
New Zealand	Soil Limit (SL)	100	NZWWA (2003)
	Eco-SGV (EC_{30} aged) – sensitive soil (Agricultural land use)[h]	130–150	Cavanagh and Munir (2016)
	Eco-SGV (EC_{30} aged) – tolerant soil (Agricultural land use)[i]	190–210	
USA	Eco-SSL plants[j]	70	EPA (2007)
	Eco-SSL soil invertebrates[j]	80	
	Eco-SSL avian[j]	28	
	Eco-SSL mammalian[k]	49	

[a] Significant toxic effects; [b]ISQG: Interim sediment quality guideline values; [c]The value beyond which further soil testing is needed in the area; [d]In other than an area referred to in industrial, storage, or transport area if the concentration of one substance or several substances exceeds 150 mg/kg; [e]In an area used as an industrial, storage or transport area or as other corresponding areas if the Cu concentration exceeds the prescribed upper guideline value; [f]The maximum level of pollution above which it is likely to harm human health or the environment, as well as the level which must be achieved after remediation; [g]The level at or above which functional characteristics of soil and ground are significantly disrupted or pollution poses a direct threat to human health or the environment; [h]Eco-SGVs: soil guideline values for the protection of ecological receptors. Recent soils were chosen to represent a sensitive soil (Low SOM and CEC and thus may be expected to have greater availability of contaminants); [i]Allophanic soils were chosen to represent tolerant soils (High sorption capacity for contaminants, and thus would be expected to exhibit lower toxicity) and [j]The Ecological Soil Screening Level (Eco-SSL) derivation process is used to derive a set of risk-based Eco-SSLs that are frequently of ecological concern for plants and animals at hazardous waste sites.

stipulated an allowable maximum of 8 kg/ha/year, reduced to 6 kg/ha/year in 2006 (IFOAM, 2018). In 2015, the European Union included Cu-based fungicides among the candidate substances for substitution, later imposing stricter limits (4 kg Cu/ha/year) to reduce the risk of increasing Cu accumulation in soils (European Commission, 2018). Fagnano et al. (2020) estimated that a dose of 4 kg Cu/ha diluted in the tilled soil layer (0.3 m) could correspond to a 1.0 mg/kg Cu/year increase. In the current rules for organic production systems, the European Community permits the application of up to 30 kg Cu/ha every 8 years (3.75 kg Cu/ha/year, Miotto et al., 2017). A meta-analysis on Cu ecotoxicity concluded that applying 4 kg Cu/ha/year should not substantially modify soil biological quality and functions (Karimi et al., 2021).

Plant protection products containing Cu are acceptable in organic farming at lower application rates. However, the amounts applied have been restricted in Europe since 2006 to a maximum of 28 kg/ha over 7 years (European Commission, 2018, 2021). Copper is included in the EU's list of candidates for substitution (European Commission, 2011). Elsewhere, regulatory controls of plant protection products containing Cu in different crops and against individual pathogens are set nationally. Cu is widely used in organic production systems in the USA under the USDA National Organics Program (Organic Materials Review Institute (OMRI), 2018). Organic certification is listed for basic inorganic Cu formulations (Cu sulphate, Cu oxychloride) used for plant disease control. However, restrictions apply; copper must be 'used in a manner that minimises accumulation in soil'.

In the current national authorisation of Cu compounds in Germany, the application of Cu is restricted to 3 kg Cu/ha/year (4 kg Cu/ha/year for hops), although Kuehne et al. (2013) found lesser annual amounts of Cu were used in conventional grapes (0.8 kg/ha), hops (1.7 kg/ha), potatoes (0.8 kg/ha), and apples (1.4 kg/ha). Elsewhere, there are limitations of Cu usage in agricultural fields in Brazil of 3 kg Cu/ha/year, in organic agriculture in East Africa of 8 kg Cu/ha/year. In Austria, the organic farming association (Bio Austria) sets levels that are crop-specific: 3 kg Cu/ha/year in wine growing and fruit, 2 kg Cu/ha/year in arable crops, and 4 kg Cu/ha/year for hops (IFOAM, 2018). The total Cu input in organic farming in Australia (since 2002) should not exceed 8 kg Cu/ha/year. In New Zealand, the acceptable level of Cu application is 8 kg Cu/ha/year in conventional orchards and 3 kg Cu/ha/year in organic orchards in kiwifruit farms (Kiwifruit Vine Health, 2019).

The usage of copper in organic farming across Europe is complex due to overlapping EU regulations and variations in nationally authorised plant protection practices, crops, diseases for which it may be used, and allowable quantity. Significant discrepancies exist between the maximum amount permitted and actual usage; for example, research in Switzerland, France, and Germany highlighted that organic farmers often use substantially less than the maximum allowed amounts, ranging from 3% to 80%, depending on the crop. This reflects a high awareness among farmers about the need to minimise copper use (Kuehne et al., 2017; Speiser et al., 2018; Tamm et al., 2022). The variability in copper usage across European regions or producers often depends on climatic conditions, cultivars, or production systems, although annual copper limits are not always adhered to (Katsoulas et al., 2020; van Bruggen et al., 2016).

14.6 Copper Chemistry and Accumulation in Soil

The bioavailability and mobility in soils of different formulations of applied Cu vary to some extent, for example, in the order: Cu oxychloride ($3Cu(OH)_2{\cdot}CuCl_2$) >Cu nitrate ($Cu(NO_3)_2{\cdot}4H_2O$) > Bordeaux mixture ($CuSO_4/Ca(OH)_2$ (Wang et al., 2020). Copper in soils exists in various chemical forms, including ionic, complex, and precipitated, with its mobility and biological availability influenced by soil characteristics such as texture, organic matter, and pH (Flemming & Trevors, 1989). The total amount of Cu in soil can be divided into a pool of sorbed copper in the solid phase and a pool in the liquid phase, with some equilibrium between the two. In solution, copper exists either complexed with organic or mineral species or as free Cu^{2+} ions. The small, labile fraction of free copper is often used as a proxy for bioavailable copper. However, the total amount of copper represents the pool that can become available to biota (Sereni et al., 2023).

Soil parameters such as organic matter content, pH, ionic strength, and dissolved organic carbon significantly influence copper speciation (Degryse et al., 2009; Mondaca et al., 2015). Soil organic matter plays a significant role in copper immobilisation, as it binds copper ions, reducing solubility and mobility (Arias et al., 2006; Karlsson et al., 2006; Komárek et al., 2010). Copper immobilisation in soils occurs through its adsorption by clay minerals, organic fractions, and humic substances that form stable complexes, reducing availability (Sereni et al., 2023). The fate of organic substance-bound copper depends on the nature of the organic molecules, with less soluble fractions forming very stable complexes and soluble fractions retaining copper in solution (Arias et al., 2006; Karlsson et al., 2006; Komárek et al., 2010). The formation of complexes with humic and fulvic acids plays a crucial role in soil Cu retention (Komárek et al., 2010). Consequently, in soils rich in organic substances and humification, Cu is less likely to remain in the soluble phase, reducing its availability to plants, thus lowering phytotoxicity risk (La Torre et al., 2018). Absorption, precipitation, and complexation reactions are also influenced by pH levels: lowering pH to below 6.5 increases copper solubility and mobility. In alkaline soils, copper tends to form insoluble precipitates, whereas in acidic soils, copper solubility increases, making it more available to plants (Janik et al., 2009; La Torre et al., 2018; Mozaffari et al., 1996). The nature of these complexes and their stability are crucial in determining the environmental impact of Cu, underscoring the importance of managing soil pH and organic matter to mitigate copper toxicity in agricultural settings. Maintaining soil Cu less soluble forms reduces its availability to plants and lowers the risk of phytotoxicity (La Torre et al., 2018).

14.7 Transfer to Food Products and Formulations of Cu

Copper accumulation in plant tissues can significantly impair plant growth and physiology, disrupting nutrient uptake, leading to deficiencies in essential nutrients like phosphorus and potassium, altering chloroplast structure and pigment composition, and impacting photosynthetic efficiency, plant health, and productivity (Ambrosini et al., 2018; Cambrollé et al., 2015; Hossain et al., 2020; Yuan et al., 2013).

Copper application as a foliar spray tends to remain on plant surfaces without absorption or else reach the soil. However, concentrations in edible plant parts can exceed safe consumption thresholds through surface contamination or uptake. Copper uptake from soil to plant roots is frequently recorded, but its onward transfer via the xylem to above-ground and edible plant parts is generally highly restricted. Xylem loading controls the movement of Cu ions during plant growth (Printz et al., 2016). Typically, uncontaminated plants contain concentrations ranging from 1 to 50 mg Cu/kg (dry weight), with toxicity symptoms appearing when levels exceed 15–20 mg/kg in above-ground tissues (Beni & Rossi, 2009; NAS (National Academy of Sciences), 1977; Xiong & Wang, 2005; Yruela, 2005). Studies indicate varied correlations between soil copper content and its concentration in edible plant parts. Often, there is no significant correlation between copper content in the edible parts of crops and its total or bioavailable concentrations in soil (Agrelli et al., 2017), although there are exceptions (e.g., Beygi & Jalali, 2019; Fagnano et al., 2020). There are regulatory acceptable limits for Cu concentration in food products that vary between countries; for example, in Europe, Cu residues in fruit should not exceed 5 mg/kg, but there is a higher maximum in Switzerland of 15 mg/kg (Kurnik et al., 2012).

The most common inorganic formulations applied a spray to crops are copper hydroxide, copper sulphate, and copper oxychloride solutions. Organic pesticide formulations, where copper is complexed with organic molecules, require lower application rates than traditional inorganic products. These are partially systemic, potentially allowing for a higher accumulation of copper within the tissues of treated plants that may result in copper concentrations in harvested fruits exceeding the maximum EU residue limits [MRLs] (Kurnik et al., 2012). Copper-based compounds such as octanoate, gluconate, and ureate can frequently result in levels exceeding the legal MRL in apples at harvest, often more so than inorganic formulations. Copper levels in apple fruits have been reported to surpass the EC legal MRL of 5 mg/kg fresh mass, reaching up to 21 mg/kg fresh mass; in Switzerland, the limit is 15 mg/kg fresh mass. Considering the water content of apples, which ranges from 80% to 85%, these values correspond to approximately 105–140 mg/kg dry mass (Rasooli Sharabiani et al., 2021). Recent development of copper

TABLE 14.5
Organic Formulations of Cu Pesticides, Including the Use of Nanotechnology

Formulation	Manufacturer	Chemical form of Copper	G Cu^{2+}/kg
Atempo	Nidorf DE	Cu-octanoate complex	19.0
CinkarnaEDTA	Cinkarna d.o.o. SLO	Cu-Ca oxychloride-EDTA chelate	32.0
Cinkarna AminoT	Cinkarna d.o.o. SLO	Cu-Ca oxychloride-titanium dioxide-amino complex	27.0
CinkarnaEDTAN	Cinkarna d.o.o. SLO	Cu-Ca oxychloride-EDTA chelate Na	39.3
Cinkarna Amino	Cinkarna d.o.o. SLO	Cu-Ca oxychloride amino complex	27.5
Coptrel 500	Yara Vita TM GB	Cu-oxide-urea complex	330.0
Copper Protein	Nova Prot GmbH DE	Cu-hydroxide-protein complex	18.0
Cuprablau Z	Cinkarna d.o.o. SLO	Cu-Ca oxychloride	350.0
Cuprablau ZU	Cinkarna d.o.o. SLO	Cu-Ca oxychloride Na	350.0
Kupro 190 SC	Agro Ruse SLO	Cu-oxosulphate	190.0
Labicuper	Macasa S.L.ES	Cu-gluconate complex	65.0
Peptiram 5	Sicit 2000 S.p.a.l	Cu-S-peptide complex	50.0
Protex-CU	Area Europe BV NL	Cu-S water soluble	60.0

N[a]-Formulation with copper in the form of nanoparticles (1–5 nm).
Source: Kah et al. (2022).

formulations utilising nanotechnology (Table 11.5) to enhance copper's availability and effectiveness of Cu significantly impacts its short-term mobility and effectiveness in agricultural applications (Kah et al., 2022). Different copper formulations exhibit markedly different effects on bacterial and fungal growth, highlighting the significance of selecting appropriate copper compounds based on specific agricultural needs and environmental considerations (Vázquez-Blanco et al., 2020).

14.8 Conclusion

The persistent use of copper-based fungicides in agriculture and horticulture has led to highly elevated concentrations and intolerable accumulation in many production soils. This has long-term ramifications for soil health and environmental sustainability, with potential consequences to human health. Critical issues associated with copper accumulation in soil include detrimental effects on plant roots, microbial diversity, invertebrate communities, and ecosystem functionality. Biological adaptation and acclimation to elevated soil copper, maintenance of high soil pH, and soil organic matter provide a natural resilience to contamination of soils that buffers the impact and mitigates toxicity of Cu, but this provides a limited and unsatisfactory safeguard for the future. The metal is strongly associated with soil organic matter, the dynamics of which drive ecosystem functionality. The decomposition of Cu-contaminated soil organic matter both determines and is determined by the amounts of Cu, its mobility, and toxicity. Restricted mobility of Cu within plants generally means that concentrations are not enhanced in food products, although organic formulations and Cu nanoparticles may counter this protection.

Regulatory efforts have been instrumental in curbing excessive copper use, yet challenges remain due to variations in enforcement and compliance across different regions. As the food production sectors have moved towards environmentally safer and more sustainable practices, developing and adopting alternative disease management strategies is paramount. Reducing copper dependency is essential to ensure the long-term health of agricultural landscapes. Various tools and strategies to reduce reliance on copper-based plant protection products include disease-resistant crop varieties, cultural measures, better disease forecasting models, decision support systems, and alternative plant protection products. However, the complete elimination of copper fungicides currently seems unlikely. These complexities underscore the ongoing challenge within organic farming to balance effective disease management with environmental sustainability and health considerations. Finding alternatives to Cu and innovative agricultural practices remains critical in striving towards reduced dependence on Cu, aligning with the principles of organic farming, and maintaining high crop productivity. Possible alternative substances that offer disease control without copper's environmental downsides can complement or be substituted for copper in disease management, including biocontrol agents, plant extracts, and naturally derived substances such as chitosan and laminarin.

Acknowledgements

This chapter of the book is adapted from Dasom Jeon's PhD thesis. The authors are grateful to Lincoln University for grant funding to support Dasom Jeon's PhD stipend and operational funds.

REFERENCES

Adrees, M., Ali, S., Rizwan, M., Ibrahim, M., Abbas, F., Farid, M., Zia-ur-Rehman, M., Irshad, M. K., & Bharwana, S. A. (2015). The effect of excess copper on growth and physiology of important food crops: a review. *Environmental Science and Pollution Research*, *22*(11), 8148–8162. https://doi.org/10.1007/s11356-015-4496-5

Agrelli, D., Adamo, P., Cirillo, T., Duri, L. G., Duro, I., Fasano, E., Ottaiano, L., Ruggiero, L., Scognamiglio, G., & Fagnano, M. (2017). Soil versus plant as indicators of agroecosystem pollution by potentially toxic elements. *Journal of Plant Nutrition and Soil Science*, *180*(6), 705–719. https://doi.org/10.1002/jpln.201700050

Aikpokpodion, P., Lajide, L., Festus Aiyesanmi, A., Lajide, L., & Aiyesanmi, A. (2010). Heavy metals contamination in fungicide treated Cocoa plantations in Cross River State, Nigeria. *Journal of Agricultural and Environmental Sciences*, *8*(3), 268–274. https://www.researchgate.net/publication/228360668

Allen, S. E., Grimshaw, H. M., Parkinson, J. A., & Quarmby, C. (1974). *Chemical Analysis of Ecological Material*. Oxford: Blackwell Scientific.

Alloway, B. J. (2013). *Heavy Metals in Soils: Trace Metals and Metalloids in Soils and their Bioavailability*. In B. J. Alloway (Ed.), Vol. 22. Netherlands: Springer Netherlands. https://doi.org/10.1007/978-94-007-4470-7

Ambrosini, V. G., Rosa, D. J., Bastos de Melo, G. W., Zalamena, J., Cella, C., Simão, D. G., Souza da Silva, L., Pessoa dos Santos, H., Toselli, M., Tiecher, T. L., & Brunetto, G. (2018). High copper content in vineyard soils promotes modifications in photosynthetic parameters and morphological changes in the root system of 'Red Niagara' plantlets. *Plant Physiology and Biochemistry*, *128*, 89–98. https://doi.org/10.1016/j.plaphy.2018.05.011

Aminah, T. S., Murray, B. M., Karl, J. C., Joshua, A. P., Paul, E. C., & Quirine, K. (2020). *Agronomy Fact Sheet Series of Copper (Cu) (Issue Fact sheet #113)*. Virginia: Copper Development Association Inc.

Arena, M., Auteri, D., Barmaz, S., Bellisai, G., Brancato, A., Brocca, D., Bura, L., Byers, H., Chiusolo, A., Court Marques, D., Crivellente, F., De Lentdecker, C., Egsmose, M., Erdos, Z., Fait, G., Ferreira, L., Goumenou, M., Greco, L., Ippolito, A., & Villamar-Bouza, L. (2018). Peer review of the pesticide risk assessment of the active substance copper compounds copper(I), copper(II) variants namely copper hydroxide, copper oxychloride, tribasic copper sulfate, copper(I) oxide, Bordeaux mixture. *EFSA Journal*, *16*(1), e05152. https://doi.org/10.2903/j.efsa.2018.5152

Arias, M., Pérez-Novo, C., López, E., & Soto, B. (2006). Competitive adsorption and desorption of copper and zinc in acid soils. *Geoderma*, *133*(3–4), 151–159. https://doi.org/10.1016/j.geoderma.2005.07.002

Aspelin, A. L., & Grube, A. H. (1999). *Pesticide Industry Sales and Usage: 1996 and 1997 Market Estimates. 733-R-99-001*. Washington, DC: Office of Pesticide Programs, USEPA.

AsureQuality. (2018). *Asurequality Organic Standard Version 7*.

Australian and New Zealand Environment and Conservation Council (ANZECC). (1992). *Australian and New Zealand Guidelines for the Assessment and Management of Contaminated Sites*.

Australian and New Zealand Environment and Conservation Council (ANZECC). (2000). *Australian and New Zealand Guidelines for Fresh and Marine Water Quality* (Vol. 1). https://www.waterquality.gov.au/sites/default/files/documents/anzecc-armcanz-2000-guidelines-vol1.pdf

Baker, D. E. (1990). Copper. In B. J. Alloway (Ed.), Heavy Metals in Soils. Wiley; Chichester.

Beni, C., & Rossi, G. (2009). Conventional and organic farming: estimation of some effects on soil, copper accumulation and wine in a central Italy vineyard. *Agrochimica*, *53*(3), 147–159.

Bereswill, R., Golla, B., Streloke, M., & Schulz, R. (2012). Entry and toxicity of organic pesticides and copper in vineyard streams: Erosion rills jeopardise the efficiency of riparian buffer strips. *Agriculture, Ecosystems and Environment*, *146*(1), 81–92. https://doi.org/10.1016/j.agee.2011.10.010

Besnard, E., Chenu, C., & Robert, M. (2001). Influence of organic amendments on copper distribution among particle-size and density fractions in Champagne vineyard soils. *Environmental Pollution*, *112*(3), 329–337. https://doi.org/10.1016/S0269-7491(00)00151-2

Beygi, M., & Jalali, M. (2019). Assessment of trace elements (Cd, Cu, Ni, Zn) fractionation and bioavailability in vineyard soils from the Hamedan, Iran. *Geoderma*, *337*, 1009–1020. https://doi.org/10.1016/j.geoderma.2018.11.009

Bolan, N., Adriano, D., Mani, S., & Khan, A. (2003). Adsorption, complexation, and phytoavailability of copper as influenced by organic manure. *Environmental Toxicology and Chemistry*, *22*(2), 450–456. https://doi.org/10.1897/1551-5028(2003)022<0450:ACAPOC>2.0.CO;2

Bortoluzzi, E. C., Korchagin, J., Moterle, D. F., Santos, D. R., & Caner, L. (2019). Accumulation and precipitation of Cu and Zn in a centenarian vineyard. *Soil Science Society of America Journal*, *83*(2), 492–502. https://doi.org/10.2136/sssaj2018.09.0328

Brun, L. A., Maillet, J., Richarte, J., Herrmann, P., & Remy, J. C. (1998). Relationships between extractable copper, soil properties and copper uptake by wild plants in vineyard soils. *Environmental Pollution*, *102*(2–3), 151–161. https://doi.org/10.1016/S0269-7491(98)00120-1

Cabinet Regulation No. 804. Republic of Latvia. (2005). *Regulations Regarding Quality Standards for Soil and Ground.* https://faolex.fao.org/docs/pdf/lat190875ENG.pdf

Cambrollé, J., García, J. L., Figueroa, M. E., & Cantos, M. (2015). Evaluating wild grapevine tolerance to copper toxicity. *Chemosphere*, *120*, 171–178. https://doi.org/10.1016/j.chemosphere.2014.06.044

Canadian Council of Ministers of the Environment. (2007). *Canadian Soil Quality Guidelines for the Protection of Environmental and Human Health (Update 7.0).* https://support.esdat.net/Environmental%20Standards/canada/soil/rev_soil_summary_tbl_7.0_e.pdf

Cao, X., Luo, J., Wang, X., Chen, Z., Liu, G., Khan, M. B., Kang, K. J., Feng, Y., He, Z., & Yang, X. (2020). Responses of soil bacterial community and Cd phytoextraction to a Sedum alfredii-oilseed rape (Brassica napus L. and Brassica juncea L.) intercropping system. *Science of the Total Environment*, *723*, 138152. https://doi.org/10.1016/j.scitotenv.2020.138152

Carlon, C. (2007). *Derivation Methods of Soil Screening Values in Europe. A Review and Evaluation of National Procedures Towards Harmonization.* Europe: European Commission.

Cavanagh, J. A. E., & Munir, K. (2016). *Development of soil guideline values for the protection of ecological receptors (Eco-SGVs): Technical document.* www.landcareresearch.co.nz

Chaignon, V., Sanchez-Neira, I., Herrmann, P., Jaillard, B., & Hinsinger, P. (2003). Copper bioavailability and extractability as related to chemical properties of contaminated soils from a vine-growing area. *Environmental Pollution*, *123*(2), 229–238. https://doi.org/10.1016/S0269-7491(02)00374-3

Chander, K., Dyckmans, J., Joergensen, R., Meyer, B., & Raubuch, M. (2001). Different sources of heavy metals and their long-term effects on soil microbial properties. *Biology and Fertility of Soils*, *34*(4), 241–247. https://doi.org/10.1007/s003740100406

Coelho, F. C., Squitti, R., Ventriglia, M., Cerchiaro, G., Daher, J. P., Rocha, J. G., Rongioletti, M. C. A., & Moonen, A. C. (2020). Agricultural use of copper and its link to Alzheimer's disease. *Biomolecules*, *10*(6), 1–19. https://doi.org/10.3390/biom10060897

Cruz, F. J. R., Ferreira, R. L. da C., Conceição, S. S., Lima, E. U., Neto, C. F. de O., Galvão, J. R., Lopes, S. da C., & Viegas, I. de J. M. (2022). Copper toxicity in plants: nutritional, physiological, and biochemical aspects. Edited by Josphert Ngui Kimatu. IntechOpen. doi: 10.5772/intechopen.105212

Degryse, F., Smolders, E., & Parker, D. R. (2009). Partitioning of metals (Cd, Co, Cu, Ni, Pb, Zn) in soils: concepts, methodologies, prediction and applications - a review. *European Journal of Soil Science*, *60*(4), 590–612. https://doi.org/10.1111/j.1365-2389.2009.01142.x

Dickinson, N. M., & Lepp, N. W. (1985). A model of retention and loss of fungicide-derived copper in different-aged stands of coffee in Kenya. *Agriculture, Ecosystems and Environment*, *14*(1–2), 15–23. https://doi.org/10.1016/0167-8809(85)90081-7

Eijsackers, H., Beneke, P., Maboeta, M., Louw, J. P. E., & Reinecke, A. J. (2005). The implications of copper fungicide usage in vineyards for earthworm activity and resulting sustainable soil quality. *Ecotoxicology and Environmental Safety*, *62*(1), 99–111. https://doi.org/10.1016/j.ecoenv.2005.02.017

Elmer, W. H., & White, J. C. (2016). The use of metallic oxide nanoparticles to enhance growth of tomatoes and eggplants in disease infested soil or soilless medium. *Environmental Science: Nano*, *3*, 1072–1079. https://doi.org/https://doi.org/10.1039/C6EN00146G

EPA. (2007). *Ecological Soil Screening Levels for Copper, Interim Final* (Issue February). https://www.epa.gov/chemical-research/interim-ecological-soil-screening-level-documents

European Commission. (2011). *Consolidated text: Commission Implementing Regulation (EU) No 540/2011 of* 25 May 2011 *implementing Regulation (EC) No 1107/2009 of the European Parliament and of the Council as regards the list of approved active substances (Text with EEA relevance).* Official Journal of the European Union. https://eur-lex.europa.eu/legal-content/EN/TXT/?uri=CELEX%3A02011R0540-20230701&qid=1690342352168

European Commission. (2018). *Commission Implementing Regulation (EU) 2018/1981 of 13* December 2018 *renewing the approval of the active substances copper compounds, as candidates for substitution, in accordance with Regulation (EC) No 1107/2009 of the European Parliament and of the Co.* Official Journal of the European Union. https://data.europa.eu/eli/reg_ impl/2018/1981/oj

European Commission. (2021). *Consolidated text: Commission Implementing Regulation (EU)* 2021/1165 of 15 July 2021 *authorising certain products and substances for use in organic production and establishing their lists (Text with EEA relevance).* Official Journal of the European Union. https://eur-lex.europa.eu/eli/reg_impl/2021/1165/oj

European Food Safety Authority (EFSA). (2006). *Tolerable upper intake level on vitamins and minerals.* https://www.efsa.europa.eu/sites/default/files/efsa_rep/blobserver_assets/ndatolerableuil.pdf

Fagnano, M., Agrelli, D., Pascale, A., Adamo, P., Fiorentino, N., Rocco, C., Pepe, O., & Ventorino, V. (2020). Copper accumulation in agricultural soils: Risks for the food chain and soil microbial populations. *Science of the Total Environment*, *734*, 139434. https://doi.org/10.1016/j.scitotenv.2020.139434

Flemming, C. A., & Trevors, J. T. (1989). Copper toxicity and chemistry in the environment: a review. *Water, Air, and Soil Pollution*, *44*(1–2), 143–158. https://doi.org/10.1007/BF00228784

Flores-Vélez, L. M., Ducaroir, J., Jaunet, A. M., & Robert, M. (1996). Study of the distribution of copper in an acid sandy vineyard soil by three different methods. *European Journal of Soil Science*, *47*(4), 523–532. https://doi.org/10.1111/j.1365-2389.1996.tb01852.x

Gallagher, D. L., Johnston, K. M., & Dietrich, A. M. (2001). Fate and transport of copper-based crop protectants in plasticulture runoff and the impact of sedimentation as a best management practice. *Water Research*, *35*(12), 2984–2994. https://doi.org/10.1016/S0043-1354(00)00594-7

Helling, B., Reinecke, S. A., & Reinecke, A. J. (2000). Effects of the fungicide copper oxychloride on the growth and reproduction of Eisenia fetida (Oligochaeta). *Ecotoxicology and Environmental Safety*, *46*(1), 108–116. https://doi.org/10.1006/eesa.1999.1880

Hossain, M. S., Abdelrahman, M., Tran, C. D., Nguyen, K. H., Chu, H. D., Watanabe, Y., Hasanuzzaman, M., Mohsin, S. M., Fujita, M., & Tran, L. S. P. (2020). Insights into acetate-mediated copper homeostasis and antioxidant defense in lentil under excessive copper stress. *Environmental Pollution*, *258*, 113544. https://doi.org/10.1016/j.envpol.2019.113544

Hsu, H. W., Bondy, S. C., & Kitazawa, M. (2018). Environmental and dietary exposure to copper and its cellular mechanisms linking to Alzheimer's disease. *Toxicological Sciences*, *163*(2), 338–345. https://doi.org/10.1093/toxsci/kfy025

IFOAM. (2018). *Strategy for the Minimisation of Copper in Organic Farming in Europe*. https://eur-lex.europa.eu/LexUriServ/LexUriServ.do?uri=CONSLEG:1991R2092:20080514:EN:PDF

Janik, L. J., Forrester, S. T., & Rawson, A. (2009). The prediction of soil chemical and physical properties from mid-infrared spectroscopy and combined partial least-squares regression and neural networks (PLS-NN) analysis. *Chemometrics and Intelligent Laboratory Systems*, *97*(2), 179–188. https://doi.org/10.1016/j.chemolab.2009.04.005

Jeon, D. (2024). *Copper Contamination of Fruit Orchard Soils: Biotic Impacts*. Lincoln University: Lincoln.

Jouquet, P., Dauber, J., Lagerlöf, J., Lavelle, P., & Lepage, M. (2006). Soil invertebrates as ecosystem engineers: Intended and accidental effects on soil and feedback loops. *Applied Soil Ecology*, *32*(2), 153–164. https://doi.org/10.1016/j.apsoil.2005.07.004

Kabata-Pendias, A. (2010). *Trace Elements in Soils and Plants*. CRC Press: Boca Raton, FL. https://doi.org/10.1201/b10158

Kabata-Pendias, A., & Szteke, B. (2015). *Trace Elements in Abiotic and Biotic Environments* (1st ed.). CRC Press: Boca Raton, FL.

Kah, M., Navarro, D., Schenkeveld, W., Kookana, R. S., Kirby, J. K., Santra, S., & Ozcan, A. (2022). Fate of copper in soil: Effect of agrochemical (nano)formulations and soil properties. *Environmental Science: Nano*, *9*(2), 653–662. https://doi.org/10.1039/D1EN00213A

Karimi, B., Masson, V., Guilland, C., Leroy, E., Pellegrinelli, S., Giboulot, E., Maron, P. A., & Ranjard, L. (2021). Ecotoxicity of copper input and accumulation for soil biodiversity in vineyards. *Environmental Chemistry Letters*, *19*(3), 2013–2030. https://doi.org/10.1007/s10311-020-01155-x

Karlsson, T., Persson, P., & Skyllberg, U. (2006). Complexation of copper(II) in organic soils and in dissolved organic matter - EXAFS evidence for chelate ring structures. *Environmental Science and Technology*, *40*(8), 2623–2628. https://doi.org/10.1021/es052211f

Katsoulas, N., Løes, A. K., Andrivon, D., Cirvilleri, G., de Cara, M., Kir, A., Knebl, L., Malińska, K., Oudshoorn, F. W., Willer, H., & Schmutz, U. (2020). Current use of copper, mineral oils and sulphur for plant protection in organic horticultural crops across 10 European countries. *Organic Agriculture*, *10*, 159–171. https://doi.org/10.1007/s13165-020-00330-2

Kiwifruit Vine Health. (2019). *KVH Information Sheet. September*. www.kvh.org.nz/spray_products

Komárek, M., Čadková, E., Chrastný, V., Bordas, F., & Bollinger, J. C. (2010). Contamination of vineyard soils with fungicides: A review of environmental and toxicological aspects. *Environment International*, *36*(1), 138–151. https://doi.org/10.1016/j.envint.2009.10.005

Kovačič, G. R., Lešnik, M., & Vršič, S. (2013). An overview of the copper situation and usage in viticulture. *Bulgarian Journal of Agricultural Science*, *19*(1), 50–59.

Kuehne, S., Bieberich, L., Piorr, H., & Landzettel, C. (2013). Moglichkeiten zur Reduktion kupferhaltiger Pflanzenschutzmittel fur den Oko-Kartoffelanbau. *Kartoffelbau*, *6*, 31–33.

Kuehne, S., Roßberg, D., Röhrig, P., Von Mehring, F., Weihrauch, F., Kanthak, S., Kienzle, J., Patzwahl, W., Reiners, E., & Gitzel, J. (2017). The use of copper pesticides in Germany and the search for minimization and replacement strategies. *Organic Farming*, *3*(1), 0301006. https://doi.org/10.12924/of2017.03010066

Kurnik, V., Gaberšek, V., Unuk, T., Tojnko, S., Vogrin, A., Vajs, S., & Lešnik, M. (2012). Einfluss neuartiger Kupferformulierungen auf den Kupfergehalt in der Apfelfrucht. *Erwerbs-Obstbau*, *54*(4), 161–170. https://doi.org/10.1007/s10341-012-0172-9

La Torre, A., Iovino, V., & Caradonia, F. (2018). Copper in plant protection: Current situation and prospects. In *Phytopathologia Mediterranea*, *57*(2), 201–236. https://doi.org/10.14601/Phytopathol_Mediterr-23407

Lepp, N. W., Dickinson, N. M., & Ormand, K. L. (1984). Distribution of fungicide-derived copper in soils, litter and vegetation of different aged stands of coffee (Coffea arabica L.) in Kenya. *Plant and Soil*, *77*(2–3), 263–270. https://doi.org/10.1007/BF02182929

Li, X., Meng, D., Li, J., Yin, H., Liu, H., Liu, X., Cheng, C., Xiao, Y., Liu, Z., & Yan, M. (2017). Response of soil microbial communities and microbial interactions to long-term heavy metal contamination. *Environmental Pollution*, *231*, 908–917. https://doi.org/10.1016/j.envpol.2017.08.057

Linder, M. C., & Hazegh-Azam, M. (1996). Copper biochemistry and molecular biology. *The American Journal of Clinical Nutrition*, *63*(5), 797S–811S. https://doi.org/10.1093/ajcn/63.5.797

López-Bucio, J., Cruz-Ramírez, A., & Herrera-Estrella, L. (2003). The role of nutrient availability in regulating root architecture. *Current Opinion in Plant Biology*, *6*(3), 280–287. https://doi.org/10.1016/S1369-5266(03)00035-9

Loska, K., & Wiechuła, D. (2003). Application of principal component analysis for the estimation of source of heavy metal contamination in surface sediments from the Rybnik Reservoir. *Chemosphere*, *51*, 723–733.

Melendez, M. V., Heckman, J. R., Murphy, S., & D'amico, F. (2020). New Jersey farm soil copper levels resulting from copper fungicide applications. *HortTechnology*, *30*(2), 268–272. https://doi.org/10.21273/HORTTECH04494-19

Merrington, G., Rogers, S. L., & Van Zwieten, L. (2002). The potential impact of long-term copper fungicide usage on soil microbial biomass and microbial activity in an avocado orchard. *Australian Journal of Soil Research*, *40*(5), 749–759. https://doi.org/10.1071/SR01084

Ministry of the Environment (MEF). (2007). Threshold and guideline values for harmful substances in soil. Government Decree on the Assessment of Soil Contamination and Remediation Needs. Finland: Ministry of the Environment. https://faolex.fao.org/docs/pdf/fin113198.pdf

Miotto, A., Ceretta, C. A., Girotto, E., Trentin, G., Kaminski, J., De Conti, L., Moreno, T., Elena, B., & Brunetto, G. (2017). Copper accumulation and availability in sandy, acid, vineyard soils. *Communications in Soil Science and Plant Analysis*, *48*(10), 1167–1183. https://doi.org/10.1080/00103624.2017.1341908

Mirlean, N., Baisch, P., & Medeanic, S. (2009). Copper bioavailability and fractionation in copper-contaminated sandy soils in the wet subtropics (Southern Brazil). *Bulletin of Environmental Contamination and Toxicology*, *82*(3), 373–377. https://doi.org/10.1007/s00128-008-9620-5

Mirlean, N., Roisenberg, A., & Chies, J. O. (2007). Metal contamination of vineyard soils in wet subtropics (southern Brazil). *Environmental Pollution*, *149*(1), 10–17. https://doi.org/10.1016/j.envpol.2006.12.024

Mitchell, K., Trakal, L., Sillerova, H., Avelar-González, F. J., Guerrero-Barrera, A. L., Hough, R., & Beesley, L. (2018). Mobility of As, Cr and Cu in a contaminated grassland soil in response to diverse organic amendments; a sequential column leaching experiment. *Applied Geochemistry*, *88*, 95–102. https://doi.org/10.1016/j.apgeochem.2017.05.020

Mondaca, P., Neaman, A., Sauvé, S., Salgado, E., & Bravo, M. (2015). Solubility, partitioning, and activity of copper-contaminated soils in a semiarid region. *Journal of Plant Nutrition and Soil Science*, *178*(3), 452–459. https://doi.org/10.1002/jpln.201400349

Morgan, R. K., & Johnston, H. (1991). The accumulation of copper in a New Zealand orchard soil. *Journal of the Royal Society of New Zealand*, *21*(4), 323–327. https://doi.org/10.1080/03036758.1991.10420830

Morgan, R. K., & Taylor, E. (2004). Copper accumulation in vineyard soils in New Zealand. *Environmental Sciences*, *1*(2), 139–167. https://doi.org/10.1080/15693430512331342602

Morris, M. C., Evans, D. A., Tangney, C. C., Bienias, J. L., Schneider, J. A., Wilson, R. S., & Scherr, P. A. (2006). Dietary copper and high saturated and trans fat intakes associated with cognitive decline. Archives of *Neurology*, *63*(8), 1085. https://doi.org/10.1001/archneur.63.8.1085

Mozaffari, M., Alva, A. K., & Chen, E. Q. (1996). Relation of copper extractable from soil and pH to copper content and growth of two citrus rootstocks. *Soil Science*, *161*, 786–792.

NAS (National Academy of Sciences). (1977). *Copper. Committee on Medical and Biologic effects of environmental pollutants.* National Research Council. Washington, DC: National Academy of Science.

National Environment Protection Council (NEPC). (2011). *Schedule B5c - Soil quality guidelines for Arsenic, Chromium (III), Copper, DDT, Lead, Naphtalene, Nickel and Zinc, Australia* (Issue April). https://www.nepc.gov.au/sites/default/files/2022-09/schedule-b5c-guideline-soil-quality-guidelines-sep10.pdf

New Zealand Water & Waste Association (NZWWA). (2003). *Guidelines for the Safe Application of Biosolids to Land in New Zealand* (Issue August).

Oorts, K. (2013). *Heavy Metals in Soils*. In B. J. Alloway (Ed.), Vol. 22, pp. 367–394. Netherlands: Springer. https://doi.org/10.1007/978-94-007-4470-7

Organic Materials Review Institute (OMRI). (2018). *Crop Products List (Updated* July 29, *2023)*. https://www.omri.org/us-list

Peña, N., Antón, A., Kamilaris, A., & Fantke, P. (2018). Modeling ecotoxicity impacts in vineyard production: Addressing spatial differentiation for copper fungicides. *Science of the Total Environment*, *616–617*, 796–804. https://doi.org/10.1016/j.scitotenv.2017.10.243

Pietrzak, U., & McPhail, D. C. (2004). Copper accumulation, distribution and fractionation in vineyard soils of Victoria, Australia. *Geoderma*, *122*(2-4 SPEC. IIS.), 151–166. https://doi.org/10.1016/j.geoderma.2004.01.005

Printz, B., Lutts, S., Hausman, J. F., & Sergeant, K. (2016). Copper trafficking in plants and its implication on cell wall dynamics. *Frontiers in Plant Science*, *7*(May), 1–16. https://doi.org/10.3389/fpls.2016.00601

Rasooli Sharabiani, V., Kaveh, M., Abdi, R., Szymanek, M., & Tanaś, W. (2021). Estimation of moisture ratio for apple drying by convective and microwave methods using artificial neural network modeling. Scientific Reports, 11(1), 9155. https://doi.org/10.1038/s41598-021-88270-z

Ruyters, S., Salaets, P., Oorts, K., & Smolders, E. (2013). Copper toxicity in soils under established vineyards in Europe: A survey. *Science of the Total Environment*, *443*, 470–477. https://doi.org/10.1016/j.scitotenv.2012.11.001

Sauvé, S., Hendershot, W., & Allen, H. E. (2000). Solid-solution partitioning of metals in contaminated soils: Dependence on pH, total metal burden, and organic matter. *Environmental Science and Technology*, *34*(7), 1125–1131. https://doi.org/10.1021/es9907764

Schramel, O., Michalke, B., & Kettrup, A. (2000). Study of the copper distribution in contaminated soils of hop fields by single and sequential extraction procedures. *Science of the Total Environment*, *263*(1–3), 11–22. https://doi.org/10.1016/S0048-9697(00)00606-9

Sereni, L., Guenet, B., & Lamy, I. (2023). Mapping risks associated with soil copper contamination using availability and bio-availability proxies at the European scale. *Environmental Science and Pollution Research*, *30*(8), 19828–19844. https://doi.org/10.1007/s11356-022-23046-0

Shen, X.-L., Yu, J.-H., Zhang, D.-F., Xie, J.-X., & Jiang, H. (2014). Positive relationship between mortality from Alzheimer's disease and soil metal concentration in Mainland China. *Journal of Alzheimer's Disease*, *42*(3), 893–900. https://doi.org/10.3233/JAD-140153

Speiser, B., Schärer, H.-J., & Tamm, L. (2018). Direct plant protection in organic farming. In Improving Organic Crop Cultivation (pp. 303–322). Cambridge; Burleigh Dodds Science Publishing.

Squitti, R., Ghidoni, R., Simonelli, I., Ivanova, I. D., Colabufo, N. A., Zuin, M., Benussi, L., Binetti, G., Cassetta, E., Rongioletti, M., & Siotto, M. (2018). Copper dyshomeostasis in Wilson disease and Alzheimer's disease as shown by serum and urine copper indicators. *Journal of Trace Elements in Medicine and Biology*, *45*, 181–188. https://doi.org/10.1016/j.jtemb.2017.11.005

Squitti, R., Lupoi, D., Pasqualetti, P., Dal Forno, G., Vernieri, F., Chiovenda, P., Rossi, L., Cortesi, M., Cassetta, E., & Rossini, P. M. (2002). Elevation of serum copper levels in Alzheimer's disease. *Neurology*, *59*(8), 1153–1161. https://doi.org/10.1212/WNL.59.8.1153

Squitti, R., Pasqualetti, P., Dal Forno, G., Moffa, F., Cassetta, E., Lupoi, D., Vernieri, F., Rossi, L., Baldassini, M., & Rossini, P. M. (2005). Excess of serum copper not related to ceruloplasmin in Alzheimer disease. *Neurology*, *64*(6), 1040–1046. https://doi.org/10.1212/01.WNL.0000154531.79362.23

Tamm, L., Thuerig, B., Apostolov, S., Blogg, H., Borgo, E., Corneo, P. E., Fittje, S., De Palma, M., Donko, A., Experton, C., Marín, É. A., Morell Pérez, Á., Pertot, I., Rasmussen, A., Steinshamn, H., Vetemaa, A., Willer, H., & Herforth-Rahmé, J. (2021). Copper use in Organic Agriculture in Twelve European Countries. www.BLW.admin.ch

Tamm, L., Thuerig, B., Apostolov, S., Blogg, H., Borgo, E., Corneo, P. E., Fittje, S., de Palma, M., Donko, A., Experton, C., Marín, É. A., Pérez, Á. M., Pertot, I., Rasmussen, A., Steinshamn, H., Vetemaa, A., Willer, H., & Herforth-Rahmé, J. (2022). Use of copper-based fungicides in organic agriculture in twelve European countries. Agronomy, *12*(3), 12030673. https://doi.org/10.3390/agronomy12030673

Toselli, M., Baldi, E., Marcolini, G., Malaguti, D., Quartieri, M., Sorrenti, G., & Marangoni, B. (2009). Response of potted grapevines to increasing soil copper concentration. *Australian Journal of Grape and Wine Research*, *15*(1), 85–92. https://doi.org/10.1111/j.1755-0238.2008.00040.x

van Bruggen, A. H. C., Gamliel, A., & Finckh, M. R. (2016). Plant disease management in organic farming systems. *Pest Management Science*, *72*(1), 30–44. https://doi.org/10.1002/ps.4145

Van Zwieten, L., Rust, J., Kingston, T., Merrington, G., & Morris, S. (2004). Influence of copper fungicide residues on occurrence of earthworms in avocado orchard soils. *Science of the Total Environment*, *329*(1–3), 29–41. https://doi.org/10.1016/j.scitotenv.2004.02.014

Vasudevaraju, P., Bharathi, T., Jyothsna, T., Shamasundar, N., Rao, S., Balaraj, B., Rao, K., & Sathyanarayana Rao, T. (2010). New evidence on iron, copper accumulation and zinc depletion and its correlation with DNA integrity in aging human brain regions. *Indian Journal of Psychiatry*, *52*(2), 140. https://doi.org/10.4103/0019-5545.64590

Vavoulidou, E., Avramides, E. J., Papadopoulos, P., Dimirkou, A., Charoulis, A., & Konstantinidou-Doltsinis, S. (2005). Copper content in agricultural soils related to cropping systems in different regions of Greece. *Communications in Soil Science and Plant Analysis*, *36*(4–6), 759–773. https://doi.org/10.1081/CSS-200043367

Vázquez-Blanco, R., Arias-Estévez, M., Bååth, E., & Fernández-Calviño, D. (2020). Comparison of Cu salts and commercial Cu based fungicides on toxicity towards microorganisms in soil. *Environmental Pollution*, *257*, 113585. https://doi.org/10.1016/j.envpol.2019.113585

VROM. (2000). *Circular Values and Intervention Values for Soil Remediation Annex A: Target Values, Soil Remediation Intervention Values and Indictive Levels for Serious. Reference DBO/1999226863 Contamination.*

Wang, Q., Liu, J., Wang, Y., & Yu, H. (2015). Accumulations of copper in apple orchard soils: distribution and availability in soil aggregate fractions. *Journal of Soils and Sediments*, *15*(5), 1075–1082. https://doi.org/10.1007/s11368-015-1065-y

Wang, Q. Y., Sun, J. Y., Xu, X. J., & Yu, H. W. (2020). Distribution and availability of fungicide-derived copper in soil aggregates. *Journal of Soils and Sediments*, *20*(2), 816–823. https://doi.org/10.1007/s11368-019-02441-0

Wang, Q. Y., Zhou, D. M., & Cang, L. (2009). Microbial and enzyme properties of apple orchard soil as affected by long-term application of copper fungicide. *Soil Biology and Biochemistry*, *41*(7), 1504–1509. https://doi.org/10.1016/j.soilbio.2009.04.010

Wang, Y. P., Shi, J. Y., Wang, H., Lin, Q., Chen, X. C., & Chen, Y. X. (2007). The influence of soil heavy metals pollution on soil microbial biomass, enzyme activity, and community composition near a copper smelter. *Ecotoxicology and Environmental Safety*, *67*(1), 75–81. https://doi.org/10.1016/j.ecoenv.2006.03.007

Weng, H. X., Zhang, X. M., Chen, X. H., & Wu, N. Y. (2003). The stability of the relative content ratios of Cu, Pb and Zn in soils and sediments. *Environmental Geology*, *45*(1), 79–85. https://doi.org/10.1007/s00254-003-0859-1

Wightwick, A. M., Mollah, M. R., Partington, D. L., & Allinson, G. (2008). Copper fungicide residues in Australian vineyard soils. *Journal of Agricultural and Food Chemistry*, *56*(7), 2457–2464. https://doi.org/10.1021/jf0727950

Wyszkowska, J., Borowik, A., Kucharski, M., & Kucharski, J. (2013). Oddzia£ywanie kadmu, miedzi i cynku na roœliny, drobnoustroje i enzymy glebowe. *Journal of Elementology*, *18*(4), 769–796. https://doi.org/10.5601/jelem.2013.18.4.455

Xiong, Z. T., & Wang, H. (2005). Copper toxicity and bioaccumulation in Chinese cabbage (Brassica pekinensis Rupr.). *Environmental Toxicology*, *20*(2), 188–194. https://doi.org/10.1002/tox.20094

Yang, X. E., Peng, H. Y., Jiang, L. Y., & He, Z. L. (2005). Phytoextraction of copper from contaminated soil by Elsholtzia splendens as affected by EDTA, citric acid, and compost. *International Journal of Phytoremediation*, *7*(1), 69–83. https://doi.org/10.1080/16226510590915855

Yruela, I. (2005). Copper in plants. *Brazilian Journal of Plant Physiology*, *17*, 145–156.

Yuan, H. M., Xu, H. H., Liu, W. C., & Lu, Y. T. (2013). Copper regulates primary root elongation through PIN1-mediated auxin redistribution. *Plant and Cell Physiology*, *54*(5), 766–778. https://doi.org/10.1093/pcp/pct030

Zhang, C., Nie, S., Liang, J., Zeng, G., Wu, H., Hua, S., Liu, J., Yuan, Y., Xiao, H., Deng, L., & Xiang, H. (2016). Effects of heavy metals and soil physicochemical properties on wetland soil microbial biomass and bacterial community structure. *Science of the Total Environment*, *557–558*, 785–790. https://doi.org/10.1016/j.scitotenv.2016.01.170

15

Arsenic Pathways from the Environment to the Human Body: The Case of Rice Consumption

Thanh Son Tran and Kyoung-Woong Kim

15.1 Presence of Arsenic in the Environment

Arsenic is a naturally occurring element with the atomic number 33. Arsenic compounds have been known since ancient times, with major uses in medicine and industry long before the element was isolated. In the 4th century BCE, Aristotle referred to a substance known as sandarach, which we now recognize as the mineral realgar, a sulfide of arsenic. By the 1st century CE, both Pliny the Elder and Pedanius Dioscorides had described auripigmentum, believed to be the dyestuff orpiment (As_2S_3). The element itself might have been first observed in the 13th century by Albertus Magnus, who noted a metal-like substance when arsenicum (another name for As_2S_3) was heated with soap. However, German pharmacist Johann Schroederirst identified free arsenic in 1649 by heating its oxide with charcoal (Shikha Srivastava, 2015).

Arsenic is a naturally occurring element in the Earth's crust, ranking as the 20th most abundant (Cullen' & Reimer, 1989). It is found in air, water, and soil (Maud & Rumsby, 2008; Nriagu et al., 2007; Vishwakarma et al., 2021) at concentrations of 1.5–2 ppm (National Research Council (US) Committee on Medical and Biological Effects of Environmental Pollutants, 1977). Although arsenic is a major constituent of over 245 mineral species, including arsenides, sulfides, and sulfosalts, metallic arsenic is rarely found in its native form in nature. Arsenopyrite, a compound of iron, arsenic, and sulfur, is the most prevalent arsenic mineral. Other less common arsenic-containing minerals include orpiment, realgar, and enargite, which are arsenic sulfides. Environmental arsenic is often found in higher concentrations within sulfide deposits and is typically associated with metals like copper, lead, and silver.

Arsenic is primarily obtained as a by-product from processing copper, silver, gold, and other metal ores rather than directly from arsenic ores. Environmental regulations mandate the removal of arsenic from ores to prevent its release into the environment through gases, fluids, or solids. Copper-gold deposits in countries like the Philippines and Chile are rich in arsenic. Many nations produce arsenic as a by-product of metal ore smelting, with China leading the production, followed by Chile and Morocco. About a dozen other countries with metal smelters also contribute smaller amounts. The USA has not mined arsenic since 1985 and now imports all the metallic arsenic and arsenic compounds it utilizes. Recycling of arsenic is generally limited, except for waste from factories that produce arsenic compounds and the small-scale recovery of arsenic metal from gallium arsenide (GaAs) semiconductor scrap.

Arsenic is naturally released into the environment through volcanic eruptions and wind-blown dust. Additionally, human activities such as mining and smelting arsenic-containing ores considerably increase its environmental presence. Moreover, coal-fired power plants and waste incineration are important anthropogenic sources, releasing arsenic into the atmosphere through combustion (Masindi & Muedi, 2018). Environmental arsenic cannot be destroyed; it only changes form or binds to other particles. Arsenic may react with oxygen or other molecules in air, water, or soil, or it may be influenced by bacterial activity in soil or sediments. The transformation of arsenic affects its mobility and bioavailability in the environment. For instance, arsenic attached to fine particles can remain airborne for an extended period and travel considerable distances.

DOI: 10.1201/9781003492115-19

The soil concentrations of arsenic vary widely, typically ranging from 0.1 to 40 ppm, depending on the underlying geology and human activities. Soils overlying arsenic-rich ore deposits may contain several hundred parts per million of arsenic, but extreme cases of up to 8000 ppm have been reported (National Research Council (US) Committee on Medical and Biological Effects of Environmental Pollutants, 1977). Surface, agricultural, garden, and rhizospheric arsenic concentrations in 17 locations varied between 126 and 1600 mg/kg. Rhizospheric soil showed the maximum concentration due to plant and microorganism interactions. High concentrations of arsenic have been reported in soil in India, especially in the states of Assam, Bihar, Uttar Pradesh, and West Bengal, with levels reaching up to 41.2 mg/kg. In other countries, such as Bangladesh, Pakistan, Nepal, China, Vietnam, Iran, and South Africa, soil arsenic concentrations of 7–46, 3–266, 6.1–16.7, 9.7–110, 6–20, 7–1500, and 18.3 mg/kg, respectively, were reported (Patel et al., 2023). Furthermore, arsenic concentrations in sediments ranged from <0.5 to 30 mg/kg, increasing with increasing depth (Keesari et al., 2022).

Arsenic is highly soluble in water and often adheres to particles in water or sediment, enabling it to enter lakes, rivers, and groundwater. The arsenic concentrations generally range from 1 to 2 μg/L in natural waters, including open ocean seawater. However, in regions with volcanic rock, sulfide mineral deposits, or other natural sources, the concentrations can reach up to 12 mg/L. Elevated levels are also found near anthropogenic sources, such as mining and agrochemical manufacturing, as well as in geothermal waters, which have a mean concentration of 500 μg/L and can reach up to 25 mg/L (WHO, 2011).

Arsenic concentrations in the air vary significantly between regions. The arsenic concentrations in remote or rural areas range from 0.02 to 4 ng/m^3, whereas in urban areas, they range from 3 to 200 ng/m^3. Near industrial sources, the concentration can exceed 1000 ng/m^3 (WHO, 2011). The significant differences in concentration between regions are due to human activities because arsenic is released by combustion and typically forms highly soluble oxides. The wind carries these particles and eventually settles back to the ground through wet or dry deposition. Arsenic is naturally present in coal and oil, and power plants release arsenic into the atmosphere (Pacyna, 2006). Thus, arsenic contamination in water, soil, and air is a primary environmental and public health concern (Bhat et al., 2024).

15.2 Arsenic Emissions in Human Habitats

Arsenic exists in natural and anthropogenic forms, posing enormous global environmental, agricultural, and health challenges. In this section, we examine the sources and mobility of arsenic.

15.2.1 Natural sources

Arsenic can enter human habitats through various natural processes. The primary source of arsenic in soils is the parent rock (Smedley & Kinniburgh, 2002). Weathering of arsenic minerals and rainwater releases arsenic into the soil. However, volcanic eruptions also emit arsenic-containing gases and particles, contaminating soil.

Approximately one-third of the arsenic present in Earth's atmosphere originates from natural sources. The most important natural source is volcanic eruptions, followed by arsenic-containing vapors from solid or liquid arsenic salts at low temperatures (Hatzis & Karatolios, 2016). Volcanic eruptions may account for 20%–40% of total natural arsenic emissions, atmospheric aerosols contribute 30%–50%, and sea salt aerosols contribute $<10\%$ (Nriagu et al., 2007).

Leaching from geological formations is a primary natural source of arsenic that strongly affects its levels in groundwater (Shaji et al., 2021). Elevated concentrations of arsenic in groundwater have been reported in many countries, including Bangladesh (Meng-Hang et al., 2014), Vietnam (Stopelli et al., 2020; Winkel et al., 2011), China (Guo et al., 2014), Taiwan (Jean et al., 2010), Argentina (Litter et al., 2019), and Canada (Wang & Mulligan, 2006), primarily due to these natural processes. Areas of thermal activity or regions with rocks with a high arsenic content often have high levels of arsenic in groundwater.

15.2.2 Human Activities

Human activities contribute markedly to the emission of arsenic into the environment. Industrial uses of arsenic, including mining, metal processing, and the production and use of pesticides and fertilizers, are important anthropogenic sources. For instance, the incineration of preserved wood products treated with chromate copper arsenate releases arsenic into the environment. Furthermore, arsenic-treated lumber discarded in landfills and emissions from coal-fired power plants substantially increases environmental arsenic levels.

The primary sources of arsenic contamination from mining activities include ore processing, mine tailings, and acid mine drainage, which involve several geochemical and biological processes. During mining operations, arsenic sulfide minerals, such as arsenopyrite, realgar, and orpiment, are exposed to air and water, resulting in the oxidation of sulfides, which can produce sulfuric acid that dissolves arsenic and other heavy metals, leading to the mobilization of these metals in water bodies (Zhuang et al., 2023). The waste materials or tailings generated by the extraction of valuable minerals often contain large amounts of arsenic that can leach arsenic into soil and water over time. Acid mine drainage can also increase the leaching of arsenic from mine tailings and waste rock piles; the acidic conditions solubilize arsenic, allowing its migration into groundwater and surface water. The arsenic is then taken up by plants and aquatic organisms, entering the food chain and posing risks to human health and wildlife (Faria et al., 2023).

The crushing and grinding of ore release arsenic into the air as dust (Martin et al., 2014), and the extraction and refining of metals such as copper, gold, and lead often involve the release of arsenic. For example, arsenic can be released through ore smelting, roasting, and hydrometallurgical processes. During smelting, the ores are heated to extract metals, and arsenic can be released into the atmosphere as arsenic trioxide, a highly toxic compound. Roasting is a method of oxidizing sulfide ores that can release arsenic oxides into the environment as dust or dissolved in water (Shamsuddin, 2021). Hydrometallurgical processes use aqueous chemistry to extract metals from ores, and arsenic can be mobilized and released into the process water, potentially contaminating local water bodies if it is not managed appropriately (Shi et al., 2019).

Historically, arsenic-based pesticides were widely used in agriculture. These included compounds like lead arsenate and calcium arsenate. The use of arsenic compounds in agriculture, mainly wood preservatives, gained momentum during the 1970s. By 1980, 70% of arsenic consumption in the USA was for wood preservation. However, this practice was eventually halted because the US Environmental Protection Agency banned using the arsenic and chromium-based wood preservative known as chromated copper arsenate. Arsenical herbicides containing cacodylic acid as an active ingredient were extensively used in the USA, with purposes ranging from golf courses to cotton fields and aiding plant drying before harvesting (Bencko & Foong, 2017). Although the use of these herbicides has declined due to health and environmental concerns, residual contamination remains in soil and water (US Environmental Protection Agency, 2004). Arsenic can also be present in some commercial fertilizers, especially those derived from phosphate rock and can leach into the soil over time (Chaney, 2012).

15.3 Arsenic Pathways from the Environment to the Human Body

15.3.1 Inhalation Pathway

Arsenic exposure through inhalation primarily occurs in environments where arsenic is present in the air, often attached to particulate matter. Airborne arsenic can be a concern in areas with intensive industrial activity or where arsenic pesticides have been used. The inhalation of arsenic trioxide, particularly in the workplace, is a major risk, and 60% of the body absorbs 90% of inhaled arsenic (Sullivan & Krieger, 2001). The absorption of airborne arsenic particles via the lungs involves two critical processes: deposition and absorption. Arsenic particles are deposited onto the lung surface after inhalation and absorbed from the deposited material into the bloodstream.

In lung cancer patients exposed to arsenic in cigarette smoke, approximately 40% of the inhaled arsenic is thought to be deposited in the lungs, and 75%–85% of deposited arsenic is absorbed. Accordingly, 30%–34% of inhaled arsenic is absorbed through the lungs. In workers exposed to arsenic trioxide dust in smelters, the amount of arsenic excreted in urine (the primary route of excretion) is approximately 40%–60% of the estimated inhaled dose. Studies assessing urinary metabolites show that arsenic absorption correlates with time-weighted average arsenic air concentrations determined using personal breathing zone air samplers (Yager et al., 1997). Although the exact deposition percentage was not measured in these cases, nearly all deposited arsenic is probably absorbed. Intratracheal instillation of arsenic compounds, such as sodium arsenite, sodium arsenate, and arsenic trioxide, resulted in rapid and nearly complete clearance from the lungs (approximately 60%–90% within 1 day). However, less soluble forms, such as arsenic sulfide and lead arsenate, clear more slowly. This suggests lower absorption rates for highly insoluble inhaled arsenic (Life Systems, Inc, United States. Agency for Toxic Substances and Disease Registry, 1993).

15.3.2 Dermal Pathway

Dermal exposure is a less common exposure route than ingestion and inhalation. Certain occupational accidents, such as those involving coal-fired power plants or incinerators, as well as coal or other products that come into contact with arsenic through ash or environmental release (Baker et al., 2018), involving compounds like arsenic trichloride or arsenic acid, can cause toxic systemic effects through skin contact. Although rare, these instances highlight the potential for arsenic to enter the body via the dermal route under specific circumstances.

15.3.3 Ingestion Pathway

The primary route of arsenic entering the human body is ingestion of contaminated water and food. Arsenic concentrations exceeding permissible limits in food and drinking water are a widespread problem caused by water pollution in developed and developing nations. Exposure to elevated arsenic levels poses life-threatening risks. Water sources, including surface water, groundwater, and food sources like fish, crops, and cereals, contribute significantly to human arsenic exposure (Khosravi-Darani et al., 2022). Inorganic arsenic is highly soluble in water, making contaminated groundwater a major source of exposure. Contaminated water is used for drinking and preparing food and irrigating crops, thereby introducing arsenic into the food chain. Foods such as rice, which are often cultivated using arsenic-laden water, frequently contain elevated levels of arsenic, posing additional risks when consumed regularly.

15.4 Arsenic Ingestion via Rice Consumption

Rice is a staple food for many of the world's population. It is often grown in water-saturated conditions, making it particularly susceptible to absorbing any arsenic in the irrigation water. This section explores the pathways through which arsenic enters the human body from rice consumption and the potential health impacts.

15.4.1 Rice Cultivation

Rice (*Oryza sativa*) is an edible starchy cereal grain derived from the grass plant family (Poaceae). Archaeological evidence shows rice consumption dating back to 8000 BCE, although the exact timeline of its domestication remains debated. Rice is vital in global food security, particularly in East and Southeast Asia, where approximately half the world's population relies on it as a staple food. Most of the world's rice crop is consumed by humans rather than livestock. The countries with the highest daily rice consumption per capita include Bangladesh, the Lao People's Democratic Republic, Cambodia, Vietnam, Myanmar, Thailand, Indonesia, and the Philippines (Muthayya et al., 2014). Globally, two domesticated rice species are cultivated, *O. sativa* (Asian) and *Oryza glaberrima* (African) (Sweeney & McCouch, 2007), although *O. glaberrima* has mostly been supplanted by the higher-yielding *O. sativa*, leading to a decline in the varieties of *O. glaberrima*. Nevertheless, *O. glaberrima* still accounts for about 20% of the rice cultivated in West Africa. However, it is now seldom found in West African markets, replaced mainly by Asian strains (Linares, 2002).

Rice is predominantly cultivated in regions where flooding is used to grow crops, especially in South and Southeast Asia (Panda & Barik, 2021). Rice cultivation has shaped many cultures in China, India, and Southeast Asia. Although each region has unique methods for cultivating rice, the process generally involves the following stages. Farmers meticulously prepare the land using precision tools to create uniform grades and slopes and optimize water management. The ideal soils for rice cultivation are heavy clay and silt loam, which retain water effectively. Rice can be either directly seeded in the field or transplanted from seedbeds. Direct seeding requires more seeds but less labor, whereas transplanting is labor-intensive but uses fewer seeds.

Traditional methods involve constructing earthen levees to flood the fields, maintaining a water layer 5–10cm deep. Modern techniques, such as poly-tubing and periodic irrigation, are also used to conserve water and increase efficiency. Rice plants grow to a height of 7.6–10.2cm over approximately 120days. Farmers ensure that the fields remain flooded during this period to prevent soil erosion and improve water quality. Once mature, the fields are drained and combined to cut the rice and separate the grain from the stalks. The harvested grains are dried to reduce moisture content before milling (Manikandan, 2023).

Depending on the region, planting typically takes place between late February to May. The duration of rice cultivation ranges from 100 to 210days, depending on the variety and growing conditions. In temperate regions, the average time from sowing to harvest is 130–150days. In Vietnam, rice cultivation duration varies by region and variety, with farming divided into three main seasons: winter–spring (December to April, 120–150days), summer–autumn (May to August, 100–120days), and autumn–winter (September to December, 90–100days) (Hong Van et al., 2014).

Prolonged flooding in paddy fields can mobilize arsenic from soil minerals into the water, where rice plants absorb it through their roots. Using arsenic-contaminated groundwater for irrigation significantly increases the arsenic concentration in rice grains compared with those irrigated with arsenic-free water.

15.4.2 Pathway of Arsenic into the Human Body from Rice Consumption

Arsenic can be present in both organic and inorganic forms. The term "organic" in chemistry refers to carbon-containing compounds. In an organic arsenic compound, the arsenic atom is bonded to a carbon atom, which could be part of a sugar molecule, such as ribose. These organic arsenic compounds have a more complex structure than inorganic ones and are generally harmless. Conversely, inorganic arsenic compounds do not contain carbon, are typically simpler molecules, such as arsenic trioxide, and are highly toxic (Hughes, 2002). Natural processes, such as weathering and volcanic eruptions, and human activities, such as mining, arsenic-based pesticides, and industrial processes, increase arsenic levels in the environment.

Rice is typically grown in flooded conditions, which create an anaerobic environment in the soil. Under these conditions, arsenic becomes more mobile and bioavailable (Bradham et al., 2018). In paddy soils, both organic and inorganic forms of arsenite (As(III)) and arsenate (As(V)) are the primary arsenic species that affect soil conditions, and methylated arsenic species are also present (Li et al., 2016). In irrigation water, arsenic (both arsenate and arsenite) derived mainly from groundwater-based irrigation may temporarily sink into the topsoil and can be easily released into the water–soil system and embedded in the food chain, including rice (Akinbil & Haque, 2012; Sandhi et al., 2022). Plant roots can selectively absorb certain forms of arsenic, and all arsenic species can be transported into plant cells through specific transporter proteins (Mitra et al., 2017).

Figure 15.1 shows the uptake and translocation of arsenic in rice plants. The process depends strongly on the species of arsenic present in the rhizosphere. The presence of silicon and phosphorus in soil and pore water also affects arsenic uptake by rice plants. There are two primary pathways for arsenic uptake in rice. The first is the phosphate transport pathways, which take up arsenate because it is a phosphate analog. The second is aquaporin channels used for silicate transport in the roots, which take up arsenite because it is a silicic acid analog, as well as undissociated methylated arsenic species, such as dimethylarsinic acid and monomethyl arsenic acid. There is considerable variation in the uptake efficiency of different arsenic species. Compared with inorganic arsenic species, the uptake efficiency of methylated arsenic species is lower, and the translocation efficiency is higher. The transfer of arsenic to rice grains is more significant than in other cereals, probably due to the high soil/shoot ratio and differences in arsenic speciation and dynamics in aerobic and anaerobic soils (Islam et al., 2016a). Rice plants cultivated in arsenic-contaminated regions

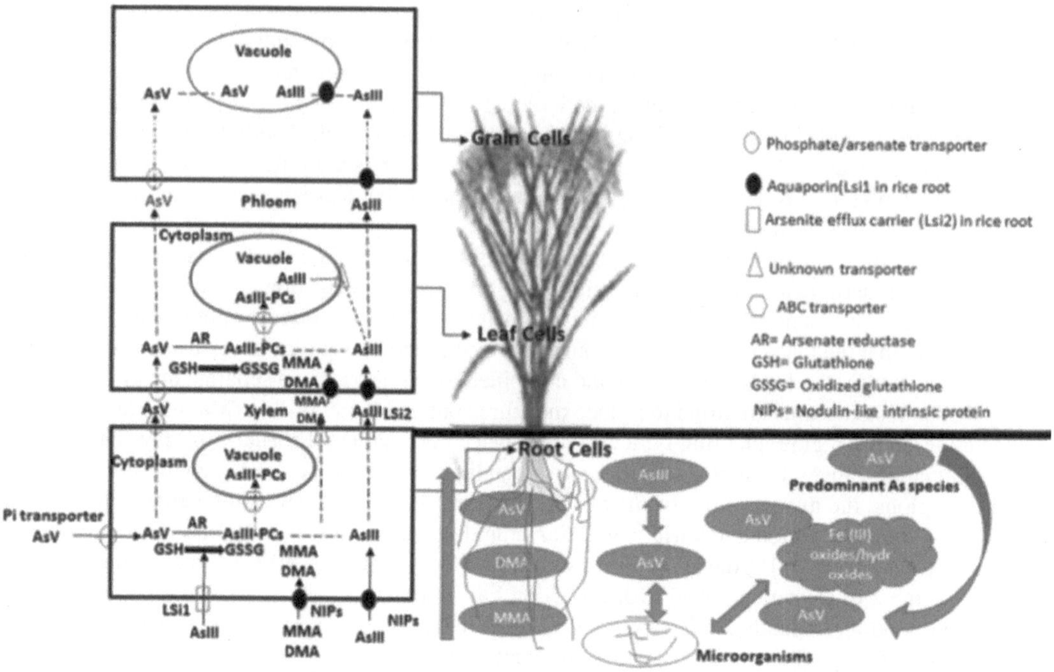

FIGURE 15.1 Possible mechanisms of arsenic uptake in rice plants (Ali et al., 2009; Islam et al., 2016b).

can accumulate up to 2 mg/kg of arsenic in their grains and up to 92 mg/kg in their vegetative parts (Shukla et al., 2023). The proportion of organic and inorganic arsenic in rice grains varies considerably based on factors such as rice variety, cultivation conditions, soil composition, and location. Inorganic arsenic, especially arsenite and arsenate, typically constitutes around 60%–80% of the total arsenic content in rice grains (Begum et al., 2024; Meharg & Zhao, 2012; Sabbagh, 2023; Singh et al., 2019).

The primary pathway of arsenic into the human body from rice consumption is through ingestion. Rice can accumulate arsenic in higher concentrations than other food crops due to its cultivation methods (Karagas et al., 2019; Rokonuzzaman et al., 2022). The amount of rice consumed per person varies hugely by region and dietary habits. On average, the average rice consumption per capita is highest in South Asia (630 g/day), followed by Southeast Asia (239 g/day) and China (200 g/day), and is much lower in South America (68 g/day) and Europe (18 g/day) (Bhavadharini et al., 2020). Once absorbed, arsenic is distributed via the bloodstream to various organs, including the liver, kidneys, spleen, and lungs. The gastrointestinal tract absorbs about 95% of ingested soluble trivalent arsenic compounds. Regular consumption of arsenic-laden rice leads to a cumulative increase in arsenic levels in the human body, which can pose serious health risks.

The daily excretion of arsenic is influenced by the level of exposure and an individual's metabolic rate. Typically, a person is exposed to around 50 μg of arsenic per day from food, water, and air sources. Most of this arsenic is eliminated through urine. Research indicates that the average daily urinary excretion of arsenic ranges from 30 to 60 μg. Studies on humans who ingested known amounts of arsenite or arsenate show that only a tiny fraction is excreted in feces, with 45%–85% being excreted in urine within 1–3 days. At low exposure levels, urinary arsenic levels generally increase linearly with intake. Overall, the body eliminates arsenic relatively quickly, with half-lives of 40–60 h in humans (Atlanta (GA): Agency for Toxic Substances and Disease Registry (US), 2007). After high-dose exposure, the elimination of arsenic is prolonged. In the case of acute poisoning, elevated levels of arsenic in urine can still be detected several weeks after exposure has ended (Kosnett, 2017). Arsenic excretion patterns were monitored in a group consuming a specific rice diet. Rice consumption led to a consistent increase in urine arsenic levels. At a steady state, approximately 40% of the arsenic from rice was excreted through urine (Meharg et al., 2014).

15.4.3 Adverse Effects of Arsenic on Human Health

Long-term exposure to arsenic through rice consumption can lead to chronic arsenic poisoning, called arsenicosis. Health effects associated with arsenic exposure include skin lesions, cardiovascular diseases, and increased risk of various cancers, particularly skin, lung, and bladder cancers. Additionally, arsenic exposure has been linked to developmental effects, neurotoxicity, and adverse pregnancy outcomes. Accordingly, arsenic exposure is a critical public health issue.

Acute arsenic poisoning results from ingesting high levels of arsenic in a short period. Symptoms usually manifest within 30 min and include severe gastrointestinal distress, characterized by abdominal pain, nausea, vomiting, and diarrhea, because of the high sensitivity of the gastrointestinal tract to arsenic, leading to inflammation and damage. Cardiovascular effects, such as hypotension (low blood pressure), arrhythmia (irregular heartbeats), and shock, can occur as arsenic interferes with cellular respiration and enzyme function. Neurological symptoms may include headache, dizziness, confusion, and convulsion, with severe cases potentially leading to coma and death (Saha et al., 1999). The lethal dose of arsenic for humans can vary, but it is estimated to range from 70 to 180 mg of ingested arsenic trioxide. This equates to approximately 2–20 mg/kg of body weight. For an average adult, this is approximately 140–1400 mg. Even a small amount, less than 1/8 teaspoon, can be fatal to a healthy adult, and a smaller amount can be deadly for children, older adults, or those with compromised health (Debnath et al., 2020).

Long-term exposure to arsenic, often through consumption of contaminated food, especially rice (Hassan et al., 2017; Rokonuzzaman et al., 2022), has more insidious effects. The International Agency for Research on Cancer classifies inorganic arsenic as a Group 1 carcinogen, indicating sufficient evidence of carcinogenicity in humans (Martinez et al., 2011). Prolonged consumption of arsenic-laden rice leads to cellular damage and genetic mutations, disrupting normal cellular processes and promoting malignant transformations. Populations with high rice consumption exhibit higher incidences of these cancers, highlighting the significant health risks associated with dietary arsenic exposure (Sanchez et al., 2020). One of the earliest signs of chronic arsenic exposure is skin changes, including hyperpigmentation, the appearance of dark patches, and the development of small, corn-like growths called keratoses. Arsenic exposure can have severe neurological effects, including peripheral neuropathy, characterized by numbness and tingling in the extremities. In children, chronic exposure can lead to delayed development and cognitive impairment. Pregnant women exposed to arsenic are at risk of fetal complications and birth defects (Tyler & Allan, 2014). Arsenic poisoning is also associated with diabetes, pulmonary diseases, and gastrointestinal problems (Muzaffar et al., 2023). The element interferes with cellular metabolism and enzyme function, leading to widespread organ damage.

Monitoring arsenic levels in rice and implementing strategies to mitigate exposure is essential for reducing its associated health risks. These strategies might include developing rice varieties that are less capable of absorbing arsenic or using water management techniques that minimize arsenic uptake. Public awareness and regulatory measures are also crucial to managing arsenic's impact in rice-consuming regions.

REFERENCES

Akinbil, O. C., & Haque, A. M. M. (2012). Arsenic contamination in irrigation water for rice production in Bangladesh: a review. *Trends in Applied Sciences Research*, *7*(5), 331–349. https://doi.org/10.3923/tasr.2012.331.349

Ali, W., Isayenkov, S. V., Zhao, F. J., & Maathuis, F. J. M. (2009). Arsenite transport in plants. *Cellular and Molecular Life Sciences*, 66(14), 2329–2339. https://doi.org/10.1007/s00018-009-0021-7

Atlanta (GA): Agency for Toxic Substances and Disease Registry (US). (2007). Toxicological Profile for Arsenic. The Agency, Atlanta, GA. https://www.ncbi.nlm.nih.gov/books/NBK591624/#ch3.s4.4

Baker, B. A., Cassano, V. A., & Murray, C. (2018). Arsenic exposure, assessment, toxicity, diagnosis, and management. *Journal of Occupational & Environmental Medicine*, 60(12), e634–e639. https://doi.org/10.1097/JOM.0000000000001485

Begum, M., Li, L., Young, E., Carey, M., Li, G., Zhu, Y. G., Banwart, S., Meharg, A. A., & Meharg, C. (2024). Fertilization enhances grain inorganic arsenic assimilation in rice. *Exposure and Health*, 16(2), 417–430. https://doi.org/10.1007/s12403-023-00563-y

Bencko, V., & Foong, F. Y. L. (2017). The history of arsenical pesticides and health risks related to the use of agent blue. *Annals of Agricultural and Environmental Medicine*, 24(2), 312–316. https://doi.org/10.26444/aaem/74715

Bhat, A., Ravi, K., Tian, F., & Singh, B. (2024). Arsenic contamination needs serious attention: an opinion and global scenario. *Pollutants*, 4(2), 196–211. https://doi.org/10.3390/pollutants4020013

Bhavadharini, B., Mohan, V., Dehghan, M., Rangarajan, S., Swaminathan, S., Rosengren, A., Wielgosz, A., Avezum, A., Lopez-Jaramillo, P., Lanas, F., Dans, A. L., Yeates, K., Poirier, P., Chifamba, J., Alhabib, K. F., Mohammadifard, N., Zatońska, K., Khatib, R., Keskinler, M. V., & Yusuf, S. (2020). White rice intake and incident diabetes: a study of 132,373 participants in 21 countries. *Diabetes Care*, 43(11), 2643–2650. https://doi.org/10.2337/dc19-2335

Bradham, K. D., Diamond, G. L., Burgess, M., Juhasz, A., Klotzbach, J. M., Maddaloni, M., Nelson, C., Scheckel, K., Serda, S. M., Stifelman, M., & Thomas, D. J. (2018). In vivo and in vitro methods for evaluating soil arsenic bioavailability: relevant to human health risk assessment. *Journal of Toxicology and Environmental Health - Part B: Critical Reviews*, 21(2), 83–114. https://doi.org/10.1080/10937404.2018.1440902

Chaney, R. (2012). Food safety issues for mineral and organic fertilizers. *Advances in Agronomy*, 117, 51–116. https://doi.org/10.1016/B978-0-12-394278-4.00002-7

Cullen', W. R., & Reimer, K. J. (1989). Arsenic speciation in the environment. Chemical reviews, 89(4), 713–764.

Debnath, D., Nath, B. D., Pervin, R., & Hossain, Md. A. (2020). Arsenic toxicity. In Metal Toxicology Handbook (pp. 373–391). CRC Press, Boca Raton, FL. https://doi.org/10.1201/9780429438004-28

Faria, M. C. da S., Hott, R. de C., Santos, M. J. dos, Santos, M. S., Andrade, T. G., Bomfeti, C. A., Rocha, B. A., Barbosa, F., & Rodrigues, J. L. (2023). Arsenic in mining areas: environmental contamination routes. *International Journal of Environmental Research and Public Health*, 20(5), 20054291. https://doi.org/10.3390/ijerph20054291

Guo, H., Wen, D., Liu, Z., Jia, Y., & Guo, Q. (2014). A review of high arsenic groundwater in Mainland and Taiwan, China: distribution, characteristics and geochemical processes. *Applied Geochemistry*, 41, 196–217. https://doi.org/10.1016/j.apgeochem.2013.12.016

Hassan, F. I., Niaz, K., Khan, F., Maqbool, F., & Abdollahi, M. (2017). The relation between rice consumption, arsenic contamination, and prevalence of diabetes in South Asia. *EXCLI Journal*, 16, 1132–1143. https://doi.org/10.17179/excli2017-222

Hatzis, G., & Karatolios, K. (2016). *Chapter 6- The Contribution of Pollutants and Environmental Chemicals in Cardiovascular Disease*. In: N. Papageorgiou (Ed.), Cardiovascular Diseases, pp. 119–136. Academic Press, New York. https://doi.org/10.1016/B978-0-12-803312-8.00006-9

Hong Van, N. P., Thi Nga, T., Arai, H., Hosen, Y., Huu Chiem, T., & Inubushi, K. (2014). Rice straw management by farmers in a triple rice production system in the Mekong delta, viet nam. *Tropical Agriculture Development*, 58(4), 155–162.

Hughes, M. F. (2002). Arsenic toxicity and potential mechanisms of action. *Toxicology Letters*, 133(1), 1–16. https://doi.org/10.1016/S0378-4274(02)00084-X

Islam, S., Rahman, M. M., Islam, M. R., & Naidu, R. (2016a). Arsenic accumulation in rice: consequences of rice genotypes and management practices to reduce human health risk. *Environment International*, 96, 139–155. https://doi.org/10.1016/j.envint.2016.09.006

Islam, S., Rahman, M. M., Islam, M. R., & Naidu, R. (2016b). Arsenic accumulation in rice: consequences of rice genotypes and management practices to reduce human health risk. *Environment International*, 96, 139–155. https://doi.org/10.1016/j.envint.2016.09.006

Jean, J.-S., Bundschuh, J., Chen, C.-J., Guo, H.-R., Liu, C.-W, Lin, T.-F., & Chen, Y.-H. (2010). *The Taiwan Crisis: A Showcase of the Global Arsenic Problem*. CRC Press, London. https://doi.org/10.1201/b10535

Karagas, M. R., Punshon, T., Davis, M., Bulka, C. M., Slaughter, F., Karalis, D., Argos, M., & Ahsan, H. (2019). Rice intake and emerging concerns on arsenic in rice: a review of the human evidence and methodologic challenges. *Current Environmental Health Reports*, 6(4), 361–372. https://doi.org/10.1007/s40572-019-00249-1

Keesari, T., Chidambaram, S., Pethaperumal, S., Kamaraj, P., Sharma, D. A., & Sinha, U. K. (2022). Arsenic distribution in sediments of multi-tier sedimentary formation of coastal Pondicherry, India – Implications on groundwater quality. *Marine Pollution Bulletin*, 174, 113193. https://doi.org/10.1016/j.marpolbul.2021.113193

Khosravi-Darani, K., Rehman, Y., Katsoyiannis, I. A., Kokkinos, E., & Zouboulis, A. I. (2022). Arsenic exposure via contaminated water and food sources. *Water (Switzerland)*, 14(12), 14121884. https://doi.org/10.3390/w14121884

Kosnett, M. J. (2017). Arsenic. In Critical Care Toxicology: Diagnosis and Management of the Critically Poisoned Patient (pp. 1639–1667). Springer International Publishing, Singapore. https://doi.org/10.1007/978-3-319-17900-1_16

Li, N., Wang, J., & Song, W.-Y. (2016). Arsenic uptake and translocation in plants. *Plant and Cell Physiology*, 57(1), 4–13. https://doi.org/10.1093/pcp/pcv143

Life Systems, Inc, United States. (1993). *Toxicological Profile for Arsenic*. Agency for Toxic Substances and Disease Registry, Atlanta, GA.

Linares, O. F. (2002). African rice (Oryza glaberrima): history and future potential. *Proceedings of the National Academy of Sciences of the United States of America*, 99(25), 16360–16365. https://doi.org/10.1073/pnas.252604599

Litter, M. I., Ingallinella, A. M., Olmos, V., Savio, M., Difeo, G., Botto, L., Farfán Torres, E. M., Taylor, S., Frangie, S., Herkovits, J., Schalamuk, I., González, M. J., Berardozzi, E., García Einschlag, F. S., Bhattacharya, P., & Ahmad, A. (2019). Arsenic in Argentina: occurrence, human health, legislation and determination. *Science of the Total Environment*, 676, 756–766. https://doi.org/10.1016/j.scitotenv.2019.04.262

Manikandan, K. (2023). *Pocket Hand Book on Rice Cultivation (Issue December)*. Thannambikkai, Coimbatore, Tamil Nadu.

Martin, R., Dowling, K., Pearce, D., Sillitoe, J., & Florentine, S. (2014). Health effects associated with inhalation of airborne arsenic arising from mining operations. *Geosciences (Switzerland)*, 4(3), 128–175. https://doi.org/10.3390/geosciences4030128

Martinez, V. D., Vucic, E. A., Becker-Santos, D. D., Gil, L., & Lam, W. L. (2011). Arsenic exposure and the induction of human cancers. *Journal of Toxicology*, 2011, 1–13. https://doi.org/10.1155/2011/431287

Masindi, V., & Muedi, K. L. (2018). Environmental contamination by heavy metals. In: E. D. M. Saleh & R. F. Aglan (Eds.), Heavy Metals, p. 7. IntechOpen, London. https://doi.org/10.5772/intechopen.76082

Maud, J., & Rumsby, P. (2008). *A Review of the Toxicity of Arsenic in Air*. Environment Agency, Bristol, UK.

Meharg, A. A., & Zhao, F.-J. (2012). Risk from arsenic in rice grain. *Arsenic & Rice*, 9789400729, 31–50. https://doi.org/10.1007/978-94-007-2947-6_3

Meharg, A. A., Williams, P. N., Deacon, C. M., Norton, G. J., Hossain, M., Louhing, D., Marwa, E., Lawgalwi, Y., Taggart, M., Cascio, C., & Haris, P. (2014). Urinary excretion of arsenic following rice consumption. *Environmental Pollution*, 194, 181–187. https://doi.org/10.1016/j.envpol.2014.07.031

Meng-Hang, Y., Yuan-Sheng, Z., Hai, H., Kun, C., Bing, L., Guang-Yuan, S., & Xue-Wei, Z. (2014). Arsenic trioxide exerts anti-lung cancer activity by inhibiting angiogenesis. *Current Cancer Drug Targets*, 14(6), 557–566. https://doi.org/10.2174/1568009614666140725090000

Mitra, A., Chatterjee, S., Moogouei, R., & Gupta, D. (2017). Arsenic accumulation in rice and probable mitigation approaches: a review. *Agronomy*, 7(4), 67. https://doi.org/10.3390/agronomy7040067

Muthayya, S., Sugimoto, J. D., Montgomery, S., & Maberly, G. F. (2014). An overview of global rice production, supply, trade, and consumption. *Annals of the New York Academy of Sciences*, 1324(1), 7–14. https://doi.org/10.1111/nyas.12540

Muzaffar, S., Khan, J., Srivastava, R., Gorbatyuk, M. S., & Athar, M. (2023). Mechanistic understanding of the toxic effects of arsenic and warfare arsenicals on human health and environment. *Cell Biology and Toxicology*, 39(1), 85–110. https://doi.org/10.1007/s10565-022-09710-8

National Research Council (US) Committee on Medical and Biological Effects of Environmental Pollutants. (1977). *Arsenic: Medical and Biologic Effects of Environmental Pollutants*. National Academies Press (US), Washington, DC. https://doi.org/10.1097/00043764-197506000-00009

Nriagu, J., Bhattacharya, P., Mukherjee, A., Bundschuh, J., Zevenhoven, R., & Loeppert, R. (2007). Arsenic in soil and groundwater: an overview. In Arsenic in Soil and Groundwater Environment - Biogeochemical Interactions, Health Effects and Remediation (pp. 3–60). Elsevier, Amsterdam, The Netherlands. https://doi.org/10.1016/S0927-5215(06)09001-1

Pacyna, J. M. (2006). *Atmospheric Emissions of Arsenic, Cadmium, Lead and Mercury from High Temperature Processes in Power Generation and Industry* (pp.69–87). John Wiley & Sons Ltd, Norway.

Panda, D., & Barik, J. (2021). Flooding tolerance in rice: focus on mechanisms and approaches. *Rice Science*, 28(1), 43–57. https://doi.org/10.1016/j.rsci.2020.11.006

Patel, K. S., Pandey, P. K., Martín-Ramos, P., Corns, W. T., Varol, S., Bhattacharya, P., & Zhu, Y. (2023). A review on arsenic in the environment: contamination, mobility, sources, and exposure. *RSC Advances*, *13*(13), 8803–8821. https://doi.org/10.1039/d3ra00789h

Rokonuzzaman, M., Li, W. C., Man, Y. B., Tsang, Y. F., & Ye, Z. (2022). Arsenic accumulation in rice: sources, human health impact and probable mitigation approaches. *Rice Science*, 29(4), 309–327. https://doi.org/10.1016/j.rsci.2022.02.002

Sabbagh, S. (2023). Arsenic contamination in rice, radiation and chemical methods of measurement, and implications for food safety. *Journal of Food Science and Technology*, 60(7), 1870–1887. https://doi.org/10.1007/s13197-022-05469-2

Saha, J. C., Dikshit, A. K., Bandyopadhyay, M., & Saha, K. C. (1999). A review of arsenic poisoning and its effects on human health. *Critical Reviews in Environmental Science and Technology*, 29(3), 281–313. https://doi.org/10.1080/10643389991259227

Sanchez, T. R., Klu, Y. A. K., Genkinger, J. M., Lacey, J. V., Chung, N. T., & Navas-Acien, A. (2020). Association between rice consumption and risk of cancer incidence in the California teachers study. *Cancer Causes & Control*, 31(12), 1129–1140. https://doi.org/10.1007/s10552-020-01350-9

Sandhi, A., Yu, C., Rahman, M. M., & Amin, M. N. (2022). Arsenic in the water and agricultural crop production system: Bangladesh perspectives. *Environmental Science and Pollution Research*, 29(34), 51354–51366. https://doi.org/10.1007/s11356-022-20880-0

Shaji, E., Santosh, M., Sarath, K. V., Prakash, P., Deepchand, V., & Divya, B. V. (2021). Arsenic contamination of groundwater: a global synopsis with focus on the Indian Peninsula. *Geoscience Frontiers*, 12(3), 101079. https://doi.org/10.1016/j.gsf.2020.08.015

Shamsuddin, M. (2021). Roasting of sulfide minerals. In M. Shamsuddin (Ed.), Physical Chemistry of Metallurgical Processes (pp. 39–68). Springer International Publishing, Singapore. https://doi.org/10.1007/978-3-030-58069-8_2

Shi, M.-Q., Chai, L.-Y., & Liang, Y.-J. (2019). Pollution source distribution of arsenic in the typical smelter. In Arsenic Pollution Control in Nonferrous Metallurgy (pp. 17–27). Springer, Singapore. https://doi.org/10.1007/978-981-13-6721-2_2

Shikha Srivastava, A. K. D. (2015). Natural bio-remediation of arsenic contamination: a short review. *Journal of Waste Water Treatment & Analysis*, 06(01), 1000186. https://doi.org/10.4172/2157-7587.1000186

Shukla, J., Narayan, S., Mishra, A., Shirke, P. A., & Kumar, M. (2023). Reduction of arsenic accumulation in rice grain by endophytic fungus Serendipita indica. *Rhizosphere*, 26, 100680. https://doi.org/10.1016/j.rhisph.2023.100680

Singh, S. P., Upadhyay, S. K., Pandey, A., & Kumar, S. (2019). Molecular approaches in plant biology and environmental challenges. In Energy, Environment, and Sustainability (pp. 1–5). Springer, Singapore. https://doi.org/10.1007/978-981-15-0690-1_1

Smedley, P. L., & Kinniburgh, D. G. (2002). A review of the source, behaviour and distribution of arsenic in natural waters. *Applied Geochemistry*, 17(5), 517–568. https://doi.org/10.1016/S0883-2927(02)00018-5

Stopelli, E., Duyen, V. T., Mai, T. T., Trang, P. T. K., Viet, P. H., Lightfoot, A., Kipfer, R., Schneider, M., Eiche, E., Kontny, A., Neumann, T., Glodowska, M., Patzner, M., Kappler, A., Kleindienst, S., Rathi, B., Cirpka, O., Bostick, B., Prommer, H., & Berg, M. (2020). Spatial and temporal evolution of groundwater arsenic contamination in the Red River delta, Vietnam: interplay of mobilisation and retardation processes. *Science of the Total Environment*, 717(February), 137143. https://doi.org/10.1016/j.scitotenv.2020.137143

Sullivan, J. B., & Krieger, G. R. (2001). *Clinical Environmental Health and Toxic Exposures*. Lippincott Williams & Wilkins, Philadelphia, PA.

Sweeney, M., & McCouch, S. (2007). The complex history of the domestication of rice. *Annals of Botany*, 100(5), 951–957. https://doi.org/10.1093/aob/mcm128

Tyler, C. R., & Allan, A. M. (2014). The effects of arsenic exposure on neurological and cognitive dysfunction in human and rodent studies: a review. *Current Environmental Health Reports*, 1(2), 132–147. https://doi.org/10.1007/s40572-014-0012-1

US Environmental Protection Agency. (2004). *Historic Arsenical Pesticide Research*. US Environmental Protection Agency, Denver, CO.

Vishwakarma, Y. K., Tiwari, S., Mohan, D., & Singh, R. S. (2021). A review on health impacts, monitoring and mitigation strategies of arsenic compounds present in air. *Cleaner Engineering and Technology*, 3, 100115. https://doi.org/10.1016/j.clet.2021.100115

Wang, S., & Mulligan, C. N. (2006). Occurrence of arsenic contamination in Canada: sources, behavior and distribution. *Science of the Total Environment*, 366(2), 701–721. https://doi.org/10.1016/j.scitotenv.2005.09.005

WHO. (2011). *Arsenic in Drinking-Water. WHO Guidelines for Drinking-Water Quality*. WHO, Geneva.

Winkel, L. H. E., Trang, P. T. K., Lan, V. M., Stengel, C., Amini, M., Ha, N. T., Viet, P. H., & Berg, M. (2011). Arsenic pollution of groundwater in Vietnam exacerbated by deep aquifer exploitation for more than a century. *Proceedings of the National Academy of Sciences*, 108(4), 1246–1251. https://doi.org/10.1073/pnas.1011915108

Yager, J. W., Hicks, J. B., & Fabianova, E. (1997). Brogan & partners airborne arsenic and urinary excretion of arsenic metabolites during boiler cleaning operations in a Slovak coal-fired power plant. *Source: Environmental Health Perspectives*, 105(8), 836–842.

Zhuang, F., Huang, J., Li, H., Peng, X., Xia, L., Zhou, L., Zhang, T., Liu, Z., He, Q., Luo, F., Yin, H., & Meng, D. (2023). Biogeochemical behavior and pollution control of arsenic in mining areas: a review. *Frontiers in Microbiology*, 14, 1043024. https://doi.org/10.3389/fmicb.2023.1043024

16

Arsenic in Rice: Existing Know-How and Potential Mitigation Appraches to Limit Its Level in Rice Grains

MD Rokonuzzaman and Wai Chin Li

16.1 Introduction

Arsenic is a potent class-one carcinogen (IARC, 2004), meaning it has no safe threshold level of exposure. In humans, it is known to induce a wide range of chronic diseases, some of which can be fatal (McClintock et al., 2012). Culturing staple foods like rice in As-affected areas is a predominant exposure source via the human diet (Meharg et al., 2009; Chowdhury et al., 2024). Naujokas et al. (2013), Podgorski and Berg (2020), and Biswas et al. (2021) estimate that about 200 million individuals across 105 developed and developing countries are at danger of As poisoning due to drinking As elevated water and rice. Every year, new countries and regions are added to the list of those from which the As contamination issue has not been previously reported. Consequently, the number of individuals afflicted continues to rise, necessitating more stringent regulations (Signes-Pastor et al., 2017). Moreover, there have been historical adjustments in worldwide food-taking behavior throughout the last 50 years due to the migration of Asian citizens to Western nations, which has dramatically boosted rice consumption (Islam et al., 2016). In addition to organic species, the presence of inorganic As (iAs) in rice makes this a much bigger concern (Wang et al., 2021). Because rice is often traded to various locations, sometimes even different continents from where it is grown, the As issue in rice has now gone worldwide, impacting more than only the countries that cultivate this crop (Saha et al., 2022).

The dissolution of As-enriched minerals is the primary natural source of As in groundwater, but human activities such as pesticide use, mining, coal combustion, and other industrial processes also contribute to this dilemma (Bundschuh et al., 2011; Singh et al., 2015). The primary focus of worldwide attention is the presence of As above the safe limit, particularly in groundwater, where a substantial portion of the worldwide population relies on it for irrigation and consumption purposes (Sarkar and Paul, 2016). Elevated levels of As are responsible for many acute and chronic human health issues (Jaishankar et al., 2014). Singh et al. (2015) have portrayed that As phytotoxicity can decrease plant yield and have adversative effects on metabolic, development, and growth processes. Approximately 50% of the global population eats rice daily. In most Asian nations, groundwater is still the primary irrigation source and is considered the staple sustenance. Consequently, the populations in these countries consume a substantial amount of As (Islam et al., 2016). About 90% of the world's rice comes from Asia (Arunakumara et al., 2013). To illustrate the point, in 2017, the world produced 735 million metric tonnes of rice, with Asia accounting for approximately 680 million metric tonnes of that total (FAO, 2018).

Consequently, mitigation strategies must be implemented in paddy agroecosystems contaminated with As to ensure safer rice grain production with minimal As content. Numerous studies have examined various mitigation strategies for their efficacy, practicality, and technological feasibility during the past few decades to lower the As content of rice. We attempt to evaluate technical agronomic practices, genetic approaches, biological, and post-harvest processing techniques to reduce the bioavailability of As in the solution of paddy soil and the quantity absorbed and deposited in rice grains. The environmental and

DOI: 10.1201/9781003492115-20

social ramifications of these technologies and their future potential and applications are also intended to be addressed in this chapter. To produce rice grains with a lower As content, we also discuss research gaps and potential avenues for further investigation.

16.2 Understanding Arsenic Contamination in Rice

16.2.1 Sources of Arsenic in Rice

Arsenic pollution exists in two main ways: natural processes and anthropogenic activities (Figure 16.1).

16.2.1.1 Natural Sources of Arsenic Pollution

Arsenic pollution in groundwater primarily comes from minerals containing As, as Zacarías-Estrada et al. (2020) noted. Mineral weathering is another crucial factor in As contamination, as minerals with high As levels can be deposited on various natural soil types, including acidic sulfate, bog, and peaty soil. Additionally, As contamination can arise from weathering products like sulfide and As-containing minerals, iron oxyhydroxide/oxide, hydrothermal water, and volcanic ash. Natural events like volcanic eruptions, marine aerosols, and wind movement also enhance As contamination (Shrivastava et al., 2015).

Arsenic contamination primarily arises from three main natural sources: hydrothermal activity, Cenozoic sediments, and mineral deposits (Masuda, 2018). Water with a high As level may be the consequence of seeping losses of the pyroclastic material, as volcanic glass dissolves at alkaline pH. The predominance of As^{5+} causes most trace elements to become mobilized as a mix of anions and oxyanions, influenced by reactions with carbonate. Natural processes can result in As accumulation ranging from 333 to 2098 μg/L, often exceeding the WHO recommended levels by 200 times (Maizel et al., 2018).

16.2.1.2 Anthropogenic Sources of Arsenic Pollution

Several human activities contribute to As pollution. These include emissions from industrial processes, burning fossil fuels, treating wood, using pesticides, fungicides, and herbicides excessively, and applying fertilizers that contain As (Shrivastava et al., 2015; Dolphen and Thiravetyan, 2019). Pesticides and wood preservatives both make heavy use of As. Arsenic-containing insecticides

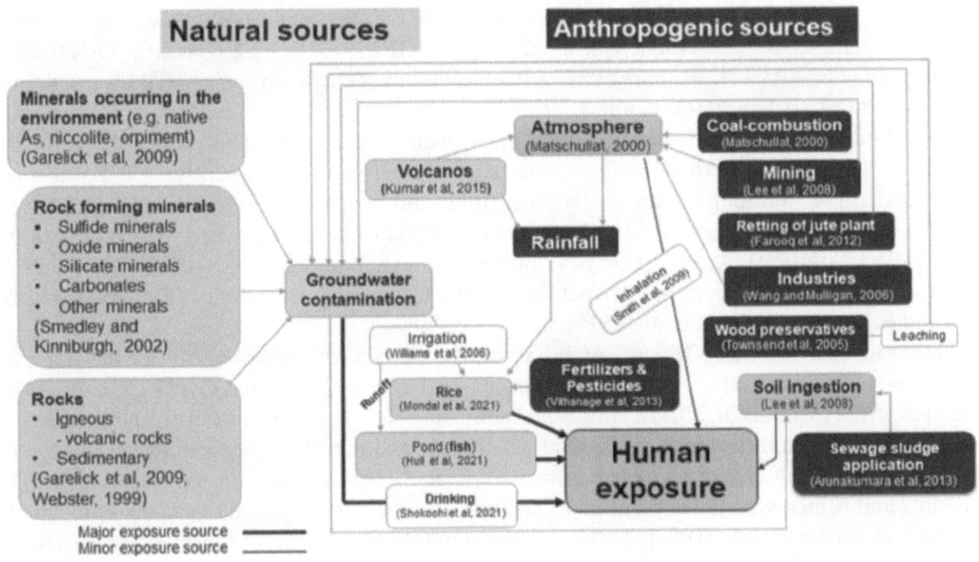

FIGURE 16.1 Sources of arsenic and possible pathways for human exposure.

have been used less recently, while As alloys are still used in semiconductors. Disproportionate groundwater extraction and mining result in As dispersion to remote areas, increasing the polluted zone's size.

Arsenic in the hot spring usually shifts to the river's surface water, which mixes with the silt from the riverbed. The hot spring releases arsenite, which microbes oxidize to generate arsenate. After being transformed, this type of arsenate eventually falls to the riverbed as a precipitate. The naturally occurring As near mining sites tends to stay localized. In contrast, human activities related to ore mining, including the release of tailings and sludge, cause As to disperse into the nearby soil and water systems (Masuda, 2018).

Consequently, greater levels of As are present in sand aquifers and in the filtration systems of river banks that lack oxygen, as observed by Ahmad et al. (2020). Within the Indian subcontinent, significant deposition occurs in fly ash generated during coal burning. Also, As particulate matter is released into the atmosphere during the smelting process of the ore. Furthermore, it has been observed by Kepel et al. (2020) that the sediments found in mine tailing locations contain As levels that are 30 times higher than those found in uncontaminated sediment. A large amount of As is also linked to deep glauconitic sand aquifers in the Late Tertiary period. Human actions, like using herbicides, antimicrobials, or growth supplements for pigs and poultry, can introduce organoarsenicals into the environment, or bacteria can generate them (Chen et al., 2015).

16.2.2 Mechanisms of Arsenic Uptake and Transport by Rice Plants

When compared in their inorganic forms, arsenite (As(III)) is far more hazardous and has different transport pathways than arsenate (As(V)). Plants can absorb arsenate because of phosphate (PHT) transporters (Cao et al., 2018). Arsenate and phosphorus use the exact transport mechanisms to move within rice plants. Arsenate enters cells via OsPHT1;1 (Kamiya et al., 2013) and OsPHT1;8 (Wu et al., 2011) transporters in rice. A study conducted by Wang et al. (2016) brought attention to the role of OsPT8 as an As(V) transporter in rice. The proteins involved in phosphorus transport, including OsPHF1 (P transporter traffic facilitator 1) and PHR2 (P starvation response 2), also influence the absorption and translocation of As(V) (Wu et al., 2011). Nonetheless, the mechanism for inorganic phosphate (Pi) transport does not enable As absorption (Xu et al., 2008). Multiple aquaglyceroporins facilitate the As species transportation, such as As(III) and undissociated methylated As, predominantly via aquaporin channels belonging to the nodulin 26-like intrinsic protein (NIP) family (Zhao et al., 2010).

The NIP class transporter OsNIP2;1 (Lsi1) is a significant facilitator of silicic acid (Si) uptake, which is essential for the absorption of As(III) (Ma et al., 2008). Other transporters involved in this process in rice include OsNIP3;2 (Bienert et al., 2008), OsNIP1;1, OsNIP2;2 (OsLsi6), OsNIP3;1 (Ma et al., 2008), and OsNIP3;3 (Katsuhara et al., 2014). In low-oxygen conditions, As(III) is the dominant form of As, and OsLsi1 facilitates its uptake into root cells, while OsLsi2 mediates its release into the xylem stele. Their specific locations within root cells ensure efficient As absorption (Ma et al., 2008). Research by Chen et al. (2017b) highlighted the significant role of OsNIP3;2 in As(III) absorption by rice lateral roots. Additional transporters identified and characterized in rice include AtNIP1;1, AtNIP1;2, AtNIP3;1 (in roots), AtNIP1;1, AtNIP1;2, AtNIP5;1 (Kamiya and Fujiwara, 2009) (in roots), AtNIP5;1, AtNIP6;1, AtNIP7;1 (in developing anthers and pollen tissues), OsNIP1;1, OsNIP2;1, OsNIP3;1, OsNIP3;2 (in anthers and suspension cells), OsLsi6, among others (Bienert et al., 2008; Kamiya and Fujiwara, 2009; Xu et al., 2015).

In addition to their roles as NIPs, research has shown that several aquaporins from the rice plasma membrane intrinsic protein (PIP) family, including OsPIP2;4, OsPIP2;6, and OsPIP2;7, can aid in the transfer of arsenite (Mosa et al., 2012). However, methylated As compound absorption and translocation processes have received little attention. Li et al. (2009) showed that the aquaporin OsLsi1, also known as NIP2;1, increased the uptake of monomethylarsonic acid (MMA) and dimethylarsinic acid (DMA) by rice roots and helped with the absorption of arsenite. Li et al. (2009) found that OsLsi1-deficient rice mutants had an approximately 50% and 80% reduction in DMA and MMA absorption, respectively.

In addition to transferring silicic acid, Ma et al. (2008) found that the OsLsi2 transporter helps load arsenite into the xylem and enables its transfer from the root to the shoot. Mawia et al. (2021) suggested that the transport of As by Lsi2 will likely decline when silicon (Si) is added. In flooded conditions, the level of arsenite was decreased with the application of Si in polished rice by 33%, attributed to a 50% reduction in As transport through the plant's vascular system (Fleck et al., 2013). Both arsenate and phosphorus (P) use the same transporters to move within rice plants. Continuous expression of the P transporter OsPht1;8 occurs in environments with adequate oxygenation, leading to increased uptake of As(V) into the xylem. According to Awasthi et al. (2017), DMA can be more readily transported to the shoots, although As(V), As(III), and MMA mostly build up in the roots.

16.2.3 Factors Influencing Arsenic Speciation, Uptake, and Bioavailability in Rice

16.2.3.1 Soil pH

There are several oxidation states that As may exist in, including –3, 0, +3, and +5; however, As is present in nature in both solid and liquid forms. Typically, it is encountered in its elemental form, As(–3), or as As(0) in nature (Wang and Mulligan, 2006). When the pH of groundwater is between 6.5 and 8.5, As activates rapidly. In paddy soils, As is usually present in pore fluids as inorganic oxyanions, mainly As(V) or As(III). Dimethylarsinic acid (DMAsV) and Monomethylarsonic acid (MMAsV) are the organic As compounds most commonly detected in this environment (Upadhyay et al., 2019). Because soils with higher pH levels have negative surface charges, which might increase the release of As(V) and As(III) into the soil, there is a strong relationship between total grain As content and soil pH. At low pH levels, the most common form of As in mammals is the corrosive form, H_3AsO_4, which occurs under oxidizing circumstances (Wang and Mulligan, 2006). Smedley and Kinniburgh (2002) state that the species of As found in paddy fields are determined by several processes, including the formation of root plaques, the reduction and oxidation of As by microbes, and the reduction of metal hydro(oxides).

16.2.3.2 Rice Genotype and Transporter Phylogeny

The primary factors affecting As absorption in rice grains are genetic differences and the interplay between environmental conditions and genotype (Islam et al., 2016). Various genotypes react differently under different circumstances. Bhattacharya et al. (2010a,b) examined the As bioaccumulation in twelve different types of rice cultivated in the Nadia District of West Bengal, India. These varieties included local types, high-yielding, and hybrids. Their study revealed that these rice varieties absorbed different levels of As. Significant accumulators included rice cultivars like IR68144-27, BRRI dhan29, BR11, Nanyangzhan, TD71, T-Aman rice, and IAPAR9, with As content varying from 1.25 to 1.835 mg/kg. In addition, consistent genetic variations in As accumulation were noted in 76 different rice types, encompassing traditional Bangladeshi varieties and improved cultivars developed by the Bangladesh Rice Research Institute (BRRI). These were observed in Sonargaon and Faridpur by Norton et al. (2009). Compared to brown bran (0.42 mg/kg), Bangladeshi landraces exhibited the highest As concentration in red bran (0.57 mg/kg). Additional genotypes exhibiting elevated levels of grain As concentration are Jagliboro, Tulsiboro, Kazlaboro, Goa Bish, and Lafai. Japonica cultivars consistently exhibited lower grain-As concentrations compared to Indica rice genotypes.

16.2.3.3 Soil Texture

The mobility of As is primarily influenced by the soil's texture, which results from charge differences on the soil's surface. This surface texture also impacts soil adsorption and desorption processes (Abedi and Mojiri, 2020). The distribution of As and particle size fractions in soil and the amounts of clay and silt are intricately correlated (Williams et al., 2011). According to Ustohalova (2011), the surface area of soils with finer textures, like clays and silts, was much larger than that of sandy or coarse soils. Pasias et al. (2013) noted that Fe oxides are found only in finer-textured soils, leading to a higher As scavenging ability in clay soils than in coarse soils.

Consequently, fine-textured soils have lower As concentrations compared to coarse-textured soils. Research indicates that As phytotoxicity is significantly higher in sandy and loamy soils than in clay soils. Furthermore, As uptake by rice is easier from loamy soils than clay or sandy soils (Pasias et al., 2013). Suda et al. (2023) found that in anaerobic soils, acid detergent-soluble organic matter (ADSOM), which mainly consists of hemicellulose and non-fiber organic matter (OM), can cause the dissolution of As. A possible explanation for rice's decreased As absorption might be the increased formation of iron plaque on rice roots due to OM amendments (Syu et al., 2019).

16.2.3.4 Iron, Phosphate, and Silicon in Soil

According to Majumdar et al. (2023), the redox chemistry of iron (Fe) dramatically influences how As behaves in soil. Gu and Deng (2007) also noted that soil amorphous ferric hydroxide and As are highly correlated. Both the As(III) and As(V) primarily cling to the surface of Fe-oxyhydroxides (FeOOH) found in soil. According to Majumdar et al. (2023), the effect of soil redox conditions on the adsorption process is significant. The availability of As in soil is heavily influenced by its physicochemical properties, including pH, OM, and irrigation regimes. P in soil greatly influences how various rice varieties absorb As. Cultivars with low internal P levels are more sensitive to arsenate, according to a study by Lihong and Guilan (2009). Reduced As buildup in rice grains can be achieved by increasing the soil's P content, as pointed out by Lu et al. (2010). Studies demonstrate that the sorption process of ferric hydroxide in soil correlates favorably with Fe, P, and As concentrations. Arsenic and phosphorus do not significantly impede one other's absorption. Iron and ferric hydroxide formation is pivotal in influencing the availability and solubility of P and As in the soil.

The mobility and absorption of As from soil minerals into rice plants decrease when Si is introduced into the soil (Zhang et al., 2020). In addition to boosting shoot growth, research has demonstrated that Si intake strengthens rice nodes, shoots, and the xylem-phloem integrity in the leaves. As a result, the translocation of As is inhibited, and the internal structure of the rice plant is improved. Seyfferth et al. (2018) found that silicic acid (H_4SiO_4), a chemically similar compound to As(III) in the form of H_3AsO_3, is present in high concentrations in the pore water of rice fields.

Silicon and As in the form of As(III) are the primary elements at the soil–water interface. These elements vie for the transporters that facilitate Si uptake in rice plants. Silicon or As(III) can be transported simultaneously depending on their respective concentrations. Besides this competitive inhibition, Si can also reduce the movement of As by binding to FeOOH surfaces, lowering the abundance of As binding sites in iron plaque and facilitating As absorption (Gang et al., 2018). Thus, the occurrence of both Si and P in soils has been demonstrated to affect As absorption in rice, according to Syu et al. (2019).

16.2.3.5 Irrigation Water Quality and Treatment Regimes

Water quality significantly affects how easily plants can absorb As from the soil, according to a study by Islam and colleagues in 2007. Overexploitation of groundwater contaminated with As (Majumder and Banik, 2019) and various water treatment methods (Islam et al., 2019) also impact its availability. Upland rice culture, which involves low natural water stagnancy, is a preferred technique for field setup. This type of cultivation is considered semi-aerobic (Majumdar, 2021). Rice grains and straws have lower levels of As when there are no aerobic conditions or flooding, but As absorption by rice plants is increased in flooded or anaerobic environments, as noted by Li and colleagues in 2009. Islam and others found in 2017 that rice growers had significantly higher As absorption rates in anaerobic conditions compared to aerobic conditions, with a tenfold difference.

The erosion of upland soil can result in the deposition of As content in lowland fields, increasing the likelihood of rice accumulating As (Barla et al., 2017). Several factors can affect how As settles in irrigated rice fields and accumulates into the rice. The factors that come into play are the soil type, the background soil As concentration, the groundwater As level used for irrigation, the depth and distance from the irrigation source, and the duration of the monsoon flooding season.

16.2.3.6 Microbial Activity in Soil

Soil and rice rhizosphere aerobic and anaerobic bacteria participate in As reduction and oxidation, As compound breakdown, and other related activities (Yang et al., 2019). The ability of microbes to oxidize arsenite to arsenate is a detoxifying mechanism that decreases the toxicity and mobility of As in flooded and non-flooded environments (Zhang et al., 2015). Jia et al. (2012) and Kumarathilaka et al. (2019) have demonstrated that microorganisms are essential in converting arsenite and arsenate into MMA, DMA, and arsines.

Begum et al. (2024) identified that bacteria from the Euryarchaeota phylum, which can reduce sulfate and produce methane, are likely vital contributors to As methylation. Additionally, the study highlights that As in rice is primarily transported and transformed by iron-reducing bacteria belonging to the Firmicutes phylum. The research also found a positive association between these bacteria and iAs and its metabolites, MMA, and DMA. According to Dai et al. (2020), there was an increase in methylated As compounds and an increase in the release of As from iron deposits when iron-reducing bacteria, notably *Clostridium sensu strictu* (from the Firmicutes phylum), were present. In addition to the factors highlighted above, fluctuations in microbiological activity across the seasons might further impact the accumulation and availability of As (Brammer and Ravenscroft, 2009).

16.3 Health Implications of Arsenic in Rice

16.3.1 Chronic Arsenic Exposure and Associated Health Risks

Numerous severe health issues can develop as a result of As exposure, including cancers of the bladder and lungs, skin conditions, gastrointestinal disturbances, cardiovascular problems, respiratory ailments, neurological disorders, hormonal imbalances such as diabetes, developmental challenges, skin diseases, and reproductive abnormalities (Oberoi et al., 2014; Sobel et al., 2020). The appearance of red flag lesions on the skin indicates acute internal injury (Chakraborti, 2011). Regrettably, researchers have been unable to develop any medication that can effectively treat chronic As toxicity (Chakraborti et al., 2018). According to Roychowdhury (2008), iAs constitute approximately 96.8% of rice's total As (tAs) content. However, Rahman et al. (2014) indicate that rice from Asian countries can have iAs levels as high as 99%. In their 2016 study in India, Kumar and colleagues demonstrated that consuming rice and vegetables had a notable effect on human health, with both foods exhibiting a health risk index (HRI) exceeding 1.

The "lifetime cancer risk" per 10,000 persons resulting from rice consumption was measured by several researchers. The rates for Taiwan, the US, Bangladesh, China, India, and Italy are 1.04, 1.30, 22.10, 15.20, 6.90, and 0.70, respectively (Meharg et al., 2009; Chen et al., 2016; Islam et al., 2016). Long-term exposure to As in Bangladesh often leads to arsenicosis, which is frequently manifested as skin lesions. This exposure has been shown to affect the health of adults and children negatively. The risk of neonatal mortality, stillbirth, preterm birth, spontaneous abortion, delayed child growth, reduced IQ, low birth weight, compromised immune system, impaired neurodevelopment, and neurotoxicity has been increased by exposure to As during pregnancy (Vahter, 2008).

However, individuals with lower protein intake are more likely to experience toxicity, as stated by Ahmad et al. (2018). Additionally, malnourished individuals are more susceptible to developing skin lesions related to As exposure, as noted by Kapaj et al. (2006). Sampson et al. (2008) noted that disorders caused by As exposure usually become visible 8–10 years after ingesting meals with high As contamination levels. However, in Cambodia, there have been instances where symptoms appeared just three years after drinking water with very high As concentrations (3500 mg/L). These cases were also linked to malnutrition and the socioeconomic status of the individuals consuming the water (Sampson et al., 2008).

16.3.2 Regulatory Standards and Guidelines

The quantity of rice eaten daily is the most important factor in determining the daily As intake from rice, as Singh et al. (2015) stated. The FAO (2004) reports that rice consumption varies widely between countries, with rates spanning from about 0.9 to 650 g/person/day. Countries in Asia, including Vietnam, Myanmar, Laos, Cambodia, and Bangladesh, typically have higher rice consumption rates than certain European and African countries. FAO (2004) extended that 46 Asian, South American, and African countries consume rice weighing over 100 g/person/day. As an illustration, Asian adults typically consume between 200 and 600 g/person/day (Zavala and Duxbury, 2008; Zhu et al., 2008; Garnier et al., 2010). The daily As intake from drinking 2 L of water at the recommended level of 10 µg As/L set by the World Health Organization is far higher than many of these figures.

Because of inadequate regulations in several developed countries, rice containing high levels of As can be sold and consumed without consumers knowing the contamination. A 2012 USFDA study, "Arsenic in Rice: Full Analytical Results from Rice/Rice Product Sampling" analyzed As levels in 165 rice products and discovered an average concentration of 222 ± 135 µg As/kg. Since rice can be a significant source of As exposure, it is crucial to establish stringent controls for As content in this staple food. Studies have conclusively shown that moderate rice consumption increases As levels in children's urine across the United States. An analysis of 229 pregnant women revealed a strong correlation between eating rice and exposure to As through water, resulting in considerably elevated As levels in their urine. A further investigation in Bangladesh revealed notable connections between rice intake, As in urine, and the development of skin abnormalities linked to As exposure. It is crucial to consider the gastrointestinal bioavailability of As from rice. However, numerous studies indicate that higher consumption of rice results in elevated exposure to As and heightened hazards.

China enforces the strictest regulations concerning As levels in rice, setting a limit of 150 µg of iAs per kg (Munera-Picazo et al., 2015). Despite these regulations, the widespread use of rice as a staple food in Asia means that it remains a significant source of As exposure, surpassing even drinking water in this regard. The FAO/WHO suggests a daily tolerable intake of 2.1 µg of As per kg of body weight. Furthermore, the FAO/WHO has established a Provisional Tolerable Weekly Intake (PTWI) of 15 µg per kg of body weight (WHO, 1989; Díaz et al., 2004; Roychowdhury, 2008). The No Adverse Effect Level (NOAEL) for chronic oral consumption is set at 1 µg per kg per day for all individuals, while the Low Adverse Effect Level (LOAEL) ranges from 10 to 100 µg per kg of body weight per day (Larsen and Berg, 2001; Díaz et al., 2004).

16.4 Existing Approaches to Limit As Loading in Rice Grains

Using modern technology to extract groundwater for irrigation purposes has led to a higher degree of exploitation, leading to a higher risk of As contamination in places already prone to As contamination. When rice is cultivated in irrigated paddy fields that utilize contaminated groundwater, As seeps into the rice via soil and water (the permissible limit for irrigation water, soil, and rice has been presented in Table 16.1). This has been documented in studies conducted by Rahman et al. (2007) and Zhao et al. (2010). Various studies have shown that diverse agricultural methods can reduce the rice As content and how much the human body absorbs. Several methods utilized to diminish the absorption of As in rice include (1) Technical agronomic practices, (2) Genetic approaches, (3) Biological approaches, and (4) Post-harvest processing techniques to minimize the uptake and movement of As.

16.4.1 Technical Agronomic Practices

16.4.1.1 Water Management and Irrigation Practices

Intermittent flooding, aerobic conditions, and alternate wetting and drying (AWD) are some water management strategies that have recently attracted interest in rice cultivation. These methods have been found to sustain high grain yields while reducing grain As levels (Hu et al., 2015). Research has shown a correlation between elevated soil As levels and the usage of groundwater polluted with As for irrigation purposes. For instance, a thorough investigation of paddy soils in Bangladesh conducted by Meharg and Rahman (2003) showed that elevated soil As levels were strongly associated with irrigation

TABLE 16.1
Permissible Limit of As for Irrigation Water, Soil, and Rice

Irrigation Water As			Soil As			Rice As		
Country/ Organization	**Permissible Limit (mg/L)**	**References**	**Country/ Organization**	**Permissible Limit (mg/kg)**	**References**	**Country/ Organization**	**Permissible Limit (mg/kg)**	**References**
Mexico	100	Bundschuh et al. (2012)	FAO	50	Pescod (1992)	WHO	0.37	WHO (2016); Suriyagoda et al. (2018)
Columbia	10	Alonso et al. (2014)	China	30	Massaquoi et al. (2015); Zhou et al. (2016)	FAO	1	Singh et al. (2015)
Venezuela	50	Bundschuh et al. (2012)	Taiwan	5.65	Chang et al. (1999)	China	0.20	Yao Y et al. (2021)
Ecuador	100	Bundschuh et al. (2012)	Vietnam	12	Phuong et al. (2008)	China ((iAs in foods)	0.15	Chakraborti et al. (2018)
Taiwan	50	Chou et al. (2016)	Argentina	20	Tarvainen et al. (2013)	WHO	0.37	WHO (2016); Suriyagoda et al. (2018)
Peru	200	Bundschuh et al. (2012)	Australia	10–20	Duxbury and Zavala (2005)	FAO	1	Singh et al. (2015)
Bolivia	50	Bundschuh et al. (2012)	Germany	200	German Federal Soil Protection Act (1998)	JFWCAC (iAs in polished rice)	0.20	EFSA (2014)
Chile	100	Bundschuh et al. (2012)	Canada	20	Duxbury and Zavala (2005)	Bangladesh	No regulation	Aziz et al. (2015)
Brazil (trees, cereals, and forages)	33	Bundschuh et al. (2012)	UK	10–20	Duxbury and Zavala (2005)	US	No regulation	Hojsak et al. (2015)
FAO	100	Chakraborti et al. (2018)	UK	50	Smith (1996)	European Union (EU)	No regulation	Hojsak et al. (2015)
WHO	<10	WHO (2004); Arain et al. (2009)	Netherlands	10–20	Duxbury and Zavala (2005)			
Argentina	100	Bundschuh et al. (2012)	Southeastern USA and Greece	40	Dudka and Miller (1999)			
			Norway	20	Hansen and Danielsberg (2009)			
			Belgium	45	Soil Remediation Act (1995)			

(Continued)

TABLE 16.1 (*Continued*)

Permissible Limit of As for Irrigation Water, Soil, and Rice

Irrigation Water As			Soil As			Rice As		
Country/ Organization	**Permissible Limit (mg/L)**	**References**	**Country/ Organization**	**Permissible Limit (mg/kg)**	**References**	**Country/ Organization**	**Permissible Limit (mg/kg)**	**References**
			Germany	50	German Federal Soil Protection Act (1998)			
			New Jersey, USA	20	Tarvainen et al. (2013)			
			U.S. Environmental Protection Agency	24	USEPA (1996)			
			European Union	20	Rahman et al. (2007)			
			Global average	10	Rahman et al. (2013)			
			Japan	15	Japan (2016)			
			Thailand	3.9	Punshon et al. (2017)			

water polluted with As and with the continuous use of tubewells. Similarly, Sarkar et al. (2012) identified a direct relationship between elevated soil As levels and irrigation with As-laden groundwater. Growing rice under flooded conditions can enhance As mobility by breaking down iron oxyhydroxides (Kim et al., 2009). Shrivastava et al. (2020) revealed that reducing As buildup in rice grains can be achieved by combining aerobic water management with alternate irrigation methods.

Because of its strong affinity for soil minerals, oxygen helps rice plants retain less As in their vegetative and grainy phases as they mature. Soil solutions render As immobile, unlike conventional flooded-field rice (Li et al., 2009). Talukder et al. (2011) witnessed that cultivating rice aerobically improves grain yield and lowers grain As content. Hu et al. (2015) also discovered a similar result under periodic puddles, aerobic or AWD settings, rather than continuous flooding. Consequently, implementing deficit irrigation techniques in rice cultivation decreases the presence of As in the soil-roots-shoot-grain pathway (Newbigging et al., 2015).

Regardless of the amount of pollution in the irrigation water, deficit irrigation can lower soil As levels (Rinklebe et al., 2016). As opposed to As(V), the mobility of As(III) is enhanced at lower redox potentials, as found by Hu et al. (2015). In aerobic and AWD settings, the decrease of As content is usually accompanied by several variables, including soil micropore As(III) levels are lower, more Fe, S, and As-oxidizing bacteria are present, less S and Fe-reducing bacteria are present, and Fe plaques binding As(III) and As(V) are formed (Deng and Kim, 2016). Moreover, research by Moreno-Jiménez et al. (2014) indicated that using sprinkler irrigation in rice cultivation lowered As content in the grains. Majumdar et al. (2024) confirmed that applying the AWD technique can reduce As bioavailability by up to 26%, and this method is effective across various geographic regions and seasonal changes.

The study also noted that the microbial community exhibited increased resilience to As and enhanced metabolic activity under anaerobic wetland conditions (AWD). In addition, Polizzotto et al. (2015) also showed that enhancing irrigation networks with hessian buildings is an efficient method for reducing As content in irrigation water before discharging on the paddy fields. The modified channels improved As removal from the solution, achieving a 7% reduction in tAs compared to a 3% reduction in the unaltered channels. The cause of this phenomenon can be linked to a rise in the duration between events and the effectiveness of particle capture. In contrast, the elimination through the creation of oxidizing components does not appear to be notably enhanced.

16.4.1.2 Seed Priming Technology

Seed priming (SP) helps to promote seed growth while preventing radicle emergence through sun drying. Various SP strategies exist, including osmopriming, nanopriming, nutri-priming (macro or micronutrients), hydropriming, hormone priming, biopriming, and using natural compounds derived from plants (Farooq et al., 2019). SP involves treating or supplementing seeds with various inorganic or organic compounds and exposure to different temperatures (Waqas et al., 2019). The SP method is straightforward, efficient, and affordable and poses minimal risk. This approach has successfully safeguarded various crops from numerous environmental stresses (Sahoo et al., 2019).

Mridha et al. (2021) proposed that the collective impact of potassium humate (K-humate) on As stress is attributed to its ability to regulate gibberellin and abscisic acid levels, as well as its capability to chelate metals through humate compounds. According to Moulick et al. (2018a, 2018b), applying selenium (Se) to rice seeds before sowing may improve seed germination and encourage the development of seedlings. Reducing As toxicity through forming a selenium pool surrounding the seed coat, assisting detoxification processes, and preserving cell membranes by maintaining total phenolics and proline levels are all part of the underlying mechanism. Restoring a balanced redox state after exposure to As also controls the absorption of the metalloid. These effects were observed in both soil-based and hydroponic environments.

Another experiment was carried out to evaluate the efficacy of SP with Zn as a possible ameliorating agent in As-stress circumstances. No substantial change was seen in the pattern of As absorption in the seedlings from the unprimed seeds to the Zn-primed ones, although the germination of the latter was enhanced (Moulick et al., 2019). Priming seeds increases the absorption and transportation of micronutrients to the aboveground portions of seedlings, boosting their development (Moulick et al., 2018a, 2019). The effects of raising Se-primed seedlings to maturity in soils polluted with and devoid of As were studied by Moulick et al. (2018a, 2018b). They observed that under identical As stress circumstances, Se-primed plants produced lower As levels in brown rice and better yield characteristics than unprimed plants of any rice. Rice plants have not been extensively studied for their ability to reduce As stress, despite SP being the most practicable technique for alleviating induced stress (Khan and Gupta, 2018).

16.4.1.3 Nanotechnology Application

The advancement and widespread use of nanotechnology in recent times have expedited the emergence of engineered nanoparticles and their pragmatic application in the agriculture fields (Wang et al., 2019a; Wu et al., 2021). Wu et al. (2021) found that incorporating nano-titanium oxide (TiO_2) into rice seedlings markedly diminishes As accumulation and alleviates the detrimental impact of oxidative stress induced by As exposure. Similarly, research by Cui et al. (2020) on single cells revealed that nanoscale silica ($nSiO_2$) can effectively reduce As uptake by lessening oxidative stress and increasing cell wall thickness by up to 77.4%. The expression of the OsLis1 and OsLis2 genes was also decreased in rice cells treated with $nSiO_2$. Additionally, Liu et al. (2014) found that nanoscale silica (Si) helps sequester As inside the cell walls of rice shoots, which is likely responsible for the reduced As levels in brown rice.

According to Huang et al. (2018), the antioxidant effects on the rice plant's aerial parts are diminished by introducing nano-Fe_3O_4 and nano-Fe^0. However, Hu et al. (2020) found that the decrease in iAs levels in rice is caused by the production of Fe plaque when Fe^0 is applied. Fe_3O_4 exhibits a strong affinity for negatively charged As particles on its surface, inhibiting As entry into its structure (Khan et al., 2020). The presence of CuO reduces the life cycle of plants, which reduces the amount of As load in rice plants, bringing the concentration below the threshold when harmful effects become apparent (0.80 µg/kg per day), as stated by Liu et al. (2018). Furthermore, the distinctive "nano-effect" of CuO on As accumulation in rice was also brought to light by Wang et al. (2019a).

According to Ma et al. (2020), Zinc ions (Zn^{2+}) and zinc oxide (ZnO) considerably decrease As content in the tissues of rice plants. This suggests that Zn^{2+} treatment might influence the OsLsil transporter. Arsenate levels dropped noticeably in roots treated with Zn^{2+} due to the rigorous function of the As reductase in the root cells to convert As(V) to As(III). Wu et al. (2020) and Yan et al. (2021) found that supplementing with Zn nutrients via ZnO treatment significantly improves germination, boosts biomass, and decreases As buildup in rice. Conversely, Wang et al. (2018) found that ZnO NPs (nanoparticles) combined with As(V) or As(III) might harm plants, possibly lowering As levels in rice plants by making As transporters less efficient.

16.4.1.4 Biochars

Pyrolysis, or the thermal breakdown of organic material under hypoxia, is the process used for producing biochar (BC) (Wang et al., 2019b). The material offers numerous advantages, such as cost-effectiveness, oxygen-rich functional groups, robustness, efficiency, a well-developed pore structure with a large surface area, and significant potential for cleaning contaminated water and soil (Che et al., 2023; Chaudhary et al., 2024). Its high porosity and surface properties make it an effective solution for remediating soils contaminated with organic and inorganic pollutants from an environmental standpoint (Cope et al., 2014). Biologically modified BC, such as Mn oxide-modified BC or Fe(III)-treated or CaO-saturated rice hulls, has been reported to have greater As soaking efficacy than BC generated solely from agricultural by-products. However, there are variances (Agrafioti et al., 2014; Wu et al., 2020). Some significant BC impregnation evidence for As-safe rice production is presented in Table 16.2.

TABLE 16.2

Some Significant Biochar Impregnation Evidence for As-Safe Rice Production

Impregnation Type	Material Source	Temperature	Characteristics/ Functional Group	Application	Results	References
Iron (Fe)	Fe_3O_4 and rice husk	400°C	As (mg/kg) = 2.83; CEC (cmol/kg) = 70.16; Pb (mg/kg) = 6.68; Cd (mg/kg) = 0.065; ratio of Fe to modified BC = 19.5%; pH = 4.90	Open-aerated field experiment	Noteworthy decrease of up to 47.3% in grain As	Yao Y et al. (2021)
	$FeSO_4 \cdot 7H_2O$, corncobs, and hen eggshells	450°C	For [a]FCB, [a]FEB, and [a]FCEB, C (%) = 62.6, 7.8, and 35.3; H (%) = 2.0, 1.6, and 1.9; Fe (%) = 5.4, 9.2, and 7.7; As (mg/kg) = 0.04, 0.73, and 0.35, N (%)= 0.57, 0.08, and 0.35, pH = 8.71, 8.65, and 8.67, respectively	Pot experiments	As level of brown rice dropped by 29%–60%, and the average As mobility receded by 56%	Islam et al. (2021a)
Zero valent iron (Fe°)	Fe°- Palm fiber	700°C	-	Pot experiments	Reduces uoto 42% of grain As	Qiao et al. (2019)
	Fe°-amended biochar	-	C–O, C = O, Fe–O and –OH	Pot experiments	Reduces >50% As in brown rice	Islam et al. (2021b)
Iron and Manganese	$Fe(NO_3)_3$, $FeSO_4$, Mn oxide and corn straw	600° C	[b]$FMBC_1$: C (%)= 67.27, N (%) = 1.83, H (%) = 2.38; pH = 9.60 [b]$FMBC_2$: C (%) = 53.82, N (%) = 1.41, H (%) = 2.22; pH = 3.17	Pot experiments	Developed Mn and Fe plaque; 2% $FMBC_2$ decreased total grain As by 68.9%–78.3%; converted As(III) to As(V)	Lin et al. (2017)

(Continued)

TABLE 16.2 (*Continued*)
Some Significant Biochar Impregnation Evidence for As-Safe Rice Production

Impregnation Type	Material Source	Temperature	Characteristics/ Functional Group	Application	Results	References
Fe-Mn-La	Corn stem	600°C	[c]$FMLBC_1$ = 25:4:1:1, [c]$FMLBC_2$ = 25:4:1:3, and [c]$FMLBC_3$ = 25:4:1:5	Laboratory	significantly reduced the As(III) in paddy soils; controls As volatilization	Lin et al. (2021)
Birnessite	rice hull	600°C	pH = 5.83±0.07 Mn (mg/kg) = 3146±47; Si (mg/kg) = 174.26±11.68; average pore diameter (nm) 6.0944; Total pore volume (mL/g) 0.1772; specific surface area (m^2/g) 116.3	Pot experiments	The rice yield was enhanced by 10%–34%; the process included As(III) oxidation; modification reduced the risk of cancer through rice intake	Kumarathilaka et al. (2021b)
Iron-Manganese-Cerium oxide	Maize stalks	600°C	[d]$FMCBC_1$= 24:2:3:4; [d]$FMCBC_2$= 24:2:3:8, and [d]$FMCBC_3$ = 24:2:3:10	Pot experiments	Reduced bioavailable forms of As; improved soil pH	Zhang G et al. (2020)
$FeSO_4·7H_2O$, $FeCl_3$, and Fe°	Rice straw	300°C	Amount of Fe loading was BC <BC-$FeCl_3$ <BC-FeOS <BC-Fe	Laboratory	Drastically lowered the $NaHCO_3$-extractable As concentrations in paddy soils	Wu et al. (2018)
Fe^0, Fe^{3+}, and Ca^{2+}	The organic portion of municipal solid waste and rice husk	300°C	-	Laboratory	Accelerate the prevention of As(V) loading; exhibited high As(V) removal	Agrafioti et al. (2014)
Fe^0 and silica sol	Palm thread	200°C–300°C	C (> 50%), Fe (2.16%), K (1.24%); Particle size (< 5.00 mm) >90%; pH =6	Field experiments	Fe-BC-Si greatly raised soil pH, which in turn reduced As availability	Pan et al. (2019)

[a] Fe-enriched corncob biochar (FCB), Fe-enriched charred eggshell (FEB), and Fe-enriched corncob-eggshell biochar (FCEB).
[b] In $FMBC_1$, weight ratio of BC:$KMnO_4$:$Fe(NO_3)_3$ was 25:4:1; and in $FMBC_2$, weight ratio of BC:$KMnO_4$:$FeSO_4$ was 18:3:1.
[c] FMLBCs with BC:Fe: Mn:La weight ratios of 25:4:1:1 ($FMLBC_1$), 25:4:1:3 ($FMLBC_2$), and 25:4:1:5 ($FMLBC_3$).
[d] The weight ratio of BC:Fe:Mn:Ce were 24:2:3:4 ($FMCBC_1$, w/w/w), 24:2:3:8 ($FMCBC_2$, w/w/w), and 24:2:3:10 ($FMCBC_3$, w/w/w), respectively.

Agrafioti et al. (2014) found that the efficiency of blocking As(V) was enhanced when using BC that had been treated with calcium (Ca^{2+}) and iron (Fe^0, Fe^{3+}) agents. Wu et al. (2018) noted that As elimination in solution is commonly achieved using iron-enriched BC due to the strong attraction between As and iron oxides. They evaluated the remediation effectiveness of BC combined with $FeCl_3$ (BC-$FeCl_3$), Fe-oxyhydroxy sulfate (BC-FeOS), and Fe^0 (BC-ZVI). The findings indicated that BC-FeOS was the most effective at immobilizing As and cleaning up soil contaminated with this element. Pan et al. (2019) observed that increasing soil pH by using BC, Fe^0, or silica sol, individually or together, facilitated the reduction of As levels in grains.

Yao et al. (2021) noted a decrement in rice As content due to its attachment to Fe plaque and crystalline hydrous oxide. Iron (Fe) plaque development on roots considerably reduced As levels in rice grains, as Islam et al. (2021a) also found. Qiao et al. (2019) investigated the synergistic effects of palm fiber and

Fe^0, which seemed to decrease As bioavailability through forming Fe plaque. Research conducted by Lin et al. (2017) indicates that the reduction in As levels in rice tissue after applying Fe-Mn oxide-amended corn straw BC (FMBC) is a result of the transformation of As from its trivalent to pentavalent form, which is made possible by the presence of Fe-Mn oxide. The acceleration of the As conversion is likely facilitated by the creation of plaques on the surfaces of the roots that contain Fe and Mn.

The As content in grains is reduced because Fe-Mn plaques limit As absorption by rice plants, as shown by Lian et al. (2020). Zhang G et al. (2020) demonstrated that adding iron-manganese-cerium oxide to BC generated from maize stalks reduces the availability of As, increases soil pH, and perhaps enhances the soil's ability to undergo reduction-oxidation reactions. Research by Lin et al. (2021) suggests that there might be a way to control the amount of As released into the air and slow down its crystallization, dissolution, and methylation by combining BC derived from corn stems with Fe-Mn-La. According to Kumarathilaka et al. (2021a), the silicon and sulfate in rice water are responsible for reducing rice BC (RBC). Furthermore, Kumarathilaka et al. (2021b) proposed that intermittent water management coupled with birnessite-modified rice husk BC (Mn-RBC) might potentially reduce grain As by oxidizing As(III) through Mn.

16.4.2 Genetic Approaches

It is crucial to identify and comprehend the As detoxification mechanisms in rice plants to select breeding cultivars and produce transgenic plants (Chen et al., 2017a). Wu et al. (2011) found that controlling and altering P transporters significantly impacted the amount of As absorption in rice plants. Reducing the amount of As is achieved by gene manipulation and mutation. Kamiya et al. (2013) observed that the overexpression lines elevated As levels in wild-type plants. The mutation in the OsPT8 transporter also decreased the absorption and concentration of AsV. According to Xu et al. (2017), the As(V) reductase enzymes, including OsHAC1;1, OsHAC1;2, and OsHAC4, are responsible for the conversion of As(V) to As(III) in the roots of rice. Foreign gene introduction decreases As buildup in grains of genetically engineered rice plants, according to research by Duan et al. (2012).

Increasing the discharge of As(III) from the roots into the soil around the rice plant can reduce As levels in the grain, as Zhao et al. (2010) pointed out. Consistent with this, Sun et al. (2018) discovered that brown rice has lower As content in its grains, straw, husks, and leaves. The increased expression of the OsNIP1 resulted in the reduction of 1 and OsNIP3;3 genes. A similar slowed or reduced migration from the root to the shoot was also documented by Singh et al. (2019). Duan et al. (2012) have discovered comparable results indicating that rice plants expressing a specific ACR3 gene, which encodes a protein responsible for removing As(III), experience improved growth and reduced As buildup in the rice grains.

The strategic application of gene editing technologies like CRISPR/Cas9 has demonstrated efficacy in investigating gene functions in rice plants (Biswas et al., 2019). Consequently, important genes involved in As absorption and transport in rice, such as OsPHT1;8, Lsi1, and Lsi2, may be selectively modified using the CRISPR/Cas system. According to Ye et al. (2017), As buildup was reduced after using CRISPR/Cas9 technology to remove the OsPT4 gene. Specifically, compared to wild-type plants, OsPT4 mutant rice plants showed a reduction in root As levels of 17%–30% and demonstrated significant resistance to AsV, indicating enhanced tolerance. Xu et al. (2017) found that modifying the OsHAC4 gene impacted rice's detoxification process and buildup of As(V). Root exodermis and epidermis are the primary sites of gene expression; however, the gene is present in the cytoplasm and the nucleus. Overexpressing OsHAC4 resulted in a notable increase in arsenate tolerance, with As concentrations dropping by 39.2%–42.1% in roots and 38.2%–40% in shoots. Overexpressing the OsGrx_C7 gene increased rice plant tolerance to As(III) and reduced its accumulation, according to Verma et al. (2020).

16.4.3 Biological Approaches

Soil microbes are essential in mineral mobilization, some metabolic processes, and As speciation in the rhizosphere (Hayat et al., 2017). The crucial function of soil microbes in the oxidation and reduction processes of As in the environment has been proven by several investigations, such as Kruger et al. (2013), Hayat et al. (2017), and Kumarathilaka et al. (2018). Some microbes can endure higher levels of As that would be detrimental to other organisms (Hayat et al., 2017). Singh et al. (2016) investigated how introducing As-tolerant bacterial strains, such as *Staphylococcus arlettae* (NBRIEAG-6), *Brevibacillus* sp. (NBRIEAG-6), and *Staphylococcus* sp. (NBRIEAG-6), affects the growth of rice in As-contaminated soils. Their findings revealed that these bacterial strains could reduce the As content in rice by about one-third.

Microorganisms can methylate the toxic form of iAs compounds to produce less toxic compounds, including dimethylarsinous acid (DMA^{III}), dimethylarsinic acid (DMA^{V}), monomethylarsonous acid (MMA^{III}), and monomethylarsonic acid (MMA^{V}) (Jia et al., 2012). This conversion occurs sequentially under aerobic (oxidizing) conditions, as Kruger et al. (2013) demonstrated. As Jia et al. (2012) demonstrated that microorganisms sequentially transformed arsenite and arsenate to methylated As species. Nevertheless, the bacteria undergo a detoxification process involving the As-resistant gene, transforming As molecules (Kruger et al., 2013). An increase in As(III) oxidizing microorganisms and a subsequent decrease in the concentration of bioavailable As in the soil solution was observed after adding NO_3 to waterlogged paddy soils (Zhang et al., 2017). Additionally, research by Zhang et al. (2015) found that *Paracoccus sp.* can serve as an electron acceptor and oxidize arsenite. This process can occur either by utilizing NO_3 under submerged conditions or using O_2 under aerobic conditions.

In rice agroecosystems, a detoxification process might reflect microorganisms converting arsenite to arsenate (Zhang et al., 2015). This transformation, carried out by bacteria that oxidize As(III), likely results in the entrapment of As(V) within the paddy soil matrix and root plaque (Dong et al., 2014). Nevertheless, the rate at which As methylation occurs can be enhanced by introducing bacteria through inoculation (bioaugmentation) and stimulating microbial activities (biostimulation). Ma et al. (2014) found that adding 2.43 tonnes of rice straw per acre in field conditions enhanced the presence of the arsM gene in paddy soils. Similarly, Chen et al. (2017c) demonstrated that As-contaminated soils could have 483.2 mg/kg of tAs eliminated per year by applying 5% rice straw and *Pseudomonas putida* KT2440. According to Yang et al. (2018), microbes are believed to have a role in releasing As from iron minerals and controlling As mobilization by managing As methylation in the rhizosphere of rice fields.

16.4.4 Post-Harvest Processing Techniques

16.4.4.1 Mechanical Methods

16.4.4.1.1 Parboiling

The process of parboiling has undergone a comprehensive investigation, and it has been determined that the extent and technique of parboiling have a noteworthy influence on the As level in rice and its nutritious components, as demonstrated by Bhattacharya (2011) and Menon et al. (2024). Chowdhury et al. (2018) found that parboiled rice grains contain higher levels of As when exposed to As-contaminated water than rice grains that have been dried using sunlight. Chowdhury et al. (2024) corroborated this result through a recent study, showing that the levels of As-accumulation, distribution, and absorption increased by 24% during the parboiling phase and by 34% during the cooking process when using As-contaminated water. Consequently, the amount of As in the cooked rice increased.

The effects of two distinct cooking methods were investigated by Menon et al. (2024), parboiled and absorbed (PBA), and excess water (EW), on the levels of various As forms and essential nutrients in parboiled, brown, and white rice. The rice samples were prepared using either clean or intentionally contaminated tap water. The findings revealed that the PBA and EW methods were similarly effective in reducing iAs in white and brown rice, achieving 54%–58% removal rates. Nevertheless, the EW approach showed superior efficacy in eliminating iAs from parboiled rice, achieving a reduction of roughly 50%, compared to the PBA method, which only reduced to around 39%. Additionally, it was discovered that cooked brown rice outperformed other rice varieties containing numerous vital nutrient elements. Furthermore, the cooking techniques employed substantially impacted the reduction of Mo, Cu, Fe, and K.

In their 2007 study, Rahman et al. examined how traditional parboiling followed by polishing impacted two varieties of rice cultivated in Bangladesh: BRRI dhan28 and BRRI hybrid dhan1. They observed that these rice types' tAs concentrations were reduced by 37.5% and 62.5%, respectively, after parboiling and polishing, for an average reduction of 50% when compared to raw rice. According to the study, parboiled rice's As concentration was lowered by discarding the soaking and boiling processes. Mwale et al. (2018) made a minor adjustment to the traditional parboiling technique in their study. Instead of using rough rice, they parboiled wholegrain or polished rice. They used a ratio of 1:10 (rice to water) and parboiled the rice for 5 min using water free of As. They discovered that this method resulted in a 44% average reduction in tAs in rice grains from various countries, including the UK, Nigeria, Myanmar, and Sri Lanka.

Rahman et al. (2019) found that iAs levels in rice were significantly reduced, with an average drop of 25% using a modified parboiling procedure. Similarly, Menon et al. (2021) conducted experiments to assess how parboiling affects As concentrations in cooked rice. Their study included three types of white rice (sushi, white pudding, and white long grain) and three types of brown rice (brown Basmati, short grain brown, and Thai brown). The rice was cleaned, soaked, and parboiled at four parts water to one part rice. The results showed that cooked brown rice reduced iAs by 54% and white rice by 73%.

16.4.4.1.2 Polishing

To produce polished (white) rice, unpolished (brown) rice's outer layers or bran are usually removed by polishing. Studies by Narukawa et al. (2014), Yim et al. (2017), and Hensawang et al. (2020) have reported a direct relationship between the quantities of As lost from the polished grain and the degree of polishing. Once again, research has shown that the extent of milling and polishing has varying effects on the concentration of As and critical elements in rice grains (Naito et al., 2015; Pedron et al., 2019). When comparing the concentration of tAs in non-polished rice to that in 50%, 30%, and 10% polished rice, Narukawa et al. (2014) discovered that the grains of polished rice contained around 8.3%, 29.9%, and 55.8% of the total concentration of As, respectively.

The effects of a step-by-step polishing process were studied by Hensawang et al. (2020). They discovered that the most significant decrease in As (26%) occurred when the rice, which had not been polished, went through the initial polishing stage. This particular phase resulted in a loss of around 5.9% of the rice's mass. Their study found that 31% of the tAs content was removed from the grain, leading to an 8.1% reduction in the mass of the unpolished rice. Pedron et al. (2019) noted that unpolished rice reduced As levels by up to 41% and a mass loss of 18.3%. Naito et al. (2015) studied tAs concentration and its different forms in popular Japanese rice varieties. The rice was subjected to polishing and storage for varying durations. They revealed that removing 5% of rice bran (by weight) decreased tAs by 16%–20% and iAs by 11%–25%. However, increasing the removal to 10% led to a more significant reduction, with tAs decreasing by 34%–39% and iAs by 30%–49%. Lower concentrations of tAs (45%–53%) and iAs (39%–61%) were observed in cooked rice after rinsing with deionized water.

Conversely, Hensawang et al. (2020) found a more substantial reduction (57%–62%) in iAs levels across all varieties (5.9%, 6.3%, 7.5%, and 8.1%) of polished rice. Meanwhile, Pedron et al. (2019) investigated the impact of varying rice polishing durations in their research. They discovered that polishing rice for 60 s, 40 s, and 20 s reduced tAs by 41.3%, 15.6%, and 13.8% compared to unpolished rice. Furthermore, washing these polished rice samples for 8 h at 35°C using a 1:2 ratio of rice to water reduced As content by 55%, 38%, and 41%. According to Yim et al. (2017), more polishing resulted in lower tAs levels in Korean rice. When comparing brown rice from two separate areas to 10% polished rice, the As content in the former was 154%–172% higher.

16.4.4.1.3 Washing and Cooking

Rice is traditionally prepared for consumption by cooking the grains after they have been washed. Multiple investigations have shown that different amounts of As may be removed after washing with As-free water (Atiaga et al., 2020; Khanam et al., 2022). Das et al. (2022) identified multiple factors that influence the As content in cooked rice, highlighting that As concentration in the cooking water significantly affects As transfer between rice and water. The study used five distinct water samples with various amounts of As to determine how As was released from cooked rice into water during cooking. Nonetheless, when the concentration of As in the water rose, the percentage of As discharged decreased. It was also shown that the variety and cultivar affected the amount of As transferred from cooked rice to water. To conduct a flow optimization experiment, Chowdhury et al. (2020) washed rice grains three times for 10 min each, maintaining a 1:2 ratio of rice to water. The grains were then soaked for an additional 10 min. After soaking, the rice was cooked using a 1:3 water-to-rice ratio, and the leftover porridge was thrown out. According to the study, using this approach to boil rice in water with a low concentration of As (<10 mg/L) significantly reduced the amount of As by almost 76%.

The amount of As in cooked rice was decreased by 24% when water with 40 mg/L of As was utilized. Nevertheless, according to Atiaga et al. (2020), any ratio of rice to water (r/w) will significantly enhance the As concentration in cooked rice if the water used for cooking is contaminated with the As. Mandal et al. (2019) revealed that the As level in cooked rice rose by 56% when washed three times and boiled in a substantial quantity of As-elevated water. According to the research by Halder et al. (2014), Mwale et al. (2018), Mandal et al. (2019), and Atiaga et al. (2020), As reduction by 30%–80% could be achieved in rice cooked with As-free water. Mihucz et al. (2010) found that washing rice three times at 2:1 water-to-rice ratio, followed by cooking it at a 6:1 water-to-rice ratio, can decrease As level in cooked rice by as much as 45%.

Halder et al. (2014) recorded a comparable decrease in As levels in several rice cultivars from West Bengal, India. They observed an average reduction of 33% (ranging from 7.5% to 66%) in As content while using a cooking ratio 1:6 (rice to water). However, they found that washing the rice using a 6:1 water-to-rice ratio had little effect on reducing As concentrations. Mwale et al. (2018) discovered that by using six times the amount of water typically used for cooking, there was a significant decrease of up to 30% in the concentration of tAs. Before boiling the rice in a considerable amount of water, Mandal et al. (2019) incorporated a soaking stage of 15–20 min. They discovered that washing resulted in an 80% decrease in tAs levels compared to uncooked rice.

According to Pedron et al. (2019), the optimum period to soak and wash rice for As removal was 8 h. They also discovered that rinsing with a small volume of water washed the As much more effectively because of the friction between the grains. Following an 8 h washing process using a 2:1 water-to-rice ratio, they found that As levels were 31% lower in the polished rice. However, several studies have consistently demonstrated that washing and cooking rice with a 6:1 water-to-rice ratio can lead to an average As reduction of 30%–35% (Halder et al., 2014; Mwale et al., 2018). According to Mandal et al. (2019) and Atiaga et al. (2020), it is possible to remove up to 80% of the As by cooking the rice in a 10:1 water-to-rice ratio and enhancing up to five times the washing frequency before cooking.

16.4.4.1.4 Continuous Flow of Percolating Water

In 2015, Carey et al. designed a method that uses a continuous flow of water to cook rice at temperatures close to boiling, effectively reducing As content to minimal or undetectable levels. They utilized two techniques for water percolation: one involved using Soxhlet reflux equipment to recycle the cooking water, allowing it to circulate through the rice grains continuously. The second approach involved using a coffee percolator in a domestic environment. In a laboratory setting, polished rice showed an average iAs removal of 59%, while wholegrain rice demonstrated a 69% reduction. The coffee percolator removed 49% of iAs from polished and wholegrain rice. However, these technologies could remove up to 85% of iAs from different rice varieties. The utilization of these devices for rice preparation appears impractical, especially for rural communities, yet it presents a concept to create specialized rice cookers for places afflicted by As. Signes-Pastor et al. (2017) studied the utilization of rice bran, a by-product known for its high As content and nutritional value, as a strategy for removing As. Results indicated that when rice brans were cooked in boiling water that did not contain As, up to 96% of the iAs were removed. The amount of iAs removed was directly related to the amount of cooking water utilized. This procedure had a minimal impact on other essential micro- and macro-nutrient concentrations.

16.4.4.2 Soaking Methods

16.4.4.2.1 Water Soaking

Chen et al. (2021) investigated how temperature influences the As concentration in various grain types, including extra-long, medium, short, and waxy grains, throughout the soaking process with a water-to-rice ratio of 10:1. When the rice was soaked at room temperature for one hour, there was a 22% reduction in iAs. Extending the soaking time to eight hours resulted in a 52% reduction. Soaking the rice for 10 min at 80°C reduced iAs by 27.5%–48.3%, except waxy rice, which only decreased by 15%. Non-waxy white rice showed an average of 40% reduction. As a result, soaking times are shortened as the temperature of the water used for soaking rises. However, further optimization of temperature-dependent soaking is necessary to shorten the soaking duration of the methods.

16.4.4.2.2 Chemical Soaking

Using vinegar solutions with concentrations of 0%, 1%, and 5% in a 1:2 vinegar-to-water ratio at 35°C, Pedron et al. (2019) investigated the effect of pH variations on As elimination. They observed that adding vinegar, which lowered the pH from 6.03 to 4.70, reduced the effectiveness of As deduction. Zhang et al. (2020) revealed that introducing acetic acid in the soaking procedure did not influence the grain-As content but led to the depletion of essential elements. Conversely, after soaking the rice for 12h in citric acid solutions ranging from 1 to 0.0001 M with pH values between 2.5 and 6.4, Pogoson et al. (2021) boiled the rice in a massive quantity of water with a water-to-rice ratio of 5:1. They discovered that, unlike the water-only control group, the inclusion of citric acid hindered the elimination of tAs in cooked rice. However, when compared to raw rice, 1/3 of tAs was eliminated by using a 1 M citric acid solution. This outcome held significance as it demonstrated that substantial quantities of DMA were eliminated under the specified pre-soaking conditions. This result remained unaltered by pre-soaking and cooking with water alone.

In addition, the researchers utilized a 1 M calcium carbonate solution to soak the rice for 12h. This process effectively neutralized the acid and resulted in the removal of up to 63% of tAs. Notably, the soaking method achieved a maximum removal of 66% and 82% iAs DMA (Pogoson et al., 2021). The reason for the reduced effectiveness of different acids in As elimination at low pH is possible because more iAs (as As(III)) are present in the bran. It is possible that the oxidation of As(III) was aided by the water reaching the deeper layers of the endosperm during the prolonged soaking. Citric acid was the only edible acid that successfully reduced arsenate levels in rice grains. The 24h soaking time required by this procedure makes it questionable for home usage.

16.5 Challenges and Prospects of the Emerging Mitigation Approaches

Most endeavors to address the problems caused by As are concentrated on cutting-edge and innovative research in technical agronomic practices, genetic approaches, biological approaches, and post-harvest processing techniques. Some worthwhile approaches must be prioritized in a more detailed appraisal of the potential of prospective mitigation strategies. In rice environments, various soil amendments, including Fe and Mn, can control the As absorption. However, high levels of these amendments might harm the microorganisms supporting rice plant growth. As a result, field-scale evaluations of microbial abundance under various soil amendment circumstances are critical for ensuring the safety and efficacy of amendments in paddy fields contaminated with As. The fate and toxicity of materials utilized in the environmental system have been the subject of few studies in the literature.

The extensive use of BCs and nanomaterials (NMs) for soil amendments offers a viable strategy for remediating As-contaminated soils. However, this approach may raise concerns about possible negative impacts on ecosystems. The long-term fate and degradation status of BCs and BCs impregnated with various chemicals and compounds in rice ecosystems have yet to be investigated. There are certain disadvantages to applying BC or impregnated or modified BC in submerged or aerobic fields. Under field conditions, the effectiveness of BC derived from different sources in mitigating As may not meet expectations. Overusing rice husk BC has been found to raise As concentrations in soil solutions in certain instances (Wu et al., 2020). The physicochemical parameters of the BC and soil characteristics may intimately connect with the varied results of BC supplementation in As-contaminated rice fields (Kumarathilaka et al., 2020).

Additionally, Khan et al. (2017) found that arsM gene regulation by some BC may enhance As volatilization in paddy soils. Therefore, it is essential to take appropriate precautions before applying BC to rice fields, as it may contain specific elemental As species (Wang et al., 2017; Sui et al., 2020). Finally, a more comprehensive investigation is still necessary to determine the method of preparation, quantity, and administration of BC to alleviate the stress of various rice-growing ecologies. Furthermore, BCs and NMs can accumulate in microbial and plant systems before entering the food chain, posing a significant threat to human health. The development of BC and NMs that are biocompatible and biodegradable will reduce their accumulation in agricultural systems in the future.

A significant challenge with using biological methods to lower As content in paddy fields is that the introduced microbes might not adapt well to the field environment and could be outcompeted by the existing microbial community. Additionally, there is a lack of information on how these microbes interact with other biological agents in the paddy soils and limited data on their availability to plants on a large scale in the field. As a result, identifying and inoculating microbes into various paddy environments is not feasible.

Regarding the post-harvest techniques, polishing reduces about 50% of the As and a substantial amount of minerals and vitamins. Washing the rice in a 2:1 water-to-rice ratio and boiling it in a substantial quantity of water, with at least a 4:1 water-to-rice ratio, can eradicate more than one-third of the As. Soaking efficiently eliminated around 30% of As, but only after at least 8 h. However, it was deemed impractical for daily use. Pre-cooking can minimize the loss of nutrients caused by polishing while efficiently eliminating As. The continuous percolating water approach effectively eliminated 49% of iAs when cooking rice, but its practicality for residential use is uncertain.

However, it is essential to overcome substantial limitations and fill existing knowledge gaps to guarantee the broad acceptance and safe application of BC, NMs, soil amendments, and biological agents for As remediation. Hence, additional field studies are necessary to assess their application's safety, dependability, efficiency, consistency, and long-term impact. Furthermore, placing greater emphasis on optimizing the dosage and ensuring these materials' safe and precise distribution would enhance their potential benefits in the fields. Additionally, the combined use of water management strategies, BC, NMs, and/or biological agents can significantly decrease the levels of methylated and inorganic As species in rice farming systems. Utilizing genetic approaches in conjunction with modern agronomic practices holds potential for further research in lowering levels of As. For instance, integrating genetically modified rice varieties with NMs or mineral-enriched soils can lower As loading in rice grains. Future research should investigate the universal applicability of these methods across different geographical areas and climatic conditions in rice paddies. Additionally, it is crucial to determine the cost-effective and most efficient strategies for decreasing As concentration in rice grains. These approaches should be thoroughly examined to determine the optimal combination.

In addition to the above points, adopting a well-coordinated multidisciplinary or transdisciplinary approach is needed to tackle the As issues. This method involves leveraging the expertise of several fields to solve specific aspects of the problem. The research administration should also prioritize managing long-term interdisciplinary or transdisciplinary international research projects, promoting awareness through a broad and integrated worldwide network, and cooperating on themes linked to As. This method addresses the limitations of adopting innovative mitigation techniques by examining the interaction and link between social, economic, biological, and agronomic aspects (Pearson et al., 2018). The policymakers must prioritize strategic research that addresses broader research concerns rather than focusing on specific areas or features, in addition to fundamental and applied research. Furthermore, it is vital to promote the widespread implementation of the developed mitigation strategies, whether used individually or in combination, to ensure their practicality for end users while minimizing any negative impact on biodiversity.

16.6 Conclusion

The toxicity of As is an increasing concern, and controlling it is a significant human health priority globally. Multiple human organs are susceptible to cancer caused by As, a potent carcinogen. The domestic and global trading of rice has transformed the issue of As contamination into an international concern. Consequently, various strategies have been devised to tackle this complex problem. This chapter offers an in-depth examination of recent progress and future possibilities in rice's most effective techniques for lowering As levels. Among the strategies developed in the past 20 years, technical agronomic practices, genetic approaches, biological approaches, and post-harvest processing techniques have demonstrated promising results in reducing As levels in paddy environments despite encountering specific challenges and limitations during their implementation. Over the years, traditional agronomic methods have been used, notwithstanding their expensive implementation, limited As removal, and toxic sludge production, among other issues (Kumar et al., 2019).

Moreover, the implementation of nanoparticles and BCs in rice paddies has decreased As the buildup of rice grains, thereby improving food safety. The percentage of As that can be removed from rice varies depending on the cultivar and type. Washing and cooking can remove 6%–80%, soaking can remove 37%–63%, polishing can remove 38%–55%, and parboiling can remove 39%–54% of As. In addition to a single mitigation technique, combining multiple approaches may have a dual impact in reducing the presence of As. Nevertheless, an exhaustive investigation is necessary to verify the effectiveness of integrated approaches in determining the optimal combinations for decreasing grain As content. Evaluating any strategy's economic effects, whether a standalone approach or a combined one, is crucial. Additionally, it is essential to grasp the characteristics of different places polluted with As and various rice varieties. The degradation status of BC in paddy fields is a matter of significant concern that necessitates obtaining a comprehensive understanding through substantial field-based study. The toxic effects of NMs on paddy soils and plants must be investigated for the sake of public health. In general, the benefits of each technique greatly surpass the drawbacks, and it is essential to focus on overcoming these limitations to reap the advantages fully.

Acknowledgments

The work described in this paper was fully supported by a fellowship award from the Research Grants Council of the Hong Kong Special Administrative Region, China (Project No. EdUHK PDFS2324-8S02).

Declaration of Competing Interest

The authors declare that they have no known competing financial interests or personal relationships that could have appeared to influence this work.

REFERENCES

Abedi, T., & Mojiri, A. (2020). Arsenic uptake and accumulation mechanisms in rice species. *Plants*, 9(2), 129.

Agrafioti, E., Kalderis, D., & Diamadopoulos, E. (2014). Ca and Fe modified biochars as adsorbents of arsenic and chromium in aqueous solutions. *Journal of Environmental Management*, 146, 444–450.

Ahmad, A., van der Wens, P., Baken, K., de Waal, L., Bhattacharya, P., & Stuyfzand, P. (2020). Arsenic reduction to < 1 μg/L in Dutch drinking water. *Environment International*, 134, 105253.

Ahmad, S. A., Khan, M. H., & Haque, M. (2018). Arsenic contamination in groundwater in Bangladesh: implications and challenges for healthcare policy. *Risk Management and Healthcare Policy*, 11, 251.

Alonso, D. L., Latorre, S., Castillo, E., & Brandão, P. F. B. (2014). Environmental occurrence of arsenic in Colombia: a review. *Environmental Pollution,* 186, 272–281.

Arain, M. B., Kazi, T. G., Baig, J. A., Jamali, M. K., Afridi, H. I., Shah, A. Q., & Sarfraz, R. A. (2009). Determination of arsenic levels in lake water, sediment, and foodstuff from selected area of Sindh, Pakistan: estimation of daily dietary intake. *Food and Chemical Toxicology,* 47(1), 242–248.

Arsenic in Rice: Full Analytical Results from Rice/Rice Product Sampling - September 2012. [http://www.fda.gov/Food/FoodSafety/FoodContaminantsAdulteration/Metals/ ucm319916.htm]

Arunakumara, K. K. I. U., Walpola, B. C., & Yoon, M.-H. (2013). Current status of heavy metal contamination in Asia's rice lands. *Reviews in Environmental Science and Bio/Technology,* 12(4), 355–377. doi:10.1007/s11157-013-9323-1

Atiaga, O., Nunes, L. M., & Otero, X. L. (2020). Effect of cooking on arsenic concentration in rice. *Environmental Science and Pollution Research*, 27(10), 10757–10765.

Awasthi, S., Chauhan, R., Srivastava, S., & Tripathi, R. D. (2017). The journey of arsenic from soil to grain in rice. *Frontiers in Plant Science*, 8, 262071.

Aziz, A., Ullah, S. M., & Ullah, M. R. (2015). Arsenic in rice grains at Sonargaon, Bangladesh. *Bangladesh Journal of Botany*, 44(1), 85–89.

Barla, A., Shrivastava, A., Majumdar, A., Upadhyay, M. K., & Bose, S. (2017). Heavy metal dispersion in water saturated and water unsaturated soil of Bengal delta region, India. *Chemosphere,* 100(168), 807–816.

Begum, M., Li, L., Young, E., Carey, M., Li, G., Zhu, Y. G., & Meharg, C. (2024). Fertilization enhances grain inorganic arsenic assimilation in rice. *Exposure and Health,* 16(2), 417–430.

Bhattacharya, K. R. (2011). *Rice Quality: A Guide to Rice Properties and Analysis.* Amsterdam, The Netherlands: Elsevier.

Bhattacharya, P., Samal, A. C., Majumdar, J., & Santra, S. C. (2010a). Arsenic contamination in rice, wheat, pulses, and vegetables: a study in an arsenic affected area of West Bengal, India. *Water, Air, & Soil Pollution,* 213(1), 3–13.

Bhattacharya, P., Samal, A. C., Majumdar, J., & Santra, S. C. (2010b). Accumulation of arsenic and its distribution in rice plant (Oryza sativa L.) in Gangetic West Bengal, India. *Paddy and Water Environment,* 8(1), 63–70.

Bienert, G. P., Thorsen, M., Schüssler, M. D., Nilsson, H. R., Wagner, A., Tamás, M. J., & Jahn, T. P. (2008). A subgroup of plant aquaporins facilitate the bi-directional diffusion of $As(OH)_3$ and $Sb(OH)_3$ across membranes. *BMC Biology*, 6, 1–15.

Biswas, S., Li, R., Yuan, Z., Zhang, D., Zhao, X., & Shi, J. (2019). Development of methods for effective identification of CRISPR/Cas9-induced indels in rice. *Plant Cell Reports*, *38*, 503–510.

Biswas, B., Chakraborty, A., Chatterjee, D., Pramanik, S., Ganguli, B., Majumdar, K. K., & Bhowmick, S. (2021). Arsenic exposure from drinking water and staple food (rice): a field scale study in rural Bengal for assessment of human health risk. *Ecotoxicology and Environmental Safety*, 228, 113012.

Brammer, H., & Ravenscroft, P. (2009). Arsenic in groundwater: a threat to sustainable agriculture in South and South-east Asia. *Environmental International,* 35, 647–654. doi: 10.1016/j.envint.2008.10.004.

Bundschuh, J., Bhattacharya, P., Sracek, O., Mellano, M. F., Ramírez, A. E., Storniolo, A. D. R., & Jean, J. S. (2011). Arsenic removal from groundwater of the Chaco-Pampean Plain (Argentina) using natural geological materials as adsorbents. *Journal of Environmental Science and Health, Part A*, 46(11), 1297–1310.

Bundschuh, J., Nath, B., Bhattacharya, P., Liu, C. W., Armienta, M. A., López, M. V. M., & Tenuta Filho, A. (2012). Arsenic in the human food chain: the Latin American perspective. *Science of the Total Environment*, 429, 92–106

Cao, Y., Sun, D., Chen, J. X., Mei, H., Ai, H., Xu, G., & Ma, L. Q. (2018). Phosphate transporter PvPht1; 2 enhances phosphorus accumulation and plant growth without impacting arsenic uptake in plants. *Environmental Science & Technology*, 52(7), 3975–3981.

Carey, M., Jiujin, X., Gomes Farias, J., & Meharg, A. A. (2015). Rethinking rice preparation for highly efficient removal of inorganic arsenic using percolating cooking water. *PLoS One*, 10(7), e0131608.

Chakraborti, D. (2011). Arsenic: occurrence in groundwater. In: Jerome ÂÂ, Nriagu O (eds) Encyclopedia of Environmental Health. Burlington: Elsevier, pp. 165–180.

Chakraborti, D., Singh, S. K., Rahman, M. M., Dutta, R. N., Mukherjee, S. C., Pati, S., & Kar, P. B. (2018). Groundwater arsenic contamination in the Ganga river basin: a future health danger. *International Journal of Environmental Research and Public Health,* 15(2), 180.

Chang, T. K., Shyu, G. S., Lin, Y. P., & Chang, N. C. (1999). Geostattstical analysis of soil arsenic content in Taiwan. *Journal of Environmental Science & Health Part A,* 34(7), 1485–1501.

Chaudhary, H., Dinakaran, J., & Rao, K. S. (2024). Comparative analysis of biochar production methods and their impacts on biochar physico-chemical properties and adsorption of heavy metals. *Journal of Environmental Chemical Engineering*, 12(3), 113003.

Che, H., Wei, G., Fan, Z., Zhu, Y., Zhang, L., Wei, Z., & Wei, L. (2023). Super facile one-step synthesis of sugarcane bagasse derived N-doped porous biochar for adsorption of ciprofloxacin. *Journal of Environmental Management*, 335, 117566.

Chen, G., Lai, B., Chen, T., Lin, H., & Mao, X. (2021). Brief soaking at above-gelatinization temperature reduces inorganic arsenic in cooked rice. *Cereal Chemistry*, 98(1), 144–153.

Chen, H. L., Lee, C. C., Huang, W. J., Huang, H. T., Wu, Y. C., Hsu, Y. C., & Kao, Y. T. (2016). Arsenic speciation in rice and risk assessment of inorganic arsenic in Taiwan population. *Environmental Science and Pollution Research*, 23, 4481–4488.

Chen, J., Bhattacharjee, H., & Rosen, B. P. (2015). ArsH is an organoarsenical oxidase that confers resistance to trivalent forms of the herbicide MSMA and the poultry growth promoter roxarsone. *Molecular Microbiology*, 96(5), 1042.

Chen, Y., Han, Y. H., Cao, Y., Zhu, Y. G., Rathinasabapathi, B., & Ma, L. Q. (2017a). Arsenic transport in rice and biological solutions to reduce arsenic risk from rice. *Frontiers in Plant Science*, 8, 245590.

Chen, Y., Sun, S. K., Tang, Z., Liu, G., Moore, K. L., Maathuis, F. J., & Zhao, F. J. (2017b). The Nodulin 26-like intrinsic membrane protein OsNIP3; 2 is involved in arsenite uptake by lateral roots in rice. *Journal of Experimental Botany*, 68(11), 3007–3016.

Chen, P., Li, J., Wang, H. Y., Zheng, R. L., & Sun, G. X. (2017c). Evaluation of bioaugmentation and biostimulation on arsenic remediation in soil through biovolatilization. *Environmental Science and Pollution Research, 24*, 21739-21749.

Chou, M.-L., Jean, J.-S., Sun, G.-X., Yang, C.-M., Hseu, Z.-Y., Kuo, S.-F., et al. (2016). Irrigation practices on rice crop production in arsenic-rich paddy soil. *Crop Science,* 56(1), 422. doi:10.2135/cropsci2015.04.0233

Chowdhury, N. R., Das, A., Joardar, M., De, A., Mridha, D., Das, R., & Roychowdhury, T. (2020). Flow of arsenic between rice grain and water: Its interaction, accumulation and distribution in different fractions of cooked rice. *Science of the Total Environment*, 731, 138937.

Chowdhury, N. R., Ghosh, S., Joardar, M., Kar, D., & Roychowdhury, T. (2018). Impact of arsenic contaminated groundwater used during domestic scale post harvesting of paddy crop in West Bengal: arsenic partitioning in raw and parboiled whole grain. *Chemosphere*, 211, 173–184.

Chowdhury, N. R., Joardar, M., Das, A., Bhattacharya, P., & Roychowdhury, T. (2024). Current opinion on the role of post-harvesting and cooking on arsenic mobility in rice grain, its surmounting risk towards human and domestic livestock with sustained management. *Current Opinion in Environmental Science & Health*, 2024, 100535.

Cope, C. O., Webster, D. S., & Sabatini, D. A. (2014). Arsenate adsorption onto iron oxide amended rice husk char. *Science of the Total Environment*, 488, 554–561.

Cui, J., Li, Y., Jin, Q., & Li, F. (2020). Silica nanoparticles inhibit arsenic uptake into rice suspension cells via improving pectin synthesis and the mechanical force of the cell wall. *Environmental Science: Nano*, 7(1), 162–171.

Dai, J., Tang, Z., Jiang, N., Kopittke, P. M., Zhao, F. J., & Wang, P. (2020). Increased arsenic mobilization in the rice rhizosphere is mediated by iron-reducing bacteria. *Environmental Pollution*, 263, 114561.

Das, A., Chowdhury, N. R., Mridha, D., Joardar, M., De, A., Majumder, S., & Roychowdhury, T. (2022). Accumulation and translocation of arsenic in rice with its distributional flow during cooking: a study in West Bengal, India. In: Niazi NK, Bibi I, Aftab T (eds) Global Arsenic Hazard: Ecotoxicology and Remediation. Cham: Springer International Publishing, pp. 159–177.

Deng, B., & Kim, E. S. (2016). Co-adsorption of trichloroethylene and arsenate by iron-impregnated granular activated carbon. *Water Environment Research*, 88(5), 394–402.

Díaz, O. P., Leyton, I., Muñoz, O., Núñez, N., Devesa, V., Súñer, M. A., ... & Montoro, R. (2004). Contribution of water, bread, and vegetables (raw and cooked) to dietary intake of inorganic arsenic in a rural village of Northern Chile. *Journal of Agricultural and Food Chemistry,* 52(6), 1773–1779.

Dolphen, R., Thiravetyan, P., 2019. Reducing arsenic in rice grains by leonardite and arsenic–resistant endophytic bacteria. *Chemosphere* 223, 448–454. https://doi.org/ 10.1016/j.chemosphere.2019.02.054.

Dong, D. T., Yamaguchi, N., Makino, T., & Amachi, S. (2014). Effect of soil microorganisms on arsenite oxidation in paddy soils under oxic conditions. *Soil Science and Plant Nutrition*, 60(3), 377–383. doi:10.108 0/00380768.2014.897924

Duan, G., Kamiya, T., Ishikawa, S., Arao, T., & Fujiwara, T. (2012). Expressing ScACR3 in rice enhanced arsenite efflux and reduced arsenic accumulation in rice grains. *Plant and Cell Physiology*, 53(1), 154–163.

Dudka, S., & Miller, W. P. (1999). Accumulation of potentially toxic elements in plants and their transfer to human food chain. *Journal of Environmental Science & Health Part B,* 34(4), 681–708.

Duxbury, J., & Zavala, Y., (2005). What are Safe Levels of Arsenic in Food and Soils. *Behavior of Arsenic in Aquifers, Soils and Plants (Conference Proceedings), International Symposium*, Dhaka.

EFSA (European Food Safety Authority). (2014). Dietary exposure to inorganic arsenic in the European population. *EFSA Journal*, 12(3), 3597.

FAO (Food and Agriculture Organization of the United Nations). (2004) https://faostat.fao.org/site/336/DesktopDefault.aspx?PageID=336

FAO (Food and Agriculture Organization of the United Nations). (2018). Rice Market Monitor (RMM). XXI(1), www.fao.org/3/i9243en/I9243EN.pdf

Farooq, M., Usman, M., Nadeem, F., ur Rehman, H., Wahid, A., Basra, S. M., & Siddique, K. H. (2019). Seed priming in field crops: potential benefits, adoption and challenges. *Crop and Pasture Science*, 70(9), 731–771.

Fleck, A. T., Mattusch, J., & Schenk, M. K. (2013). Silicon decreases the arsenic level in rice grain by limiting arsenite transport. *Journal of Plant Nutrition and Soil Science*, 176(5), 785–794.

Gang, L. I., ZHENG, M., Jianfeng, T. A. N. G., Hojae, S. H. I. M., & Chao, C. A. I. (2018). Effect of silicon on arsenic concentration and speciation in different rice tissues. *Pedosphere*, 28(3), 511–520.

Garnier, J. M., Travassac, F., Lenoble, V., Rose, J., Zheng, Y., Hossain, M. S., & van Geen, A. (2010). Temporal variations in arsenic uptake by rice plants in Bangladesh: the role of iron plaque in paddy fields irrigated with groundwater. *Science of the Total Environment,* 408(19), 4185–4193.

German Federal Soil Protection Act. (1998). Acton Protection against Harmful Changes to Soil and on Rehabilitation of Contaminated Sites. Federal Law Gazette I p. 502, (BBodSchG), Article 3. https://germanlawarchive.iuscomp.org/?p=322 (accessed on 20 March 2018)

Gu, Z., & Deng, B. (2007). Use of iron-containing mesoporous carbon (IMC) for arsenic removal from drinking water. *Environmental Engineering Science*, 24(1), 113–121.

Halder, D., Biswas, A., Šlejkovec, Z., Chatterjee, D., Nriagu, J., Jacks, G., & Bhattacharya, P. (2014). Arsenic species in raw and cooked rice: implications for human health in rural Bengal. *Science of the Total Environment*, 497, 200–208

Hansen, H. J., & Danielsberg, A. (2009). Condition classes for contaminated sites. The Environment Agency, Report TA-2533 (in Norwegian).

Hayat, K., Menhas, S., Bundschuh, J., & Chaudhary, H. J. (2017). Microbial biotechnology as an emerging industrial wastewater treatment process for arsenic mitigation: a critical review. *Journal of Cleaner Production*, 151, 427–438.

Hensawang, S., Lee, B. T., Kim, K. W., & Chanpiwat, P. (2020). Probabilistic assessment of the daily intake of microelements and toxic elements via the consumption of rice with different degrees of polishing. *Journal of the Science of Food and Agriculture*, 100(10), 4029–4039.

Hu, L., Zeng, M., Lei, M., Liao, B., & Zhou, H. (2020). Effect of zero-valent iron on arsenic uptake by rice (Oryza sativa L.) and its relationship with iron, arsenic, and phosphorus in soil and iron plaque. *Water, Air, & Soil Pollution*, 231, 1–11.

Hu, P., Ouyang, Y., Wu, L., Shen, L., Luo, Y., & Christie, P. (2015). Effects of water management on arsenic and cadmium speciation and accumulation in an upland rice cultivar. *Journal of Environmental Sciences*, 27, 225–231.

Huang, Q., Liu, Q., Lin, L., Li, F. J., Han, Y., & Song, Z. G. (2018). Reduction of arsenic toxicity in two rice cultivar seedlings by different nanoparticles. *Ecotoxicology and Environmental Safety*, 159, 261–271.

Islam, M. S., Chen, Y., Weng, L., Ma, J., Khan, Z. H., Liao, Z., & Li, Y. (2021b). Watering techniques and zero-valent iron biochar pH effects on As and Cd concentrations in rice rhizosphere soils, tissues and yield. *Journal of Environmental Sciences*, 100, 144–157.

Islam, M. S., Magid, A. S. I. A., Chen, Y., Weng, L., Ma, J., Arafat, M. Y., & Li, Y. (2021a). Effect of calcium and iron-enriched biochar on arsenic and cadmium accumulation from soil to rice paddy tissues. *Science of the Total Environment*, 785, 147163.

Islam, M., Jahiruddin, M., & Islam, S. (2007). Arsenic linkage in the irrigation water-soil-rice plant systems. *Pakistan Journal of Scientific and Industrial Research,* 50, 85–90.

Islam, S., Rahman, M. M., & Naidu, R. (2019). Impact of water and fertilizer management on arsenic bioaccumulation and speciation in rice plants grown under greenhouse conditions. *Chemosphere*, 214, 606–613.

Islam, S., Rahman, M. M., Islam, M. R., & Naidu, R. (2016). Arsenic accumulation in rice: consequences of rice genotypes and management practices to reduce human health risk. *Environment International,* 96, 139–155. doi:10.1016/j.envint.2016.09.006

Islam, S., Rahman, M. M., Islam, M. R., & Naidu, R. (2017). Effect of irrigation and genotypes towards reduction in arsenic load in rice. *Science of the Total Environment*, 609, 311–318.

Jaishankar, M., Tseten, T., Anbalagan, N., Mathew, B. B., & Beeregowda, K. N. (2014). Toxicity, mechanism and health effects of some heavy metals. *Interdisciplinary Toxicology*, 7(2), 60–72.

Jia, Y., Huang, H., Sun, G. X., Zhao, F. J., & Zhu, Y. G. (2012). Pathways and relative contributions to arsenic volatilization from rice plants and paddy soil. *Environmental Science & Technology*, 46(15), 8090–8096.

Kamiya, T., & Fujiwara, T. (2009). Arabidopsis NIP1; 1 transports antimonite and determines antimonite sensitivity. *Plant and Cell Physiology*, 50(11), 1977–1981.

Kamiya, T., Islam, R., Duan, G., Uraguchi, S., & Fujiwara, T. (2013). Phosphate deficiency signaling pathway is a target of arsenate and phosphate transporter OsPT1 is involved in As accumulation in shoots of rice. *Soil Science and Plant Nutrition*, 59(4), 580–590.

Kapaj, S., Peterson, H., Liber, K., & Bhattacharya, P. (2006). Human health effects from chronic arsenic poisoning–a review. *Journal of Environmental Science and Health, Part A*, 41(10), 2399–2428.

Katsuhara, M., Sasano, S., Horie, T., Matsumoto, T., Rhee, J., & Shibasaka, M. (2014). Functional and molecular characteristics of rice and barley NIP aquaporins transporting water, hydrogen peroxide and arsenite. *Plant Biotechnology*, 31(3), 213–219.

Kepel, B., Bodhi, W., & Tallei, T. E. (2020). Isolation and Identification of Arsenic-resistant Bacteria for Possible Application in Arsenic Bioremediation. *Pakistan Journal of Biological Sciences: PJBS*, 23(1), 63–67.

Khan, E., & Gupta, M. (2018). Arsenic–silicon priming of rice (Oryza sativa L.) seeds influence mineral nutrient uptake and biochemical responses through modulation of Lsi-1, Lsi-2, Lsi-6 and nutrient transporter genes. *Scientific Reports*, 8(1), 10301.

Khan, M. N., Mobin, M., Abbas, Z. K., AlMutairi, K. A., & Siddiqui, Z. H. (2017). Role of nanomaterials in plants under challenging environments. *Plant Physiology and Biochemistry*, 110, 194–209.

Khan, S., Akhtar, N., Rehman, S. U., Shujah, S., Rha, E. S., & Jamil, M. (2020). Biosynthesized iron oxide nanoparticles (Fe_3O_4 NPs) mitigate arsenic toxicity in rice seedlings. *Toxics*, 9(1), 2.

Khanam, R., Kulsum, P. G. P. S., Mandal, B., Hazra, G. C., & Kundu, D. (2022). The mechanistic pathways of arsenic transport in rice cultivars: soil to mouth. *Environmental Research*, 204, 111942.

Kim, K., Moon, J. T., Kim, S. H., & Ko, K. S. (2009). Importance of surface geologic condition in regulating As concentration of groundwater in the alluvial plain. *Chemosphere*, 77(4), 478–484.

Kruger, M. C., Bertin, P. N., Heipieper, H. J., & Arsène-Ploetze, F. (2013). Bacterial metabolism of environmental arsenic—mechanisms and biotechnological applications. *Applied Microbiology and Biotechnology*, 97, 3827–3841.

Kumar, K., Gupta, D., Mosa, K. A., Ramamoorthy, K., & Sharma, P. (2019). Arsenic transport, metabolism, and possible mitigation strategies in plants. *Plant-metal interactions*, 141–168.

Kumarathilaka, P., Bundschuh, J., Seneweera, S., & Ok, Y. S. (2021a). An integrated approach of rice hull biochar-alternative water management as a promising tool to decrease inorganic arsenic levels and to sustain essential element contents in rice. *Journal of Hazardous Materials*, 405, 124188.

Kumarathilaka, P., Bundschuh, J., Seneweera, S., & Ok, Y. S. (2021b). Rice genotype's responses to arsenic stress and cancer risk: The effects of integrated birnessite-modified rice hull biochar-water management applications. *Science of the Total Environment*, 768, 144531.

Kumarathilaka, P., Seneweera, S., Meharg, A., & Bundschuh, J. (2018). Arsenic speciation dynamics in paddy rice soil-water environment: sources, physico-chemical, and biological factors-a review. *Water Research*, 140, 403–414.

Kumarathilaka, P., Seneweera, S., Ok, Y. S., Meharg, A. A., & Bundschuh, J. (2020). Mitigation of arsenic accumulation in rice: An agronomical, physico-chemical, and biological approach–A critical review. *Critical Reviews in Environmental Science and Technology*, 50(1), 31–71.

Kumarathilaka, P., Seneweera, S., Ok, Y. S., Meharg, A., & Bundschuh, J. (2019). Arsenic in cooked rice foods: assessing health risks and mitigation options. *Environment International*, 127, 584–591.

Larsen, E. H., & Berg, T. (2001). Trace element speciation and international food legislation–a Codex Alimentarius position paper on arsenic as a contaminant. *Trace Element Speciation for Environment, Food and Health*, 2001, 251–260.

Li, R. Y., Ago, Y., Liu, W. J., Mitani, N., Feldmann, J., McGrath, S. P., & Zhao, F. J. (2009). The rice aquaporin Lsi1 mediates uptake of methylated arsenic species. *Plant Physiology*, 150(4), 2071–2080.

Lian, F., Liu, X., Gao, M., Li, H., Qiu, W., & Song, Z. (2020). Effects of Fe-Mn-Ce oxide–modified biochar on As accumulation, morphology, and quality of rice (Oryza sativa L.). *Environmental Science and Pollution Research*, 27, 18196–18207.

Lihong, W. A. N. G., & Guilan, D. U. A. N. (2009). Effect of external and internal phosphate status on arsenic toxicity and accumulation in rice seedlings. *Journal of Environmental Sciences*, 21(3), 346–351.

Lin, L., Gao, M., Liu, X., Qiu, W., & Song, Z. (2021). Effect of Fe–Mn–La-modified biochar composites on arsenic volatilization in flooded paddy soil. *Environmental Science and Pollution Research*, 28(36), 49889–49898.

Lin, L., Gao, M., Qiu, W., Wang, D., Huang, Q., & Song, Z. (2017). Reduced arsenic accumulation in indica rice (Oryza sativa L.) cultivar with ferromanganese oxide impregnated biochar composites amendments. *Environmental Pollution*, 231, 479–486.

Liu, J., Simms, M., Song, S., King, R. S., & Cobb, G. P. (2018). Physiological effects of copper oxide nanoparticles and arsenic on the growth and life cycle of rice (Oryza sativa japonica 'Koshihikari'). *Environmental Science & Technology*, 52(23), 13728–13737.

Liu, W. J., McGrath, S. P., & Zhao, F. J. (2014). Silicon has opposite effects on the accumulation of inorganic and methylated arsenic species in rice. *Plant and Soil*, 376, 423–431.

Lu, Y., Dong, F., Deacon, C., Chen, H. J., Raab, A., & Meharg, A. A. (2010). Arsenic accumulation and phosphorus status in two rice (Oryza sativa L.) cultivars surveyed from fields in South China. *Environmental Pollution*, 158(5), 1536–1541.

Ma, J. F., Yamaji, N., Mitani, N., Xu, X. Y., Su, Y. H., McGrath, S. P., & Zhao, F. J. (2008). Transporters of arsenite in rice and their role in arsenic accumulation in rice grain. *Proceedings of the National Academy of Sciences*, 105(29), 9931–9935.

Ma, R., Shen, J., Wu, J., Tang, Z., Shen, Q., & Zhao, F. J. (2014). Impact of agronomic practices on arsenic accumulation and speciation in rice grain. *Environmental Pollution*, 194, 217–223.

Ma, X., Sharifan, H., Dou, F., & Sun, W. (2020). Simultaneous reduction of arsenic (As) and cadmium (Cd) accumulation in rice by zinc oxide nanoparticles. *Chemical Engineering Journal*, 384, 123802.

Maizel, D., Balverdi, P., Rosen, B., Sales, A. M., & Ferrero, M. A. (2018). Arsenic-hypertolerant and arsenic-reducing bacteria isolated from wells in Tucumán, Argentina. *Canadian Journal of Microbiology*, 64(11), 876–886.

Majumdar, A. (2021). *Arsenic bio-regulation management and altered plant physiology under different irrigation regime* (Doctoral dissertation, Indian Institute of Science Education and Research Kolkata).

Majumdar, A., Upadhyay, M. K., Giri, B., Karwadiya, J., Bose, S., & Jaiswal, M. K. (2023). Iron oxide doped rice biochar reduces soil-plant arsenic stress, improves nutrient values: an amendment towards sustainable development goals. *Chemosphere*, 312, 137117.

Majumdar, A., Upadhyay, M. K., Giri, B., Yadav, P., Moulick, D., Sarkar, S., & Roychowdhury, T. (2024). Sustainable water management in rice cultivation reduces arsenic contamination, increases productivity, microbial molecular response, and profitability. *Journal of Hazardous Materials*, 466, 133610.

Majumder, S., & Banik, P. (2019). Geographical variation of arsenic distribution in paddy soil, rice and rice-based products: A meta-analytic approach and implications to human health. *Journal of Environmental Management*, 233, 184–199.

Massaquoi, L. D., Ma, H., Liu, X. H., Han, P. Y., Zuo, S.-M., & Hua, Z.-X., et al. (2015). Heavy metal accumulation in soils, plants, and hair samples: an assessment of heavy metal exposure risks from the consumption of vegetables grown on soils previously irrigated with wastewater. *Environmental Science and Pollution Research*, 22(23), 18456–18468. doi:10.1007/s11356-015-5131-1

Masuda, H. (2018). Arsenic cycling in the Earth's crust and hydrosphere: interaction between naturally occurring arsenic and human activities. *Progress in Earth and Planetary Science*, 5(1), 1–11.

Mawia, A. M., Hui, S., Zhou, L., Li, H., Tabassum, J., Lai, C., & Hu, P. (2021). Inorganic arsenic toxicity and alleviation strategies in rice. *Journal of Hazardous Materials*, 408, 124751.

McClintock, T. R., Chen, Y., Bundschuh, J., Oliver, J. T., Navoni, J., Olmos, V., & Parvez, F. (2012). Arsenic exposure in Latin America: biomarkers, risk assessments and related health effects. *Science of the Total Environment*, 429, 76–91.

Meharg, A. A., & Rahman, M. M. (2003). Arsenic contamination of Bangladesh paddy field soils: implications for rice contribution to arsenic consumption. *Environmental Science & Technology*, *37*(2), 229–234.

Meharg, A. A., Williams, P. N., Adomako, E., Lawgali, Y. Y., Deacon, C., Villada, A., et al. (2009). Geographical variation in total and inorganic arsenic content of polished (white) rice. *Environmental Science & Technology*, 43, 1612–1617.

Menon, M., Dong, W., Chen, X., Hufton, J., & Rhodes, E. J. (2021). Improved rice cooking approach to maximise arsenic removal while preserving nutrient elements. *Science of the Total Environment*, 755, 143341.

Menon, M., Nicholls, A., Smalley, A., & Rhodes, E. (2024). A comparison of the effects of two cooking methods on arsenic species and nutrient elements in rice. *Science of the Total Environment*, 914, 169653.

Mihucz, V. G., Silversmit, G., Szalóki, I., De Samber, B., Schoonjans, T., Tatár, E., & Záray, G. (2010). Removal of some elements from washed and cooked rice studied by inductively coupled plasma mass spectrometry and synchrotron based confocal micro-X-ray fluorescence. *Food Chemistry*, 121(1), 290–297.

Moreno-Jiménez, E., Meharg, A. A., Smolders, E., Manzano, R., Becerra, D., Sánchez-Llerena, J., et al. (2014). Sprinkler irrigation of rice fields reduces grain arsenic but enhances cadmium. *Science of the Total Environment*, 485–486, 468–473. doi:10.1016/j.scitotenv.2014.03.106

Mosa, K. A., Kumar, K., Chhikara, S., Mcdermott, J., Liu, Z., Musante, C., & Dhankher, O. P. (2012). Members of rice plasma membrane intrinsic proteins subfamily are involved in arsenite permeability and tolerance in plants. *Transgenic Research*, 21, 1265–1277.

Moulick, D., Santra, S. C., & Ghosh, D. (2018a). Rice seed priming with se: a novel approach to mitigate as induced adverse consequences on growth, yield and as load in brown rice. *Journal of Hazardous Materials*, 355, 187–196.

Moulick, D., Santra, S. C., & Ghosh, D. (2018b). Effect of selenium induced seed priming on arsenic accumulation in rice plant and subsequent transmission in human food chain. *Ecotoxicology and Environmental Safety*, 152, 67–77.

Moulick, D., Santra, S. C., Ghosh, D., & Panda, S. K. (2019). An assessment of efficiency of zinc priming in rice (cv. MTU-7029) during germination and early seedling growth. In: Hasanuzzaman M, Fotopoulos V (eds) Priming and Pretreatment of Seeds and Seedlings: Implication in Plant Stress Tolerance and Enhancing Productivity in Crop Plants. Singapore: Springer, pp. 495–507.

Mridha, D., Paul, I., De, A., Ray, I., Das, A., Joardar, M., & Roychowdhury, T. (2021). Rice seed (IR64) priming with potassium humate for improvement of seed germination, seedling growth and antioxidant defense system under arsenic stress. *Ecotoxicology and Environmental Safety*, 219, 112313.

Munera-Picazo, S., Cano-Lamadrid, M., Castaño-Iglesias, M. C., Burló, F., & Carbonell-Barrachina, Á. A. (2015). Arsenic in your food: potential health hazards from arsenic found in rice. *Nutrition and Dietary Supplements*, 2015, 1–10.

Mwale, T., Rahman, M. M., & Mondal, D. (2018). Risk and benefit of different cooking methods on essential elements and arsenic in rice. *International Journal of Environmental Research and Public Health*, 15(6), 1056.

Naito, S., Matsumoto, E., Shindoh, K., & Nishimura, T. (2015). Effects of polishing, cooking, and storing on total arsenic and arsenic species concentrations in rice cultivated in Japan. *Food Chemistry*, 168, 294–301.

Narukawa, T., Matsumoto, E., Nishimura, T., & Hioki, A. (2014). Determination of sixteen elements and arsenic species in brown, polished and milled rice. *Analytical Sciences*, 30(2), 245–250.

Naujokas, M. F., Anderson, B., Ahsan, H., Aposhian, H. V., Graziano, J. H., Thompson, C., & Suk, W. A. (2013). The broad scope of health effects from chronic arsenic exposure: update on a worldwide public health problem. *Environmental Health Perspectives*, 121(3), 295–302.

Newbigging, M., Paliwoda, E., & Chris, X. (2015). Rice: Reducing arsenic content by controlling water irrigation. *Journal of Environmental Sciences*, 27(4), 129–131.

Norton, G. J., Islam, M. R., Deacon, C. M., Zhao, F. J., Stroud, J. L., McGrath, S. P., & Meharg, A. A. (2009). Identification of low inorganic and total grain arsenic rice cultivars from Bangladesh. *Environmental Science & Technology*, 43(15), 6070–6075.

Oberoi, S., Barchowsky, A., & Wu, F. (2014). The global burden of disease for skin, lung, and bladder cancer caused by arsenic in food. *Cancer Epidemiology Biomarkers & Prevention,* 23(7), 1187–1194. doi:10.1158/1055-9965.epi-13-1317

Pan, D., Liu, C., Yu, H., & Li, F. (2019). A paddy field study of arsenic and cadmium pollution control by using iron-modified biochar and silica sol together. *Environmental Science and Pollution Research*, 26(24), 24979–24987.

Pasias, I. N., Thomaidis, N. S., & Piperaki, E. A. (2013). Determination of total arsenic, total inorganic arsenic and inorganic arsenic species in rice and rice flour by electrothermal atomic absorption spectrometry. *Microchemical Journal*, 108, 1–6.

Pearson, K. A., Millar, G. M., Norton, G. J., & Price, A. H. (2018). Alternate wetting and drying in Bangladesh: water-saving farming practice and the socioeconomic barriers to its adoption. *Food and Energy Security*, 7(4), e00149.

Pedron, T., Segura, F. R., Paniz, F. P., de Moura Souza, F., dos Santos, M. C., de Magalhães Júnior, A. M., & Batista, B. L. (2019). Mitigation of arsenic in rice grains by polishing and washing: evidencing the benefit and the cost. *Journal of Cereal Science*, 87, 52–58.

Pescod, M. B. (1992). Wastewater treatment and use in agriculture-FAO irrigation and drainage paper 47. Food and Agriculture Organization of the United Nations, Rome.

Phuong, N. M., Kang, Y., Sakurai, K., Iwasaki, K., Kien, C. N., Van Noi, N., & Son, L. T. (2008). Arsenic contents and physicochemical properties of agricultural soils from the Red River Delta, Vietnam. *Soil Science and Plant Nutrition,* 54(6), 846–855.

Podgorski, J., & Berg, M. (2020). Global threat of arsenic in groundwater. *Science*, 368(6493), 845–850.

Pogoson, E., Carey, M., Meharg, C., & Meharg, A. A. (2021). Reducing the cadmium, inorganic arsenic and dimethylarsinic acid content of rice through food-safe chemical cooking pre-treatment. *Food Chemistry, 338*, 127842.

Polizzotto, M. L., Birgand, F., Badruzzaman, A. B. M., & Ali, M. A. (2015). Amending irrigation channels with jute-mesh structures to decrease arsenic loading to rice fields in Bangladesh. *Ecological Engineering*, 74, 101–106.

Punshon, T., Jackson, B. P., Meharg, A. A., Warczack, T., Scheckel, K., & Guerinot, M. L. (2017). Understanding arsenic dynamics in agronomic systems to predict and prevent uptake by crop plants. *Science of the Total Environment*, 581, 209–220.

Qiao, J., Yu, H., Wang, X., Li, F., Wang, Q., Yuan, Y., & Liu, C. (2019). The applicability of biochar and zero-valent iron for the mitigation of arsenic and cadmium contamination in an alkaline paddy soil. *Biochar*, 1(2), 203–212.

Rahman, H., Carey, M., Hossain, M., Savage, L., Islam, M. R., & Meharg, A. A. (2019). Modifying the parboiling of rice to remove inorganic arsenic, while fortifying with calcium. *Environmental Science & Technology*, 53(9), 5249–5255.

Rahman, M. A., & Hasegawa, H. (2011). High levels of inorganic arsenic in rice in areas where arsenic-contaminated water is used for irrigation and cooking. *Science of the Total Environment*, 409(22), 4645–4655.

Rahman, M. A., Hasegawa, H., Rahman, M. M., Rahman, M. A., & Miah, M. A. M. (2007). Accumulation of arsenic in tissues of rice plant (Oryza sativa L.) and its distribution in fractions of rice grain. *Chemosphere*, 69(6), 942–948.

Rahman, M. A., Rahman, A., Khan, M. Z. K., & Renzaho, A. M. (2018). Human health risks and socio-economic perspectives of arsenic exposure in Bangladesh: a scoping review. *Ecotoxicology and environmental safety*, 150, 335–343.

Rahman, M. M., Asaduzzaman, M., & Naidu, R. (2013). Consumption of arsenic and other elements from vegetables and drinking water from an arsenic-contaminated area of Bangladesh. *Journal of Hazardous Materials*, 262, 1056–1063.

Rahman, M. S., Islam, M. N., Hassan, M. Z., Islam, S. A., & Zaman, S. K. (2014). Impact of water management on the arsenic content of rice grain and cultivated soil in an arsenic contaminated area of Bangladesh. *Journal of Environmental Science and Natural Resources*, 7(2), 43–46.

Rinklebe, J., Shaheen, S. M., & Yu, K. (2016). Release of As, Ba, Cd, Cu, Pb, and Sr under pre-definite redox conditions in different rice paddy soils originating from the U.S.A. and Asia. *Geoderma*, 270, 21–32.

Roychowdhury, T. (2008). Impact of sedimentary arsenic through irrigated groundwater on soil, plant, crops and human continuum from Bengal delta: special reference to raw and cooked rice. *Food and Chemical Toxicology*, 46(8), 2856–2864.

Saha, A., Sen Gupta, B., Patidar, S., & Martínez-Villegas, N. (2022). Identification of soil arsenic contamination in rice paddy field based on hyperspectral reflectance approach. *Soil Systems*, 6(1), 30.

Sahoo, S., Borgohain, P., Saha, B., Moulick, D., Tanti, B., & Panda, S. K. (2019). Seed priming and seedling pre-treatment induced tolerance to drought and salt stress: recent advances. In: Hasanuzzaman M, Fotopoulos V (eds) Priming and Pretreatment of Seeds and Seedlings: Implication in Plant Stress Tolerance and Enhancing Productivity in Crop Plants. Singapore: Springer, pp. 253–263.

Sampson, M. L., Bostick, B., Chiew, H., Hagan, J. M., & Shantz, A. (2008). Arsenicosis in Cambodia: case studies and policy response. *Applied Geochemistry*, 23(11), 2977–2986.

Sarkar, A., & Paul, B. (2016). The global menace of arsenic and its conventional remediation - a critical review. *Chemosphere*, 158, 37–49. doi:10.1016/j.chemosphere.2016.05.043

Sarkar, S., Basu, B., Kundu, C. K., & Patra, P. K. (2012). Deficit irrigation: an option to mitigate arsenic load of rice grain in West Bengal, India. *Agriculture, Ecosystems & Environment*, 146(1), 147–152.

Seyfferth, A. L., Limmer, M. A., & Dykes, G. E. (2018). On the use of silicon as an agronomic mitigation strategy to decrease arsenic uptake by rice. *Advances in Agronomy*, 149, 49–91.

Shrivastava, A., Barla, A., Majumdar, A., Singh, S., & Bose, S. (2020). Arsenic mitigation in rice grain loading via alternative irrigation by proposed water management practices. *Chemosphere*, 238, 124988.

Shrivastava, A., Ghosh, D., Dash, A., & Bose, S. (2015). Arsenic contamination in soil and sediment in India: sources, effects, and remediation. *Current Pollution Reports*, 1, 35–46.

Signes-Pastor, A. J., Carey, M., & Meharg, A. A. (2017). Inorganic arsenic removal in rice bran by percolating cooking water. *Food Chemistry*, 234, 76–80

Singh, J., Singh, V., Vineeth, T. V., Kumar, P., Kumar, N., & Sharma, P. C. (2019). Differential response of Indian mustard (Brassica juncea L., Czern and Coss) under salinity: photosynthetic traits and gene expression. *Physiology and Molecular Biology of Plants*, 25, 71–83.

Singh, N., Marwa, N., Mishra, J., Verma, P. C., Rathaur, S., & Singh, N. (2016). Brevundimonas diminuta mediated alleviation of arsenic toxicity and plant growth promotion in Oryza sativa L. *Ecotoxicology and Environmental Safety*, 125, 25–34.

Singh, R., Singh, S., Parihar, P., Singh, V. P., & Prasad, S. M. (2015). Arsenic contamination, consequences and remediation techniques: a review. *Ecotoxicology and Environmental Safety*, 112, 247–270. doi:10.1016/j.ecoenv.2014.10.009

Smedley, P., & Kinniburgh, D. (2002). A review of the source, behaviour and distribution of arsenic in natural waters. *Applied Geochemistry*, 17 (5), 517–568.

Smith, S. R. (1996). *Agricultural recycling of Sewage Sludge and the Environment*. Wallingford, UK: CAB International, p. 382.

Sobel, M. H., Sanchez, T. R., Jones, M. R., Kaufman, J. D., Francesconi, K. A., Blaha, M. J., & Navas-Acien, A. (2020). rice intake, arsenic exposure, and subclinical cardiovascular disease among us adults in MESA. *Journal of the American Heart Association*, 9(4), e015658.

Soil Remediation Act. (1995). Bodemsaneringsdecreet, Belgium 29 October 1995.

Suda, A., Baba, K., Sakurai, G., Furuya, M., & Yamaguchi, N. (2023). Enhanced dissolution of arsenic in anaerobic soils upon organic amendment application: acid detergent-soluble organic matter as a potential indicator. *Scientific Reports*, 13(1), 217.

Sui, F., Wang, J., Zuo, J., Joseph, S., Munroe, P., Drosos, M., & Pan, G. (2020). Effect of amendment of biochar supplemented with Si on Cd mobility and rice uptake over three rice growing seasons in an acidic Cd-tainted paddy from central South China. *Science of the Total Environment*, 709, 136101.

Sun, S. K., Chen, Y., Che, J., Konishi, N., Tang, Z., Miller, A. J., & Zhao, F. J. (2018). Decreasing arsenic accumulation in rice by overexpressing Os NIP 1; 1 and Os NIP 3; 3 through disrupting arsenite radial transport in roots. *New Phytologist*, 219(2), 641–653.

Suriyagoda, L. D. B., Dittert, K., & Lambers, H. (2018). Mechanism of arsenic uptake, translocation and plant resistance to accumulate arsenic in rice grains. *Agriculture, Ecosystems & Environment*, 253, 23–37.

Syu, C. H., Wu, P. R., Lee, C. H., Juang, K. W., & Lee, D. Y. (2019). Arsenic phytotoxicity and accumulation in rice seedlings grown in arsenic-contaminated soils as influenced by the characteristics of organic matter amendments and soils. *Journal of Plant Nutrition and Soil Science*, 182(1), 60–71.

Talukder, A. S. M. H. M., Meisner, C. A., Sarkar, M. A. R., & Islam, M. S. (2011). Effect of water management, tillage options and phosphorus status on arsenic uptake in rice. *Ecotoxicology and Environmental Safety*, 74(4), 834–839.

Tarvainen, T., Albanese, S., Birke, M., Poňavič, M., Reimann, C., & Team, T. G. P. (2013). Arsenic in agricultural and grazing land soils of Europe. *Applied Geochemistry*, 28, 2–10.

Upadhyay, M. K., Shukla, A., Yadav, P., & Srivastava, S. (2019). A review of arsenic in crops, vegetables, animals and food products. *Food Chemistry*, 276, 608–618.

USEPA (United States Environmental Protection Agency). (1996). Report: Recent Developments for In Situ Treatment of Metals contaminated Soils, U.S. Environmental Protection Agency, Office of Solid Waste and Emergency Response.

Ustohalova, V. (2011). Management and export of wastes: human health implications. *Encyclopedia of Environmental Health*, 2011, 603–611.

Vahter, M. (2008). Health effects of early life exposure to arsenic. *Basic & Clinical Pharmacology & Toxicology*, 102(2), 204–211.

Verma, P. K., Verma, S., Tripathi, R. D., & Chakrabarty, D., (2020). A rice glutaredoxin regulates the expression of aquaporin genes and modulate root responses to provide arsenic tolerance. *Ecotoxicology and Environmental Safety*, 195, 110471.

Wang, H., Cui, S., Ma, L., Wang, Z., & Wang, H. (2021). Variations of arsenic forms and the role of arsenate reductase in three hydrophytes exposed to different arsenic species. *Ecotoxicology and Environmental Safety*, 221, 112415.

Wang, N., Xue, X. M., Juhasz, A. L., Chang, Z. Z., & Li, H. B. (2017). Biochar increases arsenic release from an anaerobic paddy soil due to enhanced microbial reduction of iron and arsenic. *Environmental Pollution*, 220, 514–522.

Wang, P., Zhang, W., Mao, C., Xu, G., & Zhao, F. J. (2016). The role of OsPT8 in arsenate uptake and varietal difference in arsenate tolerance in rice. *Journal of Experimental Botany*, 67(21), 6051–6059.

Wang, S., & Mulligan, C. N. (2006). Occurrence of arsenic contamination in Canada: sources, behavior and distribution. *Science of the Total Environment*, 366(2–3), 701–721.

Wang, X., Sun, W., & Ma, X. (2019a). Differential impacts of copper oxide nanoparticles and Copper (II) ions on the uptake and accumulation of arsenic in rice (Oryza sativa). *Environmental Pollution*, 252, 967–973.

Wang, X., Sun, W., Zhang, S., Sharifan, H., & Ma, X. (2018). Elucidating the effects of cerium oxide nanoparticles and zinc oxide nanoparticles on arsenic uptake and speciation in rice (Oryza sativa) in a hydroponic system. *Environmental Science & Technology*, 52(17), 10040–10047.

Wang, Y., Xiao, X., Xu, Y., & Chen, B. (2019b). Environmental effects of silicon within biochar (sichar) and carbon–silicon coupling mechanisms: a critical review. *Environmental Science & Technology*, 53(23), 13570–13582.

Waqas, M., Korres, N. E., Khan, M. D., Nizami, A. S., Deeba, F., Ali, I., & Hussain, H. (2019). Advances in the concept and methods of seed priming. In: Hasanuzzaman M, Fotopoulos V (eds) Priming and Pretreatment of Seeds and Seedlings: Implication in Plant Stress Tolerance and Enhancing Productivity in Crop Plants. Singapore: Springer, pp. 11–41.

WHO (World Health Organization). (1989). Evaluation of certain food additives and contaminants (Thirty-third report of the Joint FAO/WHO Expert Committee on Food Additives). WHO Technical Report Series, No. 776, p. 64.

WHO (World Health Organization). (2004). IARC, Working Group on some drinking water disinfectants and contaminants, including arsenic, vol 84. Lyon. Monograph 1.

WHO (World Health Organization). (2016). Arsenic international programme on chemical safety. Health Impacts of Chemicals Cited November 22, 2016.

Williams, P. N., Zhang, H., Davison, W., Meharg, A. A., Hossain, M., Norton, G. J., & Islam, M. R. (2011). Organic matter solid phase interactions are critical for predicting arsenic release and plant uptake in Bangladesh paddy soils. *Environmental Science & Technology*, 45(14), 6080–6087.

Wu, C., Cui, M., Xue, S., Li, W., Huang, L., Jiang, X., & Qian, Z. (2018). Remediation of arsenic-contaminated paddy soil by iron-modified biochar. *Environmental Science and Pollution Research*, 25, 20792–20801.

Wu, F., Fang, Q., Yan, S., Pan, L., Tang, X., & Ye, W. (2020). Effects of zinc oxide nanoparticles on arsenic stress in rice (Oryza sativa L.): germination, early growth, and arsenic uptake. *Environmental Science and Pollution Research*, 27, 26974–26981.

Wu, J., Li, Z., Wang, L., Liu, X., Tang, C., & Xu, J. (2020). A novel calcium-based magnetic biochar reduces the accumulation of As in grains of rice (Oryza sativa L.) in As-contaminated paddy soils. *Journal of Hazardous Materials*, 394, 122507.

Wu, X., Hu, J., Wu, F., Zhang, X., Wang, B., Yang, Y., & Wang, X. (2021). Application of TiO_2 nanoparticles to reduce bioaccumulation of arsenic in rice seedlings (Oryza sativa L.): a mechanistic study. *Journal of Hazardous Materials*, 405, 124047.

Wu, Z., Ren, H., McGrath, S. P., Wu, P., & Zhao, F. J. (2011). Investigating the contribution of the phosphate transport pathway to arsenic accumulation in rice. *Plant Physiology*, 157(1), 498–508.

Xu, J., Shi, S., Wang, L., Tang, Z., Lv, T., Zhu, X., & Wu, Z. (2017). OsHAC4 is critical for arsenate tolerance and regulates arsenic accumulation in rice. *New Phytologist*, 215(3), 1090–1101.

Xu, W., Dai, W., Yan, H., Li, S., Shen, H., Chen, Y., & Ma, M. (2015). Arabidopsis NIP3; 1 plays an important role in arsenic uptake and root-to-shoot translocation under arsenite stress conditions. *Molecular Plant*, 8(5), 722–733.

Xu, X. Y., McGrath, S. P., Meharg, A. A., & Zhao, F. J. (2008). Growing rice aerobically markedly decreases arsenic accumulation. *Environmental Science & Technology*, 42(15), 5574–5579.

Yan, S., Wu, F., Zhou, S., Yang, J., Tang, X., & Ye, W. (2021). Zinc oxide nanoparticles alleviate the arsenic toxicity and decrease the accumulation of arsenic in rice (Oryza sativa L.). *BMC Plant Biology*, 21, 1–11.

Yang, Y. P., Zhang, H. M., Yuan, H. Y., Duan, G. L., Jin, D. C., Zhao, F. J., & Zhu, Y. G. (2018). Microbe mediated arsenic release from iron minerals and arsenic methylation in rhizosphere controls arsenic fate in soil-rice system after straw incorporation. *Environmental Pollution*, 236, 598–608.

Yang, Y., Hu, H., Fu, Q., Zhu, J., & Huang, G. (2019). Water management of alternate wetting and drying reduces the accumulation of arsenic in brown rice - as dynamic study from rhizosphere soil to rice. *Ecotoxicology and Environmental Safety*, 185, 109711. doi:10.1016/j.ecoenv.2019.109711

Yao, Y., Zhou, H., Yan, X. L., Yang, X., Huang, K. W., Liu, J., & Liao, B. H. (2021). The Fe_3O_4-modified biochar reduces arsenic availability in soil and arsenic accumulation in indica rice (Oryza sativa L.). *Environmental Science and Pollution Research*, 28, 18050–18061.

Ye, Y., Li, P., Xu, T., Zeng, L., Cheng, D., Yang, M., & Lian, X. (2017). OsPT4 contributes to arsenate uptake and transport in rice. *Frontiers in Plant Science*, 8, 2197.

Yim, S. R., Park, G. Y., Lee, K. W., Chung, M. S., & Shim, S. M. (2017) Determination of total arsenic content and arsenic speciation in different types of rice. *Food Science and Biotechnology*, 26, 293–298. doi:10.1007/ s10068-017-0039-9

Zacarías-Estrada, O. L., Ballinas-Casarrubias, L., Montero-Cabrera, M. E., Loredo-Portales, R., Orrantia-Borunda, E., & Luna-Velasco, A. (2020). Arsenic removal and activity of a sulfate reducing bacteria-enriched anaerobic sludge using zero valent iron as electron donor. *Journal of Hazardous Materials*, 384, 121392.

Zavala, Y. J., & Duxbury, J. M. (2008). Arsenic in rice: I. Estimating normal levels of total arsenic in rice grain. *Environmental Science & Technology*, 42(10), 3856–3860.

Zhang, G., Liu, X., Gao, M., & Song, Z. (2020). Effect of Fe–Mn–Ce modified biochar composite on microbial diversity and properties of arsenic-contaminated paddy soils. *Chemosphere*, 250, 126249.

Zhang, J., Zhao, S., Xu, Y., Zhou, W., Huang, K., Tang, Z., & Zhao, F. J. (2017). Nitrate stimulates anaerobic microbial arsenite oxidation in paddy soils. *Environmental Science & Technology*, 51(8), 4377–4386.

Zhang, S. Y., Zhao, F. J., Sun, G. X., Su, J. Q., Yang, X. R., Li, H., & Zhu, Y. G. (2015). Diversity and abundance of arsenic biotransformation genes in paddy soils from southern China. *Environmental Science & Technology*, 49(7), 4138–4146.

Zhang, S., Geng, L., Fan, L., Zhang, M., Zhao, Q., Xue, P., & Liu, W. (2020). Spraying silicon to decrease inorganic arsenic accumulation in rice grain from arsenic-contaminated paddy soil. *Science of the Total Environment*, 704, 135239.

Zhao, F.-J., McGrath, S. P., & Meharg, A. A. (2010). Arsenic as a food chain contaminant: mechanisms of plant uptake and metabolism and mitigation strategies. *Annual Review of Plant Biology*, 61(1), 535–559. doi:10.1146/annurev-arplant-042809-112152

Zhou, H., Yang, W.-T., Zhou, X., Liu, L., Gu, J.-F., Wang, W.-L., et al. (2016). Accumulation of heavy metals in vegetable species planted in contaminated soils and the health risk assessment. *International Journal of Environmental Research and Public Health*, 13(3), 289.

Zhu, Y. G., Sun, G. X., Lei, M., Teng, M., Liu, Y. X., & Chen, N. C., (2008). High percentage inorganic arsenic content of mining impacted and nonimpacted Chinese rice. *Environmental Science & Technology*, 42, 5008–5013.

17

Sources, Dynamics, and Mitigation Strategies of Arsenic and Cadmium and Their Co-Contamination in Paddy Systems: A Comprehensive Review

Yihan Chi, Yuanyuan Tang, and Wai Chin Li

17.1 Introduction

Arsenic (As) and cadmium (Cd), as trace elements in natural environments, are categorized into heavy metal(loid)s that have aroused global concerns in the past decades. In the earth's crust, As and Cd occur naturally with average concentrations of approximately 1.5 and 0.2 mg/kg, respectively (Chen and Costa, 2021; Marković et al., 2019). Besides the natural weathering of rocks, these elements can be introduced into human living environments through multiple industrial and agronomical activities, posing severe risks to public health (Bolan et al., 2013; Mitra et al., 2017). Despite its toxic nature, As has long been used as medicine, insecticides, herbicides, wood preservatives, and components of alloy (Zhao et al., 2009). Cadmium has been commonly used in electroplating and producing pigments, batteries, and coatings (Clemens et al., 2013). Along with growing awareness of their ecological, environmental, and health hazards, the production and utilization of As and Cd have been restricted and reduced in many European countries, the US, China, etc. Nevertheless, the As and Cd levels in aqueous and soil environments are increasing steadily, especially in some Asian countries, owing to mining and smelting activities, the applications of phosphate (P) fertilizers and manure, and the non-degradable nature of these elements (Zhao et al., 2015).

Arsenic is primarily present in sulfidic minerals such as arsenopyrite in anaerobic conditions, and the oxidative dissolution of these minerals leaches As into groundwater and surface water (Smedley and Kinniburgh, 2002). Arsenic has two oxidation states that have biological implications, i.e., As(III) and As(V), with As(III) generally considered more labile and toxic than As(V) (Garbinski et al., 2019). Inorganic As species can be transformed into monomethylarsonic acid (MMA), dimethylarsinic acid (DMA), and trimethylarsine oxide (TMAO) under more oxidized conditions by microbial methylation, resulting in lower toxicity (Somenahally et al., 2011). In natural environments, Cd can exist in various forms, including free ions, ions adsorbed onto iron (Fe), aluminum (Al), and manganese (Mn) (hydr)oxides, inorganic and organic complexes, and precipitates as phosphates, sulfides, and carbonates (Adriano, 2001). As and Cd are non-essential for most living organisms with no biological function. However, they can enter plant and microbial cells as hitchhikers through transport pathways/transporters for essential elements (Clemens and Ma, 2016). Excessive accumulation generates extensive and potential food-chain contamination.

Based on previous clinical evidence, chronic intake of As or Cd, even at very low doses, can have serious health consequences. The inorganic forms of As (iAs) are highly toxic, and oral intake of iAs can cause cancer in various parts of the human body such as skin, lungs, and bladder (Straif et al., 2009). Long-term exposure to As is also associated with a series of adverse effects such as skin lesions, diabetes, and developmental and neuro-toxicity (Cubadda et al., 2017). Cadmium is also listed as a Class I human carcinogen, similar to iAs, by the International Agency for Research on Cancer (IARC), which may increase cancer risks in the lungs, bladder, endometrium, prostate, and breast (Song et al., 2017). Prolonged dietary exposure to Cd may also significantly damage renal and skeletal systems.

DOI: 10.1201/9781003492115-21

Non-occupational exposure to iAs and Cd is almost exclusively through consuming contaminated agricultural products, with rice accounting for a substantial proportion (Suda and Makino, 2016; Upadhyay et al., 2019). Rice (*Oryza sativa*), one of the most essential food crops, is a staple food for over half of the global population, particularly in East and Southeast Asia. However, compared to other cereal crops, rice more readily uptakes As and Cd, leading to a higher accumulation of these elements in its grains (Chen and Costa, 2021; Yu et al., 2017). The higher water demand for rice cultivation further exacerbates the As and Cd contamination in paddy soils through sewage irrigation (Hu et al., 2016; Suriyagoda et al., 2018a). Moreover, the major rice-producing regions in many Asian countries, such as China, India, and Bangladesh, face elevated Cd contamination from a high geochemical background or poorly regulated mining and smelting activities (Norton et al., 2009; Zou et al., 2021).

Consequently, rice constitutes a significant dietary As and Cd exposure route in these areas, making the mitigation of As and Cd accumulation in rice imperative. Governments and organizations have imposed clear limits and guidelines for the concentrations of As and Cd in rice grains and related products. For instance, the Codex Alimentarius Commission of the United Nation's Food and Agriculture Organization and the World Health Organization (FAO/WHO), the European Union, and the Chinese government set the maximum allowable inorganic As concentration in brown rice at 0.2 mg/kg (Zhao and Wang, 2020). Safety thresholds of 0.4 and 0.2 mg/kg for the Cd concentration in rice grains have also been proposed by FAO/WHO and the Chinese government, respectively (Ma et al., 2021). Nevertheless, rice iAs and Cd concentrations exceeding the safety standards have been frequently reported in recent years, indicating that As and Cd contamination in paddy fields remains severe (Chi et al., 2018; Ge et al., 2024; Zhang et al., 2024).

Although As and Cd, the most prevalent toxic substances in paddy systems, have been extensively studied from multiple perspectives regarding their environmental behavior over the past few decades, there remains a notable lack of systematic reviews comparing the inputs and dynamics of As and Cd in paddy soils. To better prevent and mitigate As and Cd contamination in paddy fields, it is essential to comprehensively understand the similarities and differences in their pollution and transformation pathways and the key environmental factors influencing these processes. Since As and Cd often exhibit disparate chemical properties in the soil environment, selecting effective remediation techniques and amendments requires a thorough grasp of the relevant knowledge. Therefore, this chapter aims to review the input and transformation processes of As and Cd in paddy fields and discuss the critical environmental factors determining their fate and relevant mitigation strategies based on current knowledge.

In this chapter, the common and distinct sources of As and Cd in paddy soils, along with their relative contributions, will be systematically discussed. This chapter will also analyze the changes in the speciation and chemical fractions of As and Cd in paddy soils. It will depict the impact of environmental condition variations on their mobility, identifying the key factors. The fundamental mechanisms governing the transformation of As and Cd, as well as effective immobilization and remediation strategies, particularly those targeting combined contamination, will be thoroughly examined. Finally, this chapter will highlight the unresolved mechanisms in their environmental behavior and remaining challenges in contamination remediation to provide perspectives and directions for future research.

17.2 Sources of As and Cd in Paddy Fields

17.2.1 General Contamination Status of As and Cd

Arsenic levels in soil and water that exceed the safety threshold are widely detected in many countries worldwide, particularly in East and South Asia. The issue is particularly severe in India, Bangladesh, and China, with approximately 70.4, 32.0, and 2.3 million people estimated to be affected by As contamination, respectively (Raju, 2022; Sarkar and Biswajit, 2016). The major rice-producing regions in these areas suffer from elevated As contamination, and rice is frequently reported to contain risky levels of As/iAs (Ali et al., 2020; Meharg, 2004). Rice grains produced in areas with elevated As levels in groundwater and mine-impacted zones are particularly prone to

As accumulation (Williams et al., 2009, 2006). Cadmium contamination in paddy fields has also been found globally, with the contamination levels, range, and related studies being more prominent in East Asian countries, i.e., China, Japan, and Korea (Watanabe et al., 1996). Owing to the high mobility of Cd and the physiological characteristics of rice, the translocation rate of Cd from soil to rice grains is typically higher than that of most toxic trace elements (Ali et al., 2020). This potentially amplifies Cd contamination in paddy soils.

It is important to note that Cd and As are China's most common soil contaminants, according to a national survey conducted in 2014. The survey revealed that soil Cd and As contents exceeded quality standards at 7.0% and 2.7%, respectively (Zhao et al., 2015). The spatial distribution and temporal changes of As and Cd contamination in the paddy fields of China have been well documented and systematically analyzed in a range of studies (Mu et al., 2019; Wang et al. 2019; Yuan et al., 2021; Zhang et al., 2024). Data indicate that soil As and Cd contamination in southern China, the major rice cropping region, is significantly more severe than in other regions. This can be attributed to the development of mining and smelting activities (Wang et al., 2019; Yuan et al., 2021; X. Zhang et al., 2015). Since the early 1980s, the soil Cd levels in southern China have increased by over 50%, although a slight decrease in the last decade has been observed. Meanwhile, soil As levels in China have continuously increased over the past 20 years, especially in southern China. It is estimated that the proportion of paddy fields with the As concentration exceeding 20 mg/kg (the screening value for agricultural soils in China) has increased from 13.0% in 2000 to 17.1% in 2020 and will reach 18.3% by 2040 (Zhang et al., 2024). Therefore, it is imperative to identify and control the sources of As and Cd contamination to ensure safe rice production and public health.

17.2.2 Natural Sources of As and Cd

17.2.2.1 As

Arsenic can act as a substitute for sulfur (S), phosphorus (P(V)), silicate (Si(IV)), Al(III), Fe(III), and titanium (Ti(IV)) in mineral structures and composes over 200 rock-forming minerals (Punshon et al., 2017). Among these minerals, arsenopyrite (FeAsS) and arsenian pyrite ($Fe(S, As)_2$) are the most prevalent in ore deposits and serve as the primary sources of As in ore zones, with concentrations potentially reaching up to 10% As (Tabelin et al., 2018). As-containing minerals' natural weathering and biogeochemical transformation can lead to substantial release and soil contamination in adjacent and downstream areas (Punshon et al., 2017). Specifically, the reductive dissolution of As-bearing Fe oxyhydroxides dispersed in aquifers contributes primarily to the As contamination in groundwater. It should be responsible for the paddy-soil As contamination in Bangladesh and West Bengal (India) (Raju, 2022; Sarkar and Biswajit, 2016).

17.2.2.2 Cd

Unlike As, which has natural geochemical processes as an essential origin (Zhou et al., 2018), the natural sources of Cd, mainly from weathering of parental materials and volcanic activities, account for only around 10% of the total release into soils and water (Khan et al., 2017; Zou et al., 2021). In the karst region of southwestern China, Cd-rich carbonate rocks result in a higher geochemical background of Cd in Guangxi (0.267 mg/kg) and Guizhou (0.209 mg/kg) compared to the nationwide background (0.097 mg/kg) (Wen et al., 2020). Therefore, the weathering of parent materials can directly cause As and Cd contamination in paddy fields through pedogenesis and aggravate soil As contamination through groundwater.

17.2.3 Anthropogenic Sources of As and Cd

17.2.3.1 As

Compared to natural sources, anthropogenic sources are believed to contribute more significantly to As and Cd contamination in soils, especially in rice paddies. These sources are primarily associated with industrial and agricultural activities (Ali et al., 2020). Approximately 52,000–112,000 tonnes of As are discharged into the environment annually due to human activities, with 6200 tonnes attributed to smelting processes, especially the smelting of copper (Cu) (Sarkar and Biswajit, 2016). As previously discussed, high concentrations of As can be integrated into the structure of, or adsorbed onto, oxide minerals and hydrous metal oxides, such as pyrite and ferrihydrite (Fan et al., 2018; Punshon et al., 2017). Arsenic may constitute over 1% of ore and mine waste, and As sulfides are particularly susceptible to oxidation. Consequently, most mining activities lead to As release as a byproduct, resulting in high As levels from 0.01 to over 10 mg/L in mine wastewater and impacted streams (Craw and Bowell, 2014).

Coal combustion is another primary source of As contamination. Through this pathway, approximately 9000 tonnes of As was released in China in 2020. Arsenic is one of the most volatile hazardous elements in coal fly ash, mainly as calcium (Ca)-As(III, V) complexes (Shen et al., 2015). Coal around the world typically contains 0.5–80 mg/kg of As, but in certain cases, e.g., in southwestern China, the As concentration in coals can reach up to 35,000 mg/kg. In some villages in Guizhou, China, more than 3000 cases of chronic As poisoning were reported to be linked to the burning of local As-rich coals (C. Wang et al., 2018).

In the mid-20th century, in the US, arsenic was used extensively in pesticides, e.g., Ca arsenate and lead (Pb) arsenate. The leaching of these arsenicals as legacy contaminants into groundwater may explain the As contamination in some paddy fields in the US (Hudak, 2000; Majumder and Banik, 2019). It was also used in herbicides and wood preservatives and could enter soils by applying organic manure (Punshon et al., 2017). Although the sources of anthropogenic As usually vary among different regions, the contamination pathways of paddy fields mainly converge on atmospheric deposition, wastewater irrigation, and fertilization practices (Peng et al., 2019; Yi et al., 2018).

17.2.3.2 Cd

Compared with As contamination, anthropogenic sources contribute more significantly to the Cd contamination in farmland soils (Khan et al., 2017; Zhou et al., 2018). Around 5600–38,000 tonnes of Cd are released annually from anthropogenic sources, mainly mining and smelting activities, fossil fuel combustion, P fertilizers, and sewage sludge (Khan et al., 2017). In some rapidly developing countries like China, mining activities are deemed the primary anthropogenic source of Cd pollution in paddy soils (Liu et al., 2016). Most Cd in the earth's crust exists with zinc (Zn), Pb, and tungsten (W) ores as co-occurring or associated minerals, resulting in higher Cd levels in soils near Pb-Zn and W mines compared to the background levels (Shi et al., 2022). In China, Cd-containing minerals predominantly occur in the central, southwestern, and eastern regions, constituting around 88% of the proven deposits (Yuan et al., 2012). As a result, these areas overlap considerably with the major rice-producing regions in China, resulting in particularly prominent Cd contamination in paddy fields.

Some investigations claimed that 57% of Cd-contaminated paddy fields were located in mine-impacted regions (Liu et al., 2016). Additionally, the much higher irrigation water demand of double rice cropping in south China aggravated the Cd contamination in paddy fields compared to other cropping systems (Huang et al., 2019). In many water-scarce regions, treated mine wastewater is likely to be used for paddy rice irrigation, although this issue has gradually come under control in recent years (Edelstein and Ben-Hur, 2018; Zhang et al., 2015). Acid mine drainage (AMD) with heavy loads of As and Cd is inevitably generated in mining processes. Soil acidification caused by AMD irrigation further exacerbates Cd contamination in rice grains by increasing Cd mobility (Zhang et al., 2013; Zou et al., 2021). It is important to note that Cd released from mining and smelting activities typically has higher mobility and associated toxicity than other sources.

Despite the varying contributions of different sources, soil As and Cd contamination share the same primary pathways, i.e., atmospheric deposition (Peng et al., 2019; Yi et al., 2018). Cd-rich dust is discharged into the atmosphere during Zn smelting and deposited in nearby areas (Khan et al., 2017). In China, the annual input of Cd into farmland soils through atmospheric deposition is estimated to be around 1298 tonnes (Peng et al., 2019). In some mine-impacted paddy fields in central China, atmospheric deposition contributed to 76.4%–98.3% of the total input flux of Cd (Wang et al., 2021).

Additionally, the application of P fertilizer and livestock manure, as well as sewage sludge, are notable sources of Cd contamination in paddy fields (Bolan et al., 2013). Phosphate fertilizers from rock phosphate may contain excessive Cd, leading to substantial Cd accumulation in soils through long-term or intensive application. A nationwide survey estimated the average Cd concentration in P fertilizers in China to be 0.75 mg/kg. This is significantly lower than in European countries, where the average is

7.4 mg/kg (Li et al., 2020). It was reported that long-term application of phosphate fertilizer raised soil Cd levels by 1–14-fold in some paddy fields in China, Korea, and the USA (Park et al., 2021).

In some livestock breeding areas, the application of manure fertilizer can become the second most important source of Cd contamination in farmlands, reaching up to 33.8% of the total input (Peng et al., 2019; Yi et al., 2018). Sewage sludge can also be rich in Cd, especially when sewage is mixed with industrial wastewater (Singh and Agrawal, 2008). Composting sewage sludge with rice straw may increase its Cd concentration by 12%–60%, according to previous research (Cai et al., 2007). Reports claim elevated Cd levels in paddy fields and rice grains derived from sewage sludge application (Latare et al., 2014).

Paddy fields are more susceptible to As and Cd contamination due to their unique cultivation practices and geographic location. Anthropogenic sources of As and Cd in paddy fields are often linked to intensive industrial and agricultural activities. Mining and smelting are the dominant sources, with atmospheric deposition being a major contributor. Arsenic contamination differs from Cd contamination because As enrichment in paddy fields can result from anthropogenic activities and a high geochemical background.

17.3 Behaviors of As and Cd in Paddy Soils

17.3.1 Redox Transformation of Soil As in Paddy Fields

Arsenic exists in the natural environment primarily in four oxidation states, –3, 0, +3, and +5. The native states, –3 and 0, are rarely detected in paddy fields (Kumarathilaka et al., 2018). Arsenic is present in multiple inorganic and organic forms in paddy soils, in both liquid and solid phases (Zhao et al., 2010). It is well accepted that the mobility, bioavailability, and toxicity of soil As are closely related to its speciation (Sarwar et al., 2021). The inorganic forms of As consist of two oxyanions, i.e., the trivalent arsenite (As(III)) and the pentavalent arsenate (As(V)).

The organic forms MMA and DMA are detected as minor components of the soil As pool (Zhao et al., 2013; Zheng et al., 2013b). Other species of As, such as TMAO, arsenosugar, arsenobetaine, and thioarsenates, are also found in the environment but are rarely detected in paddy soils (Mondal et al., 2022). The speciation and associated bioavailability of soil As are fundamentally affected by soil redox conditions (Honma et al., 2016b).

In aerobic soil conditions, the major form of As is the less labile arsenate which is readily adsorbed to or precipitated with Fe, Mn (oxy)hydroxides, and clay minerals. When paddy soils undergo prolonged flooding, the reduced form, arsenite, predominates and is released into the soil solution with higher mobility due to its lower affinity for soil particles (Das et al., 2016; Suriyagoda et al., 2018a). Compared with arsenate, arsenite, as an analog of silicic acid, is more readily taken up by rice plants through a series of aquaporins, posing greater phytotoxicity and human health risk (Suriyagoda et al., 2018b). Arsenate enters rice roots mainly through phosphate transporters. Methylated As species generally have a lower affinity for transporters and are less available to rice plants (Zhao et al., 2009).

Microbes play a crucial role in the direct oxidation, reduction, and methylation of As species in paddy fields, mainly determining the toxicity and fate of soil As (S.-Y. Zhang et al., 2015). Microbial oxidation of As(III) to As(V) is mainly catalyzed by As(III) oxidase encoded by *aioA* and *aioB* genes, which are present in various heterotrophic and chemolithoautotrophic microbes abundant in paddy soils (Jia et al., 2014; Shi et al., 2020; Tong et al., 2019). Two microbial pathways are responsible for soil As(V) reduction: the dissimilatory or respiratory pathway mediated by the As(V) respiratory reductase (ArrA) and the detoxification/cytoplastic pathway catalyzed by cytoplasmic As(V) reductase (ArsC) (Qiao et al., 2018a; Zhang et al., 2017).

The genes encoding As(V) reductases, i.e., *ArrA* and *ArsC*, are frequently detected in *Geobacter, Shewanella, Rhizobiales,* and *Pseudomonadales*, in paddy soils (Y. Chi et al., 2022b; Zecchin et al., 2017; S.-Y. Zhang et al., 2015). Except for the historical use of methylated As compounds as pesticides or herbicides, the MMA and DMA present in paddy soils and rice grains are almost exclusively driven by soil microbes (Kaya et al., 2024; Zhao et al., 2013). This process is catalyzed by arsenite methyltransferase (ArsM), encoded by the gene *arsM*, which is widely distributed in various microbes and is high efficient under anaerobic soil conditions (Viacava et al., 2020).

17.3.2 Major Soil Components Determining the Sequestration and Release of As and Cd

17.3.2.1 Immobilization and Release of Soil As

Despite the direct transformation of As species, soil microorganisms also indirectly affect As mobility through the biogeochemical cycles of redox-sensitive elements such as Fe, Mn, and S (W. Shi et al., 2021b, 2021a). Iron oxides, hydroxides, and oxyhydroxides are pivotal in regulating soil As mobility through adsorption and co-precipitation. The oxyhydroxide minerals of Fe, including amorphous ferrihydrite and crystalline lepidocrocite (γ-FeOOH), goethite (α-FeOOH), and hematite (α-Fe_2O_3), represent a major pool of As in the bulk soil, rhizosphere, and root surfaces (as the root Fe plaque) in paddy fields (Yu et al., 2016a). The lower mobility of arsenate compared to arsenite in paddy environments can be explained by the higher affinity of As(V) for Fe oxides (Xu et al., 2016a). In this way, the microbial oxidation of aqueous Fe(II) and reductive dissolution of Fe(III) minerals are crucial in controlling the sequestration and remobilization of As in paddy soils (Honma et al., 2016b).

The biotransformation of As in soils is strongly coupled with the biogeochemical cycling of Fe. Some As(V)-reducing bacteria, e.g., *Geobacter, Shewanella putrefaciens* 200, are capable of reducing As(V) and Fe(III) concurrently, thereby markedly promoting As mobilization (Zheng et al., 2013a). Even under aerobic conditions, elevated abundances of Fe-reducing bacteria (FeRB) may lead to enhanced As release in the rice rhizosphere, promoting plant As uptake (Dai et al., 2020). Similar to Fe minerals, Mn oxides contribute to As immobilization in paddy soils (Suda and Makino, 2016). Mn oxides are strong adsorbents and oxidants of As; As(III) can be oxidized to As(V) by Mn(IV) and subsequently adsorbed by Mn(II) (X. Xu et al., 2017). Additionally, Mn minerals, such as birnessite, can delay the reductive dissolution of Fe(III) and associated As release under anoxic conditions, thereby enhancing As immobilization under redox fluctuations (Ehlert et al., 2016).

Sulfur cycling, especially microbial sulfate reduction, contributes to both the immobilization and release of As in paddy soils, though its specific effects remain controversial. Dissimilatory sulfate-reducing bacteria (SRB) facilitate the reductive dissolution of Fe(III) oxides and the release and reduction of solid phase As(V) to aqueous As(III) under soil flooding (Kumarathilaka et al., 2018; W. Shi et al., 2021a). The newly formed ferrous sulfide minerals (FeS and FeS_2) are believed to sequester substantial amounts of As, and sulfide can coprecipitate with As, forming some insoluble mineral-like phases such as realgar (AsS) and amorphous As_2O_3 (Burton et al., 2014; Fan et al., 2018).

However, some studies argue that Fe sulfides have a low affinity for arsenite, and the formation of As sulfides is inhibited by Fe(II) and Fe(III) (Kocar et al., 2010). Additionally, excessive sulfide anions in soil solution may promote the formation of thioarsenic species with weak adsorption properties, enhancing As partitioning into the aqueous phase (Zhao et al., 2017). The system is further complicated by the formation of As sulfide complexes associated with secondary Fe minerals, which facilitate As sequestration (Burton et al., 2014; W. Shi et al., 2021b; Xu et al., 2016b).

Based on current knowledge, the ultimate impact of sulfate reduction on soil As mobility probably depends on soil S supply and incubation time. Compared with S-deficit soils, abundant S can reduce ferrous release under reduced soil conditions and adsorb aqueous As. Sulfur supplementation has been reported to reduce soil As mobility and rice As uptake effectively (Jia et al., 2015). With long-term incubation, some researchers found that soil dissolved As levels displayed an "M" shape variation over 240 h with sulfate addition and were eliminated from soil solution at the end (Xu et al., 2016a). A recent investigation of multiple paddy fields also found that soil As was effectively immobilized under long-term anoxia along with rich soil S, not showing remobilization under redox fluctuations (Y. Chi et al., 2022b).

Soil organic matter (SOM), especially dissolved organic matter (DOM), is also a key determinant of As mobility in paddy soils. Depending on the type and nature, different components of soil DOM can have dual functions in the sequestration and release of As during redox fluctuations. Due to their negative charges, DOM components can competitively adsorb onto Fe (oxy)hydroxides and release As (Bauer and Blodau, 2006; Williams et al., 2011). DOM acts as electron shuttles to promote the reductive dissolution of Fe(III) oxides under soil flooding, facilitating As remobilization into soil pore water (Mladenov et al., 2015).

However, DOM can also form complexes with As in soil solution and reduce the bioavailability and root uptake of As (Pothier et al., 2020; Wang et al., 2015). The roles of DOM in the biogeochemistry of soil As are closely associated with the types and doses of DOM and environmental conditions like redox potential. For instance, labile DOM, such as amino acids and carbohydrates, can be directly used as a carbon source by reducing microbes and fuel microbial Fe(III) reduction. Recalcitrant DOM, like humic substances, primarily serves as electron shuttles and conducts complexation with As, which may involve Fe in a ternary As-Fe-DOM complex (Z. Gao et al., 2023; Wang et al., 2015). DOM may contribute to As mobilization when paddy soil is drained by inhibiting the formation and precipitation of secondary Fe(III) (oxy)hydroxides (Wu et al., 2019).

17.3.2.2 Adsorption and Remobilization of Soil Cd

The mobility and bioavailability of Cd in paddy soils are primarily determined by sorption-desorption processes, similar to those of As. As a result, they are associated with active soil factors, including Fe/Mn oxides, S, and SOM (Hussain et al., 2021; Loganathan et al., 2012). However, compared with As, cadmium, as a cation trace element, exhibits a non-variable oxidation state, occurring in paddy soils predominantly as free ions, carbonate, and inorganic and organometallic complexes (Bolan et al., 2013). This indicates that soil Cd mobility depends more on the occurrence and transformation of key soil components. Specifically, the redox transformation of soil Fe has significant implications for the immobilization of soil Cd.

It is reported that amorphous Fe oxides can effectively reduce the labile Cd, including the water-soluble and exchangeable fractions in paddy soils, and reduce Cd uptake by rice plants (Yu et al., 2016b). During the microbial reductive dissolution of Fe(III) oxides, Cd is initially released into soil solution but is subsequently adsorbed onto or coprecipitated with the newly formed secondary Fe minerals (Muehe et al., 2013b; Yuan et al., 2018). The effect of Fe(III) reduction on Cd sequestration is thought to be linked to the duration of paddy soil flooding, wherein short-term anoxia leads to Cd mobilization. In contrast, prolonged flooding causes readsorption and even incorporation of Cd into the mineral lattice (Y. Gao et al., 2023).

Nevertheless, Cd has a generally weaker binding strength than most metal cations such as Pb^{2+}, Cu^{2+}, Zn^{2+}, and Ni^{2+}, owing to its lower electronegativity and a larger hydrated ionic radius (M. Shi et al., 2021; Suda and Makino, 2016). This implies a lower affinity of Fe oxides for Cd than for other heavy metals, and the adsorption sites being more likely to face competitive occupation. In some cases, competitive adsorption can also happen between Cd^{2+} and Fe^{2+} (Fulda et al., 2013). Therefore, the effects of Fe oxides and their transformation on Cd mobility warrant more systematic study under diverse redox conditions.

Sulfur mainly contributes to Cd sequestration in paddy soils through the microbial sulfate reduction process (De Livera et al., 2011). Under soil flooding, sulfate is reduced to sulfide, forming poorly soluble CdS, which significantly reduces Cd mobility (Fulda et al., 2013). Meanwhile, Fe^{2+} produced from the reductive dissolution of Fe(III) oxides can form FeS or FeS_2 with S^{2-}, further leading to the adsorption and coprecipitation of Cd (Wang et al., 2020). Reducible S supply was reported to control Cd immobilization under anoxic conditions substantially (Fulda et al., 2013). However, oxidative dissolution occurs in these metal sulfides when paddy soil is drained, and Cd is released during this process. Sulfate can also form soluble complexes with Cd and enhance Cd mobility (Liu et al., 2022).

In this context, more abundant reactive S can increase Cd bioavailability under oxidized conditions. However, some studies claimed that the oxidative dissolution of CdS contributed much less than the desorption of Cd from Fe and Mn oxides to labile Cd upon soil drainage (J. Wang et al., 2019). It is also worth noting that the oxidation of CdS is inhibited by the presence of ZnS but is promoted by CuS through a voltaic effect, which hints at an influence of chalcophile elements on Cd transformation (Huang et al., 2021a). A recent study highlighted that hydroxyl radicals produced in the oxidation of Fe(II) sulfides may play a key role in the oxidative dissolution of CdS under soil drainage (Huang et al., 2021b).

SOM also plays a vital role in regulating Cd mobility in paddy soils (Li et al., 2015; C. Yuan et al., 2021). The immobilization effect of SOM on Cd is mainly achieved by direct complexation and by indirectly adjusting soil pH. First, SOM's abundant carboxyl and hydroxyl groups enable its efficient binding of soil Cd and formation of strong complexes (Hu et al., 2016). Increasing SOM content in paddy fields can effectively decrease bioavailable Cd fraction and reduce Cd accumulation in rice tissues (Li et al., 2022). Meanwhile, SOM can act as an electron donor, driving microbial Fe(III) and sulfate reduction

(Kappler et al., 2021). This process results in the adsorption and coprecipitation of Cd, as mentioned above. More importantly, microbial Fe and sulfate respiration results in elevated soil pH in acidic paddy fields and contributes to Cd adsorption by increasing soil negative charge and enhancing Cd hydrolysis (Loganathan et al., 2012). Additionally, Cd binding to some SOM components, such as humic acids, induces the formation of ternary complexes by adsorption onto metal oxides and clay minerals (Bolan et al., 2013; Yuan et al., 2019).

17.3.3 Environmental Factors Controlling the Mobility and Bioavailability of As and Cd in Paddy Fields

17.3.3.1 Redox Potential

In most cases, redox transformations resulting from water management practices are the dominant factor controlling soil As bioavailability in paddy fields (Li et al., 2009; Somenahally et al., 2011). Under aerobic conditions during soil drainage, As mainly exists as protonated arsenate forms, such as $H_2AsO_4^-$ and $HAsO_4^{2-}$, which readily form stable complexes and co-precipitates with anions (Majumdar et al., 2019). As(V) exhibits a higher affinity for Fe(III) (oxyhydr)oxides, abundant in aerobic paddy soils and are immobilized through adsorption (Sarwar et al., 2021). When paddy soils are under reduced conditions due to continuous flooding, the dominant species of soil As becomes As(III) in the form of $H_2ASO_3^{3-}$, which is released into soil solution with elevated bioavailability (Das et al., 2016; Kumarathilaka et al., 2018). The reductive dissolution of Fe(III) (oxyhydr)oxides as soil Eh declines also contributes to the mobilization of soil As since secondary ferrous minerals, e.g., FeS and FeS_2, exhibit a weaker ability to sequester labile As (Punshon et al., 2017; M. Shi et al., 2021; X. Xu et al., 2017).

Compared with As, soil Cd responds differently to changes in redox environment (Li et al., 2017; Moreno-Jiménez et al., 2014). Reducing conditions from soil flooding facilitates the formation of secondary metal sulfides, including CdS and FeS_X, resulting in the precipitation and adsorption of Cd (Li et al., 2021; Muehe et al., 2013a). The microbial respiration of sulfate and ferric oxides under decline Eh consumes protons, raising soil pH and immobilizing soil Cd by increasing net negative charges. On the other hand, increasing soil Eh leads to the oxidative dissolution of CdS and the formation of water-soluble $CdSO_4$ (Rinklebe et al., 2016; Wang et al., 2020). Therefore, soil Eh, as the most variable environmental factor in paddy soil, has contrasting effects on regulating the mobility and potential bioavailability of soil As and Cd (Honma et al., 2016b). This phenomenon presents significant challenges for addressing As and Cd co-contamination in paddy fields through water management, one of the most cost-effective strategies for decreasing As and Cd accumulation in rice plants (Zhao and Wang, 2020).

17.3.3.2 pH

Soil pH is a crucial factor driving the adsorption and desorption of Cd in paddy fields (Zhu et al., 2016). Like other heavy metals, Cd is typically adsorbed by surface hydroxyl groups and the negative charges of metal oxides (Colzato et al., 2018). Changes in these variable charges are considered responsible for the increases in surface negative charge under soil pH rises (Bolan et al., 1999). Higher pH also promotes the formation of hydrolyzed $CdOH^+$ species, which are preferentially adsorbed over Cd^{2+} (Palansooriya et al., 2020). Together, these mechanisms result in enhanced adsorption and immobilization of Cd under elevated soil pH. Various studies have reported sharp increases in Cd adsorption onto metal oxides when pH was increased from 6 to 9 (Honma et al., 2016b; Lackovic et al., 2003; Loganathan et al., 2012). In contrast, soil acidification decreases surface negative charge and dissolution of metal oxides, leading to Cd desorption and remobilization in paddy soils (Bolan et al., 2013; Qiu et al., 2012).

Due to the species transformation of soil As, the effects of soil pH changes on As mobility appear to be more complex than those of Cd. It has been reported that Fe (oxyhydr)oxides preferentially adsorb arsenate under acidic soil conditions, while arsenite adsorption is favored under alkaline conditions (Miretzky and Cirelli, 2010; Suriyagoda et al., 2018a). Previous studies found that increased soil pH facilitated As release and plant As uptake in paddy systems (Ahmed et al., 2011; Y. Chi et al., 2022b; Honma et al., 2016b). This can be ascribed to decreased positive charge as pH increases and surface precipitation of As(V) on amorphous Fe oxides in acidic environments (Sarwar et al., 2021; M. Shi et al., 2021). However, this effect appears prominent only in acidic paddy soils and negligible in neutral and alkaline soils (Yao et al., 2022). It is notable that pH effects on As and Cd's mobility contribute to these elements' contrasting geochemical behaviors in acidic paddy soils.

17.3.3.3 Competing Ions

The immobilization and release of As and Cd in paddy soils are largely driven by adsorption and desorption mechanisms. Other soil cations/anions may thus influence the mobility of As and Cd through competing for binding sites (Benjamin and Leckie, 1981; W. Shi et al., 2021b). It is well-established that phosphate and silicic acid are analogs to arsenate and arsenite, respectively, and compete with As for adsorbing onto Fe (oxyhydr)oxides (Kumarathilaka et al., 2018; Tao et al., 2006). Phosphate plays a key role in regulating labile As content at the soil-water interface by promoting As desorption from oxide minerals and soil particles, owing to its ubiquity and abundance in natural environments. However, such competitive adsorption appears ineffective in releasing As from Fe oxyhydroxides and sulfides or after a long adsorption period (Biswas et al., 2014; Duan et al., 2022; Han et al., 2020).

It has also been reported that phosphate addition induces microbial reduction of As(V) and a substantial As release from solid phases, mainly as As(III) (Maier et al., 2019). Interestingly, P starvation causes bacteria to enhance As(III) oxidation to As(V) as a substitute for phosphate in the biosynthesis of specific molecules (Q. Wang et al., 2018). The effect of silicic acid on soil As mobility and rice As uptake seems controversial. Although adding silicic acid resulted in higher As levels in soil solution, it remained unclear whether this is due to increases in DMA or inorganic As (Dykes et al., 2020; Kersten and Daus, 2015). It is important to note that silicic acid may also compete with arsenite for binding sites and transporters during root uptake, consequently decreasing rice As accumulation (Meharg and Meharg, 2015; Seyfferth et al., 2016). This effect often overweighs the impact of As mobilization into soil solution (Liu et al., 2014).

Competitive adsorption of Cd with other cations is also a critical mechanism determining Cd bioavailability in paddy soils. Generally, soil Cd exhibits higher solubility and bioavailability under greater ionic strength from irrigation water with higher salinity (W. Chi et al., 2022b, 2022a). This can be attributed to the competitive adsorption of other cations for limited adsorption sites, consequently releasing Cd (Loganathan et al., 2012). Anions, such as Cl^-, can form ion pairs and complexes with Cd^{2+}, increasing its mobility (Hu et al., 2016). Introducing specific ions can also alter soil pH and electrostatic potential, thus impacting the adsorption equilibrium of Cd at soil–water interfaces (Loganathan et al., 2012). Since Cd^{2+} exhibits a lower affinity for Fe/Mn oxides and soil particles, the presence of other heavy metals, e.g., Pb, facilitates the desorption of Cd in soils after sorption saturation (Jiang et al., 2012; Li et al., 2013). Therefore, as the most mobile heavy metal, Cd's co-contamination with other heavy metals, prevalent in mine-impacted paddy fields, potentially exacerbates the risks of soil Cd contamination.

17.4 Mitigation Strategies for Soil As and Cd Contamination

The existing arable land often needs to be used more efficiently to meet the increasing food demand in the context of continuous population growth globally. Paddy fields cannot be left fallow when suffering from soil As or Cd contamination. The widespread paddy system contamination in some East and Southeast Asian countries renders expensive remediation techniques, such as soil replacement and chemical leaching, less feasible. At the same time, cost-effective in-situ solidification/stabilization strategies have gradually become the focus of relevant research and application (Hu et al., 2016; Zhao and Wang, 2020). These strategies mainly include the application of soil amendments, paddy management practices, and the selection and breeding of As/Cd-exclusive rice cultivars.

17.4.1 Soil Amendments

17.4.1.1 Liming Agents

Due to the significant impact of pH on the mobility of Cd, incorporating alkaline materials to correct soil acidity (i.e., liming) is considered a direct method to mitigate Cd's bioavailability and plant uptake in paddy fields (Palansooriya et al., 2020; Rizwan et al., 2016). Traditional liming materials primarily consist of calcium (Ca) and magnesium (Mg) oxides and hydroxides, including quicklime (CaO), hydrated lime ($Ca(OH)_2$), limestone ($CaCO_3$), calcite, and dolomite (Bolan et al., 2014). Some industrial by-products, such as fly ash, steel-making slag, and cement kiln dust, have also been used to neutralize acidic farmland soils. These materials offer a low-cost remediation and an appropriate disposal method for solid wastes (Qiu et al., 2012; Shaheen and Rinklebe, 2015). Such solid wastes, typically with an alkaline pH of 8.3–13.0, can effectively raise soil pH after incorporation and increase the negative charges in variable charge soil (Hamid et al., 2019).

More Cd is adsorbed onto Fe/Mn/Al (hydr)oxides, clay minerals, and soil particles due to increased sorption sites and a higher affinity of hydrolyzed Cd species for these sites under neutral conditions, as mentioned above (Loganathan et al., 2012). The Cd immobilization effects of different liming agents can be reflected by the pH-changing ability associated with their Ca, Mg, and Si contents. For instance, the Cd stabilization effects of different coal fly ashes varied widely in various studies, which can be explained by their varying pH levels of 8.1–12.6 (Bidar et al., 2016; Y. Chi et al., 2022a). Consecutively applying fly ash with a slightly alkaline pH of 8.1 only reduced soil bioavailable Cd and Cd concentration in rice grains by 23.7% and 35.7%, respectively, with the effect persisting for only one crop season (Y. Chi et al., 2022a). By contrast, fly ash with a pH of 13.1 was reported to reduce soil available Cd by 88.6% (Qiu et al., 2012).

These liming materials can improve soil nutrient status and health by alleviating soil acidity and increasing inorganic minerals, thereby promoting crop growth (Bhaskarachary et al., 2012; Honghua He et al., 2017; Wang et al., 2016). Nevertheless, these industrial by-products often carry excessive heavy metals, such as As, Pb, Zn, and Cu, derived from their parent materials (Singh et al., 2016). This can result in secondary soil contamination, although plant uptake may be unaffected in the short term because of the liming effect. This drawback significantly limits the extensive use of by-product materials as liming agents, relegating them to a temporary solution for severely contaminated paddy fields.

The effect of liming on soil As mobility can be complex. There is concern that liming might enhance rice As accumulation, as increased soil pH is often associated with enhanced As release (Honma et al., 2016b). Recent studies indicate that the effects of liming materials on soil As dissolution might depend on Fe supply, but raising soil pH to over 7.0 led to substantial As release (Yu et al., 2023). In most cases, liming agents generally exhibit no obvious effect on soil As mobility (Chen et al., 2018; Y. Chi et al., 2022a).

17.4.1.2 Adsorbents

Various adsorbents, including metal oxides and clay minerals, are widely used to reduce the mobility of both As and Cd in paddy soils (Farrah and Pickering, 1977; Palansooriya et al., 2020). The oxides and hydrous oxides of Fe, Mn, and Al are natural adsorbents for As and Cd, which typically have large specific surface areas and surface reactivity (Bao et al., 2022; Kim et al., 2018). Among them, Fe-bearing materials, such as natural and synthesized zero-valent iron (ZVI), ferric oxides, Fe oxyhydroxides (e.g., ferrihydrite and goethite), and other binary oxides, are most extensively used as soil amendments in As- and/or Cd-contaminated paddy soils (Makino et al., 2016; Qiao et al., 2018b). Fe oxides naturally possess a high affinity for As (ASO_4^{3-} and ASO_3^{3-}) for their rich positive charges in acidic paddy soils (Miretzky and Cirelli, 2010; Wu et al., 2022). By adding Fe-rich materials, soil As is preferentially adsorbed onto Fe oxides through ligand exchange for OH^- and OH_2 groups, further forming inner-sphere complexes on the surfaces of Fe oxyhydroxides (Moulick et al., 2021; T. Yang et al., 2012). Incorporating Fe-bearing amendments may also promote the deposition of Fe (hydr)oxides under anoxic conditions, rather than increasing the reductive dissolution of Fe(III), thus facilitating As immobilization in flooded paddy fields (Yamaguchi et al., 2017). ZVI has been reported to be an effective amendment for inhibiting soil As dissolution, attributed to the As(III) oxidation on ZVI surfaces and formation of FeAsS-like species (Makino et al., 2016; Suda et al., 2018).

Applying ZVI is also considered a promising technique for mitigating Cd contamination in paddy fields. Hashimoto and Yamaguchi (2013) found that ZVI addition enhanced CdS formation under flooded paddy soils, reducing Cd concentration in soil solution and exchangeable forms. Fe-bearing materials are also commonly used for soil Cd immobilization (Hamid et al., 2019; Mubeen et al., 2023). Cadmium can be adsorbed onto pristine or secondary Fe oxides by binding or coprecipitating with surface hydroxyl groups under both aerobic and anoxic soil conditions (Muehe et al., 2013b).

However, Cd^{2+} generally has a lower affinity for Fe oxides than other heavy metals and is readily released through ion exchange by other divalent and trivalent cations in large amounts (M. Shi et al., 2021). Consequently, Fe (hydr)oxides are generally less effective in reducing Cd's bioavailability than As in paddy soils. It is also worth noting that some industrial by-products that contain Fe oxides, such as steel-making slag, are used for soil remediation as cost-effective alternatives for expensive materials like ZVI (Huaidong He et al., 2017; Honma et al., 2016a).

Some natural clays and clay minerals, such as montmorillonite, bentonite, zeolite, kaolinite, and illite, are also effective adsorbents for reducing Cd mobility in paddy soils (Sebastian and Prasad, 2014). This effect is mainly ascribed to their large specific surface area and abundant negative charge under acidic conditions (Hamid et al., 2019). Cadmium in soil solution can be adsorbed onto clay minerals through ion exchange or subjected to surface complexation via electrostatic attraction (Novikau and Lujaniene, 2022). Some clay minerals, e.g., sepiolite, can also increase soil pH, contributing to Cd immobilization (Y. Xu et al., 2017).

17.4.1.3 Complexing Agents

Organic amendments, including composts, animal manure, sewage sludge, and rice straw, have traditionally been used as a Cd stabilizer in paddy soils (Khan et al., 2017; Liu et al., 2021; Palansooriya et al., 2020). These materials are typically produced locally and are therefore easily accessible. Compared with other soil amendments, the application of these organic wastes is better accepted by farmers owing to their effects on supplementing essential nutrients and sustaining soil's long-term fertility (Palansooriya et al., 2020). Applying organic amendments can elevate soil DOM with higher molecular weight, hydrophobicity, and aromaticity (Liu et al., 2021).

Effective Cd immobilization is mainly achieved by the binding of Cd^{2+} with the reactive groups of soil DOM, including hydroxyl, carbonyl, and phenoxyl groups, resulting in the formation of insoluble complexes (Khan et al., 2017). Meanwhile, calcareous organic wastes further enhance Cd immobilization by increasing soil pH (Hu et al., 2016). In paddy soils with limited SOM, organic amendments can also improve microbial reduction of Fe(III) and sulfate by providing carbon sources, consequently raising soil pH (Dong et al., 2023; Muehe et al., 2013a; Suriyagoda et al., 2018a). Increasing SOM is also believed to enhance soil water holding capacity (Z. Yang et al., 2012), which may help paddy soils maintain a lower redox potential and reduce Cd bioavailability under intermittent flooding.

The effects of organic amendments on Cd mobility depend largely on their composition, application dose, and soil characteristics (Khan et al., 2017). Chloride ion (Cl^-) and hydrophilic DOM with lower molecular weights released from these amendments may enhance Cd bioavailability by forming soluble complexes (Cheng et al., 2007; Liu et al., 2021). Moreover, some organic amendments, especially those based on sewage sludge and animal manure, may contain excessive toxic heavy metal(loid)s and exacerbate soil contamination with long-term input (Xu et al., 2015).

On the other hand, the effects of organic amendments on the lability and plant uptake of As in paddy fields can be complex. Increased SOM theoretically enhances soil As release by forming soluble As-DOM complexes, competing with As for sorption sites, and facilitating the reductive dissolution of As-bearing Fe(III) oxides (Bao et al., 2022; W. Shi et al., 2021b; Syu et al., 2019). Higher levels of DOC in paddy soils are frequently reported to enhance As release from the solid phase into pore water and increase As accumulation in rice tissues (Norton et al., 2013; Williams et al., 2011; Xiao et al., 2017). However, there was also literature documenting effective As fixation through the organic amendment incorporation (Das et al., 2008; Majumder et al., 2021), possibly explained by the formation of insoluble As-OM complexes, especially when saturated with cations (Suriyagoda et al., 2018a). Soil As mainly interacts with SOM by forming ternary As-Fe oxides-OM complexes, which may either enhance or inhibit the dissolution of As across various soil conditions (Cai et al., 2020; Fakour et al., 2016). In this context, the role of organic amendments in As transformation and mobilization requires further investigation.

Notably, biochar application exhibits similar effects to traditional organic amendments on the mobility of As and Cd in paddy soils (Qiao et al., 2018a; R. Wang et al., 2019). Biochar has been characterized by a high pH, a large surface area, a high cation exchange capacity, and abundant oxygen-containing functional groups (X. Zhang et al., 2013). Therefore, biochar application readily immobilizes soil Cd through surface adsorption, complexation, and precipitation, which appears to be more effective than traditional organic amendments. In contrast, biochar facilitates the biotic and abiotic reduction of As(V) and Fe(III) primarily via electron shuttling and strongly adsorbs to Fe, resulting in enhanced As release (Chen et al., 2016; Vithanage et al., 2017; Wu et al., 2020).

17.4.2 Field Management

17.4.2.1 Water Management

Water management practices are straightforward and powerful measures for controlling the mobility of As and Cd in paddy soils (Zhao and Wang, 2020). As discussed above, the soil water regime largely determines the redox potential, and flooding leads to a rapid reduction in soil Eh. Meanwhile, changes in soil Eh are closely linked to pH dynamics, and soil anoxia generally brings soil pH toward neutrality by consuming protons in acidic soils while increasing P_{CO_2} in alkaline soils (Wan et al., 2018). The heading stage is considered the most important for determining grain As and Cd accumulation. Flooding for 3 weeks before and after heading markedly reduces grain Cd levels but increases As levels (Arao et al., 2009).

In contrast, intermittent flooding and sprinkler irrigation effectively mitigate rice As accumulation but increase Cd risks (Moreno-Jiménez et al., 2014; Somenahally et al., 2011). It is important to note that different water management practices sometimes arise from limited accessibility to irrigation water resources. For instance, in double rice cropping regions of southern China, precipitation predominantly occurs during the early rice growth season, and the late season requires an alternate wetting and drying (AWD) strategy. Consequently, rice grown in the late season commonly accumulates higher Cd and lower As concentrations than in the early season (Chi et al., 2018; Liu et al., 2017).

Studies have shown that by applying different water management strategies, bioavailable As exhibits a negative correlation with soil Eh and a positive correlation with soil pH, while bioavailable Cd shows the opposite (Honma et al., 2016b; Yao et al., 2022). The trade-off between As and Cd dissolution implies a theoretical "optimal Eh/pH" that can mitigate both As and Cd mobility. However, it is almost impossible to precisely maintain specific redox potential and pH due to limitations imposed by precipitation, irrigation infrastructure, and labor costs. In this context, fluctuating soil water regimes are mostly responsible for the opposing dynamics of As and Cd sequestration, dissolution, and plant uptake in paddy fields.

17.4.2.2 Intercropping and Crop Rotation

Intercropping is a traditional cropping practice used to improve nutrient utilization and pest/disease control. Intercropped plants share the same rhizosphere and are thus affected by each other's root activities. Plants that can immobilize soil As and Cd or hyperaccumulate these trace elements are suitable for intercropping with rice in contaminated paddy fields (Mubeen et al., 2023). Xiang et al. (2023) found that intercropping with wetland plants, *Pontederia cordata* and *Canna indica*, effectively reduced Cd accumulation in all tissues, including the grains. Similar Cd mitigation effects were observed when rice was intercropped with the hyperaccumulating plant *Solanum nigrum*, due to increased rhizosphere soil pH and reduced bioavailable Cd levels (Kama et al., 2023). It was also reported that intercropping of high-Cd and low-Cd cultivars reduced Cd accumulation in both cultivars because of increased soil pH and enhanced root Fe plaque (Xu et al., 2021). In these cases, intercropping also improved rice growth and grain yield, which could be explained by higher nutrient uptake efficiency and lower Cd stress. The effect of intercropping on plant As accumulation is less frequently reported, but Huang et al. (2021c) found that intercropping with water spinach (*Ipomoea aquatica*) significantly decreased the As accumulation and increased the grain yields of rice plants.

Crop rotation is another promising agronomical practice to alleviate the As and Cd contamination in paddy fields (Rizwan et al., 2016). It is essentially a phytoextraction strategy aimed at reducing the health risks posed by soil As and Cd contamination while minimizing the losses in agricultural output and farmers' income (Mubeen et al., 2023). The cultivation of some crops, such as oilseed rape (*Brassica juncea*), tobacco (*Nicotiana tabacum*), and chicory (*Cichorium intybus*), and hyperaccumulator plants, such as Chinese brake fern (*Pteris vittata*), can effectively reduce the As or Cd accumulation in rice grains in the subsequent cropping season and reduce soil As and Cd contents after consecutive cultivation (Deng et al., 2023; Liu et al., 2023; Warke et al., 2022; Yang et al., 2022; Yu et al., 2014). However, some plant species may mobilize soil Cd and increase Cd accumulation in the subsequent rice (Yu et al., 2014). Some crops, like tobacco, may suffer from Cd toxicity and exhibit yield loss and quality degradation (Yang et al., 2022). Therefore, selecting plant species for crop rotation with rice requires careful consideration, especially in highly contaminated paddy fields.

17.4.3 Cultivar Breeding

Rice cultivars vary significantly in their ability to accumulate As and Cd. According to Pinson et al. (2015), the differences in grain As and Cd concentrations can reach up to 125- and 154-fold among different rice cultivars. The abilities to accumulate As and Cd, related to multiple genes, quantitative trait loci, and plant physiological characteristics, are highly heritable (Clemens and Ma, 2016; Liu et al., 2020; Uraguchi and Fujiwara, 2013, 2012). In this context, screening and selecting cultivars with consistently lower As and/or Cd concentrations in the grains is considered a cost-effective and eco-friendly method for mitigating the food-chain contamination of As and Cd through rice consumption. Many efforts have been made in selecting and breeding As and Cd low-accumulating cultivars in the past decades (Chi et al., 2018; Duan et al., 2017; Norton et al., 2009; Yu et al., 2006; Zhang et al., 2014). However, cultivar selection is a time-consuming and labor-intensive task. Different rice cultivars' As and Cd accumulation capacities vary to different extents across various soil environments (Chi et al., 2018). Multiple field trials across different crop seasons are necessary to identify cultivars with stable low-accumulating characteristics.

The turnover rate of hybrid rice cultivars is rapid, and selected cultivars are often phased out before they can be promoted and applied. Therefore, recent studies have focused on the genetic and physiological biomarkers of low-As and low-Cd rice cultivars. For instance, radial oxygen loss, Fe plaque formation, and the development of root apoplastic barrier, which vary significantly among rice cultivars, are conducive to inhibiting As and Cd uptake and limiting the upward transport of Cd (Mei et al., 2012; Qi et al., 2020; Wang et al., 2011). Varied expression of the Cd transporter genes, such as *OsHMA3* and *Nramp5*, may explain the different levels of Cd accumulation among rice cultivars (Liu et al., 2020; Sui et al., 2018).

17.5 Future Perspectives and Research Needs

The widespread attention to the contamination of As and Cd in paddy fields has resulted in marked achievements in understanding the transformation mechanisms and developing effective mitigation techniques. However, several prominent issues and critical mechanisms still warrant further investigation. First, in some mine-impacted paddy fields, the accumulation of both As and Cd exceeds the safety threshold, resulting in dual contamination. Due to the contrasting dynamics of As and Cd under redox changes, remediation techniques often alleviate one type of contamination while exacerbating the other. One practical strategy to address this dilemma is to combine different methods, such as water management and soil amendments. However, this significantly increases remediation costs, thus limiting extensive use. A cost-effective method to simultaneously immobilize soil As and Cd is urgently needed.

Second, the sustainability of current mitigation techniques, especially soil amendments, must be verified over time. For instance, soil reacidification, as well as the consumption and migration of amendments, may diminish or even eliminate mitigation effects, necessitating repeated remediation. Furthermore, if periodic practices are necessary, the time intervals should be carefully determined to balance costs and effects.

Third, recent findings suggest that the physiological characteristics of different rice cultivars related to As and Cd accumulation are influenced by cultivar×environment interactions, implying varied performances of the low-As or low-Cd cultivars selected based on these characteristics. Therefore, future research is needed to understand how environmental and genetic characteristics collectively shape the As and Cd accumulation capacities of different rice cultivars. Advanced technologies, such as metabolomics, proteomics, spatial transcriptomics, and plant phenotyping platforms may aid relevant research.

Finally, using advanced computational methods such as machine learning offers a novel and valuable perspective for predicting As and Cd contamination in rice, as well as evaluating consequent public health risks. Machine learning algorithms can process complex, multidimensional environmental and soil characteristics. By employing predictive analytics, these models can forecast As and Cd accumulation in rice grains and facilitate a better understanding of the biogeochemical pathways that govern the mobilization and uptake of these toxic metals. This approach could greatly enhance risk assessments associated with dietary exposures and inform the development of more precise and cost-effective remediation strategies (Figures 17.1–17.3).

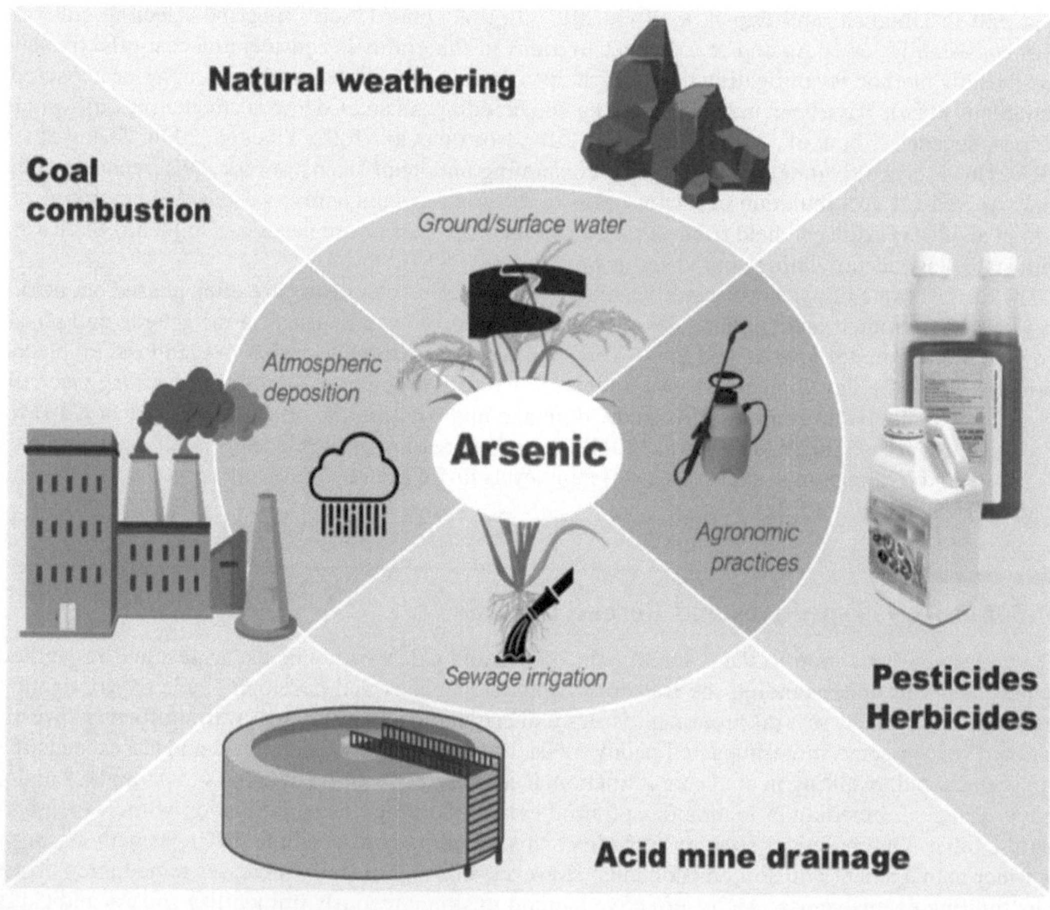

FIGURE 17.1 Common sources of As in paddy fields.

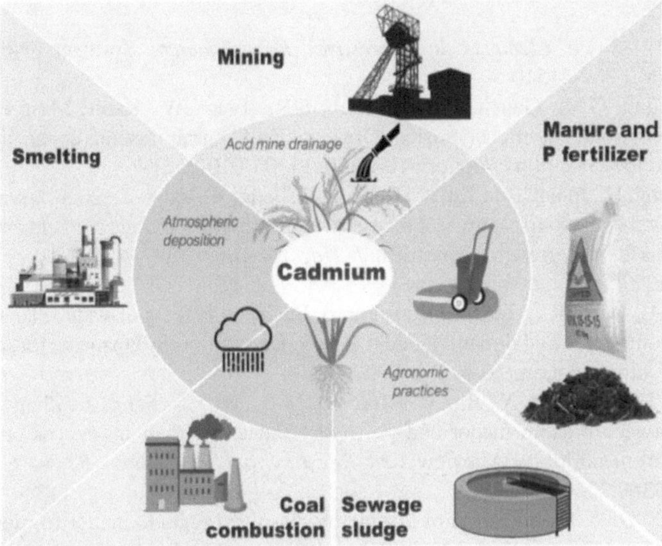

FIGURE 17.2 Common sources of Cd in paddy fields.

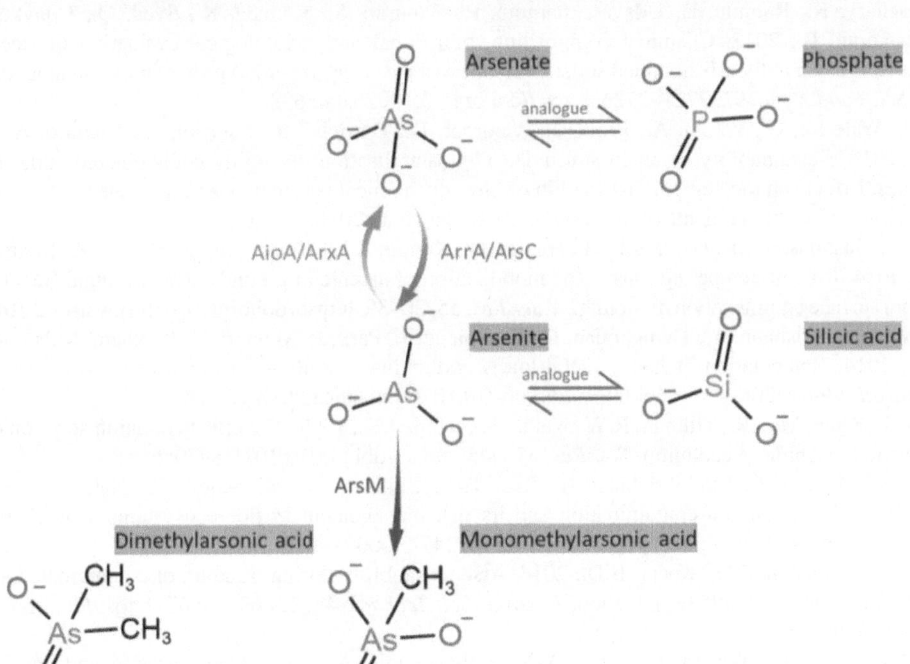

FIGURE 17.3 Species transformation of As in paddy soils.

Acknowledgments

This chapter was supported by the National Natural Science Foundation of China (41977329 and 41501355).

REFERENCES

Adriano, D.C., 2001. *Trace Elements in Terrestrial Environments.* Springer, New York. https://doi.org/10.1007/978-0-387-21510-5

Ahmed, Z.U., Panaullah, G.M., Gauch, H., McCouch, S.R., Tyagi, W., Kabir, M.S., Duxbury, J.M., 2011. Genotype and environment effects on rice (Oryza sativa L.) grain arsenic concentration in Bangladesh. *Plant Soil.* 338, 367–382. https://doi.org/10.1007/s11104-010-0551-7

Ali, W., Mao, K., Zhang, H., Junaid, M., Xu, N., Rasool, A., Feng, X., Yang, Z., 2020. Comprehensive review of the basic chemical behaviours, sources, processes, and endpoints of trace element contamination in paddy soil-rice systems in rice-growing countries. *J. Hazard. Mater.* 397, 122720. https://doi.org/10.1016/j.jhazmat.2020.122720

Arao, T., Kawasaki, A., Baba, K., Mori, S., Matsumoto, S., 2009. Effects of water management on cadmium and arsenic accumulation and dimethylarsinic acid concentrations in Japanese rice. *Environ. Sci. Technol.* 43, 9361–9367. https://doi.org/10.1021/es9022738

Bao, Y., Bolan, N.S., Lai, J., Wang, Y., Jin, X., Kirkham, M.B., Wu, X., Fang, Z., Zhang, Y., Wang, H., 2022. Interactions between organic matter and Fe (hydr)oxides and their influences on immobilization and remobilization of metal(loid)s: A review. *Crit. Rev. Environ. Sci. Technol.* 52, 4016–4037. https://doi.org/10.1080/10643389.2021.1974766

Bauer, M., Blodau, C., 2006. Mobilization of arsenic by dissolved organic matter from iron oxides, soils and sediments. *Sci. Total Environ.* 354, 179–190. https://doi.org/10.1016/j.scitotenv.2005.01.027

Benjamin, M.M., Leckie, J.O., 1981. Competitive adsorption of Cd, Cu, Zn, and Pb on amorphous iron oxyhydroxide. *J. Colloid Interface Sci.* 83, 410–419. https://doi.org/10.1016/0021-9797(81)90337-4

Bhaskarachary, K., Ramulu, P., Udayasekhararao, P., Bapurao, S., Kamala, K., Syed, Q., Udaykumar, P., Sesikeran, B., 2012. Chemical composition, nutritional and toxicological evaluation of rice (*Oryza sativa*) grown in fly ash amended soils: Evaluation of rice (*Oryza sativa*) grown in fly ash amended soils. *J. Sci. Food Agric.* 92, 2721–2726. https://doi.org/10.1002/jsfa.5698

Bidar, G., Waterlot, C., Verdin, A., Proix, N., Courcot, D., Détriché, S., Fourrier, H., Richard, A., Douay, F., 2016. Sustainability of an in situ aided phytostabilisation on highly contaminated soils using fly ashes: Effects on the vertical distribution of physicochemical parameters and trace elements. *J. Environ. Manage.* 171, 204–216. https://doi.org/10.1016/j.jenvman.2016.01.029

Biswas, A., Gustafsson, J.P., Neidhardt, H., Halder, D., Kundu, A.K., Chatterjee, D., Berner, Z., Bhattacharya, P., 2014. Role of competing ions in the mobilization of arsenic in groundwater of Bengal Basin: Insight from surface complexation modeling. *Water Res.* 55, 30–39. https://doi.org/10.1016/j.watres.2014.02.002

Bolan, N., Kunhikrishnan, A., Thangarajan, R., Kumpiene, J., Park, J., Makino, T., Kirkham, M.B., Scheckel, K., 2014. Remediation of heavy metal(loid)s contaminated soils – To mobilize or to immobilize? *J. Hazard. Mater.* 266, 141–166. https://doi.org/10.1016/j.jhazmat.2013.12.018

Bolan, N.S., Khan, M.A.R., Tillman, R.W., Naidu, R., Syers, J.K., 1999. The effects of anion sorption on sorption and leaching of cadmium. *Soil Res.* 37, 445. https://doi.org/10.1071/S97046

Bolan, N.S., Makino, T., Kunhikrishnan, A., Kim, P.-J., Ishikawa, S., Murakami, M., Naidu, R., Kirkham, M.B., 2013. Cadmium contamination and its risk management in rice ecosystems. *Adv. Agron.* 119, 183–273. https://doi.org/10.1016/B978-0-12-407247-3.00004-4

Burton, E.D., Johnston, S.G., Kocar, B.D., 2014. Arsenic mobility during flooding of contaminated soil: The effect of microbial sulfate reduction. *Environ. Sci. Technol.* 48, 13660–13667. https://doi.org/10.1021/es503963k

Cai, Q.-Y., Mo, C.-H., Wu, Q.-T., Zeng, Q.-Y., Katsoyiannis, A., 2007. Concentration and speciation of heavy metals in six different sewage sludge-composts. *J. Hazard. Mater.* 147, 1063–1072. https://doi.org/10.1016/j.jhazmat.2007.01.142

Cai, X., Wang, P., Li, Z., Li, Y., Yin, N., Du, H., Cui, Y., 2020. Mobilization and transformation of arsenic from ternary complex OM-Fe(III)-As(V) in the presence of As(V)-reducing bacteria. *J. Hazard. Mater.* 381, 120975. https://doi.org/10.1016/j.jhazmat.2019.120975

Chen, H., Zhang, W., Yang, X., Wang, P., McGrath, S.P., Zhao, F.-J., 2018. Effective methods to reduce cadmium accumulation in rice grain. *Chemosphere* 207, 699–707. https://doi.org/10.1016/j.chemosphere.2018.05.143

Chen, Q.Y., Costa, M., 2021. Arsenic: A global environmental challenge. *Annu. Rev. Pharmacol. Toxicol.* 61, 47–63. https://doi.org/10.1146/annurev-pharmtox-030220-013418

Chen, Z., Wang, Y., Xia, D., Jiang, X., Fu, D., Shen, L., Wang, H., Li, Q.B., 2016. Enhanced bioreduction of iron and arsenic in sediment by biochar amendment influencing microbial community composition and dissolved organic matter content and composition. *J. Hazard. Mater.* 311, 20–29. https://doi.org/10.1016/j.jhazmat.2016.02.069

Cheng, H., Xu, W., Liu, J., Zhao, Q., He, Y., Chen, G., 2007. Application of composted sewage sludge (CSS) as a soil amendment for turfgrass growth. *Ecol. Eng.* 29, 96–104. https://doi.org/10.1016/j.ecoleng.2006.08.005

Chi, W., Yang, Y., Liu, T., Sun, Y., Du, Y., Qin, H., Li, X., 2022a. Effects of water salinity on cadmium availability at soil–water interface: Implication for salt water intrusion. *Environ. Sci. Pollut. Res.* 29, 68892–68903. https://doi.org/10.1007/s11356-022-20606-2

Chi, W., Yang, Y., Zhang, K., Wang, P., Du, Y., Li, X., Sun, Y., Liu, T., Li, F., 2022b. Seawater intrusion induced cadmium activation via altering its distribution and transformation in paddy soil. *Chemosphere* 307, 135805. https://doi.org/10.1016/j.chemosphere.2022.135805

Chi, Y., Li, F., Tam, N.F., Liu, C., Ouyang, Y., Qi, X., Li, W.C., Ye, Z., 2018. Variations in grain cadmium and arsenic concentrations and screening for stable low-accumulating rice cultivars from multi-environment trials. *Sci. Total Environ.* 643, 1314–1324. https://doi.org/10.1016/j.scitotenv.2018.06.288

Chi, Y., Peng, L., Tam, N.F., Lin, Q., Liang, H., Li, W.C., Ye, Z., 2022a. Effects of fly ash and steel slag on cadmium and arsenic accumulation in rice grains and soil health: A field study over four crop seasons in Guangdong, China. *Geoderma*, 419, 115879. https://doi.org/10.1016/j.geoderma.2022.115879

Chi, Y., Tam, N.F., Li, W.C., Ye, Z., 2022b. Multiple geochemical and microbial processes regulated by redox and organic matter control the vertical heterogeneity of As and Cd in paddy soil. *Sci. Total Environ.* 839, 156229. https://doi.org/10.1016/j.scitotenv.2022.156229

Clemens, S., Aarts, M.G.M., Thomine, S., Verbruggen, N., 2013. Plant science: The key to preventing slow cadmium poisoning. *Trends Plant Sci.* 18, 92–99. https://doi.org/10.1016/j.tplants.2012.08.003

Clemens, S., Ma, J.F., 2016. Toxic heavy metal and metalloid accumulation in crop plants and foods. *Annu. Rev. Plant Biol.* 67, 489–512. https://doi.org/10.1146/annurev-arplant-043015-112301

Colzato, M., Alleoni, L.R.F., Kamogawa, M.Y., 2018. Cadmium sorption and extractability in tropical soils with variable charge. *Environ. Monit. Assess.* 190, 345. https://doi.org/10.1007/s10661-018-6666-7

Craw, D., Bowell, R.J., 2014. The characterization of arsenic in mine waste. *Rev. Mineral. Geochem.* 79, 473–505. https://doi.org/10.2138/rmg.2014.79.10

Cubadda, F., Jackson, B.P., Cottingham, K.L., Van Horne, Y.O., Kurzius-Spencer, M., 2017. Human exposure to dietary inorganic arsenic and other arsenic species: State of knowledge, gaps and uncertainties. *Sci. Total Environ.* 579, 1228–1239. https://doi.org/10.1016/j.scitotenv.2016.11.108

Dai, J., Tang, Z., Jiang, N., Kopittke, P.M., Zhao, F.-J., Wang, P., 2020. Increased arsenic mobilization in the rice rhizosphere is mediated by iron-reducing bacteria. *Environ. Pollut.* 263, 114561. https://doi.org/10.1016/j.envpol.2020.114561

Das, D.K., Sur, P., Das, K., 2008. Mobilisation of arsenic in soils and in rice (Oryza sativa L.) plants affected by organic matter and zinc application in irrigation water contaminated with arsenic. *Plant Soil Environ.* 54, 30–37. https://doi.org/10.17221/2778-PSE

Das, S., Chou, M.-L., Jean, J.-S., Liu, C.-C., Yang, H.-J., 2016. Water management impacts on arsenic behavior and rhizosphere bacterial communities and activities in a rice agro-ecosystem. *Sci. Total Environ.* 542, 642–652. https://doi.org/10.1016/j.scitotenv.2015.10.122

De Livera, J., McLaughlin, M.J., Hettiarachchi, G.M., Kirby, J.K., Beak, D.G., 2011. Cadmium solubility in paddy soils: Effects of soil oxidation, metal sulfides and competitive ions. *Sci. Total Environ.* 409, 1489–1497. https://doi.org/10.1016/j.scitotenv.2010.12.028

Deng, X., Wu, S., Yang, Y., Qin, Y., Huang, Q., Wu, W., Rong, X., Zeng, Q., 2023. A rice–chicory rotation pattern ensures safe grain production and phytoremediation of cadmium-contaminated paddy fields: A four-year field experiment in southern China. *Chemosphere* 322, 138192. https://doi.org/10.1016/j.chemosphere.2023.138192

Dong, H., Zeng, Q., Sheng, Y., Chen, C., Yu, G., Kappler, A., 2023. Coupled iron cycling and organic matter transformation across redox interfaces. *Nat. Rev. Earth Environ.* 4, 659–673. https://doi.org/10.1038/s43017-023-00470-5

Duan, G., Shao, G., Tang, Z., Chen, H., Wang, B., Tang, Z., Yang, Y., Liu, Y., Zhao, F.-J., 2017. Genotypic and environmental variations in grain cadmium and arsenic concentrations among a panel of high yielding rice cultivars. *Rice* 10, 9. https://doi.org/10.1186/s12284-017-0149-2

Duan, Y., Li, R., Yu, K., Zeng, G., Liu, C., 2022. Effects of geochemical and hydrodynamic transiency on desorption and transport of As in heterogeneous systems. *Sci. Total Environ.* 835, 155381. https://doi.org/10.1016/j.scitotenv.2022.155381

Dykes, G.E., Chari, N.R., Seyfferth, A.L., 2020. Si-induced DMA desorption is not the driver for enhanced DMA availability after Si addition to flooded soils. *Sci. Total Environ.* 739, 139906. https://doi.org/10.1016/j.scitotenv.2020.139906

Edelstein, M., Ben-Hur, M., 2018. Heavy metals and metalloids: Sources, risks and strategies to reduce their accumulation in horticultural crops. *Sci. Hortic.* 234, 431–444. https://doi.org/10.1016/j.scienta.2017.12.039

Ehlert, K., Mikutta, C., Kretzschmar, R., 2016. Effects of manganese oxide on arsenic reduction and leaching from contaminated floodplain soil. *Environ. Sci. Technol.* 50, 9251–9261. https://doi.org/10.1021/acs.est.6b01767

Fakour, H., Lin, T., Lo, S., 2016. Equilibrium modeling of arsenic adsorption in a ternary arsenic–iron oxide–natural organic matter system. *CLEAN – Soil Air Water* 44, 1287–1295. https://doi.org/10.1002/clen.201500962

Fan, L., Zhao, F., Liu, J., Frost, R.L., 2018. The As behavior of natural arsenical-containing colloidal ferric oxyhydroxide reacted with sulfate reducing bacteria. *Chem. Eng. J.* 332, 183–191. https://doi.org/10.1016/j.cej.2017.09.078

Farrah, H., Pickering, W., 1977. The sorption of lead and cadmium species by clay minerals. *Aust. J. Chem.* 30, 1417. https://doi.org/10.1071/CH9771417

Fulda, B., Voegelin, A., Kretzschmar, R., 2013. Redox-controlled changes in cadmium solubility and solid-phase speciation in a paddy soil As affected by reducible sulfate and copper. *Environ. Sci. Technol.* 47, 12775–12783. https://doi.org/10.1021/es401997d

Gao, Y., Tong, H., Zhao, Z., Cheng, N., Wu, P., 2023. Effects of Fe oxides and their redox cycling on Cd activity in paddy soils: A review. *J. Hazard. Mater.* 456, 131665. https://doi.org/10.1016/j.jhazmat.2023.131665

Gao, Z., Guo, H., Chen, D., Yu, C., He, C., Shi, Q., Qiao, W., Kersten, M., 2023. Transformation of dissolved organic matter and related arsenic mobility at a surface water-groundwater interface in the Hetao Basin, China. *Environ. Pollut.* 334, 122202. https://doi.org/10.1016/j.envpol.2023.122202

Garbinski, L.D., Rosen, B.P., Chen, J., 2019. Pathways of arsenic uptake and efflux. *Environ. Int.* 126, 585–597. https://doi.org/10.1016/j.envint.2019.02.058

Ge, Y., Huang, X., Lu, L., 2024. Mitigating cadmium contamination in rice: Insights from a large-scale meta-analysis of amendment effects. *Plant Soil.* https://doi.org/10.1007/s11104-024-06689-5

Hamid, Y., Tang, L., Sohail, M.I., Cao, X., Hussain, B., Aziz, M.Z., Usman, M., He, Z., Yang, X., 2019. An explanation of soil amendments to reduce cadmium phytoavailability and transfer to food chain. *Sci. Total Environ.* 660, 80–96. https://doi.org/10.1016/j.scitotenv.2018.12.419

Han, Y.-S., Park, J.-H., Min, Y., Lim, D.-H., 2020. Competitive adsorption between phosphate and arsenic in soil containing iron sulfide: XAS experiment and DFT calculation approaches. *Chem. Eng. J.* 397, 125426. https://doi.org/10.1016/j.cej.2020.125426

Hashimoto, Y., Yamaguchi, N., 2013. Chemical speciation of cadmium and sulfur K-Edge XANES spectroscopy in flooded paddy soils amended with zerovalent iron. *Soil Science Soc of Amer J.* 77, 1189–1198. https://doi.org/10.2136/sssaj2013.01.0038

He, H., Dong, Z., Peng, Q., Wang, X., Fan, C., Zhang, X., 2017. Impacts of coal fly ash on plant growth and accumulation of essential nutrients and trace elements by alfalfa (Medicago sativa) grown in a loessial soil. *J. Environ. Manage.* 197, 428–439. https://doi.org/10.1016/j.jenvman.2017.04.028

He, H., Tam, N.F.Y., Yao, A., Qiu, R., Li, W.C., Ye, Z., 2017. Growth and Cd uptake by rice (Oryza sativa) in acidic and Cd-contaminated paddy soils amended with steel slag. *Chemosphere* 189, 247–254. https://doi.org/10.1016/j.chemosphere.2017.09.069

Honma, T., Ohba, H., Kaneko, A., Nakamura, K., Makino, T., Katou, H., 2016a. Effects of soil amendments on arsenic and cadmium uptake by rice plants (*Oryza sativa* L. cv. Koshihikari) under different water management practices. *Soil Sci. Plant Nutr.* 62, 349–356. https://doi.org/10.1080/00380768.2016.1196569

Honma, T., Ohba, H., Kaneko-Kadokura, A., Makino, T., Nakamura, K., Katou, H., 2016b. Optimal soil Eh, pH, and water management for simultaneously minimizing arsenic and cadmium concentrations in rice grains. *Environ. Sci. Technol.* 50, 4178–4185. https://doi.org/10.1021/acs.est.5b05424

Hu, Y., Cheng, H., Tao, S., 2016. The challenges and solutions for cadmium-contaminated rice in China: A critical review. *Environ. Int.* 92–93, 515–532. https://doi.org/10.1016/j.envint.2016.04.042

Huang, H., Chen, H.-P., Kopittke, P.M., Kretzschmar, R., Zhao, F.-J., Wang, P., 2021a. The voltaic effect as a novel mechanism controlling the remobilization of cadmium in paddy soils during drainage. *Environ. Sci. Technol.* 55, 1750–1758. https://doi.org/10.1021/acs.est.0c06561

Huang, H., Ji, X.-B., Cheng, L.-Y., Zhao, F.-J., Wang, P., 2021b. Free radicals produced from the oxidation of ferrous sulfides promote the remobilization of cadmium in paddy soils during drainage. *Environ. Sci. Technol.* 55, 9845–9853. https://doi.org/10.1021/acs.est.1c00576

Huang, S.Y., Zhuo, C., Du, X.Y., Li, H.S., 2021c. Remediation of arsenic-contaminated paddy soil by intercropping aquatic vegetables and rice. *International Journal of Phytoremediation* 23, 1021–1029. https://doi.org/10.1080/15226514.2021.1872485

Huang, Y., Wang, L., Wang, W., Li, T., He, Z., Yang, X., 2019. Current status of agricultural soil pollution by heavy metals in China: A meta-analysis. *Sci. Total Environ.* 651, 3034–3042. https://doi.org/10.1016/j.scitotenv.2018.10.185

Hudak, P.F., 2000. Distribution and sources of arsenic in the southern high Plains Aquifer, Texas, USA. *J. Environ. Sci. Health Part A* 35, 899–913. https://doi.org/10.1080/10934520009377010

Hussain, B., Ashraf, M.N., Shafeeq-ur-Rahman, Abbas, A., Li, J., Farooq, M., 2021. Cadmium stress in paddy fields: Effects of soil conditions and remediation strategies. *Sci. Total Environ.* 754, 142188. https://doi.org/10.1016/j.scitotenv.2020.142188

Jia, Y., Bao, P., Zhu, Y.-G., 2015. Arsenic bioavailability to rice plant in paddy soil: Influence of microbial sulfate reduction. *J. Soils Sediments* 15, 1960–1967. https://doi.org/10.1007/s11368-015-1133-3

Jia, Y., Huang, H., Chen, Z., Zhu, Y.-G., 2014. Arsenic uptake by rice is influenced by microbe-mediated arsenic redox changes in the rhizosphere. *Environ. Sci. Technol.* 48, 1001–1007. https://doi.org/10.1021/es403877s

Jiang, H., Li, T., Han, X., Yang, X., He, Z., 2012. Effects of pH and low molecular weight organic acids on competitive adsorption and desorption of cadmium and lead in paddy soils. *Environ. Monit. Assess.* 184, 6325–6335. https://doi.org/10.1007/s10661-011-2422-y

Kama, R., Ma, Q., Nabi, F., Aidara, M., Huang, P., Li, Z., He, J., Diatta, S., Li, H., 2023. Hyperaccumulator Solanum nigrum L. intercropping reduced rice cadmium uptake under a high-bed and low-ditch planting system. *Plants* 12, 4027. https://doi.org/10.3390/plants12234027

Kappler, A., Bryce, C., Mansor, M., Lueder, U., Byrne, J.M., Swanner, E.D., 2021. An evolving view on biogeochemical cycling of iron. *Nat. Rev. Microbiol.* 19, 360–374. https://doi.org/10.1038/s41579-020-00502-7

Kaya, C., Uğurlar, F., Ashraf, M., Hou, D., Kirkham, M.B., Bolan, N., 2024. Microbial consortia-mediated arsenic bioremediation in agricultural soils: Current status, challenges, and solutions. *Sci. Total Environ.* 917, 170297. https://doi.org/10.1016/j.scitotenv.2024.170297

Kersten, M., Daus, B., 2015. Silicic acid competes for dimethylarsinic acid (DMA) immobilization by the iron hydroxide plaque mineral goethite. *Sci. Total Environ.* 508, 199–205. https://doi.org/10.1016/j.scitotenv.2014.11.093

Khan, M.A., Khan, S., Khan, A., Alam, M., 2017. Soil contamination with cadmium, consequences and remediation using organic amendments. *Sci. Total Environ.* 601–602, 1591–1605. https://doi.org/10.1016/j.scitotenv.2017.06.030

Kim, M., Kim, J., Kim, M., Kim, Y.-S., Nam, S.M., Moon, D.H., Hyun, S., 2018. Factors influencing As(V) stabilization in the mine soils amended with iron-rich materials. *Environ. Sci. Pollut. Res.* 25, 26757–26765. https://doi.org/10.1007/s11356-017-0044-9

Kocar, B.D., Borch, T., Fendorf, S., 2010. Arsenic repartitioning during biogenic sulfidization and transformation of ferrihydrite. *Geochim. Cosmochim. Acta* 74, 980–994. https://doi.org/10.1016/j.gca.2009.10.023

Kumarathilaka, P., Seneweera, S., Meharg, A., Bundschuh, J., 2018. Arsenic speciation dynamics in paddy rice soil-water environment: Sources, physico-chemical, and biological factors - A review. *Water Res.* 140, 403–414. https://doi.org/10.1016/j.watres.2018.04.034

Lackovic, K., Angove, M.J., Wells, J.D., Johnson, B.B., 2003. Modeling the adsorption of Cd(II) onto Muloorina illite and related clay minerals. *J. Colloid Interface Sci.* 257, 31–40. https://doi.org/10.1016/S0021-9797(02)00031-0

Latare, A.M., Kumar, O., Singh, S.K., Gupta, A., 2014. Direct and residual effect of sewage sludge on yield, heavy metals content and soil fertility under rice–wheat system. *Ecol. Eng.* 69, 17–24. https://doi.org/10.1016/j.ecoleng.2014.03.066

Li, H., Luo, N., Li, Y.W., Cai, Q.Y., Li, H.Y., Mo, C.H., Wong, M.H., 2017. Cadmium in rice: Transport mechanisms, influencing factors, and minimizing measures. *Environ. Pollut.* 224, 622–630. https://doi.org/10.1016/j.envpol.2017.01.087

Li, H., Yang, Z., Dai, M., Diao, X., Dai, S., Fang, T., Dong, X., 2020. Input of Cd from agriculture phosphate fertilizer application in China during 2006–2016. *Sci. Total Environ.* 698, 134149. https://doi.org/10.1016/j.scitotenv.2019.134149

Li, R.Y., Stroud, J.L., Ma, J.F., McGrath, S.P., Zhao, F.J., 2009. Mitigation of arsenic accumulation in rice with water management and silicon fertilization. *Environ. Sci. Technol.* 43, 3778–3783. https://doi.org/10.1021/es803643v

Li, T., Jiang, H., Yang, X., He, Z., 2013. Competitive sorption and desorption of cadmium and lead in paddy soils of eastern China. *Environ. Earth Sci.* 68, 1599–1607. https://doi.org/10.1007/s12665-012-1853-2

Li, X., Mu, L., Zhang, C., Fu, T., He, T., 2022. Effect of amendments on bioavailability of cadmium in soil-rice system: A field experiment study. *Environ. Sci. Pollut. Res.* 30, 37659–37668. https://doi.org/10.1007/s11356-022-24875-9

Li, Z., Huang, B., Huang, J., Chen, G., Zhang, C., Nie, X., Luo, N., Yao, H., Ma, W., Zeng, G., 2015. Influence of removal of organic matter and iron and manganese oxides on cadmium adsorption by red paddy soil aggregates. *RSC Adv.* 5, 90588–90595. https://doi.org/10.1039/C5RA16501F

Li, Z., Liang, Y., Hu, H., Shaheen, S.M., Zhong, H., Tack, F.M.G., Wu, M., Li, Y.-F., Gao, Y., Rinklebe, J., Zhao, J., 2021. Speciation, transportation, and pathways of cadmium in soil-rice systems: A review on the environmental implications and remediation approaches for food safety. *Environ. Int.* 156, 106749. https://doi.org/10.1016/j.envint.2021.106749

Liu, C., Gao, Z., Shang, L., Yang, C., Ruan, B., Zeng, D., Guo, L., Zhao, F., Huang, C., Qian, Q., 2020. Natural variation in the promoter of *OsHMA3* contributes to differential grain cadmium accumulation between *Indica* and *Japonica* rice. *J. Integr. Plant Biol.* 62, 314–329. https://doi.org/10.1111/jipb.12794

Liu, N., Lou, X., Li, X., Shuai, Z., Liu, H., Jiang, Z., Wei, S., 2021. Rhizosphere dissolved organic matter and iron plaque modified by organic amendments and its relations to cadmium bioavailability and accumulation in rice. *Sci. Total Environ.* 792, 148216. https://doi.org/10.1016/j.scitotenv.2021.148216

Liu, W.-J., McGrath, S.P., Zhao, F.-J., 2014. Silicon has opposite effects on the accumulation of inorganic and methylated arsenic species in rice. *Plant Soil* 376, 423–431. https://doi.org/10.1007/s11104-013-1991-7

Liu, X., Tian, G., Jiang, D., Zhang, C., Kong, L., 2016. Cadmium (Cd) distribution and contamination in Chinese paddy soils on national scale. *Environ. Sci. Pollut. Res.* 23, 17941–17952. https://doi.org/10.1007/s11356-016-6968-7

Liu, Y., Zeng, H., Zhou, H., Zhang, S., Tie, B., Peng, L., Zeng, Q., Peng, H., Luo, S., 2023. Variation of Cd and As accumulation in crops under oilseed rape–rice rotation system in response to different contaminated rice straw-return methods. *Plant Soil* 489, 309–321. https://doi.org/10.1007/s11104-023-06018-2

Liu, Y., Zhang, C., Zhao, Y., Sun, S., Liu, Z., 2017. Effects of growing seasons and genotypes on the accumulation of cadmium and mineral nutrients in rice grown in cadmium contaminated soil. *Sci. Total Environ.* 579, 1282–1288. https://doi.org/10.1016/j.scitotenv.2016.11.115

Liu, Z., Wang, Qi-qi, Huang, S., Kong, L., Zhuang, Z., Wang, Qi, Li, H., Wan, Y., 2022. The risks of sulfur addition on cadmium accumulation in paddy rice under different water-management conditions. *J. Environ. Sci.* 118, 101–111. https://doi.org/10.1016/j.jes.2021.08.022

Loganathan, P., Vigneswaran, S., Kandasamy, J., Naidu, R., 2012. Cadmium sorption and desorption in soils: A review. *Crit. Rev. Environ. Sci. Technol.* 42, 489–533. https://doi.org/10.1080/10643389.2010.520234

Ma, J.F., Shen, R.F., Shao, J.F., 2021. Transport of cadmium from soil to grain in cereal crops: A review. *Pedosphere* 31, 3–10. https://doi.org/10.1016/S1002-0160(20)60015-7

Maier, M.V., Wolter, Y., Zentler, D., Scholz, C., Stirn, C.N., Isenbeck-Schröter, M., 2019. Phosphate induced arsenic mobilization as a potentially effective in-situ remediation technique—preliminary column tests. *Water* 11, 2364. https://doi.org/10.3390/w11112364

Majumdar, A., Upadhyay, M.K., Kumar, J.S., Sheena, Barla, A., Srivastava, S., Jaiswal, M.K., Bose, S., 2019. Ultra-structure alteration via enhanced silicon uptake in arsenic stressed rice cultivars under intermittent irrigation practices in Bengal delta basin. *Ecotoxicol. Environ. Saf.* 180, 770–779. https://doi.org/10.1016/j.ecoenv.2019.05.028

Majumder, S., Banik, P., 2019. Geographical variation of arsenic distribution in paddy soil, rice and rice-based products: A meta-analytic approach and implications to human health. *J. Environ. Manage.* 233, 184–199. https://doi.org/10.1016/j.jenvman.2018.12.034

Majumder, S., Powell, M.A., Kumar Biswas, P., Banik, P., 2021. The role of agronomic factors (rice cultivation practices and soil amendments) on Arsenic fractionation: A strategy to minimise Arsenic uptake by rice, with some observations related to cadmium. *CATENA* 206, 105556. https://doi.org/10.1016/j.catena.2021.105556

Makino, T., Nakamura, K., Katou, H., Ishikawa, S., Ito, M., Honma, T., Miyazaki, N., Takehisa, K., Sano, S., Matsumoto, S., Suda, A., Baba, K., Kawasaki, A., Yamaguchi, N., Akahane, I., Tomizawa, M., Arao, T., 2016. Simultaneous decrease of arsenic and cadmium in rice (*Oryza sativa* L.) plants cultivated under submerged field conditions by the application of iron-bearing materials. *Soil Sci. Plant Nutr.* 62, 340–348. https://doi.org/10.1080/00380768.2016.1203731

Marković, J., Jović, M., Smičiklas, I., Šljivić-Ivanović, M., Onjia, A., Trivunac, K., Popović, A., 2019. Cadmium retention and distribution in contaminated soil: Effects and interactions of soil properties, contamination level, aging time and in situ immobilization agents. *Ecotoxicol. Environ. Saf.* 174, 305–314. https://doi.org/10.1016/j.ecoenv.2019.03.001

Meharg, A.A., 2004. Arsenic in rice – understanding a new disaster for South-East Asia. *Trends Plant Sci.* 9, 415–417.

Meharg, C., Meharg, A.A., 2015. Silicon, the silver bullet for mitigating biotic and abiotic stress, and improving grain quality, in rice? *Environ. Exp. Bot.* 120, 8–17. https://doi.org/10.1016/j.envexpbot.2015.07.001

Mei, X.Q., Wong, M.H., Yang, Y., Dong, H.Y., Qiu, R.L., Ye, Z.H., 2012. The effects of radial oxygen loss on arsenic tolerance and uptake in rice and on its rhizosphere. *Environ. Pollut.* 165, 109–117. https://doi.org/10.1016/j.envpol.2012.02.018

Miretzky, P., Cirelli, A.F., 2010. Remediation of arsenic-contaminated soils by iron amendments: A review. *Crit. Rev. Environ. Sci. Technol.* 40, 93–115. https://doi.org/10.1080/10643380802202059

Mitra, A., Chatterjee, S., Moogouei, R., Gupta, D., 2017. Arsenic accumulation in rice and probable mitigation approaches: A review. *Agronomy* 7, 67. https://doi.org/10.3390/agronomy7040067

Mladenov, N., Zheng, Y., Simone, B., Bilinski, T.M., McKnight, D.M., Nemergut, D., Radloff, K.A., Rahman, M.M., Ahmed, K.M., 2015. Dissolved organic matter quality in a shallow aquifer of Bangladesh: Implications for arsenic mobility. *Environ. Sci. Technol.* 49, 10815–10824. https://doi.org/10.1021/acs.est.5b01962

Mondal, S., Pramanik, K., Ghosh, S.K., Pal, P., Ghosh, P.K., Ghosh, A., Maiti, T.K., 2022. Molecular insight into arsenic uptake, transport, phytotoxicity, and defense responses in plants: A critical review. *Planta* 255, 87. https://doi.org/10.1007/s00425-022-03869-4

Moreno-Jiménez, E., Meharg, A.A., Smolders, E., Manzano, R., Becerra, D., Sánchez-Llerena, J., Albarrán, Á., López-Piñero, A., 2014. Sprinkler irrigation of rice fields reduces grain arsenic but enhances cadmium. *Sci. Total Environ.* 485–486, 468–473. https://doi.org/10.1016/j.scitotenv.2014.03.106

Moulick, D., Samanta, S., Sarkar, S., Mukherjee, A., Pattnaik, B.K., Saha, S., Awasthi, J.P., Bhowmick, S., Ghosh, D., Samal, A.C., Mahanta, S., Mazumder, M.K., Choudhury, S., Bramhachari, K., Biswas, J.K., Santra, S.C., 2021. Arsenic contamination, impact and mitigation strategies in rice agro-environment: An inclusive insight. *Sci. Total Environ.* 800, 149477. https://doi.org/10.1016/j.scitotenv.2021.149477

Mu, T., Wu, T., Zhou, T., Li, Z., Ouyang, Y., Jiang, J., Zhu, D., Hou, J., Wang, Z., Luo, Y., Christie, P., Wu, L., 2019. Geographical variation in arsenic, cadmium, and lead of soils and rice in the major rice producing regions of China. *Sci. Total Environ.* 677, 373–381. https://doi.org/10.1016/j.scitotenv.2019.04.337

Mubeen, S., Ni, W., He, C., Yang, Z., 2023. Agricultural strategies to reduce cadmium accumulation in crops for food safety. *Agriculture* 13, 471. https://doi.org/10.3390/agriculture13020471

Muehe, E.M., Adaktylou, I.J., Obst, M., Zeitvogel, F., Behrens, S., Planer-Friedrich, B., Kraemer, U., Kappler, A., 2013a. Organic carbon and reducing conditions lead to cadmium immobilization by secondary Fe mineral formation in a pH-neutral soil. *Environ. Sci. Technol.* 47, 13430–13439. https://doi.org/10.1021/es403438n

Muehe, E.M., Obst, M., Hitchcock, A., Tyliszczak, T., Behrens, S., Schröder, C., Byrne, J.M., Michel, F.M., Krämer, U., Kappler, A., 2013b. Fate of Cd during microbial Fe(III) mineral reduction by a novel and Cd-tolerant *geobacter* species. *Environ. Sci. Technol.* 47, 14099–14109. https://doi.org/10.1021/es403365w

Norton, G.J., Adomako, E.E., Deacon, C.M., Carey, A.-M., Price, A.H., Meharg, A.A., 2013. Effect of organic matter amendment, arsenic amendment and water management regime on rice grain arsenic species. *Environ. Pollut.* 177, 38–47. https://doi.org/10.1016/j.envpol.2013.01.049

Norton, G.J., Duan, G., Dasgupta, T., Islam, M.R., Lei, M., Zhu, Y., Deacon, C.M., Moran, A.C., Islam, S., Zhao, F.-J., Stroud, J.L., McGrath, S.P., Feldmann, J., Price, A.H., Meharg, A.A., 2009. Environmental and genetic control of arsenic accumulation and speciation in rice grain: Comparing a range of common cultivars grown in contaminated sites across Bangladesh, China, and India. *Environ. Sci. Technol.* 43, 8381–8386. https://doi.org/10.1021/es901844q

Novikau, R., Lujaniene, G., 2022. Adsorption behaviour of pollutants: Heavy metals, radionuclides, organic pollutants, on clays and their minerals (raw, modified and treated): A review. *J. Environ. Manage*. 309, 114685. https://doi.org/10.1016/j.jenvman.2022.114685

Palansooriya, K.N., Shaheen, S.M., Chen, S.S., Tsang, D.C.W., Hashimoto, Y., Hou, D., Bolan, N.S., Rinklebe, J., Ok, Y.S., 2020. Soil amendments for immobilization of potentially toxic elements in contaminated soils: A critical review. *Environ. Int*. 134, 105046. https://doi.org/10.1016/j.envint.2019.105046

Park, H.J., Kim, S.U., Jung, K.Y., Lee, S., Choi, Y.D., Owens, V.N., Kumar, S., Yun, S.W., Hong, C.O., 2021. Cadmium phytoavailability from 1976 through 2016: Changes in soil amended with phosphate fertilizer and compost. *Sci. Total Environ*. 762, 143132. https://doi.org/10.1016/j.scitotenv.2020.143132

Peng, H., Chen, Y., Weng, L., Ma, J., Ma, Y., Li, Y., Islam, Md.S., 2019. Comparisons of heavy metal input inventory in agricultural soils in North and South China: A review. *Sci. Total Environ*. 660, 776–786. https://doi.org/10.1016/j.scitotenv.2019.01.066

Pinson, S.R.M., Tarpley, L., Yan, W., Yeater, K., Lahner, B., Yakubova, E., Huang, X.-Y., Zhang, M., Guerinot, M.L., Salt, D.E., 2015. Worldwide genetic diversity for mineral element concentrations in rice grain. *Crop Science* 55, 294–311. https://doi.org/10.2135/cropsci2013.10.0656

Pothier, M.P., Lenoble, V., Garnier, C., Misson, B., Rentmeister, C., Poulain, A.J., 2020. Dissolved organic matter controls of arsenic bioavailability to bacteria. *Sci. Total Environ*. 716, 137118. https://doi.org/10.1016/j.scitotenv.2020.137118

Punshon, T., Jackson, B.P., Meharg, A.A., Warczack, T., Scheckel, K., Guerinot, M.L., 2017. Understanding arsenic dynamics in agronomic systems to predict and prevent uptake by crop plants. *Sci. Total Environ*. 581–582, 209–220. https://doi.org/10.1016/j.scitotenv.2016.12.111

Qi, X., Tam, N.F., Li, W.C., Ye, Z., 2020. The role of root apoplastic barriers in cadmium translocation and accumulation in cultivars of rice (*Oryza sativa* L.) with different Cd-accumulating characteristics. *Environ. Pollut*. 264, 114736. https://doi.org/10.1016/j.envpol.2020.114736

Qiao, J., Li, X., Hu, M., Li, F., Young, L.Y., Sun, W., Huang, W., Cui, J., 2018a. Transcriptional activity of arsenic-reducing bacteria and genes regulated by lactate and biochar during arsenic transformation in flooded paddy soil. *Environ. Sci. Technol*. 52, 61–70. https://doi.org/10.1021/acs.est.7b03771

Qiao, J., Liu, T., Wang, X., Li, F., Lv, Y., Cui, J., Zeng, X., Yuan, Y., Liu, C., 2018b. Simultaneous alleviation of cadmium and arsenic accumulation in rice by applying zero-valent iron and biochar to contaminated paddy soils. *Chemosphere* 195, 260–271. https://doi.org/10.1016/j.chemosphere.2017.12.081

Qiu, H., Gu, H.-H., He, E.-K., Wang, S.-Z., Qiu, R.-L., 2012. Attenuation of metal bioavailability in acidic multi-metal contaminated soil treated with fly ash and steel slag. *Pedosphere* 22, 544–553. https://doi.org/10.1016/S1002-0160(12)60039-3

Raju, N.J., 2022. Arsenic in the geo-environment: A review of sources, geochemical processes, toxicity and removal technologies. *Environ. Res*. 111782.

Rinklebe, J., Shaheen, S.M., Yu, K., 2016. Release of As, Ba, Cd, Cu, Pb, and Sr under pre-definite redox conditions in different rice paddy soils originating from the U.S.A. and Asia. *Geoderma* 270, 21–32. https://doi.org/10.1016/j.geoderma.2015.10.011

Rizwan, M., Ali, S., Adrees, M., Rizvi, H., Zia-ur-Rehman, M., Hannan, F., Qayyum, M.F., Hafeez, F., Ok, Y.S., 2016. Cadmium stress in rice: Toxic effects, tolerance mechanisms, and management: A critical review. *Environ. Sci. Pollut. Res*. 23, 17859–17879. https://doi.org/10.1007/s11356-016-6436-4

Sarkar, A., Biswajit, P., 2016. The global menace of arsenic and its conventional remediation - A critical review. *Chemosphere* 37–49.

Sarwar, T., Khan, S., Muhammad, S., Amin, S., 2021. Arsenic speciation, mechanisms, and factors affecting rice uptake and potential human health risk: A systematic review. *Environ. Technol. I*nnov. 22, 101392. https://doi.org/10.1016/j.eti.2021.101392

Sebastian, A., Prasad, M.N.V., 2014. Cadmium minimization in rice. A review. *Agron. Sustain. Dev*. 34, 155–173. https://doi.org/10.1007/s13593-013-0152-y

Seyfferth, A.L., Morris, A.H., Gill, R., Kearns, K.A., Mann, J.N., Paukett, M., Leskanic, C., 2016. Soil incorporation of silica-rich rice husk decreases inorganic arsenic in rice grain. *J. Agric. Food Chem*. 64, 3760–3766. https://doi.org/10.1021/acs.jafc.6b01201

Shaheen, S.M., Rinklebe, J., 2015. Impact of emerging and low cost alternative amendments on the (im)mobilization and phytoavailability of Cd and Pb in a contaminated floodplain soil. *Ecol. Eng*. 74, 319–326. https://doi.org/10.1016/j.ecoleng.2014.10.024

Shen, F., Liu, J., Zhang, Z., Dai, J., 2015. On-line analysis and kinetic behavior of arsenic release during coal combustion and pyrolysis. *Environ. Sci. Technol.* 49, 13716–13723. https://doi.org/10.1021/acs.est.5b03626

Shi, J., Du, P., Luo, H., Wu, H., Zhang, Y., Chen, J., Wu, M., Xu, G., Gao, H., 2022. Soil contamination with cadmium and potential risk around various mines in China during 2000–2020. *J. Environ. Manage.* 310, 114509. https://doi.org/10.1016/j.jenvman.2022.114509

Shi, K., Wang, Q., Wang, G., 2020. Microbial oxidation of arsenite: regulation, chemotaxis, phosphate metabolism and energy generation. *Front. Microbiol.* 11, 569282. https://doi.org/10.3389/fmicb.2020.569282

Shi, M., Min, X., Ke, Y., Lin, Z., Yang, Z., Wang, S., Peng, N., Yan, X., Luo, S., Wu, J., Wei, Y., 2021. Recent progress in understanding the mechanism of heavy metals retention by iron (oxyhydr)oxides. *Sci. Total Environ.* 752, 141930. https://doi.org/10.1016/j.scitotenv.2020.141930

Shi, W., Song, W., Luo, Y., Qile, G., Zheng, J., Lü, C., He, J., 2021a. Transformation pathways of arsenic species: SRB mediated mechanism and seasonal patterns. *Chemosphere* 263, 128255. https://doi.org/10.1016/j.chemosphere.2020.128255

Shi, W., Song, W., Zheng, J., Luo, Y., Qile, G., Lü, S., Lü, X., Zhou, B., Lü, C., He, J., 2021b. Factors and pathways regulating the release and transformation of arsenic mediated by reduction processes of dissimilated iron and sulfate. *Sci. Total Environ.* 768, 144697. https://doi.org/10.1016/j.scitotenv.2020.144697

Singh, P.K., Tripathi, P., Dwivedi, S., Awasthi, S., Shri, M., Chakrabarty, D., Tripathi, R.D., 2016. Fly-ash augmented soil enhances heavy metal accumulation and phytotoxicity in rice (Oryza sativa L.); A concern for fly-ash amendments in agriculture sector. *Plant Growth Regul.* 78, 21–30. https://doi.org/10.1007/s10725-015-0070-x

Singh, R.P., Agrawal, M., 2008. Potential benefits and risks of land application of sewage sludge. *Waste Manag.* 28, 347–358. https://doi.org/10.1016/j.wasman.2006.12.010

Smedley, P.L., Kinniburgh, D.G., 2002. A review of the source, behaviour and distribution of arsenic in natural waters. *Appl. Geochem.* 17, 517–568. https://doi.org/10.1016/S0883-2927(02)00018-5

Somenahally, A.C., Hollister, E.B., Yan, W., Gentry, T.J., Loeppert, R.H., 2011. Water management impacts on arsenic speciation and iron-reducing bacteria in contrasting rice-rhizosphere compartments. *Environ. Sci. Technol.* 45, 8328–8335. https://doi.org/10.1021/es2012403

Song, Y., Wang, Y., Mao, W., Sui, H., Yong, L., Yang, D., Jiang, D., Zhang, L., Gong, Y., 2017. Dietary cadmium exposure assessment among the Chinese population. *PLOS ONE* 12, e0177978. https://doi.org/10.1371/journal.pone.0177978

Straif, K., Benbrahim-Tallaa, L., Baan, R., Grosse, Y., Secretan, B., El Ghissassi, F., Bouvard, V., Guha, N., Freeman, C., Galichet, L., Cogliano, V., 2009. A review of human carcinogens—Part C: Metals, arsenic, dusts, and fibres. *Lancet Oncol.* 10, 453–454. https://doi.org/10.1016/S1470-2045(09)70134-2

Suda, A., Makino, T., 2016. Functional effects of manganese and iron oxides on the dynamics of trace elements in soils with a special focus on arsenic and cadmium: A review. *Geoderma* 270, 68–75. https://doi.org/10.1016/j.geoderma.2015.12.017

Suda, A., Yamaguchi, N., Taniguchi, H., Makino, T., 2018. Arsenic immobilization in anaerobic soils by the application of by-product iron materials obtained from the casting industry. *Soil Sci. Plant Nutr.* 64, 210–217. https://doi.org/10.1080/00380768.2017.1385371

Sui, F.-Q., Chang, J.-D., Tang, Z., Liu, W.-J., Huang, X.-Y., Zhao, F.-J., 2018. Nramp5 expression and functionality likely explain higher cadmium uptake in rice than in wheat and maize. *Plant Soil* 433, 377–389. https://doi.org/10.1007/s11104-018-3849-5

Suriyagoda, L.D.B., Dittert, K., Lambers, H., 2018a. Arsenic in rice soils and potential agronomic mitigation strategies to reduce arsenic bioavailability: A review. *Pedosphere* 28, 363–382. https://doi.org/10.1016/S1002-0160(18)60026-8

Suriyagoda, L.D.B., Dittert, K., Lambers, H., 2018b. Mechanism of arsenic uptake, translocation and plant resistance to accumulate arsenic in rice grains. *Agric. Ecosyst. Environ.* 253, 23–37. https://doi.org/10.1016/j.agee.2017.10.017

Syu, C.-H., Wu, P.-R., Lee, C.-H., Juang, K.-W., Lee, D.-Y., 2019. Arsenic phytotoxicity and accumulation in rice seedlings grown in arsenic-contaminated soils as influenced by the characteristics of organic matter amendments and soils. *J. Plant Nutr. Soil Sci.* 182, 60–71. https://doi.org/10.1002/jpln.201800337

Tabelin, C.B., Igarashi, T., Villacorte-Tabelin, M., Park, I., Opiso, E.M., Ito, M., Hiroyoshi, N., 2018. Arsenic, selenium, boron, lead, cadmium, copper, and zinc in naturally contaminated rocks: A review of their sources, modes of enrichment, mechanisms of release, and mitigation strategies. *Sci. Total Environ.* 645, 1522–1553. https://doi.org/10.1016/j.scitotenv.2018.07.103

Tao, Y., Zhang, S., Jian, W., Yuan, C., Shan, X., 2006. Effects of oxalate and phosphate on the release of arsenic from contaminated soils and arsenic accumulation in wheat. *Chemosphere* 65, 1281–1287. https://doi.org/10.1016/j.chemosphere.2006.04.039

Tong, H., Liu, C., Hao, L., Swanner, E.D., Chen, M., Li, F., Xia, Y., Liu, Y., Liu, Y., 2019. Biological Fe(II) and As(III) oxidation immobilizes arsenic in micro-oxic environments. *Geochim. Cosmochim. Acta* 265, 96–108. https://doi.org/10.1016/j.gca.2019.09.002

Upadhyay, M.K., Shukla, A., Yadav, P., Srivastava, S., 2019. A review of arsenic in crops, vegetables, animals and food products. *Food Chem.* 276, 608–618. https://doi.org/10.1016/j.foodchem.2018.10.069

Uraguchi, S., Fujiwara, T., 2012. Cadmium transport and tolerance in rice: Perspectives for reducing grain cadmium accumulation. *Rice* 5, 5. https://doi.org/10.1186/1939-8433-5-5

Uraguchi, S., Fujiwara, T., 2013. Rice breaks ground for cadmium-free cereals. *Curr. Opin. Plant Biol.* 16, 328–334. https://doi.org/10.1016/j.pbi.2013.03.012

Viacava, K., Meibom, K.L., Ortega, D., Dyer, S., Gelb, A., Falquet, L., Minton, N.P., Mestrot, A., Bernier-Latmani, R., 2020. Variability in arsenic methylation efficiency across aerobic and anaerobic microorganisms. *Environ. Sci. Technol.* 54, 14343–14351. https://doi.org/10.1021/acs.est.0c03908

Vithanage, M., Herath, I., Joseph, S., Bundschuh, J., Bolan, N., Ok, Y.S., Kirkham, M.B., Rinklebe, J., 2017. Interaction of arsenic with biochar in soil and water: A critical review. *Carbon* 113, 219–230. https://doi.org/10.1016/j.carbon.2016.11.032

Wan, Y., Camara, A.Y., Yu, Y., Wang, Q., Guo, T., Zhu, L., Li, H., 2018. Cadmium dynamics in soil pore water and uptake by rice: Influences of soil-applied selenite with different water managements. *Environ. Pollut.* 240, 523–533. https://doi.org/10.1016/j.envpol.2018.04.044

Wang, C., Liu, H., Zhang, Y., Zou, C., Anthony, E.J., 2018. Review of arsenic behavior during coal combustion: Volatilization, transformation, emission and removal technologies. *Prog. Energy Combust. Sci.* 68, 1–28. https://doi.org/10.1016/j.pecs.2018.04.001

Wang, F., Peng, L., Zhou, X., Zeng, Q., Luo, S., 2021. Typical sources of Cd to paddy fields in different contaminated areas and their impacts on Cd accumulation in topsoil and rice in Changzhutan, China. *Environ. Res.* 193, 110523. https://doi.org/10.1016/j.envres.2020.110523

Wang, G., Hu, Z., Li, S., Wang, Y., Sun, X., Zhang, X., Li, M., 2020. Sulfur controlled cadmium dissolution in pore water of cadmium-contaminated soil as affected by DOC under waterlogging. *Chemosphere* 240, 124846. https://doi.org/10.1016/j.chemosphere.2019.124846

Wang, J., Wang, P.-M., Gu, Y., Kopittke, P.M., Zhao, F.-J., Wang, P., 2019. Iron–manganese (Oxyhydro)oxides, rather than oxidation of sulfides, determine mobilization of Cd during soil drainage in paddy soil systems. *Environ. Sci. Technol.* 53, 2500–2508. https://doi.org/10.1021/acs.est.8b06863

Wang, M.Y., Chen, A.K., Wong, M.H., Qiu, R.L., Cheng, H., Ye, Z.H., 2011. Cadmium accumulation in and tolerance of rice (Oryza sativa L.) varieties with different rates of radial oxygen loss. *Environ. Pollut.* 159, 1730–1736. https://doi.org/10.1016/j.envpol.2011.02.025

Wang, P., Chen, H., Kopittke, P.M., Zhao, F.-J., 2019. Cadmium contamination in agricultural soils of China and the impact on food safety. *Environ. Pollut.* 249, 1038–1048. https://doi.org/10.1016/j.envpol.2019.03.063

Wang, Q., Kang, Y., Alowaifeer, A., Shi, K., Fan, X., Wang, L., Jetter, J., Bothner, B., Wang, G., McDermott, T.R., 2018. Phosphate starvation response controls genes required to synthesize the phosphate analog arsenate. *Environ. Microbiol.* 20, 1782–1793. https://doi.org/10.1111/1462-2920.14108

Wang, R., Wei, S., Jia, P., Liu, T., Hou, D., Xie, R., Lin, Z., Ge, J., Qiao, Y., Chang, X., Lu, L., Tian, S., 2019. Biochar significantly alters rhizobacterial communities and reduces Cd concentration in rice grains grown on Cd-contaminated soils. *Sci. Total Environ.* 676, 627–638. https://doi.org/10.1016/j.scitotenv.2019.04.133

Wang, W., Zeng, C., Sardans, J., Wang, C., Zeng, D., Peñuelas, J., 2016. Amendment with industrial and agricultural wastes reduces surface-water nutrient loss and storage of dissolved greenhouse gases in a subtropical paddy field. *Agric. Ecosyst. Environ.*, 无 231, 296–303. https://doi.org/10.1016/j.agee.2016.07.012

Wang, X., Peng, B., Tan, C., Ma, L., Rathinasabapathi, B., 2015. Recent advances in arsenic bioavailability, transport, and speciation in rice. *Environ. Sci. Pollut. Res.* 22, 5742–5750. https://doi.org/10.1007/s11356-014-4065-3

Warke, M., Sarkar, D., Zhang, Z., Neve, S., Datta, R., 2022. Human health risk mitigation from arsenic in rice by crop rotation with a hyperaccumulator plant. *Environ. Sci. Pollut. Res.* 30, 12030–12040. https://doi.org/10.1007/s11356-022-22985-y

Watanabe, T., Shimbo, S., Moon, C.-S., Zhang, Z.-W., Ikeda, M., 1996. Cadmium contents in rice samples from various areas in the world. *Sci. Total Environ.* 184, 191–196. https://doi.org/10.1016/0048-9697(96)05100-5

Wen, Y., Li, W., Yang, Z., Zhuo, X., Guan, D.-X., Song, Y., Guo, C., Ji, J., 2020. Evaluation of various approaches to predict cadmium bioavailability to rice grown in soils with high geochemical background in the karst region, Southwestern China. *Environ. Pollut.* 258, 113645. https://doi.org/10.1016/j.envpol.2019.113645

Williams, P.N., Islam, M.R., Adomako, E.E., Raab, A., Hossain, S.A., Zhu, Y.G., Feldmann, J., Meharg, A.A., 2006. Increase in rice grain arsenic for regions of bangladesh irrigating paddies with elevated arsenic in groundwaters. *Environ. Sci. Technol.* 40, 4903–4908. https://doi.org/10.1021/es060222i

Williams, P.N., Lei, M., Sun, G., Huang, Q., Lu, Y., Deacon, C., Meharg, A.A., Zhu, Y.-G., 2009. Occurrence and partitioning of cadmium, arsenic and lead in mine impacted paddy rice: Hunan, China. *Environ. Sci. Technol.* 43, 637–642. https://doi.org/10.1021/es802412r

Williams, P.N., Zhang, H., Davison, W., Meharg, A.A., Hossain, M., Norton, G.J., Brammer, H., Islam, M.R., 2011. Organic matter—solid phase interactions are critical for predicting arsenic release and plant uptake in Bangladesh paddy soils. *Environ. Sci. Technol.* 45, 6080–6087. https://doi.org/10.1021/es2003765

Wu, C., An, W., Liu, Z., Lin, J., Qian, Z., Xue, S., 2020. The effects of biochar as the electron shuttle on the ferrihydrite reduction and related arsenic (As) fate. *J. Hazard. Mater.* 390, 121391. https://doi.org/10.1016/j.jhazmat.2019.121391

Wu, K., Wu, C., Jiang, X., Xue, R., Pan, W., Li, W.-C., Luo, X., Xue, S., 2022. Remediation of arsenic-contaminated paddy field by a new iron oxidizing strain (Ochrobactrum sp.) and iron-modified biochar. *J. Environ. Sci.* 115, 411–421. https://doi.org/10.1016/j.jes.2021.08.009

Wu, X., Bowers, B., Kim, D., Lee, B., Jun, Y.-S., 2019. Dissolved organic matter affects arsenic mobility and iron(III) (hydr)oxide formation: implications for managed aquifer recharge. *Environ. Sci. Technol.* 53, 14357–14367. https://doi.org/10.1021/acs.est.9b04873

Xiang, H., Lan, N., Wang, F., Zhao, B., Wei, H., Zhang, J., 2023. An effective planting model to decrease cadmium accumulation in rice grains and plants: Intercropping rice with wetland plants. *Pedosphere* 33, 355–364. https://doi.org/10.1016/j.pedsph.2022.06.054

Xiao, A., Ouyang, Y., Li, W.C., Ye, Z., 2017. Effect of organic manure on Cd and As accumulation in brown rice and grain yield in Cd-As-contaminated paddy fields. *Environ. Sci. Pollut. Res.* 24, 9111–9121. https://doi.org/10.1007/s11356-017-8460-4

Xu, L., Wu, X., Wang, S., Yuan, Z., Xiao, F., Yang, M., Jia, Y., 2016a. Speciation change and redistribution of arsenic in soil under anaerobic microbial activities. *J. Hazard. Mater.* 301, 538–546. https://doi.org/10.1016/j.jhazmat.2015.09.030

Xu, L., Wu, X., Wang, S., Yuan, Z., Xiao, F., Yang, M., Jia, Y., 2016b. Speciation change and redistribution of arsenic in soil under anaerobic microbial activities. *J. Hazard. Mater.* 301, 538–546. https://doi.org/10.1016/j.jhazmat.2015.09.030

Xu, X., Chen, C., Wang, P., Kretzschmar, R., Zhao, F.-J., 2017. Control of arsenic mobilization in paddy soils by manganese and iron oxides. *Environ. Pollut.* 231, 37–47. https://doi.org/10.1016/j.envpol.2017.07.084

Xu, Y., Feng, J., Li, H., 2021. How intercropping and mixed systems reduce cadmium concentration in rice grains and improve grain yields. *J. Hazard. Mater.* 402, 123762. https://doi.org/10.1016/j.jhazmat.2020.123762

Xu, Y., Liang, X., Xu, Y., Qin, X., Huang, Q., Wang, L., Sun, Y., 2017. Remediation of heavy metal-polluted agricultural soils using clay minerals: A review. *Pedosphere* 27, 193–204. https://doi.org/10.1016/S1002-0160(17)60310-2

Xu, Y.G., Yu, W.T., Ma, Q., Zhou, H., 2015. Potential risk of cadmium in a soil-plant system as a result of long-term (10 years) pig manure application. *Plant Soil Environ.* 61, 352–357. https://doi.org/10.17221/100/2015-PSE

Yamaguchi, N., Ohkura, T., Hikono, A., Yamaguchi, H., Hashimoto, Y., Makino, T., 2017. Effects of iron amendments on the speciation of arsenic in the rice rhizosphere after drainage. *Soils* 1, 6. https://doi.org/10.3390/soils1010006

Yang, T., Chen, M.-L., Liu, L.-H., Wang, J.-H., Dasgupta, P.K., 2012. Iron(III) modification of *Bacillus subtilis* membranes provides record sorption capacity for arsenic and endows unusual selectivity for As(V). *Environ. Sci. Technol.* 46, 2251–2256. https://doi.org/10.1021/es204034z

Yang, Y., Wang, T., Li, Y., Wang, M., Chen, W., Dai, Y., 2022. Mitigating cadmium contamination of rice soils supporting tobacco–rice rotation in southern China: Win–win or lose–lose? *J. Hazard. Mater.* 425, 128052. https://doi.org/10.1016/j.jhazmat.2021.128052

Yang, Z., Xu, M., Zheng, S., Nie, J., Gao, J., Liao, Y., Xie, J., 2012. Effects of long-term winter planted green manure on physical properties of reddish paddy soil under a double-rice cropping system. *J. Integr. Agric.* 11, 655–664. https://doi.org/10.1016/S2095-3119(12)60053-7

Yao, B.-M., Wang, S.-Q., Xie, S.-T., Li, G., Sun, G.-X., 2022. Optimal soil Eh, pH for simultaneous decrease of bioavailable Cd, As in co-contaminated paddy soil under water management strategies. *Sci. Total Environ.* 806, 151342. https://doi.org/10.1016/j.scitotenv.2021.151342

Yi, K., Fan, W., Chen, J., Jiang, S., Huang, S., Peng, L., Zeng, Q., Luo, S., 2018. Annual input and output fluxes of heavy metals to paddy fields in four types of contaminated areas in Hunan Province, China. *Sci. Total Environ.* 634, 67–76. https://doi.org/10.1016/j.scitotenv.2018.03.294

Yu, G., Zheng, W., Wang, W., Dai, F., Zhang, Z., Yuan, Y., Wang, Q., 2017. Health risk assessment of Chinese consumers to Cadmium via dietary intake. *J. Trace Elem. Med. Biol.* 44, 137–145. https://doi.org/10.1016/j.jtemb.2017.07.003

Yu, H., Wang, J., Fang, W., Yuan, J., Yang, Z., 2006. Cadmium accumulation in different rice cultivars and screening for pollution-safe cultivars of rice. *Sci. Total Environ.* 370, 302–309. https://doi.org/10.1016/j.scitotenv.2006.06.013

Yu, H., Zhong, D., Zeng, H., Huang, B., Wang, X., Peng, B., Xing, B., 2023. Can simultaneous immobilization of arsenic and cadmium in paddy soils be achieved by liming? Environ. *Sci. Pollut. Res.* 30, 73677–73687. https://doi.org/10.1007/s11356-023-27536-7

Yu, H.-Y, Li, F.-B., Liu, C.-S., Huang, W., Liu, T.-X., Yu, W.-M., 2016a. Iron redox cycling coupled to transformation and immobilization of heavy metals: Implications for paddy rice safety in the red soil of South China. *Adv. Agron.* 6, 279–317. https://doi.org/10.1016/bs.agron.2015.12.006

Yu, H.-Y., Liu, C., Zhu, J., Li, F., Deng, D.-M., Wang, Q., Liu, C., 2016b. Cadmium availability in rice paddy fields from a mining area: The effects of soil properties highlighting iron fractions and pH value. *Environ. Pollut.* 209, 38–45. https://doi.org/10.1016/j.envpol.2015.11.021

Yu, L., Zhu, J., Huang, Q., Su, D., Jiang, R., Li, H., 2014. Application of a rotation system to oilseed rape and rice fields in Cd-contaminated agricultural land to ensure food safety. *Ecotoxicol. Environ. Saf.* 108, 287–293. https://doi.org/10.1016/j.ecoenv.2014.07.019

Yuan, C., Li, F., Cao, W., Yang, Z., Hu, M., Sun, W., 2019. Cadmium solubility in paddy soil amended with organic matter, sulfate, and iron oxide in alternative watering conditions. *J. Hazard. Mater.* 378, 120672. https://doi.org/10.1016/j.jhazmat.2019.05.065

Yuan, C., Li, Q., Sun, Z., Sun, H., 2021. Effects of natural organic matter on cadmium mobility in paddy soil: A review. *J. Environ. Sci.* 104, 204–215. https://doi.org/10.1016/j.jes.2020.11.016

Yuan, C., Liu, T., Li, F., Liu, C., Yu, H., Sun, W., Huang, W., 2018. Microbial iron reduction as a method for immobilization of a low concentration of dissolved cadmium. *J. Environ. Manage.* 217, 747–753. https://doi.org/10.1016/j.jenvman.2018.04.023

Yuan, S., Xiao, X., Guo, C., 2012. Regional distribution of cadmium minerals and risk assessment for potential cadmium pollution of soil in China. *J. Environ. Poll. Cont.* 34, 51–56. https://doi.org/10.15985/j.cnki.1001-3865.2012.06.021

Yuan, X., Xue, N., Han, Z., 2021. A meta-analysis of heavy metals pollution in farmland and urban soils in China over the past 20 years. *J. Environ. Sci.* 101, 217–226. https://doi.org/10.1016/j.jes.2020.08.013

Zecchin, S., Corsini, A., Martin, M., Cavalca, L., 2017. Influence of water management on the active root-associated microbiota involved in arsenic, iron, and sulfur cycles in rice paddies. *Appl. Microbiol. Biotechnol.* 101, 6725–6738. https://doi.org/10.1007/s00253-017-8382-6

Zhang, C., Wu, P., Tang, C., Tao, X., Han, Z., Sun, J., Liu, H., 2013. The study of soil acidification of paddy field influenced by acid mine drainage. *Environ. Earth Sci.* 70, 2931–2940. https://doi.org/10.1007/s12665-013-2351-x

Zhang, H., Zhang, X., Li, T., Huang, F., 2014. Variation of cadmium uptake, translocation among rice lines and detecting for potential cadmium-safe cultivars. *Environ. Earth Sci.* 71, 277–286. https://doi.org/10.1007/s12665-013-2431-y

Zhang, S., Zhang, J., Niu, L., Chen, Q., Zhou, Q., Xiao, N., Man, J., Ma, J., Wei, C., Zhang, S., Luo, Y., Yao, Y., 2024. Escalating arsenic contamination throughout Chinese soils. *Nat. Sustain.* https://doi.org/10.1038/s41893-024-01341-7

Zhang, S.-Y., Williams, P.N., Luo, J., Zhu, Y.-G., 2017. Microbial mediated arsenic biotransformation in wetlands. *Front. Environ. Sci. Eng.* 11, 1. https://doi.org/10.1007/s11783-017-0893-y

Zhang, S.-Y., Zhao, F.-J., Sun, G.-X., Su, J.-Q., Yang, X.-R., Li, H., Zhu, Y.-G., 2015. Diversity and abundance of arsenic biotransformation genes in paddy soils from Southern China. *Environ. Sci. Technol.* 49, 4138–4146. https://doi.org/10.1021/acs.est.5b00028

Zhang, X., Chen, D., Zhong, T., Zhang, X., Cheng, M., Li, X., 2015. Assessment of cadmium (Cd) concentration in arable soil in China. *Environ. Sci. Pollut. Res.* 22, 4932–4941. https://doi.org/10.1007/s11356-014-3892-6

Zhang, X., Wang, H., He, L., Lu, K., Sarmah, A., Li, J., Bolan, N.S., Pei, J., Huang, H., 2013. Using biochar for remediation of soils contaminated with heavy metals and organic pollutants. *Environ. Sci. Pollut. Res.* 20, 8472–8483. https://doi.org/10.1007/s11356-013-1659-0

Zhao, F.J., Ma, J.F., Meharg, A.A., McGrath, S.P., 2009. Arsenic uptake and metabolism in plants. *New Phytol.* 181, 777–794. https://doi.org/10.1111/j.1469-8137.2008.02716.x

Zhao, F.-J., Ma, Y., Zhu, Y.-G., Tang, Z., McGrath, S.P., 2015. Soil contamination in China: Current status and mitigation strategies. *Environ. Sci. Technol.* 49, 750–759. https://doi.org/10.1021/es5047099

Zhao, F.-J., McGrath, S.P., Meharg, A.A., 2010. Arsenic as a food chain contaminant: Mechanisms of plant uptake and metabolism and mitigation strategies. *Annu. Rev. Plant Biol.* 61, 535–559. https://doi.org/10.1146/annurev-arplant-042809-112152

Zhao, F.-J., Wang, P., 2020. Arsenic and cadmium accumulation in rice and mitigation strategies. *Plant Soil* 446, 1–21. https://doi.org/10.1007/s11104-019-04374-6

Zhao, F.-J., Zhu, Y.-G., Meharg, A.A., 2013. Methylated arsenic species in rice: Geographical variation, origin, and uptake mechanisms. *Environ. Sci. Technol.* 47, 3957–3966. https://doi.org/10.1021/es304295n

Zhao, Z., Wang, S., Jia, Y., 2017. Effect of sulfide on As(III) and As(V) sequestration by ferrihydrite. *Chemosphere* 185, 321–328. https://doi.org/10.1016/j.chemosphere.2017.06.134

Zheng, M.-Z., Li, G., Sun, G.-X., Shim, H., Cai, C., 2013b. Differential toxicity and accumulation of inorganic and methylated arsenic in rice. *Plant Soil* 365, 227–238. https://doi.org/10.1007/s11104-012-1376-3

Zheng, R., Sun, G., Zhu, Y., 2013a. Effects of microbial processes on the fate of arsenic in paddy soil. *Chin. Sci. Bull.* 58, 186–193. https://doi.org/10.1007/s11434-012-5489-0

Zhou, Y., Niu, L., Liu, K., Yin, S., Liu, W., 2018. Arsenic in agricultural soils across China: Distribution pattern, accumulation trend, influencing factors, and risk assessment. *Sci. Total Environ.* 616–617, 156–163. https://doi.org/10.1016/j.scitotenv.2017.10.232

Zhu, H., Chen, C., Xu, C., Zhu, Q., Huang, D., 2016. Effects of soil acidification and liming on the phytoavailability of cadmium in paddy soils of central subtropical China. *Environ. Pollut.* 219, 99–106. https://doi.org/10.1016/j.envpol.2016.10.043

Zou, M., Zhou, S., Zhou, Y., Jia, Z., Guo, T., Wang, J., 2021. Cadmium pollution of soil-rice ecosystems in rice cultivation dominated regions in China: A review. *Environ. Pollut.* 280, 116965. https://doi.org/10.1016/j.envpol.2021.116965

Section V

Impacts on Human Health and Development

18

Cancer, Kidney, and Bone Diseases Caused by Long-Term Consumption of Crops Containing High Cadmium Levels

Anwen Xiao, Zhihong Ye, and Yichun Li

18.1 Introduction

As a non-essential metal, cadmium (Cd) does not have a physiological function in the human body. In marine diatoms, Cd is a key component of Cd-containing carbonic anhydrase, which participates in the assimilation of inorganic carbon (Lane et al., 2005). Cd is also a typical representative of heavy metals because of its toxicity and considerable contamination (pollution) area. Cd causes persistent damage following entry into the human body because of its long half-life and lack of degradability.

Cd from dietary sources accounts for 96.5% of the total Cd exposure of the general population. By contrast, food intake accounts for 62.5% of the total Cd exposure for smokers, indicating that dietary Cd intake is the primary pathway of Cd intake (Zhu et al., 2024). Cd exposure can only be attributed to dietary intake for the population who do not smoke or have occupational exposure (Charkiewicz et al., 2023). However, completely avoiding dietary Cd intake is challenging. According to the Joint FAO/WHO Expert Committee on Food Additives (JECFA), Cd-contaminated food is a problem in more than 30 countries and regions in Asia, Africa, America, and Europe, affecting billions of people worldwide (JECFA, 2023).

The average dietary Cd intake range from various nationalities is 4.6–34.6 μg/day (Zhao et al., 2023). Although the average dietary Cd intake of the population in most countries is lower than the provisional tolerable weekly intake (PTWI) set by the European Food Safety Authority (EFSA) [2.5 μg/kg bw/week, 21.4 μg/day (60 kg)] and JECFA [25 μg/kg bw/month, 50 μg/day (60 kg)], chronic Cd exposure at low levels may result in an elevated risk of cancer (Åkesson et al., 2008; Julin et al., 2012; Zhao et al., 2023). Notably, considerable disparities exist in the levels of dietary Cd intake among geographical regions within a country. Dietary Cd intake by populations living in Cd-contaminated regions may be higher than the PTWI, especially in mine, smelting, and metal recycling areas (Horiguchi et al., 2004; Ngo et al., 2012; Zhu et al., 2016). Compared with other foods, cereals, cereal-based products, and vegetables have been consistently identified as the principal sources of dietary Cd intake (JECFA, 2023).

Due to the propensity for long-term Cd intake and the long half-life of Cd (highly persistent toxicity), the health effects of consuming Cd-contaminated foods must be seriously considered. Once Cd enters the human body, it exhibits a high propensity for retention, particularly within the kidneys and liver, where it can accumulate over extended periods, continuously damaging the body. It is difficult for the human body to remove Cd, and its half-life is 10–30 years (Genchi et al., 2020). The Cd content in the body may increase continuously owing to long-term (persistent) dietary Cd intake and the long half-life of Cd, which may result in more vital damage to the human body.

Low-level but long-term dietary Cd exposure causes kidney and bone injury and may result in cancer (Clemens et al., 2013; Tinkov et al., 2018). Cd has been designated as a human group 1 carcinogen, and Cd exposure is known to be associated with gastrointestinal tract, prostate, kidney, and nasopharynx cancers (Genchi et al., 2020; WHO, 2019). Cd exposure may also result in liver dysfunction, testicular

DOI: 10.1201/9781003492115-23

injury, and adrenal and hematopoietic system damage, although kidney and bone diseases and cancer are the primary diseases caused by Cd exposure (Clemens et al., 2013; Tinkov et al., 2018).

Dietary Cd intake (especially Cd-contaminated crops) is the primary pathway of human Cd intake, and Cd-contaminated crops have affected billions of people. Dietary Cd exposure can cause cancer, kidney, and bone diseases. However, comprehensive and systematic descriptions of Cd-contaminated crops' relationship with cancer, kidney, and bone diseases are rare. Dietary recommendations for reducing Cd exposure and damage are scattered and unsystematic. The objectives of this chapter are to systematically reveal the health risks associated with the consumption of Cd-contaminated crops and their pathogenic mechanisms, to summarize strategies to protect humans from dietary Cd exposure and damage, and to provide some related dietary recommendations. In this chapter, the intake of Cd and its fate in the human body are introduced, and then typical examples, epidemiological evidence, and pathogenic mechanisms of three Cd-induced diseases (cancer, kidney, and bone disease) are mentioned. Finally, dietary recommendations based on three strategies are provided, with the aim to reduce Cd exposure and damage.

18.2 Cd Intake and Fate in the Human Body

18.2.1 Principal Pathways of Human Cd Exposure

18.2.1.1 Crop Consumption, Especially Rice

Excluding occupational exposure and tobacco smoke inhalation, dietary Cd is the principal source of Cd exposure in humans. There are three main types of human Cd exposure: diet, respiration, and skin contact. The Cd level in the air is generally below 0.01 $\mu g/m^3$, and the principal source of air exposure to Cd is tobacco use, with respiratory Cd intake considered mainly related to smoking (Reyes-Hinojosa et al., 2019; Taha et al., 2018). Dietary Cd ingestion accounts for 96.5% of the total Cd exposure of the general public (Zhu et al., 2024), indicating that Cd-contaminated food is the predominant source of human Cd exposure.

Compared with other foods, cereals, and cereal-based products are the primary sources of dietary Cd exposure. In China, more than 67.0% of dietary Cd intake is derived from cereals, with rice and wheat being the main sources (Song et al., 2017; Zhao et al., 2022b). After surveying more than 30 countries and regions covering Asia, Africa, America, and Europe, the JECFA evaluated the estimates of dietary Cd exposure at their 73rd meeting and reported that cereals, cereal products, and vegetables were consistently identified as major Cd sources (JECFA, 2023).

Dietary Cd exposure is contingent upon two principal factors: the Cd concentration in the food and the amount consumed. This suggests that the diet structure can also influence the primary contributing factors to dietary Cd exposure. At the national level, the average levels of Cd in rice, cereals, potatoes, vegetables, and fruits were determined to be approximately equivalent (about 0.02 mg/kg) (Zhao et al., 2023). Considering the consumption rate, cereals and grains are the principal sources of dietary Cd exposure in numerous countries. Cereals and cereal products are staple foods in the United States and many European countries. In the US, cereals and bread accounted for the most significant dietary Cd intake, accounting for approximately 34.0% (Kim et al., 2018).

Among European countries, grains and their products were identified as the most significant contributors, accounting for 26.6% of dietary Cd intake (EFSA, 2012). In the UK, the "miscellaneous cereals" and bread food groups were the primary sources of dietary Cd intake (FSA, 2015). Rice is the staple food in numerous Asian countries and represents the most significant source of dietary Cd. Specifically, in South Korea, Japan, China, and Bangladesh, rice accounts for 31.0%, 44.0%, 56.0%, and 58.0% of dietary Cd intake, respectively (Al-Rmalli et al., 2012; Kim and Wolt, 2011; Song et al., 2017; Watanabe et al., 2000).

18.2.1.2 Spatial Arrangement and Chemical Speciation of Cd within Rice Grain

Bioaccessible Cd in rice may accurately represent potential consumption risks more than total Cd content (Praveena and Omar, 2017; Zhuang et al., 2016). The total dietary Cd intake (external exposure dose) is not the dose that reaches the tissue after digestion and absorption, resulting in toxic effects (internal exposure dose) (Wang, 2019). Only a fraction of Cd is absorbed upon consumption of Cd-contaminated food, not the entire amount present in the food (Mao et al., 2022). The concepts of oral bioavailability and oral bioaccessibility were proposed to describe the risk of Cd-contaminated foods better. The oral bioavailability of Cd in food refers to the Cd fraction taken up and transported into the systemic circulation (Li et al., 2019; Oomen et al., 2002; Ruby et al., 1999). The bioaccessible Cd fraction in food represents the portion that dissolves within the gastrointestinal environment and is thus available for assimilation (Li et al., 2019; Oomen et al., 2002; Ruby et al., 1999). Therefore, a more precise and realistic human health risk evaluation of dietary Cd exposure can be obtained by considering the bioaccessibility and bioavailability of Cd in food rather than the total Cd concentration in the food (Xiao et al., 2024; Yao et al., 2021).

The rice substrate mentioned in this chapter comprises all components of rice except Cd and its physical and chemical properties (nutritional composition and their physical and chemical properties), which can affect Cd bioaccessibility and bioavailability. The rice substrate will influence Cd bioaccessibility and bioavailability, with significant differences among different cultivars (Lin et al., 2022; Yao et al., 2021). A previous study reported a significant positive correlation between total Cd levels and bioaccessible Cd fractions, suggesting that elevated Cd levels in grains are consistently associated with higher Cd bioaccessibility (Liu et al., 2021a). However, other studies have revealed that elevated total Cd levels in rice grains do not invariably increase bioaccessible Cd (Tavarez et al., 2023). These inconsistent results may be related to the different rice substrates used across studies.

It is crucial to recognize that rice Cd bioaccessibility and bioavailability can vary significantly depending on the assay methods used, and there are no consistent assay methods, leading to inconsistent results regarding Cd bioaccessibility and bioavailability in rice (Li et al., 2019; Xiao et al., 2024). The Cd bioavailability of rice, as determined by mouse liver and kidney uptake, varied from 21.2% to 67.5%. In contrast, when assessed using an *in vitro* digestion method, the Cd bioavailability of rice ranged from 7.5% to 39.7%, with an average value of 23.9% (Deng et al., 2022; Xiao et al., 2024). Moreover, cooked rice grains' in vitro Cd bioaccessibility was reported as 50.0% (Peng et al., 2020).

The spatial arrangement and chemical speciation of Cd in rice grains remain largely unknown but are crucial factors for assessing the bioavailability of Cd to humans. Some studies have shown that the distribution of Cd in rice grains is not uniform, with Cd predominantly accumulated in the brain, cortex, and embryonic parts of rice (Feng et al., 2022a; Huo et al., 2016). However, another research found two distinct patterns of Cd distribution: one with a non-uniform distribution (concentrated in the aleurone stratum and external starchy endosperm) and the other with a uniform distribution (Gu et al., 2020). Differences in Cd distribution within rice grains are attributed to temporal changes in the availability of Cd from soil porewater throughout the grain-filling stage (Gu et al., 2020).

Cd is generally unevenly distributed in rice grains (Huo et al., 2016). In brown rice, the Cd concentration was found to be highest in the outermost layer (0.52 mg/kg) and lowest in the innermost layer (0.29 mg/kg), which mirrors the distribution pattern of protein content within the rice grains (Feng et al., 2019). Moreover, the alkali-extracted proteins and enzyme-extracted proteins (ERPs) exhibited Cd concentrations of 2.70 and 1.85 mg/kg, respectively, which were significantly higher than those in rice powder (0.30 mg/kg), suggesting that proteins are the primary components binding to Cd (Feng et al., 2019).

Scanning electron microscopy revealed that ERPs and prolamins displayed compact, lamellar structures, suggesting that most Cd is associated with the protein surface (Feng et al., 2019). In rice grains, most Cd (66.0%–92.0%) is likely complexed with proteins rich in thiol groups, with the residual Cd found in the form of Cd-carboxyl compounds and Cd-histidine (Gu et al., 2020).

Compared with brown rice, the degree of milling of polished rice is higher, resulting in more rice germs and cortexes rich in protein, fat, vitamins, and minerals being removed. The Cd absorption rate of cooked polished rice was higher than that of cooked brown rice, indicating that the Cd bioaccessibility of cooked polished rice is higher than that of cooked brown rice (Liu et al., 2021b). The primary reason for this reduction in Cd absorption may be attributed to the higher content of macromolecules in brown rice, such as dietary fiber (Liu, 2023). These macromolecules can slow or prevent the complete release of Cd from the substrate. In addition, they may inhibit digestive enzyme reactions by binding with Cd, thereby reducing intestinal absorption of the metal. Concurrently, dietary fiber can stimulate intestinal peristalsis, leading to increased bowel movements and facilitating the excretion of Cd through feces (Liu, 2023).

18.2.2 Fate of Cd in the Human Body (Absorption, Accumulation, Excretion)

18.2.2.1 Absorption (Duodenum and Jejunum) and Transport (Bloodstream) of Cd

Diet-derived Cd is absorbed by the intestinal tract, with the duodenum and jejunum being the principal sites of Cd absorption (Figure 18.1) (Genchi et al., 2020; Khan et al., 2022). Kinetic studies have shown that dietary Cd is absorbed into duodenal enterocytes (Liu et al., 2021b).

The cellular transport systems for Cd remain elusive, although Fe and Zn transporters play crucial roles in Cd transportation (Cirovic and Satarug, 2024; Himeno and Fujishiro, 2021). As an apical Fe transporter, divalent metal transporterl (DMT1) is a proton-coupled and membrane potential-sensitive transport protein that can transport various divalent cations (Cirovic and Satarug, 2024). DMT1 has been identified at the apical membrane of enterocytes and is recognized for its significant role in Cd absorption (Branca et al., 2018; Cirovic and Satarug, 2024). The basolateral Fe exporter, metal transporter protein 1 (MTP1), is located at the basolateral membrane of enterocytes (Cirovic and Satarug, 2024; Fujiwara et al., 2020). After being absorbed into enterocytes by DMT1, Cd, and Fe are conveyed into the systemic circulation via the duodenal Fe exporter, MTP1 (Cirovic and Satarug, 2024; Fujiwara et al., 2020). DMT1 expression peaks in the proximal duodenum and gradually diminishes along the gastrointestinal tract, reaching its lowest levels in the distal colon. This gradient in DMT1 expression could explain why the duodenum is the primary site of Cd absorption (Zalups and Ahmad, 2003).

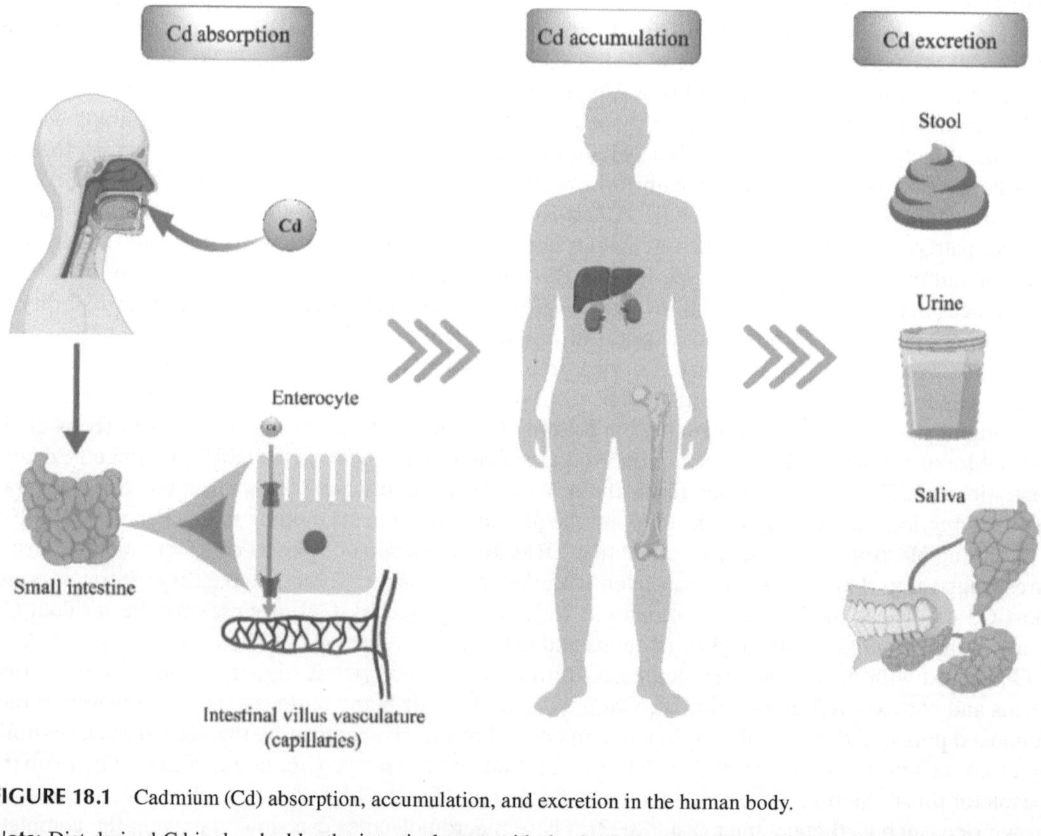

FIGURE 18.1 Cadmium (Cd) absorption, accumulation, and excretion in the human body.

Note: Diet-derived Cd is absorbed by the intestinal tract, with the duodenum and jejunum being the principal sites of Cd absorption. After entering the body, Cd accumulates mainly in the liver and kidney, resulting in cancer and kidney and bone diseases. The excretion of Cd from the body occurs slowly, with Cd exhibiting a biological half-life ranging from 10 to 30 years in humans. The main Cd excretion pathways are fecal, urinary, and salivary excretion, with fecal excretion being the primary route. Cd in the systemic circulation is primarily excreted via the urine, but this excretion pathway harms the kidneys.

In addition to Fe transporters, the Zrt-/Irt-like protein (ZIP) Zn transporters, ZIP8 and ZIP14, are also localized to the apical membrane of enterocytes; they cannot distinguish Zn^{2+}, Fe^{2+}, Mn^{2+}, and Cd^{2+} and have also been shown to be responsible for Cd transport into cells (Fujishiro and Himeno, 2019; Himeno and Fujishiro, 2021; Hu, 2021). ZIP8 and ZIP14 are responsible for Cd cellular uptake, whereas ZnT10 is involved in Cd cellular excretion (Himeno and Fujishiro, 2021).

After absorption by the intestine, dietary Cd is transported into the portal blood circulation (Branca et al., 2018; Cirovic and Satarug, 2024). The blood Cd concentration (BCd) indicates the extent of Cd exposure, particularly in cases of acute exposure. Some studies have suggested that, in individuals exposed to Cd, the BCd may vary from slightly above 0 μM up to 0.05 μM (Branca et al., 2018; Shi et al., 2023). Furthermore, human BCd significantly varies depending on age, sex, diet, and residential environment (Agency for Toxic Substances and Disease Registry USA, 2017). In the bloodstream, Cd is combined with the plasma protein albumin and is first conveyed into the liver via erythrocytes and albumin (Cirovic and Satarug, 2024; Genchi et al., 2020; Khan et al., 2022). Cd stimulates the synthesis of metallothionein (MT) proteins within hepatic cells. Cd binds to these MT proteins, forming Cd–MT complexes, which are transported to the kidneys via the bloodstream (Khan et al., 2022; Syed et al., 2023). Kinetic studies clarified that ^{109} Cd persisted in the duodenum after a single meal for up to 16 days. For the 64-day "chase" period, a portion of the ^{109} Cd from the turnover pool was redistributed, with some moving to the liver and kidneys (Chaney et al., 2004).

18.2.2.2 Cd Accumulation Mainly Occurs in the Liver and Kidney

In humans, Cd has a propensity to accumulate predominantly in the liver, kidneys, and bone, causing irreversible damage to these organs (Figure 18.1) (Genchi et al., 2020; Wang et al., 2021). Initially, Cd is concentrated in the liver, forming complexes with MT, which are then transported to other organs (Syed et al., 2023). Entry of Cd into the body results in Cd "enterohepatic circulation." The intestinal tract absorbs Cd from the diet, which is then conveyed to the liver via the blood. In the liver, Cd binds with glutathione (GSH) and cysteine, forming Cd-GSH and Cd-cysteine-binding compounds, which can be excreted into the bile and subsequently enter the intestine, where they may be reabsorbed by intestinal epithelial cells, a process known as enterohepatic circulation (Wang et al., 2021; Zhai et al., 2019).

The kidneys are the primary target organ for Cd accumulation and can reabsorb Cd. Cd toxicity can cause hepatic cell death, prompting the release of Cd–MT into the bloodstream. Cd–MT is conveyed to the kidneys, where it is filtered by glomeruli and subsequently reabsorbed by renal tubules, contributing to renal Cd accumulation (Wang et al., 2021). The renal proximal tubule, extending from the glomerulus to Henle's loop, is segmented into three distinct segments: S1 to S3. Within the S1 and S2 segments, Cd–MT is mainly reabsorbed by endocytosis (Fujishiro and Himeno, 2019). However, a portion of Cd^{2+} ions is released into the lumen of the proximal tubule from segments S1 and S2 or filtered by the glomerulus. These Cd^{2+} ions can be reabsorbed by epithelial cells in segment S3 of the proximal tubule, probably via ZIP8, which is highly expressed in the S3 segment (Fujishiro and Himeno, 2019; Himeno and Fujishiro, 2021). It is noteworthy that Cd dissociates into a free ionic form (more toxic than binding Cd) after being reabsorbed (Wang et al., 2021). Subsequently, the free ionic form of Cd recombines with the MT generated by the epithelial cells of the renal tubules and then further decomposes, resulting in significant damage to renal tubular cells (Khan et al., 2022; Wang et al., 2021). The presence of Cd in the kidneys impairs the reabsorption function of epithelial cells, ultimately resulting in proteinuria (Khan et al., 2022; Wang et al., 2021).

18.2.2.3 Excretion of Cd (Half-Life: 10–30 years)

The excretion of Cd from the body occurs slowly, with fecal, urinary, and salivary excretion being the main Cd excretion pathways (Figure 18.1) (Genchi et al., 2020; Liu et al., 2021b). Fecal excretion can remove Cd from the gastrointestinal tract but cannot remove Cd from the systemic circulation. Cd in the systemic circulation is excreted via urine, saliva, and milk (Genchi et al., 2020). Approximately 0.001%–0.005% of the body's Cd load is eliminated daily, with Cd exhibiting a biological half-life ranging from 10 to 30 years within the human body (Cirovic and Satarug, 2024; Shi et al., 2023; Wang et al., 2021).

Fecal excretion is the primary route of Cd excretion, the efficiency of which is influenced by the food substrate (Liu et al., 2021b). In a study on rats, Cd excreted via feces in the brown rice group (15.0%–55.1%) was found to be higher than that in the polished rice group (21.6%–27.4%), owing to the higher dietary fiber in brown rice promoting fecal excretion (Liu et al., 2021b).

Cd in the systemic circulation is primarily excreted via the urine, but this excretion pathway harms the kidneys. The excreted Cd originates from damaged renal tubular epithelial cells, and Cd excretion indicates ongoing tissue damage (Satarug et al., 2023). In the kidneys, Cd tends to accumulate preferentially in the proximal tubular epithelial cells, and upon cell death, Cd is released and subsequently excreted in the urine; thus, the urine Cd concentration (UCd) reflects renal Cd accumulation (long-term accumulated Cd burden on the body) and the current toxicity of Cd (Satarug, 2023; Shi et al., 2023). For postmenopausal women in the US, the mean corrected UCd concentration has been reported to be 0.73 μg/g creatinine; furthermore, high corrected UCd (>0.62 μg/mg creatinine) was positively correlated with cancer mortality (Shi et al., 2023). Moreover, genetics can influence the UCd through a quantitative trait locus (QTL) located on chromosome 16, which is heritable and can be attributed to genetic variation in ABCC1 (Grau-Perez et al., 2021). In summary, the UCd is a biomarker of Cd exposure and potential harm, and higher urinary excretion of Cd correlates with an increased health risk associated with Cd exposure.

The ATP-binding cassette (ABC) transporter is a primary cellular efflux pathway for Cd and Cd complexes, particularly ABCB6 and ABCC1/MRP1. It has been proven that Cd efflux occurs via ABCB6 and ABCC1/MRP1 (Rakvács et al., 2019; Thévenod and Lee, 2024; Tian et al., 2017). According to a QTL on chromosome 16, the genetic variation in ABCC1 could explain the difference in UCd levels among American Indians (Grau-Perez et al., 2021). Cd efflux via ABCB6 and ABCC1/MRP1 occurs when Cd binds to the tripeptide GSH (Thévenod and Lee, 2024). However, for other ABC transporters, only weak evidence of Cd efflux exists (Thévenod and Lee, 2024; Tian et al., 2017). For detoxification purposes, further studies are warranted to clarify the mechanisms of Cd efflux by other ABC transporters (including the specific chemical forms of Cd transported) (Thévenod and Lee, 2024). In addition to ABC transporters, ZnT1 may be related to Cd excretion, requiring further studies to clarify its effects on ZnT1 and its mechanisms (Satarug et al., 2021).

18.3 Diseases Caused by Long-Term Consumption of Crops with High Cd Levels

18.3.1 Cancer

18.3.1.1 "Cancer Villages" in China

Cd has been identified as a group I human carcinogen, and Cd exposure may lead to gastrointestinal tract, prostate, urinary bladder, kidney, and nasopharynx cancers (Genchi et al., 2020; WHO, 2019). A previous report indicated that Shangba (Wengyuan County, Guangdong Province, PR China) is severely polluted by Cd through both riverine and soil contamination, so much so that the area has been referred to as a "cancer village" by nongovernmental organizations (Larson, 2014). The cancer rates of the villagers in this village are higher than those of other villages owing to Cd-contaminated food (Jiang et al., 2021). In villages contaminated by Cd, the primary dietary Cd intake was from Cd-contaminated rice, contributing approximately 55.0% of the dietary Cd intake (Song et al., 2017; Zhao et al., 2023).

Some regions in other countries have high cancer rates, including Tsushima Island, Nagasaki, Japan; Shipham and Hutton villages in Somerset, UK; and the village of Gusum, Sweden, which may be due to Cd contamination of food (Arisawa et al., 2001; Elliott et al., 2000; Helmfrid et al., 2012). Moreover, the high consumption of Cd-containing rice poses significant health risks to Sri Lankans (Samarajeewa, 2022). In Cuttack district, India, the total carcinogenic risk posed by dietary Cd exceeded the threshold value (10^{-4}), indicating a potential cancer risk for both adults and children; for children, the oncogenic risk from Cd in rice alone was above this threshold, suggesting that Cd-contaminated rice poses a significant cancer risk to children (Mohanty et al., 2023).

18.3.1.2 Dietary Cd Intake Increases Cancer Risks, Mainly Gastrointestinal and Breast Cancers (Epidemiological Evidence)

Gastrointestinal cancers encompass a range of malignancies, including liver, pancreatic, esophagus, stomach, colon, and rectum cancers, while colorectal cancer is the most prevalent form. Dietary Cd intake may cause gastrointestinal cancers (Tavakoli Pirzaman et al., 2023) and breast cancer.

Cd has been linked to esophageal cancer. In Kazakhstan, a high incidence of esophageal cancer has been linked to both inhalation and oral Cd exposure (Mamyrbayev et al., 2016). Moreover, chronic Cd exposure has been associated with the development of esophageal squamous cell cancer, and Cd can also increase the chemoresistance of tumor cells (Chen et al., 2022). Within various age groups of Chinese community populations, soil Cd levels were positively correlated with an elevated risk of mortality from esophageal cancer (Jiang et al., 2021).

Gastric cancer is among the most common malignancies globally. Research indicates a correlation between gastric cancer and Cd exposure; elevated BCd levels have been consistently observed in individuals diagnosed with gastrointestinal malignancies (Ostadrahimi et al., 2017), suggesting that BCd is a risk factor for gastrointestinal cancer. Increased BCd levels potentially facilitate the initiation and development of these malignancies (Lin et al., 2018). Indeed, in Korea, a significant positive association has been observed between dietary Cd intake and the incidence of early stage gastric cancer (Kim et al., 2019). In addition, soil Cd levels are positively correlated with an increased risk of gastric cancer-related mortality (Jiang et al., 2021).

Colorectal cancer, also referred to as adenocarcinoma, is one of the deadliest forms of cancer worldwide. According to the data from Golestan Province, Iran, Cd exposure is a risk factor for colon cancer, with a >60.0% increased risk for both women and men (Kiani et al., 2021).

A significant association has been identified between Cd exposure and an elevated risk of pancreatic cancer. Studies have shown that higher pancreatic Cd levels are correlated with an increased risk of the disease, with a precise dose-response relationship (Djordjevic et al., 2019). Cd exposure is mainly involved in the progression of exocrine pancreatic cancer; individuals with high Cd levels ($\geq$ 0.59 μg/g) have demonstrated a significantly elevated risk of malignancy compared with others (Camargo et al., 2019; Djordjevic et al., 2019). Moreover, the Cd concentration in malignant pancreatic tissue (1.27–18.64 μg/g) was found to be significantly higher than that in the control (0.27–2.50 μg/g) (Djordjevic et al., 2019). Furthermore, the Cd concentration in tumor tissue has been reported to be approximately three times higher than that in adjacent non-oncological tissue (Djordjevic et al., 2019). Cd can cause renal tubular dysfunction, which has been significantly correlated with mortality from pancreatic carcinoma, suggesting that exposure to Cd increases the mortality risk of pancreatic cancer (Nogawa et al., 2022; Satarug et al., 2023).

Hepatocellular carcinoma is a common type of liver malignancy correlated with Cd exposure (Petrick et al., 2020). A previous meta-analysis revealed significantly elevated Cd levels in patients with liver malignancy compared with healthy individuals, indicating that Cd accumulation could substantially contribute to the neoplastic transformation of liver cells (Zhang et al., 2023a). Cd exposure can also affect the mortality of liver cancer. Indeed, in China, there is a positive association between soil Cd concentrations and age-standardized mortality rates for liver cancer (Jiang et al., 2021).

Cd exposure may also result in breast cancer; however, whether dietary Cd intake can cause breast cancer or not remains unclear. An analysis of 55 breast tissue samples revealed that cancerous tissues contained nearly four times more Cd than the adjacent non-cancerous tissues (Anđelković et al., 2021). This finding was further confirmed by research showing higher Cd levels in scalp hair and toenail samples from women diagnosed with breast malignancy compared with healthy individuals (Jouybari et al., 2018). A previous meta-analysis suggested that increased Cd exposure is related to a higher risk of breast cancer among women. However, uncertainties remain regarding the relative risk of non-dietary exposure and potential confounding factors such as the constituents of tobacco smoke (Flórez-García et al., 2023).

While some research has found that dietary Cd intake was not conclusively linked to an elevated risk of breast cancer in postmenopausal women, this does not entirely discount the potential for dietary Cd exposure to promote breast cancer risk (Cirovic and Satarug, 2024; Filippini et al., 2020; Flórez-García et al., 2023). Interestingly, when UCd excretion, an indicator of chronic cumulative exposure (Cd body burden), was considered, Cd exposure emerged as a notably high-risk contributor to breast malignancy

(Larsson et al., 2015; Lin et al., 2016). The lack of a link between dietary Cd intake and breast cancer risk is not surprising given that the bioavailability of Cd, and thus its potential impact is significantly influenced by the absorption rate rather than just the dietary intake levels (Cirovic and Satarug, 2024).

18.3.1.3 Cellular and Molecular Mechanisms of Cd Carcinogenesis

Although Cd is recognized as a potent human carcinogen, the precise mechanisms through which it induces carcinogenesis have not been fully elucidated. Compared with other carcinogenic metals, Cd is classified as a weak mutagen, which is attributed to its low affinity for DNA and its relatively low mutagenicity; this suggests that the tumorigenic effects of Cd may be mediated through pathways that do not directly involve genetic mutation (Charkiewicz et al., 2023; Zhao et al., 2022a). The main carcinogenic mechanisms of Cd include (1) the production of reactive oxygen species (ROS), leading to the induction of oxidative stress; (2) alterations in apoptotic pathways; (3) the induction of epigenetic modifications; and (4) a synergistic combination of these influences involving various pathways (Figure 18.2) (Balali-Mood et al., 2021; Genchi et al., 2020; Tavakoli Pirzaman et al., 2023). Cd can also promote solid tumor growth and metastasis by promoting angiogenesis and decreasing cellular adhesion ability (Tavakoli Pirzaman et al., 2023; Tran et al., 2022).

i. **Reactive oxygen species (ROS)**

The genotoxic and cytotoxic effects of Cd largely stem from its ability to produce ROS, including hydrogen peroxide (H_2O_2), hydroxyl radicals (OH), and superoxide radicals ($O_2^{\bullet-}$) (Collin, 2019; Khan et al., 2022). Cd does not generate ROS directly but instead facilitates their production

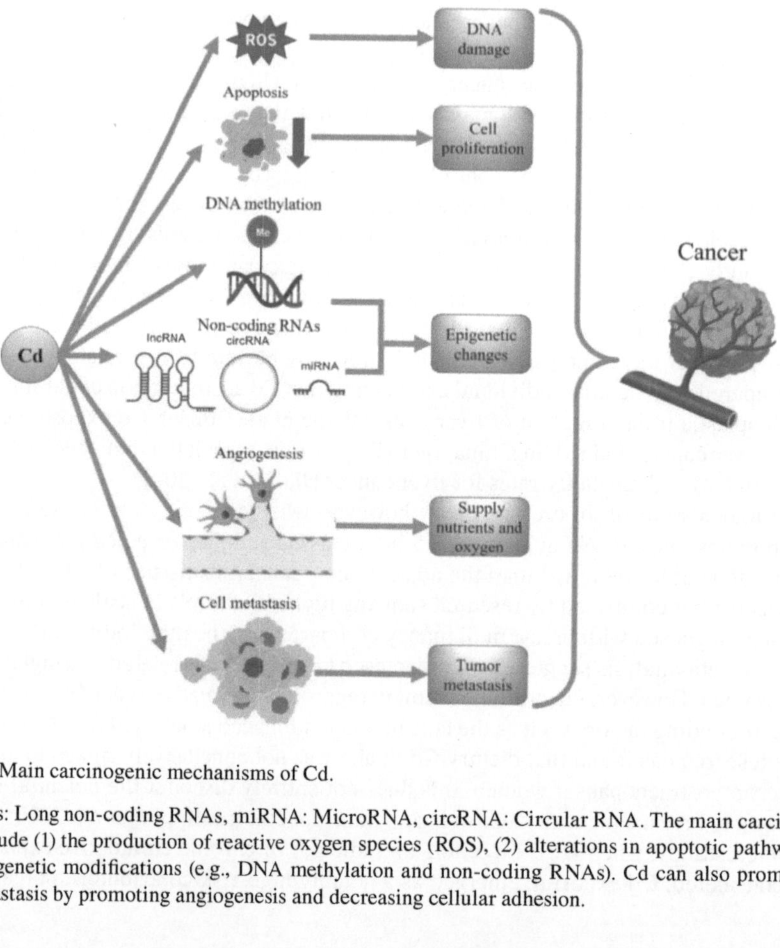

FIGURE 18.2 Main carcinogenic mechanisms of Cd.

NOTE: lncRNAs: Long non-coding RNAs, miRNA: MicroRNA, circRNA: Circular RNA. The main carcinogenic mechanisms of Cd include (1) the production of reactive oxygen species (ROS), (2) alterations in apoptotic pathways, and (3) the induction of epigenetic modifications (e.g., DNA methylation and non-coding RNAs). Cd can also promote solid tumor growth and metastasis by promoting angiogenesis and decreasing cellular adhesion.

through a catalytic role (Peana et al., 2022; Tavakoli Pirzaman et al., 2023; Unsal et al., 2020). The mechanisms by which Cd produces ROS are as follows: (1) Cd can deplete glutathione and protein-associated thiol groups, leading to increased generation of ROS due to diminished antioxidant capacity; and (2) mitochondria serve as a significant target for Cd toxicity, and Cd initially interacts with protein thiols located within the mitochondrial membrane, altering mitochondrial permeability transition and inhibiting the mitochondrial respiratory chain reaction, subsequently leading to the production of ROS (Nemmiche, 2017; Unsal et al., 2020).

ROS can inflict damage in several forms, including elevated lipid peroxidation, DNA damage, and protein oxidation, all of which are associated with cancer development (Unsal et al., 2020). ROS can severely damage DNA, which, if unrepaired, can lead to genetic mutations. Incorporating incorrect nucleotides, coupled with the failure of the DNA polymerase's proofreading mechanism, can result in replication errors (Khan et al., 2022). ROS also has the potential to damage proteins that serve as critical factors in the DNA replication process, leading to replication-born errors (Khan et al., 2022). Over time, ROS accumulation can create an imbalance between the continuous generation of ROS and the capacity of biological systems to neutralize these reactive intermediates or repair the ensuing cellular damage, resulting in oxidative stress (Tavakoli Pirzaman et al., 2023). Cd-induced oxidative stress harms different body tissues and organs, such as the gastrointestinal, cardiovascular, and skeletal systems, and may result in cancer (Collin, 2019; Khan et al., 2022; Pizzino et al., 2017). For instance, Cd-induced oxidative stress is closely related to the etiology of pancreatic cancer, while oxidative stress can cause inflammation and act as a precursor to oncogenesis (Djordjevic et al., 2019; Tavakoli Pirzaman et al., 2023).

ii. **Changes in apoptotic pathways**

Maintaining cellular integrity is crucial, and apoptosis plays a vital role in this process. During apoptosis, damaged cells are either repaired or eliminated systematically to preserve the organism's health (Tavakoli Pirzaman et al., 2023). Cd can suppress caspase activity, inhibiting apoptosis (Tavakoli Pirzaman et al., 2023). The inhibition of apoptosis increases the proliferation of deficient cells and promotes tumor development. Cd also upregulates CDK6 expression in human esophageal cells, which results in hyperphosphorylation of PKM2, inhibiting rate-limiting glycolytic enzymes. This inhibition directs a portion of carbon toward the pentose phosphate pathway, leading to the generation of NADPH and glutathione, both of which are crucial for ROS neutralization and ultimately prevent apoptosis (Tavakoli Pirzaman et al., 2023; Yang et al., 2020). Epithelial cell lines with apoptotic resistance can accumulate mutants, resulting in esophageal cancer (Tavakoli Pirzaman et al., 2023; Yang et al., 2020).

iii. **Epigenetic changes**

Epigenetics encompass dynamic and reversible alterations to genomes that do not alter the underlying DNA sequences. These modifications typically include DNA methylation, chromatin reorganization, and the action of non-coding RNAs. Such epigenetic regulation is crucial for maintaining appropriate gene expression profiles. Disruptions in these epigenetic processes can result in perturbed cellular functions and, in some cases, the onset of cancerous changes. As a weak mutagen, Cd mainly regulates gene expression by affecting epigenetic modifications and ultimately induces cancer formation, development, and angiogenesis (Charkiewicz et al., 2023; Zhao et al., 2022a). Recently, an increasing number of studies have underscored the pivotal role of epigenetic mechanisms in cancers induced by Cd exposure (Zhao et al., 2022a). However, the contribution of epigenetic mechanisms to Cd-induced carcinogenesis and angiogenesis remains largely enigmatic, and further research is urgently needed.

DNA methylation is a pivotal epigenetic mechanism implicated in Cd-induced carcinogenesis. DNA methylation, which typically involves the covalent attachment of methyl groups to the 5-position of the cytosine pyrimidine ring, is derived from S-adenosylmethionine (Jin et al., 2011). This modification generally results in gene silencing because 5-methylcytosine can prevent transcription factors from attaching to DNA and/or induce chromatin remodeling, thereby reducing DNA accessibility (Jaenisch and Bird, 2003). Given their central roles in the regulation of gene expression, DNA methylation and demethylation are implicated in a myriad of cellular processes, including cell proliferation, cell cycle regulation, and apoptosis (Lu et al., 2020).

Consequently, dysregulation of DNA methylation and demethylation is related to a spectrum of diseases and cancers (Lu et al., 2020). A previous study in mice demonstrated that dietary Cd intake induced hypermethylation at the promoter regions of key genes, including caspase-8 (critical in apoptosis) and interleukin-1β (Il-1β, critical in inflammation), downregulating their expression, indicating that dietary Cd intake may contribute to cancer development by affecting DNA methylation (Zhang et al., 2023b).

Non-coding RNAs also serve as key factors in the development of Cd-induced cancer. In the human genome, a substantial portion is transcribed into non-protein-coding RNA molecules, collectively known as non-coding RNAs. These non-coding RNAs are primarily classified into small non-coding RNAs (200 nucleotides) and long non-coding RNAs (lncRNAs; exceeding 200 nucleotides) (Yan and Bu, 2021). Within this category, microRNA (miRNA), lncRNA, and circular RNA (circRNA) have been identified as key regulators of gene expression; therefore, these non-coding RNAs are extensively implicated in the origination and development of a range of human cancers (Yan and Bu, 2021).

Recent studies have demonstrated that non-coding RNAs are frequently dysregulated in cells following Cd exposure and are critical players in Cd-induced cancer (Zhao et al., 2022a). For example, Cd can promote the expression of MT1DP, which is recruited to chromatin, where it attaches to SMARCAL1, thereby competitively inhibiting the interaction between SMARCAL1 and replication protein A complexes; this results in heightened replication stress and DNA injury and contributes to Cd-induced oncogenesis and tumor development (Feng et al., 2022b; Zhao et al., 2022a).

iv. **Angiogenesis**

The growth and metastasis of solid tumors are contingent on angiogenesis, which provides the necessary oxygen (O_2) and nutrients to the tumor and facilitates the removal of tumor metabolites (Carmeliet and Jain, 2011). Without neovascularization, a tumor cannot exceed a diameter of approximately 2 mm (Tavakoli Pirzaman et al., 2023). However, the effect of Cd on tumor angiogenesis remains poorly studied (Zhao et al., 2022a).

The current research suggests that the effect of Cd on tumor angiogenesis is dose-dependent (Zhao et al., 2022a). High concentrations of Cd (> 10 μM) may induce apoptosis in endothelial cells and reduce tumor angiogenesis (Jung et al., 2008; Kim et al., 2012), whereas lower concentrations of Cd (1–10 μM) enhance tumor blood vessel formation (Jing et al., 2012; Kim et al., 2012; Liu et al., 2015; Zhang et al., 2016). This pro-angiogenic effect is mediated through the stimulation of signaling pathways such as PKB/Akt (Liu et al., 2006), NF-κB (Dong et al., 2014), and MAPKs (Kim et al., 2012), which triggers endothelial cell activity and promotes tumor blood vessel formation (Wei et al., 2017).

Cd has been shown to enhance the tube-forming capacity of primary human umbilical vein endothelial cells through the downregulation of miR-101 (Che et al., 2021). The reduction in miR-101 leads to the upregulation of cyclooxygenase-2 and the induction of endoplasmic reticulum stress, which, in turn, increases the production of vascular endothelial growth factor, contributing to aberrant angiogenesis (Che et al., 2021). Notably, the mechanisms by which epigenetic alterations and non-coding RNAs mediate the impact of Cd on blood vessel formation are poorly understood; thus, further research is imperative to elucidate these processes (Zhao et al., 2022a).

v. **Cell metastasis**

Cell metastasis is closely related to tumor metastasis, invasion, and cancer-related mortality. It is estimated that approximately 90.0% of cancer-related fatalities stem from cancer cell metastasis, which is often initiated by epithelial-mesenchymal transition (EMT) in the early stages of cancer (Tran et al., 2022). During EMT, epithelial cells transform, losing their cellular polarity and tight junctions while acquiring migratory and invasive properties. The EMT plays a crucial role in cell–cell and cell–matrix adhesion, maintenance of cell polarity, tissue development, regeneration of injured tissue, and preservation of tissue homeostasis (Lamouille et al., 2014; Lecuit and Le Goff 2007; Marconi et al., 2021). However, in the context of cancer, neoplastic epithelial cells can misregulate EMT, thereby enhancing the mobility and invasiveness of cancer cells (Tiwari et al., 2012).

Recent investigations have revealed that Cd can disrupt the expression of miRNAs that are pivotal in modulating the expression of genes implicated in EMT, which promote the progression of cancer (Tran et al., 2022). For instance, oncogenic miRNAs, such as miR-221 and miR-155, are upregulated in Cd-induced pancreatic ductal adenocarcinoma; this upregulation leads to increased expression of SNAIL and ZEB1 proteins, whereas miR-126, Wnt-11, and E-cadherin are downregulated, resulting in Cd-induced EMT and oncogenesis (Mortoglou et al., 2022).

18.3.2 Kidney Disease

18.3.2.1 Kidneys, as a Principal Target Organ for Cd Toxicity

Within the age bracket of 30–60 years, Cd in the kidney accounts for 60.0% of the total Cd in the body (Thévenod and Lee, 2024). Compared with the other 50 mammalian species, human kidneys have been observed to exhibit the highest rates of Cd accumulation, and the Cd concentration in female kidneys is more than double that in male kidneys (Koizumi et al., 2008). Moreover, compared with other organs, the kidneys are considered to be affected most by Cd toxicity (Charkiewicz et al., 2023).

Cd in the kidneys can be partially excreted through urine. Cd excretion is an indicator of ongoing tissue injury, and the excreted Cd is thought to originate from injured renal tubular epithelial cells (Satarug et al., 2023). Three stages of UCd excretion can be distinguished: (1) Cd accumulates in the renal cortex through binding to MT, and the UCd is proportional to the kidney Cd level, reflecting historical exposure levels; (2) with sustained high Cd exposure leading to the depletion of MT, the UCd becomes indicative of both current and historical Cd exposure; (3) renal tubular damage decreases the Cd reabsorption ability, leading to a significant increase in urinary excretion of Cd, which reflects ongoing Cd exposure and Cd excretion from the renal system (Charkiewicz et al., 2023).

The severity of kidney damage caused by Cd is contingent on the degree of Cd–MT accumulation in the organ. The primary tubules are predominantly affected, with early manifestations including the onset of proteinuria (Charkiewicz et al., 2023). Proteinuria, a symptom with potentially chronic persistence, is often accompanied by other urinary abnormalities, such as glycosuria, aminoaciduria, elevated excretion of Ca and P, and increased creatinine concentration, and then progresses to overt kidney disease, such as chronic kidney disease (CKD) and toxic nephropathy (Lech and Sadlik, 2017). Furthermore, if exposure to Cd is both elevated and prolonged, or if it coincides with other triggers, kidney impairment may advance to end-stage kidney disease and death (Wang et al., 2021).

CKD, a progressive syndrome characterized by significant morbidity and mortality, affects approximately 8.0%–16.0% of the global population (De Nicola and Zoccali, 2016; Kalantar-Zadeh et al., 2021). CKD is clinically defined by an estimated glomerular filtration rate (GFR) $\leq$ 60 mL/min 1.73 m^{-2} or by an elevated urinary albumin-to-creatinine ratio exceeding 30 mg/g for a minimum of 3 months (Glassock et al., 2017). The hallmark histological features of CKD include interstitial tissue fibrosis, interstitial mononuclear cell infiltration, and renal tubule atrophy (Mohanty et al., 2023). In clinical trials, the efficacy of CKD treatments is assessed based on the ability to mitigate the rate of GFR decline (Kalantar-Zadeh et al., 2021; Levey et al., 2014; Satarug et al., 2023).

18.3.2.2 Cd Exposure Increases the Risk of CKD and Kidney Failure (Epidemiological Evidence)

Chronic Cd exposure correlates with a decreased GFR (an essential indicator for diagnosing CKD), increasing the risk of developing CKD and exacerbating its progression (Smereczański and Brzóska, 2023; Zang et al., 2019). In CKD endemic areas, a WHO study showed high Cd concentrations in soils and vegetables and human tissues (Samarajeewa, 2022). Low-level environmental Cd exposure has also been shown to be correlated with a decline in GFR among participants in multiple cycles in the U.S. National Health and Nutrition Examination Survey (NHANES), which covers 17 years from 1999 to 2016 (Satarug et al., 2023).

Elevated concentrations of UCd and BCd are related to lower GFR and higher albumin excretion, indicating that Cd is associated with CKD. A Spanish epidemiological study revealed that even low UCd

(0.27 μg/g creatinine) increased the risk of albuminuria by 58.0% (Grau-Perez et al., 2017). Furthermore, in the NHANES from 2009 to 2012, UCd > 0.22 μg/L was linked to increased albumin excretion among participants (Zhu et al., 2019). Similarly, in Shanghai, UCd > 1.72 μg/g creatinine was linked to higher albumin excretion in Cd-exposed residents (Feng et al., 2022c). In Korean epidemiological studies, BCd (1.74 μg/L, 2.08 μg/L) was correlated with a low estimated GFR (Chung et al., 2014; Myong et al., 2012). The BCd levels linked to a low estimated GFR in the U.S. population were approximately one-fourth of those of the Korean population (Lin et al., 2014; Madrigal et al., 2019). Moreover, dietary Cd intake may increase the mortality from kidney failure. In a prospective cohort study of individuals residing in a Cd-impacted region, comprising 1169 males and 1433 females, a lifetime Cd intake of 1 g or more, half the permissible limit set by regulatory authorities, was correlated with an elevated risk of mortality attributed to kidney failure, particularly in female participants (49.0%) (Nishijo et al., 2020).

18.3.2.3 Cellular and Molecular Mechanisms of Kidney Disease (e.g., Kidney Tubular Impairment)

Within the renal system, Cd–MT is filtered by glomeruli and subsequently reabsorbed by renal tubules (Wang et al., 2021). Following reabsorption, Cd is liberated as free ions, exhibiting higher toxicity than that of bound Cd. These liberated Cd ions subsequently bind to newly synthesized MT produced by epithelial cells of renal tubules and further decompose, causing substantial damage to these cells. This process impairs the ability of epithelial cells to reabsorb substances, resulting in the development of proteinuria (Yang and Shu, 2015). A slight elevation in the urinary excretion of low-molecular-weight proteins is typically reversible; however, an elevation of an order of magnitude > 1 leads to permanent renal tubular damage, which may advance to overt kidney disease (WHO, 2019).

Currently, the molecular mechanisms underlying Cd-induced renal toxicity are believed to primarily include oxidative stress, mitochondrial dysfunction, apoptosis, DNA methylation changes, and Ca^{2+} homeostasis disruption, which are independent and interrelated (Figure 18.3) (Wang et al., 2021). Cd can induce renal cells to produce a large amount of ROS and reduce their ability to clear ROS (reducing

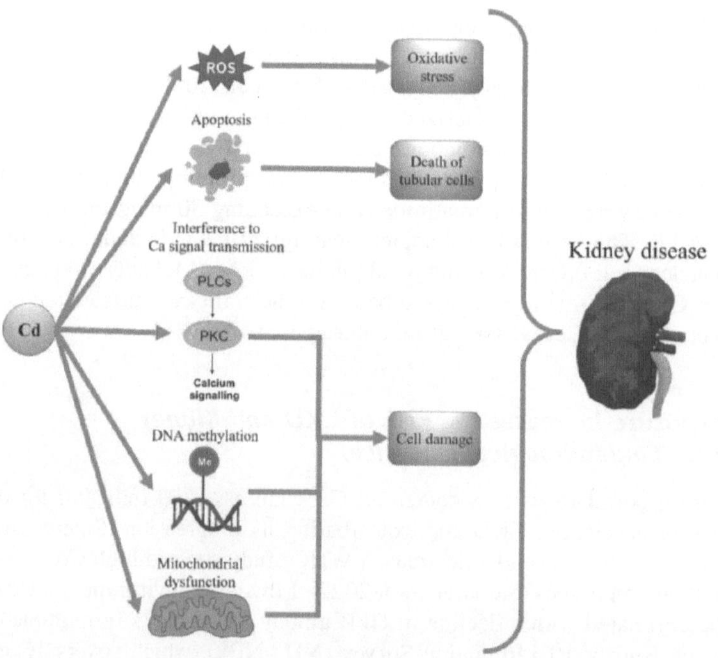

FIGURE 18.3 Main mechanism of Cd-induced renal toxicity.

Note: The mechanisms underlying Cd-induced renal toxicity primarily include oxidative stress, apoptosis, interference with Ca-related signal transmission, DNA methylation changes, and mitochondrial dysfunction.

antioxidant enzyme activity by replacing metals [Fe and Cu] and binding to thiol groups of antioxidant enzymes), causing renal injury (Jin et al., 2018; Satarug et al., 2017). Cd can also induce apoptosis by inducing the release of apoptotic factors, pro-apoptotic proteins (p53), and other factors, while tubular cells, which contain numerous mitochondria, are greatly dependent on cellular self-digestion to sustain cellular homeostasis, rendering them particularly vulnerable to Cd-induced apoptosis (Chen et al., 2016; Lee et al., 2016; Satarug, 2018).

Through Ca^{2+} channels, Cd^{2+} can enter the cell to replace Ca^{2+} and activate related molecular mechanisms that interfere with Ca-related signal transmission, leading to cell damage (Jin et al., 2018; Wang et al., 2021). Cd can also cause chronic kidney injury by disrupting iron metabolism rather than by the typical ferroptosis pathway (Zhang et al., 2024).

18.3.3 Bone Disease

18.3.3.1 "Itai-Itai Disease" in Japan - A Typical Example of Cd-Induced Bone Disease

"Itai-itai disease" (Japanese; "it hurts it hurts") is the first typical disease that was discovered to be related to Cd, as well as being the most representative and well-known. In Toyama Prefecture, "itai-itai disease" was first recorded before 1931 (Reyes-Hinojosa et al., 2019). In the late 1960s, an epidemiologic study disclosed that Zn mining adjacent to the Jinzu River was the source of Cd contamination and that dietary Cd ingestion (Cd contaminated rice) resulted in the "itai-itai disease" (Nogawa et al., 2017; Reyes-Hinojosa et al., 2019; Zhao et al., 2023). The impact of Cd has persisted, and by the end of 2000, 20 new cases of "itai-itai disease" were identified and acknowledged by the Toyama government (Reyes-Hinojosa et al., 2019).

"Itai-itai disease" is recognized as the most severe form of chronic Cd poisoning, and it is characterized by intense bone pain, especially in individuals with spine and joint injuries (Genchi et al., 2020; Reyes-Hinojosa et al., 2019). "Itai-itai disease" presents a spectrum of symptoms, including reduced bone mineralization, serious skeletal decalcification, an increased incidence of fractures and deformities in long bones, osteomalacia, osteoporosis, and severe bone pain (Aoshima, 2016; Baba et al., 2014; Nishijo et al., 2017). The condition often progresses over several years, with patients initially experiencing pain in the dorsal and femoral regions that gradually extend throughout the body, culminating in bed confinement due to recurrent bone fractures (Reyes-Hinojosa et al., 2019). Other complications, such as coughing, anemia, and renal failure, also exist, which, if untreated, can be fatal (Genchi et al., 2020).

The dietary Cd intake can also lead to other musculoskeletal disorders, including osteoporosis and osteoarthritis (OA) (Reyes-Hinojosa et al., 2019). Osteoporosis is characterized by reduced bone mineral density (BMD), which increases fracture risk (Hernlund et al., 2013; Ma et al., 2021). OA is a degenerative disease characterized by a multitude of anatomical and physiological changes within joint tissues, including joint cartilage degradation, bone remodeling, and the formation of osteophytes (Allen et al., 2022).

18.3.3.2 Cd Exposure Increases Osteoporosis, OA, and Fracture Risks (Epidemiological Evidence)

Dietary Cd intake (Cd-contaminated rice) can affect BMD and osteoporosis. An increase in dietary Cd intake is correlated with both the magnitude and the percentage decrease in BMD (Chen et al., 2009). Regarding highly contaminated areas, a positive association was detected between the decline in BMD and both UCd and BCd. Moreover, in the high-Cd-exposure group, the prevalence of osteoporosis increased in connection with the UCd and BCd levels, both in males and females (Chen et al., 2009). The Cd levels that correlate with a lower bone mass range from 0.72 to 1.35 μg/L in blood and range from 2.14 to 3.99 μg/g of creatinine in urine. BMD is related to BCd, which leads to osteoporosis, while UCd is related to a reduction in bone mass (Chen et al., 2013a, b). However, a previous meta-analysis showed that UCd was a suitable indicator of osteoporosis, whereas BCd is not (Li et al., 2021). In the Mae Sot District of Thailand, previous research revealed that elevated dietary Cd intake (Cd-contaminated rice) was related to an elevated risk of osteopathy, mainly affecting Ca resorption, leading to renal tubular dysfunction and accelerating bone resorption (Nambunmee et al., 2010). The Cd concentration of the hip joint in people with fractures (0.85 μg/g) exceeded that of patients with OA (0.64 μg/g), suggesting that higher Cd concentrations result in more severe musculoskeletal diseases (Brodziak-Dopierała et al., 2015).

OA may also be related to Cd exposure. Patients with OA have been shown to exhibit elevated Cd concentrations in the tibia compared with those who had experienced a knee injury. An elevated Cd concentration was also observed in the hip joint of patients with degenerative changes (Lanocha-Arendarczyk et al., 2015). Furthermore, there was a notable difference in Cd concentration between men and women with OA in the tibia (0.016 mg/kg in men, 0.014 mg/kg in women), femur (0.017 mg/kg in men, 0.013 mg/kg in women), and meniscus (0.008 mg/kg in men, 0.007 mg/kg in women) (Roczniak et al., 2017).

18.3.3.3 Pathogenic Mechanisms of Bone Disease

Cd can cause bone disease via two mechanisms: indirect and direct effects on bone cells (Figure 18.4). Originally, the impact of Cd on bones was considered indirect. Cd exposure may induce changes in the skeletal system through its inhibitory effects on primary renal tubular cells, and this inhibition can reduce the efficiency of vitamin D metabolism, consequently impairing Ca absorption (Charkiewicz et al., 2023; WHO, 2019). Cd also inhibits the reabsorption of Ca and P, which can ultimately result in bone demineralization (Charkiewicz et al., 2023; WHO, 2019). Furthermore, Cd can replace Ca and other divalent ions (Mg, Fe, Zn, and Cu) and interfere with their metabolism in cells, disturbing their cascades, ultimately causing bone demineralization, osteomalacia, and osteoporosis and affecting other vital organs (Charkiewicz et al., 2023; Syed et al., 2023).

In addition to indirectly affecting bone, Cd has direct osteotoxic effects. The predominant mechanism of Cd-induced injury is believed to involve the direct suppression of bone formation and stimulation of bone resorption, which is achieved by affecting the activity of osteoblasts and osteoclasts (the key cells responsible for bone formation and breakdown, respectively) and substituting Cd for Ca in the

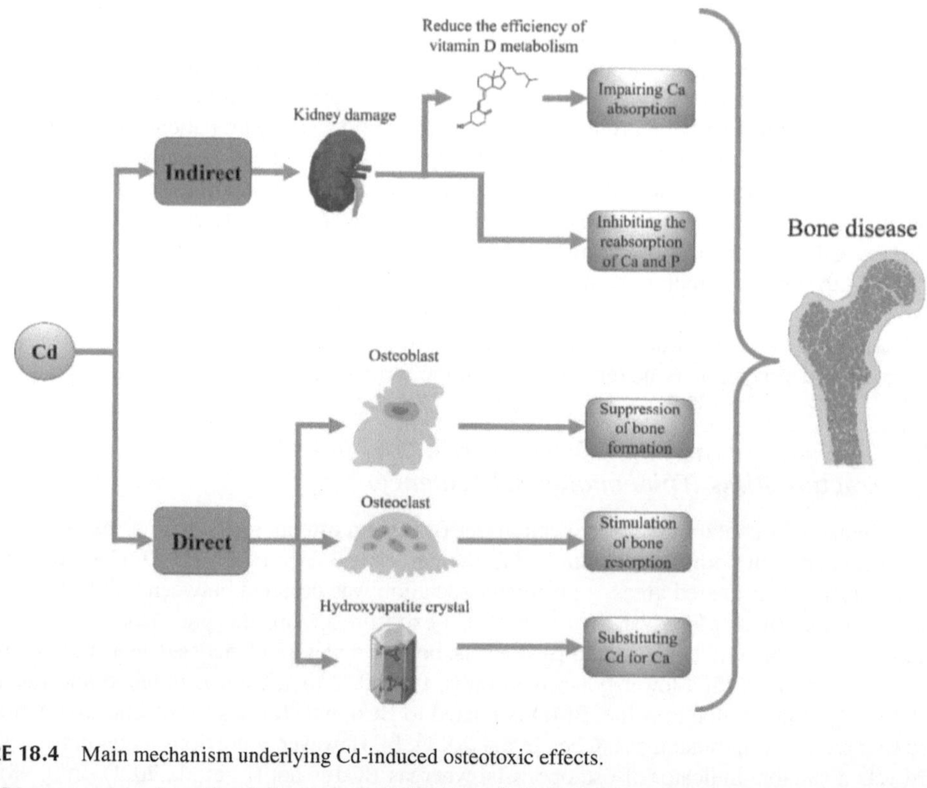

FIGURE 18.4 Main mechanism underlying Cd-induced osteotoxic effects.

Note: Cd exposure may harm the skeletal system indirectly by inhibiting primary renal tubular cells, which can reduce the efficiency of vitamin D metabolism, thereby impairing Ca absorption. Cd also inhibits Ca and P reabsorption, resulting in bone demineralization. The predominant mechanism of direct Cd-induced osteotoxic effects is the suppression of bone formation and stimulation of bone resorption, which are achieved by affecting the activity of osteoblasts and osteoclasts and substituting Cd for Ca in the hydroxyapatite crystals.

hydroxyapatite crystals (Reyes-Hinojosa et al., 2019). Cd exposure inhibits osteoblast and osteoclast differentiation by drastically reducing P2X7 expression (Ma et al., 2021). ROS may also be involved in Cd-induced osteoporosis. ROS plays a critical role in bone remodeling, and overproduction of ROS by osteoclasts can enhance bone resorption (Reyes-Hinojosa et al., 2019). Runt-related transcription factor 2 (RUNX2) mRNA serves as a pivotal regulatory element in the differentiation of bone-forming cells and bone formation, while also controlling the expression of critical bone matrix proteins. Cd-induced oxidative stress harms osteoblasts by reducing the expression of RUNX2 mRNA (Choi and Han, 2015; Reyes-Hinojosa et al., 2019).

18.4 Dietary Recommendations for Reducing Cd Exposure and Damage

Three strategies are employed to protect humans from damage caused by dietary Cd exposure: (1) decreasing dietary Cd intake, (2) reducing Cd absorption capacity in the intestine, and (3) counteracting Cd-induced harm (Figure 18.5). Some foods may reduce Cd damage through multiple strategies.

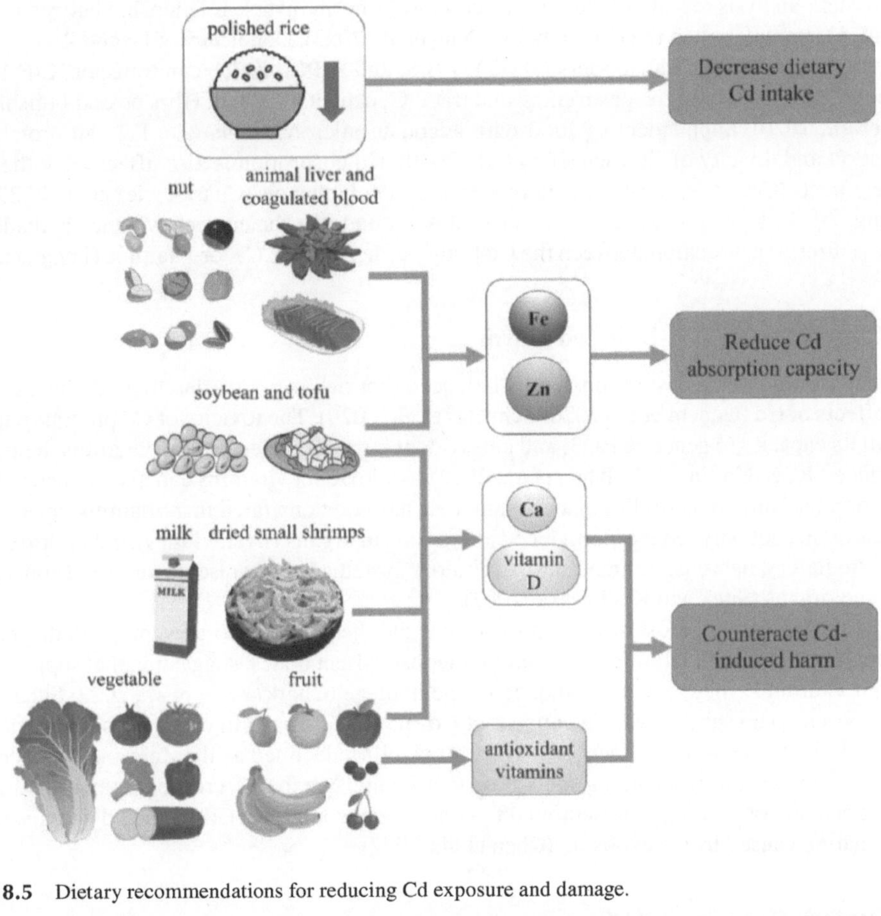

FIGURE 18.5 Dietary recommendations for reducing Cd exposure and damage.

Note: (1) Decreasing dietary Cd intake: consuming foods from areas without Cd contamination; adjusting dietary structure by reducing rice consumption; switching to other staple foods or vegetables and fruits; and replacing polished rice with brown rice. (2) Reducing the Cd absorption capacity of the intestine by consuming foods rich in Fe and Zn, such as nuts, animal liver and coagulated blood, and soybean and its products. (3) Counteracting Cd-induced harm by consuming food that is rich in antioxidant vitamins, such as fruits and vegetables, which can decrease the ROS produced by Cd; and consuming milk and milk products, dried small shrimps, soybean and its products, and some fruits and vegetables, all of which can supplement the Ca and vitamin D lost as a result of Cd toxicity.

18.4.1 Decreasing Dietary Cd Intake

Given the long half-life of Cd in humans and the lack of a therapeutically efficacious chelation treatment, the most feasible preventive strategy is to reduce Cd absorption rates (Cirovic and Satarug, 2024). The habitual intake of foods with elevated Cd levels is a significant factor contributing to the body's Cd burden, underscoring the need to reduce the intake of foods with high Cd concentrations to mitigate potential health risks (Cirovic and Satarug, 2024). To reduce dietary Cd intake, it is necessary to consume foods (especially staple foods, such as rice and wheat) from areas without Cd contamination. Moreover, in terms of adjusting the dietary structure, for people who consume rice as their staple food, it is advisable to reduce rice consumption and switch to other staple foods or switch to vegetables and fruits. The Cd accumulation capacity of rice is more substantial than that of other crops, suggesting that lower rice consumption may decrease dietary Cd intake. Given its lower Cd absorption rate, cooked brown rice is safer than cooked polished rice (Liu et al., 2021b).

18.4.2 Reducing Cd Absorption Capacity in the Intestine

A balanced diet containing certain bioelements (especially Fe and Zn) can prevent Cd absorption. A previous meta-analysis revealed robust evidence supporting an inverse relationship between Zn and Fe status and Cd accumulation in the human body (Peng et al., 2023). Deficiency of Fe and Zn may increase the expression of Fe and Zn transporters (DMT1, ZIP8, and ZIP14, which can transport Cd^{2+}) and lead to enhanced Cd absorption in the gastrointestinal tract (Genchi et al., 2020; Himeno and Fujishiro, 2021; Schaefer et al., 2020). Supplementing food with adequate micronutrients (Zn, Fe) can protect against the absorption and toxicity of Cd (Schaefer et al., 2020). Cd absorption is also affected by the levels of essential elements (Ca, Se, P), vitamins, antioxidants, and polyphenols (Charkiewicz et al., 2023; Cirovic and Satarug, 2024). However, a previous meta-analysis found that the current evidence is inadequate to establish a definitive association between the Cd body burden and Se, Ca, or vitamins (Peng et al., 2023).

18.4.3 Counteracting Cd-Induced Harm

In addition to inhibiting Cd absorption, a well-balanced diet rich in antioxidant ingredients can mitigate the toxic effects of Cd (Genchi et al., 2020; Schaefer et al., 2020). The toxicity of Cd predominantly originates from its capacity to generate ROS, and antioxidant ingredients (especially vitamins) help decrease the Cd-induced ROS (Collin, 2019; Khan et al., 2022). Antioxidant vitamins can also reduce Cd concentrations in organs. Numerous studies in animal models have demonstrated that vitamins, including E, C, A, and β-carotene, not only reduce Cd-induced alterations in organs (liver, kidney) and multiple systems, including circulatory, nervous, immune, and respiratory systems but can also decrease Cd concentrations in tissues and organs (Sitek and Kozłowska, 2022).

Supplementing substances lost due to Cd toxicity is another strategy to prevent Cd damage. Cd may cause a deficiency in Ca and vitamin D and simultaneously affect bones, suggesting that supplementation with Ca and vitamin D may reduce Cd-induced bone damage (Charkiewicz et al., 2023; Schaefer et al., 2020). Moreover, some substances can counteract Cd-induced damage in other ways. A previous study demonstrated that the cytotoxic effects of Cd on cell viability declined as the Zn^{2+} and Mn^{2+} concentrations increased in the culture solutions, indicating that Zn and Mn may decrease Cd toxicity (Tang et al., 2023). Furthermore, omega-3 polyunsaturated fatty acids can counteract Cd-mediated damage by reducing inflammation caused by Cd exposure (Chen et al., 2022).

18.4.4 Dietary Recommendations

It is necessary to decrease the consumption of crops containing high Cd levels. It is also advisable to encourage the consumption of natural food products rich in antioxidant vitamins and essential minerals (especially Fe and Zn), which can help mitigate the risk of Cd (Sitek and Kozłowska, 2022). Vegetables and fruits serve as vital dietary sources of essential minerals and vitamins, and consuming adequate amounts of these foods can help to counteract the harmful effects of Cd by elevating the body's levels of these crucial nutrients (Wang et al., 2021). Note that any vegetables and fruits consumed should be Cd-free.

18.5 Conclusion

Cd does not have a physiological function (non-essential metal) and is a persistent toxic substance. The primary route of human Cd exposure is through Cd-contaminated crop consumption, especially rice and wheat. After consuming a Cd-contaminated crop, the Cd is absorbed by the intestinal tract and transported into the portal and systemic circulation. In the bloodstream, Cd is first transported into the liver, where a portion of the Cd remains concentrated. The remaining Cd is transported into the kidneys and other organs via blood circulation. Cd primarily accumulates in the liver and kidneys.

Cd can indirectly produce ROS, increasing lipid peroxidation, DNA damage, and protein oxidation. The accumulation of ROS will impose oxidative stress, damaging different body tissues and organs and resulting in diseases such as cancer and kidney and bone diseases. As a human carcinogen, Cd can also interfere with apoptosis and regulate gene expression by affecting epigenetic modifications, which ultimately induce cancer formation, development, and angiogenesis. Moreover, kidney tubules can be damaged by Cd-induced oxidative stress, mitochondrial dysfunction, apoptosis, and DNA methylation, resulting in proteinuria. As the damage from Cd increases, irreversible tubular dysfunction develops, which progresses to overt kidney disease (CKD, end-stage renal disease). Cd-induced kidney damage also interferes with Ca, P, and vitamin D metabolism, which contributes to decreased BMD and increased fracture risk. Cd can also reduce bone formation and increase bone resorption by directly acting on bone cells, resulting in musculoskeletal diseases ("itai-itai disease," osteoporosis, and OA). It is necessary to decrease the consumption of crops containing high Cd levels to protect humans from Cd damage. Moreover, adequate amounts of Cd-free vegetables, fruits, and other food products rich in antioxidant vitamins and essential minerals (especially Fe and Zn) can help reduce Cd absorption and counteract its harmful effects.

Acknowledgments

The Special Fund for Scientific Innovation Strategy-Construction of High-Level Academy of Agriculture Science (R2024YJ-YB1003) financially supported this work. We wish to thank Dr. Yang Zhijian (Department of Joint Surgery, First Affiliated Hospital, SunYat-Sen University) for his suggestion regarding Figure 18.1 and Figure 18.4. We wish to thank Dr. Chen Jiahong (Guangdong Provincial Key Laboratory of Food Quality and Safety, College of Food Science, South China Agricultural University) for her suggestion regarding the meaning of rice substrate. The figures in this chapter were created with BioGDP.com, and some material comes from the following websites: https://90sheji.com; https://ian.umces.edu/media-library; https://www.51yuansu.com; https://616pic.com; https://www.ztupic.com. We thank LetPub (www.letpub.com.cn) for its linguistic assistance during the preparation of this manuscript.

REFERENCES

Agency for Toxic Substances and Disease Registry USA. 2017. "*Toxicological Profile for Cadmium*." Department of Health and Humans Services, Public Health Service, Centers for Disease Control, Atlanta, GA, USA.

Åkesson, A., Julin, B., and Wolk, A. 2008. "Long-term dietary cadmium intake and postmenopausal endometrial cancer incidence: A population-based prospective cohort study." *Cancer Research* 68(15): 6435–6441.

Allen, K. D., Thoma, L. M., and Golightly, Y. M. 2022. "Epidemiology of osteoarthritis." *Osteoarthritis and Cartilage* 30(2): 184–195.

Al-Rmalli, S. W., Jenkins, R. O., and Haris, P. I. 2012. "Dietary intake of cadmium from Bangladeshi foods." *Journal of Food Science* 77(1): T26–33.

Arisawa, K., Nakano, A., Saito, H., Liu, X. J., Yokoo, M., Soda, M., Koba, T., Takahashi, T., and Kinoshita, K. 2001. "Mortality and cancer incidence among a population previously exposed to environmental cadmium." *International Archives of Occupational and Environmental Health* 74: 255–262.

Anđelković, M., Djordjevic, A. B., Miljaković, E. A., Javorac, D., Čolaković, N., Oprić, S., Petričević, S., Granić, M., Kotur-Stevuljević, J., Antonijević, B., and Bulat, Z. 2021. "Cadmium tissue level in women diagnosed with breast cancer—A case control study." *Environmental Research* 199: 111300.

Aoshima, K. 2016. "Itai-itai disease: Renal tubular osteomalacia induced by environmental exposure to cadmium—historical review and perspectives." *Soil Science and Plant Nutrition* 62(4): 319–326.

Baba, H., Tsuneyama, K., Kumada, T., Aoshima, K., and Imura, J. 2014. "Histopathological analysis for osteomalacia and tubulopathy in itai-itai disease." *The Journal of Toxicological Sciences* 39(1): 91–96.

Balali-Mood, M., Naseri, K., Tahergorabi, Z., Khazdair, M. R., and Sadeghi, M. 2021. "Toxic mechanisms of five heavy metals: Mercury, lead, chromium, cadmium, and arsenic." *Frontiers in Pharmacology* 12: 643972.

Branca, J. J. V., Morucci, G., and Pacini, A. 2018. "Cadmium-induced neurotoxicity: Still much ado." *Neural Regeneration Research* 13(11): 1879–1882.

Brodziak-Dopierała, B., Kwapuliński, J., Sobczyk, K., and Wiechuła, D. 2015. "Analysis of the content of cadmium and zinc in parts of the human hip joint." *Biological Trace Element Research* 163: 73–80.

Camargo, J., Pumarega, J. A., Alguacil, J., Sanz-Gallén, P., Gasull, M., Delclos, G. L., Amaral, A. F. S., and Porta, M. 2019. "Toenail concentrations of trace elements and occupational history in pancreatic cancer." *Environment International* 127: 216–225.

Carmeliet, P., and Jain, R. K. 2011. "Molecular mechanisms and clinical applications of angiogenesis." *Nature* 473(7347): 298–307.

Chaney, R. L., Reeves, P. G., Ryan, J. A., Simmons, R. W., Welch, R. M., and Scott Angle, J. 2004. "An improved understanding of soil Cd risk to humans and low cost methods to phytoextract Cd from contaminated soils to prevent soil Cd risks." *Biometals* 17(5): 549–553.

Charkiewicz, A. E., Omeljaniuk, W. J., Nowak, K., Garley, M., and Nikliński, J. 2023. "Cadmium toxicity and health effects—a brief summary." *Molecules* 28(18): 6620.

Che, L., Wu, Z. L., Huang, L. Y., Wu, J. S., Du, Z. B., Lin, J. X., Su, Y. H., Chen, X. X., Lin, Z. N., and Lin, Y. C. 2021. "MicroRNA-101 inhibits cadmium-induced angiogenesis by targeting cyclooxygenase-2 in primary human umbilical vein endothelial cells." *Biochemical Pharmacology* 189: 114192.

Chen, J., Zhou, Z., Lin, X., Liao, J., Zhang, Y., Xie, B., Huang, Y., and Peng, L. 2022. "Environmental cadmium exposure promotes the development, progression and chemoradioresistance of esophageal squamous cell carcinoma." *Frontiers in Cell and Developmental Biology* 10: 792933.

Chen, X., Gan, C., Zhu, G., and Jin, T. 2013a. "Benchmark dose for estimation of cadmium reference level for osteoporosis in a Chinese female population." *Food and Chemical Toxicology* 55: 592–595.

Chen, X., Li, J., Cheng, Z., Xu, Y., Wang, X., Li, X., Xu, D., Kapron, C. M., and Liu, J. 2016. "Low dose cadmium inhibits proliferation of human renal mesangial cells via activation of the JNK pathway." *International Journal of Environmental Research and Public Health* 13(10): 990.

Chen, X., Wang, G., Li, X., Gan, C., Zhu, G., Jin, T., and Wang, Z. 2013b. "Environmental level of cadmium exposure stimulates osteoclasts formation in male rats." *Food and Chemical Toxicology* 60: 530–535.

Chen, X., Zhu, G., Jin, T., Åkesson, A., Bergdahl, I. A., Lei, L., Weng, S. F., and Liang, Y. 2009. "Changes in bone mineral density 10 years after marked reduction of cadmium exposure in a Chinese population." *Environmental Research* 109(7): 874–879.

Chen, Z., Lu, Q., Wang, J., Cao, X., Wang, K., Wang, Y., Wu, Y., and Yang, Z. 2022. "The function of omega-3 polyunsaturated fatty acids in response to cadmium exposure." *Frontiers in Immunology* 13: 1023999.

Choi, W. J., and Han, S. H. 2015. "Blood cadmium is associated with osteoporosis in obese males but not in non-obese males: The Korea National Health and Nutrition Examination Survey 2008–2011." *International Journal of Environmental Research and Public Health* 12(10): 12144–12157.

Chung, S., Chung, J. H., Kim, S. J., Koh, E. S., Yoon, H. E., Park, C. W., Chang, Y. S., and Shin, S. J. 2014. Blood lead and cadmium levels and renal function in Korean adults. *Clinical and Experimental Nephrology* 18: 726–734.

Cirovic, A., and Satarug, S. 2024. "Toxicity tolerance in the carcinogenesis of environmental cadmium." *International Journal of Molecular Sciences* 25(3): 1851.

Clemens, S., Aarts, M. G. M, Thomine, S., and Verbruggen, N. 2013. "Plant science: The key to preventing slow cadmium poisoning." *Trends in Plant Science* 18(2): 92–99.

Collin, F. 2019. "Chemical basis of reactive oxygen species reactivity and involvement in neurodegenerative diseases." *International Journal of Molecular Sciences* 20(10): 2407.

De Nicola, L., and Zoccali, C. 2016. "Chronic kidney disease prevalence in the general population: Heterogeneity and concerns." *Nephrology Dialysis Transplantation* 31(3): 331–335.

Deng, Z., Yao, L., Fang, M., Wu, Y., and Gong, Z. 2022. "Health risk assessment using in vitro simulation in assessing bioavailability of cadmium in rice from main producing areas across China." *Journal of Food Science* 87(6): 2766–2774.

Djordjevic, V. R., Wallace, D. R., Schweitzer, A., Boricic, N., Knezevic, D., Matic, S., Grubor, N., Kerkez, M., Radenkovic, D., Bulat, Z., Antonijevic, B., Matovic, V., and Buha, A. 2019. "Environmental cadmium exposure and pancreatic cancer: Evidence from case control, animal and *in vitro* studies." *Environment International* 128: 353–361.

Dong, F., Zhou, X., Li, C., Yan, S., Deng, X., Cao, Z., Li, L., Tang, B., Allen, T. D., and Liu, J. 2014. "Dihydroartemisinin targets VEGFR2 via the NF-κB pathway in endothelial cells to inhibit angiogenesis." *Cancer Biology & Therapy* 15(11): 1479–1488.

EFSA. 2012. "Cadmium dietary exposure in the European population." *EFSA Journal* 10: 1–37.

Elliott, P., Arnold, R., Cockings, S., Eaton, N., Järup, L., Jones, J., Quinnc, M., Rosatoc, M., Thorntonb, I., Toledanoa, M., Tristan, E., and Wakefield, J. 2000. "Risk of mortality, cancer incidence, and stroke in a population potentially exposed to cadmium." *Occupational and Environmental Medicine* 57(2): 94–97.

Feng, W., Dong, T., Li, K., Wang, T., Chen, Z., and Wang, R. 2019. "Characterization of binding behaviors of Cd^{2+} to rice proteins." *Food Chemistry* 275: 186–192.

Feng, W., Fan, D., Li, K., Wang, T., Zhang, H., Zhou, X., Wan, J., and Wang, R. 2022a. "Removal of cadmium from rice grains by acid soaking and quality evaluation of decontaminated rice." *Food Chemistry* 371: 131099.

Feng, W., Qi, Z., Dong, Z., Liu, W., Xu, M., Gao, M., and Liu, S. 2022b. "LncRNA MT1DP promotes cadmium-induced DNA replication stress by inhibiting chromatin recruitment of SMARCAL1." *Science of the Total Environment* 807: 151078.

Feng, X., Zhou, R., Jiang, Q., Wang, Y., and Yu, C. 2022c. "Analysis of cadmium accumulation in community adults and its correlation with low-grade albuminuria." *Science of the Total Environment* 834: 155210.

Filippini, T., Torres, D., Lopes, C., Carvalho, C., Moreira, P., Naska, A., Kasdagli, M. I., Malavolti, M., Orsini, N., and Vinceti, M. 2020. "Cadmium exposure and risk of breast cancer: A dose-response meta-analysis of cohort studies." *Environment International* 142: 105879.

Flórez-García, V. A., Guevara-Romero, E. C., Hawkins, M. M., Bautista, L. E., Jenson, T. E., Yu, J., and Kalkbrenner, A. E. 2023. "Cadmium exposure and risk of breast cancer: A meta-analysis." *Environmental Research* 219: 115109.

FSA. 2015. "Total diet study: Metals and other elements" [online] (https://www.food.gov.uk/research/research-projects/total-diet-study-metals-and-other-elements#:~:text=The%20Total%20Diet%20Study%20(TDS,or%20nutritional%20quality%20of%20food).

Fujishiro, H., and Himeno, S. 2019. "New insights into the roles of ZIP8, a cadmium and manganese transporter, and its relation to human diseases." *Biological and Pharmaceutical Bulletin* 42(7): 1076–1082.

Fujiwara, Y., Lee, J. Y., Banno, H., Imai, S., Tokumoto, M., Hasegawa, T., Seko, Y., Nagase, H., and Satoh, M. 2020. "Cadmium induces iron deficiency anemia through the suppression of iron transport in the duodenum." *Toxicology Letters* 332: 130–139.

Genchi, G., Sinicropi, M. S., Lauria, G., Carocci, A., and Catalano, A. 2020. "The effects of cadmium toxicity." *International Journal of Environmental Research and Public Health* 17(11): 3782.

Glassock, R. J., Warnock, D. G., and Delanaye, P. 2017. "The global burden of chronic kidney disease: Estimates, variability and pitfalls." *Nature Reviews Nephrology* 13(2): 104–114.

Grau-Perez, M., Pichler, G., Galan-Chilet, I., Briongos-Figuero, L. S., Rentero-Garrido, P., Lopez-Izquierdo, R., Navas-Acien, A., Weaver, V., García-Barrera, T., Gomez-Ariza, J. L., Martín-Escudero, J. C., Chaves, F. J., Redon, J., and Tellez-Plaza, M. 2017. "Urine cadmium levels and albuminuria in a general population from Spain: A gene-environment interaction analysis." *Environment International* 106: 27–36.

Grau-Perez, M., Voruganti, V. S., Balakrishnan, P., Haack, K., Goessler, W., Franceschini, N., Redón, J., Cole, S. A., Navas-Acien, A., and Tellez-Plaza, M. 2021. "Genetic variation and urine cadmium levels: ABCC1 effects in the strong heart family study." *Environmental Pollution* 276: 116717.

Gu, Y., Wang, P., Zhang, S., Dai, J., Chen, H. P., Lombi, E., Howard, D. L., Ent, A., Zhao, F. J., and Kopittke, P. M. 2020. "Chemical speciation and distribution of cadmium in rice grain and implications for bioavailability to humans." *Environmental Science & Technology* 54(19): 12072–12080.

Helmfrid, I., Berglund, M., Löfman, O., and Wingren, G. 2012. "Health effects and exposure to polychlorinated biphenyls (PCBs) and metals in a contaminated community." *Environment International* 44: 53–58.

Hernlund, E., Svedbom, A., Ivergård, M., Compston, J., Cooper, C., Stenmark, J., McCloskey, E. V., Jonsson, B., and Kanis, J. A. 2013. "Osteoporosis in the European Union: Medical management, epidemiology and economic burden: A report prepared in collaboration with the International Osteoporosis Foundation (IOF) and the European Federation of Pharmaceutical Industry Associations (EFPIA)." *Archives of Osteoporosis* 8: 1–115.

Himeno, S., and Fujishiro, H. 2021. "Roles of zinc transporters that control the essentiality and toxicity of manganese and cadmium." *Yakugaku Zasshi: Journal of the Pharmaceutical Society of Japan* 141(5): 695–703. (in Japanese)

Horiguchi, H., Oguma, E., Sasaki, S., Miyamoto, K., Ikeda, Y., Machida, M., and Kayama, F. 2004. "Dietary exposure to cadmium at close to the current provisional tolerable weekly intake does not affect renal function among female Japanese farmers." *Environmental Research* 95(1): 20–31.

Hu, J. 2021. "Toward unzipping the ZIP metal transporters: Structure, evolution, and implications on drug discovery against cancer." *The FEBS Journal* 288(20): 5805–5825.

Huo, Y., Du, H., Xue, B., Niu, M., and Zhao, S. 2016. "Cadmium removal from rice by separating and washing protein isolate." *Journal of Food Science* 81(6): T1576–T1584.

Jaenisch, R., and Bird, A. 2003. "Epigenetic regulation of gene expression: How the genome integrates intrinsic and environmental signals." *Nature Genetics* 33(3): 245–254.

JECFA. 2023. "Safety evaluation of certain contaminants in food: Prepared by the ninety-first meeting of the Joint FAO/WHO Expert Committee on Food Additives (JECFA)." 1–56 (https://www.who.int/publications/i/item/9789240060760).

Jiang, A., Gong, L., Ding, H., and Wang, M. 2021. "Cancer mortality and long-term environmental exposure of cadmium in contaminated community based on a third retrospective cause of death investigation of residents living in the Guangdong Province from 2004 to 2005." *Biological Trace Element Research* 199(12): 4504–4515.

Jin, B., Li, Y., and Robertson, K. D. 2011. "DNA methylation: Superior or subordinate in the epigenetic hierarchy?" *Genes & Cancer* 2(6): 607–617.

Jin, Y. Y., Zhou, R., and Kuang, X. Y. 2018. "Research progress on renal injury induced by cadmium." *Journal of Environmental and Occupational Medicine* 35(2): 180–184.

Jing, Y., Liu, L. Z., Jiang, Y., Zhu, Y., Guo, N. L., Barnett, J., Rojanasakul, Y., Agani, F., and Jiang, B. H. 2012. "Cadmium increases HIF-1 and VEGF expression through ROS, ERK, and AKT signaling pathways and induces malignant transformation of human bronchial epithelial cells." *Toxicological Sciences* 125(1): 10–19.

Jouybari, L., Saei Ghare Naz, M., Sanagoo, A., Kiani, F., Sayehmiri, F., Sayehmiri, K., and Hasanpour Dehkordi, A. 2018. "Toxic elements as biomarkers for breast cancer: A meta-analysis study." *Cancer Management and Research* 2018: 69–79.

Julin, B., Wolk, A., Bergkvist, L., Bottai, M., and Akesson, A. 2012. "Dietary cadmium exposure and risk of postmenopausal breast cancer: A population-based prospective cohort study." *Cancer Research* 72(6): 1459–1466.

Jung, Y. S., Jeong, E. M., Park, E. K., Kim, Y. M., Sohn, S., Lee, S. H., Baik, E. J., and Moon, C. H. 2008. "Cadmium induces apoptotic cell death through p38 MAPK in brain microvessel endothelial cells." *European Journal of Pharmacology* 578(1): 11–18.

Kalantar-Zadeh, K., Jafar, T. H., Nitsch, D., Neuen, B. L., and Perkovic, V. 2021. "Chronic kidney disease." *The Lancet* 398(10302): 786–802.

Khan, Z., Elahi, A., Bukhari, D. A., and Rehman, A. 2022. "Cadmium sources, toxicity, resistance and removal by microorganisms-A potential strategy for cadmium eradication." *Journal of Saudi Chemical Society* 26(6): 101569.

Kiani, B., Hashemi Amin, F., Bagheri, N., Bergquist, R., Mohammadi, A. A., Yousefi, M., Roshandel, G., Beirami, S., and Hoseini, B. 2021. "Association between heavy metals and colon cancer: An ecological study based on geographical information systems in North-Eastern Iran." *BMC Cancer* 21, 1–12.

Kim, H., Lee, J., Woo, H. D., Kim, D. W., Choi, I. J., Kim, Y. I., and Kim, J. 2019. "Association between dietary cadmium intake and early gastric cancer risk in a Korean population: A case–control study." *European Journal of Nutrition* 58, 3255–3266.

Kim, J., Lim, W., Ko, Y., Kwon, H., Kim, S., Kim, O., Park, G., Choi, H., and Kim, O. 2012. "The effects of cadmium on VEGF-mediated angiogenesis in HUVECs." *Journal of Applied Toxicology* 32(5): 342–349.

Kim, K., Melough, M. M., Vance, T. M., Noh, H., Koo, S. I., and Chun, O. K. 2018. "Dietary cadmium intake and sources in the US." *Nutrients* 11(1): 2.

Kim, M., and Wolt, J. D. 2011. "Probabilistic risk assessment of dietary cadmium in the South Korean population." *Food Additives & Contaminants: Part A* 28(1): 62–70.

Koizumi, N., Murata, K., Hayashi, C., Nishio, H., and Goji, J. 2008. "High cadmium accumulation among humans and primates: Comparison across various mammalian species—a study from Japan." *Biological Trace Element Research* 121, 205–214.

Lamouille, S., Xu, J., and Derynck, R. 2014. "Molecular mechanisms of epithelial–mesenchymal transition." *Nature Reviews Molecular Cell Biology* 15(3): 178–196.

Lane, T. W., Saito, M. A., George, G. N., Pickering, I. J., Prince, R. C., and Morel, F. M. 2005. "Biochemistry: A cadmium enzyme from a marine diatom." *Nature* 435(7038): 42–42.

Lanocha-Arendarczyk, N., Kosik-Bogacka, D. I., Prokopowicz, A., Kalisinska, E., Sokolowski, S., Karaczun, M., Zietek, P., Podlasińska, J., Pilarczyk, B., Tomza-Marciniak, A., Baranowska-Bosiacka, I., Gutowska, I., Safranow, K., and Chlubek, D. 2015. "The effect of risk factors on the levels of chemical elements in the tibial plateau of patients with osteoarthritis following knee surgery." *BioMed Research International* 2015: 650282.

Larson, C. 2014. "Environmental science China gets serious about its pollutant-laden soil." *Science* 343(6178): 1415–1416.

Larsson, S. C., Orsini, N., and Wolk, A. 2015. "Urinary cadmium concentration and risk of breast cancer: A systematic review and dose-response meta-analysis." *American Journal of Epidemiology* 182(5): 375–380.

Lech, T., and Sadlik, J. K. 2017. "Cadmium concentration in human autopsy tissues." *Biological Trace Element Research* 179: 172–177.

Lecuit, T., and Le Goff, L. 2007. "Orchestrating size and shape during morphogenesis." *Nature* 450(7167): 189–192.

Lee, J. Y., Tokumoto, M., Fujiwara, Y., Hasegawa, T., Seko, Y., Shimada, A., and Satoh, M. 2016. "Accumulation of p53 via down-regulation of UBE2D family genes is a critical pathway for cadmium-induced renal toxicity." *Scientific Reports* 6(1): 21968.

Levey, A. S., Inker, L. A., Matsushita, K., Greene, T., Willis, K., Lewis, E., Zeeuw, D. D., Cheung, A. K., and Coresh, J. 2014. "GFR decline as an end point for clinical trials in CKD: A scientific workshop sponsored by the National Kidney Foundation and the US Food and Drug Administration." *American Journal of Kidney Diseases* 64(6): 821–835.

Li, D., Lin, H., Zhang, M., Meng, J., Hu, L., and Yu, B. 2021. "Urine cadmium as a risk factor for osteoporosis and osteopenia: A meta-analysis." *Frontiers in Medicine* 8: 648902.

Li, H. B., Li, M. Y., Zhao, D., Li, J., Li, S. W., Juhasz, A. L., Basta, N. T., Luo, Y. M., and Ma, L. Q. 2019. "Oral bioavailability of As, Pb, and Cd in contaminated soils, dust, and foods based on animal bioassays: A review." *Environmental Science & Technology* 53(18): 10545–10559.

Lin, J., Zhang, F., and Lei, Y. 2016. "Dietary intake and urinary level of cadmium and breast cancer risk: A meta-analysis." *Cancer Epidemiology* 42: 101–107.

Lin, Q., Hamid, Y., Yin, X., Hussain, B., He, Z., and Yang, X. 2022. "Screening of low-Cd accumulating early rice cultivars coupled with phytoremediation and agro-production: Bioavailability and bioaccessibility tests." *Science of the Total Environment* 844: 157143.

Lin, X., Peng, L., Xu, X., Chen, Y., Zhang, Y., and Huo, X. 2018. "Connecting gastrointestinal cancer risk to cadmium and lead exposure in the Chaoshan population of Southeast China." *Environmental Science and Pollution Research* 25: 17611–17619.

Lin, Y. S., Ho, W. C., Caffrey, J. L., and Sonawane, B. 2014. "Low serum zinc is associated with elevated risk of cadmium nephrotoxicity." *Environmental Research* 134: 33–38.

Liu, F., Wang, B., Li, L., Dong, F., Chen, X., Li, Y., Dong, X., Wada, Y., Kapron, C. M., and Liu, J. 2015. "Low-dose cadmium upregulates VEGF expression in lung adenocarcinoma cells." *International Journal of Environmental Research and Public Health* 12(9): 10508–10521.

Liu, L. Z., Hu, X. W., Xia, C., He, J., Zhou, Q., Shi, X., Fang, J., Jiang, B. H. 2006. "Reactive oxygen species regulate epidermal growth factor-induced vascular endothelial growth factor and hypoxia-inducible factor-1α expression through activation of AKT and P70S6K1 in human ovarian cancer cells." *Free Radical Biology and Medicine* 41(10): 1521–1533.

Liu, P., Xiao, W., Wang, K., Yang, Z., and Wang, L. 2021a. "Bioaccessibility of Cd and its correlation with divalent mineral nutrients in locally grown rice from two provinces in China." *Biological Trace Element Research* 2021: 1–10.

Liu, R., He, Q., Wu, Y., Wang, J., and Fu, J. 2021b. "24 h absorption and excretion profiles of cadmium from contaminated cooked brown rice and white rice in female rats." *Journal of Cereal Science* 102: 103300.

Liu, R. Y. 2023. "Study on the absorption, accumulation, excretion and damage of cadmium in staple rice matrices." *Central South University of Forestry and Technology* 2023: 1–13.

Lu, Y., Chan, Y. T., Tan, H. Y., Li, S., Wang, N., and Feng, Y. 2020. "Epigenetic regulation in human cancer: The potential role of epi-drug in cancer therapy." *Molecular Cancer* 19: 1–16.

Ma, Y., Ran, D., Cao, Y., Zhao, H., Song, R., Zou, H., Gu, J., Yuan, Y., Bian, J., Zhu, J., and Liu, Z. 2021. The effect of P2X7 on cadmium-induced osteoporosis in mice. *Journal of Hazardous Materials* 405: 124251.

Madrigal, J. M., Ricardo, A. C., Persky, V., and Turyk, M. 2019. "Associations between blood cadmium concentration and kidney function in the US population: Impact of sex, diabetes and hypertension." *Environmental Research* 169: 180–188.

Mamyrbayev, A., Djarkenov, T., Dosbayev, A., Dusembayeva, N., Shpakov, A., Umarova, G., Drobchenko, Y., Kunurkulzhayev, T., Zhaylybaev, M., and Isayeva, G. 2016. "The incidence of malignant tumors in environmentally disadvantaged regions of Kazakhstan." *Asian Pacific Journal of Cancer Prevention: APJCP* 17(12): 5203.

Mao, P., Wu, J., Li, F., Sun, S., Huang, R., Zhang, L., Mo, J., Li, Z., and Zhuang, P. 2022. "Joint approaches to reduce cadmium exposure risk from rice consumption." *Journal of Hazardous Materials* 429: 128263.

Marconi, G. D., Fonticoli, L., Rajan, T. S., Pierdomenico, S. D., Trubiani, O., Pizzicannella, J., and Diomede, F. 2021. "Epithelial-mesenchymal transition (EMT): The type-2 EMT in wound healing, tissue regeneration and organ fibrosis." *Cells* 10(7): 1587.

Mohanty, S., Nayak, R. K., Jena, B., Padhan, K., Mohapatra, K. K., Sahoo, S. K., Dash, P. K., Das, J., Behera, S. K., Sahu, A., Nayak, J. K., Padhan, S., and Datta, D. 2023. "Heavy metal contamination in rice, pulses, and vegetables from CKDu-endemic areas in Cuttack district, India: A health risk assessment." *Frontiers in Environmental Science* 11: 1248373.

Mortoglou, M., Buha Djordjevic, A., Djordjevic, V., Collins, H., York, L., Mani, K., Valle, E., Wallace, D., and Uysal-Onganer, P. 2022. "Role of microRNAs in response to cadmium chloride in pancreatic ductal adenocarcinoma." *Archives of Toxicology* 96: 1–19.

Myong, J. P., Kim, H. R., Baker, D., and Choi, B. 2012. Blood cadmium and moderate-to-severe glomerular dysfunction in Korean adults: Analysis of KNHANES 2005–2008 data. *International Archives of Occupational and Environmental Health* 85: 885–893.

Nambunmee, K., Honda, R., Nishijo, M., Swaddiwudhipong, W., Nakagawa, H., and Ruangyuttikarn, W. 2010. "Bone resorption acceleration and calcium reabsorption impairment in a Thai population with high cadmium exposure." *Toxicology Mechanisms and Methods* 20(1): 7–13.

Nemmiche, S. 2017. "Oxidative signaling response to cadmium exposure." *Toxicological Sciences* 156(1): 4–10.

Ngo, D. M., Hough, R. L., Le, T. T., Nyberg, Y., Le, B. M., Nguyen, C. V., Nguyen, M. K., and Ingrid, O. 2012. "Assessing dietary exposure to cadmium in a metal recycling community in Vietnam: Age and gender aspects." *Science of the Total Environment* 416: 164–171.

Nishijo, M., Nakagawa, H., Suwazono, Y., Nogawa, K., and Kido, T. 2017. "Causes of death in patients with Itai-itai disease suffering from severe chronic cadmium poisoning: A nested case–control analysis of a follow-up study in Japan." *BMJ Open* 7(7): e015694.

Nishijo, M., Nogawa, K., Suwazono, Y., Kido, T., Sakurai, M., and Nakagawa, H. 2020. "Lifetime cadmium exposure and mortality for renal diseases in residents of the cadmium-polluted Kakehashi River Basin in Japan." *Toxics* 8(4): 81.

Nogawa, K., Sakurai, M., Ishizaki, M., Kido, T., Nakagawa, H., and Suwazono, Y. 2017. "Threshold limit values of the cadmium concentration in rice in the development of itai-itai disease using benchmark dose analysis." *Journal of Applied Toxicology* 37(8): 962–966.

Nogawa, K., Watanabe, Y., Sakuma, S., Sakurai, M., Nishijo, M., Ishizaki, M., Morikawa, Y., Kido, T., Nakagawa, H., and Suwazono, Y. 2022. "Renal tubular dysfunction and cancer mortality in the Japanese general population living in cadmium-non-contaminated areas." *Journal of Applied Toxicology* 42(9): 1458–1466.

Oomen, A. G., Hack, A., Minekus, M., Zeijdner, E., Cornelis, C., Schoeters, G., Verstraete, W., Wiele, T. V., Wragg, J., Rompelberg, C. J. M., Adriënne, J. A. M. S., and Van Wijnen, J. H. 2002. "Comparison of five in vitro digestion models to study the bioaccessibility of soil contaminants." *Environmental Science & Technology* 36(15): 3326–3334.

Ostadrahimi, A., Payahoo, L., Somi, M. H., Hashemzade, S. H., Esfahani, A., Asgharijafarabadi, M., Mobasseri, M., Samadi, N., Faraji, S., and KhajeBishak, Y. 2017. "The association between blood cadmium levels and the risk of gastrointestinal cancer in Tabriz, northwest of Iran." *Polish Annals of Medicine* 24(2): 133–137.

Peana, M., Pelucelli, A., Chasapis, C. T., Perlepes, S. P., Bekiari, V., Medici, S., and Zoroddu, M. A. 2022. Biological effects of human exposure to environmental cadmium. *Biomolecules* 13(1): 36.

Peng, L., Chen, J., Chen, L., Ding, W., Gao, Y., Wu, Y., and Xiong, Y. L. 2020. "Effect of degree of milling on the cadmium in vitro bioaccessibility in cooked rice." *Journal of Food Science* 85(11): 3756–3763.

Peng, X., Li, C., Zhao, D., and Huang, L. 2023. "Associations of micronutrients exposure with cadmium body burden among population: A systematic review." *Ecotoxicology and Environmental Safety* 256: 114878.

Petrick, J. L., Florio, A. A., Znaor, A., Ruggieri, D., Laversanne, M., Alvarez, C. S., Ferlay, J., Valery, P. C., Bray, F., and McGlynn, K. A. 2020. "International trends in hepatocellular carcinoma incidence, 1978–2012." *International Journal of Cancer* 147(2): 317–330.

Pizzino, G., Irrera, N., Cucinotta, M., Pallio, G., Mannino, F., Arcoraci, V., Squadrito, F., Altavilla, D., and Bitto, A. 2017. "Oxidative stress: Harms and benefits for human health." *Oxidative Medicine and Cellular Longevity* 2017: 8416763.

Praveena, S. M., and Omar, N. A. 2017. "Heavy metal exposure from cooked rice grain ingestion and its potential health risks to humans from total and bioavailable forms analysis." *Food Chemistry* 235: 203–211.

Rakvács, Z., Kucsma, N., Gera, M., Igriczi, B., Kiss, K., Barna, J., Kovács, D., Vellai, T., Bencs, L., Reisecker, J. M., Szoboszlai, N., and Szakács, G. 2019. "The human ABCB6 protein is the functional homologue of HMT-1 proteins mediating cadmium detoxification." *Cellular and Molecular Life Sciences* 76, 4131–4144.

Reyes-Hinojosa, D., Lozada-Pérez, C. A., Cuevas, Y. Z., López-Reyes, A., Martínez-Nava, G., Fernández-Torres, J., Olivos-Meza, A., Landa-Solis, C., Gutiérrez-Ruiz, M. C., Castillo, E. R. D., and Martínez-Flores, K. 2019. "Toxicity of cadmium in musculoskeletal diseases." *Environmental Toxicology and Pharmacology* 72: 103219.

Roczniak, W., Brodziak-Dopierała, B., Cipora, E., Jakóbik-Kolon, A., Kluczka, J., and Babuśka-Roczniak, M. 2017. "Factors that affect the content of cadmium, nickel, copper and zinc in tissues of the knee joint." *Biological Trace Element Research* 178: 201–209.

Ruby, M. V., Schoof, R., Brattin, W., Goldade, M., Post, G., Harnois, M., Mosby, D. E., Casteel, S. W., Berti, W., Carpenter, M., Edwards, D., Cragin, D., and Chappell, W. 1999. "Advances in evaluating the oral bioavailability of inorganics in soil for use in human health risk assessment." *Environmental Science & Technology* 33(21): 3697–3705.

Samarajeewa, U. 2022. "Heavy metals and food safety in Sri Lanka: A review." *Journal of the National Science Foundation of Sri Lanka* 50(3): 541–562.

Satarug, S. 2012. "Long-term exposure to cadmium in food and cigarette smoke, liver effects and hepatocellular carcinoma." *Current Drug Metabolism* 13(3): 257–271.

Satarug, S. 2018. "Dietary cadmium intake and its effects on kidneys." *Toxics* 6(1): 15.

Satarug, S. 2023. "Is environmental cadmium exposure causally related to diabetes and obesity?" *Cells* 13(1): 83.

Satarug, S., Garrett, S. H., Somji, S., Sens, M. A., and Sens, D. A. 2021. "Zinc, zinc transporters, and cadmium cytotoxicity in a cell culture model of human urothelium." *Toxics* 9(5): 94.

Satarug, S., Vesey, D. A., and Gobe, G. C. 2017. "Kidney cadmium toxicity, diabetes and high blood pressure: The perfect storm." *The Tohoku Journal of Experimental Medicine* 241(1): 65–87.

Satarug, S., Vesey, D. A., Gobe, G. C., and Phelps, K. R. 2023. "Estimation of health risks associated with dietary cadmium exposure." *Archives of Toxicology* 97(2): 329–358.

Schaefer, H. R., Dennis, S., and Fitzpatrick, S. 2020. "Cadmium: Mitigation strategies to reduce dietary exposure." *Journal of Food Science* 85(2): 260–267.

Shi, J. W., Fan, D. X., and Li, M. Q. 2023. "The relationship between cadmium exposure and mortality in postmenopausal females: A cohort study of 2001–2018 NHANES." *Nutrients* 15(21): 4604.

Sitek, A., and Kozłowska, L. 2022. "The role of well-known antioxidant vitamins in the prevention of cadmium-induced toxicity." *International Journal of Occupational Medicine and Environmental Health* 35(4): 367.

Smereczański, N. M., and Brzóska, M. M. 2023. "Current levels of environmental exposure to cadmium in industrialized countries as a risk factor for kidney damage in the general population: A comprehensive review of available data." *International Journal of Molecular Sciences* 24(9): 8413.

Song, Y., Wang, Y., Mao, W., Sui, H., Yong, L., Yang, D., Jiang, D., Zhang, L., and Gong, Y. 2017. "Dietary cadmium exposure assessment among the Chinese population." *PLoS One* 12(5): e0177978.

Syed, M. H., Rubab, S. A., Abbas, S. R., Qutaba, S., Mohd Zahari, M. A., and Abdullah, N. 2023. "Effects of cadmium acetate contaminated drinking water on vital organs: A histopathological and biochemical study." *Journal of Biochemical and Molecular Toxicology* 37(8): e23382.

Taha, M. M., Mahdy-Abdallah, H., Shahy, E. M., Ibrahim, K. S., and Elserougy, S. 2018. "Impact of occupational cadmium exposure on bone in sewage workers." *International Journal of Occupational and Environmental Health* 24(3–4): 101–108.

Tang, Y., Wang, H., Wang, Q., and Wang, X. 2023. "Amelioration of cadmium cytotoxicity to human cells by nutrient cation contents and the building of a biotic ligand model." *Ecotoxicology and Environmental Safety* 253: 114690.

Tavakoli Pirzaman, A., Ebrahimi, P., Niknezhad, S., Vahidi, T., Hosseinzadeh, D., Akrami, S., Ashrafi, A. M., Velayatimehr, M. M., Hosseinzadeh, R., and Kazemi, S. 2023. "Toxic mechanisms of cadmium and exposure as a risk factor for oral and gastrointestinal carcinomas." *Human & Experimental Toxicology* 2023: 42.

Tavarez, M., Grusak, M. A., and Sankaran, R. P. 2023. "The effect of exogenous cadmium and zinc applications on cadmium, zinc and essential mineral bioaccessibility in three lines of rice that differ in grain cadmium accumulation." *Foods* 12(21): 4026.

Thévenod, F., and Lee, W. K. 2024. "Cadmium transport by mammalian ATP-binding cassette transporters." *BioMetals* 2024: 1–23.

Tian, J., Hu, J., Chen, M., Yin, H., Miao, P., Bai, P., and Yin, J. 2017. "The use of mrp1-deficient (Danio rerio) zebrafish embryos to investigate the role of Mrp1 in the toxicity of cadmium chloride and benzo [a] pyrene." *Aquatic Toxicology* 186, 123–133.

Tinkov, A. A., Filippini, T., Ajsuvakova, O. P., Skalnaya, M. G., Aaseth, J., Bjørklund, G., Gatiatulina, E. R., Popova, E. V., Nemereshina, O. N., Huang, P. T, Vinceti, M., and Skalny, A. V. 2018. "Cadmium and atherosclerosis: A review of toxicological mechanisms and a meta-analysis of epidemiologic studies." *Environmental Research* 162, 240–260.

Tiwari, N., Gheldof, A., Tatari, M., and Christofori, G. 2012. "EMT as the ultimate survival mechanism of cancer cells." *Seminars in Cancer Biology* 22(3): 194–207.

Tran, F., Lee, E., Cuddapah, S., Choi, B. H., and Dai, W. 2022. "MicroRNA–gene interactions impacted by toxic metal(oid)s during EMT and carcinogenesis." *Cancers* 14(23): 5818.

Unsal, V., Dalkıran, T., Çiçek, M., and Kölükçü, E. 2020. "The role of natural antioxidants against reactive oxygen species produced by cadmium toxicity: A review." *Advanced Pharmaceutical Bulletin* 10(2): 184.

Wang, M., Chen, Z., Song, W., Hong, D., Huang, L., and Li, Y. 2021. "A review on cadmium exposure in the population and intervention strategies against cadmium toxicity." *Bulletin of Environmental Contamination and Toxicology* 106: 65–74.

Wang, Y. X. 2019. "Study on the bioaccessibility and in vivo and vitro bioavailability of cadmium in rice." *Wuhan Polytechnic University* 2019: 1–18. (in Chinese)

Watanabe, T., Zhang, Z. W., Moon, C. S., Shimbo, S., Nakatsuka, H., Matsuda-Inoguchi, N., Higashikawa, K., and Ikeda, M. 2000. "Cadmium exposure of women in general populations in Japan during 1991–1997 compared with 1977–1981." *International Archives of Occupational and Environmental Health* 73(1): 26–34.

Wei, T., Jia, J., Wada, Y., Kapron, C. M., and Liu, J. 2017. "Dose dependent effects of cadmium on tumor angiogenesis." *Oncotarget* 8(27): 44944–44959.

WHO. 2019. "Preventing disease through healthy environments: Exposure to cadmium: A major public health concern." *World Health Organization* 2019: 1–6.

Xiao, W., Yang, Y., Tang, N., Huang, X., Zhang, Q., Zhao, S., Chen, D., Guo, B., Zhao, Z., Jiang, Y., and Ye, X. 2024. "Innovative accumulative risk assessment of co-exposure to Cd, As, and Pb in contaminated rice based on their in vivo bioavailability and in vitro bioaccessibility." *Science of the Total Environment* 912: 168922.

Yan, H., and Bu, P. 2021. "Non-coding RNA in cancer." *Essays in Biochemistry* 65(4): 625–639.

Yang, J., Guo, T., Liang, X., Zhai, Y., Cheng, Y., Sun, H., Cui, Y., and Cheng, X. 2020. "Cadmium inhibits apoptosis of human esophageal epithelial cells by upregulating CDK6." *Ecotoxicology and Environmental Safety* 205: 111146.

Yao, L., Wang, Y., Deng, Z., Wu, Q., Fang, M., Wu, Y., and Gong, Z. 2021. "Study on the bioaccessibility and bioavailability of Cd in contaminated rice in vitro and in vivo." *Journal of Food Science* 86(8): 3730–3742.

Yang, H., and Shu, Y. 2015. "Cadmium transporters in the kidney and cadmium-induced nephrotoxicity." *International Journal of Molecular Sciences* 16(1): 1484–1494.

Zalups, R. K., and Ahmad, S. 2003. "Molecular handling of cadmium in transporting epithelia." *Toxicology and Applied Pharmacology* 186(3): 163–188.

Zang, Y., Devleesschauwer, B., Bolger, P. M., Goodman, E., and Gibb, H. J. 2019. "Global burden of late-stage chronic kidney disease resulting from dietary exposure to cadmium, 2015." *Environmental Research* 169: 72–78.

Zhai, Q., Liu, Y., Wang, C., Zhao, J., Zhang, H., Tian, F., Lee, Y. K., and Chen, W. 2019. "Increased cadmium excretion due to oral administration of Lactobacillus plantarum strains by regulating enterohepatic circulation in mice." *Journal of Agricultural and Food Chemistry* 67(14): 3956–3965.

Zhang, K., Long, M., Dong, W., Li, J., Wang, X., Liu, W., Huang, Q., Ping, Y., Zou, H., Song, R., Liu, G., Ran, D., and Liu, Z. 2024. "Cadmium induces kidney iron deficiency and chronic kidney injury by interfering with the iron metabolism in rats." *International Journal of Molecular Sciences* 25(2): 763.

Zhang, L., Wang, Y., Li, T., Zhuo, W., and Zhu, Y. 2023a. "Elevated serum and hair levels of cadmium as a risk factor for liver carcinoma: A meta-analysis." *Nutrition and Cancer* 75(6): 1438–1447.

Zhang, H., Li, L., Wang, Y., Dong, F., Chen, X., Liu, F., Xu, D., Yi, F., Kapron, C. M., and Liu, J. 2016. "NF-κB signaling maintains the survival of cadmium-exposed human renal glomerular endothelial cells." *International Journal of Molecular Medicine* 38(2): 417–422.

Zhang, S. N., Xie, W. Y., Zhai, Z. Q., Chen, C., Zhao, F. J., and Wang, P. 2023b. "Dietary intake of household cadmium-contaminated rice caused genome-wide DNA methylation changes on gene/hubs related to metabolic disorders and cancers." *Environmental Pollution* 327, 121553.

Zhao, D., Wang, P., and Zhao, F. J., 2023. "Dietary cadmium exposure, risks to human health and mitigation strategies." *Critical Reviews in Environmental Science and Technology* 53(8): 939–963.

Zhao, L., Islam, R., Wang, Y., Zhang, X., and Liu, L. Z. 2022a. "Epigenetic regulation in chromium-, nickel-and cadmium-induced carcinogenesis." *Cancers* 14(23): 5768.

Zhao, X., Shao, Y., Ma, L., Shang, X. H., Zhao, Y. F., and Wu, Y. N. 2022b. "Exposure to lead and cadmium in the Sixth Total Diet Study—China, 2016–2019." *China CDC Weekly* 4(9): 176–179.

Zhu, H., Tang, X., Gu, C., Chen, R., Liu, Y., Chu, H., Zhang, Z. 2024. "Assessment of human exposure to cadmium and its nephrotoxicity in the Chinese population." *Science of the Total Environment* 918: 170488.

Zhu, P., Liang, X. X., Wang, P., Wang, J., Gao, Y. H., Hu, S. G., Huang, Q., Huang, R., Jiang, Q., Wu, S. X., Li, Z. X., Ling, H. T., Xu, Y. H., Wu, N. Y., Zou, F., and Yang, X. F. 2016. "Assessment of dietary cadmium exposure: A cross-sectional study in rural areas of south China." *Crop Science* 62: 284–290.

Zhu, X. J., Wang, J. J., Mao, J. H., Shu, Q., and Du, L. Z. 2019. "Relationships of cadmium, lead, and mercury levels with albuminuria in US adults: Results from the national health and nutrition examination survey database, 2009–2012." *American Journal of Epidemiology* 188(7): 1281–1287.

Zhuang, P., Zhang, C., Li, Y., Zou, B., Mo, H., Wu, K., Wu, J., and Li, Z. 2016. "Assessment of influences of cooking on cadmium and arsenic bioaccessibility in rice, using an in vitro physiologically-based extraction test." *Food Chemistry* 213: 206–214.

19

Effects of Exposure to Low and High Concentrations of Trace Elements on Male Infertility

Leonid Perelomov, Irina Perelomova, Maria Gertsen, and Yulia Venevtseva

19.1 Introduction

Human fertility is a complex phenomenon that depends on variables, including genetic, physiological, and environmental influences (Jurewicz et al., 2009, Mendiola et al., 2009, Wong and Cheng, 2011). Infertility can affect countless people and have a significant impact on their lives and families. According to the World Health Organization, approximately 17.5% of adults worldwide have infertility, or about one in six people. Infertility rates are similar in high-, middle-, and low-income countries, with approximately 17.8% of people in high-income countries experiencing infertility and approximately 16.5% in low- and middle-income countries. This means that infertility is a significant problem throughout the globe, regardless of economic status (WHO, 2023).

Recently, there has been a pressing issue studying environmental factors' impact on reproductive health, including those related to lifestyle, living conditions, and professional activities. It is worth noting that the study of this intimate process as human fertility is constrained by ethical considerations and methodological challenges, mainly due to the specificity and differences in reproductive physiology among various mammalian species that serve as model organisms (Perelomov et al., 2016).

Microelements are chemical elements concentrated in environmental objects below 10^{-3}%. This includes a wide range of metals, transition metals, and non-metals with various chemical properties. Due to this purely quantitative definition, some elements can be macro- and micronutrients, depending on the specific biological system.

Microelements, from the perspective of their nutritional significance for humans, can be divided into three main groups: essential elements, those that are likely essential, and potentially toxic elements. An element is considered critical to an organism if it is an integral part of organic structures performing vital functions or if reducing its exposure below a certain level consistently decreases physiologically significant function (WHO, 1996). However, as known, essential microelements can cause toxic effects at high concentrations, while potentially toxic elements may be necessary at specific levels for normal physiological processes. This is in line with the principle of the ecological optimum. Recommendations for the intake of microelements, primarily through food, are presented as safe consumption ranges that the human body has adapted to during evolution. Deviating from these ranges, whether decreasing or increasing intake, can lead to changes in various body systems' physiological functions and, thus, to adverse health outcomes, including reproductive health problems.

This chapter examines the impact of deficiency and excess of trace elements on the functioning of the male reproductive system. Brief information on the physiology of the male reproductive system is provided, and the main indicators of its health are shown. The role of zinc, copper, iodine, selenium, manganese, molybdenum, and iron as micronutrients necessary for preventing the development of infertility is considered. Particular attention is paid to the toxic effect of high concentrations of microelements on indicators of male reproductive health: disorders in the endocrine system (the "hypothalamus-pituitary-testicles" axis) and the formation of the male genital tract, a decrease in the quantity and quality of sperm, the occurrence of genetic mutations.

DOI: 10.1201/9781003492115-24

19.2 Male Reproductive Health Indicators

Male sexual differentiation and maturation are driven by androgens formed in the fetal testes. Androgens are typically steroids, responsible for the development and maintenance of male characteristics. These hormones control the activation of primary male reproductive organs and the development of secondary male sex characteristics (Tue Nguyen et al., 2018).

The main integrative center of the functional reproductive system in men is the hypothalamus. It receives information from the central nervous system and the testes and regulates the formation and secretion of gonadotropin-releasing hormone (GnRH). GnRH is secreted in a pulsatile fashion and acts as a necessary link in stimulating the synthesis and release of gonadotropins from the pituitary gland, such as luteinizing hormone (LH) and follicle-stimulating hormone (FSH) (Figure 19.1).

In the testes, these hormones bind to specific receptors on Leydig and Sertoli cells. Leydig cells are located in the connective tissue area between the seminiferous tubules. Their primary function is to produce testosterone and steroidal intermediates from cholesterol via a series of enzymatic pathways under the control of LH. The FSH acts on Sertoli cells on the basement membrane of the seminiferous tubules and is connected by tight junctions. These cells form a continuous and complete lining within the tubular wall and establish the blood-testis barrier. They provide nourishment to the developing sperm under spermatogenesis. Sertoli cells have several other functions, including destroying defective sperm cells, secretion of fluid that helps transport sperm into the epididymis, and release of hormone inhibin that helps regulate FSH and sperm production (Figure 19.1).

Testosterone inhibits GnRH, LH, and FSH secretion, while inhibin B inhibits FSH secretion via feedback inhibition mechanisms. Prolactin release also inhibits GnRH's release from the hypothalamus, decreasing LH and FSH release from the anterior pituitary (Raut et al., 2019). Details of endocrine control of testicular function in adults are shown in Table 19.1.

The growth, development, and functions of organs of the male reproductive system are regulated by the action of the androgens, mediated via the androgen receptors (ARs). The main types of androgens include testosterone, dihydrotestosterone (DHT), dehydroepiandrosterone (DHEA), androstenedione, androstenediol, and androsterone. Testosterone, the primary androgen, is synthesized and secreted primarily in the Leydig cells in males, which are stimulated by LHs. Testosterone plays a vital role in developing the reproductive system in male fetuses and stimulating secondary male sexual characteristics, including body hair, male body habitus, and increased muscle growth. Of all the testosterone entering the bloodstream,

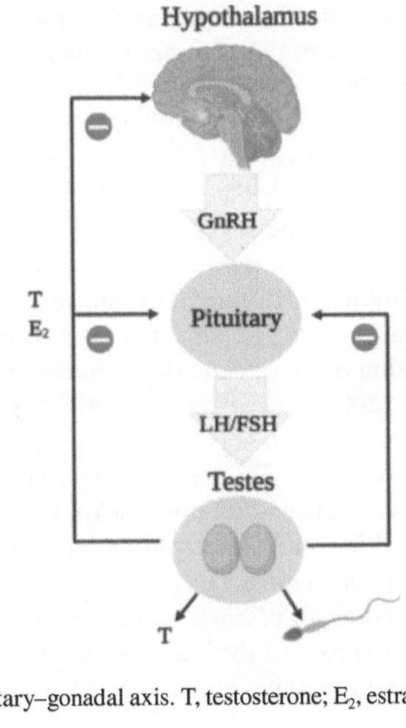

FIGURE 19.1 The hypothalamus–pituitary–gonadal axis. T, testosterone; E_2, estradiol. (Modified from Sikka and Naz, 2002).

TABLE 19.1

Endocrine Control of Testicular Function in Adult Men

Hormone	Source	Major Target	Direct or Indirect Effect on Target (s)
Luteinizing hormone (LH)	Anterior pituitary	Leydig cells	Stimulate steroidogenesis (testosterone production)
Follicle-stimulating hormone (FSH)	Anterior pituitary	Sertoli cells Sertoli and/or germ cells	Stimulate protein synthesis (e.g. androgen-binding protein) Maturation of spermatids into spermatozoa (spermiogenesis)
Testosterone	Leydig cells	Male accessory glands	Maintain structure and function
		Hypothalamus and pituitary	Negative feedback control on the release of FSH and LH
Estradiol	Leydig cells	Anterior pituitary	Negative feedback control on the release of FSH and LH
Inhibin	Sertoli cells	Anterior pituitary	Negative feedback control on the release of FSH

Source: Overstreet and Blazak (1983).

98% is bound and weakly active. Only 2% of free testosterone significantly stimulates various organs and systems in the body (Sagalov, 2006). DHT is made from testosterone by the enzyme 5α-reductase in the prostate, testes, hair follicles, and adrenal glands. DHEA is made in the adrenal cortex from cholesterol and is a primary precursor for estrogens. Androstenedione is an androgenic hormone produced by the testes and adrenal cortex. This hormone is converted to testosterone and other androgens. Androstenediol is a precursor for testosterone and has weak androgenic effects. Androsterone is a byproduct of androgen decay, and it also has minor masculinizing effects and significantly less testosterone activity.

ARs are activated by binding to androgens in the cytoplasm and then translocated into the nucleus and bind androgen response elements. ARs are expressed in many organs, including the testes, prostate, pituitary, hypothalamus, and liver. In the presence of androgens, ARs function as DNA-binding transcription factors that control gene expression. The other functions of ARs include facilitating the recruitment of other DNA-binding proteins (Sagalov, 2006), inducing rapid activation of kinase-signaling cascades, and modulating intracellular calcium levels (Boron and Boulpaep, 2011). These interactions between androgens and ARs are pivotal in ensuring proper spermatogenesis, spermatozoids survival, and functions (Tue Nguyen et al., 2018).

Spermatogenesis is a complex process involving the division of spermatogonia, stem cells capable of reproducing similar cells or producing daughter cells that differentiate into spermatocytes. These spermatocytes undergo further division, generating cell lines that eventually become spermatids and spermatozoa. There are six stages in the development of the germinal epithelium. Sperm development from the first stage to the sixth is one cycle. Each cycle lasts approximately 16 days in humans, and the transformation from early spermatogonia to mature sperm takes 4.6 cycles. Thus, spermatogenesis in humans lasts about 74 days (Sagalov, 2006).

The germinal epithelium, which consists of four to eight layers of cells, forms the walls of the seminiferous tubules and constitutes up to 85%–90% of the testicular volume. It contains spermatogenic cells at different stages of development and Sertoli cells that support them. Spermatogonia and young spermatocytes are located closer to the basal membrane of the seminiferous tubules. As they mature, they move to the upper layers and deep into the lumen of the tubules. Approximately 100–200 million sperm are produced in the testis daily in an adult male (Boron and Boulpaep, 2011). However, only a significantly smaller number of them enter the ejaculate due to partial death during development in the testis and vas deferens. Maturation and accumulation of these cells occur in the latter, particularly in the epididymis (Boron and Boulpaep, 2011, Sagalov, 2006).

In addition to sperm maturation, the epididymis is a storage site for male reproductive cells. Estimates suggest that the extratesticular reservoir of sperm contains approximately 440 million cells, with more than 50% located in the caudal region of the epididymis. Sperm from the epididymis's tail enter the vas deferens, which are moved by peristaltic contractions of the duct walls, propelling through the ejaculatory ducts and into the urethra under ejaculation. During this process, the secretions from the seminal vesicles and prostate gland enter the posterior urethra, contributing to the overall composition of semen.

During ejaculation, the muscles at the base of the bladder contract, and the external sphincter relaxes, allowing sperm to move down the urethra. This is made possible by the rhythmic contractions of the perineal and bulbourethral muscles. Most of the seminal fluid is produced in the seminal vesicles, with smaller amounts produced by the prostate. The seminal vesicles contribute to the nutrient supply for sperm, providing fructose and other substances like prostaglandins and clotting factors. The primary function of seminal plasma is thought to be creating a buffer for the acidic environment in the vagina. After ejaculation, a lot of sperm is formed, which is broken down by enzymes in the prostate within 20–30 min. Prostate secretions add zinc ions, phospholipids, and other substances to the semen.

Therefore, the following indicators can be considered for assessing male reproductive health in the context of toxic microelement exposure:

1. The status of the hypothalamus–pituitary–testis axis, which regulates hormonal function;
2. The condition of the reproductive tract condition where sperms are produced and mature;
3. Quality and quantity of sperm, including concentration in seminal fluid and motility;
4. Genetic material, including gene, chromosome, and genomic mutations.

The pathogenesis of male infertility remains poorly understood. It is generally accepted today that various etiological factors, including unknown ones in the idiopathic form, trigger similar pathogenetic processes that ultimately decrease sperm quality and count (Pizent et al., 2012). There are some exceptions, such as specific forms of infertility characterized by specific pathogens, in particular, excretory disorders that make it difficult for sperm to travel through the genital tract, as well as severe hormonal imbalances that inhibit spermatogenesis sharply.

Oxidative stress is currently considered to be one of the most significant pathogenic mechanisms for the development of male infertility. This condition arises from the abnormal accumulation of molecules that contain oxygen in an unreduced form, known as reactive oxygen species (ROS).

Another important pathogenetic mechanism for the development of male infertility is the increase in apoptosis, a programmed cell death process. Usually, there is a balance between pro- and anti-apoptotic factors, which is disrupted under the influence of the abovementioned etiological factors. Increased apoptosis in the testicular tissue has been observed in various forms of male infertility, including varicocele, inflammatory processes, and hormonal disorders (Aitken and Koppers, 2011, Sinha Hikim and Swerdloff, 1999).

For a long time, another mechanism for the development of infertility has been actively studied—the formation of anti-sperm antibodies. These antibodies can occur as a result of various factors, such as vasectomy, obstruction of the genital tract, trauma, testicular torsion, tumor, or infection. However, they have been found in sick and healthy men (Bohring and Krause, 2003). Therefore, the role of these antibodies in the development of infertility remains unclear.

Among medical professionals, the classification of male infertility proposed by Tiktinsky et al. is widely used (Tiktinsky et al., 2010). This classification is easy to understand and covers most potential causes of male infertility.

1. Secretory infertility:
   a. Primary testicular failure (congenital and acquired);
   b. Secondary testicular failure:
      - Central origin (hypothalamic-pituitary region),
      - Disconnection (endocrine and other organs).
2. Excretere infertility
   a. Diseases and malformations of the urethra;
   b. Diseases and defects of accessory sex glands (inflammation of the accessory glands);
   c. Excretere-obstructive infertility;
   d. Apermatism.
3. Immune infertility
4. Combined infertility
5. Relative infertility

19.3 Trace Elements as Micro-Nutrients for the Development of the Endocrine System and Semen Quality

An essential element was initially defined as one whose dietary deficiency consistently and adversely changed a biological function (growth, development, maturation, etc.) from optimal. This change was reversible or preventable by physiological amounts of the element (Nielsen, 2003). However, based on this definition, many elements were classified as essential in the case of small changes in physiological or biochemical functions under their action. Sometimes, such changes result from a suboptimal function, and sometimes, they result from a pharmacological or toxic action in the body, including an effect on intestinal microorganisms. As a result, if the element deficiency does not lead to significant changes in the body's functioning or its death, now many researchers consider an element essential when it has a well-known biochemical function. Thus, there is no universally accepted list of vital trace elements.

Trace elements have five known functions in living organisms (Nielsen, 2003):

1. In close association with enzymes, some elements are integral to catalytic centers at which the biochemical reactions necessary for life occur. Working in concert with a protein and frequently with another organic coenzyme, an element attracts a substrate molecule and facilitates its conversion into a specific final product.
2. Some elements donate or accept electrons in reduction or oxidation reactions. In addition to generating and utilizing metabolic energy, redox reactions frequently involve the chemical transformation of molecules.
3. Some elements, particularly iron, bind, transport, and release oxygen within living organisms.
4. Some elements have structural roles, imparting stability and spatial structure to important biological molecules.
5. Some elements have regulatory roles. These elements control essential biological processes through such actions as inhibiting enzymatic reactions, facilitating the binding of molecules to receptor sites on cell membranes, altering the structure or ionic nature of membranes to prevent or allow specific molecules or ions to enter a cell, and inducing genes to express themselves, resulting in the formation of proteins involved in life processes (Nielsen, 2003).

WHO considers 21 trace elements essential from the point of view of nutrition and health (WHO, 1996). Crucial elements include I, Zn, Se, Cu, Mo, Cr, Fe, and Co. In contrast, possibly essential elements include Mn, Si, Ni, B, and V. Eight microelements are potentially toxic, with some performing critical biological functions (F, Pb, Cd, Hg, As, Al, Li, and Sn) (WHO, 1996). Microelements (or trace elements) are usually defined as elements that are required in amounts between 1 and 100 mg/day by adults and, as already said, make up less than 0.01% of the total body weight (Mehri, 2020).

To confirm the essentiality of an element, they are guided by four basic principles (WHO, 1996):

1. A dietary deprivation of the element in some model organisms results in a changed biological structure, function, or tissue composition that is reversible or preventable by an intake of an apparent physiological amount of the element.
2. The element fills the need at physiological concentrations for a known in vivo biochemical action to proceed *in vitro.*
3. The element is a component of biologically essential molecules in some life forms.
4. The element has an essential function in lower life forms.

An element is considered to have strong circumstantial support for essentiality if all four types of evidence exist (Nielsen, 2003). With the development of science, the list of essential elements is increasingly expanding. For example, B has recently been found to be required to complete the life cycle of fish and frogs, and the biochemical function of Cr has been identified.

Results from animal studies suggest that the most likely micronutrients to affect human reproductive health are zinc, copper, iodine, and selenium (WHO, 1996). Currently, the list of microelements that affect the reproductive system of men is more comprehensive than those specified by WHO. This is because of new research data and the therapeutic function of some microelements. According to various sources, such elements may include Mn (Mirnamniha et al., 2019, Wirth and Mijal, 2010), Mo (Zhai et al., 2013), and Fe (Żak et al., 2020).

19.3.1 Zinc

Zinc (Zn) is the second trace element regarding human quantity after iron. Its positive effect on male reproduction is known and is described in many review papers (Maciejewski et al., 2022, Mishra and Acharya, 2004, Vickram et al., 2021).

Zinc's important biochemical and physiological role is due to its being part of many enzymes (more than 200) and stabilizing the molecular structure of subcellular constituents and membranes. Zinc synthesizes and degrades proteins, nucleic acids, carbohydrates, and lipids. It has recently been shown to play an essential role in transcription and translation, thus, in the processes of genetic expression. Zinc's participation in such basic activities explains its essentiality in all life forms. It plays a vital role in cell proliferation, differentiation, and metabolic activity (Mehri, 2020, WHO, 1996).

Zinc is one of the most significant elements in the male sperm. In semen, Zn is secreted mainly from the prostate (Brito et al., 1986). Its concentration in the testes is higher than in other tissues and comparable with that in the kidneys and liver (Cheah and Yang, 2011). Zinc's concentration in seminal plasma and cells usually increases compared to pathologies. Kothari and Chaudhari (2016) observed decreased Zn's concentrations in the seminal plasma of individuals suffering from asthenoteratozoospermia, oligoasthenoteratozoospermia, and azoospermia compared to individuals with normozoospermia. A systematic review of the literature showed that semen Zn's levels in infertile men were significantly lower than those in healthy men. Zinc supplementation significantly increased semen volume, sperm motility, and the percentage of normal sperm morphology (Zhao et al., 2016). Colagar et al. (2009) reported that fertile individuals have significantly higher seminal plasma Zn concentrations than infertile subjects. Investigations indicated that semen Zn concentration positively correlates with sperm motility and is inverse to the number of immotile spermatozoids (Khan et al., 2011). Zinc deficiency leads to low sperm quality (primarily, low sperm motility) via several mechanisms, including deficiency of sex hormones, oxidative stress, DNA damage, apoptosis, impaired spermatogenesis, and deficiency of Zn-finger proteins.

The most significant amount of Zn is found in germ cells. Zn is also detectable in Leydig cells, late-type B spermatogonia, and spermatids. A lack of Zn is noted in interstitial tissue and sterol cells. At the cellular level, 50% of the Zn amount accumulates in the cytosol, a little less is in the nucleus, and the least accumulates in cell membranes (Kambe et al., 2015). Some works have shown the existence of Zn in sperm flagella (Yamaguchi et al., 2009). Zinc is needed for the proper functioning of the HPG axis. The synthesis of testosterone in the Leydig cell, and thus the serum androgens level, depends on Zn since the microelement is a cofactor of the 5α reductase, transforming testosterone into active DHT (Hunt et al., 1992).

The most crucial function of Zn in sperm physiology is its participation in spermatogenesis (Fallah et al., 2018). The zinc ion is associated with major events in the acquisition of fertilization ability by spermatozoa, including motility, capacitation, and acrosomal exocytosis. To fertilization, sperm must reside in the female reproductive tract for several hours. During this time, they undergo a series of biochemical and motility changes collectively called capacitation, allowing the spermatozoon to interact with the oocyte, undergo acrosomal exocytosis, and penetrate the egg (Allouche-Fitoussi and Breitbart, 2020). Zinc affects both sperm quality and function.

Zinc deficiency is associated with increased apoptosis and decreased sperm count and motility. A deficiency of Zn has been shown to cause an inflammatory response and increased pro-apoptotic signaling (Bax, caspase-3) in germ cells. At the same time, anti-apoptotic signals are reduced (Bcl-2) (Omu et al., 2015). Zinc deficiency is associated with a decreased expression of Zip6 and Zip10 and an altered structure of the seminal tubules with abnormal germ epithelium, independent of systemic levels of Zn and testosterone. Deficiency of the element in Leydig cells may also indirectly contribute to impaired spermatogenesis (Kumari et al., 2011).

Zinc is necessary for testicular development, normal function of the testes, as well as epididymis and prostate function (Croxford et al., 2011). Zinc deficiency is associated with hypogonadism, seminiferous tubular contraction, loss of testicular mass, Leydig cell damage, and inadequate development of secondary sexual characteristics in humans (Kumar et al., 2018). Zinc also improved DNA methylation, chromatin integrity, and increased spermatogonia cells in the model study (Prasad, 2013). The element is a cofactor for DNA replication, transcription and packaging, protein biosynthesis, cell differentiation, and proliferation, essential for sperm formation (Chia et al., 2000, Croxford et al., 2011). Recent works have shown that the increased level of ROS and oxidative stress caused by the element deficiency is another significant reason for low sperm quality and male infertility (Colagar et al., 2009).

It serves as an antioxidant through several mechanisms. First, it is a primary cofactor for Cu/Zn-superoxide dismutase (SOD), which has antioxidative properties for sperm survival and helps sperm survive and function (Colagar and Marzony, 2009). Zinc also reduces hydroxyl radical (OH) production because it competes with Cu and Fe for membrane binding sites (Michalśka-Mosiej et al., 2016). Zinc can also bind to SH groups in proteins and inhibits protein oxidation by occupying binding sites for Cu in lipids and DNA (Ebisch et al., 2007). It also has an anti-inflammatory function by reducing the expression of TNF-α, an inflammatory cytokine, and overexpression of interleukins (IL-2 and IL-4) as antiinflammatory cytokines in human semen (Omu et al., 2015).

These numerous data suggest the exceptional role of semen Zn physiological concentrations in men's fertility.

19.3.2 Copper

Copper (Cu) is a microelement necessary for the growth and development of all living organisms. It is the body's third most abundant trace element, after Fe and Zn. In an organism, this element can exist in two oxidation states: as a cuprous (Cu^{+}) and as a cupric (Cu^{2+}) ion (Veldhuis et al., 2009). Redox properties of Cu and its capacity to participate in one-electron transfer reactions determine the biological activity of the element. Due to its redox properties, Cu is a cofactor for more than 30 essential enzymes. Many Cu-dependent enzymes catalyze reactions of fundamental metabolic processes involved in energy or antioxidant metabolism (Herman et al., 2020, Mirnamniha et al., 2019). Copper deficiency leads to the impaired function of enzymes responsible for respiration, detoxifying oxygen free radicals, connective tissue formation, neurotransmitter synthesis, pigment production, iron metabolism, etc. It also contributes to the redox system and protects sperm cells against oxidative damage.

Copper is a main component of Cu/Zn-superoxide dismutase, and therefore, decreased levels of Cu can negatively affect SOD activity. For example, Abdul-Rasheed (2010) observed a significant decrease of Cu in the seminal plasma of azoospermic patients, which was associated with reduced activity of SOD and a higher risk of oxidative stress. Conversely, using the same redox activity, copper can generate an extremely harmful hydroxyl radical through the Fenton reaction (Van den Berghe and Klomp, 2010), which may cause oxidative damage to proteins and nucleic acids, lipid peroxidation, and enzyme inhibition.

In reproductive biology and vertebrates, Cu can affect peptide and protein function both in the gonads and pituitary. Copper acts at the pituitary receptors, which control the release of LH (Jockenhövel et al., 1990, Slivkova et al., 2009). In the pituitary, copper binds to the vital reproductive peptides GnRH I and neurokinin B, modifying their structure and function. In the male gonads, Cu plays a role in testosterone production and the formation of sperm morphology (Peacey et al., 2020).

Outside of the pituitary, Cu can have effects directly on the gonads, and it is well-known that the element can play a role in spermatogenesis and sperm motility (Roy et al., 2014). Several proteins involved in Cu homeostasis, including the import pump Ctrl, are expressed in germ cells and mature spermatozoa, confirming Cu's role in both the development and then function of sperm cells (Kuo et al., 2006, Ogorek et al., 2017, Steiger et al., 2010).

Results of numerous human and animal studies indicate that both testis structure and the process of spermatogenesis (production, maturation, motility, and fertilizing capacity of the spermatozoa) are affected in response to increased Cu concentration (Chen et al., 2020). Strong evidence shows that different Cu-deficient mammalian males produce ejaculates of lower volume, with lower spermatozoa concentration, poorer spermatozoa motility, and changed morphology. For example, Saleh et al. (2008) found that the mean level of seminal plasma Cu in azoospermic patients was significantly lower than in healthy individuals. Another study reported a significant positive relationship between seminal plasma Cu and semen volume (Akinloye et al., 2011). Some studies reported a significant positive relationship between serum Cu levels and sperm motility (Machal et al., 2002). Histological analyses have demonstrated that the seminiferous tubules of Cu-deficient animals are less developed. This was mainly explained by the inactivation of the function of the Sertoli cells (Tvrda et al., 2015, Van Niekerk and Van Niekerk, 1989). Studies using the mouse model showed that in adult males, Cu concentration in the testes (0.94 μg/g wet tissue) is lower in comparison with other organs such as the brain (1.5 μg/g), kidney (4.4 μg/g), and liver (4.94 μg/g) (Higdon et al., 2001).

To deal with the dual nature of Cu, resulting in simultaneous usefulness and toxicity, tightly regulated molecular mechanisms have evolved and operated in organisms to ensure its bioavailability and to reduce its toxicity, basically by a complex network of proteins involved in the regulation of copper uptake, transport, utilization, storage, and excretion (Higdon et al., 2001).

19.3.3 Iodine

Iodine (I) is a microelement, mainly essential in the biosynthesis of thyroid hormones, which are responsible for basal metabolic rate and, thus, the growth and development of organisms. As with most micronutrients, the daily requirements for I change at different life stages, generally ranging between 90 and 290 μg (Jenzer and Sadeghi, 2016).

The effects of iodine as a trace element associated with fertility have been well studied in the female body. Numerous studies show that both hypo- and hyperthyroidism can cause menstrual abnormalities, increased risk of miscarriage, and possible long-term effects in offspring. This relation with infertility is due to alterations in the hypothalamic–pituitary–gonadal axis (Trokoudes et al., 2006). Its role in the functioning of the male reproductive system has been poorly studied. Scientists have identified thyroid hormone receptors directly on cells within the testes, indicating that thyroid hormones significantly affect their growth and development (Wagner et al., 2008). It is thought that thyroid hormone receptors are located on the Sertoli cells of the testes, and it is believed that triiodothyronine (T3) binds directly to these receptors (Cooke, 1991). Identifying thyroid hormone receptors in Sertoli cells and the findings that thyroid hormones influence the growth and development of the testicles increased the interest in these hormones' role in spermatogenesis and male fertility (Partal-Lorente et al., 2017, Peacey et al., 2020).

Some studies have shown that iodine can have other extra-thyroid effects. The presence of iodine transporters (sodium/iodide symporters) has been described in germ and Leydig cells (Micali et al., 2013, Russo et al., 2011) and pendrin in Sertoli cells (Lacroix et al., 2001). These transporters suggest that iodine may affect male gonads and sperm quality independent of thyroid hormone levels.

The study of Manna et al. (2001) showed that an acute addition of T3 appeared to increase LH receptors and steroidogenesis of Leydig cells. However, a substantial reduction in these levels was observed due to chronic treatment. T3 has been seen to directly stimulate basal testosterone generation, which could result from its action on Leydig cells (Maran et al., 2000). The effect of T3 on Leydig cell steroidogenesis appears to be through Sertoli cell paracrine factors or gonadotropins. Thyroid hormones seem to impact testes maturation by inhibiting the proliferation of Sertoli cells while stimulating differentiation (Orth et al., 1988, Rajender et al., 2011).

Hypothyroidism negatively indirectly affects male fertility by increasing body weight and thus slowing down the metabolism. Generally, due to hypothyroidism, the Leydig and Sertoli cells are less stimulated to differentiate into mature cells, negatively affecting spermatogenesis (Krassas and Pontikides, 2004). This increases the testes size in pre-pubertal males, and if the conditions remain, extensive damage could be done to the adult testes (Wajner et al., 2009). In the same way, this could decrease the number of mature, viable, and functional cells for spermatogenesis later in life. Recent investigations of the association between I intake, and semen quality among fertile men in China showed that men with deficient or excess iodine levels had a 5% higher semen volume relative to those with optimal iodine intake (Sun et al., 2020).

Thus, the influence of iodine on the male reproductive system is significant, but it is based on various subtle mechanisms that require further study.

19.3.4 Selenium

Selenium (Se) is included in selenoproteins, which functions as redox regulators, antioxidants, in immunoglobin production, lymphocyte proliferation, and as an anticancer and antiviral element in the body (Kieliszek and Błażejak, 2013). Moreover, Se is considered an element that protects against Cd poisoning (Maciejewski et al., 2022). In nature, Se exists in inorganic (selenate and selenite) and organic (selenomethionine and selenocysteine) forms, representing good dietary sources of the element. The recommended dietary dose of Se may differ depending on geographical area and selenium level in the diet, but it is usually 55 μg (Farag et al., 2021). Deficiency of Se can adversely affect enzyme activities that depend on the element, i.e., iodothyronine deiodinases, selenoprotein W, and glutathione peroxidases (Higdon et al., 2001).

Selenium is also another significant element that is essential for the maintenance of male fertility. Studies in rats have revealed that male infertility develops after several generations of feeding a Se-deficient diet (Behne et al., 1982). With a mild deficiency, Se accumulates in the testes preferentially found in the midpiece of spermatozoa, which contains mitochondria. Developmental studies in rats showed changes associated with Se deficiency, e.g., changes in the morphology of spermatids and spermatozoa and, finally, a complete absence of mature germinal cells (Behne et al., 1996, Calvin et al., 1987).

Shalini and Bansal (2005) reported that diet deficiency of Se causes increased oxidative stress and negatively affects spermatogenesis. The critical role of Se in spermatogenesis is mainly mediated by selenoproteins, including phospholipid hydroperoxide glutathione peroxidase (PHGPx), and selenoprotein P. PHGPx is the most abundant expressed selenoprotein by testicular germ cells which provides an essential link between Se, sperm quality, and male fertility (Ursini et al., 1999). Selenoproteins glutathione peroxidases (GPx) GPx1, GPx3, mGPx4, cGPx4, and GPx5 protect against oxidative damage to sperm during the maturation process, while selenoproteins, such as mGPx4 and snGPx4, act as structural components of mature sperm (Ahsan et al., 2014, Maciejewski et al., 2022). Aitken and Koppers (2011) and Akinloye et al. (2005) assumed that Se may positively influence Leydig cells, thus influencing testosterone secretion and, consequently, spermatogenesis.

Several authors investigated the effect of Se supplementation on sperm parameters quality and fertility potency in men with abnormal sperm parameters. Morbat et al. (2018) have shown that Se addition (50 μg/day for 3 months) significantly increases the sperm count, motility, viability, and normal morphology, as well as the ejaculate volume of infertile men. Selenium supplementation was also associated with increased serum Mg, FSH, testosterone, and glutathione levels, while malondialdehyde (MDA) content was significantly decreased. These data suggest the importance of Se for spermatogenesis and its antioxidant properties in mitigating oxidative stress. In a double-blind clinical trial study, Se addition (100 μg/day for 3 months) significantly increased sperm motility in infertile men with reduced sperm motility (Scott et al., 1998). Keskes Ammar et al. (2003) (Keiding et al., 1994) showed that vitamin E and Se supplementation significantly decreased malondialdehyde but increased glutathione concentrations in the seminal plasma of infertile men.

Normal gonadal function can be disrupted by hypothyroidism. Hypothyroidism disturbs testicular function directly and centrally by affecting the hypothalamic–pituitary–testicular axis with unclear mechanisms. By affecting thyroid iodide supply, Se could relieve these disturbances (El-Shaer et al., 2024).

There are currently no reported data describing a direct effect of Se on GnRH; however, there is potentially an interaction with LH. Selenium administration was shown by Kheradmand et al. (2019) to reverse the ability of the synthetic glucocorticoid dexamethasone to lower serum LH levels. Although this could result from selenium antagonizing the cellular effects of dexamethasone by reducing ROS, Se treatment alone caused an upward trend in serum LH, suggesting that Se may promote LH production or release its own. In support of this hypothesis, other researchers proposed that Se can upregulate GnRH receptor (GnRHR) activity in the anterior pituitary to increase LH synthesis (Lukusa and Lehloenya, 2017).

19.3.5 Manganese

Manganese (Mn) is a cofactor for enzymes involved in protein and energy metabolism, antioxidant action, and mucopolysaccharide synthesis. These enzymes include the metalloenzymes: manganese-dependent superoxide dismutase, pyruvate carboxylase, and arginase, and the manganese-activated enzymes:

phosphoenolpyruvate carboxykinase, glycosyl transferases, glutamine synthetase, and farnesyl pyrophosphate synthetase (Nielsen, 2003).

Data published to date indicate that Mn is critical for the normal function of human reproductive organs, sperm motility, and fertilization at low concentrations. Manganese is a component of several enzymes, particularly Mn^{2+}-SOD, which protects cells against free radicals and oxidative damage. Tajaddini et al. (2014) showed that Mn protects mice epididymal sperm parameters and testis structure treated with formaldehyde due to its antioxidant properties. Manganese supplementation maintains the thiol level by reducing oxidative stress in human spermatozoa (Bansal and Kaur, 2009). Cheema et al. (2009) indicated that Mn could improve semen quality and, subsequently, the success rate of in vitro fertilization and artificial insemination. Manganese is also a potent stimulator for sperm motility by stimulating adenylate cyclase activity. Therefore, the absence of Mn can inhibit the enzyme systems required for sperm motility. Lafond et al. (1988) reported that men with lower sperm counts had lower Mn mean levels in their seminal plasma. Another previous study showed that normospermic infertile patients had higher serum Mn contents compared to those with oligospermia and azoospermia (Adejuwon et al., 1996). Li et al. (2012) showed that a decreased seminal plasma Mn could negatively affect seminal fluid volume, sperm normal morphology, fructose level, and white blood cell counts. These findings indicate a potential role of Mn in the evaluation of infertile males.

19.3.6 Molybdenum

Molybdenum (Mo) functions as an essential enzyme cofactor. In humans, only four enzymes requiring Mo have been identified: sulfite oxidase, xanthine oxidoreductase, aldehyde oxidase, and mitochondrial amidoxime-reducing component (mARC) (Novotnt and Peterson, 2018). Molybdoenzymes oxidize and detoxify various pyrimidines, purines, and pteridines; catalyze the transformation of hypoxanthine to xanthine and xanthine to uric acid; and catalyze the conversion of sulfite to sulfate. A high sulfur amino acid intake could be a stressor that would help induce signs of molybdenum deficiency in humans. In addition, all molybdoenzymes can detoxify xenobiotic compounds (Nielsen, 2003).

Investigation of the effects of Mo on sperm parameters and testicular oxidative stress using adult mice showed that the sperm parameters, including the epididymis index, sperm motility, sperm count, and morphology, increased by a moderate dose of Mo (25 mg/L). In addition, the changes in sperm parameters were accompanied by changes in the superoxide dismutase activities, the glutathione peroxidase activities, and the malondialdehyde levels in testes (Zhai et al., 2013). Using artificially obtained germ-free mice, Liu et al. (2019) found that Mo significantly reversed the thinner seminiferous tubules and disappeared tubule and germ cells. Molybdenum also normalized previously abnormal levels of testosterone, estradiol, LH, superoxide dismutase, lactate dehydrogenase, and malondialdehyde. Molybdenum-treated sterile mice obtained offspring with a typical number and gender ratio. These results suggested that Mo at optimal doze had a significant therapeutic effect on reproductive dysfunction in sterile mice. Its mechanism could be repairing damaged testicular structures, regulating abnormal reproductive hormone levels, decreased oxidative stress, and resistance to cell apoptosis (Liu et al., 2019). Using the example of 313 men aged 50–75 years, a direct correlation was shown between the content of Mo and testosterone in the blood (Rotter et al., 2015). Thus, Mo may be a new candidate for treating male infertility.

19.3.7 Iron

Iron (Fe) is essential in many physiological processes, including DNA metabolism, oxygen transport, and cellular energy generation. Deregulated iron metabolism, which results in iron overload or deficiency, is observed in many diseases (Roemhild et al., 2021).

Both low and high iron concentrations reduce the quality of spermatogenesis. For instance, in the studies referring to men with iron deficiency anemia, there was a doubling of sperm count and an improvement of all semen parameters after intravenous iron supplementation (Soliman et al., 2014). In addition, a significant role is observed in maintaining iron homeostasis for proper testicular function (Griffin et al., 2005, Leichtmann-Bardoogo et al., 2012). Authors employed model objects multi-omic analysis, including transcriptomics, metabolomics, and microbiomics, to comprehensively dissect the impact of iron deficiency on spermatogenesis. The findings were verified in detail using western blot, immunofluorescence, immunohistochemistry, quantitative reverse transcription polymerase chain reaction (qRT-PCR), and other techniques. Microbiomic analysis revealed altered gut microbiota in iron-deficient mice, and functional predictive analysis showed that gut microbiota can regulate spermatogenesis. The transcriptomic data indicated that iron deficiency directly alters the expression of meiosis-related genes. Transcriptome data also revealed that iron deficiency indirectly regulates spermatogenesis by affecting hormone synthesis, findings confirmed by metabolomics data, western blot, and immunofluorescence. The data elucidate that iron deficiency impairs spermatogenesis and increases the risk of male infertility by affecting hormone synthesis (Zhang et al., 2022).

19.4 Toxic Effects of High Trace Element Concentrations on Indicators of Male Reproductive Health

Any chemical compounds (including microelements) that alter physiological control processes and affect the normal functioning of a component of the reproductive system can potentially cause infertility. The effect may be mild or severe, and the duration may vary from transient dysfunction to permanent damage. This can occur either by direct chemical action of the agent or indirectly via the metabolic products formed during the biotransformation processes. However, the exposure required to produce toxic effects is rarely achieved in normal human life except in the case of chronic industrial or agricultural impacts or acute poisoning as a result of industrial accidents or accidental household poisonings. Significant bioconcentration of trace amounts of toxic elements entering food webs is also possible (Harad, 1978, Hunt et al., 1992).

Trace elements bind to androgen and estrogen receptors, blocking the action of androgens and modulating the action of estrogens, acting as endocrine-disrupting compounds (Choi et al., 2004). In addition, heavy metals could further affect fertility by inducing ROS, apoptosis, local necrosis, and immunosuppression up to mutagenic stimuli (Bhardwaj et al., 2021).

Possible mechanisms of action of microelements leading to damage to the reproductive system and reproduction are shown in Figure 19.2.

High doses of microelements can cause infertility in men, acting on different elements of the reproductive system.

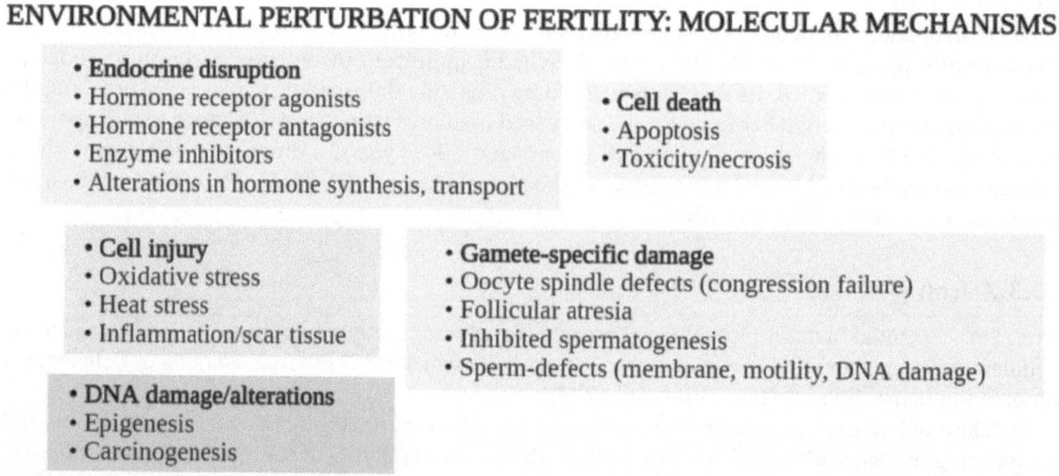

FIGURE 19.2 Mechanisms of action of microelements leading to damage to the reproductive system and reproduction. (Based on Phillips, 2015).

19.4.1 Disorders in the Endocrine System—The Hypothalamus–Pituitary–Testicular Axis

The hypothalamic–pituitary–testis axis is one of the trace elements target, which results in the alteration of the physiological function of the axis. Mainly, during the steroidogenesis process, trace elements disrupt androgen production. This could be mediated by receptors or by direct effects on gene transcription (López-Botella et al., 2021).

Many andrological pathologies seen in adults, including infertility, actually arose at a younger age due to the strong susceptibility and vulnerability of male gonads to external insults, starting from gestation age and during all growth phases. Three main phases are particularly susceptible to subsequent normal testis development and function: the intrauterine, neonatal, and puberty. However, even during infancy, when the testes are apparently "sleeping," damaging causes with permanent effects on testicular function can occur. Trace elements are among the chemicals that disrupt the endocrine system and cause changes in the sexual and reproductive organs and their functions. These elements are fascinating, given the solid experimental evidence of effects on hormone nuclear receptors, hypothalamus–pituitary–testis axis, and direct action on spermatogenesis and steroidogenesis (Di Nisio et al., 2023).

The most studied is the effect on the hypothalamus–pituitary–testicular axis of lead (Pb), one of the most common toxic trace elements. Environmental and occupational exposure to Pb may adversely affect the hypothalamic–pituitary–testicular axis, impairing the induction of spermatogenesis. Dysfunction at the reproductive axis, namely, testosterone suppression, is most susceptible and irreversible during pubertal development. Lead poisoning also appears to directly impair the process of spermatogenesis as well as sperm function (Gandhi et al., 2017). A clinical study on boys undergoing sexual development showed that marginally increased blood Pb levels might delay puberty and ultimately limit final testicular volume (Staessen et al., 2001). The effect is potentially due to impaired GnRH mRNA expression at low levels of long-term lead exposure (Sokol et al., 2002). Lead may also disturb hypothalamic GnRH release by arresting prostaglandin E2 synthesis or norepinephrine-stimulated release of prostaglandin E2 (Sokol et al., 2002).

A study looking at moderate occupational exposure to lead found that serum LH decreased with increasing duration of lead exposure (Ng et al., 1991), ultimately impacting testosterone production (Alexander et al., 1998) in Leydig cells. Lead may act at the hypothalamic–pituitary unit by interfering with the calcium-dependent secondary messenger systems that regulate LH release from secretory granule storage (Ronis et al., 1996); this disrupts the hormonal feedback loop and serves as an explanation for an increase in stored LH β (causing vacuolization of gonadotropic cells). At the testicular level, lead impairs testicular LH receptor concentrations and steroidogenesis at the pubertal stage (Ronis et al., 1996).

Decreased FSH levels have also been observed in moderately exposed lead workers (Gustafson et al., 1989). In contrast, increased (Ng et al., 1991) or (Assennato et al., 1987) FSH levels have also been reported. This disparity in reporting could be attributed to differing concentrations of Pb, varying exposure duration, and the physiological state of the reproductive axis and the testes. The metabolic function of Sertoli cells may also be impacted by lead exposure during pubertal development (Nathan et al., 1992). In auto-mechanics exposed to heavy metals, there was a negative correlation between lead and testosterone levels in the blood (Chikezie et al., 2017).

Using animal models, it was shown that Cd could also affect the activity of the hypothalamic–pituitary–testicular axis by acting at the hypothalamus (Antonio et al., 1999), the pituitary (Lafuente et al., 1999), the testes (Laskey and Phelps, 1991) and/or the sex accessory organs (Klinefelter and Hess, 1998). However, the mechanisms involved are not fully understood.

In addition to the indicated microelements, the following elements' high concentrations have a toxic effect on the different parts of the hypothalamus-pituitary-testicular axis and disrupt spermatogenesis: Hg (Davis et al., 2001), Ni (Adedara et al., 2019, Yang et al., 2021), As (Jana et al., 2006, Sun et al., 2016), Cu (Liao et al., 2019), Mn (Adedara et al., 2017, Nkpaa et al., 2018), rare earth elements (Chen et al., 2015, Lee and Park, 2022), and others. Thus, all cationic trace elements, including essential trace elements, at high concentrations harm the endocrine functions of this axis and spermatogenesis.

Recently, several studies have appeared in which the joint action of two cationic microelements-metals is studied simultaneously (Pillai et al., 2003, Adedara et al., 2017, 2019). Results indicated that co-exposure to nickel and zinc significantly abolished nickel-mediated diminution of antioxidant defense mechanisms while diminishing levels of reactive oxygen and nitrogen species and lipid peroxidation in the exposed rats' hypothalamus, epididymis, and testes. Additionally, co-exposure to zinc abated nickel-mediated diminutions in LH, FSH, serum, and intra-testicular testosterone with concomitant enhancement of sperm production and quality (Adedara et al., 2019). It should be noted that Zn was used in physiologically optimal concentrations in these experiments. At the same time, the significant decreases in the antioxidant status in the brain, testes, and epididymis, as well as in the circulatory concentrations of FSH, LH, and testosterone, were similar following separate or combined exposure of rats to As and Mn (Adedara et al., 2017).

Iodine-containing thyroid hormones have significantly impacted the male reproductive tract, spermatogenesis, and fertility. A clear link exists between thyroid hormones, testicular development, and spermatogenesis. It has been shown that hyperthyroidism causes marked changes in the gonadotropin axis and dramatically affects sperm function (Abalovich et al., 1999).

New data have been accumulated in the scientific literature in recent years, allowing an adequate risk assessment of Se concerning human health, including unexpected toxic effects of Se exposure (Vinceti et al., 2017). The primary adverse effects in laboratory animals exposed to inorganic selenium salts or to selenium-containing amino acids are various illnesses, including reproductive, although doses causing these effects are generally several times higher than normal daily Se intake (Gebreeyessus and Zewge, 2019). In humans, blood levels of triiodothyronine decreased in response to increased dietary selenium for 3 months and longer at intakes several times higher than normal intake. However, the hormone levels remained within the normal range. Toxic effects of Se observed in livestock were growth reduction, disturbance of the reproductive axis in males and females, and intrauterine resorption of fetuses (Assennato et al., 1987). High selenium intake reduced expression of the AR, DHCR24 (24-dehydrocholesterol reductase), and ABCC4 (ATP-binding cassette sub-family C member 4) (Legg et al., 2008). In several species, there is a high accumulation of Se in the pituitary (more than in the brain) (Thorlacius-Ussing and Danscher, 1985, Behne et al., 1989, Drasch et al., 2000). Studies observed that Se in drinking water (2.5–15 ppm) or ip injection of 5–20 mg/kg produced dose-dependent Se deposits in lysosomes and secretory granules, mainly in gonadotroph cells of the adenohypophysis (Behne et al., 1989, Thorlacius-Ussing and Jensen, 1988). A study following a cohort of North American men with no known fertility issues reported no change in sperm quality following 300 µg/day of Se supplementation (Hawkes et al., 2009).

Thus, the data presented indicate that microelements (cations and anions) can both normalize at optimal concentrations and disrupt at high concentrations the endocrine status of men along the hypothalamus-pituitary-testes axis. However, many mechanisms of their action remain not fully understood.

19.4.2 Disturbances in the Formation of the Male Reproductive Tract

Physical, chemical, and hormonal factors often cause abnormalities in testicular development during prenatal development. However, it should be noted that functional disorders that are detected during puberty or later in life may not always be easily linked to exposure to chemicals in the womb, leading to an underestimation of the impact of chemical exposures on reproductive tract development. Additionally, the severity of potential abnormalities in reproductive system development cannot be accurately predicted or assessed based on data from studies of chemicals' effects on adult organisms, as these organisms are more differentiated and can tolerate higher chemical loads.

The authors (Angenard et al., 2011) conducted a study on the effect of uranyl acetate on human gonads in culture over 7–12 weeks during the first trimester of pregnancy. They found embryonic testicles dysfunctional under a 0.1 mM uranium salt solution. Furthermore, human embryonic gonads appeared more sensitive to uranium than mouse gonads.

Testes are the direct target of a lot of toxicants. Animal experimentation has demonstrated lead toxicity may account for the decreased weight of testes, seminal vesicles, epididymis, and ventral prostate (Ronis et al., 1996); this may be paralleled with chronic lead exposure affecting accessory gland development beginning in childhood. Animal studies have further established the effect of lead on aberrant testicular tissue morphology (Ait Hamadouche et al., 2013), testicular germ cell death (Adhikari et al., 2001), interstitial edema (Ait Hamadouche et al., 2013), seminiferous tubule degeneration (Graca et al.,

2004), and abnormal seminal cytology (Cullen et al., 1993). Developmental exposure has also revealed limitations in the number, size, and shape of seminiferous tubules and spermatogonia (Garu et al., 2011). Other microelements can also cause changes in the reproductive tract. For example, Cd can affect the activity of the sex accessory organs (Klinefelter and Hess, 1998). The number of Sertoli cells significantly limits the maximum number of sperm produced (Sagalov, 2006). These cells begin to appear in the developing testes at around 6–7 weeks of gestation, differentiate at 8 weeks, and reach their peak number between 13 and 16 weeks of intrauterine development (Waters and Trainer, 1996). Therefore, any influences during this prenatal period, especially during these specific stages, that inhibit the growth of Sertoli cell numbers will likely result in decreased activity of spermatogenesis later in adult life. The toxic trace elements' effect on the Sertoli and Leydig cells can reduce seminiferous epithelium (Manfo et al., 2015). In addition, the affectation of the spermatogenesis process can occur through the alteration of the adhesion between germ cells or Sertoli cells (Reis et al., 2015).

There are significantly fewer studies using human subjects due to modern improvements in working conditions and ethical issues. Testicular biopsies from two adult male lead smelter workers in California with serious poisonings showed peritubular fibrosis, hypospermatogenesis, and Sertoli cell vacuolation, suggesting a direct testicular effect of Pb (Braunstein et al., 1978).

The studies' results confirm that trace elements in high concentrations can affect the male reproductive system, both at the hypothalamic-pituitary level and directly at the testicular level, causing various pathological changes.

19.4.3 Decrease in Sperm Quantity and Quality

Semen quality analysis is still considered the main prognostic sign of male infertility and an indicator of human reproductive health in general and testicular function in particular. Semen quality is frequently assessed using the WHO criteria by the following main parameters: semen volume, sperm count, and sperm motility (including progressive sperms (sperm tend to move in straight lines or large circles), nonprogressive sperms (sperm do not move in straight in lines, or they swim very small circles, and immotile sperms) (WHO, 2010). The parameters of healthy, borderline, and pathological human sperm are shown in Table 19.2.

TABLE 19.2
Parameters of Normal, Borderline, and Pathological Human Sperm

Semen Characteristic	Units	Normal	Borderline	Pathological
Volume	mL	2.0–6.0	1.5–2.0	< 1.5
pH	pH units	> 7.2	7.0–7.2	< 7.0
Sperm concentration	10^6/mL	> 2.0	10–20	< 10
Total sperm count	10^6/ejaculate	> 80	20–80	< 20
Sperm motility (0.5–2.0 h after ejaculation)	Scale 0–4	> 50	35–50	<35
Progression (at 37°C)	% motile	3 or 4	2	< 2
Straight-line velocity	µm/s	> 25	20–25	< 20
Sperm morphology	% normal	≥ 50	35–50	< 35
Head defects	per 100 sperm	< 35	35–50	> 50
Midpiece defects	per 100 sperm	≤ 20	21–25	> 25
Tail defects	per 100 sperm	≤ 20	21–25	> 25
Sperm vitality	% live	> 75	50–75	< 50
WBC	10^6/mL	0	< 1	> 1
Fructose (total)	mg %	> 100.0	50–100	< 50
Antisperm antibody	% immunobead (IgG, IgA, or IgM) binding to motile spermatozoa	0–20	20–50	> 50

Source: Adapted from Mortimer (1994).

At the same time, it should be noted that the border of spermatozoa in normothermia is periodically revised toward its reduction. In addition, sperm morphology is assessed. It is important to note that the body of a healthy man produces a significant proportion of morphologically abnormal sperm (Sharp, 1994), i.e., up to 50% of sperm may be immobile and/or irregular in shape. The striking heterogeneity of sperm morphology in human sperm (compared to virtually all other mammalian species) has long been discussed among physicians and biologists.

The findings of an analytical study (Carlsen et al., 1992) conducted on a large sample of normal men show that from 1938 to 1990, sperm concentration decreased from 113 million/mL to 66 million/ml (1.7 times). During the same period, the average ejaculate volume decreased slightly—from 3.4 to 2.75 mL. Subsequent analysis of results obtained at the same institution over different periods revealed a clear downward trend in sperm concentrations throughout the 20th century (Keiding et al., 1994). Some authors believe that the decrease in sperm concentration observed over the past decades may be caused by increased iodine consumption in the United States and some European countries (Sakamoto et al., 2004). An indirect confirmation of this relationship can be the fact that patients with hyperthyroidism have been found to have significantly compromised sperm parameters, including low motility, low ejaculate volume, low sperm concentration, and aberrant sperm morphology (Sengupta et al., 2022).

In our opinion, this decrease may be due to the general process of polychemicalization of the biosphere, in which the human body becomes part of cycles of elements, the scale of which has increased significantly compared to usual evolutionary values. So, men occupationally exposed to lead have exhibited decreased sperm concentration density per unit of seminal fluid, oligospermia, hypospermia, as well as teratozoospermia (e.g., widened, rounded, or shortened), and immature sperm (Vigeh et al., 2011). High trace element concentrations in the environment can also have deleterious effects on sperm quality, which is thought to be related to oxidative stress (Abdul-Rasheed, 2010, Lettieri et al., 2020). Subacute exposure to heavy metals and metalloids (As, Cd, Pb, and Hg) at concentrations <0.1 mg/L initiates a dyshomeostasis of Ca, Cu, Fe, and Zn that disturbs sperm parameters. These alterations worsen with prolonged exposure and higher doses (Machado-Neves, 2022).

In a meta-analysis, men with occupational exposure to lead had significantly higher blood lead levels, lower sperm counts, and poorer sperm motility than men without occupational exposure to Pb (Balachandar et al., 2020). Also, among men from infertile couples and without occupational exposure to Pb, seminal Pb levels have been negatively associated with sperm counts (Wu et al., 2012). Pizzol et al. (2021) also noted that Pb exposure was associated with reduced sperm motility and sperm morphology. A recent meta-analysis of 11 studies reported significantly higher seminal levels of cadmium among infertile than normal men (Zhang et al., 2019).

In addition to impaired sperm count and morphology, decreased sperm function is another potential consequence of lead toxicity (Taha et al., 2013). The generation of ROS may increase lipid peroxidation in sperm plasma membranes, causing a decline in fluidity, which is essential for motility, integrity, and viability (Oliveira et al., 2009). Lipid peroxides are capable of inducing DNA damage and reducing fertility during semen storage. The sperm plasma membranes are rich in polyunsaturated fatty acids containing double bonds vulnerable to free radical attack and initiation of the lipid peroxidation cascade, which is characterized by an autocatalytic, self-propagating reaction, giving rise to cell dysfunction accompanied by the loss of membrane functions and integrity. Decreased membrane functionality hinders binding with homologous and heterologous zona pellucida (i.e., sperm-oocyte fusion) and the ability to undergo acrosomal exocytose (Bansal and Bilaspuri, 2010).

19.4.4 The Influence of Microelements on the Process of Fertilization and Changes in Genetic Materials

The increased content of microelements directly in the sperm greatly influences the fertilization processes and changes in the genetic materials. The authors found that increased concentrations of trace elements in the seminal fluid may affect the egg and be sorbed into the vagina since its epithelium is permeable to a wide range of organic and inorganic compounds (Benziger and Edelson, 1983). Among the harmful mechanisms of the influence of microelements on male gametes can be both a direct effect on the genetic material (gene, chromosomal, and genomic mutations) and epigenetic changes (DNA methylation, chromatin remodeling, RNA interference (at the RNA level), protein prionization, inactivation of X-chromosomes in female).

Sperm DNA may be damaged by endogenous and exogenous factors and endanger the fetus's health (Simon et al., 2013) and, therefore, the evolutionary significance of fertility. The proportion of genetic disorders among the causes of male infertility is relatively low. Still, among men with a complete absence of sperm in their ejaculate, these disorders are quite common, accounting for up to 20% of cases (Gamidov et al., 2009).

It is known that genetic disorders, such as aneuploidy and structural abnormalities of chromosomes, can cause complete or partial inhibition of spermatogenesis. Autosomal translocations occur 10 times more frequently in sterile men than in fertile ones, and frequent deletions of the Y chromosome (Yq) have also been observed. It has been suggested that microdeletions or mutations in the azoospermia factor (AZF) locus may lead to impaired spermatogenesis in men with a normal karyotype (Bourgeron et al., 1997).

In recent years, research has suggested that severe disorders of spermatogenesis, miscarriages, and the occurrence of severe developmental defects can be caused by an increase in the size of pericentromeric heterochromatin when the body is exposed to certain unfavorable environmental factors, such as microelements and metals (Erickson et al., 1984).

At the same time, Fallah et al. (2018) found a moderate statistically significant negative correlation between parameters of sperm chromatin structure (DNA fragmentation index and DNA stainability) and sperm motility. These were found to be relatively independent predictors of male infertility. The experiments also confirmed the possible disconnect between DNA damage and sperm motility (Farag et al., 2021). He assessed the effects of factors such as culture medium and centrifugation on the chromosomes of immotile human sperm using intracytoplasmic sperm injection (ICSI) in mouse oocytes. The results showed that immotile sperm did not have significantly more DNA damage than motile sperm. However, the DNA of immotile sperm appeared to be more sensitive to damage caused by different culture media (Watanabe, 2004).

A study of paternal occupation as a risk factor for Down syndrome (aneuploidy, trisomy 21) was conducted in British Columbia (Olshan et al., 1989). The study analyzed 1008 cases of live births with Down syndrome during the period 1952–1973, taking into account maternal age correction. It showed that sick children were most often born to janitors, mechanics, farm workers, metalworkers, sheet metal workers, and sawmills and food industry workers. The disadvantage of this study was that it only studied live births with the pathology, while the mutation could also lead to fetal death. Additionally, the study did not include the mothers' occupation and the influence of other factors. Similar studies on the impact of occupation on Down syndrome incidence have been performed, but no correlation has been found (Cohen et al., 1977). Polednak and Janerich (1983) found an increased likelihood of Down syndrome among soldiers, while (Erickson et al., 1984) found no similar relationship among Vietnam War veterans.

A study of the effects of oxidative stress caused by hydrogen peroxide and iron (II) sulfate on individual nuclear genes and mitochondrial DNA in human sperm did not show significant damage to the genetic material when it interacted with reduced iron at any concentration (Sawyer et al., 2003).

At the same time, a study by Xu et al. (2003) found that cadmium in the seminal fluid can cause oxidative damage to DNA in non-smoking men, which is counteracted by selenium. Another study (Ghareeb and Sarhan, 2014) found data on oxidative damage to sperm DNA from exposure to heavy metals. Winder (1989) analyzed data on the possibility of chromosomal abnormalities in human somatic cells (in vivo) from exposure to high lead levels. Thirteen experiments confirmed the occurrence of chromosomal mutations, while two studies did not find evidence of chromosome changes. A study of Swedish children living near a lead smelter found no differences in sister chromatid sequences despite blood lead levels up to 63 μg/dL (Dalpra et al., 1983). A study by Schmid et al. (2013) showed that the sperm from 65 to 80-year-old non-smokers contained more zinc and copper than the sperm from younger men. In the older group, there was also a decrease in sperm motility and increased frequency of DNA fragmentation and chromosomal damage.

Thus, in vitro studies do not provide a clear answer regarding the relationship between exposure to trace metals and the occurrence of mutations. Nevertheless, the fact that there is a higher incidence of spontaneous abortion among women expecting a child whose father's occupation is associated with lead exposure remains (Al-Hakkak et al., 1986).

An example of epigenetic changes is the modification of numerous proteins located in the perinuclear space of sperm nuclei, the function of which is unknown. There is a hypothesis that these proteins can serve as receptors and provide a mechanism for the local distribution of genes within the nucleus (Bellvé et al., 1992). Any direct interaction of elevated trace element concentrations with these proteins can alter their structure and function and significantly affect offspring. It is noted that epigenetic chromosomal changes in the paternal pronucleus can affect the normal development of the placenta (Friedler, 1993). Epigenetic changes can occur both during gametogenesis and after fertilization before forming a zygote (Surani et al., 1990).

Many of the toxic effects we have discussed may be ascribed to the lead-induced generation of ROS (e.g., hydrogen peroxides and superoxides), which causes oxidative stress. ROS halts the synthesis of sulfhydryl antioxidants and prevents enzymatic reactions (Vigeh et al., 2011). Lead has a high affinity to sulfhydryl residues in proteins, thus acting as a nonspecific enzyme inhibitor and arresting their antioxidant activity (Ait Hamadouche et al., 2013). As oxidative stress disturbs chromatin condensation, it has been suggested that the under-protamination of sperm DNA could automatically lead to DNA fragmentation (Henkel et al., 2010). A study performed on Swiss mice found that Pb-induced ROS was associated with a significantly reduced sperm count and increased sperm abnormalities. Intraperitoneal injections of lead acetate, used to induce lead poisoning in the mice, led to an increase in lipid peroxidation potential, an indicator of oxidative stress, in the testes of the study mice compared to the control group. According to the study, the significant decrease in sperm count may be attributed to lead-induced ROS damage on the polyunsaturated fatty acids of cell membranes of germ cells and, subsequently, the spermatozoa and mature sperms (Mishra and Acharya, 2004). A computer-assisted semen analysis demonstrated that excessive ROS in semen damaged sperm concentration, motility, and other sperm motion parameters (Takeshima et al., 2017).

19.5 Conclusions

The effect of microelements on the body and the reproductive system of men follows the law of ecological optimum. This law applies regardless of the specific microelements being considered based on their essentiality. High doses of essential elements such as Zn, Cu, and Se can lead to toxic effects, while toxic trace elements are potentially necessary for the body in small quantities. Many microelements' physiological functions have not been thoroughly studied. The intake of these elements into the body through food must be within the necessary and safe ranges that the human body has evolutionarily adapted. Going beyond these ranges, whether decreasing or increasing intake, can lead to changes in physiological functions of different body systems and, consequently, deterioration of health, including reproductive health.

From the above review, it is clear that often, there is a lack of clarity and consistency in the existing ideas about the impact of micronutrients on the development of male reproductive dysfunction. This is often due to methodological limitations of studies conducted on model organisms, cells, or small sample sizes (clinical patients or a limited number of workers). Additionally, the lack of data is significantly influenced by ethical considerations.

Human reproductive health is primarily affected by deficiencies in Zn, Cu, I, and Se, according to the experts from WHO (1996). Additionally, modern data suggests that deficiencies in manganese, molybdenum, iron, and other micronutrients harm reproductive health.

The effect of high concentrations of trace elements is manifested in the state of several systems: the hormonal system, the hypothalamus–pituitary–testis axis, the male reproductive tract, sperm quantity and quality, and damage to genetic material (presence of gene, chromosome, and genomic mutations). It seems that the toxic effect of trace elements on reproductive function only manifests itself with sufficiently high exposure doses and prolonged exposure. Of particular significance is the impact of other damaging factors.

The endocrine system, along with the immune and nervous systems, is one of the three main systems that ensure the body's ability to react to environmental factors. Any disturbance, including chemical exposure, can cause tension in these systems, which can lead to problems with sexual function (from spermatogenesis to sexual intercourse). This can include the production of non-viable or genetically defective gametes, as well as the inability to conceive. However, men with unfavorable reproductive indicators still have the possibility of successful fertilization, embryo development, and fetal health due to developed reparative systems. Changes only under the influence of high doses of trace elements, as well as during long-term exposure, the presence of protective and reparative systems against chemical damage indicate the special evolutionary significance of the reproductive system. The system responsible for the inheritance of the main species characteristics is under special protection of the organism.

It is also important to remember that the impact on reproductive health depends not only on the concentration and exposure to the micronutrients but also on individual sensitivity, as well as the number of gametes produced and frequency of sexual activity with the potential for fertilization.

More research is needed to develop effective methods for protecting the human reproductive system from the harmful effects of toxic microelements. Studies should focus on identifying the specific mechanisms of damage caused by individual microelements, their combined effects, and other factors that may contribute to reproductive health issues.

Acknowledgments

The authors thank the Ministry of Education of the Russian Federation for financial support within the framework of the State assignment for 2025. No. 073-00065-25-01 from 18.03.2025

REFERENCES

Abalovich, M., O. Levalle, R. Hermes, H. Scaglia, C. Aranda, C. Zylbersztein, A. Oneto, D. Aquilano, S. Gutierrez. 1999. "Hypothalamic-pituitary-testicular axis and seminal parameters in hyperthyroid males." *Thyroid* 9(9): 857–863. doi:10.1089/thy.1999.9.857

Abdul-Rasheed, O. F. 2010. "Association between seminal plasma copper and magnesium levels with oxidative stress in Iraqi infertile men." *Oman Med J* 25(3): 168–172. doi:10.5001/omj.2010.51

Adedara, I. A., M. A. Abiola, A. N. Adegbosin, A. A. Odunewu, E. O. Farombi. 2019. "Impact of binary waterborne mixtures of nickel and zinc on hypothalamic-pituitary-testicular axis in rats." *Chemosphere* 237: 124501. doi:10.1016/j.chemosphere.2019.124501

Adedara, I. A., A. O. Abolaji, I. O. Awogbindin, E. O. Farombi. 2017. "Suppression of the brain-pituitary-testicular axis function following acute arsenic and manganese co-exposure and withdrawal in rats." *J Trace Elem Med Biol* 39: 21–29. doi:10.1016/j.jtemb.2016.07.001

Adejuwon, C. A., A. O. Ilesanmi, E. O. Ode, K. S. Akinlade. 1996. "Biophysical and biochemical analysis of semen in infertile Nigerian males." *Afr J Med Med Sci* 25(3): 217–219.

Adhikari, N., N. Sinha, R. Narayan, D. K. Saxena. 2001. "Lead-induced cell death in testes of young rats." *J Appl Toxicol* 21(4): 275–277. doi:10.1002/jat.754

Ahsan, U., Z. Kamran, I. Raza, S. Ahmad, W. Babar, M. H. Riaz, Z. Iqbal. 2014. "Role of selenium in male reproduction - a review." *Anim Reprod Sci* 146(1–2): 55–62. doi:10.1016/j.anireprosci.2014.01.009

Hamadouche, N. A., S. Nesrine, A. Abdelkeder. 2013. "Lead toxicity and the hypothalamic-pituitary-testicular axis." *Not Sci Biol* 5(1): 1–6. doi:10.15835/nsb518038

Aitken, J. R., A. J. Koppers. 2011. "Apoptosis and DNA damage in human spermatozoa." *Asian J Androl* 13: 36–42. doi:10.1038/aja.2010.68

Akinloye, O., A. O. Arowololu, O. B. Shittu, C. A. Adejuwon, B. Osotimehin. 2005. "Selenium status of idiopathic infertile Nigerian males." *Biol Trace Elem Res* 104(1): 9–18. doi:10.1385/BTER:104:1:009

Akinloye, O., F. M. Abbiyesuku, O. O. Oguntibeju, A. O. Arowojolu, E. J. Truter. 2011. "The impact of blood and seminal plasma zinc and copper concentrations on spermogram and hormonal changes in infertile Nigerian men." *Reprod Biol Endocrinol* 11(2):83–98. doi:10.1016/s1642-431x(12)60047-3

Al-Hakkak, Z. S, H. A. Hamamy, A. M. B. Murad, A. F. Hussaln. 1986. "Chromosome aberrations in workers at a storage battery plant in Iraq." *Mutat Res* 171(1): 53–60. doi:10.1016/0165–1218(86)90008-x

Alexander, B. H., H. Checkoway, E. M. Faustman, C. van Netten, C. H. Muller, T. G. Ewers. 1998. "Contrasting associations of blood and semen lead concentrations with semen quality among lead smelter workers." *Am J Ind Med* 34(5): 464–469. doi:10.1002/(sici)1097–0274(199811)34:4<464::aid-ajim20>3.0.co;2–z

Allouche-Fitoussi, D., H. Breitbart. 2020. "The role of zinc in male fertility." *Int J Mol Sci* 21(20): 7796. doi:10.3390/ijms21207796

Angenard G., V. Muczynski, H. Coffigny, C. Duquenne, R. Frydman, R. Habert, G. Livera, V. Rouiller-Fabre. 2011. "In vitro effects of Uranium on human fetal germ cells." *Reprod Toxicol* 31(4): 470–476. doi:10.1016/j.reprotox.2010.12.058

Antonio, M., I. Corpas, M. Leret. 1999. "Neurochemical changes in newborn rat's brain after gestational cadmium and lead exposure." *Toxicol Lett* 104(1–2): 1–9. doi:10.1016/s0378–4274(98)00125-8

Assennato, G., C. Paci, M. E. Baser, R. Molinini, R. G. Candela, B. M. Altamura, R. Giorgino. 1987. "Sperm count suppression without endocrine dysfunction in lead-exposed men." *Arch Environ Health* 42(2): 124–127. doi:10.1080/00039896.1987.9935808

Balachandar, R., B. S. Bagepally, R. Kalahasthi, M. Haridoss. 2020. "Blood lead levels and male reproductive hormones: a systematic review and meta-analysis." *Toxicology* 443: 152574. doi:10.1016/j.tox.2020.152574

Bansal, A. K., A. R. Kaur. 2009. "Cooperative functions of manganese and thiol redox system against oxidative stress in human spermatozoa." *J Hum Reprod Sci* 2(2): 76–80. doi:10.4103/0974–1208.57227

Bansal, A. K., G. S. Bilaspuri. 2010. "Impacts of oxidative stress and antioxidants on semen functions." *Vet Med Int* 2010: 686137. doi:10.4061/2011/686137

Behne, D., H. Hilmert, S. Scheid, H. Gessner, A. Kyriakopoulos, W. Elger. 1989. "Studies on new selenoproteins and specific selenium target tissues". In Wendel A, ed. Selenium in Biology and Medicine. Berlin, Germany: Springer-Verlag, pp. 14–20.

Behne, D., H. Weiler, A. Kyriakopoulos. 1996. "Effects of selenium deficiency on testicular morphology and function in rats." *J Reprod Fertil* 106(2): 291–297. doi:10.1530/jrf.0.1060291

Behne, D., T. Hofer, R. Berswordt-Wallrabe, W. Elger. 1982. "Selenium in the testis of the rat: studies on its regulation and its importance for the organism." *J Nutr* 112: 1682–1687. doi:10.1093/jn/112.9.1682

Bellvé, A. R., R. Chandrika, Y. S. Martinova, A. H. Barth. 1992. "The perinuclear matrix as a structural element of the mouse sperm nucleus." *Biol Reprod* 47(3):451–65. doi:10.1095/biolreprod47.3.451

Benziger, D. P., J. Edelson. 1983. "Absorption from the vagina." *Drug Metab Rev* 14(2): 137–168. doi:10.3109/03602538308991387

Bhardwaj, J. K., A. Paliwal, P. Saraf. 2021. "Effects of heavy metals on reproduction owing to infertility." *J Biochem Mol Toxicol* 35(8): e22823. doi:10.1002/jbt.22823

Bohring, C., W. Krause. 2003. "Immune infertility: towards a better understanding of sperm (auto)-immunity. The value of proteomic analysis." *Hum Reprod* 18(5): 915–924. doi:10.1093/humrep/deg207

Boron, W. F., E. L. Boulpaep. 2011. *Medical Physiology*. 2nd edition. Saunders, Elsevier: Philadelphia, US. 1352 p.

Bourgeron, T., S. Barbaux, K. McElreavey, M. Fellous. 1997. "Chromosome Y et spermatogenese." *Contracept Fertil Sexual* 25(7–8): 620–625. (In French)

Braunstein, G. D., J. Dahlgren, D. L. Loriaus. 1978. "Hypogonadism in chronically lead-poisoned men." *Infertility* 1(1): 33–51.

Brito, M., J. Figueroa, J. C. Vera, P. Cortés, R. Hott, L. O. Burzio. 1986. "Phosphoproteins are structural components of bull sperm outer dense fiber." *Gamete Res* 15(4): 327–336. doi:10.1002/mrd.1120150406

Calvin, H. I., K. Grosshans, S. R. Musicant-Shikora, S. I. Turner. 1987. "A developmental study of rat sperm and testis selenoproteins." *J Reprod Fertil* 81(1): 1–11. doi:10.1530/jrf.0.0810001

Carlsen, E., A. Giwercman, N. Keiding, N. E. Skakkebaek. 1992. "Evidence for decreasing quality of semen during past 50 years." *BMJ* 305(6854): 609–613. doi:10.1136/bmj.305.6854.609

Cheah, Y., W. Yang. 2011. "Functions of essential nutrition for high quality spermatogenesis." *Adv Biosci Biotechnol* 2: 182–197. doi:10.4236/abb.2011.24029

Cheema, R. S., A. K. Bansal, G. S. Bilaspuri. 2009. "Manganese provides antioxidant protection for sperm cryopreservation that may offer new consideration for clinical fertility." *Oxid Med Cell Longev* 2(3): 152–159. doi:10.4161/oxim.2.3.8804

Chen, H., Z. Kang, N. Qiao, G. Liu, K. Huang, X. Wang, C. Pang, Q. Zeng, Z. Tang, Y. Li. 2020. "Chronic copper exposure induces hypospermatogenesis in mice by increasing apoptosis without affecting testosterone secretion." *Biol Trace Elem Res* 195(2): 472–480. doi:10.1007/s12011-019-01852-x

Chen, J., H. J. Xiao, T. Qi, D. L. Chen, H. M. Long, S. H. Liu. 2015. "Rare earths exposure and male infertility: the injury mechanism study of rare earths on male mice and human sperm." *Environ Sci Pollut Res Int* 22: 2076–2086.

Chia, S. E., C. N. Ong, L. H. Chua, L. M. Ho, S. K. Tay. 2000. "Comparison of zinc concentrations in blood and seminal plasma and the various sperm parameters between fertile and infertile men." *J Androl* 21(1): 53–57.

Chikezie, I. C., M. A. Charles-Davies, A. M. Balogun, S. U. Okoli. 2017. "Effects of endocrine disrupting heavy metals on pituitary and gonadal hormones in normal weight automechanics in Ibadan, Nigeria." *Afr J Biomed Res* 20(1):25–35.

Choi, S. M., S. D. Yoo, B. M. Lee. 2004. "Toxicological characteristics of endocrine-disrupting chemicals: developmental toxicity, carcinogenicity, and mutagenicity." *J Toxicol Environ Health B Crit Rev* 7(1): 1–24. doi:10.1080/10937400490253229

Cohen, M. M., M. Levy, M. Eliakim. 1977. "A cytogenic evaluation of long-term colchicine therapy in the treatment of Familial Mediterranean Fever (FMF)." *Am J Med Sci* 274(2): 147–152. doi:10.1097/00000441-197709000-00005

Colagar, A. H., E. T. Marzony. 2009. "Ascorbic Acid in human seminal plasma: determination and its relationship to sperm quality." *J Clin Biochem Nutr* 45(2): 144–149. doi:10.3164/jcbn.08–251

Colagar, A. H., E. T. Marzony, M. J. Chaichi. 2009. "Zinc levels in seminal plasma are associated with sperm quality in fertile and infertile men." *Nutr Res* 29(2): 82–88. doi:10.1016/j.nutres.2008.11.007

Cooke, P. S. 1991. "Thyroid hormones and testis development: a model system for increasing testis growth and sperm production." *Ann N Y Acad Sci* 637: 122–132. doi:10.1111/j.1749–6632.1991.tb27305.x

Croxford, T. P., N. H. McCormick, S. L. Kelleher. 2011. "Moderate zinc deficiency reduces testicular Zip6 and Zip10 abundance and impairs spermatogenesis in mice." *J Nutr* 141(3): 359–365. doi:10.3945/jn.110.131318

Cullen, C., A. Singh, A. Dykeman, D. Rice, W. Foster. 1993. "Chronic lead exposure induces ultrastructural alterations in the monkey seminal vesicle." *J Submicrosc Cytol Pathol* 25(1): 127–135.

Dalpra, L., G. Tibilletti, G. Nocera, L. Giulotto, V. Auriti, V. Camelli, G. Simoni. 1983. "SCE analysis in children exposed to led emission from a smelting plant." *Mutat Res* 120(4): 249–256. doi:10.1016/0165–7992(83)90097-0

Davis, B. J., H. C. Price, R. W. O'Connor, R. Fernando, A. S. Rowland, D. L. Morgan. 2001. "Mercury vapor and female reproductive toxicity." *Toxicol Sci* 59(2): 291–296. doi:10.1093/toxsci/59.2.291

Di Nisio, A., C. Corsini, C. Foresta. 2023. "Environmental impact on the hypothalamus-pituitary-testis axis." *Environ Endocrinol Endocrine Disruptors Endocrinol* 2023: 207–238. https://doi.org/10.1007/978-3-030-39044-0_8

Drasch, G., D. S. Mail, C. Schlosser, G. Roider. 2000. "Content of non mercury-associated selenium in human tissues." *Biol Trace Elem Res* 77(3): 219–230. doi:10.1385/bter:77:3:219

Ebisch, I. M. W., C. M. Thomas, W. H. Peters, D. D. Braat, R. P. Steegers-Theunissen. 2007. "The importance of folate, zinc and antioxidants in the pathogenesis and prevention of subfertility." *Hum Reprod Update* 13(2): 163–174. doi:10.1093/humupd/dml054

El-Shaer, R. A. A., S. Ibrahim, P. M. Hewady, M. M. Atef, O. S. El-Deeb, Y. M. Hafez, R. S. Amer, J. A. El-Hameed El-Sharnoby, N. A. A. Hashish, M. M. Awad. 2024. "Selenium protects against nesfatin-1 modulation of the hypothalamic-pituitary-testicular axis during hypothyroidism in male rats". *Physiol Rep* 12(2): e15923. doi:10.14814/phy2.15923

Erickson, J. D., J. Mulinaire, P. W. McClain, T. G. Fitch, L. M. James, A. B. McClearn, M. J. Jr Adams. 1984. "Vietnam veteran's risk for fathering babies with birth defects." *J Am Med Assoc* 252(7): 903–912.

Fallah, A., A. Mohammad-Hasani, A. H. Colagar. 2018. "Zinc is an essential element for male fertility: a review of Zn roles in men's health, germination, sperm quality, and fertilization." *J Reprod Infertil* 19(2): 69–81.

Farag, M. A., S. Hamouda, S. Gomaa, A. A. Agboluaje, M. L. M. Hariri, S. M. Yousof. 2021. "Dietary micronutrients from zygote to senility: updated review of minerals' role and orchestration in human nutrition throughout life cycle with sex differences." *Nutrients* 13(11): 3740. doi:10.3390/nu13113740.

Friedler. G. 1993. "Developmental toxicology: malt-mediated effects." In Paul M, ed. Occupational and Environmental Reproductive Hazards: A Guide for Clinicians. Baltimore, US: Williams and Wilkins, pp. 52–59.

Gamidov, S. I., R. I. Ovchinnikov, D. V. Shcherbakov, R. A. Thagapsoeva. 2009. "Sovremennye aspekty patogeneza, diagnostiki i lecheniya muzhskogo besplodiya." *Kremlyovskaya medicina. Klinicheskij vestnik* 2: 26–30.

Gandhi, J., R. J. Hernandez, A. Chen, N. L. Smith, Y. R. Sheynkin, G. Joshi, S. A. Khan. 2017. "Impaired hypothalamic-pituitary-testicular axis activity, spermatogenesis, and sperm function promote infertility in males with lead poisoning." *Zygote* 25(2): 103–110. doi:10.1017/S0967199417000028

Garu, U., R. Sharma, I. Barber. 2011. "Effect of lead toxicity on developing testis of mice." *Int J Pharmaceutic Sci Res* 2: 2403–2407.

Gebreeyessus, G. D., F. A. Zewge. 2019. "A review on environmental selenium issues." *SN Appl Sci* 1: 55. doi:10.1007/s42452-018-0032–9

Ghareeb, D. A., E. M. E. Sarhan. 2014. "Role of oxidative stress in male fertility and idiopathic infertility: causes and treatment." *J Diagn Tech Biomed Anal* 2: 1–12. doi:10.4172/jdtba.1000110

Graca, A., J. Ramalho-Santos, M. de Lourdes Pereira. 2004. "Effect of lead chloride on spermatogenesis and sperm parameters in mice." *Asian J. Androl* 6(3): 237–241.

Griffin, K. P., D. T. Ward, W. Liu, G. Stewart, I. D. Morris, C. P. Smith. 2005. "Differential expression of divalent metal transporter DMT1 (Slclla2) in the spermatogenic epithelium of the developing and adult rat testis." *Am J Physiol Cell Physiol* 288(1): C176–C184. doi:10.1152/ajpcell.00061.2004

Gustafson, A., P. Hedner, A. Schutz, S. Skerfving. 1989. "Occupational lead exposure and pituitary function." *Int Arch Occup Environ Health* 61(4): 277–281. doi:10.1007/BF00381426

Harad, M. 1978. "Congenital Minamata disease: intrauterine methylmercury poisoning." *Teratology* 18(2): 285–288. doi:10.1002/tera.1420180216

Hawkes, W. C., Z. Alkan, K. Wong. 2009. "Selenium supplementation does not affect testicular selenium status or semen quality in north American men." *Journal of Andrology* 30(5): 525–533. doi:10.2164/jandrol.108.006940

Henkel, R., H. S. Bastiaan, S. Schüller, I. Hoppe, W. Starker, R. Menkveld. 2010. "Leucocytes and intrinsic ROS production may be factors compromising sperm chromatin condensation status." *Andrologia* 42(2): 69–75. doi:10.1111/j.1439-0272.2009.00967.x. PMID: 20384795

Herman, S., P. Lipiński, M. Ogórek, R. Starzyński, P. Grzmil, A. Bednarz, M. Lenartowicz. 2020. "Molecular regulation of copper homeostasis in the male gonad during the process of spermatogenesis." *Int J Mol Sci.* 21(23):9053. doi:10.3390/ijms21239053

Higdon, J., V. Drake, B. Delage, P. Tsuji. 2001. Selenium. Available online: https://lpi.oregonstate.edu/mic/minerals/selenium (accessed on 24 May 2024).

Hunt, C. D., P. E. Johnson, J. Herbel, L. K. Mullen. 1992. "Effects of dietary zinc depletion on seminal volume and zinc loss, serum testosterone concentrations, and sperm morphology in young men." *Am J Clin Nutr* 56: 148–157. doi:10.1093/ajcn/56.1.148

Jana, K., S. Jana, P. K. Samanta. 2006. "Effects of chronic exposure to sodium arsenite on hypothalamo-pituitary-testicular activities in adult rats: possible an estrogenic mode of action." *Reprod Biol Endocrinol* 4: 9. doi:10.1186/1477–7827-4–9

Jenzer, H., L. Sadeghi. 2016. "Iodine: biochemistry, deficiency and application in clinical nutrition." *Can J Clin Nutr.*5(1): 1–8. doi:10.14206/canad.j.clin.nutr.2017.01.01

Jockenhövel, F., M. Bals-Pratsch, H. P. Bertram, E. Nieschlag. 1990. "Seminal lead and copper in fertile and infertile men." *Andrologia* 22(6): 503–511.

Jurewicz, J., W. Hanke, M. Radwan, J. P. Bonde. 2009. "Environmental factors and semen quality." *Int J Occup Med Environ Health* 22(4): 305–329. doi:10.2478/v10001-009-0036-1.

Kambe, T.; T. Tsuji, A. Hashimoto, N. Itsumura. 2015. "The physiological, biochemical, and molecular roles of zinc transporters in zinc homeostasis and metabolism." *Physiol Rev* 95(3): 749–784. doi:10.1152/physrev.00035.2014

Keiding, N, A. Giwercman, E. Carlsen, N. E. Skakkebaek. 1994. "Falling sperm quality." *BMJ* 309(6947): 131. doi:10.1136/bmj.309.6947.131b

Keskes Ammar, L., N. Feki Chakroun, T. Rebai, Z. Sahnoun, H. E. A. Ghozzi. 2003. "Sperm oxidative stress and the effect of an oral vitamin E and selenium supplement on semen quality in infertile men." *Arch Androl* 49: 83–94. doi:10.1080/01485010390129269.

Khan, M. S., S. Zaman, M. Sajjad, M. Shoaib, G. Gilani. 2011. "Assessment of the level of trace element zinc in seminal plasma of males and evaluation of its role in male infertility." *Int J Appl Basic Med Res* 1(2): 93–96. doi:10.4103/2229-516X.91152

Kheradmand, N., R. Kamkar, M. Moshajjari, M. Baazm. 2019. "Effect of selenium and pentoxifylline on expression of CATSPER1 and 2 genes and FSH/LH levels in treated mice by dexamethasone." *Andrologia* 51(6): e13279. doi:10.1111/and.13279

Kieliszek, M.; S. Błażejak. 2013. "Selenium: significance, and outlook for supplementation." *Nutrition* 29(5): 713–718. doi:10.1016/j.nut.2012.11.012

Klinefelter, G. R., R. A. Hess. 1998. "Toxicology of the male excurrent ducts and accessory glands." In Korach KS, ed. Reproductive and Developmental Toxicology. New York: Marcel Dekker, pp. 553–591.

Kothari, R. P., A. R. Chaudhari. 2016. "Zinc levels in seminal fluid in infertile males and its relation with serum free testosterone." *J Clin Diag Res* 10(5): CC05–CC08. doi:10.7860/JCDR/2016/14393.7723

Krassas, G. E., N. Pontikides. 2004. "Male reproductive function in relation with thyroid alterations." *Best Pract Res Clin Endocrinol Metab* 18(2): 183–195. doi:10.1016/j.beem.2004.03.003

Kumar, S., Mishra, V., Thaker, R., Gor, M., Perumal, S., Joshi, P., Sheth, H., Shaikh, I., Gautam, A. K., Verma, Y. 2018. "Role of environmental factors & oxidative stress with respect to in vitro fertilization outcome." *Indian J Med Res* 148(1): S125–S133. doi:10.4103/ijmr.IJMR_1864_17

Kumari, D., N. Nair, R. S. Bedwal. 2011. "Effect of dietary zinc deficiency on testes of Wistar rats: morphometric and cell quantification studies." *J Trace Elem Med Biol* 25: 47–53. doi:10.1016/j.jtemb.2010.11.002

Kuo, Y. M., A. A. Gybina, J. W. Pyatskowit, J. Gitschier, J. R. Prohaska. 2006. "Copper transport protein (Ctr1) levels in mice are tissue specific and dependent on copper status." *J Nutr* 136(1): 21–26. doi:10.1093/jn/136.1.21

Lacroix, L., C. Mian, B. Caillou, M. Talbot, S. Filetti, M. Schlumberger, J. M. Bidart. 2001. "Na(+)/I(−) symporter and Pendred syndrome gene and protein expressions in human extra-thyroidal tissues." *Eur J Endocrinol* 144(3): 297–302. doi:10.1530/eje.0.1440297

Lafond, J. L., B. Sele, A. Favier. 1988. "Concentration of selected metals in normal and pathological human seminal plasma." *J Trace Elem Electrolytes Health Dis* 2(1): 19–21.

Lafuente, A., N. Marquez, S. Piquero, A. I. Esquifino. 1999. "Cadmium affects the episodic luteinizing hormone secretion in male rats: possible age-dependent effects." *Toxicol Lett* 104(1–2): 27–33. doi:10.1016/s0378-4274(98)00349-x

Laskey, J. W., P. V. Phelps. 1991. "Effect of cadmium and other metal cations on 'in vitro' Leydig cell testosterone production." *Toxicol Appl Pharmacol* 108(2): 296–306. doi:10.1016/0041-008x(91)90119-y

Lee, W. Y., H. J. Park. 2022. "Toxicity of cerium oxide nanoparticles on neonatal testicular development in mouse organ culture." *Reprod Toxicol* 111: 120–128. doi:10.1016/j.reprotox.2022.05.014

Legg, R. L., J. R. Tolman, C. T. Lovinger, E. D. Lephart, K. D. Setchell, M. J. Christensen. 2008. "Diets high in selenium and isoflavones decrease androgen-regulated gene expression in healthy rat dorsolateral prostate." *Reproductive Biology and Endocrinology* 6(1): 57. doi:10.1186/1477-7827-6-57

Leichtmann-Bardoogo, Y., L. A. Cohen, A. Weiss, B. Marohn, S. Schubert, A. Meinhardt, E. G. Meyron-Holtz. 2012. "Compartmentalization and regulation of iron metabolism proteins protect male germ cells from iron overload." *Am J Physiol Endocrinol Metab* 302(12): E1519–1530. doi:10.1152/ajpendo.00007.2012

Lettieri, G., G. D'Agostino, E. Mele, C. Cardito, R. Esposito, A. Cimmino, A. Giarra, M. Trifuoggi, S. Raimondo, T. Notari, F. Febbraio, L. Montano, M. Piscopo. 2020. "Discovery of the involvement in DNA oxidative damage of human sperm nuclear basic proteins of healthy young men living in polluted areas." *Int J Mol Sci* 21(12): 4198. doi:10.3390/ijms21124198.

Li, Y., J. Wu, W. Zhou, E. Gao. 2012. "Effects of manganese on routine semen quality parameters: results from a population-based study in China." *BMC Publ Health* 12:919. doi:10.1186/1471-2458-12-919

Liao, J., F. Yang, H. Chen, W. Yu, Q. Han, Y. Li, L. Hu, J. Guo, J. Pan, Z. Liang, Z. Tang. 2019. "Effects of copper on oxidative stress and autophagy in hypothalamus of broilers." *Ecotoxicol Environ Safety* 185: 109710. doi:10.1016/j.ecoenv.2019.109710

Liu, F. J., W. Y. Dong, H. Zhao, X. H. Shi, Y. L. Zhang. 2019. "Effect of molybdenum on reproductive function of male mice treated with busulfan." *Theriogenology* 126: 49–54. doi:10.1016/j.theriogenology.2018.12.002

López-Botella, A., I. Velasco, M. Acién, P. Sáez-Espinosa, J. L. Todolí-Torró, R. Sánchez-Romero, M. J. Gómez-Torres. 2021. "Impact of heavy metals on human male fertility-an overview." *Antioxidants (Basel)* 10(9): 1473. doi:10.3390/antiox10091473

Lukusa, K., K. Lehloenya. 2017. "Selenium supplementation improves testicular characteristics and semen quality of Saanen bucks." *Small Rumin Res.* 151: 52–58. doi:10.1016/j.smallrumres.2017.04.016

Machado-Neves, M. 2022. "Effect of heavy metals on epididymal morphology and function: an integrative review." *Chemosphere* 291(2): 133020. doi:10.1016/j.chemosphere.2021.133020

Machal, L., G. Chládek, E. Straková. 2002. "Copper, phosphorus and calcium in bovine blood and seminal plasma in relation to semen quality." *J Anim Feed Sci* 11(3): 425–435. doi:10.22358/jafs/67893/2002

Maciejewski, R., E. Radzikowska-Büchner, W. Flieger, K. Kulczycka, J. Baj, A. Forma, J. Flieger. 2022. "An overview of essential microelements and common metallic nanoparticles and their effects on male fertility." *Int J Environ Res Public Health* 19(17): 11066. doi:10.3390/ijerph191711066

Manfo, F. P. T., E. A. Nantia, P. P. Mathur. 2015. "Effect of environmental contaminants on mammalian testis." *Curr Mol Pharmacol* 7: 119–135. doi:10.2174/1874467208666150126155420

Manna, P. R., J. Kero, M. Tena-Sempere, P. Pakarinen, D. M. Stocco, I. T. Huhtaniemi. 2001. "Assessment of mechanisms of thyroid hormone action in mouse Leydig cells: regulation of the steroidogenic acute regulatory protein, steroidogenesis, and luteinizing hormone receptor function." *Endocrinology* 142(1): 319–331. doi:10.1210/endo.142.1.7900

Maran, R. R., J. Arunakaran, M. M. Aruldhas. 2000. "T3 directly stimulates basal and modulates LH induced testosterone and oestradiol production by rat Leydig cells in vitro." *Endocr J* 47(4): 417–428. doi:10.1507/endocrj.47.417

Mehri, A. 2020. "Trace elements in human nutrition (II) - an update." *Int J Prev Med* 11: 2. doi:10.4103/ijpvm.IJPVM_48_19

Mendiola, J, A. M. Torres-Cantero, A. Agarwal. 2009. "Lifestyle factors and male infertility: an evidence-based review." *Arch Med Sci* 5(1): 3.

Micali, S., V. Maggisano, A. Cesinaro, M. Celano, A. Territo, L. R. Bonetti, M. Sponziello, M. Migaldi, M. Navarra, G. Bianchi, S. Filetti, D. Russo. 2013. "Sodium/iodide symporter is expressed in the majority of seminomas and embryonal testicular carcinomas." *J Endocrinol* 216(2): 125–133. doi:10.1530/JOE-12-0495

Michalska-Mosiej, M., K. Socha, J. Soroczynska, E. Karpinska, B. Lazarczyk, M. H. Borawska. 2016. "Selenium, zinc, copper, and total antioxidant status in the serum of patients with chronic tonsillitis." *Biol Trace Element Res* 173(1): 30–34. doi:10.1007/s12011-016-0634-2

Mishra, M, U. R. Acharya. 2004. "Protective action of vitamins on the spermatogenesis in lead-treated Swiss mice." *J Trace Elem Med Biol* 18(2): 173–178. doi:10.1016/j.jtemb.2004.03.007

Mirnamniha, M., F. Faroughi, E. Tahmasbpour, P. Ebrahimi, A. Beigi Harchegani. 2019. "An overview on role of some trace elements in human reproductive health, sperm function and fertilization process." *Rev Environ Health* 34(4): 339–348. doi:10.1515/reveh-2019-0008

Morbat, M. M., A. M. Hadi, D. H. Hadri. 2018. "Effect of selenium in treatment of male infertility." *Exp Tech Urol Nephrol* 1: 1–4.

Mortimer, D. 1994. "*Practical Laboratory Andrology*", Oxford, UK: Oxford University Press, 393 p.

Nathan, E., H. F. Huang, L. Pogach, W. Giglio, J. D. Bogden, J. Seebode. 1992. "Lead acetate does not impair secretion of Sertoli cell function marker proteins in the adult Sprague Dawley rat." *Arch Environ Health* 47(5): 370–375. doi:10.1080/00039896.1992.9938377

Ng, T. P., H. H. Goh, Y. L. Ng, H. Y. Ong, C. N. Ong, K. S. Chia, S. E. Chia, J. Jeyaratnam. 1991. "Male endocrine functions in workers with moderate exposure to lead." *Br J Indust Med* 48(7): 485–491. doi:10.2307/27727282

Nielsen, F. H. 2003. "Trace elements." In Caballero B, ed. Encyclopedia of Food Sciences and Nutrition. London, England: Academic Press, pp. 5820–5828. doi:10.1016/b0-12-227055-x/01204-9

Nkpaa, K. W., B. A. Amadi, I. A. Adedara, M. O. Wegwu, E. O. Farombi. 2018. "Ethanol exacerbates manganese - induced functional alterations along the hypothalamic-pituitary-gonadal axis of male rats." *Neurosci Lett* 684: 47–54. doi:10.1016/j.neulet.2018.07.007

Novotny, J. A., C. A. Peterson. 2018. "Molybdenum." *Adv Nutr* 9(3): 272–273. doi:10.1093/advances/nmx001

Ogorek, M., M. Lenartowicz, R. Starzynski, A. Jonczy, R. Staron, A. Doniec, W. Krzeptowski, A. Bednarz, O. Pierzchala, P. Lipinski, Z. Rajfur, Z. Baster, P. Gibas-Tybur, P. Grzmil. 2017. "Atp7a and Atp7b regulate copper homeostasis in developing male germ cells in mice." *Metallomics* 9(9): 1288–1303. doi:10.1039/c7mt00134g

Oliveira, H., M. Spano, C. Santos, M. de L. Pereira. 2009. "Lead chloride affects sperm motility and acrosome reaction in mice: lead affects mice sperm motility and acrosome reaction." *Cell Biol Toxicol* 25(4): 341–353. doi:10.1007/s10565-008-9088-4

Olshan, A. F., P. A. Baird, K. Teschke. 1989. "Paternal occupational exposures and the risk of down syndrome." *Am J Hum Genet* 44(5): 646–651.

Omu, A. E., M. K. Al-Azemi, M. Al-Maghrebi, C. T. Mathew, F. E. Omu, E. O. Kehinde, J. T. Anim, M. A. Oriowo, A. Memon. 2015. "Molecular basis for the effects of zinc deficiency on spermatogenesis: an experimental study in the Sprague-dawley rat model." *Indian J Urol* 31: 57–64. doi:10.4103/0970-1591.139570

Orth, J. M., G. L. Gunsalus, A. A. Lamperti. 1988. "Evidence from sertoli cell-depleted rats indicates that spermatid number in adults depends on numbers of Sertoli cells produced during perinatal development." *Endocrinology* 122(3): 787–794. doi:10.1210/endo-122-3-787

Overstreet, J. W., Blazak, W. F. 1983. "The biology of human male reproduction: an overview". *Am J Ind Med* 4(1–2): 5–15.

Partal-Lorente, A. B., V. Maldonado-Ezequiel, L. Martinez-Navarro, I. Herrera-Contreras, C. Gutierrez-Repiso, E. García-Fuentes, P. Santiago-Fernandez. 2017. "Iodine is associated to semen quality in men who undergo consultations for infertility." *Reprod Toxicol* 73: 1–7. doi:10.1016/j.reprotox.2017.07.020

Peacey, L., C. Peacey, A. Gutzinger, C. E. Jones. 2020. "Copper(II) binding by the earliest vertebrate gonadotropin-releasing hormone, the type II isoform, suggests an ancient role for the metal." *Int J Mol Sci* 21(21): 7900. doi:10.3390/ijms21217900

Perelomov, L. V., I. V. Perelomova, U. L. Venevtseva. 2016. "The toxic effects of trace elements on male reproductive health." *Human Physiol* 42: 454–462. doi:10.1134/S0362119716030130

Phillips, K. P. 2015. "Chapter 1- Perceptions of environmental risks to fertility." In Schreier S, ed. Handbook of Fertility. London, England: Academic Press, pp. 3–17. doi:10.1016/B978-0-12-800872-0.00001-9

Pillai, A., L. Priya, S. Gupta. 2003. "Effects of combined exposure to lead and cadmium on the hypothalamic–pituitary axis function in proestrous rats." *FCT* 41(3):379–384.

Pizent, A., B. Tariba, T. Zivkovic. 2012. "Reproductive toxicity of metals in men." *Arch Hig Rada Toksikol* 63(1): 35–46. doi:10.2478/10004-1254-63-2012-2151

Pizzol, D., C. Foresta, A. Garolla, J. Demurtas, M. Trott, A. Bertoldo, L. Smith. 2021. "Pollutants and sperm quality: a systematic review and meta-analysis." *Environ Sci Pollut Res Int* 28(4): 4095–4103. doi:10.1007/s11356-020-11589-z

Polednak, A. P., D. T. Janerich. 1983. "Uses of available record systems in epidemiological studies of reproductive toxicology." *Am J Ind Med* 4(1–2): 329–348.

Prasad, A. S. 2013. "Discovery of human zinc deficiency: its impact on human health and disease." *Adv Nutr* 4: 176–190. doi:10.3945/an.112.003210

Rajender, S., M. G. Monica, L. Walter, A. Agarwal. 2011. "Thyroid spermatogenesis and male infertility." *Front Biosci* 3(3): 843–855. doi:10.2741/e292

Raut, S., S. Deshpande, N. H. Balasinor. 2019. "Unveiling the role of prolactin and its receptor in male reproduction." *Horm Metab Res* 51(4): 215–219. doi:10.1055/a-0859-1144

Reis, M. M. S., A. C. Moreira, M. Sousa, P. Mathur, P. F. Oliveira, M. G. Alves. 2015. "Sertoli cell as a model in male reproductive toxicology: advantages and disadvantages." *J Appl Toxicol* 35: 870–883. doi:10.1002/jat.3122

Roemhild, K, F. von Maltzahn, R. Weiskirchen, R. Knüchel, S. von Stillfried, T. Lammers. 2021. "Iron metabolism: pathophysiology and pharmacology." *Trends Pharmacol Sci* 42(8): 640–656. doi:10.1016/j.tips.2021.05.001

Ronis, M. J., T. M. Badger, S. J. Shema, P. K. Roberson, F. Shaikh. 1996. "Reproductive toxicity and growth effects in rats exposed to lead at different periods during development." *Toxicol Appl Pharmacol* 136(2): 361–371. doi:10.1006/taap.1996.0044

Rotter, I., D. I. Kosik-Bogacka, B. Dołęgowska, K. Safranow, M. Kuczyńska, M. Laszczyńska. 2015. "Analysis of the relationship between the blood concentration of several metals, macro- and micronutrients and endocrine disorders associated with male aging." *Environ Geochem Health* 38(3): 749–761. doi:10.1007/s10653-015-9758-0

Roy, D., S. Dey, G. C. Majumder, D. Bhattacharyya. 2014. "Copper: a biphasic regulator of caprine sperm forward progression." *Syst Biol Reprod Med* 60(1): 52–57. doi:10.3109/19396368.2013.848243

Russo, D., A. Scipioni, C. Durante, E. Ferretti, L. Gandini, V. Maggisano, D. Paoli, A. Verrienti, G. Costante, A. Lenzi, S. Filetti. 2011. "Expression and localization of the sodium/iodide symporter (NIS) in testicular cells." *Endocrine* 40(1): 35–40. doi:10.1007/s12020-011-9469-y

Sakamoto, K. Q., M. Ishizuka, A. Kazusaka, S. Fujita. 2004. "Iodine intake as a possible cause of discontinuous decline in sperm counts: a re-evaluation of historical and geographic variation in semen quality." *Jpn J Vet Res* 52(2): 85–94.

Sagalov, A. V. 2006. Ambulatorno-poliklinicheskaya andrologiya. Nizhnij Novgorod, Russia: Izdatel'stvo Nizhegorodskoj gosudarstvennoj medicinskoj akademii. 240 p. (In Rus)

Saleh, B. O. M., N. K. Hussain, A. Y. Majid, B. Thabet, K. A. Fadhil. 2008. "Status of zinc and copper concentrations in seminal plasma of male infertility and their correlations with various sperm parameters." *Iraq Postgrad Med J* 7(1): 76–80.

Sawyer, D. E., B. G. Mercer, A. M. Wiklendt, R. J. Aitken. 2003. "Quantitative analysis of gene-specific DNA damage in human spermatozoa." *Mut Res* 529(1–2): 21–34. doi:10.1016/s0027-5107(03)00101-5

Schmid, T. E., P. G. Grant, F. Marchetti, R. H. Weldon, B. Eskenazi, A. J. Wyrobek. 2013. "Elemental composition of human semen is associated with motility and genomic sperm defects among older men." *Human Reprod* 28(1): 274–282. doi:10.1093/humrep/des321

Scott, R., A. MacPherson, R. W. Yates, B. Hussain, J. Dixon. 1998. "The effect of oral selenium supplementation on human sperm motility." *Br J Urol* 82: 76–78. doi:10.1046/j.1464-410x.1998.00683.x

Sengupta, P., S. Dutta, I. R. Karkada, S. V. Chinni. 2022. "Endocrinopathies and male infertility." *Life* 12: 10. doi:10.3390/life12010010

Shalini, S., M. P. Bansal. 2005. "Role of selenium in regulation of spermatogenesis: involvement of activator protein 1." *Biofactors* 23(3): 151–162. doi:10.1002/biof.5520230304

Sharp, R. M. 1994. "Regulation of spermatogenesis." In Knobil E, Neill J, ed. The Physiology of Reproduction, 2nd ed. New York: Raven, pp. 1363–1434.

Sikka S. C., R. K. Naz. 2002. "Endocrine toxicology: male reproduction". In Derelanko MJ, Hollinger MA, eds. Handbook of Toxicology, 2nd ed. New York: CRC Press LLC.

Simon, L., I. Proutski, M. Stevenson, D. Jennings, J. McManus, D. Lutton, S. E. Lewis. 2013. "Sperm DNA damage has a negative association with live-birth rates after IVF." *Reprod Biomed Online* 26(1): 68–78. doi:10.1016/j.rbmo.2012.09.019

Sinha Hikim, A. P., R. S. Swerdloff. 1999. "Hormonal and genetic control of germ cells apoptosis in the testis." *Rev Reprod* 4(1): 38–47. doi:10.1530/ror.0.0040038

Slivkova, J., M. Popelkova, P. Massanyi, S. Toporcerova, R. Stawarz. 2009. "Concentration of trace elements in human semen and relation to spermatozoa quality." *J Environ Sci Health A Tox Hazard Subst Environ Eng* 44(4):370–375. doi:10.1080/10934520802659729

Sokol, R. Z., S. Wang, Y. J. Y. Wan, F. Z. Stanczyk, E. Gentzschein, R. E. Chapin. 2002. "Long-term, low-dose lead exposure alters the gonadotropin-releasing hormone system in the male rat." *Environ Health Perspect* 110(9): 871–874. doi:10.1289/ehp.0211087

Soliman, A., M. Yassin, V. de Sanctis. 2014. "Intravenous iron replacement therapy in eugonadal males with iron-deficiency anemia: effects on pituitary gonadal axis and sperm parameters; A pilot study." *Indian J Endocrinol Metab* 18(3): 310–316. doi:10.4103/2230-8210.131158

Staessen, J. A., T. Nawrot, E. D. Hond, L. Thijs, R. Fagard, K. Hoppenbrouwers, G. Koppen, V. Nelen, G. Schoeters, D. Vanderschueren, E. van Hecke, L. Verschaeve, R. Vlietinck, H. A. Roels. 2001. "Renal function, cytogenetic measurements, and sexual development in adolescents in relation to environmental pollutants: a feasibility study of biomarkers." *Lancet* 357(9269): 1660–1669. doi:10.1016/s0140-6736(00)04822-4

Steiger, D., M. Fetchko, A. Vardanyan, L. Atanesyan, K. Steiner, M. L. Turski, D. J. Thiele, O. Georgiev, W. Schaner. 2010. "The Drosophila copper transporter Ctr1C functions in male fertility." *J Biol Chem* 285: 17089–17097. doi:10.1074/jbc.M109.090282

Sun, H., P. Xiang, J. Luo, H. Hong, H. Lin, H. B. Li, L. Q. Ma. 2016. "Mechanisms of arsenic disruption on gonadal, adrenal and thyroid endocrine systems in humans: a review." *Environ Inter* 95: 61–68. doi:10.1016/j.envint.2016.07.020

Sun, Y., C. Chen, G. G. Liu, M. Wang, C. Shi, G. Yu, F. Lv, N. Wang, S. Zhang. 2020. "The association between iodine intake and semen quality among fertile men in China." *BMC Public Health* 20: 461. doi:10.1186/s12889-020-08547-2

Surani, M. A., N. D. Allen, S. C. Barton, R. Fundele, S. K. Howlett, M. L. Norris, W. Reik. 1990. "Developmental consequences of imprinting of parental chromosomes by DNA methylation." *Philos. Trans R Soc Lond B Biol Sci* 326(1235):313–27. doi:10.1098/rstb.1990.0014

Taha, E. A., S. K. Sayed, N. M. Ghandour, A. M. Mahran, M. A. Saleh, M. M. Amin, R. Shamloul. 2013. "Correlation between seminal lead and cadmium and seminal parameters in idiopathic oligoasthenozoospermic males." *Cent European J Urol* 66(1): 84–92. doi:10.5173/ceju.2013.01.art28

Tajaddini, S., S. Ebrahimi, B. Behnam, M. Bakhtiyari, M. T. Joghataei, M. Abbasi, M. Amini, S. Amanpour, M. Koruji. 2014. "Antioxidant effect of manganese on the testis structure and sperm parameters of formalin-treated mice." *Andrologia* 46(3): 246–253. doi:10.1111/and.12069

Takeshima, T., Y. Yumura, K. Yasuda, H. Sanjo, S. Kuroda, H. Yamanaka, A. Iwasaki. 2017. "Inverse correlation between reactive oxygen species in unwashed semen and sperm motion parameters as measured by a computer-assisted semen analyzer." *Asian J Androl* 19(3): 350–354. doi:10.4103/1008-682X.173933.

Thorlacius-Ussing, O., Danscher, G. 1985. "Selenium in the anterior pituitary of rats exposed to sodium selenite: light and electron microscopic localization." *Toxicol Appl Pharmacol* 81(1): 67–74. doi:10.1016/0041-008x(85)90121-8

Thorlacius-Ussing, O., F. T. Jensen. 1988. "Selenium in the anterior pituitary of the rat after a single injection of 75Se sodium selenite." *Biol Trace Elem Res* 15: 277–287. doi:10.1007/BF02990144

Tiktinsky, O. L., S. N. Kalinina, V. V. Mihajlichenko. 2010. *Andrologiya*. Moscow, Russia: Medicinskoe informacionnoe agentstvo. 576 p. (In Rus)

Trokoudes, K. M., N. Skordis, M. K. Picolos. 2006. "Infertility and thyroid disorders." *Curr Opin Obst Gynecol* 18(4): 446–451. doi:10.1097/01.gco.0000233941.89919.31

Tue Nguyen, H. M., D. J. Heidenberg, S. C. Sikka. 2018. "Androgen receptor modulators: the impact of environment and lifestyle choices on reproduction." *Bioenviron Issues Affect Men's Reprod Sexual Health* 2018: 277–292. doi:10.1016/b978-0-12-801299-4.00017-7

Tvrda, E., R. Peer, S. C. Sikka, A. Agarwal. 2015. "Iron and copper in male reproduction: a double-edged sword." *J Assist Reprod Genet* 32(1): 3–16. doi:10.1007/s10815-014-0344-7

Ursini, F., S. Heim, M. Kiess, M. Maiorino, A. Roveri. 1999. "Dual function of the selenoprotein PHGPx during sperm maturation. *Science* 285(5432): 1393–1396. doi:10.1126/science.285.5432.1393

Van den Berghe, P. V. E., L. W. J. Klomp. 2010. "Posttranslational regulation of copper transporters." *J Biol Inorg Chem* 15: 37–46. doi:10.1007/s00775-009-0592-7

Van Niekerk, F. E.; C. H. Van Niekerk. 1989. "The influence of experimentally induced copper deficiency on the fertility of rams. II. Macro- and microscopic changes in the testes." *J S Afr Vet Assoc* 60(1): 32–35.

Veldhuis, N. A., A. P. Gaeth, R. B. Pearson, K. Gabriel, J. Camakaris. 2009. "The multi-layered regulation of copper translocating P-type ATPases." *Biometals* 22(1): 177–190. doi:10.1007/s10534-008-9183-2

Vickram, S., K. Rohini, S. Srinivasan, D. N. Veenakumari, K. Archana, K. Anbarasu, P. Jeyanthi, S. Thanigaivel, G. Gulothungan, N. Rajendiran, P. S. Srikumar. 2021. "Role of zinc (Zn) in human reproduction: a journey from initial spermatogenesis to childbirth." *Int J Mol Sci* 22(4): 2188. doi:10.3390/ijms22042188

Vigeh, M., D. R. Smith, P. C. Hsu. 2011. "How does lead induce male infertility?" *Iran J Reprod Med* 9(1):1–8.

Vinceti, M., T. Filippini, S. Cilloni, A. Bargellini, A. V. Vergoni, A. Tsatsakis, M. Ferrante. 2017. "Health risk assessment of environmental selenium: emerging evidence and challenges (Review)." *Mol Med Rep* 15(5): 3323–3335. doi:10.3892/mmr.2017.6377

Wagner, M. S., S. M. Wajner, A. L. Maia. 2008. "The role of thyroid hormone in testicular development and function." *J Endocrinol* 199(3): 351–365. doi:10.1677/JOE-08–0218

Wajner, S. M., M. S. Wagner, A. L. Maia. 2009. "Clinical implications of altered thyroid status in male testicular function." *Arq Bras Endocrinol Metabol* 53(8): 976–982. doi:10.1590/s0004-27302009000800011

Watanabe, S. 2004. "Frequent structural chromosome aberrations in immotile human sperm exposed to culture media." *Human Reprod* 19(4): 940–947. doi:10.1093/humrep/deh148

Waters, B. L., T. D. Trainer. 1996. "Development of the human fetal testis." *Ped Path Lab Med* 16(1): 9–23.

Winder, C. 1989. "Reproductive and chromosomal effect of occupational exposure to lead on the male." *Reprod Toxicol* 3(4): 221–233. doi:10.1016/0890-6238(89)90016-6

Wirth, J. J., R. S. Mijal. 2010. "Adverse effects of low level heavy metal exposure on male reproductive function." *Syst Biol Reprod Med* 56(2): 147–167. doi:10.3109/19396360903582216

Wong, E. W. P., C. Y. Cheng. 2011. "Impacts of environmental toxicants on male reproductive dysfunction." *Trends Pharmacol Sci* 32: 290. doi:10.1016/j.tips.2011.01.001

World Health Organization. 1996. *"Trace Elements in Human Nutrition and Health."* Geneva: World Health Organization.

World Health Organization. 2023. *"Infertility Prevalence Estimates, 1990–2021."* Geneva: World Health Organization.

World Health Organization DoRHa R. 2010. *"WHO Laboratory Manual for the Examination and Processing of Human Semen. 5th ed."* Geneva: World Health Organization.

Wu, H. M., D. T. Lin-Tan, M. L. Wang, H. Y. Huang, C. L. Lee, H. S. Wang, Y. K. Soong, J. L. Lin. 2012. "Lead level in seminal plasma may affect semen quality for men without occupational exposure to lead." *Reprod Biol Endocrinol* 10: 91. doi:10.1186/1477-7827-10-91

Xu, D. X., H. M. Shen, Q. X. Zhu, L. Chua, Q. N. Wang, S. E. Chia, C. N. Ong. 2003. "The associations among semen quality, oxidative DNA damage in human spermatozoa and concentrations of cadmium, lead and selenium in seminal plasma." *Mut Res* 534(1–2): 155–163. doi:10.1016/s1383-5718(02)00274-7

Yamaguchi, S, C. Miura, K. Kikuchi, F. T. Celino, T. Agusa, S. Tanabe, T. Miura. 2009. "Zinc is an essential trace element for spermatogenesis." *Proc Natl Acad Sci U S A* 106(26): 10859–10864. doi:10.1073/pnas.0900602106

Yang, Y., Z. Zuo, Z. Yang, H. Yin, L. Wei, J. Fang, H. Guo, H. Cui, P. Ouyang, X. Chen, J. Chen, Y. Geng, Z. Chen, C. Huang, Y. Zhu, W. Liu. 2021. "Nickel chloride induces spermatogenesis disorder by testicular damage and hypothalamic-pituitary-testis axis disruption in mice." *Ecotoxicol Environ Safety* 225: 112718. doi:10.1016/j.ecoenv.2021.112718

Żak, K., M. Leziak, S. Niedobylski, M. Piwonski, K. Laszczak. 2020. "An impact of selected micronutrients on fertility." *J Education Health Sport Online* 10(9) 216–227. doi:10.12775/JEHS.2020.10.09.023

Zhai, X. W., Y. L. Zhang, Q. Qi, Y. Bai, X. L. Chen, L. J. Jin, X. G. Ma, R. Z. Shu, Z. J. Yang, F. J. Liu. 2013. "Effects of molybdenum on sperm quality and testis oxidative stress." *Syst Biol Reprod Med* 59(5): 251–255. doi:10.3109/19396368.2013.791347

Zhang, F. L., S. Yuan, P. Y. Dong, H. H. Ma, M. de Felici, W. Shen, X. F. Zhang. 2022. "Multi-omics analysis reveals that iron deficiency impairs spermatogenesis by gut-hormone synthesis axis." *Ecotoxicol Environ Safety* 248: 114344. doi.org/10.1016/j.ecoenv.2022.114344

Zhang, Y., S. Li, S. Li. 2019. "Relationship between cadmium content in semen and male infertility: a meta-analysis." *Environ Sci Pollut Res Int* 26(2): 1947–1953. doi:10.1007/s11356-018-3748-6

Zhao, J., X. Dong, X. Hu, L. Wang, B. Sun, Q. Wang, L. Li. 2016. "Zinc levels in seminal plasma and their correlation with male infertility: a systematic review and meta-analysis." *Sci Rep* 6: 22386. doi:10.1038/srep22386

20

Endocrine Disrupting Chemicals: Sources, Mechanisms, and Their Effects on Maternal, Fetal, and Early Childhood Health

Yujing Chen, Mengwei Yuan, Li-Zi Lin, Lihang Yu, Yemeng Wang, Kenneth Chow, Wen-Jing Deng, and Ming Kei Chung

LIST OF ABBREVIATIONS

Abbreviation	Definition
AGD	Anogenital distance
BDE-100	2,2′,4,4′,6-Pentabromodiphenyl ether
BDE-153	2,2′,4,4′,5,5′-Hexabromodiphenyl ether
BDE-47	2,2′,4,4′-Tetrabromodiphenyl ether
BDE-99	2,2′,4,4′,5-Pentabromodiphenyl ether
BMI	Body mass index
BPA	Bisphenol A
DBP	Dibutyl phthalate
DDD	Dichlorodiphenyldichloroethane
DDE	Dichlorodiphenyldichloroethylene
DDMS	Dichlorodiphenylchloroethane
DDMU	Dichlorodiphenylchloroethylene
DDTs	Dichlorodiphenyltrichloroethanes
DEHP	Diethylhexyl phthalate
DEP	Diethyl phthalate
DHT	Dihydrotestosterone
DMP	Dimethyl phthalate
DOHaD	Developmental origins of health and disease
EDCs	Endocrine-disrupting chemicals
GDM	Gestational diabetes mellitus
HDP	Hypertension during pregnancy
HPA	Hypothalamic-pituitary-adrenal
IGF	Insulin-like growth factor
IQ	Intelligence quotient
IRS	Indoor residual spraying
PBDEs	Polybrominated diphenyl ethers
PCOS	Polycystic ovary syndrome
PFASs	Per- and polyfluoroalkyl substances
PFHxS	Perfluorohexane sulfonic acid
PFNA	Perfluorononanoic acid
PFOA	Perfluorooctanoic acid
PFOS	Perfluorooctane sulfonic acid

DOI: 10.1201/9781003492115-25

20.1 Introduction

Endocrine-disrupting chemicals (EDCs) are a group of exogenous chemicals capable of interfering with the hormonal system of an organism (Kahn et al., 2020). EDCs affect the hormonal system through four significant mechanisms, including (1) mimicking and acting like endogenous hormones, (2) blocking endogenous hormones from regulating the endocrine pathways, (3) affecting the production, degradation, and storage of hormones in the body, and (4) modifying the sensitivity of body cells to sense the hormonal signals (O. Yang et al., 2015). EDCs comprise diverse natural and man-made chemicals (Chung et al., 2018; van der Meer et al., 2021). Human exposure to EDCs can occur through contact with various media and products such as air, water, soil, diet, clothing, and personal care products (Y. Chen et al., 2022; Chung et al., 2020; Chung et al., 2006; Chung et al., 2020; Metcalfe et al., 2022).

The endocrine system communicates through chemical messengers (i.e., hormones), which are released by glands and travel to target organs via the circulatory system. The endocrine system is critical during pregnancy and early childhood because significant changes in gene expressions and modulation of biochemical pathways are associated with rapid growth and development. Based on the theory of developmental origins of health and disease (DOHaD) (Gluckman et al., 2016), minute and seemingly insignificant changes during early life development could affect the long-term growth trajectories and susceptibility and be decisive to the onset of diseases in later life stages. Since hormones regulate other biological processes, their blood concentrations are typically low, in the range of pmol/L (Jane Ellis et al., 2003). Therefore, real-world EDC exposure scenarios are highly relevant to the development of human health and diseases because these exposures are characterized by low-dose and long-term repeated patterns (Chung et al., 2021).

Different approaches can be used to study the biological impacts of EDCs (Table 20.1). EDCs have been extensively studied through animal and cellular studies, revealing their potential to cause toxicity at various levels (Gore et al., 2015; Sabuz Vidal et al., 2021). On the other hand, many epidemiological investigations have shown the associations between EDC exposures and health and disease endpoints (Ho et al., 2022; YANG et al., 2006). While these studies have established a sound foundation regarding the

TABLE 20.1

Study Approaches to Investigate the Biological Effects of EDCs

Study Approach	Description	Advantages	Disadvantages
Epidemiological Studies	Observational studies on human populations that investigate the association between chemical exposures and health outcomes, including cohort, case-control, and cross-sectional studies.	Direct relevance to human health; high potential to demonstrate meaningful adverse health effects of EDCs in susceptible populations, such as reproductive disorders in women and developmental issues in children.	Susceptible to confounding factors and bias; expensive and time consuming; ethical considerations limit experimental control; potential exposure misclassification.
In Vivo Studies (Animal Studies)	Experimental studies using animals to investigate the biological effects, mechanisms of action, and dose-response relationships of chemicals.	Provides a way to study toxicology without experimenting on humans; allows for the study of complex interaction within organisms; enables examination of various toxicities of EDCs, such as neurotoxins, thyrotoxins, and developmental toxicity.	Ethical considerations; time-consuming; high cost; may not directly correlate with humans due to species differences.
In Vitro Studies (Cell Culture Studies)	Experimental studies that rely on growing living cells in a controlled environment to examine biological effects of chemicals.	Allows better precision; cost-effective; reduces the usage of organisms, which improves ethical considerations; easier reproducibility; enables the study of molecular effects, e.g., EDCs can bind to hormone receptors, disrupting normal cellular signaling, and gene expression.	Vulnerable to misleading results from artificial environment; limited as it does not provide the full dynamics of an organism.

Source: Information is adapted from Arango MT, et al. (https://www.ncbi.nlm.nih.gov/books/NBK459464/) and Committee on Toxicity Testing and Assessment of Environmental Agents, et al. (https://doi.org/10.17226/11523).

Abbreviations: EDCs, Endocrine-disrupting chemicals.

metabolic influence of EDCs and their impacts on non-communicable diseases (Haverinen et al., 2021; Kumar et al., 2020; Lind & Lind, 2018; Tang-Péronard et al., 2011; Wan et al., 2022), there remains a critical data gap concerning prenatal and maternal exposures and the subsequent consequences. Specifically, the impacts of EDCs on maternal, fetal, and child health are less understood. Research has shown that exposure during pregnancy can affect neurodevelopment and lead to adverse outcomes in children (Pocock et al., 1994), yet comprehensive studies focusing on these vulnerable populations are limited. Therefore, we aim to provide an overview of current evidence on how early life exposures to EDCs could influence long-term health outcomes and highlight important areas for further research efforts.

Many chemical classes are known to be EDCs or potential EDCs. These included but were not limited to metals, heavy metals, metalloids, phytoestrogens, phenols, phthalates, polybrominated chemicals, and per- and polyfluoroalkyl substances (PFASs). These chemicals vary widely regarding water solubility, environmental persistence, and bioaccumulation potential. In this chapter, we first discuss selected EDCs, focusing on their sources, environmental fates, and levels in different media. We also cover the molecular mechanisms of potential effects of EDCs obtained from cellular and animal studies. Then we transition to the epidemiological evidence of exposure risks in mothers, fetuses, and early childhood. We focus on six classes of the most commonly encountered EDCs (Table 20.2) that are of high public health relevance because of their known hazards and exposure characteristics in the human population (Table 20.3) (Louis et al., 2013): phthalates, phenols (as bisphenol A, BPA), heavy metals (as lead), pesticides (dichlorodiphenyltrichloroethanes, DDTs), polybrominated diphenyl ethers (PBDEs), and PFASs. To aid the understanding and interpretation of the readers, we only discuss one to a few representative compounds for each class. Many health outcomes are associated with EDCs, such as birth outcomes, neurodevelopment, and obesity (Figure 20.1) (Kahn et al., 2020). We specifically discuss the impacts of EDC exposures on women during pregnancy and fetal growth. Additionally, we cover the early childhood period, which spans from infancy to age eight (Kumari et al., 2019).

20.2 Commonly Encountered Endocrine-Disrupting Chemicals

20.2.1 Phthalates

Phthalates are phthalic acid esters primarily utilized as plasticizers to enhance plastics' flexibility, transparency, durability, and longevity (Y. Guo et al., 2011; Monneret, 2017). The most commonly detected phthalates in the environment include dimethyl phthalate (DMP), diethyl phthalate (DEP), dibutyl phthalate (DBP), and diethylhexyl phthalate (DEHP) (X. Gao et al., 2019).

Phthalates are usually derived from the leaching process of plastics, leading to water, sediment, and air contamination, with the highest concentrations found in sediment (Cui et al., 2024; X. Gao et al., 2019). Humans are exposed to phthalates through various pathways, including drinking water, eating food, and breathing (N. Li et al., 2024; C. Yang et al., 2020). Exposure to air accounts for only 1%, while ingestion of contaminated water and food makes up over 90% (Franco et al., 2007). The detected concentrations of phthalates in food range from 0 to 19,000 ng/g (J. Wang et al., 2023). Phthalates have been detected in human urine at concentrations ranging from 0.13 to 2.14 ng/mL (N. Li et al., 2021), meconium at 0.18–51.82 ng/g (J. Guo et al., 2020), and blood at 0.03–2.27 ng/mL (Hwa et al., 2022).

The presence of phthalate pollution in the environment poses various environmental hazards, such as disrupting soil bacterial communities and ecological functions, accumulating in crops (Q.-Q. Zhang et al., 2021), and disturbing the ecological balance in aquatic systems (Xie et al., 2023). Exposure to phthalates is progressively impacting human health. There is a significant correlation between phthalates and a range of health problems, such as cardiovascular diseases, reproductive dysfunctions, impaired physical development in fetuses and children, neurological damage, respiratory ailments, and urinary system disorders (Eales et al., 2022; Mariana & Cairrao, 2020).

The structures of both phthalates and steroid hormones include a benzene ring, which is vital to binding these contaminants to hormone receptors (Y. Li et al., 2023). Moreover, phthalates with varying side chains can stimulate and block receptors simultaneously (Weng et al., 2023). The hydrophobicity of phthalates plays a crucial role in the interaction between plasticizers and proteins (Zhou et al., 2012).

TABLE 20.2
Overview of Representative Compounds in Selected EDC Classes

EDC Classes	Description	Common Name	IUPAC Name	CAS Number	Formula	Structure	Molecular Weight	Solubility in Water
Phthalates	Class of organic compounds known as esters of phthalic acid	E.g., Dimethyl phthalate (DMP)	1,2-dimethyl benzene-1,2-dicarboxylate	131-11-3	$C_{10}H_{10}O_4$	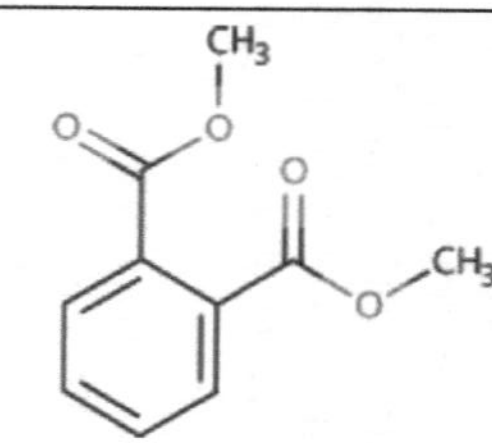	194.184 g/mol	0.43 mg/mL at 20°C
Bisphenol A (BPA)	Class of organic compounds known as bisphenols	Bisphenol A	4-[2-(4-hydroxyphenyl) propan-2-yl]phenol)	80-05-7	$C_{15}H_{16}O_2$	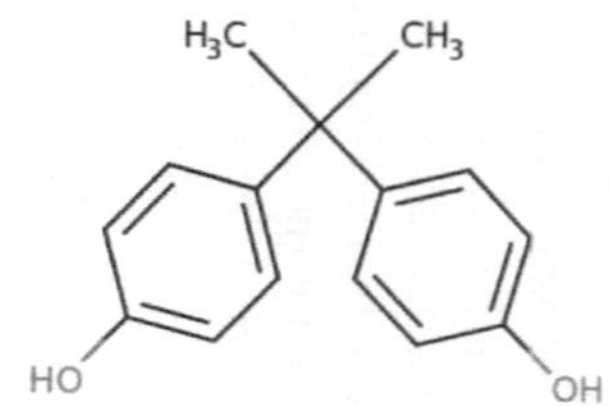	228.29 g/mol	0.12 mg/mL at 25°C
Lead	A naturally occurring metallic element	Lead	λ²-lead(2+) ion	7439-92-1	Pb	Pb^{2+}	207.2 g/mol	–t
Dichlorodiphenyltrichloroethanes (DDTs)	Class of organic compounds known as organochlorine pesticides	E.g., Dichlorodiphenyltrichloroethane (DDT)	1-chloro-4-[2,2,2-trichloro-1-(4-chlorophenyl) ethyl]benzene	50-29-3	$C_{14}H_9Cl_5$	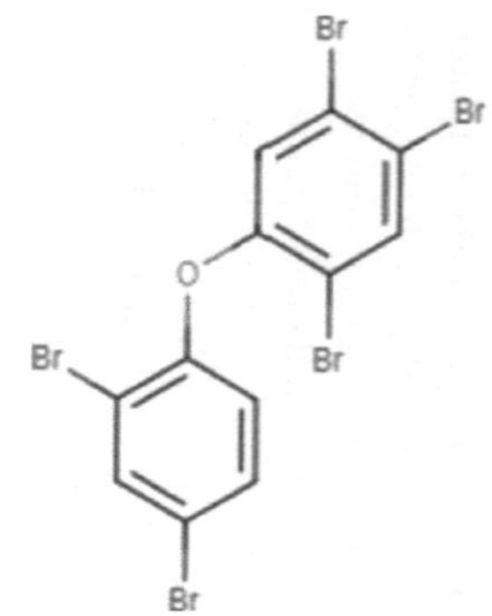	354.5 g/mol	25 μg/L at 25°C

(Continued)

TABLE 20.2 (*Continued*)

Overview of Representative Compounds in Selected EDC Classes

EDC Classes	Description	Common Name	IUPAC Name	CAS Number	Formula	Structure	Molecular Weight	Solubility in Water
Polybrominated diphenyl ethers (PBDEs)	Class of organic compounds known as bromodiphenyl ethers	E.g., 2,2′,4,4′,5-Pentabromodiphenyl ether (BDE-99)	1,2,4-tribromo-5-(2,4-dibromophenoxy) benzene	60348-660-9	$C_{12}H_5Br_5O$	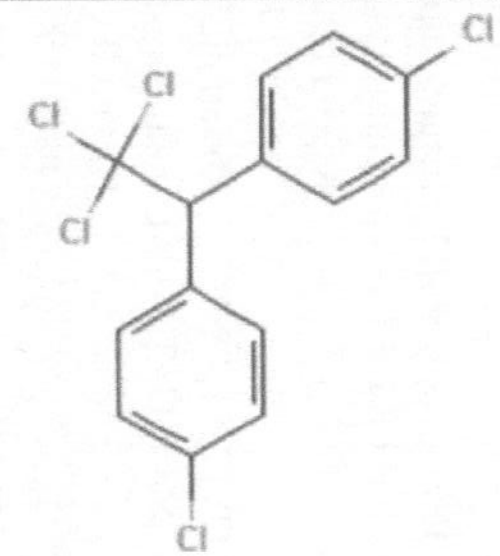	564.688 g/mol	9.0e-10 mg/mL at 20°C
Per- and polyfluoroalkyl substances (PFASs)	Class of organic compounds having fully (per) or partly (poly) fluorinated carbon chains with various functional groups	E.g., Perfluorooctane sulfonic acid (PFOS)	1-octanesulfonicacid,perfluoro-	1763-23-1	$C_8HF_{17}O_3S$	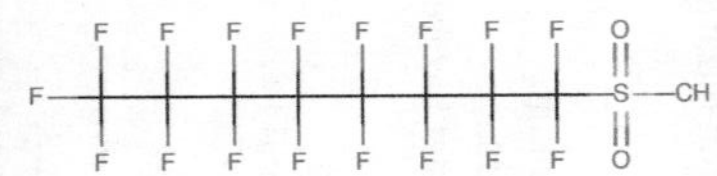	500.13 g/mol	0.68 mg/mL at 25°C

Source: Chemical information is obtained from the Human Metabolome Database (https://hmdb.ca/) and PubChem (https://pubchem.ncbi.nlm.nih.gov/)

Abbreviations: EDC, Endocrine-disrupting chemical; IUPAC, International Union of Pure and Applied Chemistry; CAS, Chemical Abstracts Service.

TABLE 20.3
EDC-Induced Hazards and Their Exposure Characteristics in the Human Population

		Gene Interactions[b]		Exposure Characteristics[b]	
EDC Classes	Overall Hazards[a]	Top Interacting Genes	Significant Enriched Pathway	Medium	Detection Frequency
Phthalates	Irritant, Health hazard, Environmental hazard	ABCC3, ABCC4, ABCG2; ABCC2; ACR	Metabolic pathways; KEGG ID: hsa01100	Urine	44.0%–98.6%
Bisphenol A (BPA)	Corrosive, Irritant, Health hazard, Environmental hazard	INS, ESR1, ESR2, MAPK1, MAPK3	Insulin signaling pathway; KEGG ID: hsa04910	Urine	0.64%–100%
Lead	Health hazard, Environmental hazard	CYP1A1, TNF, MT1, CAT, ALAD	Metabolic pathways; KEGG ID: hsa01100	Blood	0.4%–100%
Dichlorodiphenyltrichloroethanes (DDTs)	Acute Toxic, Irritant, Health Hazard, Environmental Hazard	AR, ESR1, CASP3, CYP19A1, NR1I3	Steroid hormone biosynthesis; KEGG ID: hsa00140	Serum	0%–100%
Polybrominated diphenyl ethers (PBDEs)	Acute Toxic, Health Hazard, Environmental Hazard	CYP1A1, CYP3A23-3A1, TTR, CXCL8, CYP2B1	Thyroid hormone synthesis; KEGG ID: hsa04918	Blood	17.7%–64.5%
Per- and polyfluoroalkyl substances (PFASs)	Irritant, Health hazard, Environmental hazard	PPARA, CASP3, ACOX1, BAX, BCL2	Retinol metabolism; KEGG ID: hsa00770	Serum	99.9%–100%

[a] Information from the CompTox Chemicals Dashboard (https://comptox.epa.gov/dashboard/).
[b] Information from the Comparative Toxicogenomics Database (https://ctdbase.org/).
Abbreviations: EDC, Endocrine-disrupting chemical; KEGG, Kyoto Encyclopedia of Genes and Genomes.

20.2.2 Bisphenol A (BPA)

BPA is a chemical substance extensively used in manufacturing polycarbonate plastics and epoxy resins (B. Wang et al., 2018). Found in everyday items like water bottles and food containers (Van Woerden et al., 2019), BPA is notable for its endocrine-disrupting properties, which enable it to interact with estrogen receptors and potentially affect human health (Shafei et al., 2018).

BPA does not occur naturally. Over 6.8 million tonnes of BPA are produced worldwide (Vom Saal & Welshons, 2014), and its widespread use in consumer products has made it ubiquitous. Despite attempts to mitigate unintended BPA exposures, the detection of BPA reached as high as 17.2 mg/L in hazardous waste landfill leachates in Japan and 12 μg/L in effluents in the U.S. (Corrales et al., 2015). Epidemiological studies indicate that urinary BPA is detectable in 93% of Americans (Calafat et al., 2008), with the highest levels found in infants and children (Seachrist et al., 2016). BPA was also found in the placenta tissue, cord blood, and fetal serum, which verifies its transmission from mother to fetus (Di Donato et al., 2017).

BPA rapidly dissolves under aerobic conditions with a half-life of 4.6 days. However, the potential for BPA accumulation is enhanced in marine environments, exhibiting a half-life over 30 days (Tarafdar et al., 2022). Previous studies have shown that BPA, as a xenoestrogen, may exert serious health effects in humans and wildlife, resulting in damage to tissues and organs of the body (Shafei et al., 2018; Tarafdar et al., 2022). Additionally, BPA may also cause or exacerbate human health issues such as obesity, diabetes, and heart disease (Michałowicz, 2014).

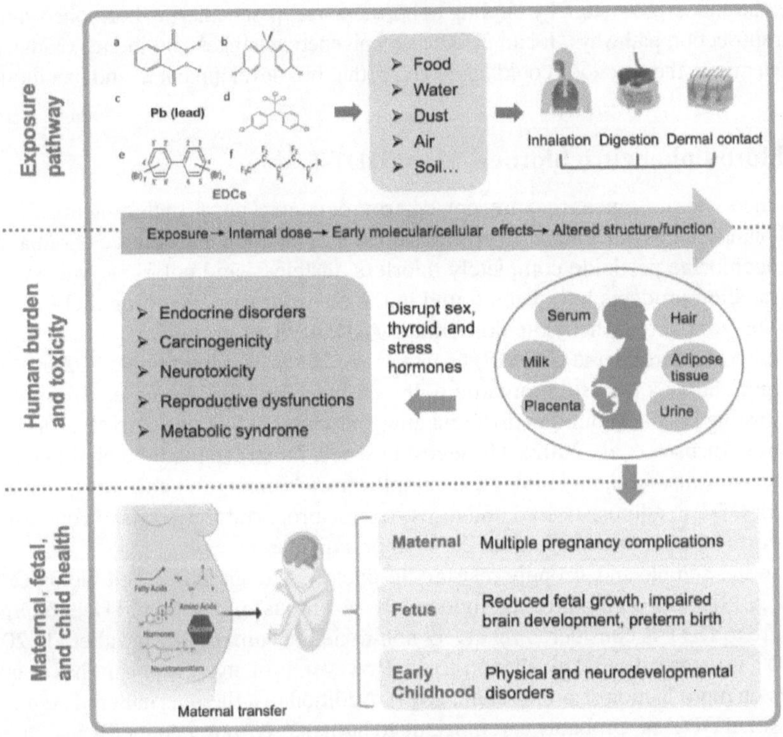

FIGURE 20.1 Endocrine disrupting chemicals (EDCs): the pathways of exposure and impacts on maternal, fetal, and early childhood health. (Created with Canva.com (@sketchify, @sketchifyedu, @nounproject). Additional elements were adapted from Starnes, H. M. et al. (2022) under a CC BY license.)

BPA triggers its biological effects through interaction with estrogen receptors, where it competes with estradiol and acts either as an agonist or antagonist through estrogen receptor-dependent signaling pathways (Y. Ma et al., 2019). This interaction leads to the inappropriate activation or inhibition of genes involved in processes such as reproductive health, which may further increase the risk of related disorders (Rubin, 2011).

20.2.3 Lead

Lead is a naturally occurring metallic element utilized in various industrial and consumer applications for an extended period. It is also considered a significant public health concern, as even minimal exposure can result in substantial adverse effects, particularly in young children. Ongoing efforts are directed toward further reducing lead exposure in the environment and human population (Wani et al., 2015).

In environmental monitoring, lead is frequently quantified in various samples, including soil, dust, water, and air. Commonly employed analytical techniques for lead include atomic absorption spectrometry and inductively coupled plasma mass spectrometry. Blood lead level measurement is the gold standard for biomonitoring, providing an integrated estimation of recent lead exposures (ATSDR, 2021a). In contrast, bone lead levels can provide information about long-term exposures. Hair and nail analysis can also indicate past lead exposures, although these matrices may be more susceptible to external contamination.

Lead contamination in soil and water can significantly affect ecosystem function and wildlife health. Lead disrupts plant growth and can be bioaccumulated in animals along the food chain (Ahamed & Siddiqui, 2007). In humans, particularly in children, lead exposure can result in severe health issues, including cognitive impairment, developmental delays, and behavioral problems. In adults, it can cause cardiovascular issues, kidney damage, and reproductive problems. Lead is particularly harmful due to its ability to accumulate in the body over time, making even low levels of exposure hazardous (Boskabady et al., 2018).

Lead affects the endocrine system by binding to hormone receptors, altering hormone synthesis, and interfering signal transduction pathways. It can also cause epigenetic changes and induce oxidative stress (Kahn et al., 2020). Over time, these actions could lead to reproductive, developmental, and metabolic health issues.

20.2.4 Dichlorodiphenyltrichloroethanes (DDTs)

Pesticides are chemical substances that prevent, destroy, or control pests, including insects, fungi, rodents, and unwanted plant species that can damage crops during cultivation and storage (Abubakar et al., 2020). DDT, an organochlorine pesticide completely odorless, tasteless, and colorless, was synthesized in the 1940s. Its degradation products have been found in the environment, including dichlorodiphenyldichloroethylene (DDE), dichlorodiphenyldichloroethane (DDD), dichlorodiphenylchloroethylene (DDMU), and dichlorodiphenylchloroethane (DDMS) (Afrin et al., 2023; Burkhard et al., 2023).

DDT is categorized as a persistent organic pollutant because of its strong environmental stability. It results in long-lasting residues found in different environmental media that also serve as ongoing sources of DDT exposure (Turusov et al., 2002). These residues accumulate in the food chain and have been consistently detected in soils, food, and biological samples from humans and animals (Turusov et al., 2002). Accumulation of DDT in adipose tissues and its estrogenic properties have raised concerns regarding the potential long-term adverse effects (Janjua, 2021) in organisms.

DDT is less water soluble and strongly resistant to photolytic degradation and biological and chemical actions (WHO, 2019). DDT is classified as "moderately hazardous" by the World Health Organization, and indirect exposure to DDT is considered relatively non-toxic for humans (Agarwal et al., 2012). Incidental exposure to DDT is regarded as relatively non-toxic. However, prolonged exposure has been linked to detrimental effects on reproduction (Loomis et al., 2015). Additionally, the International Agency for Research on Cancer classified DDT as "probably carcinogenic to humans" (Group 2A) in 2015 (Zhuang et al., 2012).

Minor structural changes of DDT analogs in the number and position of chlorine and phenolic hydroxyl moiety can cause pronounced effects on binding modes and binding affinities (Luccio-Camelo & Prins, 2011). Various analogs of DDT exhibited distinct effects on the conformational changes of the ligand-binding domain of estrogen receptor α and ER-related receptor γ, particularly regarding the orientation of helix-12. This sheds light on the diverse mechanisms by DDT analogs disrupt the endocrine system through interactions with different target receptors (Luccio-Camelo & Prins, 2011). In terms of its role as a disruptor of the androgen receptor, both of DDT's isomers (p,p′-DDT and o,p′-DDT) and its metabolite, p,p′-DDE, have been demonstrated to decrease the binding of dihydrotestosterone (DHT) to the androgen receptor in *in vivo* studies (Luccio-Camelo & Prins, 2011). Additionally, p,p′-DDE inhibited DHT-induced transcriptional activation *in vitro* (Luccio-Camelo & Prins, 2011).

20.2.5 Polybrominated Diphenyl ethers (PBDEs)

PBDEs are a class of synthetic chemicals used as chemical flame retardants to inhibit or resist the spread of fire. The structure of PBDEs includes two benzene rings, each connected by an ether bond to a central oxygen atom. PBDEs comprise 209 possible congeners, with 2,2′,4,4′-Tetrabromodiphenyl ether (BDE-47), 2,2′,4,4′,5-Pentabromodiphenyl ether (BDE-99), 2,2′,4,4′,6-Pentabromodiphenyl ether (BDE-100), and 2,2′,4,4′,5,5′-Hexabromodiphenyl ether (BDE-153) most commonly detected in human and environmental samples (Frederiksen et al., 2009).

Due to the leach-out capability, PBDEs have become ubiquitous and widely detected in ambient air particles, indoor dust, food, soils, and marine and water sediments (Deng et al., 2007; Z. Wu et al., 2020; Yu et al., 2023). Major routes of human exposure include inhalation, ingestion, and dermal contact (Johnson-Restrepo & Kannan, 2009). Among infants, diet accounts for a large proportion of the total intake of PBDEs (Fromme et al., 2016), while indoor dust is a significant source of inhaled PBDEs for toddlers and adults (Ni et al., 2013). Human biomonitoring studies have consistently shown the presence of PBDEs in serum, breast milk, and placental tissues. During pregnancy, the average concentrations of PBDEs are 3.1–112.8 ng/g (lipid weight) in umbilical cord serum, 2.6–66.8 ng/g (lipid weight) in breast milk, and about 17.6 ng/g (lipid weight) in the placenta (Z. Wu et al., 2020).

PBDE congeners are known to bioaccumulate in flora and fauna and biomagnify in the food chain, posing an ecological risk to the natural environment (Chai et al., 2019) and wildlife (D. Chen & Hale, 2010). PBDEs also exert extensive toxicity, acting as neurotoxins, thyrotoxins, estrogenic toxins, and carcinogens (Costa & Giordano, 2011; X. Wang et al., 2021). Human observational studies have shown that exposure to PBDEs is associated with endocrine dysfunction, developmental neurotoxicity, and reproductive disorders (Y. Gao et al., 2016; Sheikh & Beg, 2020). Pregnant women and neonates are particularly vulnerable due to the high risk of exposure and high sensitivity to PBDE toxicity.

PBDEs can act as agonists for estrogen and thyroid receptors, as well as antagonists for androgen (Nesan & Kurrasch, 2020). PBDEs and their hydroxylated (HO-) and methoxylated (MeO-) metabolites exhibited overall similarity of binding pattern and molecular interactions with thyroid hormone transport proteins (Sheikh & Beg, 2020). In addition, PBDEs and hydroxylated PBDEs were found to have antagonistic activity toward the glucocorticoid receptor, which can promote inflammation and cellular aging (Zota et al., 2018).

20.2.6 Per- and Polyfluoroalkyl Substances (PFASs)

PFASs are chemicals widely used in surface coatings and protectants for their unique surfactant properties (Z. Wang et al., 2021). Legacy PFASs have fully (per) or partly (poly) fluorinated carbon chains with various functional groups, categorized as short or long chains based on the length of the fluorinated carbon chain, and the most commonly studied PFASs include perfluorooctanoic acid (PFOA), perfluorooctane sulfonic acid (PFOS), perfluorohexane sulfonic acid (PFHxS), and perfluorononanoic acid (PFNA) (Evich et al., 2022).

Human exposure to PFASs occurs through dietary and non-dietary ingestion, inhalation, and dermal absorption (Holder et al., 2023), with the significance of each pathway varying globally (Sunderland et al., 2019). Dietary exposure through food and drink is predominant, accounting for over 90% of the total intake for common PFASs like PFOA and PFOS (Poothong et al., 2020). Background concentrations in drinking water can reach several hundred ng/L, while hotspot areas can reach a few mg/L (Kurwadkar et al., 2022). Detected concentrations in food chains range from pg/g to ng/g (Wee & Aris, 2023). Spatiotemporal exposure to PFASs varies considerably, with daily intake estimated at 0.01–1.7 ng/kg/day. Over the past two decades, most regions have noted decreasing trends in PFOS, PFHxS, and PFOA exposures, except for long-chain PFASs in East Asia (Fan et al., 2022). Globally, average serum concentrations were estimated at 6.3 ± 8.3 ng/mL for PFOS, 1.8 ± 2.1 ng/mL for PFHxS, 1.6 ± 1.5 ng/mL for PFOA, and 0.9 ± 0.6 ng/mL for PFNA (Fan et al., 2022).

The persistence, bioaccumulation, and toxicity of PFASs have made them a group of significant environmental contaminants (Evich et al., 2022). They are commonly detected in soil–water–plant–atmosphere ecosystems and are closely linked to human activity (Mei et al., 2021; Sima & Jaffé, 2021; Q. Wang et al., 2020). Research shows that even small concentrations can harm aquatic animals, amphibians, and insects, with significant accumulation in organisms and food webs (Brunn et al., 2023). Exposure to PFASs has been shown to affect human health (ATSDR, 2021b). A recent umbrella review shows evidence that certain PFASs were associated with the incidence of low birth weight, low efficiency of the tetanus vaccine, and low triglyceride levels. However, the proof of other health outcomes remains unclear (Xing et al., 2024).

The widespread industrial use of PFASs, with their long half-life in the human body and extreme environmental persistence, leads to chronic exposure for the general population during different developmental stages (Mokra, 2021). Both long- and short-chain PFASs would impact steroid hormone precursor levels (L. Li et al., 2024). Short-chain PFASs, especially PFHxS, often exhibit similar or higher endocrine-disrupting potential than long-chain PFASs (Mokra, 2021). *In vivo* and *in vitro* studies report that PFASs can bind to nuclear receptors, such as estrogen (Villeneuve et al., 2023), androgen (Tachachartvanich et al., 2022), and thyroid (Coperchini et al., 2021) hormone receptors, and alter steroidogenesis.

20.3 Endocrine Disruption and Maternal Health

Exposure to specific phthalates during pregnancy has been associated with adverse effects on maternal health (Weng et al., 2023). Studies indicate that phthalate exposures can disrupt crucial thyroid hormone receptor signaling necessary for pregnancy (W. Wu et al., 2022). Certain phthalates are putative agonists of regulators of glucose homeostasis and lipid metabolism (H. Wu et al., 2021) and also cause changes in prenatal circulating testosterone levels (Sathyanarayana et al., 2014). Exposure to phthalates during pregnancy may therefore be associated with increased maternal blood pressure (Han et al., 2020), long-term weight gain (Rodríguez-Carmona et al., 2019), and the development of type 2 diabetes mellitus and metabolic syndrome (Shoshtari-Yeganeh et al., 2019). Furthermore, phthalate exposures may increase the risk of unexplained recurrent miscarriages and premature birth (Aimuzi et al., 2022; Boss et al., 2018), as well as increase the risk of postpartum stress and depression (L. Hu et al., 2023).

The reproductive system is an essential target of BPA. Studies have identified that BPA can increase estradiol, progesterone, serum luteinizing hormone, and testosterone levels and could decrease serum cortisol concentrations (Y. Ma et al., 2019). Given that changes in sex hormones affect the endometrial thickness of the uterus, research has demonstrated a positive correlation between urine BPA levels and endometrial thickness in women under 37 years old and a negative correlation in women above 37 years old (Mínguez-Alarcón et al., 2015). In patients with polycystic ovary syndrome (PCOS), serum BPA levels are higher than those found in healthy women, indicating that BPA might contribute to the occurrence of PCOS (Vahedi et al., 2016). In addition, BPA levels are associated with spontaneous preterm birth, preterm premature rupture of membranes, and recurrent miscarriage (Y. Ma et al., 2019).

Lead acts as an endocrine disruptor, interfering with hormone receptors, altering hormone synthesis, and inducing oxidative stress, disrupting the hormonal balance essential for a healthy pregnancy (Rivera-Núñez et al., 2021). During pregnancy, lead exposures disrupt thyroid hormone levels. Altered thyroid function in pregnant women can lead to adverse outcomes such as miscarriage, preterm labor, and impaired fetal development (Lamb et al., 2008). Lead exposure is found to be associated with elevated levels of stress hormones, which in turn can affect pregnancy outcomes (Braun et al., 2014). In addition, lead exposure is associated with higher blood pressure, which can complicate pregnancy and increase the risk of pre-eclampsia (Blondeau et al., 2011).

Exposure to DDT has been suggested to have a connection with spontaneous abortion, which is the most prevalent adverse outcome during pregnancy, as indicated by early studies and findings from animal models (Saxena et al., 1980). Another study found a significant association between DDT exposure and maternal diastolic blood pressure (Siddiqui et al., 2002). A recent study reported several metabolic pathways, including serine and fatty acid metabolism, were associated with DDT and DDE exposures in maternal serum samples (X. Hu et al., 2020).

Gestational exposure to PBDEs has been associated with multiple adverse pregnancy outcomes, with effects on thyroid functions being the most well-characterized. Various congeners of PBDEs (e.g., BDE-47, -100, -153, and -99) have been suggested to decrease thyroid hormone levels (e.g., triiodothyronine, thyroxine, and thyroid-stimulating hormone levels), contributing to a higher prevalence of hypothyroidism in pregnant women (Gomes et al., 2024). Hypothyroidism is a risk factor for gestational diabetes mellitus (GDM) (Gong et al., 2016), and this supports the observed dose-response relationship between PBDEs and GDM risk (X. Liu et al., 2018). Studies also indicate that PBDEs were associated with decreased reproductive hormones and follicle disruption (Y. Gao et al., 2016). Women with detectable BDE-153 in follicular fluid have a higher risk of failed embryo implantation and early pregnancy loss (Johnson et al., 2012). Furthermore, maternal serum PBDE levels were related to the increased pro-inflammatory cytokines TNF-a and IL-6 (Zota et al., 2018), leading to pregnancy implications such as gestational hypertension and preeclampsia (Gomes et al., 2024).

Exposure to PFASs is associated with altered reproductive hormone levels in females, indicating potential adverse effects on maternal health (L. Li et al., 2024). Specifically, PFHxS and PFNA may disrupt androgenic and estrogenic pathways in pregnancy (Rivera-Núñez et al., 2023). Some examples of endocrine problems during pregnancy include GDM and hyper- or hypo-thyroidism. A recent review found that higher PFOA concentrations increase the risk of GDM (J. Wang et al., 2022). Changes in maternal thyroid-stimulating hormone are positively associated with exposure to PFOS and PFOA during pregnancy, though results for other thyroid hormones are inconclusive (L. Zhang et al., 2023). Although only 1%–2% of hypertension during pregnancy (HDP) cases are due to endocrine disorders, exposure to legacy PFASs (PFOA, PFOS, PFHxS) is linked to an increased risk of preeclampsia, and PFOS is further associated with HDP (C.-Y. Hu et al., 2023). Increased PFAS exposure is associated with reduced fertility in women, characterized by a lower fecundability odds ratio and a higher odds ratio for infertility (W. Wang et al., 2023). PFAS exposure may increase perceived stress and depression during pregnancy among pregnant women, raising the risk of adverse health outcomes (Eick et al., 2023).

20.4 Endocrine Disruption and Fetal Health

In addition to the impacts on fetuses through exposure to phthalates in the mother's body or sperm (Monneret, 2017), phthalates can also directly expose the fetus through the placenta. The presence of phthalates has been detected in umbilical cord blood. In early pregnancy, urine concentrations of phthalate biomarkers were associated with reductions in fetal growth, such as birth length and weight, head circumference, abdominal circumference, femur length, and changes in reproductive hormone levels (Ferguson et al., 2016; Stevens et al., 2023). Preterm birth and decreased anogenital distance (AGD) are the two outcomes related to phthalate exposure (Marie et al., 2015). Male and female newborns appear to be affected by different phthalates, suggesting gender-specific effects of phthalates (Katsikantami et al., 2016).

Exposure to BPA during pregnancy can potentially lead to adverse health outcomes for the fetus, affecting its growth, development, and long-term health. Research has linked prenatal BPA exposure to a range of effects, such as low birth weight and height and abnormal neurodevelopment problems and behaviors (Jensen et al., 2019; Pinney et al., 2017; Z. Wang et al., 2019). The fetus is particularly susceptible to neurotoxins and generally has higher levels of BPA than adults. It has been found that BPA can cross the placenta and directly enter the fetal bloodstream (Vom Saal & Welshons, 2014). BPA levels in placental tissue have been measured up to 273.9 ng/g (Troisi et al., 2014) and in cord blood up to 9.2 ng/g (Y. J. Lee et al., 2008), confirming fetal exposure through maternal transfer. Additionally, studies indicate that maternal BPA exposure correlates with a gender-specific impact on shorter AGD in boys at 12 months (Sun et al., 2018); girls who have been exposed to higher BPA levels during the first trimester also have shorter AGD (Barrett et al., 2017).

Researchers have shown that lead can cross the placental barrier and disrupt the delicate hormonal balance in the fetus (Caserta et al., 2013). This exposure can interfere with the thyroid hormone levels, which are crucial for brain development, potentially leading to cognitive impairments and developmental delays.

In addition, lead mimics the presence of calcium and interferes with parathyroid hormone regulation, which is vital for fetal bone development (Christensen et al., 2016). Disrupted calcium homeostasis can result in skeletal abnormalities and reduced fetal growth. Furthermore, lead exposure is linked to altered cortisol levels, which can impact the developing fetal hypothalamic–pituitary–adrenal (HPA) axis, influencing stress response and metabolic function postnatally (Rossi-George et al., 2011). Oxidative stress induced by lead exposures can damage placental function, reducing the fetus's nutrient and oxygen supply (Goyer, 1990). Subsequently, this can lead to low birth weight, preterm birth, and increased susceptibility to chronic diseases later in life.

The permeability of DDT through the placental barrier has been established. The concentration of pollutants in the cord serum reflects the portion passed through the placental barrier, likely close to the actual concentration in newborns that could harm fetal health (C. Xu et al., 2017). A cross-sectional study suggested that low *in utero* DDT exposure is linked to reduced birth weight, birth height, crown-heel length, and head circumference (Al-Saleh et al., 2012). For DDE, a degradation product of DDT, the adjusted odds ratios for preterm births exhibited a consistent and statistically significant rise with escalating concentrations of serum DDE (Longnecker et al., 2001). Mothers with elevated levels of DDE tended to discontinue breastfeeding their children at an earlier stage compared to those with lower exposure levels. This shorter duration of breastfeeding has been associated with increased infant mortality (Cupul-Uicab et al., 2008).

Fetuses can be exposed to PBDEs through maternal-fetal transfer, as lower brominated PBDEs can easily pass through and remain in the placenta (Zhao et al., 2013). PBDE exposures have been associated with increased levels of placental insulin-like growth factor (IGF), a fetal growth hormone needed for normal physical development (X. Xu et al., 2013). Abnormal gene expression or absence of IGF-2 can inhibit intrauterine growth and development (Cianfarani, 2012). Thyroid function and maternal transfer of glucocorticoids are also fundamental for fetal brain development and physical growth (Derakhshan et al., 2020; Nesan & Kurrasch, 2020). The disruptive effect of PBDEs on these endocrine-related mechanisms may imply the increased odds of adverse birth outcomes. Epidemiological studies have linked *in-utero* PBDE exposures to decreased birth lengths, birth weight, head circumferences (cephalic perimeter), fetal growth restriction, as well as preterm birth (Derakhshan et al., 2020; Gomes et al., 2024; The Consortium on Thyroid and Pregnancy—Study Group on Preterm Birth et al., 2019).

PFASs accumulate in the placenta; some can cross the placental barrier, accumulating in the fetus through maternal-fetal circulation (D. Ma et al., 2022; Y. Wu et al., 2023). Some studies suggest PFASs cross the placenta via passive diffusion, leading to continuous prenatal exposure (D. Ma et al., 2022). Free PFASs pass more easily than protein-bound ones because of their smaller size. It is unclear if binding affinity to transporters depends on their structure, but PFASs may enter fetal blood circulation through active transport with specific proteins (Y. Wu et al., 2023). Recent reviews suggest that prenatal exposure to PFOS, PFOA, and PFNA is linked to an increased risk of preterm birth (X. Gao et al., 2021), while PFNA exposure is associated with lower birth weight (Xing et al., 2024). Specifically, prenatal PFOA exposure negatively affects fetal head biometric parameters in women experiencing higher stress during pregnancy (Peterson et al., 2022). Lower folate status in mothers appears to increase susceptibility to adverse birth outcomes related to PFAS exposures, suggesting potential preventive measures against PFAS effects on fetal growth (B. Zhang et al., 2024).

20.5 Endocrine Disruption and Early Childhood Health

EDCs can affect neurodevelopment by targeting various hormonal axes elements (Figure 20.2). Some *in vitro* and *in vivo* studies suggest that phthalates may impact early life physical and neurodevelopment processes through anti-androgenic effects, anti-thyroid hormone activity, and adrenal imbalance (Apau et al., 2020; Bornehag & Nanberg, 2010; Hlisníková et al., 2021). The impact of phthalates on the body weight of preschool children is currently debated. Some studies suggest that high exposure levels may lead to lower infant weight and higher rates of obesity at ages 3–4 (Ferguson et al., 2022). Others suggest phthalates may affect muscle development, resulting in lower body mass index (BMI) in children (D.-W. Lee et al., 2022). Prenatal exposure to di-n-butyl phthalate is negatively correlated with children's psychomotor development (Qian et al., 2019). Phthalates may increase the risk of developing psychological disorders, especially attention-deficit hyperactivity disorder (Praveena et al., 2020). Furthermore, phthalates may increase the risk of asthma (Navaranjan et al., 2021) and thyroid function impairment (Morgenstern et al., 2017) in children.

BPA could disrupt the HPA axis and affect children's neural and psychological development (Y. Ma et al., 2019). Prenatal exposure to BPA could cause sexually dimorphic adverse effects on the HPA axis of the offspring, which have been associated with abnormal behaviors, disrupted neurodevelopment, and asthma in children. Behavioral changes such as depression, anxiety, sleep disorder, and hyperactivity were commonly observed (Rochester, 2013; the APrON Study Team et al., 2017). However, postnatal BPA exposure in boys has no association with these symptoms (Perera et al., 2016). BPA levels were also correlated with idiopathic central precocious puberty in school-aged girls (Y. Chen et al., 2018). However, higher BPA concentrations might induce later puberty in girls and earlier puberty in boys, indicating a gender-dependent effect (Berger et al., 2018).

Neurological and behavioral impacts from lead exposure are often highlighted in children. Lead primarily affects the nervous system, which could lead to cognitive and behavioral disorders. Studies have shown that lead exposure in children can result in decreased intelligence quotient (IQ), impaired cognitive functioning, learning disabilities, and conditions such as attention deficit hyperactivity disorder (Goodlad et al., 2013). These neurodevelopmental issues arise because lead interferes with synapse formation, neurotransmitter release, and neural cell migration. Additionally, lead disrupts the function of the thyroid gland, affecting hormone production and regulation, which is essential for brain development and overall growth and development (Canfield et al., 2004). Lead exposure not only affects children's neurodevelopment but also impacts the normal secretion of sex hormones, thereby adversely affecting pubertal development, as evidenced by a delay in the onset of puberty and abnormalities in the development of the reproductive organs (Pan et al., 2023). This endocrine imbalance may lead to the development of feminized features in boys or masculinized features in girls (Meeker, 2012).

Exposure to DDT may have persistent detrimental effects on the growth and development of children. Studies indicate that elevated levels of o,p'-DDT in maternal serum were associated with a nearly fourfold higher risk of breast cancer development in daughters (Cohn et al., 2015). It is suspected that DNA methylation is a potential molecular mechanism implicated in the development of breast cancer from prenatal DDT exposures (H.-C. Wu et al., 2020). Some countries still use DDT in indoor residual spraying (IRS) programs to control malaria, and children from IRS areas who experienced prenatal exposure to DDT and potentially DDE have been found to have increased odds of wheezing (Huq et al., 2020).

PBDE exposures may result in long-lasting health impacts on children, with particular concern for neurodevelopment in early life. PBDEs can affect multiple neurodevelopmental processes by disrupting thyroid hormone homeostasis and have potential neuroendocrine effects to alter neurotransmitter uptake and release (Vuong et al., 2018). Experimental and epidemiological studies have extensively documented PBDE-related neurotoxicity and behavioral alterations, including attention deficit, lower IQ, increased externalizing impulsive behaviors, and poor executive function (Y. Hu et al., 2022; J. Li et al., 2020). Meanwhile, although inconsistent findings have been noted, maternal PBDEs were related to postnatal physical growth. Some studies found positive associations of prenatal and lactational exposure to PBDEs with weight and height z-scores in infancy and preschool children (Y. Hu et al., 2022; J. Li et al., 2020), with a more pronounced association among boys. Conversely, others suggested an inverse association with BMI in 7-year-old girls (Erkin-Cakmak et al., 2015). Further studies are needed to explore the growth trajectory concerning PBDE exposures in children and elucidate the underlying sex-specific changes in the hormonal environment.

Newborns may encounter PFASs through breastfeeding and the indoor environment, with peak concentrations typically before age two due to cumulative exposure (Liew et al., 2018). Hence, children are vulnerable to PFAS exposure during prenatal, postnatal, and childhood stages (Starnes et al., 2022). Childhood exposure affects glucolipid metabolism in children and adolescents, with PFOS playing a significant role (Zheng et al., 2024). Despite study heterogeneity, pooled evidence suggests positive associations between prenatal PFAS exposure and childhood BMI and waist circumference and stronger negative associations between childhood PFAS exposures and childhood BMI (Frigerio

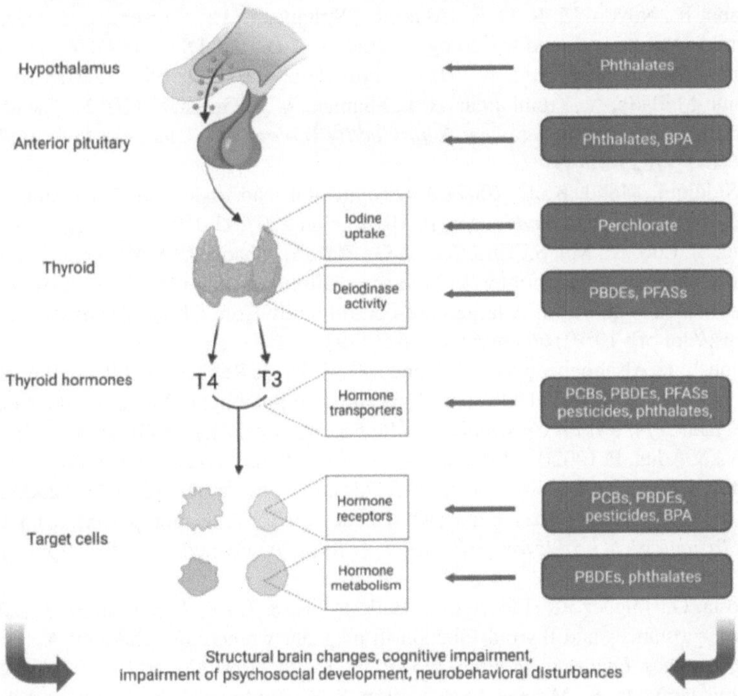

FIGURE 20.2 Effects of endocrine-disrupting chemicals on hormonal axes and their consequences on neurodevelopment, exemplified on thyroid hormone regulation. Abbreviations: BPA, bisphenol A; PBDEs, polybrominated diphenyl ethers; PCBs, polychlorinated biphenyls; PFASs, per- and polyfluoroalkyl substances; T3, triiodothyronine; T4, thyroxine. (Reproduced from Özel F, et al. (doi:10.1111/dmcn.15551) under a CC BY license.)

et al., 2023). Childhood PFHxS exposure may disrupt adolescents' energy balance and body composition, though the relationship with fat content is unclear (Lin et al., 2022). While brain development depends on hormones that PFASs might disrupt (Ballesteros et al., 2017), studies on PFAS exposure and neurodevelopmental outcomes have reported inconsistent or contradictory findings (X. Gao et al., 2023; D. Liu et al., 2024).

20.6 Conclusion

We reviewed six commonly found EDCs in humans and discussed their sources and levels in different media and biofluids, as well as their environmental fates and human health implications. Molecular and cellular toxicities of EDCs have been reported in the literature, and numerous epidemiological studies have shown their health risks to maternal, fetal, and early childhood health. Some man-made chemicals are long known to harm humans and are being phased out by various industries (i.e., legacy chemicals). Future research should focus on newer and less known emerging chemicals because significant knowledge gaps exist. Although the use and release of certain EDCs to the environment, such as lead and DDTs, have dramatically reduced in recent decades, human exposure to EDCs is almost unavoidable because of their general persistence and wide application in various consumer products in our society. Meanwhile, continuous monitoring of EDCs in different media is still critical to protecting human health, and developing safer chemical alternatives could be a long-term solution.

REFERENCES

Abubakar, Y., Tijjani, H., Egbuna, C., Adetunji, C. O., Kala, S., Kryeziu, T. L., Ifemeje, J. C., & Patrick-Iwuanyanwu, K. C. (2020). Pesticides, History, and Classification. In *Natural Remedies for Pest, Disease and Weed Control* (pp. 29–42). Elsevier, Amsterdam, The Netherlands. https://doi.org/10.1016/B978-0-12-819304-4.00003-8

Afrin, Mst. F., Kabir, E., Noyon, M. R. O. K., Akter, N., Sultana, T., Un Nayeem, J., & Uzzaman, M. (2023). Spectrochemical, biological, and toxicological studies of DDT, DDD, and DDE: An in-silico approach. *Informatics in Medicine Unlocked*, *39*, 101254. https://doi.org/10.1016/j.imu.2023.101254

Agarwal, A., Aponte-Mellado, A., Premkumar, B. J., Shaman, A., & Gupta, S. (2012). The effects of oxidative stress on female reproduction: A review. *Reproductive Biology and Endocrinology*, *10*(1), 49. https://doi.org/10.1186/1477-7827-10-49

Ahamed, M., & Siddiqui, Mohd. K. J. (2007). Environmental lead toxicity and nutritional factors. *Clinical Nutrition*, *26*(4), 400–408. https://doi.org/10.1016/j.clnu.2007.03.010

Aimuzi, R., Huang, S., Luo, K., Ma, S., Huo, X., Li, G., Tian, Y., Zhang, J., & Yu, Y. (2022). Levels and health risks of urinary phthalate metabolites and the association between phthalate exposure and unexplained recurrent spontaneous abortion: A large case-control study from China. *Environmental Research*, *212*, 113393. https://doi.org/10.1016/j.envres.2022.113393

Al-Saleh, I., Al-Doush, I., Alsabbaheen, A., Mohamed, G. E. D., & Rabbah, A. (2012). Levels of DDT and its metabolites in placenta, maternal and cord blood and their potential influence on neonatal anthropometric measures. *Science of the Total Environment*, *416*, 62–74. https://doi.org/10.1016/j.scitotenv.2011.11.020

Apau, J., Sefah, W., & Adua, E. (2020). Human contact with phthalates during early life stages leads to weight gain and obesity. *Cogent Chemistry*, *6*(1), 1815273. https://doi.org/10.1080/23312009.2020.1815273

ATSDR. (2021a). *Toxic Substance Portal*. https://wwwn.cdc.gov/TSP/substances/ToxSubstance.aspx?toxid=22

ATSDR. (2021b). *Toxicological profile for perfluoroalkyls*. https://wwwn.cdc.gov/TSP/ToxProfiles/ToxProfiles.aspx?id=1117&tid=237

Ballesteros, V., Costa, O., Iñiguez, C., Fletcher, T., Ballester, F., & Lopez-Espinosa, M.-J. (2017). Exposure to perfluoroalkyl substances and thyroid function in pregnant women and children: A systematic review of epidemiologic studies. *Environment International*, *99*, 15–28. https://doi.org/10.1016/j.envint.2016.10.015

Barrett, E. S., Sathyanarayana, S., Mbowe, O., Thurston, S. W., Redmon, J. B., Nguyen, R. H. N., & Swan, S. H. (2017). First-trimester urinary bisphenol A Concentration in relation to anogenital distance, an androgen-sensitive measure of reproductive development, in infant girls. *Environmental Health Perspectives*, *125*(7), 077008. https://doi.org/10.1289/EHP875

Berger, K., Eskenazi, B., Kogut, K., Parra, K., Lustig, R. H., Greenspan, L. C., Holland, N., Calafat, A. M., Ye, X., & Harley, K. G. (2018). Association of prenatal urinary concentrations of phthalates and bisphenol A and pubertal timing in boys and girls. *Environmental Health Perspectives*, *126*(9), 097004. https://doi.org/10.1289/EHP3424

Blondeau, B., Joly, B., Perret, C., Prince, S., Bruneval, P., Lelièvre-Pégorier, M., Fassot, C., & Duong Van Huyen, J.-P. (2011). Exposure in utero to maternal diabetes leads to glucose intolerance and high blood pressure with no major effects on lipid metabolism. *Diabetes & Metabolism*, *37*(3), 245–251. https://doi.org/10.1016/j.diabet.2010.10.008

Bornehag, C. G., & Nanberg, E. (2010). Phthalate exposure and asthma in children. *International Journal of Andrology*, *33*(2), 333–345. https://doi.org/10.1111/j.1365-2605.2009.01023.x

Boskabady, M., Marefati, N., Farkhondeh, T., Shakeri, F., Farshbaf, A., & Boskabady, M. H. (2018). The effect of environmental lead exposure on human health and the contribution of inflammatory mechanisms, a review. *Environment International*, *120*, 404–420. https://doi.org/10.1016/j.envint.2018.08.013

Boss, J., Zhai, J., Aung, M. T., Ferguson, K. K., Johns, L. E., McElrath, T. F., Meeker, J. D., & Mukherjee, B. (2018). Associations between mixtures of urinary phthalate metabolites with gestational age at delivery: A time to event analysis using summative phthalate risk scores. *Environmental Health*, *17*(1), 56. https://doi.org/10.1186/s12940-018-0400-3

Braun, J. M., Wright, R. J., Just, A. C., Power, M. C., Tamayo, Y., Ortiz, M., Schnaas, L., Hu, H., Wright, R. O., & Tellez-Rojo, M. M. (2014). Relationships between lead biomarkers and diurnal salivary cortisol indices in pregnant women from Mexico City: A cross-sectional study. *Environmental Health*, *13*(1), 50. https://doi.org/10.1186/1476-069X-13-50

Brunn, H., Arnold, G., Körner, W., Rippen, G., Steinhäuser, K. G., & Valentin, I. (2023). PFAS: Forever chemicals—persistent, bioaccumulative and mobile. Reviewing the status and the need for their phase out and remediation of contaminated sites. *Environmental Sciences Europe*, *35*(1), 20. https://doi.org/10.1186/s12302-023-00721-8

Burkhard, L. P., Elonen, G. E., & Mount, D. R. (2023). Review of DDT, DDE, DDD, DDMU and DDMS toxicity data for organisms used in estuarine and marine sediment toxicity tests. *Bulletin of Environmental Contamination and Toxicology*, *110*(6), 115. https://doi.org/10.1007/s00128-023-03754-0

Calafat, A. M., Ye, X., Wong, L.-Y., Reidy, J. A., & Needham, L. L. (2008). Exposure of the U.S. population to bisphenol A and 4- *tertiary* -octylphenol: 2003–2004. *Environmental Health Perspectives*, *116*(1), 39–44. https://doi.org/10.1289/ehp.10753

Canfield, R. L., Gendle, M. H., & Cory-Slechta, D. A. (2004). Impaired neuropsychological functioning in lead-exposed children. *Developmental Neuropsychology*, *26*(1), 513–540. https://doi.org/10.1207/s15326942dn2601_8

Caserta, D., Graziano, A., Lo Monte, G., Bordi, G., & Moscarini, M. (2013). Heavy metals and placental fetal-maternal barrier: A mini-review on the major concerns. *European Review for Medical and Pharmacological Sciences*, *17*(16), 2198–2206.

Chai, M., Ding, H., Shen, X., & Li, R. (2019). Contamination and ecological risk of polybrominated diphenyl ethers (PBDEs) in surface sediments of mangrove wetlands: A nationwide study in China. *Environmental Pollution*, *249*, 992–1001. https://doi.org/10.1016/j.envpol.2019.02.044

Chen, D., & Hale, R. C. (2010). A global review of polybrominated diphenyl ether flame retardant contamination in birds. *Environment International*, *36*(7), 800–811. https://doi.org/10.1016/j.envint.2010.05.013

Chen, Y., Wang, Y., Ding, G., Tian, Y., Zhou, Z., Wang, X., Shen, L., & Huang, H. (2018). Association between bisphenol a exposure and idiopathic central precocious puberty (ICPP) among school-aged girls in Shanghai, China. *Environment International*, *115*, 410–416. https://doi.org/10.1016/j.envint.2018.02.041

Chen, Y., Yang, J., Yao, B., Zhi, D., Luo, L., & Zhou, Y. (2022). Endocrine disrupting chemicals in the environment: Environmental sources, biological effects, remediation techniques, and perspective. *Environmental Pollution*, *310*, 119918. https://doi.org/10.1016/j.envpol.2022.119918

Christensen, P. S., Bonde, J. P., Bungum, L., Giwercman, A., Toft, G., Jönsson, B. A. G., & Specht, I. O. (2016). Environmental cadmium and lead exposure and anti-Müllerian hormone in pregnant women. *Reproductive Toxicology*, *61*, 114–119. https://doi.org/10.1016/j.reprotox.2016.03.047

Chung, M. K., Cheung, K. C., & Wong, M. H. (2020). Pollutants in Hong Kong soils: As, Cd, Cr, Cu, Hg, Pb and Zn. *bioRxiv*, 2020.02.16.951558. https://doi.org/10.1101/2020.02.16.951558

Chung, M. K., Hu, R., Cheung, K. C., & Wong, M. H. (2006). Pollutants in Hong Kong soils: Polycyclic aromatic hydrocarbons. *Chemosphere*, *67*, 464–473.

Chung, M. K., Hu, R., Cheung, K. C., & Wong, M. H. (2020). Pollutants in Hong Kong soils: Organochlorine pesticides and polychlorinated biphenyls. *bioRxiv*, 2020.02.16.951541. https://doi.org/10.1101/2020.02.16.951541

Chung, M. K., Kannan, K., Louis, G. M., & Patel, C. J. (2018). Toward capturing the exposome: Exposure biomarker variability and coexposure patterns in the shared environment. *Environmental Science & Technology*, *52*(15), 8801–8810. https://doi.org/10.1021/acs.est.8b01467

Chung, M. K., Rappaport, S. M., Wheelock, C. E., Nguyen, V. K., van, der M. T. P., Miller, G. W., Vermeulen, R., & Patel, C. J. (2021). Utilizing a biology-driven approach to map the exposome in health and disease: An essential investment to drive the next generation of environmental discovery. *Environmental Health Perspectives*, *129*(8), 085001. https://doi.org/10.1289/EHP8327

Cianfarani, S. (2012). Insulin-like growth factor-II: New roles for an old actor. *Frontiers in Endocrinology*, 3, 118. https://doi.org/10.3389/fendo.2012.00118

Cohn, B. A., La Merrill, M., Krigbaum, N. Y., Yeh, G., Park, J.-S., Zimmermann, L., & Cirillo, P. M. (2015). DDT exposure in utero and breast cancer. *The Journal of Clinical Endocrinology and Metabolism*, *100*(8), 2865–2872. https://doi.org/10.1210/jc.2015-1841

Coperchini, F., Croce, L., Ricci, G., Magri, F., Rotondi, M., Imbriani, M., & Chiovato, L. (2021). Thyroid disrupting effects of old and new generation PFAS. *Frontiers in Endocrinology*, *11*, 612320. https://doi.org/10.3389/fendo.2020.612320

Corrales, J., Kristofco, L. A., Steele, W. B., Yates, B. S., Breed, C. S., Williams, E. S., & Brooks, B. W. (2015). Global assessment of bisphenol A in the environment: Review and analysis of its occurrence and bioaccumulation. *Dose-Response*, *13*(3), 155932581559830. https://doi.org/10.1177/1559325815598308

Costa, L. G., & Giordano, G. (2011). Is decabromodiphenyl ether (BDE-209) a developmental neurotoxicant? *NeuroToxicology*, *32*(1), 9–24. https://doi.org/10.1016/j.neuro.2010.12.010

Cui, J., Bai, R., Ding, W., Liu, Q., Liu, Q., He, W., Yan, C., & Li, Z. (2024). Potential agricultural contamination and environmental risk of phthalate acid esters arrived from plastic film mulching. *Journal of Environmental Chemical Engineering*, *12*(1), 111785. https://doi.org/10.1016/j.jece.2023.111785

Cupul-Uicab, L. A., Gladen, B. C., Hernández-Ávila, M., Weber, J.-P., & Longnecker, M. P. (2008). DDE, a degradation product of DDT, and duration of lactation in a highly exposed area of Mexico. *Environmental Health Perspectives*, *116*(2), 179–183. https://doi.org/10.1289/ehp.10550

Deng, W. J., Zheng, J. S., Bi, X. H., Fu, J. M., & Wong, M. H. (2007). Distribution of PBDEs in air particles from an electronic waste recycling site compared with Guangzhou and Hong Kong, South China. *Environment International*, *33*(8), 1063–1069. https://doi.org/10.1016/j.envint.2007.06.007

Derakhshan, A., Peeters, R. P., Taylor, P. N., Bliddal, S., Carty, D. M., Meems, M., Vaidya, B., Chen, L., Knight, B. A., Ghafoor, F., Popova, P. V., Mosso, L., Oken, E., Suvanto, E., Hisada, A., Yoshinaga, J., Brown, S. J., Bassols, J., Auvinen, J., & Korevaar, T. I. M. (2020). Association of maternal thyroid function with birthweight: A systematic review and individual-participant data meta-analysis. *The Lancet Diabetes & Endocrinology*, *8*(6), 501–510. https://doi.org/10.1016/S2213-8587(20)30061-9

Di Donato, M., Cernera, G., Giovannelli, P., Galasso, G., Bilancio, A., Migliaccio, A., & Castoria, G. (2017). Recent advances on bisphenol-A and endocrine disruptor effects on human prostate cancer. *Molecular and Cellular Endocrinology*, *457*, 35–42. https://doi.org/10.1016/j.mce.2017.02.045

Eales, J., Bethel, A., Galloway, T., Hopkinson, P., Morrissey, K., Short, R. E., & Garside, R. (2022). Human health impacts of exposure to phthalate plasticizers: An overview of reviews. *Environment International*, *158*, 106903. https://doi.org/10.1016/j.envint.2021.106903

Eick, S. M., Barr, D. B., Brennan, P. A., Taibl, K. R., Tan, Y., Robinson, M., Kannan, K., Panuwet, P., Yakimavets, V., Ryan, P. B., Liang, D., & Dunlop, A. L. (2023). Per- and polyfluoroalkyl substances and psychosocial stressors have a joint effect on adverse pregnancy outcomes in the Atlanta African American maternal-child cohort. *Science of the Total Environment*, *857*, 159450. https://doi.org/10.1016/j.scitotenv.2022.159450

Erkin-Cakmak, A., Harley, K. G., Chevrier, J., Bradman, A., Kogut, K., Huen, K., & Eskenazi, B. (2015). In utero and childhood polybrominated diphenyl ether exposures and body mass at age 7 years: The CHAMACOS study. *Environmental Health Perspectives*, *123*(6), 636–642. https://doi.org/10.1289/ehp.1408417

Evich, M. G., Davis, M. J. B., McCord, J. P., Acrey, B., Awkerman, J. A., Knappe, D. R. U., Lindstrom, A. B., Speth, T. F., Tebes-Stevens, C., Strynar, M. J., Wang, Z., Weber, E. J., Henderson, W. M., & Washington, J. W. (2022). Per- and polyfluoroalkyl substances in the environment. *Science*, *375*(6580), eabg9065. https://doi.org/10.1126/science.abg9065

Fan, X., Tang, S., Wang, Y., Fan, W., Ben, Y., Naidu, R., & Dong, Z. (2022). Global exposure to per- and polyfluoroalkyl substances and associated burden of low birthweight. *Environmental Science & Technology*, *56*(7), 4282–4294. https://doi.org/10.1021/acs.est.1c08669

Ferguson, K. K., Bommarito, P. A., Arogbokun, O., Rosen, E. M., Keil, A. P., Zhao, S., Barrett, E. S., Nguyen, R. H. N., Bush, N. R., Trasande, L., McElrath, T. F., Swan, S. H., & Sathyanarayana, S. (2022). Prenatal phthalate exposure and child weight and adiposity from in utero to 6 years of age. *Environmental Health Perspectives*, 130(4), 047006. https://doi.org/10.1289/EHP10077

Ferguson, K. K., Meeker, J. D., Cantonwine, D. E., Chen, Y.-H., Mukherjee, B., & McElrath, T. F. (2016). Urinary phthalate metabolite and bisphenol A associations with ultrasound and delivery indices of fetal growth. *Environment International*, *94*, 531–537. https://doi.org/10.1016/j.envint.2016.06.013

Franco, A., Prevedouros, K., Alli, R., & Cousins, I. T. (2007). Comparison and analysis of different approaches for estimating the human exposure to phthalate esters. *Environment International*, *33*(3), 283–291. https://doi.org/10.1016/j.envint.2006.10.001

Frederiksen, M., Vorkamp, K., Thomsen, M., & Knudsen, L. E. (2009). Human internal and external exposure to PBDEs – A review of levels and sources. *International Journal of Hygiene and Environmental Health*, *212*(2), 109–134. https://doi.org/10.1016/j.ijheh.2008.04.005

Frigerio, G., Ferrari, C. M., & Fustinoni, S. (2023). Prenatal and childhood exposure to per-/polyfluoroalkyl substances (PFASs) and its associations with childhood overweight and/or obesity: A systematic review with meta-analyses. *Environmental Health*, *22*(1), 56. https://doi.org/10.1186/s12940-023-01006-6

Fromme, H., Becher, G., Hilger, B., & Völkel, W. (2016). Brominated flame retardants – Exposure and risk assessment for the general population. *International Journal of Hygiene and Environmental Health*, *219*(1), 1–23. https://doi.org/10.1016/j.ijheh.2015.08.004

Gao, X., Li, J., Wang, X., Zhou, J., Fan, B., Li, W., & Liu, Z. (2019). Exposure and ecological risk of phthalate esters in the Taihu Lake basin, China. *Ecotoxicology and Environmental Safety*, *171*, 564–570. https://doi.org/10.1016/j.ecoenv.2019.01.001

Gao, X., Ni, W., Zhu, S., Wu, Y., Cui, Y., Ma, J., Liu, Y., Qiao, J., Ye, Y., Yang, P., Liu, C., & Zeng, F. (2021). Per- and polyfluoroalkyl substances exposure during pregnancy and adverse pregnancy and birth outcomes: A systematic review and meta-analysis. *Environmental Research*, *201*, 111632. https://doi.org/10.1016/j.envres.2021.111632

Gao, X., Zuo, Q., Fu, X., Song, L., Cen, M., & Wu, J. (2023). Association between prenatal exposure to per- and polyfluoroalkyl substances and neurodevelopment in children: Evidence based on birth cohort. *Environmental Research*, *236*, 116812. https://doi.org/10.1016/j.envres.2023.116812

Gao, Y., Chen, L., Wang, C., Zhou, Y., Wang, Y., Zhang, Y., Hu, Y., Ji, L., Shi, R., Cui, C., Ding, G., Jin, J., & Tian, Y. (2016). Exposure to polybrominated diphenyl ethers and female reproductive function: A study in the production area of Shandong, China. *Science of the Total Environment*, *572*, 9–15. https://doi.org/10.1016/j.scitotenv.2016.07.181

Gluckman, P. D., Buklijas, T., & Hanson, M. A. (2016). Chapter 1- The developmental origins of health and disease (DOHaD) concept: Past, present, and future. In C. S. Rosenfeld (Ed.), *The Epigenome and Developmental Origins of Health and Disease* (pp. 1–15). Academic Press, New York. https://www.sciencedirect.com/science/article/pii/B9780128013830000013

Gomes, J., Begum, M., & Kumarathasan, P. (2024). Polybrominated diphenyl ether (PBDE) exposure and adverse maternal and infant health outcomes: Systematic review. *Chemosphere*, *347*, 140367. https://doi.org/10.1016/j.chemosphere.2023.140367

Gong, L.-L., Liu, H., & Liu, L.-H. (2016). Relationship between hypothyroidism and the incidence of gestational diabetes: A meta-analysis. *Taiwanese Journal of Obstetrics and Gynecology*, *55*(2), 171–175. https://doi.org/10.1016/j.tjog.2016.02.004

Goodlad, J. K., Marcus, D. K., & Fulton, J. J. (2013). Lead and attention-deficit/hyperactivity disorder (ADHD) symptoms: A meta-analysis. *Clinical Psychology Review*, *33*(3), 417–425. https://doi.org/10.1016/j.cpr.2013.01.009

Gore, A. C., Chappell, V. A., Fenton, S. E., Flaws, J. A., Nadal, A., Prins, G. S., Toppari, J., & Zoeller, R. T. (2015). Executive summary to EDC-2: The endocrine society's second scientific statement on endocrine-disrupting chemicals. *Endocrine Reviews*, *36*(6), 593–602. https://doi.org/10.1210/er.2015-1093

Goyer, R. A. (1990). Transplacental transport of lead. *Environmental Health Perspectives*, *89*, 101–105. https://doi.org/10.1289/ehp.9089101

Guo, J., Wu, M., Gao, X., Chen, J., Li, S., Chen, B., & Dong, R. (2020). Meconium exposure to phthalates, sex and thyroid hormones, birth size and pregnancy outcomes in 251 mother–infant pairs from Shanghai. *International Journal of Environmental Research and Public Health, 17*(21), 7711. https://doi.org/10.3390/ijerph17217711

Guo, Y., Wu, Q., & Kannan, K. (2011). Phthalate metabolites in urine from China, and implications for human exposures. *Environment International, 37*(5), 893–898. https://doi.org/10.1016/j.envint.2011.03.005

Han, X., Li, J., Wang, Y., Xu, S., Li, Y., Liu, H., Zhou, Y., Zhao, H., Fang, J., Cai, Z., & Xia, W. (2020). Association between phthalate exposure and blood pressure during pregnancy. *Ecotoxicology and Environmental Safety, 189*, 109944. https://doi.org/10.1016/j.ecoenv.2019.109944

Haverinen, E., Fernandez, M. F., Mustieles, V., & Tolonen, H. (2021). Metabolic syndrome and endocrine disrupting chemicals: An overview of exposure and health effects. *International Journal of Environmental Research and Public Health*, 18(24), Article 24. https://doi.org/10.3390/ijerph182413047

Hlisníková, H., Petrovičová, I., Kolena, B., Šidlovská, M., & Sirotkin, A. (2021). Effects and mechanisms of phthalates' action on neurological processes and neural health: A literature review. *Pharmacological Reports, 73*(2), 386–404. https://doi.org/10.1007/s43440-021-00215-5

Ho, V., Pelland-St-Pierre, L., Gravel, S., Bouchard, M. F., Verner, M.-A., & Labrèche, F. (2022). Endocrine disruptors: Challenges and future directions in epidemiologic research. *Environmental Research, 204*, 111969. https://doi.org/10.1016/j.envres.2021.111969

Holder, C., DeLuca, N., Luh, J., Alexander, P., Minucci, J. M., Vallero, D. A., Thomas, K., & Cohen Hubal, E. A. (2023). Systematic evidence mapping of potential exposure pathways for per- and polyfluoroalkyl substances based on measured occurrence in multiple media. *Environmental Science & Technology, 57*(13), 5107–5116. https://doi.org/10.1021/acs.est.2c07185

Hu, C.-Y., Qiao, J.-C., Gui, S.-Y., Xu, K.-X., Dzhambov, A. M., & Zhang, X.-J. (2023). Perfluoroalkyl and polyfluoroalkyl substances and hypertensive disorders of pregnancy: A systematic review and meta-analysis. *Environmental Research, 231*, 116064. https://doi.org/10.1016/j.envres.2023.116064

Hu, L., Mei, H., Feng, H., Huang, Y., Cai, X., Xiang, F., Chen, L., & Xiao, H. (2023). Exposure to bisphenols, parabens and phthalates during pregnancy and postpartum anxiety and depression symptoms: Evidence from women with twin pregnancies. *Environmental Research, 221*, 115248. https://doi.org/10.1016/j.envres.2023.115248

Hu, X., Li, S., Cirillo, P., Krigbaum, N., Tran, V., Ishikawa, T., La Merrill, M. A., Jones, D. P., & Cohn, B. (2020). Metabolome wide association study of serum DDT and DDE in pregnancy and early postpartum. *Reproductive Toxicology, 92*, 129–137. https://doi.org/10.1016/j.reprotox.2019.05.059

Hu, Y., Lu, Q., Huang, C., Gao, Y., Tian, Y., Fan, L., & Liu, S. (2022). Associations between prenatal exposure to polybrominated diphenyl ethers and physical growth in a seven year cohort study. *Chemosphere, 303*, 135049. https://doi.org/10.1016/j.chemosphere.2022.135049

Huq, F., Obida, M., Bornman, R., Di Lenardo, T., & Chevrier, J. (2020). Associations between prenatal exposure to DDT and DDE and allergy symptoms and diagnoses in the Venda health examination of mothers, babies and their environment (VHEMBE), South Africa. *Environmental Research, 185*, 109366. https://doi.org/10.1016/j.envres.2020.109366

Hwa, H., Peng, F., Ting, T., Chen, H., Chan, H., Yang, D., Chen, P., Kuo, Y., & Chen, P. (2022). Monitoring phthalates in maternal and cord blood: Implications for prenatal exposure and birth outcomes. *Environmental Toxicology and Chemistry, 41*(3), 715–725. https://doi.org/10.1002/etc.5280

Jane Ellis, M., Livesey, J. H., & Evans, M. J. (2003). Hormone stability in human whole blood. *Clinical Biochemistry, 36*(2), 109–112. https://doi.org/10.1016/S0009-9120(02)00440-X

Janjua, M. R. S. A. (2021). Prediction and understanding: Quantum chemical framework of transition metals enclosed in a B_{12} N_{12} inorganic nanocluster for adsorption and removal of DDT from the environment. *Inorganic Chemistry, 60*(14), 10837–10847. https://doi.org/10.1021/acs.inorgchem.1c01760

Jensen, T. K., Mustieles, V., Bleses, D., Frederiksen, H., Trecca, F., Schoeters, G., Andersen, H. R., Grandjean, P., Kyhl, H. B., Juul, A., Bilenberg, N., & Andersson, A.-M. (2019). Prenatal bisphenol A exposure is associated with language development but not with ADHD-related behavior in toddlers from the odense child cohort. *Environmental Research, 170*, 398–405. https://doi.org/10.1016/j.envres.2018.12.055

Johnson, P. I., Altshul, L., Cramer, D. W., Missmer, S. A., Hauser, R., & Meeker, J. D. (2012). Serum and follicular fluid concentrations of polybrominated diphenyl ethers and in-vitro fertilization outcome. *Environment International, 45*, 9–14. https://doi.org/10.1016/j.envint.2012.04.004

Johnson-Restrepo, B., & Kannan, K. (2009). An assessment of sources and pathways of human exposure to polybrominated diphenyl ethers in the United States. *Chemosphere*, *76*(4), 542–548. https://doi.org/10.1016/j.chemosphere.2009.02.068

Kahn, L. G., Philippat, C., Nakayama, S. F., Slama, R., & Trasande, L. (2020). Endocrine-disrupting chemicals: Implications for human health. *The Lancet Diabetes & Endocrinology*, *8*(8), 703–718. https://doi.org/10.1016/S2213-8587(20)30129-7

Katsikantami, I., Sifakis, S., Tzatzarakis, M. N., Vakonaki, E., Kalantzi, O.-I., Tsatsakis, A. M., & Rizos, A. K. (2016). A global assessment of phthalates burden and related links to health effects. *Environment International*, *97*, 212–236. https://doi.org/10.1016/j.envint.2016.09.013

Kumar, M., Sarma, D. K., Shubham, S., Kumawat, M., Verma, V., Prakash, A., & Tiwari, R. (2020). Environmental endocrine-disrupting chemical exposure: Role in non-communicable diseases. *Frontiers in Public Health*, 8, 553850. https://doi.org/10.3389/fpubh.2020.553850

Kumari, A., Sanijwan, S., & Sheoran, S. (2019). Early childhood and long term development: A systematic review. *Indian Journal of Health and Wellbeing*, *10*(1–3), 48–50.

Kurwadkar, S., Dane, J., Kanel, S. R., Nadagouda, M. N., Cawdrey, R. W., Ambade, B., Struckhoff, G. C., & Wilkin, R. (2022). Per- and polyfluoroalkyl substances in water and wastewater: A critical review of their global occurrence and distribution. *Science of the Total Environment*, *809*, 151003. https://doi.org/10.1016/j.scitotenv.2021.151003

Lamb, M. R., Janevic, T., Liu, X., Cooper, T., Kline, J., & Factor-Litvak, P. (2008). Environmental lead exposure, maternal thyroid function, and childhood growth. *Environmental Research*, *106*(2), 195–202. https://doi.org/10.1016/j.envres.2007.09.012

Lee, D.-W., Lim, H.-M., Lee, J.-Y., Min, K.-B., Shin, C.-H., Lee, Y.-A., & Hong, Y.-C. (2022). Prenatal exposure to phthalate and decreased body mass index of children: A systematic review and meta-analysis. *Scientific Reports*, 12(1), 8961. https://doi.org/10.1038/s41598-022-13154-9

Lee, Y. J., Ryu, H.-Y., Kim, H.-K., Min, C. S., Lee, J. H., Kim, E., Nam, B. H., Park, J. H., Jung, J. Y., Jang, D. D., Park, E. Y., Lee, K.-H., Ma, J.-Y., Won, H.-S., Im, M.-W., Leem, J.-H., Hong, Y.-C., & Yoon, H.-S. (2008). Maternal and fetal exposure to bisphenol A in Korea. *Reproductive Toxicology*, *25*(4), 413–419. https://doi.org/10.1016/j.reprotox.2008.05.058

Li, J., Ma, W., Zhao, Y., Jin, Y., Xiao, X., Ge, W., Shi, H., & Zhang, Y. (2020). Lactational exposure of polybrominated diphenyl ethers and its association with infant developmental measurements. *Journal of Hazardous Materials*, *388*, 122031. https://doi.org/10.1016/j.jhazmat.2020.122031

Li, L., Guo, Y., Ma, S., Wen, H., Li, Y., & Qiao, J. (2024). Association between exposure to per- and perfluoroalkyl substances (PFAS) and reproductive hormones in human: A systematic review and meta-analysis. *Environmental Research*, *241*, 117553. https://doi.org/10.1016/j.envres.2023.117553

Li, N., Liu, J., Ying, G., Lee, J. C.-K., Leung, T. F., Covaci, A., & Deng, W.-J. (2024). Endocrine disrupting chemicals in children's and their parents' urine: Is the exposure related to the Chinese and Western lifestyle? *International Journal of Hygiene and Environmental Health*, *259*, 114383. https://doi.org/10.1016/j.ijheh.2024.114383

Li, N., Ying, G.-G., Hong, H., Tsang, E. P. K., & Deng, W.-J. (2021). Plasticizer contamination in the urine and hair of preschool children, airborne particles in kindergartens, and drinking water in Hong Kong. *Environmental Pollution*, *271*, 116394. https://doi.org/10.1016/j.envpol.2020.116394

Li, Y., Yang, H., He, W., & Li, Y. (2023). Human endocrine-disrupting effects of phthalate esters through adverse outcome pathways: A comprehensive mechanism analysis. *International Journal of Molecular Sciences*, *24*(17), 13548. https://doi.org/10.3390/ijms241713548

Liew, Z., Goudarzi, H., & Oulhote, Y. (2018). Developmental exposures to perfluoroalkyl substances (PFASs): An update of associated health outcomes. *Current Environmental Health Reports*, *5*(1), 1–19. https://doi.org/10.1007/s40572-018-0173-4

Lin, L.-Z., Cai, L., Liu, Z.-Y., Gao, J., Zhou, Y., Zeng, X.-Y., Ou, Y., Dong, G.-P., Dong, P.-X., Wu, Q.-Z., Chu, C., Wu, L.-Y., Liang, L.-X., Qin, S.-J., Zeng, X.-W., Zhao, X., & Dong, G.-H. (2022). Exposure to per- and polyfluoroalkyl substances and body composition in US adolescents aged 12–18 years: An analysis of data from the national health and nutrition examination surveys 2011–2018. *Hygiene and Environmental Health Advances*, *3*, 100009. https://doi.org/10.1016/j.heha.2022.100009

Lind, P. M., & Lind, L. (2018). Endocrine-disrupting chemicals and risk of diabetes: An evidence-based review. *Diabetologia*, *61*(7), 1495–1502. https://doi.org/10.1007/s00125-018-4621-3

Liu, D., Yan, S., Liu, Y., Chen, Q., & Ren, S. (2024). Association of prenatal exposure to perfluorinated and polyfluoroalkyl substances with childhood neurodevelopment: A systematic review and meta-analysis. *Ecotoxicology and Environmental Safety*, *271*, 115939. https://doi.org/10.1016/j.ecoenv.2024.115939

Liu, X., Zhang, L., Li, J., Meng, G., Chi, M., Li, T., Zhao, Y., & Wu, Y. (2018). A nested case-control study of the association between exposure to polybrominated diphenyl ethers and the risk of gestational diabetes mellitus. *Environment International*, *119*, 232–238. https://doi.org/10.1016/j.envint.2018.06.029

Longnecker, M. P., Klebanoff, M. A., Zhou, H., & Brock, J. W. (2001). Association between maternal serum concentration of the DDT metabolite DDE and preterm and small-for-gestational-age babies at birth. *The Lancet*, *358*(9276), 110–114. https://doi.org/10.1016/S0140-6736(01)05329-6

Loomis, D., Guyton, K., Grosse, Y., Ghissasi, F. E., Bouvard, V., Benbrahim-Tallaa, L., Guha, N., Mattock, H., & Straif, K. (2015). Carcinogenicity of lindane, DDT, and 2,4-dichlorophenoxyacetic acid. *The Lancet Oncology*, *16*(8), 891–892. https://doi.org/10.1016/S1470-2045(15)00081-9

Louis, G. M. B., Sundaram, R., Schisterman, E. F., Sweeney, A. M., Lynch, C. D., Gore-Langton, R. E., Maisog, J., Kim, S., Chen, Z., & Barr, D. B. (2013). Persistent environmental pollutants and couple fecundity: The life study. *Environmental Health Perspectives*, *121*(2), 231–236. https://doi.org/10.1289/ehp.1205301

Luccio-Camelo, D. C., & Prins, G. S. (2011). Disruption of androgen receptor signaling in males by environmental chemicals. *The Journal of Steroid Biochemistry and Molecular Biology*, *127*(1–2), 74–82. https://doi.org/10.1016/j.jsbmb.2011.04.004

Ma, D., Lu, Y., Liang, Y., Ruan, T., Li, J., Zhao, C., Wang, Y., & Jiang, G. (2022). A critical review on transplacental transfer of per- and polyfluoroalkyl substances: Prenatal exposure levels, characteristics, and mechanisms. *Environmental Science & Technology*, *56*(10), 6014–6026. https://doi.org/10.1021/acs.est.1c01057

Ma, Y., Liu, H., Wu, J., Yuan, L., Wang, Y., Du, X., Wang, R., Marwa, P. W., Petlulu, P., Chen, X., & Zhang, H. (2019). The adverse health effects of bisphenol A and related toxicity mechanisms. *Environmental Research*, *176*, 108575. https://doi.org/10.1016/j.envres.2019.108575

Mariana, M., & Cairrao, E. (2020). Phthalates implications in the cardiovascular system. *Journal of Cardiovascular Development and Disease*, *7*(3), 26. https://doi.org/10.3390/jcdd7030026

Marie, C., Vendittelli, F., & Sauvant-Rochat, M.-P. (2015). Obstetrical outcomes and biomarkers to assess exposure to phthalates: A review. *Environment International*, *83*, 116–136. https://doi.org/10.1016/j.envint.2015.06.003

Meeker, J. D. (2012). Exposure to environmental endocrine disruptors and child development. *Archives of Pediatrics & Adolescent Medicine*, 166(10), 241. https://doi.org/10.1001/archpediatrics.2012.241

Mei, W., Sun, H., Song, M., Jiang, L., Li, Y., Lu, W., Ying, G.-G., Luo, C., & Zhang, G. (2021). Per- and polyfluoroalkyl substances (PFASs) in the soil–plant system: Sorption, root uptake, and translocation. *Environment International*, *156*, 106642. https://doi.org/10.1016/j.envint.2021.106642

Metcalfe, C. D., Bayen, S., Desrosiers, M., Muñoz, G., Sauvé, S., & Yargeau, V. (2022). An introduction to the sources, fate, occurrence and effects of endocrine disrupting chemicals released into the environment. *Environmental Research*, *207*, 112658. https://doi.org/10.1016/j.envres.2021.112658

Michałowicz, J. (2014). Bisphenol A – Sources, toxicity and biotransformation. *Environmental Toxicology and Pharmacology*, *37*(2), 738–758. https://doi.org/10.1016/j.etap.2014.02.003

Mínguez-Alarcón, L., Gaskins, A. J., Chiu, Y.-H., Williams, P. L., Ehrlich, S., Chavarro, J. E., Petrozza, J. C., Ford, J. B., Calafat, A. M., & Hauser, R. (2015). Urinary bisphenol A concentrations and association with *in vitro* fertilization outcomes among women from a fertility clinic. *Human Reproduction*, *30*(9), 2120–2128. https://doi.org/10.1093/humrep/dev183

Mokra, K. (2021). Endocrine disruptor potential of short- and long-chain perfluoroalkyl substances (PFASs)—A synthesis of current knowledge with proposal of molecular mechanism. *International Journal of Molecular Sciences*, *22*(4), 2148. https://doi.org/10.3390/ijms22042148

Monneret, C. (2017). What is an endocrine disruptor? *Comptes Rendus Biologies*, 340(9–10), 403–405. https://doi.org/10.1016/j.crvi.2017.07.004

Morgenstern, R., Whyatt, R. M., Insel, B. J., Calafat, A. M., Liu, X., Rauh, V. A., Herbstman, J., Bradwin, G., & Factor-Litvak, P. (2017). Phthalates and thyroid function in preschool age children: Sex specific associations. *Environment International*, *106*, 11–18. https://doi.org/10.1016/j.envint.2017.05.007

Navaranjan, G., Diamond, M. L., Harris, S. A., Jantunen, L. M., Bernstein, S., Scott, J. A., Takaro, T. K., Dai, R., Lefebvre, D. L., Azad, M. B., Becker, A. B., Mandhane, P. J., Moraes, T. J., Simons, E., Turvey, S. E., Sears, M. R., Subbarao, P., & Brook, J. R. (2021). Early life exposure to phthalates and the development of childhood asthma among Canadian children. *Environmental Research*, *197*, 110981. https://doi.org/10.1016/j.envres.2021.110981

Nesan, D., & Kurrasch, D. M. (2020). Gestational exposure to common endocrine disrupting chemicals and their impact on neurodevelopment and behavior. *Annual Review of Physiology*, *82*(1), 177–202. https://doi.org/10.1146/annurev-physiol-021119-034555

Ni, K., Lu, Y., Wang, T., Kannan, K., Gosens, J., Xu, L., Li, Q., Wang, L., & Liu, S. (2013). A review of human exposure to polybrominated diphenyl ethers (PBDEs) in China. *International Journal of Hygiene and Environmental Health*, *216*(6), 607–623. https://doi.org/10.1016/j.ijheh.2013.02.002

Pan, K., Tu, R., Cai, Z., Huang, Y., & Zhang, C. (2023). Association of blood lead with estradiol and sex hormone-binding globulin in 8–19-year-old children and adolescents. *Frontiers in Endocrinology*, *14*, 1096659. https://doi.org/10.3389/fendo.2023.1096659

Perera, F., Nolte, E. L. R., Wang, Y., Margolis, A. E., Calafat, A. M., Wang, S., Garcia, W., Hoepner, L. A., Peterson, B. S., Rauh, V., & Herbstman, J. (2016). Bisphenol A exposure and symptoms of anxiety and depression among inner city children at 10–12 years of age. *Environmental Research*, *151*, 195–202. https://doi.org/10.1016/j.envres.2016.07.028

Peterson, A. K., Eckel, S. P., Habre, R., Yang, T., Faham, D., Amin, M., Grubbs, B. H., Farzan, S. F., Kannan, K., Robinson, M., Lerner, D., Al-Marayati, L. A., Walker, D. K., Grant, E. G., Breton, C. V., & Bastain, T. M. (2022). Detected prenatal perfluorooctanoic acid (PFOA) exposure is associated with decreased fetal head biometric parameters in participants experiencing higher perceived stress during pregnancy in the MADRES cohort. *Environmental Advances*, 9, 100286. https://doi.org/10.1016/j.envadv.2022.100286

Pinney, S. E., Mesaros, C. A., Snyder, N. W., Busch, C. M., Xiao, R., Aijaz, S., Ijaz, N., Blair, I. A., & Manson, J. M. (2017). Second trimester amniotic fluid bisphenol A concentration is associated with decreased birth weight in term infants. *Reproductive Toxicology*, *67*, 1–9. https://doi.org/10.1016/j.reprotox.2016.11.007

Pocock, S. J., Smith, M., & Baghurst, P. (1994). Environmental lead and children's intelligence: A systematic review of the epidemiological evidence. *BMJ*, *309*(6963), 1189–1197. https://doi.org/10.1136/bmj.309.6963.1189

Poothong, S., Papadopoulou, E., Padilla-Sánchez, J. A., Thomsen, C., & Haug, L. S. (2020). Multiple pathways of human exposure to poly- and perfluoroalkyl substances (PFASs): From external exposure to human blood. *Environment International*, *134*, 105244. https://doi.org/10.1016/j.envint.2019.105244

Praveena, S. M., Munisvaradass, R., Masiran, R., Rajendran, R. K., Lin, C.-C., & Kumar, S. (2020). Phthalates exposure and attention-deficit/hyperactivity disorder in children: A systematic review of epidemiological literature. *Environmental Science and Pollution Research*, *27*(36), 44757–44770. https://doi.org/10.1007/s11356-020-10652-z

Qian, X., Li, J., Xu, S., Wan, Y., Li, Y., Jiang, Y., Zhao, H., Zhou, Y., Liao, J., Liu, H., Sun, X., Liu, W., Peng, Y., Hu, C., Zhang, B., Lu, S., Cai, Z., & Xia, W. (2019). Prenatal exposure to phthalates and neurocognitive development in children at two years of age. *Environment International*, *131*, 105023. https://doi.org/10.1016/j.envint.2019.105023

Rivera-Núñez, Z., Ashrap, P., Barrett, E. S., Watkins, D. J., Cathey, A. L., Vélez-Vega, C. M., Rosario, Z., Cordero, J. F., Alshawabkeh, A., & Meeker, J. D. (2021). Association of biomarkers of exposure to metals and metalloids with maternal hormones in pregnant women from Puerto Rico. *Environment International*, *147*, 106310. https://doi.org/10.1016/j.envint.2020.106310

Rivera-Núñez, Z., Kinkade, C. W., Khoury, L., Brunner, J., Murphy, H., Wang, C., Kannan, K., Miller, R. K., O'Connor, T. G., & Barrett, E. S. (2023). Prenatal perfluoroalkyl substances exposure and maternal sex steroid hormones across pregnancy. *Environmental Research*, *220*, 115233. https://doi.org/10.1016/j.envres.2023.115233

Rochester, J. R. (2013). Bisphenol A and human health: A review of the literature. *Reproductive Toxicology*, *42*, 132–155. https://doi.org/10.1016/j.reprotox.2013.08.008

Rodríguez-Carmona, Y., Cantoral, A., Trejo-Valdivia, B., Téllez-Rojo, M. M., Svensson, K., Peterson, K. E., Meeker, J. D., Schnaas, L., Solano, M., & Watkins, D. J. (2019). Phthalate exposure during pregnancy and long-term weight gain in women. *Environmental Research*, *169*, 26–32. https://doi.org/10.1016/j.envres.2018.10.014

Rossi-George, A., Virgolini, M. B., Weston, D., Thiruchelvam, M., & Cory-Slechta, D. A. (2011). Interactions of lifetime lead exposure and stress: Behavioral, neurochemical and HPA axis effects. *NeuroToxicology*, *32*(1), 83–99. https://doi.org/10.1016/j.neuro.2010.09.004

Rubin, B. S. (2011). Bisphenol A: An endocrine disruptor with widespread exposure and multiple effects. *The Journal of Steroid Biochemistry and Molecular Biology*, *127*(1–2), 27–34. https://doi.org/10.1016/j.jsbmb.2011.05.002

Sabuz Vidal, O., Deepika, D., Schuhmacher, M., & Kumar, V. (2021). EDC-induced mechanisms of immunotoxicity: A systematic review. *Critical Reviews in Toxicology*, *51*(7), 634–652. https://doi.org/10.1080/10408444.2021.2009438

Sathyanarayana, S., Barrett, E., Butts, S., Wang, C., & Swan, S. H. (2014). Phthalate exposure and reproductive hormone concentrations in pregnancy. *Reproduction*, *147*(4), 401–409. https://doi.org/10.1530/REP-13-0415

Saxena, M. C., Siddiqui, M. K. J., Bhargava, A. K., Seth, T. D., Krishnamurti, C. R., & Kutty, D. (1980). Role of chlorinated hydrocarbon pesticides in abortions and premature labour. *Toxicology*, *17*(3), 323–331. https://doi.org/10.1016/0300-483X(80)90013-X

Seachrist, D. D., Bonk, K. W., Ho, S.-M., Prins, G. S., Soto, A. M., & Keri, R. A. (2016). A review of the carcinogenic potential of bisphenol A. *Reproductive Toxicology*, *59*, 167–182. https://doi.org/10.1016/j.reprotox.2015.09.006

Shafei, A., Ramzy, M. M., Hegazy, A. I., Husseny, A. K., EL-hadary, U. G., Taha, M. M., & Mosa, A. A. (2018). The molecular mechanisms of action of the endocrine disrupting chemical bisphenol A in the development of cancer. *Gene*, *647*, 235–243. https://doi.org/10.1016/j.gene.2018.01.016

Sheikh, I. A., & Beg, M. A. (2020). Structural studies on the endocrine-disrupting role of polybrominated diphenyl ethers (PBDEs) in thyroid diseases. *Environmental Science and Pollution Research*, *27*(30), 37866–37876. https://doi.org/10.1007/s11356-020-09913-8

Shoshtari-Yeganeh, B., Zarean, M., Mansourian, M., Riahi, R., Poursafa, P., Teiri, H., Rafiei, N., Dehdashti, B., & Kelishadi, R. (2019). Systematic review and meta-analysis on the association between phthalates exposure and insulin resistance. *Environmental Science and Pollution Research*, *26*(10), 9435–9442. https://doi.org/10.1007/s11356-019-04373-1

Siddiqui, M. K., Nigam, U., Srivastava, S., Tejeshwar, D. S., & Chandrawati, R. (2002). Association of maternal blood pressure and hemoglobin level with organochlorines in human milk. *Human & Experimental Toxicology*, *21*(1), 1–6. https://doi.org/10.1191/0960327102ht198oa

Sima, M. W., & Jaffé, P. R. (2021). A critical review of modeling poly- and perfluoroalkyl substances (PFAS) in the soil-water environment. *Science of the Total Environment*, *757*, 143793. https://doi.org/10.1016/j.scitotenv.2020.143793

Starnes, H. M., Rock, K. D., Jackson, T. W., & Belcher, S. M. (2022). A critical review and meta-analysis of impacts of per- and polyfluorinated substances on the brain and behavior. *Frontiers in Toxicology*, 4, 881584. https://doi.org/10.3389/ftox.2022.881584

Stevens, D. R., Rosen, E. M., Van Wickle, K., McNell, E. E., Bommarito, P. A., Calafat, A. M., Botelho, J. C., Sinkovskaya, E., Przybylska, A., Saade, G., Abuhamad, A., & Ferguson, K. K. (2023). Early pregnancy phthalates and replacements in relation to fetal growth: The human placenta and phthalates study. *Environmental Research*, *229*, 115975. https://doi.org/10.1016/j.envres.2023.115975

Sun, X., Li, D., Liang, H., Miao, M., Song, X., Wang, Z., Zhou, Z., & Yuan, W. (2018). Maternal exposure to bisphenol A and anogenital distance throughout infancy: A longitudinal study from Shanghai, China. *Environment International*, *121*, 269–275. https://doi.org/10.1016/j.envint.2018.08.055

Sunderland, E. M., Hu, X. C., Dassuncao, C., Tokranov, A. K., Wagner, C. C., & Allen, J. G. (2019). A review of the pathways of human exposure to poly- and perfluoroalkyl substances (PFASs) and present understanding of health effects. *Journal of Exposure Science & Environmental Epidemiology*, *29*(2), 131–147. https://doi.org/10.1038/s41370-018-0094-1

Tachachartvanich, P., Singam, E. R. A., Durkin, K. A., Furlow, J. D., Smith, M. T., & La Merrill, M. A. (2022). In Vitro characterization of the endocrine disrupting effects of per- and poly-fluoroalkyl substances (PFASs) on the human androgen receptor. *Journal of Hazardous Materials*, *429*, 128243. https://doi.org/10.1016/j.jhazmat.2022.128243

Tang-Péronard, J. L., Andersen, H. R., Jensen, T. K., & Heitmann, B. L. (2011). Endocrine-disrupting chemicals and obesity development in humans: A review. *Obesity Reviews*, *12*(8), 622–636. https://doi.org/10.1111/j.1467-789X.2011.00871.x

Tarafdar, A., Sirohi, R., Balakumaran, P. A., Reshmy, R., Madhavan, A., Sindhu, R., Binod, P., Kumar, Y., Kumar, D., & Sim, S. J. (2022). The hazardous threat of bisphenol A: Toxicity, detection and remediation. *Journal of Hazardous Materials*, *423*, 127097. https://doi.org/10.1016/j.jhazmat.2021.127097

The APrON Study Team, Giesbrecht, G. F., Ejaredar, M., Liu, J., Thomas, J., Letourneau, N., Campbell, T., Martin, J. W., & Dewey, D. (2017). Prenatal bisphenol A exposure and dysregulation of infant hypothalamic-pituitary-adrenal axis function: Findings from the APrON cohort study. *Environmental Health*, *16*(1), 47. https://doi.org/10.1186/s12940-017-0259-8

The Consortium on Thyroid and Pregnancy—Study Group on Preterm Birth, Korevaar, T. I. M., Derakhshan, A., Taylor, P. N., Meima, M., Chen, L., Bliddal, S., Carty, D. M., Meems, M., Vaidya, B., Shields, B., Ghafoor, F., Popova, P. V., Mosso, L., Oken, E., Suvanto, E., Hisada, A., Yoshinaga, J., Brown, S. J., & Peeters, R. P. (2019). Association of thyroid function test abnormalities and thyroid autoimmunity with preterm birth: A systematic review and meta-analysis. *JAMA*, *322*(7), 632. https://doi.org/10.1001/jama.2019.10931

Troisi, J., Mikelson, C., Richards, S., Symes, S., Adair, D., Zullo, F., & Guida, M. (2014). Placental concentrations of bisphenol A and birth weight from births in the Southeastern U.S. *Placenta*, *35*(11), 947–952. https://doi.org/10.1016/j.placenta.2014.08.091

Turusov, V., Rakitsky, V., & Tomatis, L. (2002). Dichlorodiphenyltrichloroethane (DDT): Ubiquity, persistence, and risks. *Environmental Health Perspectives*, *110*(2), 125–128. https://doi.org/10.1289/ehp.02110125

Vahedi, M., Saeedi, A., Poorbaghi, S. L., Sepehrimanesh, M., & Fattahi, M. (2016). Metabolic and endocrine effects of bisphenol A exposure in market seller women with polycystic ovary syndrome. *Environmental Science and Pollution Research*, *23*(23), 23546–23550. https://doi.org/10.1007/s11356-016-7573-5

van der Meer, T. P., Chung, M. K., van Faassen, M., Makris, K. C., van Beek, A. P., Kema, I. P., Wolffenbuttel, B. H. R., van Vliet-Ostaptchouk, J. V., & Patel, C. J. (2021). Temporal exposure and consistency of endocrine disrupting chemicals in a longitudinal study of individuals with impaired fasting glucose. *Environmental Research*, *197*, 110901. https://doi.org/10.1016/j.envres.2021.110901

Van Woerden, I., Bruening, M., Montresor-López, J., & Payne-Sturges, D. C. (2019). Trends and disparities in urinary BPA concentrations among U.S. emerging adults. *Environmental Research*, *176*, 108515. https://doi.org/10.1016/j.envres.2019.05.046

Villeneuve, D. L., Blackwell, B. R., Cavallin, J. E., Collins, J., Hoang, J. X., Hofer, R. N., Houck, K. A., Jensen, K. M., Kahl, M. D., Kutsi, R. N., Opseth, A. S., Santana Rodriguez, K. J., Schaupp, C., Stacy, E. H., & Ankley, G. T. (2023). Verification of in vivo estrogenic activity for four per- and polyfluoroalkyl substances (PFAS) identified as estrogen receptor agonists via new approach methodologies. Environmental Science & Technology, *57*(9), 3794–3803. https://doi.org/10.1021/acs.est.2c09315

Vom Saal, F. S., & Welshons, W. V. (2014). Evidence that bisphenol A (BPA) can be accurately measured without contamination in human serum and urine, and that BPA causes numerous hazards from multiple routes of exposure. *Molecular and Cellular Endocrinology*, *398*(1–2), 101–113. https://doi.org/10.1016/j.mce.2014.09.028

Vuong, A. M., Yolton, K., Dietrich, K. N., Braun, J. M., Lanphear, B. P., & Chen, A. (2018). Exposure to polybrominated diphenyl ethers (PBDEs) and child behavior: Current findings and future directions. *Hormones and Behavior*, *101*, 94–104. https://doi.org/10.1016/j.yhbeh.2017.11.008

Wan, M. L. Y., Co, V. A., & El-Nezami, H. (2022). Endocrine disrupting chemicals and breast cancer: A systematic review of epidemiological studies. *Critical Reviews in Food Science and Nutrition*, *62*(24), 6549–6576. https://doi.org/10.1080/10408398.2021.1903382

Wang, B., Zhou, W., Zhu, W., Chen, L., Wang, W., Tian, Y., Shen, L., & Zhang, J. (2018). Associations of female exposure to bisphenol A with fecundability: Evidence from a preconception cohort study. *Environment International*, *117*, 139–145. https://doi.org/10.1016/j.envint.2018.05.003

Wang, J., Weng, X., Liu, S., Zhang, H., Zhu, Q., Fu, Q., Wang, T., Liao, C., & Jiang, G. (2023). Occurrence, fate, and reduction measures of phthalates in the cooking process: A review. *Environment & Health*, *1*(5), 300–314. https://doi.org/10.1021/envhealth.3c00091

Wang, J., Zhang, J., Fan, Y., Li, Z., Tao, C., Yan, W., Niu, R., Huang, Y., Xu, Q., Wang, X., Xu, Q., Han, L., & Lu, C. (2022). Association between per- and polyfluoroalkyl substances and risk of gestational diabetes mellitus. *International Journal of Hygiene and Environmental Health*, *240*, 113904. https://doi.org/10.1016/j.ijheh.2021.113904

Wang, Q., Ruan, Y., Lin, H., & Lam, P. K. S. (2020). Review on perfluoroalkyl and polyfluoroalkyl substances (PFASs) in the Chinese atmospheric environment. *Science of the Total Environment*, *737*, 139804. https://doi.org/10.1016/j.scitotenv.2020.139804

Wang, W., Hong, X., Zhao, F., Wu, J., & Wang, B. (2023). The effects of perfluoroalkyl and polyfluoroalkyl substances on female fertility: A systematic review and meta-analysis. *Environmental Research*, *216*, 114718. https://doi.org/10.1016/j.envres.2022.114718

Wang, X., Hales, B. F., & Robaire, B. (2021). Effects of flame retardants on ovarian function. *Reproductive Toxicology*, *102*, 10–23. https://doi.org/10.1016/j.reprotox.2021.03.006

Wang, Z., Buser, A. M., Cousins, I. T., Demattio, S., Drost, W., Johansson, O., Ohno, K., Patlewicz, G., Richard, A. M., Walker, G. W., White, G. S., & Leinala, E. (2021). A new OECD definition for per- and polyfluoroalkyl substances. *Environmental Science & Technology*, *55*(23), 15575–15578. https://doi.org/10.1021/acs.est.1c06896

Wang, Z., Liang, H., Tu, X., Yuan, W., Zhou, Z., Jin, L., Miao, M., & Li, D.-K. (2019). Bisphenol A and pubertal height growth in school-aged children. *Journal of Exposure Science & Environmental Epidemiology*, *29*(1), 109–117. https://doi.org/10.1038/s41370-018-0063-8

Wani, A. L., Ara, A., & Usmani, J. A. (2015). Lead toxicity: A review. *Interdisciplinary Toxicology*, *8*(2), 55–64. https://doi.org/10.1515/intox-2015-0009

Wee, S. Y., & Aris, A. Z. (2023). Revisiting the "forever chemicals", PFOA and PFOS exposure in drinking water. *Npj Clean Water*, *6*(1), 57. https://doi.org/10.1038/s41545-023-00274-6

Weng, X., Zhu, Q., Liao, C., & Jiang, G. (2023). cumulative exposure to phthalates and their alternatives and associated female reproductive health: Body burdens, adverse outcomes, and underlying mechanisms. *Environmental Science & Technology*, *57*(22), 8189–8212. https://doi.org/10.1021/acs.est.3c00823

WHO. (2019). *The WHO Recommended Classification of Pesticides by Hazard and Guidelines to Classification*, 2019 edition. https://www.who.int/publications/i/item/9789240005662

Wu, H., Just, A. C., Colicino, E., Calafat, A. M., Oken, E., Braun, J. M., McRae, N., Cantoral, A., Pantic, I., Pizano-Zárate, M. L., Tolentino, M. C., Wright, R. O., Téllez-Rojo, M. M., Baccarelli, A. A., & Deierlein, A. L. (2021). The associations of phthalate biomarkers during pregnancy with later glycemia and lipid profiles. *Environment International*, *155*, 106612. https://doi.org/10.1016/j.envint.2021.106612

Wu, H.-C., Cohn, B. A., Cirillo, P. M., Santella, R. M., & Terry, M. B. (2020). DDT exposure during pregnancy and DNA methylation alterations in female offspring in the child health and development study. *Reproductive Toxicology*, *92*, 138–147. https://doi.org/10.1016/j.reprotox.2019.02.010

Wu, W., Ma, Z., Yang, F., Wu, P., Zhang, D., Zeng, R., Sun, D., & Cao, L. (2022). Urinary phthalate metabolites in pregnant women: Occurrences, related factors, and association with maternal hormones. *Environmental Science and Pollution Research*, *29*(22), 33372–33382. https://doi.org/10.1007/s11356-022-18590-8

Wu, Y., Bao, J., Liu, Y., Wang, X., & Qu, W. (2023). A review on per- and polyfluoroalkyl substances in pregnant women: Maternal exposure, placental transfer, and relevant model simulation. *Toxics*, *11*(5), 430. https://doi.org/10.3390/toxics11050430

Wu, Z., He, C., Han, W., Song, J., Li, H., Zhang, Y., Jing, X., & Wu, W. (2020). Exposure pathways, levels and toxicity of polybrominated diphenyl ethers in humans: A review. *Environmental Research*, *187*, 109531. https://doi.org/10.1016/j.envres.2020.109531

Xie, Z., Zhang, X., Wu, J., & Wu, Y. (2023). Risk assessment of phthalate metabolites accumulated in fish to the Indo-Pacific humpback dolphins from their largest habitat. *Science of the Total Environment*, *876*, 163094. https://doi.org/10.1016/j.scitotenv.2023.163094

Xing, W.-Y., Sun, J.-N., Liu, F.-H., Shan, L.-S., Yin, J.-L., Li, Y.-Z., Xu, H.-L., Wei, Y.-F., Liu, J.-X., Zheng, W.-R., Zhang, Y.-Y., Song, X.-J., Liu, K.-X., Liu, J.-C., Wang, J.-Y., Jia, M.-Q., Chen, X., Li, X.-Y., Liu, C., & Wu, Q.-J. (2024). Per- and polyfluoroalkyl substances and human health outcomes: An umbrella review of systematic reviews with meta-analyses of observational studies. *Journal of Hazardous Materials*, *472*, 134556. https://doi.org/10.1016/j.jhazmat.2024.134556

Xu, C., Yin, S., Tang, M., Liu, K., Yang, F., & Liu, W. (2017). Environmental exposure to DDT and its metabolites in cord serum: Distribution, enantiomeric patterns, and effects on infant birth outcomes. *Science of the Total Environment*, *580*, 491–498. https://doi.org/10.1016/j.scitotenv.2016.11.196

Xu, X., Yekeen, T. A., Xiao, Q., Wang, Y., Lu, F., & Huo, X. (2013). Placental IGF-1 and IGFBP-3 expression correlate with umbilical cord blood PAH and PBDE levels from prenatal exposure to electronic waste. *Environmental Pollution*, *182*, 63–69. https://doi.org/10.1016/j.envpol.2013.07.005

Yang, C., Harris, S. A., Jantunen, L. M., Kvasnicka, J., Nguyen, L. V., & Diamond, M. L. (2020). Phthalates: Relationships between air, dust, electronic devices, and hands with implications for exposure. *Environmental Science & Technology*, *54*(13), 8186–8197. https://doi.org/10.1021/acs.est.0c00229

Yang, M., Park, M. S., & Lee, H. S. (2006). Endocrine disrupting chemicals: Human exposure and health risks. *Journal of Environmental Science and Health, Part* C, *24*(2), 183–224. https://doi.org/10.1080/10590500600936474

Yang, O., Kim, H. L., Weon, J.-I., & Seo, Y. R. (2015). Endocrine-disrupting chemicals: Review of toxicological mechanisms using molecular pathway analysis. *Journal of Cancer Prevention*, *20*(1), 12–24. https://doi.org/10.15430/JCP.2015.20.1.12

Yu, X., Liu, B., Yu, Y., Li, H., Li, Q., Cui, Y., & Ma, Y. (2023). Polybrominated diphenyl ethers (PBDEs) in household dust: A systematic review on spatio-temporal distribution, sources, and health risk assessment. *Chemosphere*, *314*, 137641. https://doi.org/10.1016/j.chemosphere.2022.137641

Zhang, B., Wang, Z., Zhang, J., Dai, Y., Ding, J., Guo, J., Qi, X., Wu, C., & Zhou, Z. (2024). Prenatal exposure to per- and polyfluoroalkyl substances, fetal thyroid function, and intelligence quotient at 7 years of age: Findings from the Sheyang Mini Birth Cohort Study. *Environment International*, *187*, 108720. https://doi.org/10.1016/j.envint.2024.108720

Zhang, L., Liang, J., & Gao, A. (2023). Contact to perfluoroalkyl substances and thyroid health effects: A meta-analysis directing on pregnancy. *Chemosphere*, *315*, 137748. https://doi.org/10.1016/j.chemosphere.2023.137748

Zhang, Q.-Q., Ma, Z.-R., Cai, Y.-Y., Li, H.-R., & Ying, G.-G. (2021). Agricultural plastic pollution in China: Generation of plastic debris and emission of phthalic acid esters from agricultural films. *Environmental Science & Technology*, *55*(18), 12459–12470. https://doi.org/10.1021/acs.est.1c04369

Zhao, Y., Ruan, X., Li, Y., Yan, M., & Qin, Z. (2013). Polybrominated diphenyl ethers (PBDEs) in aborted human fetuses and placental transfer during the first trimester of pregnancy. *Environmental Science & Technology*, *47*(11), 5939–5946. https://doi.org/10.1021/es305349x

Zheng, Q., Yan, W., Gao, S., & Li, X. (2024). The effect of PFAS exposure on glucolipid metabolism in children and adolescents: A meta-analysis. *Frontiers in Endocrinology*, *15*, 1261008. https://doi.org/10.3389/fendo.2024.1261008

Zhou, X.-M., Lü, W.-J., Su, L., Shan, Z.-J., & Chen, X.-G. (2012). Binding of phthalate plasticizers to human serum albumin in vitro: A multispectroscopic approach and molecular modeling. *Journal of Agricultural and Food Chemistry*, *60*(4), 1135–1145. https://doi.org/10.1021/jf204380r

Zhuang, S., Zhang, J., Wen, Y., Zhang, C., & Liu, W. (2012). Distinct mechanisms of endocrine disruption of DDT-related pesticides toward estrogen receptor α and estrogen-related receptor γ. *Environmental Toxicology and Chemistry*, *31*(11), 2597–2605. https://doi.org/10.1002/etc.1986

Zota, A. R., Geller, R. J., Romano, L. E., Coleman-Phox, K., Adler, N. E., Parry, E., Wang, M., Park, J.-S., Elmi, A. F., Laraia, B. A., & Epel, E. S. (2018). Association between persistent endocrine-disrupting chemicals (PBDEs, OH-PBDEs, PCBs, and PFASs) and biomarkers of inflammation and cellular aging during pregnancy and postpartum. *Environment International*, *115*, 9–20. https://doi.org/10.1016/j.envint.2018.02.044

21

Exploring Endocrine Disruptors: Implication for Child Health Development and Global Policies

Ziying Li, Xin Chen, Lingcheng Su, and Yi Yang

21.1 Introduction

Endocrine-disrupting chemicals (EDCs) are defined as exogenous chemicals that interfere with the synthesis, secretion, transport, binding, action, metabolism, and elimination of natural hormones in the human and animal body, leading to dysfunction in homeostasis, reproduction, development, and behaviour (Gore et al., 2015). Thousands of chemicals with diverse chemical structures are potential EDCs, including natural chemicals produced by plants and animals and synthetic chemicals widely used for domestic, industrial, and agricultural purposes (Liao & Kannan, 2013, 2014).

EDCs occur in various environmental matrices, including air, water, soil, and vegetation. Many EDCs are volatile and can also be transported and deposited, spreading over a long geographical range through the grasshopper effect (Rizzo et al., 2023). Humans can be exposed to EDCs through various routes, including airborne inhalation, ingestion, and dermal absorption. For children (from birth to age 12) and adolescents (age 13–18) who are in a critical period of growth and development, exposure to widespread EDCs through contact with consumer products, such as personal care products, toys, processed foods, and other childcare products, is of great public health concern (He et al., 2018; Venkatesan & Halden, 2015). Given the fact that children have relatively larger body surface area and are still undergoing development (Braun, 2017; Ramírez et al., 2022), exposure of children and adolescents to EDCs is expected to be high, rendering this critical group particularly vulnerable.

EDCs are readily detected in children's blood serum (Xu et al., 2018) and urine (Perera et al., 2012). Concentration of EDCs and their metabolites in blood and urine can serve as an excellent biomarker to estimate the chemical exposure period/concentration, intake amount, and health risk. EDCs are capable of disrupting the endocrine system by (1) mimicking or antagonising endogenous hormones and affecting their action, (2) inhibiting or stimulating hormone secretion processes, (3) interfering with or altering hormonal, metabolic pathways, and (4) altering hormone receptor binding processes, resulting in impaired growth, development and function (Ramírez et al., 2022). Exposure to EDCs may lead to various diseases, including obesity, asthma, and neuro-developmental disorders in children and adolescents (Paciência et al., 2019). However, the underlying mechanisms of this are not yet fully understood. More than 300 billion USD have been spent on EDC-related diseases, equivalent to more than 2% of the U.S. gross domestic product (GDP) (Attina et al., 2016). European Union (EU) estimated that the median cost of 157 billion euros per year has been allocated to treat diseases caused by EDCs, equivalent to 1.23% of the EU GDP, of which a total of 42,400 cases and 1.54 billion euros have been associated with childhood obesity (Trasande et al., 2015). Due to their adverse effects, banning and restricting use of EDCs in the production of consumer products (e.g., toys, personal care products, and food package materials) have already been implemented by many countries and organisations including the U.S., Canada, China, Japan, Korea, and EU (CPSC, 2008).

DOI: 10.1201/9781003492115-26

21.2 Sources and Children Exposure Levels of Common EDCs

21.2.1 Classification and Sources of EDCs

Thousands of chemicals can be defined as EDCs due to their ability to bind to receptors or affect enzymes and transport proteins in the human body (Gore et al., 2015). EDCs can be divided into two broad categories according to their properties and applications. Natural chemicals, including hormones, can be found in animals, e.g., estrone (E1) and 17β-estradiol (E2), or plants, e.g., genistein. These chemicals are used for pharmaceutical purposes or are added as dietary supplements (Liao & Kannan, 2014). Synthetic chemicals include industrial solvents and their by-products, plastics additives, plasticisers, pesticides, herbicides, fungicides, pharmaceutical agents, disinfectants, and flame retardants, which are produced for industrial and agricultural purposes (Liao & Kannan, 2013, 2014).

Most chemicals found in pesticides, herbicides, and fungicides fall under the category of persistent organic pollutants, with six classes of EDCs being commonly used and known for their harmful effects on human health. These classes include bisphenol analogues (BPA), flame retardants (FRs), polycyclic aromatic hydrocarbons (PAHs), polychlorinated biphenyls (PCBs), phthalates (PEs), perfluoroalkyl substances (PFAS), and personal care products (PCPs), which are frequently employed as additives, plasticisers, surfactants, disinfectants, and are commonly found in electronic devices as flame retardants, thus contributing to their pervasive presence in the environment.

Children and adolescents can be continuously exposed to EDCs emitted from various sources and pathways (Figure 21.1). Exposure to EDCs released from consumer products, which commonly act as indoor pollutants and can attach to dust particles, poses a significant concern as they enter the human body through inhalation (Paciência et al., 2019). Prenatal exposure through placental transport, newborns and infants are exposed through breastfeeding is an essential source of EDC exposure to children. Additionally, bottle and plastic toy use may be a potential source of exposure for breastfeeding infants, with diet being the main source of exposure (Fromme et al., 2009). EDCs like BPA and phthalates,

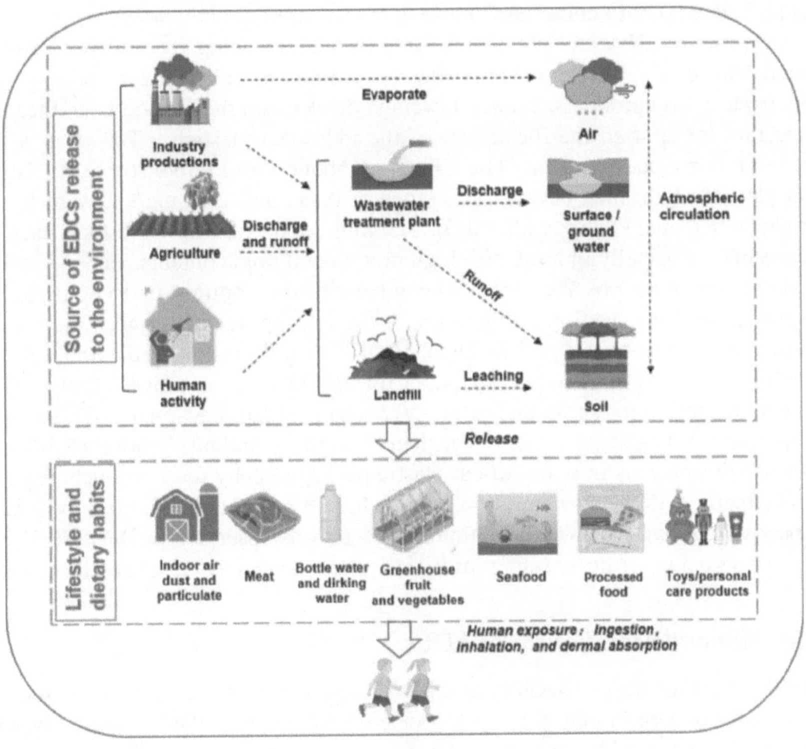

FIGURE 21.1 The origins of EDCs released into the environment and the pathways of children's exposure.

leached from food containers and baby products such as plastic baby teethers and feeding bottles, are commonly ingested orally (Braun, 2017). Children and adolescents, being a high-risk group due to their frequent hand-to-mouth activities, are particularly susceptible to EDC exposure from consumer products, including personal care items, which can also be absorbed through the skin (Cilluffo et al., 2022; Paciência et al., 2019).

21.2.2 Exposure Level, Half-Life, and Metabolism of EDCs

21.2.2.1 EDC Level in Food and Daily Necessities

Uptake, accumulation, and translocation of EDCs by different plants, especially vegetables and fruits, could potentially threaten human health. Lu et al. (2013) reported that BPA was detectable in the range of 0.2–9.0 µg/kg in all vegetables and fruits (e.g., lettuce, tomato, pumpkin, potato, carrot, citrus, apple, and strawberry) collected from U.S. local commercial sources, illustrating a wide occurrence of BPA in fresh food. In addition, EDCs were also detected in other foodstuffs, such as fish, meat, and egg. It was also reported BPA in non-canned foodstuffs of animal origin samples in France, including meat, poultry, game, offal, delicatessen meats, fish, crustaceans, and molluscs, and BPA concentrations ranging between 0.09 and 60.19 µg/kg, with the highest concentration found in poultry and game (mean: 4.13 µg/kg) (Gorecki et al., 2017). Additionally, Kuzukiran et al. (2018) investigated various EDCs in hen eggs (battery-produced, free-range, and organic). The study showed that the eggs were contaminated by more than one chemical, particularly phthalates, PCBs, and PBDEs, with concentrations ranging between 2.6–83.6, <0.43–10.5, and <0.38–0.7 µg/kg, respectively. Higher phthalates and PBDE levels were observed in battery-produced eggs, while organic eggs exhibited higher PCB concentrations. In recent years, there has been growing concern about the contamination of phthalates in soil, vegetables, and fruits due to the rapid growth of the greenhouse vegetable industry worldwide and the high demand for the plastic film as agricultural soil cover. Wang et al. (2015) indicated that the concentrations of total phthalates ranging between 510 and 7160 µg/kg in plastic film greenhouse vegetables, with DBP, DEHP, and di-*n*-octyl phthalate (DnOP) comprising more than 90% of total phthalates.

Children are exposed to BPA and phthalates daily through leaching from food containers made of plastic or coated with epoxy liners, including but not limited to baby feeding bottles, plastic drinking bottles, microwave food products, canned beverage drinks, and metal food containers (Duan et al., 2020). In 2011, Taiwan reported the illegal use of the industrial plasticiser DEHP as a food additive to replace palm oil as a clouding agent. The China Alcoholic Drinks Industry Association (CADIA) confirmed that plasticising agents, between the range of 0.495 and 2.32 mg/kg, were found in almost all alcohol products in China (Yang et al., 2013). The migration of EDCs from food packages resulted in EDCs in foodstuffs, especially in food with high protein and lipid contents. The migration of phthalates (diisobutyl phthalate) was observed from silicon baby feeding bottles, up to 80 µg/kg, while levels of BPA from polyamide baby feeding bottles were 1–1005 µg/kg. Higher concentrations of BPA were reported in canned foods (mean: 15.1 µg/kg; range: <0.01–146 µg/kg) compared with food stored in glass (mean: 0.64 µg/kg; range: <0.01–4.16 µg/kg), paper (mean: 0.84 µg/kg; range: <0.01–2.37 µg/kg), and plastic containers (mean:0.95 µg/kg; range: <0.01–20.8 µg/kg) (Liao & Kannan, 2013). García-Córcoles et al. (2018) reported that bisphenol B (BPB), bisphenol S (BPS), and bisphenol P (BPP) were the most abundant among the bisphenols in ready-to-eat plastic packaged baby food products (e.g., fruits, meat, fish, and milk) in Spain, with the concentration ranging between <0.3 and 8.5, <0.3 and 49.2, and <1 and 7.7 µg/kg, respectively. However, lower detection frequency was observed for BPA (20%), with concentration ranging from <0.3 to 1.1 µg/kg, suggesting that other bisphenols are gradually replacing BPA.

21.2.2.2 Half-Life and Metabolism of EDCs

Understanding the metabolism of EDCs is crucial for conducting epidemiological studies on human health. Certain EDCs undergo rapid degradation or elimination from the human body within a short timeframe, ranging from several hours to days. However, some EDCs can persist in body fat and remain stored for years. This variability in metabolic behaviour underscores the importance of considering the diverse kinetics and longevity of EDCs when assessing their impact on human health (Bansal et al., 2018). Some EDCs have glucuronide and sulphate conjugates, such as BPA and TCS. Metabolite concentrations differ in children and adult urine samples. Based on urinary metabolite concentrations, daily intake level, and dietary exposure of adults and children can be estimated; and such estimates serve as references to establish regulations (Kamrin, 2009; Xu et al., 2018).

The half-life of EDCs is influenced by their physicochemical properties and the surrounding natural environment. Studies have reported that the biological half-life of PCBs is around 10 years (Yang et al., 2018); 5.4, 8.5, and 3.8 years for PFOS, perfluorohexanesulfonate (PFHS), and PFOA, respectively (Olsen et al., 2007); weeks to months for PBDEs (Xu et al., 2018); hours to days for FRs (He et al., 2018); less than 24 h for TCS and BPA (Kim & Chevrier, 2020). Repeated exposure to EDCs through consumer products can result in a constant but low-level circulation in the human body, even though some EDCs have short biological half-lives and do not tend to accumulate.

21.3 EDCs and Children's Development

The wide occurrence of EDCs in industrial or food products poses a significant risk to human health, especially for children and adolescents, due to their diet, behaviour, physiology, and development. Exposure to EDCs during early childhood may disturb neurodevelopment, pubertal development, foetal growth, and early reproductive tract development and increase the risk of behavioural disorders and cognitive deficits in later life stages (Lee & Choi, 2023). Hormone-driven epigenetic responses could also be altered by EDCs, leading to long-term health problems, including metabolite disorder, asthma, obesity, and behavioural disorders (Table 21.1).

TABLE 21.1
Diseases Associated with EDCs for Children and Adolescents

EDCs	Mechanism	Disease-Associated	References
FRs	Bind to thyroid transport proteins; change the activity of thyroid hormone receptor; regulate the conversion of testosterone and oestradiol; alter mRNA expression of the progesterone receptor, androgen receptor, and oestrogen receptor; active peroxisome proliferator-activated receptors (obesogen); increase reactive oxygen species formation.	Poorer intellectual performance; decreased cognitive and motor function; behavioural problems; Poorer intellectual performance and memory deterioration; poorer social skills.	Gibson et al. (2018); Gorecki et al. (2017); Hamra et al. (2019); He et al. (2018); Kuzukiran et al. (2018); Xu et al. (2018)
PAHs	Affect serum superoxide dismutase and serum glutathione peroxidase; Mitochondrial oxidative stress; DNA damage; DNA methylation, histone modification, long interspersed nuclear element activation, and non-coding RNA, specifically microRNA (miRNA) alterations.	Neurodevelopment; cardiovascular diseases; reproduction-related effects; and endocrine outcomes; respiratory morbidity in decreasing lung function; attention-deficit/hyperactivity disorder; behavioural problems; asthma; and hypertension.	Cilluffo et al. (2022); Lee and Choi (2023); Ramírez et al. (2022); Rundle et al. (2012)
PCBs	Interfere with thyroid homeostasis; bind to nuclear proteins and DNA in the liver; affect the Ca^{2+} homeostasis in neuronal cell; interact with the oestrogen receptor.	Attention deficit hyperactivity disorder; autism spectrum disorder; childhood obesity.	Kuzukiran et al. (2018); Neugebauer et al. (2015); Yang et al. (2018)
PEs	Interfere with androgens, thyroid hormones, and glucocorticoids; bind to peroxisome proliferator-activated receptors (obesogen); inhibit insulin secretion; promote or suppress the T helper 2 (Th2) and Th17 phenotypes.	Asthma; childhood allergies; childhood obesity; higher systolic blood pressure; learning and behavioural problems; poorer intellectual performance.	Braun (2017); Duan et al. (2020); Haggerty et al. (2021); Paciência et al. (2019); Zarean et al. (2019)

(Continued)

TABLE 21.1 (*Continued*)
Diseases Associated with EDCs for Children and Adolescents

EDCs	Mechanism	Disease-Associated	References
PFAS	Bind to peroxisome proliferator-activated receptors (obesogen) and thyroid transport proteins; interfere with glucocorticoids.	Attention deficit hyperactivity disorder; autism spectrum disorder; childhood obesity; immune suppression.	Braun (2017); Duan et al. (2020); Fisher et al. (2020); Forns et al. (2020)
PPCPs	Affect mitochondrial function; disrupt proper calcium signalling; alter zinc homeostasis; affect immunological parameters; Androgen antagonist by inhibiting aromatase activity; bind to oestrogen receptor; inhibit sulfation of oestrogens through inhibition of sulfotransferase enzymes; affect mitochondrial function; Bind to nuclear and membrane oestrogen receptors, thyroid, glucocorticoid, and peroxisome proliferator-activated receptors; interact with steroidogenic enzymes, exert epigenetic modifications.	Asthma and allergic diseases; decreased cognitive function; Asthma and allergic diseases; childhood obesity; attention deficit hyperactivity disorder; anxiety and depression behaviours; childhood obesity; poorer language skills; learning problems.	Kim and Chevrier (2020); Olsen et al. (2007); Quirós-Alcalá et al. (2018, 2019); Skogheim et al. (2021)

21.3.1 Metabolic Interference and Obesity

Multiple studies have demonstrated a causal correlation between exposure concentrations of EDCs and the development of obesity in model organisms. In humans, observational studies have provided data suggesting that many of the EDCs known to contribute to obesity are associated with a higher prevalence of obesity, particularly during critical developmental stages. Because EDCs may contribute to obesity, either alone or by increasing susceptibility to other factors, such as a high-fat diet (Braun, 2017). Children and adolescents may be more susceptible to the effects of EDCs due to their higher water consumption per unit weight compared to adults, as well as their larger body surface area, increased hand-to-mouth activity, higher ventilation rates, and greater intestinal absorption (Braun, 2017). In a meta-analysis of multiple studies, BPA and phthalates consistently showed associations with general and abdominal obesity in both children and adults, with some evidence suggesting age and gender dependence. In contrast, exposure to BPA in adults and 2,5-dichlorophenol in children were significantly associated with overweight, obesity, and increased waist circumference (Ribeiro et al., 2020). Maternal smoking during pregnancy has been robustly associated with an elevated risk of childhood obesity, as supported by compelling epidemiological data. This causal relationship can potentially be attributed to the increased likelihood of exposure to certain substances, such as BPA, due to smoking (Skogheim et al., 2021; Sunman et al., 2019).

Another study suggested that some PCPs, such as benzophenone-3 (BP-3) level, might be a risk factor for Insulin Resistance (IR) and obesity in children. It indicated that obesity mediated the relationship between BP-3 and IR in children (Li et al., 2022). Exposure to low molecular weight phthalate metabolites is more likely to cause obesity in male children and adolescents. In contrast, obesity in adults is relatively more susceptible to high molecular weight phthalate metabolites (Zarean et al., 2019). Similar results were also investigated in PAH exposure. Prenatal exposure to confined PAHs resulted in high BMI Z-scores in infants, suggesting obesity at 5 years of age and fat accumulation at 7 years of age. Pregnant women who were monitored for PAH exposure in controlled environments were exposed to PAHs of 21% and 25% at 5 and 7 years of age, respectively. It is hypothesised that the increase in foetal adiposity during this period, without changing lean body mass, may contribute to the bioaccumulation of PAHs (Rundle et al., 2012).

Children and adolescents still in the developmental stages exhibit increased susceptibility to the effects of EDCs. The level of cytochrome P450 involved in the phase I response in the livers of children and adolescent foetuses is 30%–60% of the adult value compared to adults (Fernandez et al., 2011). In addition, EDCs can stimulate adipocyte differentiation by binding to or activating peroxisome proliferator-activated receptors (PPARα or PPARγ), resulting in childhood obesity (Braun, 2017). Significantly, EDCs can induce mitochondrial dysfunction in insulin resistance, which is another possible mechanism for childhood obesity (Li et al., 2022).

21.3.2 Behavioural, Learning Issues, and Neurodevelopmental Effects

Among the various neurobehavioural disorders, autism spectrum disorders (ASDs) and attention-deficit/hyperactivity disorder (ADHD) have received more and more attention because they are one of the major global public health problems, impose substantial economic burdens on society, and potentially related to chemical exposure in the environment (Forns et al., 2020).

Previous studies have demonstrated that exposure to EDCs during pregnancy affects children's attention and behaviour. Prenatal exposure to polychlorinated dibenzo-p-dioxins and polychlorinated dibenzo-parafurans (PCDD/Fs) and PCBs affect attention performance in school-aged children, with higher levels of exposure to PCDD/Fs and PCBs being associated with a higher number of omission errors on attention tests in their offspring (Neugebauer et al., 2015). Similarly, prenatal exposure to BPA was found to be associated with changes in children's behaviour, with higher levels of BPA exposure associated with higher levels of emotional reactivity and intensity of aggressive behaviour in boys. At the same time, girls would score lower on a variety of Behavioural Syndrome ratings, including anxiety/depression and aggression, suggesting that BPA's effects on children's behaviour are sex-specific (Perera et al., 2012).

In addition, a cross-sectional analysis of data from a prospective birth cohort study involving 205 adolescents, it was discovered a significant association between exposure to specific phthalates and an elevated risk of ADHD-related behaviours. Notably, the strongest associations were observed for phthalates exhibiting antiandrogenic activity (Shoaff et al., 2019). Another study showed that increasing levels of antiandrogenic phthalate metabolites in urine were associated with increased maladaptive behaviours and decreased adaptive skills, with a doubling of these metabolite concentrations linked to a higher risk of externalising behaviour during adolescence. Although the associations were generally stronger in males, no statistically significant sex differences were observed (Shoaff et al., 2019). A previous study explored the impact of maternal exposure (27–38 weeks gestation) to NP and BPA on the neurodevelopment of children aged 2–6 years. The findings suggested that exposure to NP and BPA throughout prenatal and childhood stages has different impacts on neurodevelopment, with no discernible gender-specific effects (Wang et al., 2024).

In another study of publicly available data from a high-risk cohort, it was found that exposure to six different EDCs during specific trimesters of pregnancy had an impact on neurodevelopmental outcomes in children at age 3, with varying effects observed more prominently in females, highlighting the need for further research on EDC mixtures and the timing of exposure during pregnancy to understand their impact on child health better (Oskar et al., 2024). Moreover, children from 3 to 14 years exposed to EDCs during their critical neurodevelopmental windows have been associated with the risk of autistic traits, including the EDCs of phthalates (Haggerty et al., 2021), PFAS (Skogheim et al., 2021), FRs (Hamra et al., 2019), and PCPs (Quirós-Alcalá et al., 2018, 2019).

Neurotoxicity of EDCs in children may involve a variety of mechanisms, including direct interactions with nuclear hormone receptors (e.g., oestrogen receptor, androgen receptor, and thyroid receptor) (Ramírez et al., 2022). For example, impaired iodine uptake during development can affect thyroid hormone synthesis, bioavailability, function, and metabolism, leading to impaired neurocognitive development, among others. Early exposure to EDCs with hormone-disrupting properties (e.g., PBDEs and PCBs) has been implicated as a risk factor for increased incidence of developmental disorders. In addition, flame retardants can cause changes in the levels, transport, and receptors of certain neurotransmitters (e.g., dopamine, serotonin, glutamate, and γ-aminobutyric acid) EDCs can also inhibit or activate signalling molecules (e.g., the nuclear factor kappa b), which disrupts neuroimmune responses (Quoc et al., 2022). Sex hormones play an essential role in neurodevelopment. The effects of different sex hormones on cognitive traits, including memory, learning, and structural organisation of the brain, have been demonstrated (Sunman et al., 2019). EDCs interference of sex hormones by EDCs during brain development may lead to neurodevelopmental deterioration (Özel & Rüegg, 2023). EDCs can also be associated with changes in thyroid hormone levels, and their effects on neurocognitive disorders (e.g., ADHD) have been investigated. Still, the relationship to disorders such as autism remains inconclusive.

Neurodevelopmental processes in infancy and adolescence are critical periods of vulnerability to the harmful effects of EDCs because of the high level of neuroendocrine activity during this period (GolesTanzadeh et al., 2020). However, molecular mechanisms suggest that genes associated with

ASDs are expressed in the neocortex before birth and are involved in pathways related to excessive cell proliferation and impaired developmental processes. From this, exposure to EDCs before birth can also reveal adverse effects on neurodevelopment (Gibson et al., 2018; Hamra et al., 2019). Thus, it is indisputable that neurodevelopmental effects in children and adolescents were affected by prenatal and developmental exposure to EDCs.

21.3.3 Reproductive System Development

21.3.3.1 Male

The well-documented decline in male reproductive health has been extensively studied and reported. Studies have highlighted the diminishing serum testosterone levels and declining sperm quality observed among adolescents (Lokeshwar et al., 2021). Proper hormonal control is essential for testicular development and function, making them susceptible to disruption by exogenous hormones. Consistent with these findings, a recent study showed a negative correlation between exposure to EDCs in various types of environments and anal distance in young boys (Rodprasert et al., 2021). This indicates that EDC exposure can potentially impact the development of the male reproductive system.

A meta-analysis revealed an association between the levels of MEHP, MEHHP, MEOHP, and MECPP (metabolites of DEHP) in maternal urine and an increased risk of shortened anoscrotal and anopenile distance in male offspring (Zarean et al., 2019). A previous study indicated a positive correlation between the risk of cryptorchidism in Denmark and the levels of several congeners in different chemical groups, i.e., the levels of four FRs were significantly higher in the group of cryptorchid boys than in the control group (Krysiak-Baltyn et al., 2012). Furthermore, boys exhibiting symptoms related to reproductive abnormalities have shown correlations between their exposure levels to EDCs, such as BPA and phthalates, in amniotic fluid, cord blood, or serum samples and the levels of reproductive hormones during early developmental stages (Sunman et al., 2019).

Studies have also found that exposure to phthalates (exceptionally high molecular weight phthalates such as DEHP and DiNP), as well as certain phenols are associated with hypospadias and other genital malformations (Fisher et al., 2020), shorter penile size (Sunman et al., 2019), decreased anogenital distance (Bornehag et al., 2015), hormonal abnormalities suggesting impairment of Leydig cells (Muerköster et al., 2020), and testicular volumes (Hart et al., 2018).

21.3.3.2 Female

The menstrual cycle and fertility ability of females are very sensitive to hormonal imbalances, and changes in endocrine function due to exposure to EDCs can lead to several abnormalities during critical periods and stages of life cycle development, especially during childhood and adolescence. Bisphenol A is one of the most representative EDCs with endocrine-disrupting effects globally and is closely associated with female reproductive health, such as causing menstrual irregularities, impaired fertility, and endometriosis (Gore et al., 2015). In utero, neonatal exposure to BPA causes long-term adverse effects (Newbold et al., 2007).

DEHP is a notable EDC capable of inducing adverse effects. Therefore, it is crucial to consider the exposure risk to DEHP during essential developmental periods. Prenatal exposure to DEHP in humans is associated with shorter gestation periods. In a study with a sample size of 28 (68%) patients with menstrual disorders, it was noted that the patients' exposure to significantly high levels of phthalates [dimethyl, diethyl, dibutyl, and di(2-ethylhexyl)] and their major metabolite, mono(2-ethylhexyl) phthalate. Shortened gestation (or preterm labour) is related to long-term maternal health outcomes such as increased risk of type 2 diabetes and cardiovascular disease. In addition, it poses direct risks to both mother and baby (Tanz et al., 2019). In addition, an increased rate of miscarriage has been reported in women exposed to high-concentration phthalates in occupational settings (Toft et al., 2012). Delayed puberty occurs in U.S. and Danish girls with higher urinary concentrations of phthalate metabolites, including DEHP (Wolff et al., 2014). There have also been studies showing an association between early puberty/precocious puberty and exposure to phthalates (GolesTanzadeh et al., 2020), as well as a possible correlation between exposure to plasticisers with estrogenic and antiandrogenic activity and causing premature gynecomastia (Kamrin, 2009).

Different EDCs pose different bioaccumulation patterns. PCB 153 congeners are more likely to accumulate in the follicular wall than non-ortho PCB 126 congeners. 71%, 71.4%, and 30.4% of the total PCB 153 exposure were in small, medium, and large follicles, respectively (Gregoraszczuk & Ptak, 2013). EDCs can act at any level of the 'HPG-peripheral tissue' endocrine axis through various mechanisms and can also directly affect puberty-specific genes or HPG pathways. Indirectly, EDCs can act as inducers to alter metabolic programs during foetal and early childhood development, leading to changes in metabolism and peripheral hormones associated with the onset of puberty (GolesTanzadeh et al., 2020).

21.3.4 Immune System

Exposure to EDCs is associated with altered immune function, usually by suppressing immunity and thereby increasing susceptibility to infection or by enhancing the immune response, leading to inflammatory, allergic, or autoimmune diseases. EDCs may contribute to the pathogenesis of androgen receptors (AR) by exerting combined effects on the immune system. There are various receptors on the surface of immune cells, such as oestrogen receptor, oestrogen-associated receptor, toll-like receptor, and peroxisome proliferator-activated receptor γ, all of which can affect the development, lifespan, and function of the immune system, and the EDCs activate the immune cells and synthesise the inflammatory mediators by binding to these receptors, thus regulating the T helper 1 cell (Th1)/T helper 2 cell (Th2) function, increasing immunoglobulin expression levels and further contributing to anaphylactic reactions (Nowak et al., 2019). These disturbances can bring about a host of immune system disorders in the early period of human development. Children with higher exposure to PFOS were more likely to develop the common cold, gastroenteritis, and otitis media (Goudarzi et al., 2017). In another work, children with higher in-utero and exposure concentration to PFOS had a lower antibody response to routine childhood immunisations and an increased risk of having antibody levels below those needed for long-term protection (Ghassabian et al., 2022).

Exposure of pregnant females to varying levels of BPA has been shown to adversely affect the immune system of offspring (Bansal et al., 2018). It is also known that EDCs may have paracrine effects on Th2 cell differentiation, Th2 cytokine production, and elevated levels of immunoglobulin E, which has been associated with allergic asthma (Bornehag et al., 2015; Paciência et al., 2019). Only a few epidemiological studies have demonstrated the correlation between EDC exposure and human cytokine responses and have reported that higher prenatal exposure to phenol is also related to elevated circulating IL-6 in mothers and their offspring (Ghassabian et al., 2022; Kelley et al., 2019). EDCs can induce oxidative stress, disrupt the intestinal microbiome, and interfere with circadian rhythms EDCs can induce oxidative stress, disrupt the gut microbiome, and interfere with circadian rhythms, which may lead to immune dysfunction and trigger an inflammatory response (Bansal et al., 2018).

21.3.5 Long-Term Exposure Risk, Children Health, and Sustainable Development Implication

The impact of EDCs extends beyond the individuals directly exposed to them, as they can induce epigenetic effects with far-reaching consequences for the offspring and the entire population. The epigenetic effects of EDC exposure depend largely on the physicochemical properties of chemicals, time of exposure, dose, studied endpoints, etc. Importantly, behavioural and neurological disorders (e.g., psychomotor, neurobehavioural, and cognitive function), obesity, and asthma are some of the common epigenetic effects induced by EDCs. Although EDCs can pose a significant long-term threat to humans and wildlife, the specific mechanisms by which EDCs alter these processes and the window of sensitivity to EDCs remain unknown (Robaire et al., 2022). Epigenetics can alter DNA methylation patterns, microRNA profile, and histone modification.

However, it is important to note that the existing studies primarily rely on animal models and lack data on human beings (Rizzo et al., 2023). Therefore, more epidemiological research was suggested to focus on this aspect and prioritise human studies to validate and confirm the findings observed in model animals. For example, transgenerational epigenetic effects were observed following exposure to EDCs (e.g., the antiandrogenic compound phenazine and the estrogenic compound methoxychlor) during the assessment of gonadal sex in rats (Rizzo et al., 2023). It has also been demonstrated that the epigenetic and phenotypic effects of EDC exposure on the testes propagate at least to the F3 generation in the male germline. However, these offspring are not directly exposed to EDCs (Van Cauwenbergh et al., 2020). The F2 and F3 generations have been shown to have a comparatively low but significant incidence of pregnancy abnormalities, including uterine haemorrhage (~9% incidence) and mammary cancers (~2% incidence). Yet, the epigenetic basis for these abnormalities remains unknown (Nilsson et al., 2008). Although the cause of this impact and the sexual dimorphism of epigenetic germline transmission are unclear, they may be related to the nature of the initial injury and/or the duration of exposure. Therefore, it is reasonable to believe that exposure during the critical period of growth and development in early childhood is more likely to produce such alterations.

To reveal the mechanism of the early processes of inheritance, it would be valuable to conduct intergenerational, multigenerational, and transgenerational human studies to investigate the epigenetic effects of EDC on germ cells. This would reveal the potential transgenerational effects of exposure to EDCs, which would be assessed for the sustainability of the human population as a whole (Van Cauwenbergh et al., 2020). Furthermore, there is a need for a more comprehensive approach to toxicity testing of new chemicals before they are introduced to the market. This approach would ensure a thorough evaluation of their potential epigenetic effects and overall safety profiles, contributing to better protection of human health and the environment. Research is urgently required to unravel the effects of long-term exposure to low doses of EDCs, because exposure to EDCs during the critical developmental period could increase the risks of childhood diseases and cast long-term adverse effects on subsequent generations.

21.4 Relevant Regulatory Policies for EDC in Children's Products

EDCs are restricted because evidence suggests their harmful effects on humans. The first and most crucial step to reduce the human burden is to eliminate or phase out the production of these chemicals in commercial products. Many EDCs, such as BPA, TCS, phthalates, PBDEs, parabens, and PCBs, have been regulated in consumer products, especially baby products by various international organisations or agencies and individual countries (Figure 21.2).

BPA was first synthesised and widely used in the late 19th century. After being considered as a synthetic oestrogen in the early 20th century, Canada was the first country to declare BPA as 'dangerous substances', and Health Canada regulated its use in infant bottles in 2008 (European Parliament, 2011). Afterwards, more countries started to restrict or phase out baby feeding bottles containing BPA. In 2010, Denmark was the first European country to limit the use of BPA in baby feeding bottles after the risk assessment conducted by Danish DTU food institutes (European Parliament, 2011). France also banned the addition of BPA in the manufacture, import, and export of feeding bottles in 2010 (Özdal & Yeşilcubuk, 2014). After the publication of the European Food Safety Authority (EFSA) scientific opinion, European Commission decided to restrict the use of BPA in the manufacture of infant feeding bottles in 2011 (European Parliament, 2011). Austria restricted the use of BPA in pacifiers and teething rings, and Sweden restricted its use in varnish and food packaging coatings for children aged 0–3 in 2011 and 2013, respectively (ECHA, 2014). Moreover, the Health Ministry in China regulated the production of BPA-containing infant feeding containers (Mahamuni & Shrinithivihahshini, 2017); the U.S. Food and Drug Administration (USFDA) also banned the use of BPA in baby feeding bottles and children's drinking cups in the same year. France banned BPA in all food packages starting from 1 January 2015. EU agreed to restrict the use of BPA in thermal paper in 2016, and the mandate became effective in January 2020 (European Parliament, 2016). EU identified BPA as an endocrine disruptor for human health and the environment in the coming year (ECHA, 2017a). However, the Australian and New Zealand governments did not regulate the use of BPA; instead, they introduced a voluntary phase-out of the use of BPA in polycarbonate baby feeding bottles. In 2018, EU banned using BPA in sippy cups (European Parliament, 2018). BPA in thermal paper has recently and gradually been replaced by BPS (ECHA, 2020). In 2021, EFSA re-evaluated the risks of BPA to public health in foodstuffs (EFSA, 2021).

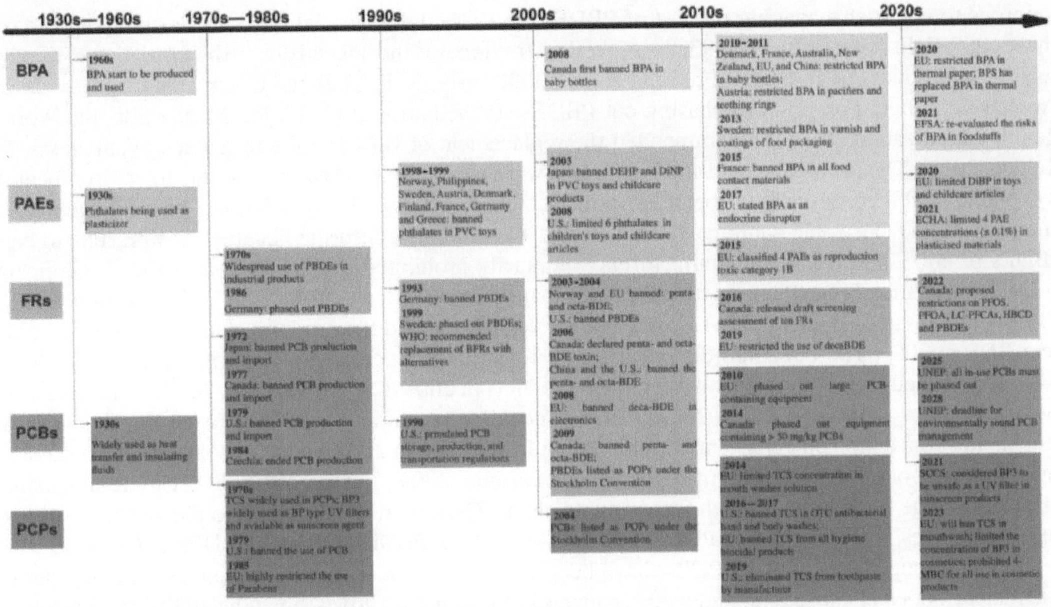

FIGURE 21.2 History of EDC regulations – Timeline of significant legislations in the 20th and 21st centuries (BPA: European Parliament, 2011, 2016; Özdal & Yeşilcubuk, 2014; ECHA, 2014; Mahamuni & Shrinithivihahshini, 2017; ECHA, 2020; EFSA, 2021; PAEs: ECHA, 2015, 2018, 2021; Kamrin, 2009; McCombie et al., 2017; Yang et al., 2013; FRs: ECJ, 2008; European Parliament, 2003; Fromme et al., 2009; Venkatesan & Halden, 2015; PCBs: European Commission, 2023; Melymuk et al., 2022; UN Environment Program, 2024; PCPs: European Commission, 2022; European Parliament, 2014; Lee et al., 2019; SCCS, 2021, 2022; USFDA, 2016, 2017, 2020).

The wide use of phthalates as a plasticiser, as reported in the news and publications, raised concerns (McCombie et al., 2017). Many countries, such as Austria (banned in January 1999), Denmark (banned in April 1999), Finland (banned in November 1999), France (banned in July 1999), Germany (banned in August 1999), Greece (banned in January 1999), Norway (banned in December 1998), Sweden (banned in September 1998), and the Philippines (banned in October 1998) have prohibited the use of phthalates in soft vinyl toys (Kamrin, 2009). Six phthalates, DEHP, DiDP, DiNP, BzBP, DnOP, and DBP, have been banned in toys that can be placed in the mouth in Europe (Commission Decision 1999/815/EC of 7 December 1999). In 2011, the Taiwan Food and Drug Administration reported that plasticisers (DEHP and DiNP) were illegally added to clouding agents used in foods and beverages (Yang et al., 2013). After 2015, DEHP, DnBP, BzBP, and DiBP, classified as reproduction toxic category 1B, could not be used without the permission of the EU (ECHA, 2015). Moreover, the amendment (Directive 2005/84/EC) stated that EU member states should limit the use of DEHP, DBP, and BzBP in all toys and childcare articles, as well as the use of DiNP, DnOP, and DiDP in toys that could be placed in the mouth to 0.1% by January 2017 (McCombie et al., 2017). In the U.S., the contents of BBP, DBP, and DEHP should not be higher than 0.1% in children's toys and childcare articles, and the concentration of DINP, DIDP, and DnOP should not be more than 0.1% in toys or childcare article that can be placed in the mouth (CPSC, 2008). In August 2003, the Japanese Minister for Health, Labour, and Welfare banned the production of PVC toys and childcare products containing DiNP and DEHP. However, there are no standards for phthalates in China. Still, the nation's Certification and Accreditation Administration requires that toys be subjected to inspection and certification review in accordance with the international standard ISO8124.1:2002. Recently, ECHA reported that DiBP concentration should be equal to or not greater than 0.1% by weight of the plasticised materials after 2020 (ECHA, 2018). Moreover, ECHA-restricted concentration of BBP, DBP, DEHP, and DiBP should not be higher than 0.1% by weight individually or in combination with any plasticised material in articles used by consumers or in indoor areas (ECHA, 2021).

The widespread use of PBDEs in flame retardants consumer products and in packaging and solid elastomers, has increased since 1970 (Prevedouros et al., 2004). Germany and Sweden were the first two countries to consider managing the use of PBDEs. In 1986, the industrial users in Germany agreed to phase out PBDEs voluntarily. In 1993, the official restriction was introduced due to the formation of dioxin during the production of PBDEs (Fromme et al., 2009). In 1999, the National Chemicals Inspectorate in Sweden submitted a proposal to phasing out PBDEs (Prevedouros et al., 2004). Meanwhile, the World Health Organization (WHO) recommended the replacement of BFRs with alternatives (Venkatesan & Halden, 2015). The penta- and octa-BDE mixtures were banned in Europe in 2004 due to environmental and public health concerns (European Parliament, 2003). Afterwards, in 2008, EU banned deca-BDE in electrical and electronic equipment (ECJ, 2008). In the U.S., California became the first state to ban PBDEs in 2003, and the U.S. manufacturers voluntarily prohibited the production of these chemicals in December 2004. Since then, many states have banned penta- and octa-BDE in 2006 (Fromme et al., 2009; Venkatesan & Halden, 2015). Furthermore, Norway banned the production and use of penta- and octa-BDE in 2003. The Canadian Environmental Protection Act declared penta- and octa-BDE toxic in 2006 and banned the production and use of new products in 2009 (Canada Gazette, 2009). China banned the use of penta- and octa-BDE in 2006. PBDEs, including penta- and octa-BDE, were listed under the Stockholm Convention on Persistent Organic Pollutants (POPs) in 2009. Canada released a draft screening assessment of ten organic FRs in 2016 (Canada Gazette, 2009). In December 2018, Environment and Climate Change Canada and Health Canada published a Consultation Document to phase out the use of toxic substances, including PFOS, PFOA, LC-PFCAs, HBCD, PBDEs, DP, and DBDPE (ECHA, 2017b).

Other common EDCs, such as PCBs, have been widely used as heat transfer fluids, insulating fluids for transformers and capacitors, and wood, paper, plastic or ink additives since the 1930s. PCB production and imports were banned in Japan, Canada, and the U.S. in 1972, 1977, and 1979, respectively (Melymuk et al., 2022). PCB production ended in Czechia in 1984 (Melymuk et al., 2022). The U.S. promulgated PCB storage, production, and transportation regulations in 1990. Afterwards, PCB was listed as POPs under the Stockholm Convention in 2004. European countries had to phase out all large PCB-containing equipment in 2010. In 2014, all equipment containing > 50 mg/kg PCBs must be phased out in Canada (European Commission, 2023). The Stockholm Convention provides the elimination of PCBs by 2025 and ensures environmentally sound waste management of PCB-containing liquids and PCB-contaminated equipment by 2028 (UN Environment Program, 2024).

PCPs, including TCS, BP-3, 4-MBC, and parabens, were banned by many countries since the 1970s. TCS was developed in the 1960s and has been widely used in personal care products since the 1970s. Standards have been developed to manage TCS in food contact materials and cosmetics. In Japan, the maximum allowable content of TCS for sterilising and preserving cosmetics was 0.3%. In the U.S. and Korea, TCS's maximum allowable permissible content was 0.3% for soap, deodorant, and rinse-off products (Lee et al., 2019). In 2014, EU limited TCS to a maximum concentration of 0.2% in mouth washes solutions and 0.3% in special cosmetic products, such as toothpaste, hand and body soaps, and face powders (European Parliament, 2014). In September 2016, TCS was banned by the USFDA in over-the-counter (OTC) antibacterial hand and body washes, such as hand and liquid soaps (USFDA, 2016), and hospital products by the end of 2018 (USFDA, 2017). However, TCS can still be applied to toothpaste, hand sanitiser, and mouthwash. Afterwards, EU banned TCS from all human hygiene biocidal products in 2017, and manufacturers in the U.S. voluntarily eliminated TCS from toothpaste in 2019 planned to be banned in mouthwash by EU (SCCS, 2022). Iso- propyl, iso-butyl, benzyl, and pentyl parabens are banned in cosmetics products (European Parliament, 2014). However, the USFDA classified parabens as 'safe' compounds (USFDA, 2020). SCCS considered BP3 unsafe as a UV filter at a maximum concentration of 6% in sunscreen in 2021 (SCCS, 2021). The maximum concentration of BP3 allowed in cosmetics was restricted by EU (6% in face products, hand products, and lip products; 2.2% in body products; 0.5% in other products); 4-MBC should be prohibited for all use in cosmetic products (European Commission, 2022).

Still, studies on reference values of risk assessment and metabolic mechanisms for many EDCs remain unknown. Alternatives that show similar chemical properties as the banned EDCs were used to manufacture consumer products. However, no regulatory policies have been developed for these alternatives, considered EDCs. Children and adolescents are vulnerable groups because of their routine and regular use of consumer products. Governments and organisations need to pay more attention to low-dose exposure to EDCs and alternatives at the stage of children's development.

21.5 Conclusion

EDCs are of increasing concern in public health around the world due to their widespread presence and long-term harmful effects on human health, particularly on the hormonal systems of children. Even at very low concentrations of exposure, EDCs can have significant impacts. These chemicals are widely available in all aspects of human life, such as in various industrial processes, food, and consumer products. These are major sources of exposure for children and adolescents through inhalation, ingestion, and dermal absorption. Through previous discussions, it has been demonstrated that exposure to EDCs is associated with adverse effects on children's health, including neurodevelopmental disorders (e.g., ADHD), metabolic interference (e.g., obesity), immune system effect (e.g., asthma), and disruptions in reproductive system development (e.g., early or delayed sexual maturation). Furthermore, exposure to EDCs during critical periods of early development can result in transgenerational effects, impacting subsequent generations who have never been directly exposed to these chemicals. The epigenetic and transgenerational effects tend to substantially influence the sustainable development of the population as a whole.

Biomonitoring studies conducted in different countries have consistently shown that children and adolescents are regularly exposed to a mixture of EDCs with varying levels of risk. However, the limited availability of reference values and dose-related health risk information has hindered the progress of exposure assessment for EDCs and their metabolites, which is still in its early stages.

To address these issues, future research should focus on identifying and evaluating exposure pathways, investigating exposure levels, and assessing the adverse effects and public health risks related to EDCs and their metabolites in children and adolescents. Recognising the potential adverse effects of EDCs on children and adolescents, many countries have imposed restrictions on the use of certain EDCs in toys and childcare articles. Nevertheless, the lack of consistent regulatory standards and toxicological assessments across different regions allows EDCs to be present in childcare articles on the market. By allocating more resources to these research aspects, it will benefit to establish a robust scientific evident that will inform the development of comprehensive and effective legislation focused on ensuring consumer product safety.

REFERENCES

Attina, T. M., Hauser, R., Sathyanarayana, S., Hunt, P. A., Bourguignon, J.-P., Myers, J. P., DiGangi, J., Zoeller, R. T., & Trasande, L. (2016). Exposure to endocrine-disrupting chemicals in the USA: A population-based disease burden and cost analysis. *The Lancet Diabetes & Endocrinology*, 4(12), 996–1003. https://doi.org/10.1016/S2213-8587(16)30275-3

Bansal, A., Henao-Mejia, J., & Simmons, R. A. (2018). Immune system: An emerging player in mediating effects of endocrine disruptors on metabolic health. *Endocrinology*, 159(1), 32–45. https://doi.org/10.1210/en.2017-00882

Bornehag, C.-G., Carlstedt, F., Jönsson, B. A., Lindh, C. H., Jensen, T. K., Bodin, A., Jonsson, C., Janson, S., & Swan, S. H. (2015). Prenatal phthalate exposures and anogenital distance in Swedish boys. *Environmental Health Perspectives*, 123(1), 101–107. https://doi.org/10.1289/ehp.1408163

Braun, J. M. (2017). Early-life exposure to EDCs: Role in childhood obesity and neurodevelopment. *Nature Reviews Endocrinology*, 13(3), 161–173. https://doi.org/10.1038/nrendo.2016.186

Canada Gazette. (2009). Regulations amending the organic products regulations. https://gazette.gc.ca/rp-pr/p2/2013/2013-10-09/html/sor-dors167-eng.html (accessed 09 Oct 2013)

Cilluffo, G., Ferrante, G., Murgia, N., Mancini, R., Pichini, S., Cuffari, G., Giudice, V., Tirone, N., Malizia, V., Montalbano, L., Fasola, S., Pacifici, R., Viegi, G., & La Grutta, S. (2022). Effects of polycyclic aromatic hydrocarbons on lung function in children with asthma: A mediation analysis. *International Journal of Environmental Research and Public Health*, 19(3), 1826.

CPSC (Consumer Product Safety Commission), (2008). Consumer product safety improvement act of 2008. https://www.cpsc.gov/s3fs-public/cpsia.pdf (accessed 14 Aug 2008)

Duan, C., Fang, Y., Sun, J., Li, Z., Wang, Q., Bai, J., Peng, H., Liang, J., & Gao, Z. (2020). Effects of fast food packaging plasticizers and their metabolites on steroid hormone synthesis in H295R cells. *Science of the Total Environment*, 726, 138500. https://doi.org/10.1016/j.scitotenv.2020.138500

ECHA. (2014). Annex XV restriction report. https://ECHA.europa.eu/documents/10162/8cab2a02-98dd-2aa0-d951-68dc263457a8 (accessed 06 May 2014)

ECHA. (2015). Authorisation List. https://ECHA.europa.eu/authorisation-list (accessed 21 Feb 2015)

ECHA. (2017a). The Member State Committee (MSC) unanimously agrees that bisphenol A is an endocrine disruptor. https://ECHA.europa.eu/-/msc-unanimously-agrees-that-bisphenol-a-is-an-endocrine-disruptor (accessed 16 Jun 2017)

ECHA. (2017b). Commission Regulation (EU) 2017/227. https://eur-lex.europa.eu/legal-content/EN/TXT/?uri=CELEX:32017R0227 (accessed 28 Apr 2024)

ECHA. (2018). ANNEX XVII TO REACH – Conditions of restriction: Restrictions on the manufacture, placing on the market and use of certain dangerous substances, mixtures and articles. https://ECHA.europa.eu/documents/10162/aaa92146-a005-1dc2-debe-93c80b57c5ee (accessed 18 Dec 2018)

ECHA. (2020). The use of bisphenol A and its alternatives in thermal paper in the EU during 2014–2022. https://ECHA.europa.eu/documents/10162/23294236/bpa_thermal_paper_report_2020_en.pdf/59eca269-c788-7942-5c17-3bd822d9cba0 (accessed Jun 2020)

ECHA. (2021). ECHA regulation updates limits of DEHP in FCMs. https://www.foodpackagingforum.org/news/ECHA-regulation-updates-limits-of-dehp-in-fcms (accessed 8 Dec 2021)

ECJ (European Court of Justice). (2008). Cases C-14/06 and C-295/06, judgement of the court, 1 April 2008, directive 2002/95/EC and commission decision 2005/717/ EC. https://op.europa.eu/en/publication-detail/-/publication/db746402-649a-4561-98e8-1b783be5700e/language-en (accessed 01 Apr 2008)

EFSA. (2021). Stakeholder meeting on the draft scientific opinion on re-evaluation of bisphenol A (BPA). https://www.efsa.europa.eu/en/events/stakeholder-meeting-draft-scientific-opinion-re-evaluation-bisphenol-bpa (accessed 24 Jan 2022)

European Commission. (2022). Commission Regulation (EU) 2022/1176. https://eur-lex.europa.eu/eli/reg/2022/1176 (accessed 28 Apr 2024)

European Commission. (2023). Polychlorinated biphenyls and polychlorinated terphenyls (PCBs/PCTs). https://environment.ec.europa.eu/topics/waste-and-recycling/pcbspcts_en (accessed 28 Apr 2024)

European Parliament. (2003). Directive 2003/11/EC of the European Parliament and of the council of 6 February 2003 amending for the 24th time Council Directive 76/769/EEC relating to restrictions on the marketing and use of certain dangerous substances and preparations. https://eur-lex.europa.eu/eli/dir/2003/11/oj (accessed 15 Feb 2003)

European Parliament. (2011). Commission directive 2011/8/EU of 28 January 2011 amending directive 2002/72/EC as regards the restriction of use of Bisphenol A in plastic infant feeding bottles. https://op.europa.eu/en/publication-detail/-/publication/cab230a2-7299-43f8-bbcf-3f98cb291147/language-en (accessed 28 Jan 2011)

European Parliament. (2014). Regulation (EC) No 1004/2014 of the European Parliament and of the Council. Official journal of the European Union L107/5–9. https://op.europa.eu/en/publication-detail/-/publication/4f42e643-c083-11e3-86f9-01aa75ed71a1/language-en (accessed 09 Apr 2014)

European Parliament. (2016). Commission regulation (EU) 2016/2235 of 12 December 2016, amending annex XVII to regulation (EC) No 1907/2006 of the European Parliament and of the council concerning the registration, evaluation, authorisation and restriction of chemicals (REACH) as regards bisphenol A. https://eur-lex.europa.eu/legal-content/EN/TXT/?uri=uriserv%3AOJ.L_.2016.337.01.0003.01.ENG (accessed 12 Dec 2016)

European Parliament. (2018). Commission Regulation (EU) 2018/213 on the use of bisphenol A in varnishes and coatings intended to come into contact with food and amending Regulation (EU) No 10/2011 as regards the use of that substance in plastic food contact materials. https://eur-lex.europa.eu/eli/reg/2018/213/oj (accessed 14 Feb 2018)

Fernandez, E., Perez, R., Hernandez, A., Tejada, P., Arteta, M., & Ramos, J. (2011). Factors and mECHAnisms for pharmacokinetic differences between pediatric population and adults. *Pharmaceutics*, 3(1), 53–72. https://doi.org/10.3390/pharmaceutics3010053

Fisher, B. G., Thankamony, A., Mendiola, J., Petry, C. J., Frederiksen, H., Andersson, A. M., Juul, A., Ong, K. K., Dunger, D. B., Hughes, I. A., & Acerini, C. L. (2020). Maternal serum concentrations of bisphenol A and propyl paraben in early pregnancy are associated with male infant genital development. *Human Reproduction*, 35(4), 913–928. https://doi.org/10.1093/humrep/deaa045

Forns, J., Verner, M. A., Iszatt, N., Nowack, N., Bach, C. C., Vrijheid, M., Costa, O., Andiarena, A., Sovcikova, E., Høyer, B. B., Wittsiepe, J., Lopez-Espinosa, M. J., Ibarluzea, J., Hertz-Picciotto, I., Toft, G., Stigum, H., Guxens, M., Liew, Z., & Eggesbø, M. (2020). Early life exposure to perfluoroalkyl substances (PFAS) and ADHD: A meta-analysis of nine European population-based studies. *Environmental Health Perspectives*, 128(5), 057002. https://doi.org/10.1289/EHP5444

Fromme, H., Körner, W., Shahin, N., Wanner, A., Albrecht, M., Boehmer, S., Parlar, H., Mayer, R., Liebl, B., & Bolte, G. (2009). Human exposure to polybrominated diphenyl ethers (PBDE), as evidenced by data from a duplicate diet study, indoor air, house dust, and biomonitoring in Germany. *Environment International*, 35(8), 1125–1135. https://doi.org/10.1016/j.envint.2009.07.003

García-Córcoles, M. T., Cipa, M., Rodríguez-Gómez, R., Rivas, A., Olea-Serrano, F., Vílchez, J. L., & Zafra-Gómez, A. (2018). Determination of bisphenols with estrogenic activity in plastic packaged baby food samples using solid-liquid extraction and clean-up with dispersive sorbents followed by gas chromatography tandem mass spectrometry analysis. *Talanta*, 178, 441–448. https://doi.org/10.1016/j.talanta.2017.09.067

Ghassabian, A., Vandenberg, L., Kannan, K., & Trasande, L. (2022). Endocrine-disrupting chemicals and child health. *Annual Review of Pharmacology and Toxicology*, 62(1), 573–594. https://doi.org/10.1146/annurev-pharmtox-021921-093352

Gibson, E. A., Siegel, E. L., Eniola, F., Herbstman, J. B., Factor-Litvak, P. (2018). Effects of polybrominated diphenyl ethers on child cognitive, behavioral, and motor development. *International Journal of Environmental Research and Public Health*, 15(8), 1636. https://doi.org/10.3390/ijerph15081636

GolesTanzadeh, M., Riahi, R., & Kelishadi, R. (2020). Association of phthalate exposure with precocious and delayed pubertal timing in girls and boys: A systematic review and meta-analysis. *Environmental Science: Processes & Impacts*, 22(4), 873–894. https://doi.org/10.1039/C9EM00512A

Gore, A. C., Chappell, V. A., Fenton, S. E., Flaws, J. A., Nadal, A., Prins, G. S., Toppari, J., & Zoeller, R. T. (2015). Executive summary to EDC-2: The endocrine society's second scientific statement on endocrine-disrupting chemicals. *Endocrine Reviews*, 36(6), 593–602. https://doi.org/10.1210/er.2015-1093

Gorecki, S., Bemrah, N., Roudot, A. C., Marchioni, E., Le Bizec, B., Faivre, F., Kadawathagedara, M., Botton, J., Rivière, G. (2017). Human health risks related to the consumption of foodstuffs of animal origin contaminated by bisphenol A. *Food and Chemical Toxicology*, 110, 333–339. https://doi.org/10.1016/j.fct.2017.10.045

Goudarzi, H., Miyashita, C., Okada, E., Kashino, I., Chen, C.-J., Ito, S., Araki, A., Kobayashi, S., Matsuura, H., & Kishi, R. (2017). Prenatal exposure to perfluoroalkyl acids and prevalence of infectious diseases up to 4 years of age. *Environment International*, 104, 132–138. https://doi.org/10.1016/j.envint.2017.01.024

Gregoraszczuk, E. L., & Ptak, A. (2013). Endocrine-disrupting chemicals: Some actions of POPs on female reproduction. *International Journal of Endocrinology*, 2013, 1–9. https://doi.org/10.1155/2013/828532

Haggerty, D. K., Strakovsky, R. S., Talge, N. M., Carignan, C. C., Glazier-Essalmi, A. N., Ingersoll, B. R., Karthikraj, R., Kannan, K., Paneth, N. S., & Ruden, D. M. (2021). Prenatal phthalate exposures and autism spectrum disorder symptoms in low-risk children. *Neurotoxicology and Teratology*, 83, 106947. https://doi.org/10.1016/j.ntt.2021.106947

Hamra, G. B., Lyall, K., Windham, G. C., Calafat, A. M., Sjödin, A., Volk, H., & Croen, L. A. (2019). Prenatal exposure to endocrine-disrupting chemicals in relation to autism spectrum disorder and intellectual disability. *Epidemiology*, 30(3), 418–426. https://doi.org/10.1097/EDE.0000000000000983

Hart, R. J., Frederiksen, H., Doherty, D. A., Keelan, J. A., Skakkebaek, N. E., Minaee, N. S., McLachlan, R., Newnham, J. P., Dickinson, J. E., Pennell, C. E., Norman, R. J., & Main, K. M. (2018). The possible impact of antenatal exposure to ubiquitous phthalates upon male reproductive function at 20 years of age. *Frontiers in Endocrinology*, 9, 288. https://doi.org/10.3389/fendo.2018.00288

He, C., English, K., Baduel, C., Thai, P., Jagals, P., Ware, R. S., Li, Y., Wang, X., Sly, P. D., & Mueller, J. F. (2018). Concentrations of organophosphate flame retardants and plasticizers in urine from young children in Queensland, Australia and associations with environmental and behavioural factors. *Environmental Research*, 164, 262–270. https://doi.org/10.1016/j.envres.2018.02.040

Kamrin, M. A. (2009). Phthalate risks, phthalate regulation, and public health: A review. *Journal of Toxicology and Environmental Health, Part B*, 12(2), 157–174. https://doi.org/10.1080/10937400902729226

Kelley, A. S., Banker, M., Goodrich, J. M., Dolinoy, D. C., Burant, C., Domino, S. E., Smith, Y. R., Song, P. X. K., & Padmanabhan, V. (2019). Early pregnancy exposure to endocrine disrupting chemical mixtures are associated with inflammatory changes in maternal and neonatal circulation. *Scientific Reports*, 9(1), 5422. https://doi.org/10.1038/s41598-019-41134-z

Kim, J., & Chevrier, J. (2020). Exposure to parabens and prevalence of obesity and metabolic syndrome: An analysis of the Canadian health measures survey. *Science of the Total Environment*, 713, 135116. https://doi.org/10.1016/j.scitotenv.2019.135116

Krysiak-Baltyn, K., Toppari, J., Skakkebaek, N. E., Jensen, T. S., Virtanen, H. E., Schramm, K. - W., Shen, H., Vartiainen, T., Kiviranta, H., Taboureau, O., Audouze, K., Brunak, S., & Main, K. M. (2012). Association between chemical pattern in breast milk and congenital cryptorchidism: Modelling of complex human exposures. *International Journal of Andrology*, 35(3), 294–302. https://doi.org/10.1111/j.1365-2605.2012.01268.x

Kuzukiran, O., Yurdakok-Dikmen, B., Sevin, S., Sireli, U. T., Iplikcioglu-Cil, G., & Filazi, A. (2018). Determination of selected endocrine disruptors in organic, free-range, and battery-produced hen eggs and risk assessment. *Environmental Science and Pollution Research*, 25(35), 35376–35386. https://doi.org/10.1007/s11356-018-3400-5

Lee, J. D., Lee, J. Y., Kwack, S. J., Shin, C. Y., Jang, H.-J., Kim, H. Y., Kim, M. K., Seo, D.-W., Lee, B.-M., & Kim, K.-B. (2019). Risk assessment of triclosan, a cosmetic preservative. *Toxicological Research*, 35(2), 137–154. https://doi.org/10.5487/TR.2019.35.2.137

Lee, K. J., & Choi, K. (2023). Non-carcinogenic health outcomes associated with polycyclic aromatic hydrocarbons (PAHs) exposure in humans: An umbrella review. *Exposure and Health*, 15(1), 95–111.

Li, L., Xu, S., & Lian, Q. (2022). The mediating function of obesity on endocrine-disrupting chemicals and insulin resistance in children. *Journal of Pediatric Endocrinology and Metabolism*, 35(9), 1169–1176. https://doi.org/10.1515/jpem-2022-0354

Liao, C., & Kannan, K. (2013). Concentrations and profiles of bisphenol A and other bisphenol analogues in foodstuffs from the United States and their implications for human exposure. *Journal of Agricultural and Food Chemistry*, 61(19), 4655–4662. https://doi.org/10.1021/jf400445n

Liao, C., & Kannan, K. (2014). A survey of alkylphenols, bisphenols, and triclosan in personal care products from China and the United States. *Archives of Environmental Contamination and Toxicology*, 67(1), 50–59. https://doi.org/10.1007/s00244-014-0016-8

Lokeshwar, S. D., Patel, P., Fantus, R. J., Halpern, J., Chang, C., Kargi, A. Y., & Ramasamy, R. (2021). Decline in serum testosterone levels among adolescent and young adult men in the USA. *European Urology Focus*, 7(4), 886–889. https://doi.org/10.1016/j.euf.2020.02.006

Lu, J., Wu, J., Stoffella, P. J., & Wilson, P. C. (2013). Analysis of bisphenol A, nonylphenol, and natural estrogens in vegetables and fruits using gas chromatography–tandem mass spectrometry. *Journal of Agricultural and Food Chemistry*, 61(1), 84–89. https://doi.org/10.1021/jf304971k

Mahamuni, D., & Shrinithivihahshini, N. D. (2017). Need for regulatory policies in India, on the use of bisphenol A in food contact plastic containers. *Current Science*, 113(05), 861. https://doi.org/10.18520/cs/v113/i05/861-868

McCombie, G., Biedermann, S., Suter, G., & Biedermann, M. (2017). Survey on plasticizers currently found in PVC toys on the Swiss market: Banned phthalates are only a minor concern. *Journal of Environmental Science and Health, Part A*, 52(5), 491–496. https://doi.org/10.1080/10934529.2016.1274176

Melymuk, L., Blumenthal, J., Sáňka, O., Shu-Yin, A., Singla, V., Šebková, K., Pullen Fedinick, K., & Diamond, M. L. (2022). Persistent problem: Global challenges to managing PCBs. *Environmental Science & Technology*, 56(12), 9029–9040. https://doi.org/10.1021/acs.est.2c01204

Muerköster, A.-P., Frederiksen, H., Juul, A., Andersson, A.-M., Jensen, R. C., Glintborg, D., Kyhl, H. B., Andersen, M. S., Timmermann, C. A. G., & Jensen, T. K. (2020). Maternal phthalate exposure associated with decreased testosterone/LH ratio in male offspring during mini-puberty. Odense child cohort. *Environment International*, 144, 106025. https://doi.org/10.1016/j.envint.2020.106025

Neugebauer, J., Wittsiepe, J., Kasper-Sonnenberg, M., Schöneck, N., Schölmerich, A., & Wilhelm, M. (2015). The influence of low level pre- and perinatal exposure to PCDD/Fs, PCBs, and lead on attention performance and attention-related behavior among German school-aged children: Results from the Duisburg birth cohort study. *International Journal of Hygiene and Environmental Health*, 218(1), 153–162. https://doi.org/10.1016/j.ijheh.2014.09.005

Newbold, R. R., Jefferson, W. N., & Padilla-Banks, E. (2007). Long-term adverse effects of neonatal exposure to bisphenol A on the murine female reproductive tract. *Reproductive Toxicology*, 24(2), 253–258. https://doi.org/10.1016/j.reprotox.2007.07.006

Nilsson, E. E., Anway, M. D., Stanfield, J., & Skinner, M. K. (2008). Transgenerational epigenetic effects of the endocrine disruptor vinclozolin on pregnancies and female adult onset disease. *Reproduction*, 135(5), 713–721. https://doi.org/10.1530/REP-07-0542

Nowak, K., Jabłońska, E., & Ratajczak-Wrona, W. (2019). Immunomodulatory effects of synthetic endocrine disrupting chemicals on the development and functions of human immune cells. *Environment International*, 125, 350–364. https://doi.org/10.1016/j.envint.2019.01.078

Olsen, G. W., Mair, D. C., Reagen, W. K., Ellefson, M. E., Ehresman, D. J., Butenhoff, J. L., & Zobel, L. R. (2007). Preliminary evidence of a decline in perfluorooctanesulfonate (PFOS) and perfluorooctanoate (PFOA) concentrations in American red cross blood donors. *Chemosphere*, 68(1), 105–111. https://doi.org/10.1016/j.chemosphere.2006.12.031

Oskar, S., Balalian, A. A., & Stingone, J. A. (2024). Identifying critical windows of prenatal phenol, paraben, and pesticide exposure and child neurodevelopment: Findings from a prospective cohort study. *Science of the Total Environment*, 920, 170754. https://doi.org/10.1016/j.scitotenv.2024.170754

Özdal, T., & Yeşilcubuk, N. Ş. (2014). Toxicity of bisphenol-A: Effects on health and regulations. *International Journal of Agricultural and Biosystems Engineering*, 8(6), 553–557. https://doi.org/10.5281/ZENODO.1092904

Özel, F., & Rüegg, J. (2023). Exposure to endocrine-disrupting chemicals and implications for neurodevelopment. *Developmental Medicine & Child Neurology*, 65(8), 1005–1011. https://doi.org/10.1111/dmcn.15551

Paciência, I., Cavaleiro Rufo, J., Silva, D., Martins, C., Mendes, F., Farraia, M., Delgado, L., De Oliveira Fernandes, E., Padrão, P., Moreira, P., Severo, M., Barros, H., & Moreira, A. (2019). Exposure to indoor endocrine-disrupting chemicals and childhood asthma and obesity. *Allergy*, 74(7), 1277–1291. https://doi.org/10.1111/all.13740

Perera, F., Vishnevetsky, J., Herbstman, J. B., Calafat, A. M., Xiong, W., Rauh, V., & Wang, S. (2012). Prenatal bisphenol A exposure and child behavior in an inner-city cohort. *Environmental Health Perspectives*, 120(8), 1190–1194. https://doi.org/10.1289/ehp.1104492

Prevedouros, K., Jones, K. C., & Sweetman, A. J. (2004). Estimation of the production, consumption, and atmospheric emissions of pentabrominated diphenyl ether in Europe between 1970 and 2000. *Environmental Science & Technology*, 38(12), 3224–3231. https://doi.org/10.1021/es049711d

Quirós-Alcalá, L., Buckley, J. P., & Boyle, M. (2018). Parabens and measures of adiposity among adults and children from the U.S. general population: NHANES 2007–2014. *International Journal of Hygiene and Environmental Health*, 221(4), 652–660. https://doi.org/10.1016/j.ijheh.2018.03.006

Quirós-Alcalá, L., Hansel, N. N., McCormack, M. C., & Matsui, E. C. (2019). Paraben exposures and asthma-related outcomes among children from the US general population. *Journal of Allergy and Clinical Immunology*, 143(3), 948–956. https://doi.org/10.1016/j.jaci.2018.08.021

Quoc, Q. L., Thi Bich, T. C., Kim, S.-H., Ryu, M. S., Park, H.-S., & Shin, Y. S. (2022). Mono-n-butyl phthalate regulates nuclear factor erythroid 2–related factor 2 and nuclear factor kappa B pathway in an ovalbumin-induced asthma mouse model. *Food and Chemical Toxicology*, 166, 113171. https://doi.org/10.1016/j.fct.2022.113171

Ramírez, V., Gálvez-Ontiveros, Y., González-Domenech, P. J., Baca, M. Á., Rodrigo, L., & Rivas, A. (2022). Role of endocrine disrupting chemicals in children's neurodevelopment. *Environmental Research*, 203, 111890. https://doi.org/10.1016/j.envres.2021.111890

Ribeiro, C. M., Beserra, B. T. S., Silva, N. G., Lima, C. L., Rocha, P. R. S., Coelho, M. S., Neves, F. D. A. R., & Amato, A. A. (2020). Exposure to endocrine-disrupting chemicals and anthropometric measures of obesity: A systematic review and meta-analysis. *BMJ Open*, 10(6), e033509. https://doi.org/10.1136/bmjopen-2019-033509

Rizzo, R., Bortolotti, D., Rizzo, S., & Schiuma, G. (Eds.). (2023). Endocrine disruptors, epigenetic changes, and transgenerational transmission. In Environment Impact on Reproductive Health: A Translational Approach. Springer International Publishing, Berlin. https://doi.org/10.1007/978-3-031-36494-5

Robaire, B., Delbes, G., Head, J. A., Marlatt, V. L., Martyniuk, C. J., Reynaud, S., Trudeau, V. L., & Mennigen, J. A. (2022). A cross-species comparative approach to assessing multi- and transgenerational effects of endocrine disrupting chemicals. *Environmental Research*, 204, 112063. https://doi.org/10.1016/j.envres.2021.112063

Rodprasert, W., Toppari, J., & Virtanen, H. E. (2021). Endocrine disrupting chemicals and reproductive health in boys and men. *Frontiers in Endocrinology*, 12, 706532. https://doi.org/10.3389/fendo.2021.706532

Rundle, A., Hoepner, L., Hassoun, A., Oberfield, S., Freyer, G., Holmes, D., Reyes, M., Quinn, J., Camann, D., Perera, F., & Whyatt, R. (2012). Association of childhood obesity with maternal exposure to ambient air polycyclic aromatic hydrocarbons during pregnancy. *American Journal of Epidemiology*, 175(11), 1163–1172. https://doi.org/10.1093/aje/kwr455

SCCS. (2021). Opinion on Benzophenone-3 (CAS No 131–57-7, EC No 205–031–5). https://health.ec.europa.eu/latest-updates/sccs-final-opinion-benzophenone-3-2021-04-15_en (accessed 28 Apr 2024)

SCCS. (2022). Safety of Triclocarban and Triclosan as substances with potential endocrine disrupting properties in cosmetic products. https://health.ec.europa.eu/publications/safety-triclocarban-and-triclosan-substances-potential-endocrine-disrupting-properties-cosmetic_en (accessed 28 Apr 2024)

Shoaff, J. R., Calafat, A. M., Schantz, S. L., & Korrick, S. A. (2019). Endocrine disrupting chemical exposure and maladaptive behavior during adolescence. *Environmental Research*, 172, 231–241. https://doi.org/10.1016/j.envres.2018.12.053

Skogheim, T. S., Weyde, K. V. F., Aase, H., Engel, S. M., Surén, P., Øie, M. G., Biele, G., Reichborn-Kjennerud, T., Brantsæter, A. L., Haug, L. S., Sabaredzovic, A., Auyeung, B., & Villanger, G. D. (2021). Prenatal exposure to per-and polyfluoroalkyl substances (PFAS) and associations with attention-deficit/hyperactivity disorder and autism spectrum disorder in children. *Environmental Research*, 202, 111692. https://doi.org/10.1016/j.envres.2021.111692

Sunman, B., Yurdakök, K., Kocer-Gumusel, B., Özyüncü, Ö., Akbıyık, F., Balcı, A., Özkemahlı, G., Erkekoğlu, P., & Yurdakök, M. (2019). Prenatal bisphenol A and phthalate exposure are risk factors for male reproductive system development and cord blood sex hormone levels. *Reproductive Toxicology*, 87, 146–155. https://doi.org/10.1016/j.reprotox.2019.05.065

Tanz, L. J., Stuart, J. J., Williams, P. L., Missmer, S. A., Rimm, E. B., James-Todd, T. M., & Rich-Edwards, J. W. (2019). Preterm delivery and maternal cardiovascular disease risk factors: The nurses' health study II. *Journal of Women's Health*, 28(5), 677–685. https://doi.org/10.1089/jwh.2018.7150

Toft, G., Jönsson, B. A. G., Lindh, C. H., Jensen, T. K., Hjollund, N. H., Vested, A., & Bonde, J. P. (2012). Association between pregnancy loss and urinary phthalate levels around the time of conception. *Environmental Health Perspectives*, 120(3), 458–463. https://doi.org/10.1289/ehp.1103552

Trasande, L., Zoeller, R. T., Hass, U., Kortenkamp, A., Grandjean, P., Myers, J. P., DiGangi, J., Bellanger, M., Hauser, R., Legler, J., Skakkebaek, N. E., & Heindel, J. J. (2015). Estimating burden and disease costs of exposure to endocrine-disrupting chemicals in the European Union. *The Journal of Clinical Endocrinology & Metabolism*, 100(4), 1245–1255. https://doi.org/10.1210/jc.2014-4324

UN Environment Program. (2024). Toward elimination of PCBs. https://www.unep.org/topics/chemicals-and-pollution-action/pollution-and-health/persistent-organic-pollutants-pops-5 (accessed 28 Apr 2024)

USFDA. (2016). FDA issues final rule on safety and effectiveness of antibacterial soaps. https://www.fda.gov/news-events/press-announcements/fda-issues-final-rule-safety-and-effectiveness-antibacterial-soaps (accessed 02 Sep 2016)

USFDA. (2017). Safety and effectiveness for health care antiseptics; topical antimicrobial drug products for over-the-counter human use. https://www.federalregister.gov/documents/2017/12/20/2017-27317/safety-and-effectiveness-of-health-care-antiseptics-topical-antimicrobial-drug-products-for (accessed 19 Dec 2017)

USFDA. (2020). Select committee on GRAS substances (SCOGS) database. https://www.hfpappexternal.fda.gov/scripts/fdcc/index.cfm?set=SCOGS&sort=Sortsubstance&order=ASC&startrow=1&type=basic&search=paraben (accessed 31 Jul 2020)

Van Cauwenbergh, O., Di Serafino, A., Tytgat, J., & Soubry, A. (2020). Transgenerational epigenetic effects from male exposure to endocrine-disrupting compounds: A systematic review on research in mammals. *Clinical Epigenetics*, 12(1), 65. https://doi.org/10.1186/s13148-020-00845-1

Venkatesan, A., & Halden, R. (2015). Effective strategies for monitoring and regulating chemical mixtures and contaminants sharing pathways of toxicity. *International Journal of Environmental Research and Public Health*, 12(9), 10549–10557. https://doi.org/10.3390/ijerph120910549

Wang, J., Chen, G., Christie, P., Zhang, M., Luo, Y., & Teng, Y. (2015). Occurrence and risk assessment of phthalate esters (PAEs) in vegetables and soils of suburban plastic film greenhouses. *Science of the Total Environment*, 523, 129–137. https://doi.org/10.1016/j.scitotenv.2015.02.101

Wang, Y., Ji, S., Chen, X., Liu, S., Zhao, J., & Long, C. (2024). Butylated hydroxyanisole controls anthracnose on postharvest citrus fruit by inducing autophagosome formation in Colletotrichum gloeosporioides. *Postharvest Biology and Technology*, 210, 112740. https://doi.org/10.1016/j.postharvbio.2023.112740

Wolff, M. S., Teitelbaum, S. L., McGovern, K., Windham, G. C., Pinney, S. M., Galvez, M., Calafat, A. M., Kushi, L. H., Biro, F. M., & on behalf of the Breast Cancer and Environment Research Program. (2014). Phthalate exposure and pubertal development in a longitudinal study of US girls. *Human Reproduction*, 29(7), 1558–1566. https://doi.org/10.1093/humrep/deu081

Xu, B., Wu, M., Wang, M., Pan, C., Qiu, W., Tang, L., & Xu, G. (2018). Polybrominated diphenyl ethers (PBDEs) and hydroxylated PBDEs in human serum from Shanghai, China: A study on their presence and correlations. *Environmental Science and Pollution Research*, 25(4), 3518–3526. https://doi.org/10.1007/s11356-017-0709-4

Yang, C., Lee, H. K., Kong, A. P. S., Lim, L. L., Cai, Z., & Chung, A. C. K. (2018). Early-life exposure to endocrine disrupting chemicals associates with childhood obesity. *Annals of Pediatric Endocrinology & Metabolism*, 23(4), 182–195. https://doi.org/10.6065/apem.2018.23.4.182

Yang, J., Hauser, R., & Goldman, R. H. (2013). Taiwan food scandal: The illegal use of phthalates as a clouding agent and their contribution to maternal exposure. *Food and Chemical Toxicology*, 58, 362–368. https://doi.org/10.1016/j.fct.2013.05.010

Zarean, M., Keikha, M., Feizi, A., Kazemitabaee, M., & Kelishadi, R. (2019). The role of exposure to phthalates in variations of anogenital distance: A systematic review and meta-analysis. *Environmental Pollution*, 247, 172–179. https://doi.org/10.1016/j.envpol.2019.01.026

Index

Note: **Bold** page numbers refer to tables and *italic* page numbers refer to figures.